Subcellular Biochemistry

Volume 28

Cholesterol:
Its Functions and Metabolism in Biology and Medicine

SUBCELLULAR BIOCHEMISTRY

SERIES EDITOR

J. ROBIN HARRIS, Institute of Zoology, University of Mainz, Mainz, Germany

ASSISTANT EDITORS

H. J. HILDERSON, University of Antwerp, Antwerp, Belgium
B. B. BISWAS, University of Calcutta, Calcutta, India

Recent Volumes in This Series:

Volume 19 **Endocytic Components: Identification and Characterization**
Edited by J. J. M. Bergeron and J. R. Harris

Volume 20 **Mycoplasma Cell Membranes**
Edited by S. Rottem and I. Kahane

Volume 21 **Endoplasmic Reticulum**
Edited by N. Borgese and J. R. Harris

Volume 22 **Membrane Biogenesis**
Edited by A. H. Maddy and J. R. Harris

Volume 23 **Physicochemical Methods in the Study of Biomembranes**
Edited by Herwig J. Hilderson and Gregory B. Ralston

Volume 24 **Proteins: Structure, Function, and Engineering**
Edited by B. B. Biswas and Siddhartha Roy

Volume 25 **Ascorbic Acid: Biochemistry and Biomedical Cell Biology**
Edited by J. Robin Harris

Volume 26 ***myo*-Inositol Phosphates, Phosphoinositides, and Signal Transduction**
Edited by B. B. Biswas and Susweta Biswas

Volume 27 **Biology of the Lysosome**
Edited by John B. Lloyd and Robert W. Mason

Volume 28 **Cholesterol: Its Functions and Metabolism in Biology and Medicine**
Edited by Robert Bittman

A Continuation Order Plan is available for this series. A continuation order will bring delivery of each new volume immediately upon publication. Volumes are billed only upon actual shipment. For further information please contact the publisher.

Subcellular Biochemistry

Volume 28

Cholesterol: Its Functions and Metabolism in Biology and Medicine

Edited by

Robert Bittman
Queens College of The City University of New York
Flushing, New York

SPRINGER SCIENCE+BUSINESS MEDIA, LLC

The Library of Congress cataloged the first volume of this title as follows:

Sub-cellular biochemistry.
London, New York, Plenum Press.
v. illus. 23 cm. quarterly.
Began with Sept. 1971 issue. Cf. New serial titles.
1. Cytochemistry—Periodicals. 2. Cell organelles—Periodicals.
QH611.S84 574.8'76 73-643479

ISSN 0306-0225
ISBN 978-1-4613-7707-8 ISBN 978-1-4615-5901-6 (eBook)
DOI 10.1007/978-1-4615-5901-6

This series is a continuation of the journal *Sub-Cellular Biochemistry,* Volumes 1 to 4 of which were published quarterly from 1972 to 1975

Originally published by Plenum Press in 1997
Softcover reprint of the hardcover 1st edition 1997

Contributors

Ashok K. Batta Department of Medicine and the Liver Center, University of Medicine and Dentistry—New Jersey Medical School, Newark, New Jersey 07103

Erik A. L. Biessen Division of Biopharmaceutics, Leiden-Amsterdam Center for Drug Research, Sylvius Laboratory, University of Leiden, 2300 RA Leiden, The Netherlands

Robert Bittman Department of Chemistry and Biochemistry, Queens College of The City University of New York, Flushing, New York 11367–1597

Ingemar Björkhem Department of Clinical Chemistry and Department of Medicine and Center for Nutrition and Toxicology, Huddinge University Hospital, Karolinska Institute, S-141 86 Huddinge, Sweden

E. Joan Blanchette-Mackie Laboratory of Cell Biochemistry and Biology, National Institutes of Diabetes and Digestive and Kidney Diseases, National Institutes of Health, Bethesda, Maryland 20892

Li-Chuan Chen Clinical Pharmacology Branch, National Cancer Institute, National Institutes of Health, Bethesda, Maryland 20892

Peter S. Coleman The Boston Biomedical Research Institute, Laboratory of Metabolic Regulation, Boston, Massachusetts 02114

Peter A. Edwards Departments of Biological Chemistry and Medicine and the Molecular Biology Institute, University of California at Los Angeles, Los Angeles, California 90095

Ronald P. J. Oude Elferink Department of Gastrointestinal and Liver Diseases, Academic Medical Center, Amsterdam, The Netherlands

Ellen Roy Elias Division of Clinical Genetics, Department of Pediatrics, New England Medical Center, Tufts University School of Medicine, Boston, Massachusetts 02111

Johan Ericsson Departments of Biological Chemistry and Medicine and the Molecular Biology Institute, University of California at Los Angeles, Los Angeles, California 90095

Albert K. Green Department of Gastrointestinal and Liver Diseases, Academic Medical Center, Amsterdam, The Netherlands

De-an Guo Departments of Chemistry and Biochemistry, Texas Tech University, Lubbock, Texas 79409

Akira Honda Department of Medicine and the Liver Center, University of Medicine and Dentistry—New Jersey Medical School, Newark, New Jersey 07103

Mira Irons Division of Clinical Genetics, Department of Pediatrics, New England Medical Center, Tufts University School of Medicine, Boston, Massachusetts 02111

Simon M. Jackson Departments of Biological Chemistry and Medicine and the Molecular Biology Institute, University of California at Los Angeles, Los Angeles, California 90095

Zhonghua Jia Departments of Chemistry and Biochemistry, Texas Tech University, Lubbock, Texas 79409

William J. Johnson Department of Biochemistry, Allegheny University of the Health Sciences, Philadelphia, Pennsylvania 19129

Tanetoshi Koyama Institute for Chemical Reaction Science, Tohoku University, Katahira 2 chome, Aoba-ku, Sendai 980–77, Japan

Howard S. Kruth Section of Experimental Atherosclerosis, National Heart, Lung, and Blood Institute, National Institutes of Health, Bethesda, Maryland 20892

Folkert Kuipers Groningen Institute for Drug Studies, Laboratory of Nutrition and Metabolism, Academic Hospital Groningen, Groningen, The Netherlands

Laura Liscum Department of Physiology, Tufts University School of Medicine, Boston, Massachusetts 02111

Monica Lopez Departments of Chemistry and Biochemistry, Texas Tech University, Lubbock, Texas 79409

Erik Lund Department of Clinical Chemistry and Department of Medicine and Center for Nutrition and Toxicology, Huddinge University Hospital, Karolinska Institute, S-141 86 Huddinge, Sweden

Anil T. Mangla Departments of Chemistry and Biochemistry, Texas Tech University, Lubbock, Texas 79409

Federico Martínez Department of Biochemistry, Faculty of Medicine, UNAM, Mexico

W. David Nes Departments of Chemistry and Biochemistry, Texas Tech University, Lubbock, Texas 79409

Gene C. Ness Department of Biochemical and Molecular Biology, College of Medicine, University of South Florida Medical School, Tampa, Florida 33612

S. Derrek Nichols Departments of Chemistry and Biochemistry, Texas Tech University, Lubbock, Texas 79409

Lien Nguyen Department of Medicine and the Liver Center, University of Medicine and Dentistry—New Jersey Medical School, Newark, New Jersey 07103

Kyozo Ogura Institute for Chemical Reaction Science, Tohoku University, Katahira 2 chome, Aoba-ku, Sendai 980–77, Japan

Peter G. Pentchev National Institute of Neurological Disorders and Stroke, Developmental and Metabolic Neurology Branch, National Institutes of Health, Bethesda, Maryland 20892

Michael C. Phillips Department of Biochemistry, Allegheny University of the Health Sciences, Philadelphia, Pennsylvania 19129

George H. Rothblat Department of Biochemistry, Allegheny University of the Health Sciences, Philadelphia, Pennsylvania 19129

Mats Rudling Department of Clinical Chemistry and Department of Medicine and Center for Nutrition and Toxicology, Huddinge University Hospital, Karolinska Institute, S-141 86 Huddinge, Sweden

Hiroshi Sagami Institute for Chemical Reaction Science, Tohoku University, Katahira 2 chome, Aoba-ku, Sendai 980–77, Japan

Gerald Salen Research Service and Medical Service, Department of Veterans Affairs Medical Center, East Orange, New Jersey 07019 and Department of Medicine and the Liver Center, University of Medicine and Dentistry—New Jersey Medical School, Newark, New Jersey 07103

Laura Sepp-Lorenzino Memorial Sloan-Kettering Cancer Center, Program in Cell Biology and Genetics and Department of Medicine, New York, New York 10021

Nicholas J. Severs National Heart and Lung Institute, Imperial College of Science, Technology, and Medicine, Royal Brompton Hospital, London, England

Sarah Shefer Department of Medicine and the Liver Center, University of Medicine and Dentistry—New Jersey Medical School, Newark, New Jersey 07103

Leo A. J. M. Sliedregt Division of Biopharmaceutics, Leiden-Amsterdam Center for Drug Research, Sylvius Laboratory, University of Leiden, 2300 RA Leiden, The Netherlands

J. Peter Slotte Department of Biochemistry and Pharmacy, Åbo Akademi University, FIN 20520 Turku, Finland

G. Jackson Snipes Departments of Neuropathology, Neurology, and Neurosurgery, Montreal Neurological Institute, McGill University, Montreal, Quebec H3A 2B4, Canada

Jerome F. Strauss III Department of Obstetrics and Gynecology, University of Pennsylvania School of Medicine, Philadelphia, Pennsylvania 19104

Ueli Suter Department of Cell Biology, Swiss Federal Institute of Technology, ETH-Hönggerberg, 8093 Zürich, Switzerland

G. S. Tint Research Service and Medical Service, Department of Veterans Affairs Medical Center, East Orange, New Jersey 07019 and Department of Medicine and the Liver Center, University of Medicine and Dentistry—New Jersey Medical School, Newark, New Jersey 07103

Theo J. C. Van Berkel Division of Biopharmaceutics, Leiden-Amsterdam Center for Drug Research, Sylvius Laboratory, University of Leiden, 2300 RA Leiden, The Netherlands

Henkjan J. Verkade Groningen Institute for Drug Studies, Laboratory of Nutrition and Metabolism, Academic Hospital Groningen, Groningen, The Netherlands

Guorong Xu Department of Medicine and the Liver Center, University of Medicine and Dentistry—New Jersey Medical School, Newark, New Jersey 07103

Wen Zhou Departments of Chemistry and Biochemistry, Texas Tech University, Lubbock, Texas 79409

Preface

The topics in this volume cover cholesterol from A to Z. Its role as a normal metabolic constituent of biological membranes and the pathological consequences of its metabolic mismanagement are described in this book. Cholesterol is required for normal cell growth and proper membrane structure and function. However, unregulated accumulation of cholesterol is cytotoxic, and a failure to maintain homeostasis of the sterol results in a number of pathological states. Among the diseases discussed in this volume are gallstone disease, atherosclerosis, Schnyder's corneal crystalline dystrophy, tumor proliferation, Niemann–Pick C disease, sitosterolemia, and the Smith–Lemli–Opitz syndrome. Therefore, an appropriate subtitle to this volume, as suggested by Peter G. Pentchev, is "Cholesterol: The Good, the Bad, and the Ugly."

The first four chapters discuss different aspects of the synthesis of sterols from isoprenoid precursors. In Chapter 1, Jackson *et al.* review the roles of isoprenoids formed in the cholesterol biosynthetic pathway on many different cell functions, including cell signaling, intracellular vesicle movement, and targeting. They also review the ability of sterols to regulate the activity of key proteins, such as the enzymes acyl cholesterol:acyl-coenzyme A transferase (ACAT), cholesterol 7α-hydroxylase, 3-hydroxy-3-methylglutaryl-coenzyme A reductase (HMG-CoA reductase), and the sterol regulatory element (SRE-1) binding proteins (SREBPs), which function in sterol-regulated transcription.

Björkhem *et al.* discuss possible mechanisms for the steroid-dependent coordinate regulation of cholesterol 7α-hydroxylase and HMG-CoA reductase in Chapter 2. Two important regulators of HMG-CoA reductase *in vivo* are the flux of cholesterol and bile acids. The most important regulator of cholesterol 7α-hydroxylase in the liver is the flux of bile acids returning to the liver via enterohepatic circulation.

The biosynthesis of prenyl diphosphates is reviewed in Chapter 3, with an emphasis on the catalytic functions of key enzymes. Ogura *et al.* discuss the prenyltransferases, with emphasis on farnesyl diphosphate (FPP) synthase and geranylgeranyl diphosphate (GGPP) synthase. Also reviewed are medium- and long-chain prenyl diphosphate synthases, which catalyze the synthesis of prenyl products with chain lengths longer than C_{20}.

Chapters 4 and 5 deal with enzymes that affect two specific features of the sterol molecule, i.e., the nature of the 24-alkyl group and the presence of additional unsaturation in the B ring. The biochemistry of inhibitors that interfere with sterol biosynthesis is described by Guo *et al.* in Chapter 4. Particular emphasis is placed on the chemistry of sterol biosynthesis inhibitors that block the introduction of an alkyl group at C_{24} of the sterol side chain and in the function of the biomethylation enzyme (*S*)-adenosyl-L-methionine:Δ^{24}-sterol methyl transferase (SMT).

An adequate supply of cholesterol from the mother to the offspring is critical during gestation, and interference with normal cholesterol biosynthesis and transport results in hindered development. The Smith–Lemli–Opitz or RSH syndrome, a frequently fatal birth defect, is discussed by Tint *et al.* in Chapter 5. The defect in this disease is in 7-dehydrocholesterol reductase. This biosynthetic cholesterol defect, which is most pronounced in the brain, was discovered in 1993.

Cholesterol is the most abundant sterol found in mammalian cells. The influence of cholesterol on the properties of lipid bilayers have been studied by sophisticated biophysical techniques for more than three decades. Chapter 6 reviews findings that indicate that hydrophobic mismatch in the effective lengths of the cholesterol and phospholipid molecules contribute to the lateral organization of the lipid bilayer.

In Chapter 7, Snipes and Suter discuss the postulated role of cholesterol in the function of the nervous system, i.e., in contributing to the ability of myelin to act as an electrical insulator. The brain contains an unusually high amount of cholesterol relative to other organs; about 10% of the dry weight of the whole brain is cholesterol, and approximately 70% of the cholesterol in the whole brain is incorporated into myelin. Inhibitors of cholesterol synthesis have the ability to delay the onset and limit the extent of myelination. Defects in cholesterol biosynthesis and metabolism are involved in a variety of pathologies that affect the nervous system, including Alzheimer's disease, the Smith–Lemli–Opitz syndrome, and tellurium toxicity to the peripheral nervous system.

Steroidogenic tissues (adrenals, gonads, and placenta) transform cholesterol into hormones. Steroid hormone synthesis is increased in response to trophic stimulation, and steroidogenic acute regulatory protein, StAR, appears to mediate the process by enhancing the mitochondrial conversion of cholesterol into pregnenolone. In Chapter 8, Martínez and Strauss review the regulation of mitochondrial cholesterol metabolism and the mechanisms of intramitochondrial sterol movement in steroidogenic cells.

How is cholesterol homeostasis achieved, i.e., how does a normal cell handle fluctuations in cholesterol content? In Chapter 9, Johnson *et al.* describe the mechanisms involved in balancing the fluxes of cholesterol into and out of the cell and the rate of *de novo* biosynthesis of cholesterol. The chapter reviews the trafficking of cholesterol from intracellular sites and the plasma membrane, where efflux to external acceptors occurs, as well as the cholesterol ester cycle, in which cholesterol can move between pools of free cholesterol and esterified cholesterol. In Chapter 10, Slotte discusses the effects of alterations in membrane content on lipid homeostasis in cells. Particular emphasis is placed on sphingomyelin–cholesterol interactions, the effects of sphingomyelin on the metabolism of exogenously derived cholesterol, and the effects of sphingomyelin in the plasma membrane on the distribution of cholesterol in the membranes of cells. In Chapter 11, Kuipers *et al.* discuss the hepatobiliary pathway, which is important in maintaining cholesterol homeostasis *in vivo.* In this pathway, cholesterol is removed via secretion from the liver; hypersecretion of cholesterol into bile results in crystallization of excess cholesterol and initiation of gallstone formation. Biliary lipid secretion may be regulated at the bile canalicular membrane. The roles of vesicular lipid transport, phosphatidylcholine transport protein, and sterol carrier protein in hepatobiliary transport of cholesterol are discussed.

Crystalline forms of cholesterol are also found in atherosclerotic lesions. In Chapter 12, Kruth discusses factors that influence how plasma cholesterol is taken up and accumulated by the vessel wall, including the trapping of lipoprotein cholesterol in macrophage foam cells, and also discusses the pathways by which lipids can leave the vessel wall.

Coleman *et al.* discuss the cholesterol–cancer connection in Chapter 13. A loss of the normal feedback downregulation of cholesterogenesis in tumor cells results in a high flux of intermediates of the cholesterol biosynthetic pathway. Moreover, high cholesterol content appears to stimulate the activity of the mitochondrial citrate transport protein, resulting in enhanced citrate export to the cytosol, formation of cytosolic acetyl-CoA, and subsequent high levels of cholesterol synthesis. Prenylated Ras-mediated signal transduction pathways are implicated in cell proliferation. The use of inhibitors of isoprenoid biogenesis and protein prenylation to give agents that have potential antitumor activity is discussed.

The defect in intracellular cholesterol transport responsible for Niemann–Pick C disease is reviewed by Pentchev *et al.* in Chapter 14. In this lipid storage disease, cholesterol and sphingomyelin accumulate in reticuloendothelial tissues, and this lipid accumulation leads to formation of cholesterol crystals in lysosomes. Niemann–Pick C fibroblasts display an altered cholesterol distribution in the Golgi, suggesting that a Golgi protein involved in lysosomal cholesterol trafficking is defective in this disease.

In Chapter 15, Salen *et al.* provide a detailed account of the rare inherited lipid storage disease sitosterolemia, with an emphasis on the characteristic clinical and biochemical abnormalities that take place. In sitosterolemia, the levels of plant sterols and 5α-stanols are elevated in the plasma and tissues. In contrast to the lipid storage disease cerebrotendinous xanthomatosis (CTX), where increased amounts of cholestanol are deposited in the brain because of apparent damage to the blood–brain barrier, in sitosterolemia the blood–brain barrier remains intact. A deficiency in the mass of HMG-CoA reductase is the reason for the reduced level of cholesterol in sitosterolemia, which is caused by the very low levels of HMG-CoA reductase mRNA that are available for enzyme translation. The catalytic activity of the small amount of the HMG-CoA reductase that is formed is normal.

In Chapter 16, Severs reviews the principal cholesterol-binding agents that have been used for cytochemical detection of cholesterol. Pitfalls in the application of filipin and saponins to localize cholesterol in membranes and protein-rich domains by ultrastructural methods are discussed.

In Chapter 17, Biessen *et al.* review recent efforts made for pharmacological interference in the atherogenic process. A disadvantage of the statins is that by interfering with mevalonate synthesis at the step catalyzed by HMG-CoA reductase, they also affect processes that require the isoprene intermediates produced in the cholesterol pathway. The authors review studies with inhibitors that block later stages of the isoprenoid biosynthetic pathway, such as inhibitors of squalene synthase, 2,3-oxidosqualene:lanosterol cyclase, and ACAT. Another strategy to lower serum cholesterol/triglyceride levels is reviewed; this involves drugs that stimulate the catabolism of atherogenic lipoproteins, such as ansamycins and galactose-terminated glycolipids. The use of antioxidants in the prevention of oxidation of low-density lipoproteins, a key initial event in the atherogenesis process, is also reviewed.

Each chapter contains many citations to the very recent literature, reflecting the activity in the rapidly advancing field of the biological and health-related functions of cholesterol. Although the scope of the research covered here is very broad, many important topics could not be covered in this volume because of space limitations. Nevertheless, it is hoped that the chapters presented here will challenge us to increase our understanding of the diverse functions of cholesterol in biology and medicine.

Robert Bittman

Contents

Chapter 2

Coordinate Regulation of Cholesterol 7α-Hydroxylase and HMG-CoA Reductase in the Liver

Ingemar Björkhem, Eric Lund, and Mats Rudling

Chapter 3

Polyprenyl Diphosphate Synthases

Kyozo Ogura, Tanetoshi Koyama, and Hiroshi Sagami

Chapter 5

The Smith–Lemli–Opitz Syndrome: A Potentially Fatal Birth Defect Caused by a Block in the Last Enzymatic Step in Cholesterol Biosynthesis

G. S. Tint, Ashok K. Batta, Guorong Xu, Sarah Shefer, Akira Honda, Mira Irons, Ellen Roy Elias, and Gerald Salen

Chapter 6

Has Nature Designed the Cholesterol Side Chain for Optimal Interaction with Phospholipids?

Robert Bittman

Chapter 7

Cholesterol and Myelin

G. Jackson Snipes and Ueli Suter

Chapter 8

Regulation of Mitochondrial Cholesterol Metabolism

Federico Martínez and Jerome F. Strauss III

Chapter 9

Lipoproteins and Cellular Cholesterol Homeostasis

William J. Johnson, Michael C. Phillips, and George H. Rothblat

Chapter 14

Biological Implications of the Niemann–Pick C Mutation

Peter G. Pentchev, E. Joan Blanchette-Mackie, and Laura Liscum

Chapter 16

Cholesterol Cytochemistry in Cell Biology and Disease

Nicholas J. Severs

Chapter 17

Approaches for the Design of Novel Anti-Atherogenic Compounds

Erik A. L. Biessen, Leo A. J. M. Sliedregt, and Theo J. C. Van Berkel

Chapter 1

Signaling Molecules Derived from the Cholesterol Biosynthetic Pathway

Simon M. Jackson, Johan Ericsson, and Peter A. Edwards

1. INTRODUCTION

The isolation and characterization of many of the enzymes involved in the synthesis of cholesterol represented a major *tour de force* in the 1950s and 1960s (Mead *et al.*, 1986; Porter and Spurgeon, 1981). These early studies led to the realization that the cyclic cholesterol molecule was derived from a polyisoprenoid, which in turn was shown to be composed of repeating five-carbon branched-chain building blocks derived from isopentenyl diphosphate (Rilling, 1985). The condensation of one or more isopentenyl diphosphate molecules with an allylic isoprenoid diphosphate (e.g., dimethylallyl diphosphate, geranyl diphosphate) is catalyzed by a family of related prenyl transferases that differ in their specificity for the allylic substrate (Edwards *et al.*, 1992; Chen *et al.*, 1994). The resulting linear isoprenoids are intermediates in the synthesis of a number of important compounds, including sterols, dolichols, heme a, steroid hormones, 1,25-dihydroxy vitamin D, bile acids, and geranylgeranyl diphosphate (Figure 1). Isoprenoid moieties, such as farnesyl and geranylgeranyl, are also post-translationally attached to certain proteins to produce prenylated proteins (Figure 1). Such modifications

Simon M. Jackson, Johan Ericsson, and Peter A. Edwards Departments of Biological Chemistry and Medicine and the Molecular Biology Institute, University of California at Los Angeles, Los Angeles, California 90095.

Subcellular Biochemistry, Volume 28: Cholesterol: Its Functions and Metabolism in Biology and Medicine, edited by Robert Bittman. Plenum Press, New York, 1997.

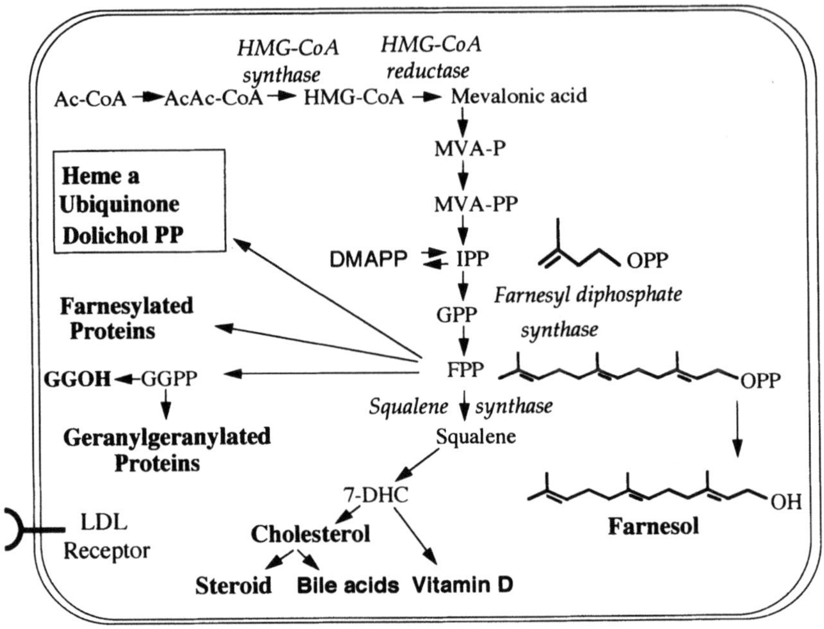

FIGURE 1. Isoprenoid synthesis and metabolism. The major products derived from isoprenoid intermediates of the cholesterol biosynthetic pathway are shown in bold. Four of the enzymes discussed in the text are shown in italics. The LDL receptor is also indicated because of its importance in the uptake of exogenous cholesterol and cellular cholesterol homeostasis. IPP, isopentenyl diphosphate; DMAPP, dimethylallyl diphosphate; Ac-CoA, acetyl coenzyme A; GGPP, geranylgeranyl diphosphate; GGOH, geranylgeraniol; 7-DHC, 7-dehydrocholesterol.

affect the localization and function of the lipidated proteins. Recent studies indicate that isoprenoid alcohols or acids may also function as novel signaling molecules.

In this chapter we propose to highlight recent studies demonstrating that isoprenoids and sterols function as signaling molecules that affect transcription, enzyme activity, and the ability of mammalian cells to respond to external stimuli. The vast literature describing the important role of thousands of isoprenoids in either insects (juvenile hormone and ecdysteroids) or plants (gibberellic acid, abscissic acid, trisporic acids, rubber, and carotenoids—which function both as plant pigments and as precursors of vitamin A and retinoic acid in mammals; Porter and Spurgeon, 1981) will be addressed briefly, if at all.

2. FARNESOIDS AND STEROLS CONTROL THE DEGRADATION OF HMG-CoA REDUCTASE

HMG-CoA reductase is a membrane-bound enzyme that was originally believed to be localized to the endoplasmic reticulum (Sabine, 1983). However, studies utilizing immunoelectron microscopy, immunofluorescence, and cell fractionation, demonstrated that the protein was localized to both peroxisomes and the endoplasmic reticulum (Keller *et al.,* 1986). These same techniques have been used to demonstrate that a number of other enzymes in the cholesterol biosynthetic pathway, including mevalonate kinase and farnesyl diphosphate synthase (Figure 1), are also localized to peroxisomes (Biardi *et al.,* 1994; Krisans *et al.,* 1994). The physiological function of the dual localization of some of these enzymes is unknown.

The initial analysis of the mammalian HMG-CoA reductase cDNA indicated that the amino acid sequence of the protein contained at least seven hydrophobic domains, which were proposed to anchor the protein in the endoplasmic reticulum (Liscum *et al.,* 1985). In contrast, the peroxisomal HMG-CoA reductase is presumed to be soluble, since it is localized to the matrix (Keller *et al.,* 1986). It is not known whether peroxisomal HMG-CoA reductase is the product of a second gene or whether it results from differential splicing or processing of a single gene. However, it seems unlikely that the soluble peroxisomal protein contains the same multiple transmembrane domains that are present in the endoplasmic reticulum-localized enzyme. Thus, the homology between the two enzymes, localized to either peroxisomes or the endoplasmic reticulum, will have to await the purification and sequence analysis of the peroxisomal protein.

In order to specifically determine the topology of the protein in the endoplasmic reticulum, Simoni and colleagues utilized antibodies directed at specific protein sequences that link the putative transmembrane domains and demonstrated that HMG-CoA reductase had eight transmembrane domains (Figure 2) (Roitelman *et al.,* 1992). It was proposed that both the amino terminus and the carboxy terminus, which contains the catalytically active site of the enzyme, must project into the cytoplasm of the cell.

The amount of HMG-CoA reductase protein in mammalian cells is highly regulated as a result of changes in a number of parameters, including altered rates of transcription of the gene, changes in both mRNA translation and stability, and in altered rates of protein degradation (Edwards and Davis, 1996). Together, these mechanisms can result in changes from 200- to 500-fold in the amount of HMG-CoA reductase protein in the livers of rats fed either hypolipidemic or hyperlipidemic diets. Insight into the factors involved in the regulated degradation of the protein originally came from studies in which cells were incubated in the presence of either excess sterols or in the presence of potent inhibitors of HMG-CoA reductase activity. These inhibitors, such as compactin and mevinolin, are structural analogues of HMG-CoA, the natural substrate of the enzyme. They are soluble and enter the cell and subsequently bind to and inactivate the enzyme (Alberts *et al.,* 1980). Indeed, the K_i for many of these competitive inhibitors approximates 10^{-9} M, as compared to the K_m of 10^{-6} M for HMG-CoA. Thus, the affinity of the enzyme for these inhibitors is 1000-fold greater than for the natural substrate. Cells incubated in the presence of these inhibitors have decreased

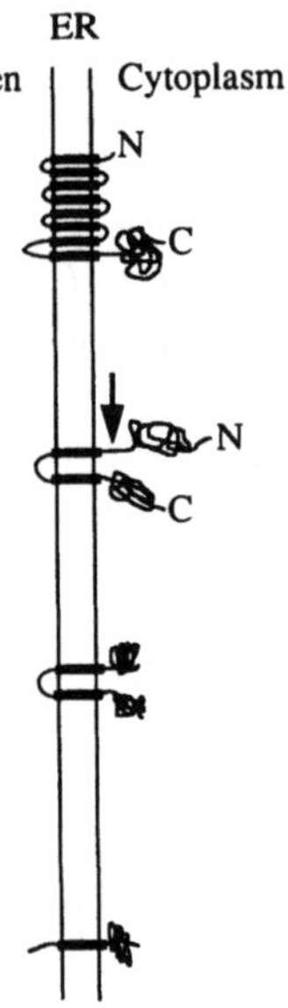

Protein	Status of the cell: Sterol-loaded	Status of the cell: Sterol-depleted
HMG-CoA-reductase	Increased protein degradation	Decreased protein degradation
SREBP	Decreased proteolytic cleavage	Increased proteolytic cleavage
ACAT	Increased enzyme activity	Decreased enzyme activity
Cholesterol 7α-hydroxylase	Increased enzyme activity	Decreased enzyme activity

FIGURE 2. The effect of alterations in cellular sterol levels on four proteins in the endoplasmic reticulum. The indicated topology of HMG-CoA reductase and sterol regulatory element binding protein (SREBP) has been reported. The topology of acyl-CoA:acyl cholesterol transferase (ACAT) and cholesterol 7α-hydroxylase is based on the published sequences of these proteins. N and C represent the amino and carboxyl terminis of the proteins, respectively. The arrow indicates the site at which SREBP is proteolytically cleaved to release the amino terminal, transcriptionally active, 68-kDa fragment.

rates of cholesterol synthesis as a result of inhibition of HMG-CoA reductase activity. Surprisingly, addition of these inhibitors, either to cells in culture or to animal diets, results in an increased mass of HMG-CoA reductase protein. In part, this increase in HMG-CoA reductase protein results from the increased stability of the HMG-CoA reductase protein; the half-life of the enzyme approximates 11 h in cells incubated in the presence of mevinolin, as compared to 4 h in cells incubated in the absence of the inhibitor (Edwards *et al.,* 1983; Goldstein and Brown, 1990). It was proposed that inhibition of HMG-CoA reductase activity deprived the cells of one or more compounds that were both normally derived from mevalonic acid (the product of the reaction catalyzed by HMG-CoA reductase) (Figure 1) and were required for normal degradation of the enzyme.

In contrast to the results with mevinolin, addition of either oxysterols or high levels of mevalonic acid to cells results in enhanced degradation of the HMG-CoA reductase protein; under these latter conditions the half-life can be as short as 45 min (Edwards *et al.,* 1983; Goldstein and Brown, 1990). These changes in protein stability are dependent on the multiple transmembrane domains of the enzyme since deletion of these sequences results in the synthesis of a truncated, soluble, catalytically active enzyme that is extremely stable (Gil *et al.*, 1985). In contrast, a chimeric protein containing the eight transmembrane domains of HMG-CoA reductase fused to β-galactosidase was posttranslationally regulated by sterols in a manner indistinguishable from that of the full length endogenous HMG-CoA reductase (Skalnik *et al.*, 1988). Taken together, these studies demonstrate that the eight transmembrane domains are required for the regulated degradation of the enzyme.

The sea urchin HMG-CoA reductase protein also contains multiple transmembrane domains that have high sequence homology with the hamster enzyme (Woodward *et al.*, 1988). However, the stability of the sea urchin enzyme in intact cells is unaffected by sterols, compactin, or mevinolin. Recently, Kumagai *et al.* (1995) made hybrid genes that encoded different combinations of hamster and sea urchin HMG-CoA reductase transmembrane domains. Plasmids encoding these hybrid proteins were transfected into mammalian cells. The rate of degradation of the hybrid HMG-CoA reductase proteins were then determined in cells incubated in the presence or absence of sterols. These studies demonstrated that the second membrane-spanning domain of the hamster enzyme is specifically required for the regulated degradation of the protein.

The identification of the signal(s) that stimulates degradation of HMG-CoA reductase has been elusive. Addition of mevalonic acid to either the diet of rats or to cells in culture resulted in decreased enzyme activity and increased degradation of HMG-CoA reductase protein (Edwards *et al.*, 1977, 1983). Nakanishi *et al.* (1988) proposed that both a sterol and non-sterol derivative of mevalonic acid (Figure 1) were required for the rapid degradation of the enzyme. Subsequent studies assessing the effect of inhibitors of specific enzymes of cholesterol biosynthesis on enzyme degradation (Correll and Edwards, 1994; Giron *et al.*, 1994) led to the proposal that the accelerated degradation of HMG-CoA reductase requires both a sterol (possibly an oxysterol) and a derivative of farnesyl diphosphate. Consistent with this proposal, Bradfute and Simoni (1994) demonstrated that the addition of derivitives of farnesyl diphosphate (farnesyl acetate or ethyl farnesyl ether) to cells also resulted in an increased rate of degradation of the enzyme.

More recently, Correll *et al.* (1994) demonstrated that the addition of farnesol, the dephosphorylated form of farnesyl diphosphate, to permeabilized cells resulted in a four-fold increase in the rate of degradation of the HMG-CoA reductase protein. Since other isoprenoid alcohols were ineffective in the assay, it was proposed that farnesol is a key intermediate normally required for the rapid destruction of the protein. However, at this time there is no information on the mechanism by which farnesol might stimulate the degradation of the protein. It is possible that the interaction of farnesol and/or a sterol with HMG-CoA reductase alters the protein conformation and results in the exposure of an epitope that is critical for degradation. Alternatively, farnesol, together with excess sterol, may accumulate in the endoplasmic reticulum and result in an altered conformation of the enzyme and exposure of a critical epitope that signals degradation of HMG-CoA reductase. In either case, HMG-CoA reductase would function as a sensor of cellular sterol levels; in the presence of high levels of sterols and farnesol in the endoplasmic reticulum, the enzyme would be degraded and the synthesis of sterols and isoprenoids would decrease until normal levels of cellular sterols were attained. Low levels of cellular sterols and farnesol would result in increased stability of HMG-CoA reductase and, hence, increased synthesis of sterols.

Interestingly, a microsomal phosphatase specific for farnesyl diphosphate has recently been described (Bansal and Vaidya, 1994). It is not known if this phosphatase is also regulated by the sterol status of the cell, thus regulating the production of farnesol. Farnesol has always been considered to be a dead-end product whose fate was excretion from the cell or body. However, recent studies have demonstrated that addition of radioactive farnesol to cells results in the appearance of radioactive farnesylated proteins, as well as other products of farnesyl diphosphate metabolism (Crick *et al.*, 1995; Fliesler and Keller, 1995). The farnesyl:protein transferase that transfers farnesyl to specific proteins is known to require farnesyl diphosphate as its substrate. Farnesol is not a substrate in this reaction (Casey, 1992; Brown and Goldstein, 1993). Thus, the incorporation of radioactive farnesol into proteins could easily be explained if the cells contain a kinase that phosphorylates the added isoprenoid alcohol. Such a kinase has not been identified in mammalian cells but was recently reported to be present in *Botryococcus braunii* (Inoue *et al.*, 1994).

These results implicate farnesol as a new and critical signaling molecule that is required, with an unidentified sterol, for the regulated degradation of HMG-CoA reductase. The purification of new enzymes that can either dephosphorylate farnesyl diphosphate or phosphorylate the isoprenoid alcohol would help to establish these proposed new pathways of isoprenoid metabolism.

The current evidence suggests that degradation of HMG-CoA reductase occurs in the endoplasmic reticulum. This conclusion is based upon the finding that the degradation of the protein was unaffected when cells were incubated with Brefeldin A, a drug that blocks the movement of proteins out of the endoplasmic reticulum (Chun *et al.*, 1990). In addition, the enzyme is degraded in permeabilized cells at rates similar to those seen in intact cells (Meigs and Simoni, 1992; Giron *et al.*, 1994) and by a process that is stimulated by farnesol (Correll *et al.*, 1994). The absence of cytosol and GTP-and ATP-generating systems in these permeabilized cells also precludes the transport of vesicles containing HMG-CoA reductase from the endoplasmic reticulum to an alternative membrane. Thus, the regulated degradation of HMG-CoA reductase occurs in a membrane compartment, the endoplasmic reticulum, that is involved in the degradation of a number of other proteins. These latter proteins include the alpha subunit of the T-cell receptor, apoprotein B, and abnormal proteins (Klausner and Sitia, 1990; Inoue and Simoni, 1992; Du *et al.*, 1994; Bonifacino and Lippincott-

Schwartz, 1991). This degradative pathway in the endoplasmic reticulum is distinct from the pathways characterized in proteosomes involving ubiquination or in lysosomes.

The degradation of HMG-CoA reductase is also regulated in the yeast *S. cerevisiae,* consistent with this being a highly conserved process (Hampton and Rine, 1994). The degradation of the yeast enzyme (HMG1p) is blocked when cells are incubated with mevinolin or other inhibitors of yeast sterol (ergosterol) biosynthesis, in a manner similar to that observed with mammalian cells. However, the impermeable nature of the yeast cell wall precludes studies in which sterols, mevalonic acid, or other potentially active compounds are added to the cells.

3. PROTEIN PRENYLATION

Protein prenylation is the term given to the covalent attachment of farnesyl or geranylgeranyl, the 15- or 20-carbon isoprenoids, respectively, to a cysteine at or near the carboxy terminal of specific proteins (Casey, 1992, 1995; Glomset and Farnsworth, 1994; Maltese, 1994). The thioether bond is maintained throughout the life of the protein and thus differs from the reversible attachment of palmitate or myristate to specific amino acids on other lipidated proteins. Protein modification by geranylgeranyl is far more common than by farnesyl. Three distinct enzymes, farnesyl transferase, geranylgeranyl transferase I, and geranylgeranyl transferase II have been identified that transfer the 15-carbon or 20-carbon isoprenoids to recipient proteins (Brown and Goldstein, 1993). Farnesyl transferase transfers farnesyl to proteins that terminate in CAAX, where C is a cysteine, A is an aliphatic amino acid, and X is methionine, serine, or alanine. Geranylgeranyl transferase I transfers geranylgeranyl to proteins that also terminate in CAAX, but where X is leucine or phenylalanine. Both of these enzymes can prenylate peptides that are as short as four amino acids, implying that the tetrapeptide is all that is required for enzyme recognition. Geranylgeranyl transferase II, or Rab prenyl transferase, transfers one or two geranylgeranyl moieties to proteins that terminate in XCXC or XXCC, where X can be any amino acid. However, other amino acid sequences, as yet unidentified, are involved in the recognition of the target protein by this enzyme.

The initial attachment of farnesyl or geranylgeranyl to the cysteine of proteins that terminate in CAAX is followed by a series of modifications that include proteolytic removal of the AAX and carboxymethylation of the resulting carboxy-terminal cysteine (Figure 3). The number of proteins that are known to be modified by prenylation has grown rapidly since the discovery that the product of the oncogene *ras* was modified by farnesylation. These include the nuclear prelamin A and lamin B; the γ subunit of trimeric G proteins; many small G proteins, including members of the rab, rac, and rho families; rhodopsin kinase; and fungal

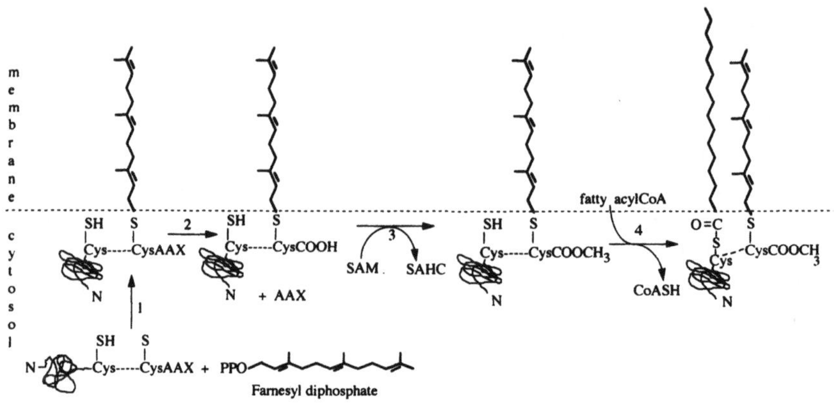

FIGURE 3. Farnesylation and palmitoylation of ras. Enzymes involved in this series of reactions include: farnesyl transferase (1); the protease that removes the three carboxy terminal amino acids, AAX (2); the methyl transferase, which requires *S*-adenosylmethionine (SAM) to methylate the C-terminal cysteine, with the concomitant release of *S*-adenosyl homocysteine (SAHC) (3); palmitoyl transferase (4) is involved in the modification of an upstream cysteine of some, but not all, ras proteins. The first reaction is cytosolic, but all subsequent reactions are associated with membrane fractions. The association of the lipidated protein with membranes is illustrated.

mating pheromones such as α-factor (Casey, 1992, 1995; Glomset and Farnsworth, 1994; Maltese, 1994; Wedegaertner *et al.,* 1995). These prenylated proteins are involved in many diverse cellular functions.

Figure 1 shows that the early part of the cholesterol biosynthetic pathway must be intact for the synthesis of both farnesyl diphosphate and geranylgeranyl diphosphate. Thus, it is not surprising that addition of either compactin or mevinolin (or other potent inhibitors of either HMG-CoA reductase or early enzymatic reactions in the cholesterol biosynthetic pathway) to cells results in the inhibition of protein prenylation and, after one cell division, cell death. Under such conditions, prenylated proteins that are normally membrane bound remain soluble and unmodified. Of particular interest was the early observation that the oncogenic ras protein lost its transforming characteristics when prenylation was blocked by either treatment of cells with inhibitors of HMG-CoA reductase or by mutation of the carboxy-terminal cysteine of ras; under either condition, ras was recovered as a soluble protein and it was no longer associated with the membrane fraction of the cell (Hancock *et al.,* 1989; Schafer *et al.,* 1989; Casey *et al.,* 1989).

Mutagenesis studies that involved altering different carboxy terminal amino acids of ras demonstrated that the anchoring of ras to membranes required prenylation of the carboxy-terminal cysteine and either palmitoylation of an upstream cysteine or the presence of an upstream polybasic region (Hancock *et al.,* 1990). Together, these different regions and/or lipid modifications are thought to enhance the association of the ras protein with the correct membrane. Since activated ras is found in a large number of human cancers, a concerted effort has been made to identify specific inhibitors of either the farnesyl transferase enzyme or of ras farnesylation *per se* (Gibbs *et al.,* 1994). Such a drug might be expected to be beneficial in the treatment of cancers that require activated ras for unregulated cell growth and division.

It is not clear whether the lipid prenyl group attached to the carboxy-terminal cysteine of specific proteins has more than one function. One function certainly involves targeting the prenylated protein to a specific cellular compartment. However, different prenylated proteins may be localized to the plasma membrane, endoplasmic reticulum, Golgi membrane, or vesicles, while some appear to be soluble and not membrane-associated. The mechanism involved in targeting different prenylated proteins to specific cellular compartments is not known. Such differential targeting might require the presence of specific receptors in different membranes. Consistent with this hypothesis, Thissen and Casey (1993) identified a high affinity binding site in microsomal membranes that was recognized by prenylated proteins. However, certain prenylated proteins, such as rab proteins, also cycle between membranes and the cytoplasm as a result of specific interactions with soluble proteins (Glomset and Farnsworth, 1994; Pfeiffer *et al.,* 1995). Guanidine nucleotide dissociation inhibitor (GDI) has been identified as one such soluble protein (Ullrich *et al.,* 1994; Pfeiffer *et al.,* 1995). GDI appears to act as a chaperone, and sequesters the hydrophobic prenyl group from the aqueous environment during the translocation of rab proteins between different membrane compartments (Figure 4).

In summary, farnesyl and geranylgeranyl diphosphates are now known to play critical roles in many cellular functions as a result of their attachment to a diverse group of proteins involved in cell signaling, intracellular vesicle movement, and targeting.

4. CHOLESTEROL AND OXYSTEROLS

Many years ago it was established that cholesterol is the most abundant sterol in mammalian cells. Many studies have demonstrated that cholesterol associates with membrane phospholipids, resulting in increased packing of the acyl chains, reduced mobility of membrane lipids, and reduced lateral diffusion of some membrane proteins. It is now realized that there is an unequal distribution and concentration of cholesterol within the different membranes of the mammalian cell. Cholesterol concentration within the endoplasmic reticulum, the site of synthesis of the cholesterol molecule *per se,* and Golgi membranes is extremely low. In contrast, the plasma membrane contains relatively high concentrations of this sterol. There are domains within the plasma membrane, called caveolae, that appear to be particularly rich in sterols. These differences in cholesterol concentration in different membranes are important. As discussed below, changes in the sterol

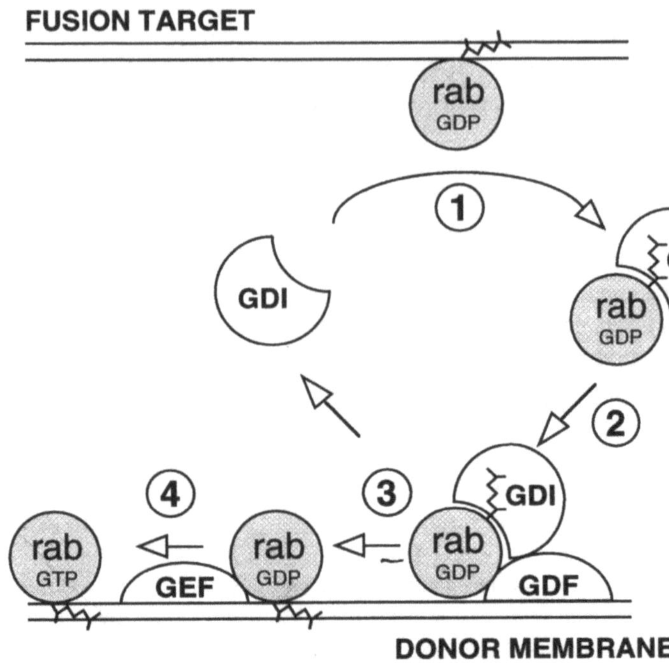

FIGURE 4. A model to explain the transfer of prenylated proteins between different membranes. Geranylgeranylated rab in the GTP-bound form is involved in targeting membrane vesicles from the donor membrane to the fusion target or acceptor membrane. Following fusion of the vesicles with the membrane, GDP-bound rab is released to the fusion target membrane. Soluble GDP dissociation inhibitor (GDI) then displaces the rab-GDP complex from the membrane to form a soluble GDI-rab-GDP complex (step 1). A rab-specific GDI displacement factor (GDF) present on the donor membranes is thought to recognize the soluble rab-GDI complex (step 2). Subsequently, the rab-GDP associates with the membrane and the GDI is released into the cytosol (step 3). Guanine exchange factor (GEF) then catalyzes the exchange of GTP for the GDP bound to rab (step 4). The GTP-bound rab can then be involved in another round of vesicle movement from the donor to acceptor membranes. (Figure courtesy of Dr. Suzanne Pfeffer.)

concentration within these membranes can have significant effects on the activities of specific enzymes and on membrane function.

4.1. Proteolytic Cleavage of Sterol Regulatory Element Binding Protein (SREBP); the Role of Sterols

Cellular cholesterol homeostasis is maintained as a result of the coordinate regulation of both the cholesterol biosynthetic pathway and the LDL receptor pathway (Figure 1). Many studies have demonstrated that the transcription of HMG-CoA synthase, HMG-CoA reductase, farnesyl diphosphate synthase, squalene synthase, and the LDL receptor genes decrease when cells are exposed to high levels of oxysterols and cholesterol. The transcription of these same genes increases when cells are deprived of cholesterol, following incubation of cells in media depleted of sterols, or in the presence of competitive inhibitors of HMG-CoA reductase (Goldstein and Brown, 1990; Spear *et al.,* 1994; Guan *et al.,* 1995; Edwards and Davis, 1996).

Goldstein and Brown and colleagues identified a 10 base-pair element in the promoters of the HMG-CoA synthase and the LDL receptor genes that is required for the sterol-regulated transcription. This element was termed sterol regulatory element 1 (SRE-1). Subsequently, a protein of 68 kDa was identified (SREBP) that bound to the SRE-1 in a specific manner and resulted in increased transcription of reporter gene constructs containing promoters derived from the LDL receptor or HMG-CoA synthase genes (Wang *et al.,* 1993; Yokoyama *et al.,* 1993). There are two human SREBP genes (SREBP-1 and SREBP-2) that encode, as a result of differential splicing, 3 distinct SREBPs. The rat homologue to SREBP-1 was independently cloned as a result of its binding to an E-box motif and termed ADD-1, due to its potential role as an *a*dipocyte *d*etermination and *d*ifferentiation factor (Tontonoz *et al.,* 1993).

SREBP-1 was shown to be synthesized as a 125-kDa precursor that was localized to the endoplasmic reticulum. This protein contains a basic helix-loop-helix-leucine zipper domain in its amino terminus and is anchored to the endoplasmic reticulum as a result of two hydrophobic domains that are located toward the carboxy terminal (Figure 2). A specific proteolytic cleavage between the transmembrane domains and the basic helix-loop-helix-leucine zipper domain results in the release of a mature, transcriptionally active, 68-kDa protein fragment (Wang *et al.,* 1994). This fragment is targeted to the nucleus, where it binds to SRE-1 and stimulates transcription of the LDL receptor and HMG-CoA synthase genes. Recent studies have shown that SREBP also regulates the transcription of the farnesyl diphosphate synthase gene following binding to a novel element termed SRE-3 (Ericsson *et al.,* 1995). Recent studies that utilize reporter genes under the control of promoter fragments derived from either the LDL receptor or farnesyl diphosphate synthase genes, demonstrate that SREBP is incapable of stimulating transcription in the absence of a second factor; increased transcription was shown to require SREBP and either Sp1 or NF-Y transcription factors (Sanchez *et al.,* 1995; Jackson *et al.,* 1995).

The SREBP-mediated enhanced transcription of genes requires that the 68-kDa mature SREBP be cleaved from the 125-kDa precursor and that it enter the nucleus and bind one of three specific *cis* elements (SRE-1, SRE-3, or an E-box motif) (Wang *et al.,* 1993; Tontonoz *et al.,* 1993; Ericsson *et al.,* 1995). These latter elements have limited identity (Table I). The cleavage of the precursor 125-kDa SREBP is regulated by the sterol status of the cell; the precursor protein is not cleaved in cells that are sterol-loaded and under these conditions

Table I
Sterol Regulatory Element Binding Protein Binds to Three Different DNA Motifs

Motif	Nucleotide sequence	Gene
SRE-1	ATCACCCCAC	LDL receptor
SRE-3	CTCACACGAG	FPP synthase
E-box	ATCACGTGA	Fatty acid synthase
Consensus	nTCACnnnAn	

transcription of genes involved in cholesterol synthesis or the endocytosis of LDL is low. In contrast, the cleavage and release of the 68-kDa fragment of SREBP is enhanced in cells that are deprived of sterols (Wang *et al.*, 1994). At this time it is not known how sterols regulate the proteolytic cleavage of SREBP. Indeed, it is not known whether the active sterol is cholesterol or an oxysterol. Excess sterols or oxysterols in the endoplasmic reticulum may activate a membrane-bound protease, result in the translocation of a protease from a different cellular localization to the endoplasmic reticulum, or alter the conformation of SREBP or the membrane and result in the exposure of a critical cleavage domain. It is intriguing that excess cellular sterols prevent the proteolytic cleavage of SREBP but result in the enhanced degradation of HMG-CoA reductase (Figure 2). Together, these two proteolytic events, which both occur in the endoplasmic reticulum, play important roles in the maintenance of cellular cholesterol homeostasis.

In summary, the transcription of a number of important genes is regulated by the nuclear localization of the mature 68-kDa SREBP. The release of this unique transcription factor from the endoplasmic reticulum is under the control of a sterol-regulated proteolytic cleavage of the precursor 125-kDa SREBP. The mechanism by which sterols regulate this cleavage event is likely to add one more interesting twist to the picture of cellular cholesterol homeostasis that has developed over the last 30 years.

4.2. Cholesterol Activation of Enzyme Activity

The endoplasmic reticulum contains at least two other enzymes (acyl-CoA:cholesterol acyl transferase (ACAT) and cholesterol 7α-hydroxylase) that are both involved in cholesterol metabolism and are regulated by the sterol content of the membrane (Figures 1 and 2). ACAT is found in all mammalian cells and transfers a fatty acid (usually oleic acid) to the 3β-hydroxyl group of cholesterol. Cholesterol 7α-hydroxylase is expressed only in the liver, consistent with its role as the rate-limiting enzyme involved in the biosynthesis of bile acids from cholesterol (see Chapter 2 by Björkhem *et al.*). The substrate for both enzymes is cholesterol. The activity of both enzymes is enhanced when the amount of cholesterol within the endoplasmic reticulum increases, consistent with substrate activation of both enzymes (Figure 2). Substrate activation appears to be the major mechanism involved in the regulation of ACAT (Cheng *et al.,* 1995). However, excess hepatic cellular sterols not only activate 7α-hydroxylase activity but also result in increased cholesterol 7α-hydroxylase mRNA levels (Noshiro *et al.,* 1990; Jelinek *et al.,* 1990), presumably via increased transcription of the gene. These studies indicate that the cholesterol concentration within the endoplasmic reticulum, or within domains in this membrane, plays an important role in the regulation of the activity of at least two enzymes that metabolize this sterol.

4.3. 27-Hydroxycholesterol

It is not known whether cholesterol *per se* or an oxygenated sterol is involved in the regulation of various cellular events, including cleavage of SREBP. Earlier studies by Kandutsch *et al.* (1978) demonstrated that addition of oxysterols, but not cholesterol, to cells resulted in dramatic decreases in the activity of HMG-CoA reductase. The high potency of exogenously added oxysterols, as compared to cholesterol, may be the result of a specific requirement for the former or it may result from the enhanced rate of entry of oxysterols into the cell as a result of their increased solubility.

Recent studies have demonstrated that cells oxidize LDL-cholesterol that enters cells via the LDL receptor to 27-hydroxycholesterol (Axelson and Larsson, 1995). 27-Hydroxycholesterol is reported to be synthesized in sufficient quantities to downregulate HMG-CoA reductase. This decrease in HMG-CoA reductase activity was prevented when cells deficient in cholesterol 27-hydroxylase, as a result of a mutation in the corresponding gene, were incubated with LDL (Axelson and Larsson, 1995). These results indicate that intracellular oxidation of exogenously added cholesterol may be important in the generation of regulatory molecules of increased potency. However, patients with *cerebrotendinous xanthomatosis,* who are defective in the cholesterol 27-hydroxylase enzyme (Cali *et al.,* 1991; see Chapter 5 by Tint *et al.*), have not been reported to exibit defective regulation of cholesterol biosynthesis. Other investigators have proposed that 27-hydroxycholesterol has a role in the elimination of cholesterol from macrophages as part of the reverse cholesterol transport process from peripheral tissue back to the liver (Björkhem *et al.,* 1994; see Chapter 2 by Björkhem *et al.*). Clearly, the exact role of either 27-hydroxycholesterol or other oxysterols in various regulatory events within the cell has not yet been definitively established.

4.4. Activation of Oocyte Meiosis by Sterols

A meiosis-inducing, diffusible compound has been detected during meiosis in germ cells of cultured fetal mouse testes (Josso and Picard, 1986). Recently,

Byskov *et al.* (1995) isolated two closely related C_{29} sterols from human follicular fluid of women and adult bull testis that both induce meiosis in naked mouse oocytes. These preliminary studies are consistent with a novel role for sterols in oocyte meiosis of adult male germ cells.

5. APOPTOSIS AND SREBP

The precursor 125-kDa SREBP was recently shown to be cleaved by a soluble protease termed CPP-32 (Wang *et al.*, 1995). This protease has homology to interleukin cleavage enzyme (ICE/CED-3) and is thought to have a critical role in apoptosis (Tewari *et al.*, 1995; Nicholson *et al.*, 1995; Whyte and Evan, 1995). The cleavage of the 125-kDa precursor SREBP by CPP-32 was independent of sterols and occurred at a site very close to, but distinct from, the sterol-regulated cleavage site (Wang *et al.*, 1995). Further studies will be required to determine whether the cleavage of SREBP and subsequent release of a transcriptionally-active mature protein is physiologically important in apoptosis.

5.1. Apoptosis; Possible Role of Farnesol and Geranylgeraniol

A number of studies have demonstrated that the addition of 20–50 μM farnesol (Haug *et al.*, 1994; Voziyan *et al.*, 1995) or geranylgeraniol (Ohizumi *et al.*, 1995) to cells in culture results in DNA fragmentation, consistent with apoptosis. Addition of farnesol to cells is also reported to inhibit the breakdown of phosphatidylcholine into diacylglycerol and phosphocholine (Voziyan *et al.*, 1995). It is not known whether these latter changes are directly related to apoptosis.

Inhibition of HMG-CoA reductase by mevinolin is also reported to result in apoptosis (Pérez-Sala and Mollinedo, 1994), even though such treatment inhibits the biosynthesis of farnesyl diphosphate, and, presumably, farnesol. Thus, the physiological relationship between apoptosis and intracellular isoprenoid alcohol levels remains to be clearly established.

6. ISOPRENOIDS AS TRANSCRIPTIONAL REGULATORS

Ecdysone is an insect steroid hormone that controls important regulatory events leading to moulting and metamorphosis. Ecdysone binds to the ecdysone receptor, a member of the nuclear hormone receptor superfamily (Mangelsdorf *et al.*, 1994). The ecdysone receptor dimerizes with ultraspiracle, the insect homolog of the mammalian retinoid X receptor (RXR), to form a transcriptionally-active heterodimer receptor complex (Yao *et al.*, 1992). Subsequent activation of target genes results in moulting.

Recently, Forman *et al.* (1995) identified a new mammalian member of the nuclear hormone receptor superfamily, termed the farnesoid X-activated receptor (FXR). Seol *et al.* (1995) independently isolated several isoforms of RIP14, the mouse homolog of the rat FXR. Sequence analysis of FXR/RIP14 and the more than 20 members of a subgroup of the superfamily that exhibit a conserved DNA-binding domain, indicated that FXR showed the highest homology with the insect ecdysone receptor. Both FXR and RIP14 were also shown to form functional heterodimeric complexes with RXR. In order to identify an active ligand/hormone, Forman *et al.* (1995) transfected cells with plasmids encoding FXR, RXR, and a reporter gene. The reporter gene was under the control of a synthetic promoter that contained multiple copies of the ecdysone responsive element (EcRE) and was responsive to a ligand-stimulated FXR/RXR complex. Addition of either farnesol or juvenile hormone III to these cells resulted in increased expression of the reporter gene. Juvenile hormone III is an insect hormone with structural similarities to farnesol (Forman *et al.*, 1995). Other isoprenoids were ineffective and did not increase the transcription of the target reporter gene.

These studies indicate that farnesol, when added to a cell, can act as a hormone and modulate transcription of synthetic reporter genes. It is not known whether this effect is of physiological importance. Further studies should define whether farnesol normally acts as an intracellular signaling molecule or whether it diffuses from cell to cell in order to alter transcription of target genes in an FXR/RXR dependent manner. The identification of genes responsive to farnesol or related isoprenoid molecules would represent an important step in defining this novel pathway.

A recent report demonstrated that addition of 4,5-didehydrogeranylgeranoic acid (4,5-ddGGA) to cells results in increased levels of retinoic acid receptor-β (RARβ) mRNA (Araki *et al.*, 1995). Surprisingly 4,5-ddGGA was as active as all- *trans* retinoic acid. In addition, the transcription of a promoter-reporter gene containing multiple copies of a retinoic acid responsive element was increased five-fold following the addition of either geranylgeranoic acid, 4,5-dd GGA, or all- *trans* retinoic acid to the cells (Araki *et al.*, 1995). All three of these activators of transcription also bound to cellular retinoic acid-binding proteins (CRABPs), soluble proteins having no known role in transcription. These studies demonstrate that acyclic isoprenoid acids can function as retinoic acid agonists. Earlier studies by Christophe and Popjak (1961) showed that geranyl diphosphate and farnesyl diphosphate could be dephosphorylated and then oxidized by cell homogenates to produce the corresponding acids. Demonstration that these acyclic acid derivatives of geranylgeranyl diphosphate are physiologically important in the regulation of genes under the control of RAR/RXR heterodimers could be of great importance. Such studies will hopefully determine whether these acyclic retinoids bind directly to RAR/RXR or, alternatively, whether they result in the release of retinoic acid from CRABP.

7. STEROID HORMONES AND VITAMIN D

Cholesterol is a required intermediate in the synthesis of steroid hormones, including cortisol, aldosterone, testosterone, estrogen, and progesterone. The synthesis and action of these hormones is beyond the scope of this review. The reader is referred to many excellent reviews that describe the normal physiology as well as the effect of genetic mutations that result in defective synthesis and action of these hormones (Donohoue *et al.*, 1995; Marx, 1995). A number of excellent reviews on the mechanism of action of these steroid hormones, their interactions with specific soluble receptors, and the formation of functional receptor dimers that alter transcription of target genes are available (Mangelsdorf *et al.*, 1994)

7-Dehydrocholesterol is an intermediate of cholesterol biosynthesis that, in response to UV light, is converted in the skin into vitamin D or cholecalciferol. The subsequent conversion to the active hormone, 1,25-dihydroxy-vitamin D, requires the action of enzymes located in the liver and either kidney or placenta. 1,25-Dihydroxy-vitamin D functions by binding to a specific nuclear receptor, a member of the nuclear hormone superfamily, and affecting the transcription of target genes that contain vitamin D responsive elements (Mangelsdorf *et al.*, 1994; Williams, 1994; Marx, 1995). The hormone plays a critical role in controlling calcium metabolism, and defects in either the receptor or synthesis of the active hormone have deliterious effects on many body functions as a result of abnormal calcium homeostasis.

8. CHOLESTEROL-RICH AND CHOLESTEROL-POOR MEMBRANE DOMAINS

The relative amount of cholesterol in specific membranes appears to play a role not only in regulating enzyme activity (see section 4) but also in cell signaling and/or protein localization. There is evidence that cholesterol and glycolipids self-associate within membranes to form microdomains (Thompson and Tillack, 1985) and that glycosyl phosphatidylinositol (GPI)-anchored proteins partition into these domains (Anderson, 1993; Parton and Simons, 1995).

Caveolae are small flask-shaped invaginations that are found in the plasma membrane of many mammalian cells (Anderson, 1993; Parton and Simons, 1995; see Chapter 16 by Severs). Caveolae were originally observed in the electron microscope and subsequently shown to be present at relatively high levels in endothelial cells, smooth muscle cells and adipocytes. Certain small molecules appear to be concentrated in caveolae and then internalized by a process termed potocytosis. Current evidence suggests that caveolae are enriched in cholesterol, glycosphingolipids, and GPI-anchored proteins, and that they also represent a novel signaling pathway by which extracellular signals are transduced into the cell (Anderson, 1993; Parton and Simons, 1995). This latter conclusion is based upon the observation that many molecules that are known to be involved in signal transduction, including src-family kinases, are localized to caveolae.

Cholesterol, enriched in caveolae, is thought to interact with VIP-21-caveolin (Murata *et al.*, 1995), a 21-kDa integral membrane protein that is palmitoylated and localized to the cytoplasmic surface of the caveolae. The importance of this cholesterol–VIP-21-caveolin interaction can be seen when cells are treated with inhibitors of HMG-CoA reductase in order to inhibit cholesterol synthesis and thus deplete cells of cholesterol. Under these conditions the number of caveolae is decreased and the structure of VIP-21-caveolin is altered (Rothberg *et al.*, 1992).

9. CONCLUSION

As discussed in this chapter, many molecules of the cholesterol biosynthetic pathway, including isoprenoids, sterols, and further metabolites of cholesterol, can function as signaling molecules. These molecules regulate diverse cell functions, ranging from meiosis to apoptosis. In light of these observations, we should expect to find a multifactorial and intricate mode of regulation of this important metabolic pathway. A major challenge is to define these mechanisms of regulation.

ACKNOWLEDGMENTS. The work was supported by Grant HL 30568 from the National Institutes of Health and a grant from the Laubisch Fund. J. E. is the recipient of an American Heart Association, Greater Los Angeles Affiliate post-doctoral fellowship. We thank Dr Suzanne Pfeffer, Stanford University, for providing the scheme shown in Figure 4.

10. REFERENCES

Alberts, A. W., Chen, J., Kuron, G., Hunt, V., Huff, J., Hoffman, C., Rothrock, J., Lopez, M., Joshua, H., Harris, E., Patchett, A., Monaghan, R., Currie, S., Stapley, E., Albers-Schonberg, G., Hensens, O., Hirshfield, J., Hoogsteen, K., Liesch, J., and Springer, J., 1980, A highly potent competitive inhibitor of hydroxymethylglutaryl-coenzyme A reductase and a cholesterol-lowering agent, *Proc. Natl. Acad. Sci. USA* **77:**3957–3961.

Anderson, R. G. W., 1993, Caveolae: Where incoming and outgoing messengers meet, *Proc. Natl. Acad. Sci. USA* **90:**10904–10913.

Araki, H., Shidoji, Y., Yamada, Y., Moriwaki, H., and Muto, Y., 1995, Retinoid agonist activities of synthetic geranyl geranoic acid derivatives, *Biochem. Biophys. Res. Commun.* **209:**66–72.

Axelson, M., and Larsson, O., 1995, Low density lipoprotein (LDL) cholesterol is converted to 27-hydroxycholesterol in human fibroblasts, *J. Biol. Chem.* **270:**15102–15110.

Bansal, V. S., and Vaidya, S., 1994, Characterization of two distinct allyl pyrophosphatase activities from rat liver microsomes, *Arch. Biochem. Biophys.* **315:**393–399.

Biardi, L., Sreedhar, A., Zokaei, A., Vartak, N. B., Bozeat, R. L, Shackelford, J. E., Keller, G. A., and Krisans, S. K., 1994, Mevalonate kinase is predominantly localized in peroxisomes and is defective in patients with peroxisome deficiency disorders, *J. Biol. Chem.* **269:**1197–1205.

Björkhem, I., Andersson, O., Diczfalusy, U., Sevastik, B., Xiu, R. J., Duan, C., and Lund, E., 1994, Atherosclerosis and sterol 27-hydroxylase: Evidence for a role of this enzyme in elimination of cholesterol from human macrophages, *Proc. Natl. Acad. Sci. USA* **91:**8592–8596.

Bonifacino, J. S., and Lippincott-Schwartz, J., 1991, *Curr. Opin. Cell Biol.* **3:**592–600.

Bradfute, D. L., and Simoni, R. D., 1994, Non-sterol compounds that regulate cholesterogenesis, *J. Biol. Chem.* **269:**6645–6650.

Brown, M. S., and Goldstein, J. L., 1993, Mad bet for rab, *Nature* **366:**14–15.

Byskov, A. G., Andersen, C. Y., Nordholm, L. N., Thogersen, H., Gouliang, X., Wassmann, O., Andersen, J. V., Guddal, E., and Roed, T., 1995, Chemical structure of sterols that activate oocyte meiosis, *Nature* **374:**559–562.

Cali, J. J., Hsieh, C.-L., Francke, U., and Russell, D. W., 1991, Mutations in the bile acid biosynthetic enzyme sterol 27-hydroxylase underlie *cerebrotendinous xanthomatosis, J. Biol. Chem.* **266:**7779–7783.

Casey, P. J., 1992, Biochemistry of protein prenylation, *J. Lipid Res.* **33:**1731–1740.

Casey, P. J., 1995, Protein lipidation in cell signaling, *Science* **268:**221–225.

Casey, P. J., Solski, P. A., Der, C. J., and Buss, J. E, 1989, p21ras is modified by a farnesyl isoprenoid. *Proc. Natl. Acad. Sci. USA* **86:**8323–8327.

Chen, A., Kroon, P. A., and Poulter, C. D., 1994, Isoprenyl diphosphate synthases: protein sequence comparisons, a phylogenetic tree, and predictions of secondary structure, *Prot. Sci.* **3:**600–607.

Cheng, D., Chang, C. C., Qu, X., and Chang, T. Y., 1995, Activation of acyl-coenzyme A:cholesterol acyltransferase by cholesterol or by oxysterol in a cell-free system, *J. Biol. Chem.* **270:**685–695.

Christophe, J., and Popjak, G., 1961, Studies in the biosynthesis of cholesterol:XIV. The origin of prenoic acids from allylpyrophosphates in liver enzyme systems, *J. Lipid Res.* **2:**244–257.

Chun, K. T., Bar-Nun, S., and Simoni, R. D., 1990, The regulated degradation of 3-hydroxy-3-methylglutaryl-CoA reductase requires a short-lived protein and occurs in the endoplasmic reticulum, *J. Biol. Chem.* **265:**22004–22010.

Correll, C. C., and Edwards, P. A., 1994, Mevalonic acid-dependent degradation of 3-hydroxy-3-methylglutaryl-coenzyme A reductase *in vivo* and *in vitro, J. Biol. Chem.* **269:**633–638.

Correll, C. C., Ng, L., and Edwards, P. A., 1994, Identification of farnesol as the non-sterol derivitive of mevalonic acid required for the accelerated degradation of 3-hydroxy-3-methylglutaryl-coenzyme A reductase, *J. Biol. Chem.* **269:**17390–17393.

Crick, D. C., Andres, D. A., and Waechter, C. J., 1995, Farnesol is utilized for protein isoprenylation and the biosynthesis of cholesterol in mammalian cells, *Biochem. Biophys. Res. Commun.* **211:** 590–599.

Donohoue, P. A., Parker, K., and Migeon, C. J., 1995, Congenital adrenal hyperplasia, in: *Metabolic Basis of Inherited Disease* (C. R. Scriver, A. L. Beaudet, W. S. Sly, and D. Valle, eds.) Vol. VII, pp. 2929–2966, McGraw-Hill, Inc., New York.

Du, E. Z., Kurth, J., Wang, S. L., Humiston, P., and Davis, R. A., 1994, Proteolysis-coupled secretion of the N terminus of apolipoprotein B. Characterization of a transient, translocation arrested intermediate, *J. Biol. Chem.* **269:**24169–24176.

Edwards, P. A., and Davis, R. A., 1996, Regulation of sterol biosynthesis, bile acid metabolism and isoprenylation of proteins, in: *Biochemistry of Lipids, Lipoproteins and Membranes,* 3rd ed. (D. E. Vance and J. E. Vance, eds.) Elsevier, Holland, pp. 341–362.

Edwards, P. A., Popják, G., Fogelman, A. M., and Edmond, J., 1977, Control of 3-hydroxy-3-methylglutaryl coenzyme A reductase by endogenously synthesized sterols *in vitro* and *in vivo, J. Biol. Chem.* **252:**1057–1063

Edwards, P. A., Lan, S.-F., and Fogelman, A. M., 1983, Alterations in the rates of synthesis and degradation of rat liver 3-hydroxy-3-methylglutaryl coenzyme A reductase produced by cholestyramine and mevinolin, *J. Biol. Chem.* **258:**10219–10222.

Edwards, P. A., Ashby, M. N., Spear, D. H., Marrero, P. F., Joly, A., and Popjak, G., 1992, Isoprenoids-biosynthesis and role in the regulation of protein function, *Biochem. Soc. Trans.* **20**:475–480.

Ericsson, J., Jackson, S. M., Lee, B. C., and Edwards, P. A., 1995, Sterol regulatory element binding protein binds to a cis element in the promoter of the farnesyl diphosphate synthase gene, *Proc. Natl. Acad. Sci. USA* **93**:945–950.

Fliesler, S. J., and Keller, K., 1995, Metabolism of [^{3}H] farnesol to cholesterol and cholesterogenic intermediates in the living rat eye, *Biochem. Biophys. Res. Commun.* **210**:695–702.

Forman, B. M., Goode, E., Chen, J., Oro, A. E., Bradley, D. J., Perlmann, T., Noonan, D. J., Burka, L. T., McMorris, T., Lamph, W. W., Evans, R. M., and Weinberger, C., 1995, Identification of a nuclear receptor that is activated by farnesol metabolites, *Cell* **81**:687–693.

Gibbs, J. B., Oliff, A., and Kohl, N. E. 1994. Farnesyltransferase inhibitors: ras research yields a potential cancer therapeutic, *Cell* **77**:175–178.

Gil, G., Faust, J. R., Chin, D. J., Goldstein, J. L., and Brown, M. S., 1985, Membrane-bound domain of HMG CoA reductase is required for sterol-enhanced degradation of the enzyme, *Cell* **41**: 249–258.

Giron, M. D., Havel, C. M., and Watson, J. A., 1994, Mevalonate-mediated suppression of 3-hydroxy-3-methylglutaryl coenzyme A reductase function in α-toxin-perforated cells, *Proc. Natl. Acad. Sci. USA* **91**:6398–6402.

Glomset, J. A., and Farnsworth, C. C., 1994, Role of protein modification reactions in programming interactions between ras-related GTPases and cell membranes. *Annu. Rev. Cell Biol.* **10**: 181–205.

Goldstein, J. L., and Brown, M. S., 1990, Regulation of the mevalonate pathway, *Nature* **343**:425–430.

Guan, G., Jiang, G., Koch, R. L., and Shechter, I., 1995, Molecular cloning and functional analysis of the promoter of the human squalene synthase gene, *J. Biol. Chem.* **270**:21958–21965.

Hampton, R. Y., and Rine, J., 1994, Regulated degradation of HMG-CoA reductase, an integral membrane protein of the endoplasmic reticulum, in yeast, *J. Cell Biol.* **125**:299–312.

Hancock, J. F., Magee, A. I., Childs, J. E., and Marshall, C. J., 1989, All ras proteins are poly-isoprenylated but only some are palmitoylated, *Cell* **57**:1167–1177.

Hancock, J. F., Paterson, H., and Marshall, C. J., 1990, A polybasic domain or palmitoylation is required in addition to the CAAX motif to localize p21ras to the plasma membrane, *Cell* **63**:133–139.

Haug, J. S., Goldner, C. M., Yazlovitskaya, E. M., Voziyan, P. A., and Melnykovych, G., 1994, Directed cell killing (apoptosis) in human lymphoblastoid cells incubated in the presence of farnesol: effect of phosphatidylcholine, *Biochim. Biophys. Acta* **1223**:133–140.

Inoue, S., and Simoni, R. D., 1992, 3-Hydroxy-3-methylglutaryl-coenzyme A reductase and T cell receptor α subunit are differentially degraded in the endoplasmic reticulum, *J. Biol. Chem.* **267**: 9080–9086

Inoue, H., Korenaga, T., Sagami, H., Koyama, T., and Ogura, K., 1994, Phosphorylation of farnesol by a cell-free system from *botryococcus braunii, Biochem. Biophys. Res. Commun.* **200**:1036–1041.

Jackson, S. M, Ericsson, J., Osborne, T. F, and Edwards, P. A., 1995, NF-Y has a novel role in sterol-dependent transcription of two cholesterogenic genes, *J. Biol. Chem.* **270**:21445–21448.

Jelinek, D. F., Andersson, S., Slaughter, C. A., and Russell, W., 1990, Cloning and regulation of cholesterol 7α-hydroxylase, the rate-limiting enzyme in bile acid biosynthesis, *J. Biol. Chem.* **265**:8190–8197.

Josso, N., and Picard, J. Y., 1986, Anti-Mullerian hormone, *Physiol. Rev.* **66**:1038–1090.

Kandutsch, A. A., Chen, H. W., and Heiniger, H.-J., 1978, Biological activity of some oxygenated sterols, *Science* **201**:498–501.

Keller, G. A., Pazirandeh, M., and Krisans, S., 1986, 3-Hydroxy-3-methylglutaryl coenzyme A reductase localization in rat liver peroxisomes and microsomes of control and cholestyramine-treated animals: quantitative biochemical and immunoelectron microscopical analysis, *J. Cell Biol.* **103**: 875–886.

Klausner, R. D., and Sitia, R., 1990, Protein degradation in the endoplasmic reticulum, *Cell* **62:** 611–614.

Krisans, S. K., Ericsson, J., Edwards, P. A., and Keller, G. A., 1994, Farnesyl diphosphate synthase is localized in peroxisomes, *J. Biol. Chem.* **269:**14165–14169.

Kumagai, H., Chun, K. T., and Simoni, R. D., 1995, Molecular dissection of the role of the membrane domain in the regulated degradation of 3-hydroxy-3-methylglutaryl coenzyme A reductase, *J. Biol. Chem.* **270:**19107–19113.

Liscum, L., Finer-Moore, J., Stroud, R. M., Luskey, K. L., Brown, M. S., and Goldstein, J. L., 1985, Domain structure of 3-hydroxy-3-methylglutaryl coenzyme A reductase, a glycoprotein of the endoplasmic reticulum, *J. Biol. Chem.* **260:**522–530.

Maltese, W. A., 1994, Posttranslational modification of proteins by isoprenoids in mammalian cells, *FASEB J.* **4:**3319–3328.

Mangelsdorf, D. J., Umesono, K., and Evans, R. M., 1994, The retinoid receptors, in: *The Retinoids: Biology, Chemistry, and Medicine,* 2nd ed. (M. B. Sporn, A. B. Roberts, and D. S. Goodman, eds.) pp. 319–349, Raven Press, New York.

Marx, S. J., 1995, Vitamin D and other calciferols, in: *Metabolic Basis of Inherited Disease* (C. R. Scriver, A. L. Beaudet, W. S. Sly and D. Valle, eds.) Vol. VII, pp. 3091–3107, McGraw-Hill Inc, New York.

Mead, J. F, Alfin-Slater, R. B, Howton, D. R, and Popjak, G., 1986, Biosynthesis of cholesterol and related substances, in: *Lipids, Chemistry, Biochemistry and Nutrition,* pp. 295–367, Plenum Press, New York.

Meigs, T. E., and Simoni, R. D., 1992, Regulated degradation of 3-hydroxy-3-methylglutaryl-coenzyme A reductase in permeabilized cells, *J. Biol. Chem.* **267:**13547–13552.

Murata, M., Peränen, J., Schreiner, R., Wieland, F., Kurzchalia, T. V., and Simons, K., 1995, VIP21/caveolin is a cholesterol-binding protein, *Proc. Natl. Acad. Sci. USA* **92:**10339–10343.

Nakanishi, M., Goldstein, J. L., and Brown, M. S., 1988, Multivalent control of 3-hydroxy-3-methylglutaryl coenzyme A reductase. Mevalonate-derived product inhibits translation of mRNA and accelerates degradation of enzyme, *J. Biol. Chem.* **263:**8929–8937.

Nicholson, D. W., Ali, A., Thornberry, N. A., Vaillancourt, J. P., Ding, C. K., Gallant, M., Gareau, Y., Griffin, P. R., Labelle, M., Lazebnik, Y. A., Munday, N. A., Raju, S. M., Smulson, M. E., Yamin, T.-T., Yu, V. L., and Miller, D. K., 1995, Identification and inhibition of the ICE/CED-3 protease necessary for mammalian apoptosis, *Nature* **376:**37–43.

Noshiro, M., Nishimoto, M., and Okuda, K., 1990, Rat liver cholesterol 7α-hydroxylase: Pretranslational regulation for circadian rhythm, *J. Biol. Chem.* **265:**10036–10041.

Ohizumi, H., Masuda, Y., Nakajo, S., Sakai, I., Ohsawa, S., and Nakaya, K., 1995, Geranylgeraniol is a potent inducer of apoptosis in tumor cells, *J. Biochem.* **117:**11–13.

Parton, R. G., and Simons, K., 1995, Digging into caveolae, *Science* **269:**1398–1399.

Pérez-Sala, D., and Mollinedo, F., 1994, Inhibition of isoprenoid biosynthesis induces apoptosis in human promyelocytic HL-60 cells, *Biochem. Biophys. Res. Commun.* **199:**1209–1215.

Pfiffer, S. R., Dirac-Svejstrup, A. B., and Soldati, T., 1995, Rab GDP dissociation inhibitor: putting Rab GTPases in the right place. *J. Biol. Chem.* **270:**17057–17059.

Porter, J. W., and Spurgeon, S. L., 1981, *Biosynthesis of Isoprenoid Compounds,* pp. 1–534, Wiley, New York.

Rilling, H. C., 1985, The mechanism of the condensation reactions of cholesterol biosynthesis, *Biochem. Soc. Trans.* **13:**997–1003.

Roitelman, J., Olender, E. H., Bar-Nun, S., Dunn, W. A., Jr., and Simoni, R. D., 1992, Immunological evidence for eight spans in the membrane domain of 3-hydroxy-3-methylglutaryl coenzyme A reductase: implications for enzyme degradation in the endoplasmic reticulum, *J. Cell Biol.* **117:** 959–973.

Rothberg, K. G., Heuser, J. E., Donzell, W. C., Ying, Y., Glenney, J. R., and Anderson, R. G. W., 1992, Caveolin, a protein component of caveolae membrane coats, *Cell* **68:**673–682.

Sabine, J. R., 1983, in: *3-Hydroxy-3-methylglutaryl Coenzyme A Reductase* (J. R. Sabine, ed.), pp. 1–257, CRC Press, Boca Raton.

Sanchez, H. B., Yieh, L., and Osborne, T. F., 1995, Cooperation by sterol regulatory element-binding protein and Sp1 in sterol regulation of low density lipoprotein receptor gene, *J. Biol. Chem.* **270:**1161–1169.

Schafer, W. R., Kim, R., Sterne, R., Thorner, J., Kim, S., and Rine, J., 1989, Genetic and pharmacological suppression of oncogenic mutations in RAS genes of yeast and humans, *Science* **245:** 379–390.

Seol, W., Choi, H., and Moore, D. D., 1995, Isolation of proteins that interact specifically with the retinoid X receptor: two novel orphan receptors, *Mol. Endo.* **9:**72–85.

Skalnik, D. G., Narita, H., Kent, C., and Simoni, R. D., 1988, The membrane domain of HMG-CoA reductase confers endoplasmic reticulum localization and sterol-regulated degradation onto β-galactosidase, *J. Biol. Chem.* **263:**6836–6841.

Spear, D. H., Ericsson, J., Jackson, S. M., and Edwards, P. A., 1994, Identification of a 6-base pair element involved in the sterol-mediated transcriptional regulation of farnesyl diphosphate synthase, *J. Biol. Chem.* **269:**25212–25218.

Tewari, M., Quan, L. T., O'Rourke, K., Desnoyers, S., Zeng, Z., Beidler, D. R., Poirier G. G., Salvesen, G. S., and Dixit, V. M., 1995, Yama/CPP32 beta, a mammalian homolog of CED-3, is a CrmA-inhibitable protease that cleaves the death substrate poly (ADP-ribose) polymerase, *Cell* **81:** 801–809.

Thissen, J. A., and Casey, P. J., 1993, Microsomal membranes contain a high affinity binding site for prenylated peptides, *J. Biol. Chem.* **268:**13780–13783.

Thompson, T. E., and Tillack, T. W. 1985. Organization of glycosphingolipids in bilayers and plasma membranes of mammalian cells. *Annu. Rev. Biophys. Biophysical Chem.* **14:**361–386.

Tontonoz, P., Kim, J. B, Graves, R. A., and Spiegelman, B. M., 1993, ADD1: A novel helix-loop-helix transcription factor associated with adipocyte determination and differentiation, *Mol. Cell. Biol.* **13:**4753–4759.

Ullrich, O., Horiuchi, H., Bucci, C., and Zerial, M., 1994, Membrane association of Rab5 mediated by GDP-dissociation inhibitor and accompanied by GDP/GTP exchange, *Nature* **368:**157–160.

Voziyan, P. A., Haug, J. S., and Melnykovych, G., 1995, Mechanism of farnesol cytotoxicity: further evidence for the role of PKC-dependent signal transduction in farnesol-induced apoptotic cell death, *Biochem. Biophys. Res. Commun.* **212:**479–486.

Wang, X., Briggs, M. R., Hua, X., Yokoyama, C., Goldstein, J. L., and Brown, M. S., 1993, Nuclear protein that binds sterol regulatory element of LDL receptor promoter. II. Purification and characterization, *J. Biol. Chem.* **268:**14497–14504.

Wang, X., Sato, R., Brown, M. S., Hua, X., and Goldstein, J. L., 1994, SREBP-1, a membrane-bound transcription factor released by a sterol-regulated proteolysis, *Cell* **77:**53–62.

Wang, X., Pai, J. T., Wiedenfeld, E. A., Medina, J. C., Slaughter, C. A., Goldstein, J. L., and Brown, M. S., 1995, Purification of an interleukin-1β converting enzyme-related cysteine protease that cleaves sterol regulatory element-binding proteins between the leucine zipper and transmembrane domains, *J. Biol. Chem.* **270:**18044–18050.

Wedegaertner, P. B., Wilson, P. T., and Bourne, H. R. 1995. Lipid modifications of trimeric G proteins. *J. Biol. Chem.* **270:**503–506.

Whyte, M., and Evan, G. 1995. The last cut is the deepest, *Nature* **376:**17–18.

Williams, G. R., 1994, Solving the specificity puzzle, *Nature* **370:**330–331.

Woodward, H. D., Allen, J. M., and Lennarz, W. J., 1988, 3-Hydroxy-3-methylglutaryl-coenzyme A reductase of the sea urchin embryo, *J. Biol. Chem.* **263:**18411–18418.

Yao, T.-P, Segraves, W. A., Oro, A. E., McKeown, M., and Evans, R. M., 1992, Drosophila ultraspiracle modulates ecdysone receptor function via heterodimer formation, *Cell* **71:**63–72.

Yokoyama, C., Wang, X., Briggs, M. R, Admon, A., Wu, J., Hua, X., Goldstein, J. L, and Brown M. S., 1993, SREBP-1, a basic-helix-loop-helix-leucine zipper protein that controls transcription of the low density lipoprotein receptor gene, *Cell* **75:**187–197.

Chapter 2

Coordinate Regulation of Cholesterol 7α-Hydroxylase and HMG-CoA Reductase in the Liver

Ingemar Björkhem, Erik Lund, and Mats Rudling

1. INTRODUCTION

Cholesterol homeostasis in the liver is dependent upon the activity of a number of key structures: 3-hydroxy-3-methylglutaryl-CoA, HMG-CoA reductase (HMGR), acyl cholesterol acyltransferase, cholesterol 7α-hydroxylase, and the LDL receptor (for a review, see Goldstein and Brown, 1990). The regulation of all these proteins is steroid dependent. This steroid-dependent regulation has been characterized in some detail at the gene level for HMG CoA synthase and the LDL receptor, and involves sophisticated coordinate transcriptional control, with interaction of *trans*-acting factors with sterol regulatory elements (SRE-l) sequences present in the promoter regions of these genes. In addition to this transcriptional control, there may also be regulatory control at different posttranscriptional levels. The detailed mechanism behind the sterol-dependent regulation of the other key enzymes is not as well characterized at present.

Ingemar Björkhem, Erik Lund, and Mats Rudling Department of Clinical Chemistry and Department of Medicine and Center for Nutrition and Toxicology, Huddinge University Hospital, Karolinska Institute, S-141 86 Huddinge, Sweden.

Subcellular Biochemistry, Volume 28: Cholesterol: Its Functions and Metabolism in Biology and Medicine, edited by Robert Bittman. Plenum Press, New York, 1997.

Among the above-mentioned enzymes, HMG-CoA reductase and cholesterol 7α-hydroxylase are particularly important, being rate-limiting for the synthesis of cholesterol and for the degradation of cholesterol to bile acids, respectively. In contrast to all the other enzymes, cholesterol 7α-hydroxylase is expressed only in the liver, and the liver is a key organ in the maintenance of cholesterol homeostasis in the body. Figure 1 summarizes the most important hepatic enzymes involved in the flux and regulation of cholesterol and bile acids. The major part of body cholesterol is excreted into the bile in intact form or after conversion into bile acids, in proportions depending on the species. The conversion of water-insoluble cholesterol to water-soluble bile acids is thus a most important elimination pathway for cholesterol. A low capacity for conversion of cholesterol into bile acids has been recognized as an independent risk factor for coronary heart disease, and decreased bile acid synthesis may be correlated to progression of atherosclerosis and coronary mortality in patients with heterozygous hypercholesterolemia (Miettinen and Gylling, 1988).

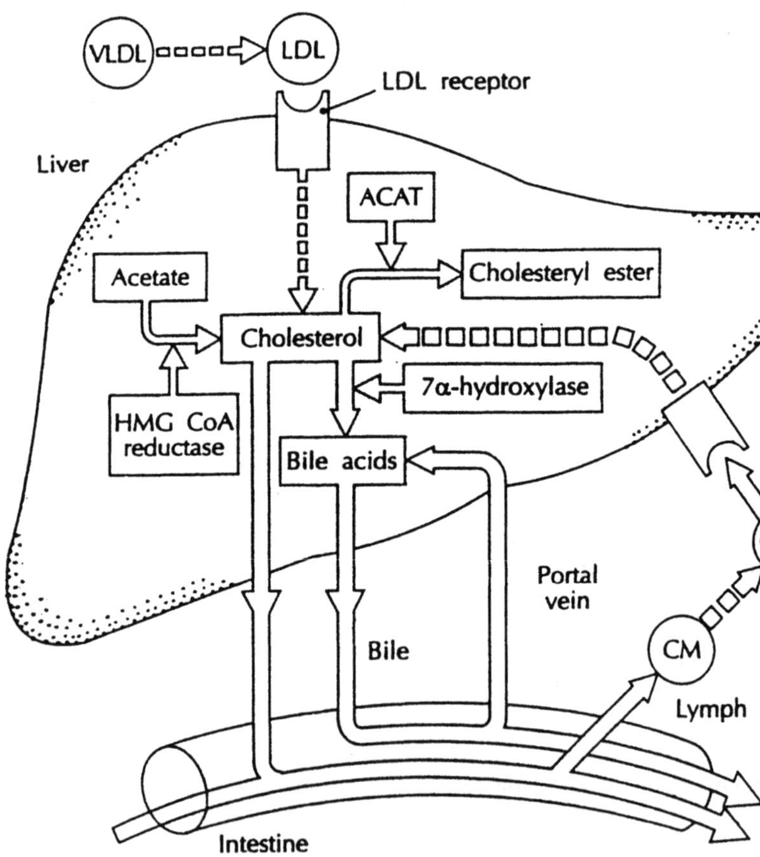

FIGURE 1. Overview of hepatic cholesterol turnover.

However, it should be pointed out that there is a newly recognized pathway for degradation of cholesterol into bile acids in which cholesterol 7α-hydroxylase is not involved (for a review, see Björkhem, 1992). Side-chain oxidized forms of cholesterol, such as 27-hydroxycholesterol, can be 7α-hydroxylated both in peripheral tissues and in the liver by a different enzyme than cholesterol 7α-hydroxylase, denoted oxysterol 7α-hydroxylase. This "acid" pathway contributes substantially to bile acid synthesis in normal subjects and may even become predominant under pathological conditions (Axelson and Sjövall, 1990). Under normal conditions, this pathway is probably less important than that involving cholesterol 7α-hydroxylase, at least under conditions in which the latter enzyme is upregulated.

Cholesterol 7α-hydroxylase is tightly regulated and under most conditions the activity of this enzyme covaries with that of the HMG-CoA reductase. Table I summarizes factors known to stimulate or suppress the two enzymes. It is evident that a primary degradation of cholesterol often leads to a compensatory increase in cholesterol synthesis. As a consequence, hypolipidemic drugs, such as cholestyramine, that stimulate degradation of cholesterol to bile acids are less effective due to the compensatory increase in the synthesis of cholesterol.

Table I
Effect of Different Treatments on HMG-CoA Reductase and Cholesterol 7α-Hydroxylase.[a,b]

Modification	HMG-CoA Reductase	Cholesterol 7α-Hydroxylase
Bile fistula and treatment with cholestyramine	+	+
Portacaval anastomosis	+	+
Lymphatic drainage	+	+
Diurnal rhythm (dark period)	+	+
Glucocorticoids	+	+
Adrenalectomy	−	−
Thyroid hormone	+	+
Thyroidectomy	−	−
Glucose after starvation	+	+
Fasting	−	−
Cholesterol feeding		
Mice, low dose	−	−
Rat, low dose	−	−
Rat, high dose	−	+
Monkeys	−	−
Rabbits	−	−

[a]For references, see Björkhem (1985) and the text.
[b]+ denotes an increase in enzyme level, − denotes a decrease in enzyme level.

It is evident that it is important to understand the regulatory links between cholesterol synthesis and degradation and that such information is essential for new therapeutic interventions.

The aim of the present review is to discuss different possible mechanisms for the coordinate regulation of HMG-CoA reductase and cholesterol 7α-hydroxylase. These two key enzymes and their regulation will be discussed first separately, and then together. In view of the vast and rapidly growing literature, this survey cannot be regarded as complete. It is inevitable that the specific interests of the authors may have influenced the selection of material. It should be mentioned that some excellent reviews covering part of the material discussed here have been published recently (Turley and Dietschy, 1988; Vlahcevic, 1991; Russell and Setchell, 1992; Twisk, 1994).

2. HMG-CoA REDUCTASE

2.1. Structure and General Properties

HMG-CoA reductase catalyzes the NADPH-dependent reduction of 3-HMG-CoA into mevalonic acid (Figure 2) and is the rate-limiting enzyme in cholesterol biosynthesis. The product mevalonate is required not only for synthesis of cholesterol but also for that of other isoprenoids such as ubiquinone, dolichol, and isopentenyl tRNA.

The enzyme is a protein of about 100 kDa. The C-terminal domain of about 60 kDa, which contains the active site, is located in the cytosol, whereas the N-terminal part of the molecule traverses the membrane of the endoplasmic reticulum several times. The membrane-bound domain is extremely hydrophobic and anchors the protein firmly to the membrane. The enzyme is regulated at the transcriptional level, posttranscriptionally by regulation of translation of mRNA, and posttranslationally by phosphorylation–dephosphorylation, as well as by controlled degradation of reductase protein.

The flux of cholesterol and bile acids are two important regulators of HMG-CoA reductase *in vivo*. In addition to cholesterol, many other sterols and nonsterol

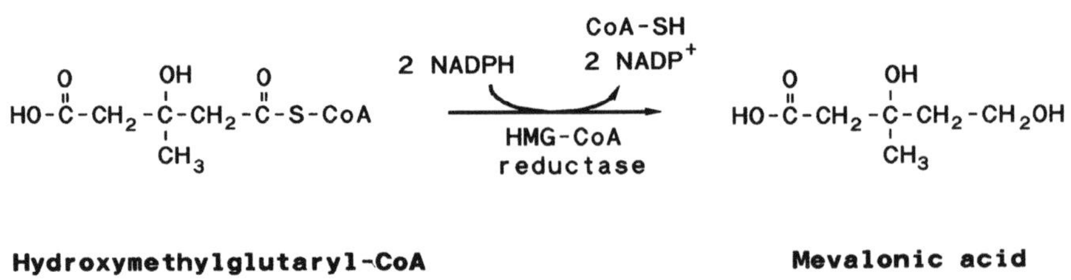

FIGURE 2. Reaction catalyzed by HMG-CoA reductase.

precursors are able to downregulate the enzyme. In intact cells its activity is suppressed by lipoprotein-derived cholesterol. Full suppression of enzyme activity is, however, achieved only if both mevalonate and sterols are present (Brown and Goldstein, 1980). Sterol-mediated regulation of HMGR occurs both at the transcriptional level (Chin *et al.*, 1982) and posttranslational level through the controlled degradation of reductase protein (Faust *et al.*, 1982; Dawson *et al.*, 1991). The latter degradation is dependent on the membrane domain of the protein, while the catalytic activity resides in the cytosolic part of the protein (Gil *et al.*, 1985). Sterol-mediated regulation of HMGR gene transcription and protein degradation operate through distinct pathways, as exemplified by a mutant cell line defective in cholesterol transport, in which LDL-induced transcriptional regulation of HMGR is defective, whereas accelerated degradation of the enzyme is normal (Dahl *et al.*, 1993). Regulation of HMGR by nonsterols and lanosterol derivatives occurs at the level of translation and enzyme degradation. The relative physiological importance of the different modes of regulation of HMGR is at present not known. Several reports support the view that exogenous cholesterol downregulates HMGR activity at the posttranscriptional rather than at the transcriptional level (Field *et al.*, 1991; Ness *et al.*, 1991; Spady *et al.*, 1992), whereas endogenous cholesterol regulates HMGR at the transcriptional level (Ness *et al.*, 1994). The diurnal variation of HMGR activity in rats is likely controlled to a large extent by the phosphorylation state of the enzyme.

Bile acids, being the final products in the degradation pathway for cholesterol, are efficient suppressors of HMG-CoA reductase *in vivo.* The suppressive effect may in part be due to the stimulatory effect of bile acids on absorption of cholesterol. It has been shown , however, that bile acids are able to also downregulate HMG-CoA reductase activity in rats with lymph fistulas, in which no cholesterol from the intestine can reach the liver (Hamprecht *et al.*, 1971, Björkhem *et al.*, 1993). The mechanism responsible for this downregulation is both transcriptional and posttranscriptional (Duckworth *et al.*, 1991; Björkhem *et al.*, 1993). Posttranscriptional regulation seems to be more important than transcriptional regulation in rats with lymph fistulas (Björkhem *et al.*, 1993).

2.2. Mechanism of Transcriptional Regulation

Transcriptional control of HMGR activity is achieved by a single sterol regulatory element (SRE-l) in the promotor of the corresponding gene (Osborne *et al.*, 1988). This element is similar to sterol regulatory elements present in the promotors of the genes for the LDL receptor and HMG-CoA synthase. There is a striking parallel in the transcriptional regulation of these structures. Although much has been revealed recently concerning the mechanisms controlling transcriptional regulation of the LDL receptor and HMG-CoA synthase, the mechanisms underlying transcriptional regulation of HMGR are still largely unclear. A candidate regulatory protein that binds to the sterol regulatory region of the HMGR gene has been identified (Osborne *et al.,* 1992), but its potential role for HMGR regulation remains to be clarified. Recent data suggest that transcriptional regulation of hepatic HMGR in intact animals is mainly exerted by endogenous sterols, whereas exogenous cholesterol regulates HMGR posttranscriptionally (Ness *et al.,* 1994).

Recently, it was reported that HMGR can be transcriptionally regulated by cyclic AMP (Bifulco *et al.,* 1995) and by the v-K-ras oncogene (Perillo *et al.,* 1995). The physiological significance of these findings is at present difficult to assess.

2.3. Translational Regulation

One class of compounds that regulates HMGR posttranscriptionally is oxygenated lanosterol derivatives (Panini *et al.,* 1992), some of which may be formed from cholesterol synthesis intermediates. 15α-Fluorolanost-7-en-3β-ol reduces the translational efficiency of the reductase mRNA, while degradation of the enzyme is unaffected (Trzaskos *et al.,* 1993). In contrast, the cholesterol synthesis intermediate 3β-hydroxylanost-8-en-32-al accelerates HMGR protein degradation in mutant cells not able to further process this compound (Leonard *et al.,* 1994). Some 15-substituted lanosterols also reduce translation and accelerate HMGR degradation (Anderson *et al.,* 1995).

2.4. Regulated Degradation of the HMG-CoA Reductase Protein

The degradation of HMGR protein is believed to be regulated both by sterols and non-sterol mevalonate-derived components (Nakanishi *et al.,* 1988). It seems likely that sterols and non-sterols operate by independent mechanisms (Roitelman and Simoni, 1992; Straka and Panini, 1995). It has, for example, been shown that mevalonate-accelerated degradation of HMGR is calcium-dependent, whereas sterol-enhanced degradation is not (Roitelman and Simoni, 1992). The nonsterol compound(s) responsible for regulated degradation is likely to be closely related to the cholesterol synthesis intermediate farnesol pyrophosphate (Parker *et al.,* 1993; Correll and Edwards, 1994; Bradfute and Simoni, 1994).

Recently, it was demonstrated that in permeabilized cells, farnesol and farnesol diphosphate accelerated the degradation of HMGR, whereas related isoprenoids such as geraniol did not. Interestingly, upon addition of the phosphatase inhibitor sodium fluoride, the effect of farnesol diphosphate on HMGR degradation was abolished, leading the authors to propose that farnesol was the mevalonate-derived nonsterol responsible for the accelerated degradation of HMGR (Correll *et al.,* 1994).

2.5. Phosphorylation

Phosphorylation, catalyzed by an AMP-activated protein kinase, inactivates HMG-CoA reductase activity (reviewed by Hardie, 1992). The phosphorylation occurs on a serine residue at a defined position in this enzyme. This reaction is not feedback-regulated by end products of the mevalonate pathway; the other modes of regulation are also not affected by the phosphorylation status of the protein (Sato *et al.*, 1993). At least in the rat, there seems to be a diurnal variation of the phosphorylation status of HMGR, leading to a high degree of phosphorylation during the light (fasting) period when cholesterol synthesis is low. Phosphorylation of HMGR is likely to be regulated by the general energy status of the cell. Thus, phosphorylation occurs as a response to increased AMP levels (hence, decreased ATP levels) in order to inhibit energy-demanding catabolic pathways. This hypothesis is supported by the fact that the AMP-activated protein kinase also inactivates acetyl-CoA carboxylase, inhibiting fatty acid biosynthesis. The recent cloning of the gene for rat and human (Aguan *et al.*, 1994) AMP-activated protein kinase (Carling *et al.*, 1994) should make possible further clarification of the role of this protein in the regulation of HMGR.

2.6. Role of Oxysterols

Oxygenated analogues of cholesterol (oxysterols) are potent suppressors of HMGR in various cell systems. It was initially suggested (Kandutsch *et al.*, 1978) that oxysterols are regulators of HMGR because of the apparent lack of suppressive effect of carefully purified cholesterol on HMGR activity or sterol synthesis in cultured cells. When added to the cell medium, oxysterols often mimic lipoproteins with respect to the effects elicited on HMGR and the LDL receptor (Chang and Limanek, 1980). The impact of LDL on HMGR activity can be abolished by ketoconazol, a cytochrome P-450 inhibitor, whereas oxysterols are effective in the presence of ketoconazol, suggesting that they undergo a cytochrome P-450-dependent intracellular formation (Gupta *et al.*, 1986; Takagi *et al.*, 1989). In view of the complicated design of these experiments and the possibility that ketoconazol may have effects on other enzymes than those containing cytochrome P-450, firm conclusions cannot be drawn.

It has been proposed that oxysterols formed intracellularly serve as important regulators, in particular, 27-hydroxycholesterol (Esterman *et al.*, 1983; Rennert *et al.*, 1990; Axelson and Larsson, 1995) and 24-hydroxycholesterol (Saucier *et al.*, 1989; Kudo *et al.*, 1989) are discussed in this context. No differences in reduction of murine hepatic HMGR activity have been found between (p.o.) administered cholesterol and analogues with reduced ability to form 24- and 27-hydroxycholesterol and other oxysterols intracellularly (Lund *et al.*, 1992; Lund and Björkhem, 1995). Patients with the rare inborn error disease cerebrotendinous xanthomatosis (CTX) lack sterol 27-hydroxylase. Fibroblasts isolated from such patients have been reported to have a normal basal activity of HMGR and a normal reponse to added LDL in the culture medium (Tint and Salen, 1982). Another group (Axelsson and Larsson, 1995) found an abnormal response to LDL added to such fibroblasts. If 27-hydroxylation of cholesterol was obligatory for the regulation of HMGR, the consequences of the enzyme deficiency for cholesterol homeostasis would probably have been far more dramatic than is the case in patients with CTX.

The intracellular transport of the oxysterol 25-hydroxycholesterol from the plasma membrane to the endoplasmic reticulum has been shown to be much more rapid than that of cholesterol (Lange *et al.*, 1995). A facilitated intracellular movement of oxysterols could possibly explain the high ability of these compounds, as compared to cholesterol, to downregulate HMGR in cells when added to the culture medium.

An oxysterol-binding protein has been identified (Taylor *et al.*, 1984), purified, and cloned (Dawson *et al.*, 1989a, 1989b; Levanon *et al.*, 1990). Upon exposure to oxysterols, this protein translocates from the cytosol to the Golgi (Ridgway *et al.*, 1992). However, conclusive evidence for the involvement of this protein in the regulation of cholesterol biosynthesis is lacking and its physiological relevance remains obscure. The possibility has been discussed that one important function of this protein may be the elimination of toxic oxysterols from cells (Björkhem *et al* 1994; Lund and Björkhem, 1995).

3. CHOLESTEROL 7α-HYDROXYLASE

3.1. Structure and General Properties

The first and rate-limiting step in the "neutral" pathway of bile acid formation is the introduction of an hydroxyl group at position C-7 in the axial (α) position (Figure 3). The cholesterol 7α-hydroxylase responsible for this reaction (EC 1.14.13.17) belongs to the large cytochrome P-450 superfamily, and is a mixed function monooxygenase located in the endoplasmic reticulum. The enzyme utilizes NADPH as an electron donor, and a protein cofactor, cytochrome P-450 reductase, for the transfer of electrons to the cytochrome.

The activity of cholesterol 7α-hydroxylase is subject to regulatory control at several different levels. Three different types of regulation have been described:

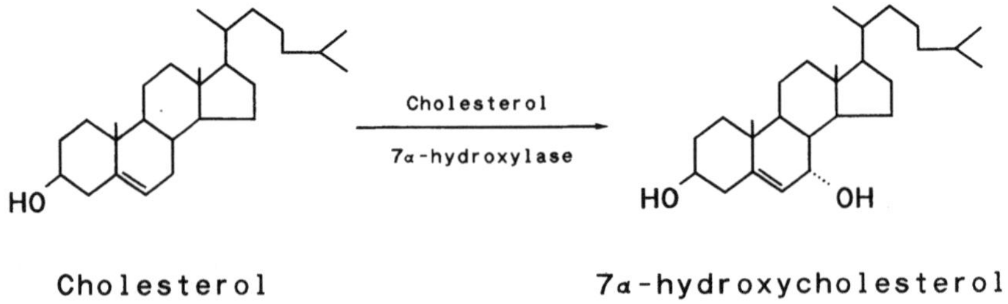

FIGURE 3. Reaction catalyzed by cholesterol 7α-hydroxylase.

long-term (e.g., feedback regulation of enzyme synthesis by bile acids), mid-term (e.g., circadian rhythm) and short-term (various hormones, drugs, and cholesterol precursors) (Sundseth and Waxman, 1990).

Concerning the role of cholesterol 7α-hydroxylase for cholesterol homeostasis, it should be emphasized that the enzyme has dual effects: control of rate of degradation of cholesterol and control of absorption of cholesterol. The direct consequences of variations in cholesterol 7α-hydroxylase activity for cholesterol absorption is not as well studied.

It is of interest to compare the changes in cholesterol 7α-hydroxylase activity with those of other key enzymes in connection with pertubations of cholesterol metabolism. Thus, interruption of bile acid enterohepatic circulation in man results in a sixfold increase in cholesterol 7α-hydroxylase activity, a fourfold increase of HMGR, and a 50% increase of LDL receptor activity (Angelin *et al.*, 1987). On the other hand, suppression of bile acid biosynthesis by chenodeoxycholic acid treatment results in a 70% reduction of cholesterol 7α-hydroxylase activity, a 50% reduction in HMGR, and a 20% reduction of LDL receptor activity. This indicates that the cholesterol 7α-hydroxylase seems to be the structure most sensitive to changes in the flux of bile acids. Recently, Spady *et al.* (1995) elegantly demonstrated the importance of cholesterol 7α-hydroxylase for cholesterol homeostasis in hamsters. Infection of the animals with a recombinant adenovirus encoding cholesterol 7α-hydroxylase was found to transiently increase hepatic microsomal cholesterol 7α-hydroxylase activity and produce a dose-dependent reduction in serum LDL-levels. The reduced cholesterol levels occurred in spite of a compensatory increase in *de novo* cholesterol synthesis in the liver.

The gene encoding cholesterol 7α-hydroxylase was recently cloned. The fact that three different groups of researchers independently pursued this cloning work (Noshiro *et al.*, 1989; Jelinek *et al.*, 1990; Li *et al.*, 1990) illustrates the importance of this enzyme and the great interest to further understand its structure, function, and regulation. The initial work concerned the rat gene. Relatively soon thereafter the same groups also presented the sequence for the human gene (Noshiro and Okuda, 1990; Cohen *et al.*, 1992; Wang and Chiang, 1993). The hamster (Crestani *et al.*, 1993) and the rabbit cDNA (Kai *et al.*, 1995) are now also available.

The gene contains six exons and five introns and encodes proteins 503 and 504 amino acids in length (rat and human, respectively) (Noshiro *et al.*, 1989; Jelinek *et al.*, 1990). The human gene has been localized to chromosome 8 (Cohen *et al.*, 1992). Analyses of the 5′-flanking region have shown that transcriptional activity requires a number of sequence elements, suggesting the involvement of several transcription factors. Initial transfection studies with promoter-reporter constructs could only detect transcription activity in HepG2 cells out of several cell lines (Molowa *et al.*, 1992). The human, rat, and hamster genes show 70–80% homology in the first 250 bases, whereas further upstream the sequences diverge (Crestani *et al.*, 1993). This indicates that the proximal promotor region may be important for regulation by physiological modulators such as bile acids, steroids, and thyroid hormones. A putative bile acid responsive element (BARE) that is conserved in the rat, hamster, and human gene has been characterized within a region of −79 to −49 nucleotides near the initiation site for transcription (Hoekman *et al.*, 1993; Chiang and Stroup, 1994; Crestani *et al.*, 1993). Three potential hepatocyte nuclear factor 3 (HNF-3) recognition sites are present in the promotor, consistent with a role of HNF-3 for cell-specific stimulation of the cholesterol 7α-hydroxylase promoter. In addition, an octanucleotide sequence has been identified in the human gene (Molowa *et al.*, 1992) that has six nucleotides identical to the consensus SRE sequence. The complex arrangement of the promoter of the enzyme requires considerably more work before we can fully understand the links between its structure and functions.

From Northern blots it has been shown that several mRNAs originate from a single gene copy (Jelinek *et al.*, 1990; Twisk *et al.*, 1993). The variable lengths are due to several polyadenylation sites present in the 3′-untranslated region, generating mRNA with variable poly-A tails. The 3′-untranslated region is rich in AU-sequences, containing AUUUA motifs and 5′-AUU-3′or 5′-UAA-3′trinucleotides, which have been linked to rapid degradation of mRNA (Shaw and Kamen, 1986; Binder *et al.*, 1989). In accordance with this, the mRNA half-life has been estimated to be 4 h (Twisk *et al.*, 1993), which corresponds to a reported half-life of 2–3 h for the enzyme (Einarsson and Johansson, 1968; Danielsson and Wikvall, 1991).

The possibility has been discussed that cholesterol 7α-hydroxylase may be modulated by phosphorylation–dephosphorylation (for reviews see Björkhem, 1985; Vlahcevic *et al.*, 1991). There is, however, no convincing experimental evidence yet, although there are potential sites for phosphorylation in the cholesterol 7α-hydroxylase protein.There is, for example, a putative Ca^{++}-calmodulin activatable protein kinase II phosphorylation sequence (Arg-Leu-Ser-Ser) that may also serve as a phosphorylation site for cAMP-dependent protein kinases.

Cytosolic factors may be of importance for the activity of cholesterol 7α-hydroxylase , and both stimulatory and inhibitory factors have been described (for a review see Vlahcevic *et al.*, 1991). While these factors modulate cholesterol 7α-hydroxylase *in vitro,* it is not known if they are of importance *in vivo.* It has been suggested that some of these factors could be of importance for delivery of cholesterol to the active site of cholesterol 7α-hydroxylase or removal of the 7α-hydroxycholesterol product by a sterol carrier protein. In any case, there are at present no data suggesting physiological regulation of cholesterol 7α-hydroxylase mediated by such factors.

3.2. Feedback Regulation of Cholesterol 7α-Hydroxylase by Bile Acids

The most important regulator of cholesterol 7α-hydroxylase activity in the liver is the flux of bile acids returning to the liver via enterohepatic circulation (for general reviews, see Björkhem 1985; Vlahcevic *et al.*, 1991; Russell and Setchell, 1992). It was shown that complete biliary diversion or administration of bile acid-binding anion-exchange polymers leads to several-fold stimulation of both cholesterol 7α-hydroxylase activity and bile acid biosynthesis. In addition, administration of bile acids to rats with biliary fistulas resulted in normalization of cholesterol 7α-hydroxylase activity and bile acid biosynthesis.

In early experiments with primary hepatocytes no direct effects of bile acids on bile acid synthesis could be demonstrated, even in concentrations higher than those found in portal blood (for a review see Twisk, 1994). It was then suggested that intact enterohepatic circulation is necessary for bile-acid-induced feedback regulation, and it was speculated that intestinal factors may be involved in the regulation of cholesterol 7α-hydroxylase. However, Kwekkeboom *et al.* (1988) and, later, several other groups demonstrated that addition of physiological concentrations of bile acids to primary cultures of pig and rat hepatocytes results in suppressed bile acid synthesis and cholesterol 7α-hydroxylase activity, mRNA, and gene transcription.

At least in the rat, hydrophobic bile acids seem to be more effective than hydrophilic bile acids in the suppression of the cholesterol 7α-hydroxylase (Heuman *et al.*, 1989). On the basis of *in vivo* experiments on rats, it was suggested that the potency to suppress is directly correlated to the hydrophobicity of the bile acid. The possibility has also been discussed that primary hydrophilic bile acids may have to be converted into more hydrophobic bile acids in the intestine to be regulatory (Hall *et al.*, 1988; Stange *et al.*,1989). Studies with primary hepatocytes of both pig and rat have, however, shown a relatively weak correlation between the ability to downregulate cholesterol 7α-hydroxylase and the hydrophobicity index (Twisk, 1994). It has been suggested that the hydroxyl groups present on the steroid backbone may form a hydrophilic microenvironment within an otherwise hydrophobic molecule, provided that the different hydroxyl groups are located in the proper position and orientation (Twisk, 1994). Creation of such an environment could be responsible for the direct or indirect selective binding of the bile acid to putative BARP or BARE, eliciting an event at the transcriptional level.

Judging from *in vivo* experiments in man, chenodeoxycholic acid seems to be a stronger suppressor of cholesterol 7α-hydroxylase than most other bile acids. A clear negative correlation has been found between the absolute as well as the relative concentration of chenodeoxycholic acid in portal blood and the activity of the cholesterol 7α-hydroxylase (Reihnér *et al.*, 1989). It was concluded from that study that the composition of individual bile acids may be more important than the concentration of total bile acids in the portal vein for the regulation of cholesterol 7α-hydroxylase activity. In a recent study on rats by Fukushima *et al.* (1995), however, no significant correlation could be found between the enzyme activity and either concentrations, composition, or overall hydrophobicity indices of portal bile acids. It should be pointed out that the composition of portal bile acids might be different from that of biliary and hepatic bile acids because efficiacy of hepatic extraction of bile acids from the portal circulation varies with their degree of hydroxylation, state of conjugation, and extent of protein binding. The contention that chenodeoxycholic acid is of particular importance for the regulation of bile acid biosynthesis in man is further supported by the fact that patients with cerebrotendinous xanthomatosis (CTX, see chapter 15 by Salen *et al.*) have a markedly reduced formation of chenodeoxycholic acid that is associated with a marked upregulation of the cholesterol 7α-hydroxylase (for a review see Björkhem and Boberg, 1994).

Under *in vivo* conditions, the flux of bile acids through the liver seems to be more important than the absolute concentration of bile acids in the bile. Ligation of the bile duct in rats leads to an unexplained stimulation of cholesterol 7α-hydroxylase activity and increased bile acid synthesis despite high concentrations of bile acids in the liver (Kinugasa *et al.*, 1981). It was recently reported that intraduodenal, but not intravenous, infusion of taurocholate downregulates cholesterol 7α-hydroxylase in rats (Pandak *et al.*, 1994a). The possibility has been discussed that an intestinal factor transported in the lymph may be of some importance for downregulation of the cholesterol 7α-hydroxylase (Åkerlund and Björkhem, 1990). Bile acids may be necessary for the absorption of this lymphatic factor from the intestine.

When cDNA and antibodies toward cholesterol 7α-hydroxylase became available, it was shown that the feeding of bile acids reduces, and cholestyramine treatment increases, cholesterol 7α-hydroxylase *in vivo* at the level of activity, protein concentration, mRNA, and transcription rate (Chiang *et al.*, 1990; Jelinek *et al.*, 1990; Li *et al.*, 1990; Pandak *et al.*, 1991).

Two different models have been discussed to explain the transcriptional repression of cholesterol 7α-hydroxylase by bile acids (Chiang and Stroup, 1994 ; Stravitz *et al.*, 1995). Bile acids may bind to a bile acid receptor, be transported to the nucleus, and then interact with the cholesterol 7α-hydroxylase promoter (a "direct" or steroid hormone-type model). Alternatively, bile acids may generate one or more extranuclear signals, which ultimately results in the covalent modification of a *trans*-acting factor required for cholesterol 7α-hydroxylase transciption (an "indirect" or second messenger-type model) (Stravitz *et al.*, 1995). At present there is little support for the direct model of regulation by bile acids.

It was recently suggested that the above BARE sequences (bile acid responsive elements) in the promoter may bind to a bile acid responsive protein, BARP. The latter protein is a positive transcription factor and will initiate transcription (Chiang and Stroup, 1994). If sufficient amounts of bile acids are present in the hepatocyte they will bind to a bile acid receptor, BAR, that thereafter may form a complex with BARP so that less BARP will bind to BARE (Figure 4). This will result in reduced expression of the gene. Further work is needed, however, to establish this mechanism and define the intermediate steps. Ramirez *et al.* (1994) have reported bile-acid-mediated repression of a reporter gene transcribed by −342 bp of the cholesterol 7α-hydroxylase 5′-flanking region. Bile-acid-dependent regulation required inclusion of a liver-specific enhancer region located 6–7 kb upstream of the start of transcription. It has been shown that activation of protein kinase C

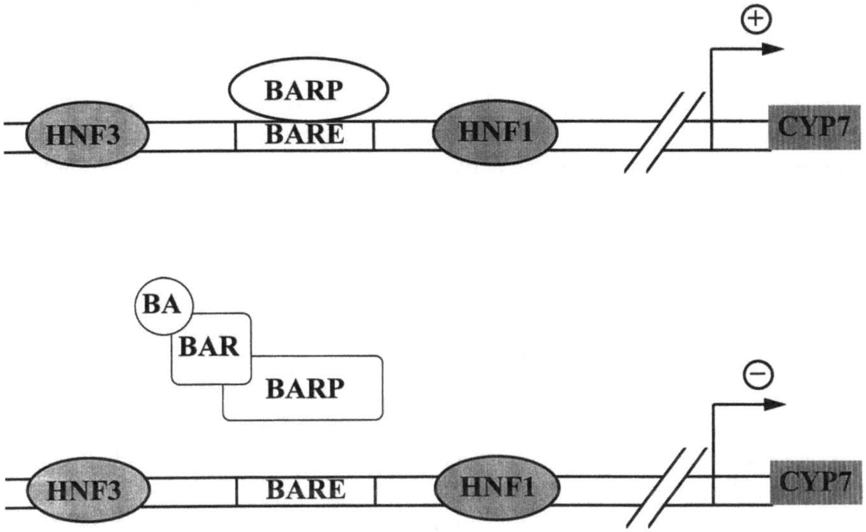

FIGURE 4. Mechanism of transcriptional regulation of cholesterol 7α-hydroxylase by bile acids according to Chiang and Stroup (1994).

results in reduced transcription and mRNA for cholesterol 7α-hydroxylase and that the suppression of cholesterol 7α-hydroxylase by bile acids may involve a protein kinase C-dependent event (Stravitz *et al.*, 1995). However, the relative importance of the protein kinase C-mediated mechanism for downregulation of cholesterol 7α-hydroxylase by bile acids cannot be assessed at present.

In addition to transcriptional regulation by bile acids, cholesterol 7α-hydroxylase may also be upregulated at a posttranscriptional level. Both in cultured rat hepatocytes (Twisk *et al.*, 1993, Hoekman *et al.*, 1993) and in some *in vivo* experiments in rats (Björkhem *et al.*, 1993), the strong suppression of cholesterol 7α-hydroxylase activity obtained with bile acids could hardly be explained by effects on transcription rate only. The possibility has been discussed that stability of mRNA corresponding to cholesterol 7α-hydroxylase may be affected by bile acids. The presence of multiple AU-rich sequences in this mRNA is compatible with a regulation at this level. There is no direct evidence for this hypothesis, however.

3.3. Regulation by Cholesterol

The possibility has been discussed that cholesterol 7α-hydroxylase activity may be regulated to some extent at the level of substrate saturation of the enzyme (Shefer *et al.*, 1981; Davis *et al.*, 1983a; Björkhem and Åkerlund, 1988; Straka *et al.*, 1990). It has been shown, however, that while different treatments of rats causes large variations in levels of enzymatic activity, these effects could not be correlated to hepatic cholesterol levels or to the degree of saturation of the cholesterol 7α-hydroxylase enzyme (Einarsson *et al.*, 1987; Einarsson *et al.*, 1989). In accordance with this, cholesterol 7α-hydroxylase was found to be nearly saturated with substrate (70–90%) under most conditions (Björkhem and Åkerlund, 1988). After upregulation of the enzyme activity by cholestyramine, there was a significant decrease in degree of saturation of the enyme, indicating some limitation for substrate under these specific conditions. When combining the cholestyramine treatment with treatment with a cholesterol synthesis inhibitor, the enzyme was found to be saturated by less than 50% (Björkhem and Åkerlund, 1988). The above results suggest that cholesterol availability is of regulatory importance for cholesterol 7α-hydroxylase only under certain conditions.

From experiments where cholesterol was delivered within liposomes to liver microsomes (cholesterol/phospholipid ratio = 0.5), Straka *et al.* (1990) suggested that cholesterol 7α-hydroxylase may be unsaturated with substrate under normal conditions also. However, since it could not be excluded that the added phospholipids may extract microsomal cholesterol or that they may have an effect per se on the enzyme, no firm conclusions can be drawn from these experiments.

Feeding rats a high cholesterol diet (2%) leads to a stimulation of cholesterol 7α-hydroxylase activity, which also occurs at the level of gene transcription (Jelinek *et al.*, 1990; Li *et al.*, 1990; Pandak *et al.*, 1991; Björkhem *et al.*, 1991). Since such a cholesterol-induced stimulation was also observed in rats with lymph fistulas, part of the stimulation may be due to a malabsorption of bile acids. In accordance with this, rats fed 2% cholesterol were shown to have a markedly increased volume of feces, increased fecal excretion of bile acids, and a reduced half-life for cholic acid (Björkhem *et al.*, 1991). Furthermore, intravenously administered liposomes, enriched in cholesterol, failed to stimulate cholesterol 7α-hydroxylase activity (Björkhem and Åkerlund, 1988). On the other hand, it was found that simultaneous administration to rats of a diet enriched in both cholesterol and bile acids led to upregulation of cholesterol 7α-hydroxylase activity and a concomitant downregulation of HMG-CoA reductase activity (Spady and Cuthbert, 1992). Since bile acids have the dual effect of stimulating cholesterol absorption and inhibiting cholesterol 7α-hydroxylase, and since high amounts of dietary cholesterol have the dual effect of downregulating HMG-CoA reductase and causing bile acid malabsorption, it is difficult to draw firm conclusions from this experiment.

When considering the natural diet of rodents, one might question the interpretation of experiments where extremely high amounts of cholesterol (0.5–5%) were used. Feeding rats a diet with 0.25% cholesterol or mice with 0.4% cholesterol results in a small downregulation rather than upregulation of cholesterol 7α-hydroxylase in both rats and mice (Björkhem *et al.*, 1991; Rudling, 1992). Such cholesterol loads may be considered to be more physiological.

It can be reasoned that because excess cholesterol can be harmful and cause disease, the balance of cholesterol across cell membranes in the liver must be subject to a delicate control. The increase of cholesterol 7α-hydroxylase enzyme activity and mRNA that follows challenge with dietary cholesterol is thus generally believed to be an adequate physiological response that should serve to reduce the cholesterol load. Since about 5% of all bile acids entering the gut escapes efficient reabsorption and since bile acids recirculate in the enterohepatic circulation (in humans, normally 6 rounds per day), it is evident that if more bile acids enter the gut this will result in a larger loss of bile acids through this pathway. However, the bile acids are designed to facilitate the absorption of dietary fat and cholesterol. Therefore, it seems somewhat puzzling that a physiological response to dietary cholesterol should lead to improvement of the prerequisites for its absorption. An alternative response could be to reduce the amount of bile acids in the intestine in order to decrease cholesterol absorption from the gut. It was recently shown that Watanabe Heritable Hyperlipidemic rabbits (WHHL-rabbits), an animal model for the human disease familial hypercholesterolemia, had levels of cholesterol 7α-hydroxylase activity only 30% of those seen in normolipidemic New Zealand White rabbits (NZW-rabbits) (Xu *et al.*, 1995). In this animal model there appears to be no compensatory increase in the excretion of bile acids, although this should be highly warranted under these conditions. In green monkeys, a cholesterol load (0.4%) seems to downregulate cholesterol 7α-hydroxylase (Rudel *et al.*, 1994). It is not known how dietary cholesterol affects human cholesterol 7α-hydroxylase.

To summarize the effect of dietary cholesterol on cholesterol 7α-hydroxylase, it seems possible that different species have different strategies for regulation of cholesterol homeostasis by the cholesterol 7α-hydroxylase and that different dietary loads may have different effects on the enzyme.

The effect of cholesterol on cholesterol 7α-hydroxylase has also been studied *in vitro*. Following incubation of primary rat liver hepatocytes with 200 μM cholesterol dissolved in β-cyclohexidin, there was no stimulation of the enzyme, although the specific activity of HMG-CoA reductase was markedly reduced (Doerner *et al.*, 1995).

It has been suggested that specific lipoproteins may stimulate cholesterol 7α-hydroxylase. β-VLDL was shown to stimulate expression of cholesterol 7α-hydroxylase activity, mRNA, and gene transcription in primary hepatocytes (Twisk, 1994). The physiological importance of this stimulation is difficult to evaluate.

Russell and Setchell (1992) have discussed the possibility that cholesterol and bile acids may have positive and negative effects via SREs or BAREs, respectively. Thus, positive SREs may outweigh negative regulation of BAREs under certain conditions. These effects may then ultimately titrate the expression of the cholesterol 7α-hydroxylase gene. Stimulation of cholesterol 7α-hydroxylase as a consequence of simultaneous administration of both cholesterol and bile acids is then consistent with positive regulation by SREs being more important than negative regulation by BAREs. As pointed out above, the gene coding for cholesterol 7α-hydroxylase has, however, not been shown to contain complete SRE sequences. A possible cholesterol responsiveness must then be mediated by as yet unidentified sequences in the gene.

The effect of cholesterol synthesis on cholesterol 7α-hydroxylase activity has also been studied. Inhibition of cholesterol synthesis by administration of a HMG-CoA reductase inhibitor has been shown to lead to a rapid downregulation of cholesterol 7α-hydroxylase activity in short- or long-term experiments with bile fistula rats (Jones *et al.,* 1993).This downregulation was shown to occur at the transcriptional level and could be prevented by infusion of mevalonate (Jones *et al.,* 1993). In view of the very high doses of inhibitor used in these experiments and the possibility of interactions other than that of HMG-CoA reductase occurring at other levels, it is difficult to draw firm conclusions. In a long-term study on rats, treatment with HMG-CoA reductase inhibitor resulted in slight inhibition of cholesterol 7α-hydroxylase, independent of whether HMG-CoA was upregulated by cholestyramine treatment or downregulated by treatment with cholesterol (Björkhem, 1986). Thus HMG-CoA reductase inhibitors may have an effect on cholesterol 7α-hydroxylase that is independent of its effect on cholesterol synthesis. It may be mentioned that long-term treatment with normal doses of HMG-CoA reductase inhibitors seems to have little or no effect on cholesterol 7α-hydroxylase activity in man (Reihnér *et al.,* 1990).

More convincing evidence for the contention that changes in cholesterol synthesis may affect cholesterol 7α-hydroxylase has come from experiments with squalestatin, a squalene synthase inhibitor that blocks the formation of cholesterol (see chapter 17 by Biessen *et al.*). When primary hepatocytes were exposed to this inhibitor, the cholesterol 7α-hydroxylase activity and mRNA levels were rapidly and markedly reduced (Doerner *et al.,* 1995). This suppressive effect of squalestatin could be reversed by addition of cholesterol to the cells. It has also been shown that inhibition of conversion of 7-dehydrocholesterol to cholesterol is associated with some downregulation of cholesterol 7α-hydroxylase in rats with chronic bile fistulas (Pandak *et al.,* 1990a). It was concluded from the results of that study that downregulation of cholesterol 7α-hydroxylase activity may occur as a result of inhibition of cholesterol synthesis at any step in the pathway.

In accordance with the contention that the rate of cholesterol synthesis may affect cholesterol 7α-hydroxylase, treatment with mevalonate has been reported to have a stimulatory effect on cholesterol 7α-hydroxylase activity and mRNA-levels in rats (Mitropoulos *et al.,* 1978; Sundseth and Waxman, 1990; Jones *et al.,* 1993). This finding has, however, not been confirmed by others (Björkhem and Åkerlund, 1988). It is possible that there are not only species differences but also strain differences in the response of cholesterol 7α-hydroxylase to mevalonate and HMG-CoA reductase inhibitors. Marked strain differences have previously been documented in the response of cholesterol 7α-hydroxylase to phenobarbital in rats (Cohen, 1975; Sudjana-Sugiaman *et al.,* 1994a), and in the response of

HMG-CoA reductase to cholesterol treatment in mice (Hwa *et al.,* 1992). Such strain differences may be responsible for many contradictory results reported by different groups.

Evidence has been presented that cholesterol 7α-hydroxylase in the upregulated state preferentially utilizes newly synthesized cholesterol (Balasubramaniam *et al.* 1979; Björkhem and Lewenhaupt, 1979). One experimental approach was to study incorporation of oxygen-18 into the different hydroxyl groups in bile acids and in cholesterol after exposing a living bile fistulated rat to oxygen-18 atmosphere (Björkhem and Lewenhaupt, 1979). The degree of labeling in the 7α-hydroxyl group reflects utilization of newly synthesized cholesterol. Under the conditions employed, about 50% of the bile acids had been formed from newly synthesized cholesterol. This does not mean, however, that newly synthesized cholesterol must be a regulator of cholesterol 7α-hydroxylase. The upregulated enzyme may consume a considerable part of adjacent preformed cholesterol and may therefore become more dependent upon newly synthesized cholesterol. This is in accordance with the finding that the degree of substrate saturation is less in rats treated with cholestyramine than in control rats (Einarsson *et al.,* 1989; Straka *et al.,* 1990).

3.4. Dietary Factors Other than Cholesterol

Starvation leads to a marked drop in cholesterol 7α-hydroxylase activity and glucose refeeding of starved rats leads to an increased activity of the enzyme (for a review see Björkhem, 1985). The level at which this regulation occurs is not known at present.

3.5. Circadian Rhythm

In the rat, the enzyme has a circadian rhythm peaking at night (Gielen *et al.,* 1969). Such a rhythm has also been shown in the mouse (Van Cantfort and Gielen, 1979), rabbit (Kai *et al.,* 1995), and man (Duane *et al.,* 1983). The circadian rhythm persists upon fasting or following the diversion of all bile by a bile fistula (Mitropoulos *et al.,* 1973). It has been shown that there are synchronous circadian rhythms of mRNA levels and activities of cholesterol 7α-hydroxylase in rabbits and rats (Kai *et al.,* 1995; Noshiro *et al.,* 1989), suggesting that transcriptional regulation here is the most likely mechanism. Previous studies have indicated that the circadian rhythm is extinguished in adrenalectomized and hypophysectomized rats (Mayer, 1974, 1976), suggesting a putative role of glucocorticoids or ACTH for the circadian rhythm. It was recently shown, however, that a diurnal rhythm is present in hypophysectomized rats, although the absolute enzymatic activity is reduced to about 30% of the activity in normal rats (Rudling *et al.,* unpublished observation). The presence of a diurnal rhythm in hypophysectomied animals argues against previously proposed causactive factor such as ACTH, glucocorticoids, and growth hormone. The pituitary dependent axis factors may instead modulate the basal expression of the enzyme.

It has been reported that the circadian rhythm of cholesterol 7α-hydroxylase mainly appears among hepatocytes close to the hepatic venules as measured by *in situ* hybridization (Berkowitz *et al.*, 1995). The exact role of this is not clear.

The developmental activity and protein expression of cholesterol 7α-hydroxylase in the rat was recently demonstrated by Smith *et al.* (1995) on rats killed one hour after the expected diurnal peak. Very low fetal levels were obtained and a transient increase was observed soon after suckling. The levels thereafter remained low until 22 days of age when a sudden eight-fold dramatic increase appeared. Enzymatic activity and protein mass were regulated in parallel. These findings are interesting when considering the recent findings by Lavery and Schibler (1993) who showed that the transcription factor DBP (albumin promoter D-site binding protein) is very important for the diurnal regulation of cholesterol 7α-hydroxylase. The presence of a DBP binding site in the rat promoter has been independently identified (Wang and Chiang, 1993). DBP accumulates to high levels in nuclei from adult rat liver only, whereas its mRNA can be found in most tissues (Mueller *et al.*, 1990). In rats, the hepatic DBP levels oscillate over about a 100-fold range during the diurnal cycle, peaking at 8 p.m. The expression of DBP is altered considerably during development, and adult oscillation commences after weaning, a period in which several factors change. There seems to be no influence by food or light. However, dexamethasone represses DBP and its mRNA in rats (Wuarin and Schibler, 1990).

It may be mentioned that at present it is not clear if HMG-CoA reductase, which also has a similar type of diurnal rhythm, might also harbor DBP-sites in its promoter.

3.6. Hormonal Effects

Of the various hormones that can regulate cholesterol 7α-hydroxylase, most previous studies deal with thyroid hormone and glucocorticoids. Both these hormones stimulate the enzyme *in vivo* in rats (for a review see Björkhem, 1985).The stimulatory effect of thyroid hormone on cholesterol 7α-hydroxylase mRNA levels *in vivo* has also been documented (Ness *et al.*, 1990). It was recently shown that the stimulatory effects of thyroid hormone and dexamethasone occur at the level of transcription, and that these effects are particularly pronounced when both hormones are added in combination to primary hepatocytes from rats (Hylemon *et al.*, 1992). These hormones have however considerably less effects on primary hepatocytes from humans (Einarsson *et al.*, unpublished observation).

Glucagon has a suppressive effect on cholesterol 7α-hydroxylase activity and bile acid synthesis in rats *in vivo* (Rudling and Angelin, 1993; Pandak *et al.*, 1995) and addition of glucagon and cAMP reduces transcription rate and mRNA levels in primary hepatocytes from rats (Hylemon *et al.*, 1992). It is therefore of interest that glucagon is known to activate protein kinase C (Tang and Houslay, 1992). Such activation has been shown to result in reduced mRNA transcription for cholesterol 7α-hydroxylase (see above).

Insulin injection in rats has previously been shown to reduce elevated cholesterol 7α-hydroxylase enzyme activity (Subbiah and Yunker, 1984). Also, when isolated hepatic microsomes were exposed to insulin the enzyme was suppressed, indicating at least in part a direct effect on the activity of the enzyme.

Recently,Twisk *et al.* (1995) showed *in vitro,* on primary rat hepatocytes, that insulin incubation could rapidly reduce total bile acid production and the activity, as well as the mRNA level, for cholesterol 7α-hydroxylase. They also showed that insulin reduced the transcriptional activity of the gene and that this inhibitory action by insulin also remained when a short (348 bp) cholesterol 7α-hydroxylase construct was used. However, *in vivo* studies on rats receiving insulin at high dose have shown that the mRNA level for cholesterol 7α-hydroxylase increases severalfold after rats are given insulin twice daily (Rudling and Angelin, 1993). There is no explanation for the different results reported. However, it must be pointed out that the *in vivo* responses may be strongly dependent on the time elapsed after the last insulin injection.

Pituitary growth hormone may have effects on cholesterol 7α-hydroxylase. Administration of this hormone to children with growth hormone deficiency resulted in an increased pool of chenodeoxycholic acid (Heubi *et al.,*1984). More work is needed, however, to establish if growth hormone has any regulatory role on cholesterol 7α-hydroxylase.

Estradiol has been reported to increase cholesterol 7α-hydroxylase activity in castrated rats (Deliconstantinos and Ramantanis, 1982) and in ovariectomized baboons (Kushwaha and Born, 1991). The level of the stimulatory effect of estradiol has not been defined. It is of interest that there is a marked sex difference in the response of HMG-CoA reductase and cholesterol 7α-hydroxylase to glycodeoxycholic acid in rabbits. Administration of bile acids led to decreased levels of the mRNA of HMG-CoA reductase and cholesterol 7α-hydroxylase in males but not in females (Xu *et al.,* 1993).

3.7. Effects of Oxysterols

Oxysterols, in particular 7α-hydroxycholesterol and 7-oxocholesterol, are competitive inhibitors of cholesterol 7α-hydroxylase (Boström, 1983, Breuer *et al.,* 1993). In a recent study, 7-oxocholesterol was infused intravenously in rats and was shown to result in upregulation of synthesis of the cholesterol 7α-hydroxylase, as shown by enzyme protein and mRNA measurements (Breuer *et al.,* 1993). The most plausible explanation is that the competitive inhibitor leads to a reduced bile acid synthesis and, as a consequence of reduced bile acid inhibition of the cholesterol 7α-hydroxylase activity, a compensatory increase in enzyme synthesis. Under normal conditions the levels of 7-oxocholesterol and other oxysterols in the liver are low, and are not likely to be of importance for the regulation of cholesterol 7α-hydroxylase activity. Under pathological conditions associated with increased lipid peroxidation there may be an accumulation of oxysterols in the liver. The above mechanism may thus be of some importance under some specific pathological conditions.

4. COORDINATE REGULATION OF HMG-CoA REDUCTASE AND CHOLESTEROL 7α-HYDROXYLASE

4.1. Distribution of the Two Enzymes

As previously mentioned, cholesterol 7α-hydroxylase and HMG-CoA reductase are located near each other in the endoplasmic reticulum. When the two enzymes are upregulated, newly synthesized cholesterol seems to be preferred by cholesterol 7α-hydroxylase. It has been shown, however, that the cholesterol synthetic enzymes, HMG-CoA reductase and HMG-CoA synthase, are localized mainly in the portal zone of the liver acinus, whereas the cholesterol 7α-hydroxylase has a pericentral localization (Twisk *et al.*, 1995). Thus the synthetic and catabolic routes of cholesterol are topographically separated, limiting linkage of both synthetic and catabolic routes to a few cells within the acinus, and suggesting a preferential utilization of preformed cholesterol under normal conditions. The distribution pattern of bile acid synthetic enzymes may change, however, upon administration of bile acid sequestrants such as colestid or cholestyramine (Twisk *et al.*, 1995), whereas expression of cholesterol synthetic enzymes remains portal under these conditions. It was suggested that utilization of certain cholesterol pools as substrate for bile acid synthesis may be dictated by relative acinar distribution, rather than the inherent preference of cholesterol 7α-hydroxylase for a particular cholesterol pool (Twisk,1994).

There must be a bile acid gradient over the acinus, and it has been reported that the bile acid concentrations differ six-fold between the periportal and pericentral parts of the acinus (Twisk, 1994). Since hepatocytes have an overcapacity to take up and excrete bile acids into the bile canaliculus, there will be a low concentration of bile acids in the pericentral area. Upregulation of the cholesterol 7α-hydroxylase by cholestyramine, for example, will result in increased expression of the enzyme over a larger part of the liver acinus. This change in distriution of the enzyme may explain the higher utilization of newly synthesized cholesterol in the upregulated state.

4.2. Consequences of a Primary Upregulation of Cholesterol 7α-Hydroxylase

In order to obtain information about the link between cholesterol 7α-hydroxylase and HMG-CoA reductase, the most simple approach would be to affect the activity of one of them and study the effect on the other. Selective inhibitors or selective stimulators must then be used. It is, however, always difficult to evaluate the degree of specificity of any inhibitor or stimulator.

Due to the fact that cholesterol 7α-hydroxylase is a cytochrome P-450, some known inductors of cytochrome P-450 may be expected to upregulate this enzyme. There are strains of rats with cholesterol 7α-hydroxylase that is stimulated by treatment with the cytochrome P-450 inducer phenobarbital (Cohen, 1975; Sudjana-Sugiaman *et al.*, 1994a). In a recent study it was shown that such primary stimulation of cholesterol 7α-hydroxylase always leads to a secondary increase in HMG-CoA reductase activity (Sudjana-Sugiaman, 1994a). Increased HMG-CoA reductase activity was found only in the liver and not in any other organ. It was suggested that the primary upregulation of cholesterol 7α-hydroxylase activity may lead to consumption of cholesterol substrate. This may lead to reduction of a pool of cholesterol that is of regulatory importance for HMG-CoA reductase. In accordance with this hypothesis there was a significant decrease in the content of free cholesterol in liver microsomes of rats responding to phenobarbital. Such decrease was not seen in nonresponding rats. It was suggested that this mechanism may be responsible for part of the covariation between the two enzymes under different conditions.

Further support for the contention that cholesterol 7α-hydroxylase activity may lead to a secondary increase in HMG-CoA reductase activity was obtained by transfection of COS-cells with a cDNA coding for human cholesterol 7α-hydroxylase (Sudjana-Sugiaman *et al.*, 1994b). This transfection resulted in a significant production of cholesterol 7α-hydroxylase enzyme and intracellular accumulation of the product, 7α-hydroxycholesterol. Presence of this enzyme activity in the nonhepatic cells was always associated with increased HMG-CoA reductase activity. The latter activity increased with the activity of the cholesterol 7α-hydroxylase

Transfection of COS-cells with cholesterol 7α-hydroxylase had no significant effect on intracellular content of cholesterol or expression of LDL-receptors. In view of the fact that a regulatory pool of cholesterol may be small in relation to the total content of cholesterol in the cell, this does not rule out the possibility that the stimulatory effect on HMG-CoA reductase is due to reduction in the size of a pool of cholesterol that is of regulatory importance. In any case, it is evident that the two key enzymes in cholesterol synthesis and degradation can also interact with each other in extrahepatic cells that are unable to produce bile acids.

4.3. Consequences of a Primary Upregulation of HMG-CoA Reductase or Increased Cholesterol Synthesis

It is difficult to obtain selective upregulation of HMG-CoA reductase. In theory, lymphatic drainage should result in depletion of cholesterol in the liver with a compensatory increase in HMG-CoA reductase. Lymphatic drainage causes upregulation of HMG-CoA reductase and a subsequent slight increase in cholesterol 7α-hydroxylase in rats (Åkerlund and Björkhem, 1990). The slight upregulation of cholesterol 7α-hydroxylase may be an effect secondary to a primary upregulation of HMG-CoA reductase, or may be due to a loss of an inhibitory factor in the lymph. A primary upregulation of HMG-CoA reductase can also be obtained after the withdrawal of a HMG-CoA reductase inhibitor. Such a treatment leads to compensatory increased enzyme synthesis because available enzyme is competitively blocked by the drug. When the drug is removed, inhibited enzyme will become active and increased enzyme activity is transiently observed. Also, during this "rebound" effect there is a slight upregulation of cholesterol 7α-hydroxylase (Björkhem, unpublished observation). This suggests that increased HMG-CoA reductase activity may lead to a secondary upregulation of cholesterol 7α-hydroxylase. This is in accordance with the demonstration that HMG-CoA reductase inhibitors as well as inhibitors, of other enzymes in cholesterol synthesis, may reduce cholesterol 7α-hydroxylase activity, and mevalonate may upregulate the enzyme. It is not known with certainty, however, if mevalonate itself or an intermediate between mevalonate and cholesterol is the stimulator (Figure 5).

4.4. Bile Acids as Common Regulators of Both HMG-CoA Reductase and Cholesterol 7α-Hydroxylase

As discussed above, bile acids are able to suppress both HMG-CoA reductase and cholesterol 7α-hydroxylase, both by transcriptional and posttranscriptional mechanisms. It has been clearly shown that the former regulation is not due to the stimulatory effect of bile acids on cholesterol absorption (Björkhem *et al.*, 1993). At present it also seems clear that the effect of bile acids on cholesterol 7α-hydroxylase is not mediated by effects on cholesterol synthesis, as previously suggested by Davis *et al.* (1983a, 1983b) on the basis of early experiments with primary hepatocytes. It is possible, however, that the effect of bile acids on the hepatic HMG-CoA reductase may be mediated by their effects on cholesterol 7α-hydroxylase. Since a primary upregulation of cholesterol 7α-hydroxylase seems to be associated with a secondary upregulation of HMG-CoA reductase both in the liver and in extrahepatic cells (see above), it seems likely that a bile-acid induced downregulation of cholesterol 7α-hydroxylase should lead to a secondary downregulation of HMG-CoA reductase. The downregulation of HMG-CoA reductase would then be due to an expanded regulatory pool of cholesterol that is also a substrate for cholesterol 7α-hydroxylase. However, further work is needed to establish this mechanism and define the regulatory pool of cholesterol.

Based on experiments in which rats were treated with cholesterol, taurocholate, and a combination of the two treatments, Shefer *et al.* (1992) concluded that mRNA levels of HMG-CoA reductase are controlled by hepatic taurocholate

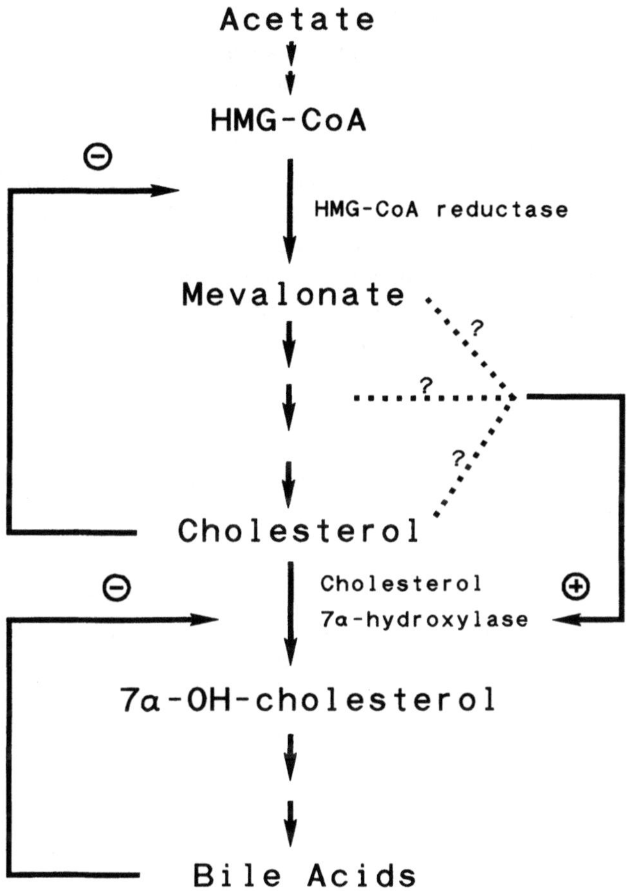

FIGURE 5. Regulation of cholesterol 7α-hydroxylase and HMG-CoA reductase at different levels by cholesterol precursors, cholesterol, and bile acids.

flux, whereas mRNA levels of cholesterol 7α-hydroxylase are controlled by the cholesterol substrate supply. In view of the fact that dietary cholesterol may have dual effects on the cholesterol 7α-hydroxlase (see above), it is difficult to draw firm conclusions from these experiments.

4.5. Oxysterols as Possible Mediators in the Coordination of the Two Enzymes

The possibility has been discussed that oxysterols may be responsible for the coupling between cholesterol 7α-hydroxylase and other key enzymes in cholesterol homeostasis. Based on experiments with Chinese hamster ovary cells transfected with a plasmid encoding rat liver cholesterol 7α-hydroxylase, Dueland *et al.* (1992) suggested that the link between cholesterol 7α-hydroxylase and the LDL receptor could be that the cholesterol 7α-hydroxylase may inactivate regulatory oxysterols (25-hydroxycholesterol). The experimental evidence for this hypothesis is weak, however. It has been clearly shown that the enzyme responsible for 7α-hydroxylation of 25-and 27-hydroxycholesterol is different from cholesterol 7α-hydroxylase (Björkhem *et al.*, 1992; Shoda *et al.*, 1993; Toll *et al.*, 1994). As pointed out above, it has not been established with certainty that oxysterols are important as regulators of cholesterol homeostasis under *in vivo* conditions (Lund and Björkhem, 1995). Furthermore, it has never been shown that introduction of a 7α-hydroxyl group in the oxysterol molecule prevents its suppressive effect on HMG-CoA reducase and other key enzymes in cholesterol homeostasis.

4.6. Phosphorylation as a Possible Mechanism for Coordination of the Two Enzymes

Scallen and Sanghvi (1983) have suggested that phosphorylation–dephosphorylation may be important for coordination of cholesterol 7α-hydroxylase, HMG-CoA reductase, and acyl CoA:cholesterol acyltransferase. It is well established that phosphorylation decreases HMG-CoA reductase and probably increases acyl-CoA:acyltransferase activity. According to Sanghvi *et al.* (1981, 1983), phosphorylation also increases the activity of cholesterol 7α-hydroxylase. If so, these changes are suitable under cholesterol deprivation, leading to increased cholesterol synthesis and decreased cholesterol degradation. The experimental support for a regulation of cholesterol 7α-hydroxylase by phosphorylation is weak, however, and contradictory results have been reported by different groups.

4.7. Concluding Remarks

It is evident that HMG-CoA reductase and cholesterol 7α-hydroxylase are controlled by a great number of dietary and hormonal factors, as well as by the flux of cholesterol and bile acids, and that this regulation involves short-term, mid-term, and long-term mechanisms. In relation to cholesterol homeostasis, long-term regulation by flux of cholesterol and bile acids seems to be most important. Figure 5 summarizes the most important features of this regulation. Under normal physiological conditions the two enzymes are affected by various treatments in the same direction, reflecting the delicate balance between synthesis and degradation of cholesterol (Table I). It seems likely that bile acids are responsible for at least part of this coordination, probably with cholesterol 7α-hydroxylase as the primary point of regulation. Under certain conditions there may be different effects on the two enzymes. While the flux of cholesterol is a most potent regulator of HMG-CoA reductase, the effect of this flux on cholesterol 7α-hydroxylase is less marked. Thus it is possible to downregulate HMG-CoA reductase by dietary or infused cholesterol, in spite of an upregulation of cholesterol 7α-hydroxylase by, for example, malabsorption of bile acids (Björkhem and Åkerlund, 1988). Normally, a malabsorption of bile acids also leads to stimulation of HMG-CoA reductase and cholesterol synthesis. A primary upregulation of HMG-CoA reductase and increased cholesterol synthesis have, however, less marked effects on cholesterol 7α-hydroxylase activity. In general, it seems as if the regulatory mechanisms have a higher potential to avoid cholesterol depletion than to avoid cholesterol excess. However, the capacity to keep a constant level of circulating cholesterol varies widely in different species, being relatively high in the rat and low in man (Dietschy *et al.*, 1993). The difficulties to produce atheromas in rats by dietary cholesterol may be related to the very high activity of the cholesterol 7α-hydroxylase in this species. From a theoretical point of view, a primary upregulation of the cholesterol 7α-hydroxylase should be an attractive approach to lower levels of circulating cholesterol in man. At present, however, anion exchange polymers are the only drugs available for this, and the effect on cholesterol degradation is counteracted by a compensatory increase in cholesterol synthesis.

ACKNOWLEDGMENTS. We thank Maud Lindblad for secretarial assistance. The work carried out in the laboratories of the authors has been supported by grants from the Swedish Medical Research Council.

5. REFERENCES

Aguan, K., Scott, J., See, C. G., and Sarkar, N. H., 1994, Characterization and chromosomal localization of the human homologue of a rat AMP-activated protein kinase-encoding gene: A major regulator of lipid metabolism in mammals, *Gene* **149:**345–350.

Åkerlund, J.-E., Björkhem, I., and Andersson, U., 1990, Studies on the regulation of cholesterol 7α-hydroxylase and HMG CoA-reductase in rat liver: effects of lymphatic drainage and ligation of the lymph duct, *J. Lipid Res.* **31:**2159–2166.

Anderson, J. A., Leonard, D. A., Cusack, K. P., and Frye, L. L., 1995, 15-Substituted lanosterols: post-transcriptional suppressors of 3-hydroxy-3-methylglutaryl coenzyme A reductase, *Arch. Biochem. Biophys.* **316:**190–196.

Angelin, B., Reihnér, E., Rudling, M., Ewerth, S., Björkhem, I., and Einarsson, K., 1987, In vitro studies of lipid metabolism in human liver, *Am. Heart J.* **113:**482–487.

Axelson, M., and Larsson, O., 1995, Low density lipoprotein (LDL) cholesterol is converted to 27-hydroxycholesterol in human fibroblasts: Evidence that 27-hydroxycholesterol can be an important intracellular mediator between LDL and the suppression of cholesterol production, *J. Biol. Chem.* **270:**15102–15110.

Axelson, M., and Sjövall, J., 1990, Potential bile acid precursors in plasma–possible indicators of biosynthetic pathways to cholic and chenodeoxycholic acids in man, *J. Steroid Biochem.* **36:**631–640.

Balasubramaniam, S., Mitropoulos, K. A., and Myant, W. B., 1979, Evidence for compartimentalization of cholesterol in rat liver microsomes, *Eur. J. Biochem.* **34:**77–83.

Berkowitz, C. M., Shen, C. S., Bilir, B. M., Guilbert, E., and Gumucio, J. J., 1995, Different hepatocytes express the cholesterol 7α-hydroxylase gene during its circadian modulation in vivo, *Hepatology* **21:**1658–1667.

Bifulco, M., Perillo, B., Saji, M., Laezza, C., Tedesco, I., Kohn, L. D., and Aloj, S. M., 1995, Regulation of 3-hydroxy-3-methylglutaryl coenzyme A reductase gene expression in FTRL-5 cells. I. Identification and characterization of a cyclic AMP responsive element in the rat reductase promoter, *J. Biol. Chem.* **270:**15231–15236.

Binder, R., Hwang, S.-P. L., Ratnasabapathy, R., and Williams, D. L., 1989, Degradation of apolipoprotein II mRNA occurs via endonucleotic cleavage at 5′-AUU-3′, 5′-UAA-3′ elements in single-stranded loop domains of the 3′-noncoding region, *J. Biol. Chem.* **264:**16910–16918.

Björkhem, I., 1985, Mechanism of bile acid biosynthesis, in: *Sterols and Bile Acids. Comprehensive Biochemistry,* Vol. 12 (H. Danielsson and J. Sjövall, eds.) pp. 231–278, Elsevier, Amsterdam.

Björkhem, I., 1986, Effects of mevinolin in rat liver. Evidence for a lack of coupling between synthesis of hydroxymethylglutaryl-CoA reductase and cholesterol 7α-hydroxylase activity, *Biochim. Biophys. Acta* **877:**43–49.

Björkhem, I., 1992, Mechanism of degradation of the steroid side chain in the formation of bile acids, *J. Lipid Res.* **33:**455–470.

Björkhem, I., and Åkerlund, J.-E., 1988, Studies on the link between HMG-CoA reductase and cholesterol 7α-hydroxylase in rat liver, *J. Lipid Res.* **29:**136–143.

Björkhem, I., and Boberg, K., 1994, Inborn errors in bile acid biosynthesis and storage of sterols other than cholesterol, in: *The Metabolic Basis of Inherited Diseases,* Seventh Ed. (C. R. Scriver, A. L. Beaudet, W. S. Sly, and D. Valle, eds.) pp. 2073–2100, McGraw Hill, New York.

Björkhem, I., and Lewenhaupt, A., 1979, Preferential utilization of newly synthesized cholesterol as substrate for bile acid synthesis, *J. Biol. Chem.* **254:**5252–5257.

Björkhem, I., Eggertsen, G., and Andersson, U., 1991, On the mechanism of stimulation of cholesterol 7α-hydroxylase by dietary cholesterol, *Biochim. Biophys. Acta* **1085:**329–335.

Björkhem, I., Nyberg, B., and Einarsson, K.,1992, 7α-Hydroxylation of 27-hydroxycholesterol in human liver microsomes, *Biochim. Biophys. Acta* **1128:**73–76.

Björkhem, I., Andersson, U., Sudjana-Sugiaman, E., Eggertsen, G., and Hylemon, P., 1993, Studies on the link between HMG-CoA reductase and cholesterol 7α-hydroxylase in lymph fistula rats: Evidence for both transcriptional and posttranscriptional mechanisms for down-regulation of the two enzymes by bile acids, *J. Lipid Res.* **34:**1497–1503.

Björkhem, I., Anderson, C., Diczfalusy, U., Sevastik, B., Xiu, R.-J., Duan, C., and Lund, E., 1994, Atherosclerosis and sterol 27-hydroxylase. Evidence for a role of this enzyme in elimination of cholesterol from human macrophages, *Proc. Natl. Acad. Sci.* USA **91:**8592–8596.

Boström, H.,1983, Binding of cholesterol to cytochrome P-450 from rabbit liver microsomes, *J. Biol. Chem.* **258:**15091–15094.

Bradfute, D. L., and Simoni, R. D., 1994, Non-sterol compounds that regulate cholesterol genesis. Analogues of farnesyl pyrophosphate reduce 3-hydroxy-3-methylglutaryl-coenzyme A reductase levels, *J. Biol. Chem.* **269:**6645–6650.

Breuer, O., Sudjana-Sugiaman, E., Eggertsen, G., Chiang, J., and Björkhem, I., 1993, Cholesterol 7α-hydroxylase is upregulated by the competitive inhibitor 7-oxocholesterol in rat liver, *Eur. J. Biochem.* **215:**705–710.

Brown, M. S., and Goldstein, J. L., 1980, Multivalent feed-back regulation of HMG CoA reductase, a control mechanism coordination isoprenoid synthesis and cell growth, *J. Lipid Res.* **21:**505–517.

van Cantfort, J., and Gielen, J. E., 1979, Comparison of rat and mouse cirdadian rhythm of cholesterol 7α-hydroxylase activity, *J. Steroid Biochem.* **10:**647–651.

Carling, D., Aguan, K., Woods, A., Verhoeven, A., Beri, R., Brennan, C., Sidebottome, C., Davison, M., and Scott, J., 1994, Mammalian AMP-activated protein kinase is homologous to yeast and plant protein kinase involved in the regulation of carbon metabolism, *J. Biol. Chem.* **269:**11442–11448.

Chang, T.-Y., and Limanek, J. S., 1980, Regulation of cytosolic acetoacetyl coenzyme A thiolase, 3-hydroxy-3-methylglutaryl coenzyme A synthase, 3-hydroxy-3-methylglutaryl coenzyme A reductase, and mevalonate kinase by low density lipoprotein and by 25-hydroxycholesterol in Chinese hamster ovary cells, *J. Biol. Chem.* **255:**7787–7795.

Chiang, J. Y. L., and Stroup, D., 1994, Identification and characterization of a putative bile acid-responsive element in cholesterol 7α-hydroxylase gene promotor, *J. Biol. Chem.* **269:**17502–17507.

Chiang, J. Y. L., Miller, W. F., and Lin, G. M., 1990, Regulation of cholesterol 7α-hydroxylase in the liver: Purification of cholesterol 7α-hydroxylase and the immunochemical evidence for the induction of cholesterol 7α-hydroxylase by cholestyramine and circadian rhythm, *J. Biol. Chem.* **265:**3889–3897.

Chin, D. J., Luskey, K. L., Faust, J. R., MacDonald, R. J., Brown, M. S., and Goldstein, J. L., 1982, Molecular cloning of 3-hydroxy-3-methylglutaryl coenzyme A reductase and evidence for regulation of its mRNA, *Proc. Natl. Acad. Sci. USA* **79:**7704–7708.

Chin, D. J., Gil, G., Faust, J. R., Goldstein, J. L., Brown, M. S., and Luskey, K. L., 1985, Sterols accelerate degradation of hamster 3-hydroxy-3-methylglutaryl coenzyme A reductase encoded by a constitutively expressed cDNA, *Mol. Cell Biol.* **5:**634–641.

Cohen, B. I., 1975, Effects of phenobarbital upon bile acid synthesis in two strains of rats, *Lipids* **10:**168–174.

Cohen, J. C., Cali, J. J., Jelinek, D. F., Mehbrabian, M., Sparkes, R. S., Lusis, A. J., Russell, D. W., and Hobbs, H. H., 1992, Cloning of the human cholesterol 7α-hydroxylase gene (CYP7) and localization to chromosome 8ql1-ql2, *Genomics* **14:**153–161.

Correll, C. C., and Edwards, P. A., 1994, Mevalonic acid-dependent degradation of 3-hydroxy-3-methylglutaryl-coenzyme A reductase *in vivo* and *in vitro*, *J. Biol. Chem.* **269:**633–638.

Correll C. C., Ng, L., and Edwards, P. A., 1994, Identification of farnesol as the non sterol derivative of mevalonic acid required for the accelerated degradation of 3-hydroxy-3-methylglutaryl-coenzyme A reductase, *J. Biol. Chem.* **269:**17390–17393.

Crestani, M., Galli, G., and Chiang, J. Y. L., 1993, Genomic cloning, sequencing, and analysis of the hamster cholesterol 7α-hydroxylase gene (CYP7), *Arch. Biochem. Biophys.* **306:**451–460.

Dahl, N. K., Guthei, W. G., and Liscum, L., 1993, Abnormal regulation of LDL-sensitive events in a cholesterol transport mutant, *J. Biol. Chem.* **268:**16979–16986.

Danielsson, H., and Wikvall, K., 1991, Evidence for a specific cytochrome P-450 with short half-life catalyzing 7α-hydroxylation of cholesterol, *Biochem. Biophys. Res. Commun.* **103:**46–51.

Davis, R. A., Hyde, P. M., Kuan, J. C. W., Malone-McNeal, M., and Archarmbault-Schexnayder, J., 1983a, Bile acid secretion by cultured rat hepatocytes. Regulation by cholesterol availability, *J. Biol. Chem.* **258:**3661–3667.

Davis, R. A., Highsmith, S. E., Malone-McNeal, M., Archambault-Schexnayder, J., Hyde, P. M., and Kuan, J-C. W., 1983b, Bile acid synthesis by cultured rat hepatocytes: Inhibition by mevinolin but not by bile acids, *J. Biol. Chem.* **258:**4079–4082.

Dawson, P. A., Ridgway, N. D., Slaughter, C. A., Brown, M. S., and Goldstein, J. L., 1989a, cDNA cloning and expression of oxysterol-binding protein, an oligomer with a potential leucine zipper, *J. Biol. Chem.* **264:**16798–16803.

Dawson P. A., van der Westhuyzen, D. R., Goldstein, J. L., and Brown, M. S., 1989b, Purification of oxysterol binding protein from hamster liver cytosol, *J. Biol. Chem.* **264:**9046–9052.

Dawson, P. A., Metherall, J. E., Ridgeway, N. D., Brown, M. S., and Goldstein, J. L., 1991, Genetic distinction between sterol-mediated transcriptional and post-transcriptional control of 3-hydroxy-3-methylglutaryl coenzyme A reductase, *J. Biol. Chem.* **266:**9128–9134.

Deliconstantinos, G., and Ramantanis, G., 1982, Evoked effects of estradiol on hepatic cholesterol 7α-hydroxylase and drug oxidase in castrated rats. *Int. J. Biochem.* **14:**811–815.

Dietschy, J. M., Turley, S. D., and Spady, D. K., 1993, Role of liver in the maintenance of cholesterol and LDL homeostasis in different animal species, including man, *J. Lipid Res.* **34:**1637–1659.

Doerner, K. C., Gurley, E. C., Vlahcevic, Z. R., and Hylemon, P. B., 1995, Regulation of cholesterol 7α-hydroxylase expression by sterols in primary hepatocyte cultures, *J. Lipid Res.* **36:**168–177.

Duane, W. C., Lewitt, D. G., and Mueller, S. M., 1983, Regulation of bile acid synthesis in man. Presence of a diurnal rhythm, *J. Clin. Invest.* **72:**1930–1936.

Duckworth, P. F., Vlahcevic, Z. R., Studer, E. J., Gurley, E. C., Heuman, D. M., Beg, Z. H., and Hylemon, P. B., 1991, Effect of hydrophobic bile acids on HMG-CoA reductase activity and mRNA levels in the rat, *J. Biol. Chem.* **266:**9413–9418.

Dueland, S., Trawick, J. D., Nenseter, M. S., MacPhee, A. A., and Davies, R. A., 1992, Expression of 7α-hydroxylase in non-hepatic cells results in liver phenotypic resistance of the LDL-receptor to cholesterol repression, *J. Biol. Chem.* **267:**22695–22698.

Einarsson, K., and Johansson, G., 1968, Effect of actinomycin D and puromycin on the conversion of cholesterol into bile acids in bile fistula rats, *FEBS Lett.* **1:**219–222.

Einarsson, K., Åkerlund, J.-E., and Björkhem, I., 1987, The pool of free cholesterol is not of major importance for regulation of the cholesterol 7α-hydroxylase activity in rat liver microsomes, *J. Lipid Res.* **28:**253–256.

Einarsson, K., Reihnér, E., and Björkhem, I., 1989, On the saturation of cholesterol 7α-hydroxylase in human liver microsomes, *J. Lipid Res.* **30:**1477–1481.

Esterman, A. L., Baum, H., Javitt, N. B., and Darlington, G. J., 1983, 26-Hydroxy-cholesterol: regulation of hydroxymethylglutaryl-CoA reductase activity in Chinese hamster ovary cell culture, *J. Lipid Res.* **24:**1304–1309.

Faust, J. R., Luskey, K. L., Chin, D. J., Goldstein, J. L., and Brown, M. S., 1982, Regulation of synthesis and degradation of 3-hydroxy-3-methylglutaryl coenzyme A reductase by low density lipoprotein and 25-hydroxycholesterol in UT-1 cells, *Proc. Natl. Acad. Sci. USA* **79:**5205–5209.

Field, F. J., Shreves, T., Fujiwara, D., Murthy, S., Albright, E., and Mathur, S. N., 1991, Regulation of gene expression and synthesis and degradation of 3-hydroxy-3-methylglutaryl coenzyme A reductase by micellar cholesterol in CaCo-2 cells, *J. Lipid Res.* **32:**1811–1821.

Fukushima, K., Ichimiya, H., Higashijima, H., Yamashita, H., Kuroki, S., Chijiiwa, K., and Tanaka, M., 1995, Regulation of bile acid synthesis in the rat: relationship between hepatic cholesterol 7α-hydroxylase activity and portal bile acids, *J. Lipid Res.* **36:**315–321.

Gielen, J., van Cantfort, J., Robaye, B., and Renson, J., 1969, Rytme circadien de la cholésterol-7α-hydroxylase chez la rat, *C. R. Acad Sci Paris* **269:**731–732.

Gil, G., Faust, J. R., Chin, D. J., Goldstein, J. L., and Brown, M. S., 1985, Membrane-bound domain of HMG CoA reductase is required for sterol-enhanced degradation of the enzyme, *Cell* **41:** 249–258.

Goldstein, J. L., and Brown, M. S., 1990, Regulation of the mevalonate pathway, *Nature* **343:**425–430.

Gupta, P. A., Sexton, R. C., and Rudney, H., 1986, Modulation of regulatory oxysterol formation and low density lipoprotein suppression of 3-hydroxy-3-methylglutaryl coenzyme A reductase activity by ketoconazole, *J. Biol. Chem.* **261:**8348–8356.

Hall, R., Kok, E., and Javitt, N. B., 1988, Bile acid synthesis: downregulation by monohydroxy bile acids, *Fed. Am. Soc. Exp. Biol.* **2:**152–156.

Hamprecht, B. R., Roscher, G., Waltinger, G., and Nussler, C., 1971, Influence of bile acids on the activity of rat liver HMG-CoA reductase. Effect of cholic acid in lymph fistula rats, *Eur. J. Biochem.* **18:**15–19.

Hardie, D. G., 1992, Regulation of fatty acid and cholesterol metabolism by the AMP-activated protein kinase, *Biochim. Biophys. Acta* **1123:**231–238.

Heubi, C. E., Burstein, S., Sperling, M. A., Gregg, D., Ravi Subbiah, M. T., and Mathews, D. E., 1984, The role of human growth hormone in the regulation of cholesterol and bile acid metabolism, *J. Clin. Endocrinol. Metab.* **57:**885–891.

Heuman, D. M., Hylemon, P. B., and Vlahcevic, Z. R., 1989, Regulation of bile acid synthesis. III. Correlation between biliary salt hydrophobicity index and the activities of enzymes regulating cholesterol and bile acid synthesis in the rat, *J. Lipid Res.* **30:**1161–1171.

Hoekmann, M. F. M., Rientjes, J. M. J., Twisk, J., Planta, R. J., Princen, H. M. G., and Mager, W. H., 1993, Transcriptional regulation of the gene encoding cholesterol 7α-hydroxylase in the rat, *Gene* **130:**217–233.

Hwa, J. J., Zollman, S., Warden, C. H., Taylor, B. A., Edwards, P. A., Fogelman, A. M., and Lusis, A. J., 1992, Genetic and dietary interactions in the regulation of HMG-CoA reductase gene expression, *J. Lipid Res.* **33:**711–725.

Hylemon, P. B., Gurley, E. C., Stravitz, R. T., Litz, J. S., Pandak, W. M., Chiang, J. Y. L., and Vlahcevic, Z. R., 1992, Hormonal regulation of cholesterol 7α-hydroxylase mRNA levels and transcriptional activity in primary rat hepatocyte cultures, *J. Biol. Chem.* **267:**16866–16871.

Jelinek, D. F., Andersson, S., Slaughter, C. A., and Russel, D. W., 1990, Cloning and regulation of cholesterol 7α-hydroxylase, the rate limiting enzyme in bile acid biosynthesis, *J. Biol. Chem.* **265:**8190–8197.

Jones, M. P., Pandak, W. M., Hylemon, P. B., and Vlahcevic, Z. R., 1993, Cholesterol 7α-hydroxylase: Evidence for regulation by mevalonate and/or metabolic products of mevalonate, *J. Lipid Res.* **34:**885–892.

Kai, M.-H., Eto, T.-A., Kondo, K.-H., Setoguchi, Y., Higashi, S., Maeda, Y., and Setogushi, T., 1995, Syncronous circadian rythms of mRNA levels and activities of cholesterol 7α-hydroxylase in the rabbit and rat, *J. Lipid Res.* **36:**367–374.

Kandutsch, A. A., Chen, H. W., and Heiniger, H.-J., 1978, Biological activity of some oxygenated sterols, *Science* **201:**498–501.

Kinugasa, T., Uchida, K., Kadowaki, M., Takase, H., Nomura, Y., and Saito, Y., 1981, Effect of bile duct ligation on bile acid metabolism in rats, *J. Lipid Res.* **22:**201–207.

Kudo, K., Emmons, G. T., Casserly, E. W., Via, D. P., Smith, L. C., Pyrek, J. S., and Schroepfer Jr., G. J., 1989, Inhibitors of sterol synthesis. Chromatography of acetate derivatives of oxygenated sterols, *J. Lipid Res.* **30:**1097–1111.

Kushwaha, R. S., and Born, K. M., 1991, Effect of estrogen and progesterone on the hepatic cholesterol 7α-hydroxylase in ovariectomized baboons, *Biochim. Biophys. Acta* **1084:**300–302.

Kwekkeboom, J., Van Voorthuizen, E. M., Princen, H. M. G., and Kempen, H. J. M., 1988, Feedback inhibition of bile acid synthesis in cultured pig hepatocytes, *Biochem. Biophys. Res. Commun.* **155:**850–856.

Lange, Y., Ye, J., and Strebel, F., 1995, Movement of 25-hydroxycholesterol from the Wasma membrane to the rough endoplasmic reticulum in cultured hepatoma cells, *J. Lipid Res.* **36:** 1092–1097.

Lavery, D. J., and Schibler, U., 1993, Circadian transcription of the cholesterol 7α-hydroxylase gene may involve the liver-enriched bZIP protein DBP, in: *Genes & Development 7,* pp. 1871–1884, Cold Spring Harbor Laboratory Press, Cold Spring Harbor.

Leonard, D. A., Kotarski, M. A., Tessiatore, J. E., Favata, M. F., and Trzaskos, J. M., 1994, Post-transcriptional regulation of 3-hydroxy-3-methylglutaryl coenzyme A reductase by 3β-hydroxy-lanost-8-en-32-al, an intermediate in the conversion of lanosterol to cholesterol, *Arch. Biochem. Biophys.* **310:**152–157.

Levanon, D., Hsieh, C.-L., Francke, U., Dawson, P. A., Ridgway, N. D., Brown, M. S., and Goldstein, J. L., 1990, cDNA cloning of human oxysterol-binding protein and localization of the gene to human chromosome 11 and mouse chromosome 19, *Genomics* **7:**65–74.

Li, Y. C., Wang, D. P., and Chiang, J. Y. L., 1990, Regulation of cholesterol 7α-hydroxylase in the liver. Cloning, sequencing and regulation of cholesterol 7α-hydroxylase mRNA, *J. Biol. Chem.* **265:** 12012–12019.

Lund, E., and Björkhem, I., 1995, Role of oxysterols in the regulation of cholesterol homeostasis: a critical evaluation, *Acc. Chem. Res.* **28:**241–249.

Lund, E., Breuer, O., and Björkhem, I., 1992, Evidence that 24- and 27-hydroxylation are not involved in the cholesterol-induced downregulation of HMG CoA reductase in mice, *J. Biol. Chem.* **267:** 25092–25097.

Mayer, D., 1974, Hormones and 7α-hydroxylation of cholesterol, in: *Advances in Bile Acid Research III: Bile Acid Meeting* (S. Matern, J. Hackenschmidt, P. Back, and W. Gerok, eds.). pp. 53–60, Schattauer Verlag, New York.

Mayer, D., 1976, The circadian rhythm of synthesis and catabolism of cholesterol, *Arch. Toxicol.* **36:**267–276.

Miettinen, T. A., and Gylling, H., 1988, Mortality and cholesterol metabolism in familial hypercholesterolemia, *Arteriosclerosis* **8:**163–167.

Mitropoulos, K. A., Balasubramaniam, S., and Myant, N. B., 1973, The effect of interruption of the enterohepatic circulation of bile acids and of cholesterol feeding on cholesterol 7α-hydroxylase in relation to the diurnal rhythm in its activity, *Biochim. Biophys. Acta* **326:**428–438.

Mitropoulos, K. A., Balasubramaniam, S., Vankatesan, S., and Reeves, B. E. A., 1978, On the mechanism for the regulation of HMG CoA reductase of cholesterol 7α-hydroxylase and of ACAT by free cholesterol, *Biochim. Biophys. Acta* **530:**99–111.

Molowa, D. T., Chen, W. S., Cimis, G. M., and Tan, C. P., 1992, Transcriptional regulation of the human cholesterol 7α-hydroxylase gene, *Biochemistry* **31:**2539–2544.

Mueller, C.R., Maire, P., and Schibler, U., 1990, DBP, a liver enriched transcriptional activator, is expressed late in ontogeny and its tissue specificity is determined posttranscriptionallly, *Cell* **61:**279–291.

Nakanishi, M., Goldstein, J. L., and Brown, M. S., 1988, Multivalent control of 3-hydroxy-3-methylglutaryl coenzyme A reductase: mevalonate-derived product inhibits translation of mRNA and accelerates degradation of enzyme. *J. Biol. Chem.* **263:**8929–8937.

Ness, G. C., Pendleton, L. C., Li, Y. C., and Chiang, J. Y. L., 1990, Effect of thyroid hormone on hepatic cholesterol 7α-hydroxylase, LDL receptor, HMG-CoA reductase, farnesyl pyrophosphate synthetase and apolipoprotein A-I mRNA levels in hypophysectomized rats, *Biochem. Biophys. Res. Commun.* **172:**1150–1156.

Ness, G. C., Keller, R. K., and Pendleton, L. C., 1991, Feed-back regulation of hepatic 3-hydroxy-3-methyl-glutaryl-CoA reductase activity by dietary cholesterol is not due to altered mRNA levels, *J. Biol. Chem.* **266:**14854–14857.

Ness, G. C., Zhao, Z., and Keller, R. K., 1994, Effect of squalene synthase inhibition on the expression of hepatic cholesterol biosynthetic enzymes, LDL receptor, and cholesterol 7α-hydroxylase, *Arch. Biochem. Biophys.* **311:**277–285.

Noshiro, M., and Okuda, K., 1990, Molecular cloning and sequence analysis of cDNA encoding human cholesterol 7α-hydroxylase, *FEBS Lett.* **268:**137–140.

Noshiro, M., Nishimoto, M., Morohashi, K.-I., and Okuda, K., 1989, Molecular cloning of cDNA for cholesterol 7α-hydroxylase from rat liver microsomes, *FEBS Lett.* **257:**97–100.

Osborne, T. F., Gil, G., Goldstein, J. L., and Brown, M. S.,1988, Operator constitutive mutation of 3-hydroxy-3-methylglutaryl coenzyme A reductase promoter abolishes protein binding to sterol regulatory element, *J. Biol. Chem.* **263:**3380–3387.

Osborne, T.F., Bennet, M., and Rhee, K., 1992, Red 25, a protein that binds specificlly to the sterol regulatory region in the promoter for HMG CoA reductase, *J. Biol. Chem.* **267:** 18973–18982.

Pandak, W. M., Heuman, D. M., Hylemon, P. B., and Vlahcevic, Z. R., 1990a, Regulation of bile acid synthesis. IV. Interrelationship between cholesterol and bile acid biosynthesis pathways, *J. Lipid Res.* **31:**79–90.

Pandak, W. M., Li, Y. C., Chiang, J. Y. L., Studer, E. J., Gurley, E. C., Heuman, D. M., Vlahcevic, Z. R., and Hylemon, P. B., 1991, Regulation of cholesterol 7α-hydroxylase mRNA and transcriptional activity by taurocholate and cholesterol in the chronic biliary diverted rat, *J. Biol. Chem.* **266:**3416–3421.

Pandak, W. M., Vlahcevic, Z. R., Heuman, D. M., Chiang, J. Y. L., and Hylemon, P. B., 1994a, Intraduodenal but not intravenous infusion of taurocholate downregulates HMG CoA reductase and cholesterol 7α-hydroxylase, *Gastroenterology* **106:**A958.

Pandak, W. M., Heuman, D. M., Compton, G., Hylemon, P. B., and Vlahcevic, Z. R., 1995, Regulation of hepatic cholesterol and bile acid biosynthesis by glucagon and enteroglucagon, *Gastroenterology* **108:**A1142.

Panini, S. R., Delate, T. A., and Sinensky, M., 1992, Post-transcriptional regulation of 3-hydroxy-3-methylglutaryl coenzyme A reductase by 24(S),25-oxidolanosterol, *J. Biol. Chem.* **267:** 12647–12654.

Parker, R. A., Pearce, B. C., Clark, R. W., Gordon, D. A., and Wright, J. J., 1993, Tocotrienols regulate cholesterol production in mammalian cells by post-transcriptional suppression of 3-hydroxy-3-methylglutaryl-coenzyme A reductase, *J. Biol Chem.* **268:**11230–11238.

Perillo, B., Tedesco, I., Laezza, C., Santillo, M., Romano, A., Aloj, S. M., and Bifulco, M., 1995, Regulation of 3-hydroxy-3-methylglutaryl coenzyme A reductase gene expression in FTRL-5 cells. II. Down-regulation by v-K-*ras* oncogene, *J. Biol. Chem.* **270:**15231–15236.

Ramirez, M., Karaoglu, D., Haro, D., Barillas, C., Bashirzadeh, R., and Gil, G., 1994, Cholesterol and bile acids regulate cholesterol 7α-hydroxylase expression at the transcriptional level in culture and in transgenic mice, *Mol. Cell Biol.* **14:**2809–2821.

Reihnér, E., Björkhem, I., Angelin, B., Ewerth, S., and Einarsson, K., 1989, Bile acid synthesis in humans: regulation of hepatic microsomal cholesterol 7α-hydroxylase activity, *Gastroenterology* **97:**1498–1505.

Reihnér, E., Rudling, M., Ståhlberg, D., Berglund, L., Ewerth, S., Björkhem, I., Einarsson, K., and Angelin, B., 1990, Influence of pravastatin, a specific inhibitor of HMG CoA reductase, on hepatic metabolism of cholesterol, *N. Engl. J. Med.* **323:**224–228.

Rennert, H., Fischer, R. T., Alvarez, J. G., Trzaskos, J. M., and Strauss III, J. F., 1990, Generation of regulatory oxysterols: 26-Hydroxylation of cholesterol by ovarian mitochondria, *Endocrinology* **127:**738–746.

Ridgway, N. D., Dawson, P. A., Ho, Y. K., Brown, M. S., and Goldstein, J. L., 1992, Translocation of oxysterol binding protein to Golgi apparatus triggered by ligand binding, *J. Cell Biol.* **116:**307–319.

Roitelman, J., and Simoni, R. D., 1992, Distinct sterol and nonsterol signals for the regulated degradation of 3-hydroxy-3-methylglutaryl-CoA reductase, *J. Biol. Chem.* **267:**25264–25273.

Rudel, L., Deckelman, C., Wilson, M., Scobey, M., and Anderson, R., 1994, Dietary cholesterol and downregulation of cholesterol 7α-hydroxylase and cholesterol absorption in African green monkeys, *J. Clin. Invest.* **93:**2463–2472.

Rudling, M., 1992, Hepatic mRNA levels for the LDL receptor and HMG-CoA reductase show coordinate regulation in vivo, *J. Lipid Res.* **33:**493–501.

Rudling, M., and Angelin, B., 1993, Stimulation of rat hepatic low density lipoprotein receptors by glucagon. Evidence of a novel regulatory mechanism in vivo, *J. Clin. Invest.* **91:**2796–2805.

Russell, D.W., and Setchell, K. D. R., 1992, Bile acid biosynthesis, *Biochemistry* **31:**4737–4739.

Sanghvi, A., Grassi, E., Warty, V., Diven, W., Wight, C., and Lester, R., 1981, Reversible activation–inactivation of cholesterol 7α-hydroxylase possibly due to phosphorylation–dephosphorylation, *Biochem. Biophys. Res. Commun.* **103:**886–892.

Sanghvi, A., Grassi, E., and Diven, W., 1983, Loss of cholesterol 7α-hydroxylase activity in vitro in the presence of bivalent metal ions and by dialysis of rat liver microsomes, *Proc. Natl. Acad. Sci. USA* **80:**2175–2178.

Sato, R., Goldstein, J. L., and Brown, M. S., 1993, Replacement of serine-871 of hamster 3-hydroxy-3-methylglutaryl-CoA reductase prevents phosphorylation by AMP-activated kinase and blocks inhibition of sterol synthesis induced by ATP depletion, *Proc. Natl. Acad. Sci. USA* **90:**9261–9265.

Saucier, S. E., Kandutsch, A. A., Gayen, A. K., Swahn, D. K., and Spencer, T. A., 1989, Oxysterol regulators of 3-hydroxy-3-methylglutaryl-CoA reductase in liver: effect of dietary cholesterol, *J. Biol. Chem.* **264:**6863–6869.

Scallen, T., and Sanghvi, A., 1983, Regulation of three key enzymes in cholesterol metabolism by phosphorylation–dephosphorylation, *Proc. Natl. Acad. Sci. USA* **80:**2477–2480.

Shefer, S., Nguyen, L. B., Salen, G., Ness, C. G., Chowdhary, I. R., Lerner, S., Batta, A. K., and Tint, G. S., 1992, Differing effects of cholesterol and taurocholate on steady state hepatic HMG-CoA reductase and cholesterol 7α-hydroxylase activities and mRNA levels in the rat, *J. Lipid Res.* **33:**1193–1200.

Shoda, J., Toll, A., Axelson, M., Pieper, F., and Sjövall, J., 1993, Formation of 7α- and 7α-hydroxylated bile acid precursors from 27-hydroxycholesterol in human liver microsomes, *Hepatology* **17:**395–403.

Smith, J. L., Lear, S. R., and Erickson, S. K., 1995, Developmental expression of elements of hepatic cholesterol metabolism in the rat, *J. Lipid Res.* **36:**641–652.

Spady, D. K., and Cuthbert, J. A., 1992, Regulation of hepatic sterol metabolism in the rat. Parallel regulation of activity and mRNA for 7α-hydroxylase but not 3-hydroxy-3-methylglutaryl-coenzyme A reductase or low density lipoprotein receptor, *J. Biol. Chem.* **267:**5584–5591.

Spady, D. K., Cuthbert, J. A., Willard, M. N., and Meidell, R. S., 1995, Adenovirus-mediated transfer of a gene encoding cholesterol 7α-hydroxylase into hamsters increases hepatic enzyme activity and reduces plasma total and low density lipoprotein cholesterol, *J. Clin. Invest.* **96:** 700–709.

Stange, E. F., Scheibner, J., and Ditschuneit, H., 1989, Role of primary and secondary bile acids as feedback inhibitors of bile acid synthesis in the rat in vivo, *J. Clin. Invest.* **84:**173–180.

Straka, M. S., and Panini, S. R., 1995, Post-transcriptional regulation of 3-hydroxy-3-methyl-glutaryl coenzyme A by mevalonate, *Arch. Biochem. Biophys.* **317:**235–243.

Straka, M. S., Junker, L. H., Zacarro, L., Zogg, D. L., Dueland, S., Everson, G. T., and Davis, R. A., 1990, Substrate stimulation of 7α-hydroxylase, an enzyme located in the cholesterol-poor ER, *J. Biol. Chem.* **265:**7145–7149.

Stravitz, R. T., Hylemon, P. B., Heuman, D. M., Hagey, L. R., Schteingart, C. D., Ton-Nu, H.-T., Hofmann, A. F., and Vlahcevic, Z. R., 1993, Transcriptional regulation of cholesterol 7α-hydroxylase mRNA by conjugated bile acids in primary cultures of raat hepatocytes, *J. Biol. Chem.* **268:**13987–13993.

Stravitz, R. T., Vlahcevic, Z. R., Gurley, E. C., and Hylemon, P. B., 1995, Repression of cholesterol 7α-hydroxylase transcription by bile acids is mediated through protein kinase C in primary cultures of rat hepatocytes, *J. Lipid Res.* **36:**1359–1369.

Subbiah, M. T., and Yunker, R. L., 1984, Cholesterol 7α-hydroxylase of rat liver: an insulin sensitive enzyme, *Biochem. Biophys. Res. Commun.* **124:**896–902.

Sudjana-Sugiaman, E., Eggertsen, G., and Björkhem, I., 1994a, Stimulation of HMG CoA reductase as a consequence of phenobarbital-induced primary stimulation of cholesterol 7α-hydroxylase in rat liver, *J. Lipid Res.* **35:**319–327.

Sudjana-Sugiaman, E., Eggertsen, G., Sjöblom, P., Maeda, Y., Okuda, K., and Björkhem, I., 1994b, Presence of cholesterol 7α-hydroxylase enzyme protein in COS-cells leads to increased HMG CoA reductase activity, *Biochem. Biophys. Res. Commun.* **202:**896–901.

Sundseth, S. S., and Waxman, D. J., 1990, Hepatic P-450 cholesterol 7α-hydroxylase. Regulation *in vivo* at the protein and mRNA level in response to mevalonate, diurnal rhythm, and bile acid feedback, *J. Biol. Chem.* **265:**15090–15095.

Takagi, K., Alvarez, J. G., Favata, M. F., Trzaskos, J. M., and Strauss III, J. F., 1989, Control of low density lipoprotein receptor gene promoter activity: ketoconazole inhibits serum lipoprotein but not oxysterol suppression of gene transcription, *J. Biol. Chem.* **264:**12352–12357.

Tang, K. Y., and Houslay, M. D., 1992. Glucagon, vasopressin and angiotensin all elicit a rapid, transient increase in hepatocyte protein kinase C activity, *Biochem J.* **283:**341–346.

Taylor, F. R., Saucier, S. E., Shown, E. P., Parish, E. J., and. Kandutsch, A. A., 1984, Correlation between oxysterol binding to a cytosolic binding protein and potency in the repression of hydroxymethylglutaryl coenzyme A reductase, *J. Biol. Chem.* **259:**12382–12387.

Tint, G., and Salen, G., 1982, Synthesis of cholesterol and its precursors but not cholestanol in cultured fibroblasts from patients with cerebrotendinous xanthomatosis, *J. Lipid Res.* **23:**597–603.

Toll, A., Wikvall, K., Sudjana-Sugiaman, E., Kondo, K. I., and Björkhem, I., 1994, 7α-Hydroxylation of 25-hydroxycholesterol in liver microsomes. Evidence that the enzyme is different from cholesterol 7α-hydroxylase, *Eur. J. Biochem.* **224**:309–316.

Trzaskos, J. M., Magolda, R. L., Favata, M. F., Fischer, R. T., Johnson, P. R., Wen, H., Ko, S. S., Leonard, D. A., and. Gaylor, J. L., 1993, Modulation of 3-hydroxy-3-methylglutaryl-CoA reductase by 15α-fluorolanost-7-en-3β-ol, *J. Biol. Chem.* **268**:22591–22599.

Turley, S. D., and Dietschy, J. M., 1988. Cholesterol metabolism and excretion, in: *The Liver, Biology and Pathology.* (I. M. Arias, W. B. Jakoby, H. Popper, D. Schachter, and D. A. Shafritz, eds.) pp. 467–492, Raven Press, New York.

Twisk, J., 1994, Regulation of cholesterol 7α-hydroxylase and sterol 27-hydroxylase in cultured rat hepatocyte, Ph.D. dissertation, Copyshop Pre-Klinische Laboratoria, Leiden, The Netherlands.

Twisk, J., Lehmann, E. M., and Princen, H. M. G., 1993, Differential feedback regulation of cholesterol 7α-hydroxylase mRNA and transcriptional activity by rat bile acids in primary monolayer cultures of rat hepatocytes, *Biochem. J.* **290**:685–691.

Twisk, J., Hoekman, M. F. M., Lehmann, E. M., Meijer, P., Mager, W. H., and Princen, H. M. G., 1994, Insulin suppresses bile acid synthesis by down-regulation of cholesterol 7α-hydroxylase and sterol 27-hydroxylase gene transcription, *Hepatology* **21**:501–510.

Twisk, J., Hoekman, M. F. M., Mager, W. H., Moorman, A. F., Boer, P. A. J., Scheja, L., Princen, H., and Gebhardt, R., 1995, Heterogenous expression of cholesterol 7α-hydroxylase and sterol 27-hydroxylase genes in the rabbit liver lobulus, *J. Clin. Invest.* **95**:1235–1243.

Vlahcevic, Z. R., Heuman, D. M., and Hylemon, P. B., 1991, Regulation of bile acid biosynthesis, *Hepatology* **13**:590–600.

Wang, D. P., and Chiang, J. Y. L., 1993, Structure and nucleotide sequences of the human cholesterol 7α-hydroxylase gene (CYP7), *Genomics.* **20**:320–323.

Wuarin, J., and Schibler, U., 1990, Expression of the liver-enriched transcriptional activator protein DBP follows a stringent circadian rhythm, *Cell* **63**:1257–1266.

Xu, G., Salen, G., Shefer, S., Batta, A. K., Ness, G. C., Nguyen, L. B., Zhao, Z., Chen, T. S., Niemann, W., and Tint, G. S., 1993, Different feedback regulation of hepatic cholesterol and bile acid synthesis by glycocholic acid in rabbits, *Gastroenterology* **105**:1192–1199.

Xu, G., Salen, G., Shefer, S., Ness, G. C., Nguyen, L. B., Parker, T. S., Chen, T. S., Zhao, Z., Donnelly, T., and Tint, G. S., 1995, Unexpected inhibition of cholesterol 7α-hydroxylase by cholesterol in New Zealand white and Watanabe heritable hyperlipiidemic rabbits, *J. Clin. Invest.* **95**: 1497–1504.

Chapter 3

Polyprenyl Diphosphate Synthases

Kyozo Ogura, Tanetoshi Koyama, and Hiroshi Sagami

1. INTRODUCTION

Isoprenoids are the most diverse and abundant compounds occurring in nature. The study of isoprenoid biosynthesis began in order to increase our understanding of the process of cholesterol biosynthesis. The discovery of isopentenyl diphosphate (IPP), made independently by Lynen *et al.* (1958) and Bloch *et al.* (1959), opened the era of enzymatic studies on the fundamental chain elongation, which starts with the condensation between IPP and dimethylallyl diphosphate (DMAPP), with elimination of inorganic pyrophosphate. This chain elongation is interesting in that the reactions are regulated to proceed consecutively and terminate precisely at discrete chain lengths according to requirements of organisms. The chain length of the products varies widely from geraniol to natural rubber. In order to understand the mechanism of these reactions, it is essential to learn how many and what kinds of enzymes are involved in such chain elongation reactions.

Farnesyl diphosphate (FPP) synthase, discovered as an enzyme responsible for squalene biosynthesis in yeast extracts by Lynen *et al.* in 1959, was the first of this class of enzymes. Since then, attempts have been made to isolate as many enzymes as possible from various organisms. To date, more than a dozen enzymes with different catalytic functions have been characterized, and they are collectively called "prenyltransferases." In a broad sense, however, prenyltransferases

---

Kyozo Ogura, Tanetoshi Koyama, and Hiroshi Sagami Institute for Chemical Reaction Science, Tohoku University, Katahira 2 chome, Aoba-ku, Sendai 980–77, Japan.

Subcellular Biochemistry, Volume 28: Cholesterol: Its Functions and Metabolism in Biology and Medicine, edited by Robert Bittman. Plenum Press, New York, 1997.

involve any enzymes that catalyze the transfer of prenyl groups to acceptors including not only IPP, but also non-isoprenoids such as aromatic compounds, proteins, etc. The term prenyltransferase is used in a narrow sense in this chapter and is limited to the prenyltransferase whose acceptor is IPP. In this case, prenyltransferase is equivalent to prenyl diphosphate synthase.

2. FPP SYNTHASE

2.1. Occurrence and Functions

The most widely occurring and best characterized prenyltransferase is FPP synthase, which occupies a particularly important branch point in the pathway of isoprenoid biosynthesis. The product, FPP, is the substrate not only for the first committed step for biosynthesis of cholesterol but also for the steps for physiologically important isoprenoids, including glycosyl carrier lipids, respiratory quinones, heme a, and prenylated proteins. The organisms that have been studied for FPP synthase include various mammalian tissues, avian liver, insects, higher plants, and eukaryotic and prokaryotic microrganisms. Homogeneously purified enzyme preparations have been obtained from human liver (Barnard and Popják, 1981), porcine liver (Yeh and Rilling, 1977), rat liver (Marrero *et al.,* 1992), chicken liver (Reed and Rilling, 1975), yeast (Eberhardt and Rilling, 1975), and *Bacillus stearothermophilus* (Koyama *et al.,* 1993). The enzyme from any organism catalyzes the condensations of IPP with DMAPP and with the resulting geranyl diphosphate (GPP) to ultimately yield all-*E*-FPP, i. e., $C_5 \rightarrow C_{10} \rightarrow C_{15}$ reactions, in the presence of Mg^{2+}, as shown in Scheme 1.

2.2. Molecular Cloning and Overexpression

FPP synthase is one of the key enzymes responsible for cholesterol synthesis and exhibits changes in the level of its activity in response to cholesterol fasting and feeding. Taking advantage of this fact, Clarke *et al.* (1987), first succeeded in isolating a cDNA for FPP synthase by using cDNA probes prepared from rat liver mRNA isolated from animals having induced and repressed levels of cholesterol synthesis.

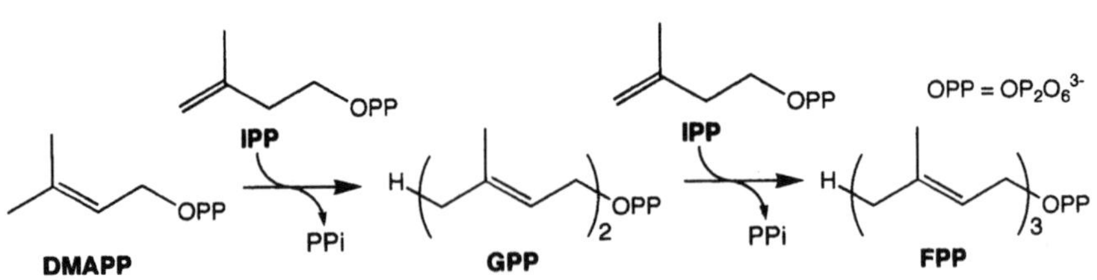

SCHEME 1. Reaction catalyzed by FPP synthase.

Thereafter, cDNAs or genomic clones encoding FPP synthase have been isolated from various organisms, including human (Sheares *et al.*, 1989), chicken (Wilkin *et al.*, 1990), yeast (Anderson *et al.*, 1989), *Escherichia coli* (Fujisaki *et al.*, 1990), *B. stearothermophilus* (Koyama *et al.*, 1993), and white lupin (Attucci *et al.*, 1995). All of these FPP synthases are homodimeric enzymes with subunits ranging in size from 32 kDa–44 kDa. Koyama *et al.* (1993) succeeded in overproducing and purifying FPP synthase from *B. stearothermophilus* in *E. coli.* In overproducing strains, the thermostable FPP synthase can account for as much as about 20% of the total cell proteins. The synthase is purified easily because of its thermostability and abundance. Heat treatment of cell-free extracts was particularly effective because the proteins of the host cells could be denatured and removed without loss of the activity of the *B. stearothermophilus* enzyme. After purification by hydrophobic chromatography followed by ion exchange (Mono-Q) chromatography, the enzyme was homogeneous (as judged by SDS-PAGE) and was crystallized (Nakane *et al.*, 1993). The subunit is composed of fewer (297) amino acid residues than those of any other prenyltransferases so far identified. The enzyme is not inactivated, even after treatment at 65 °C for 1 h. The catalytic turnover for the synthesis of FPP from GPP and IPP is 659 min^{-1} at 55 °C. Song and Poulter (1994) succeeded in overproducing and purifying a recombinant form of yeast FPP synthase by taking advantage of the addition of a C-terminal Glu-Glu-Phe α-tubulin epitope, which allowed efficient purification by immunoaffinity chromatography. Similarly, recombinant FPP synthases of rat liver (Marrero *et al.*, 1992) and avian liver (Song and Poulter, 1994) were overproduced and purified.

2.3. Crystallization and Three-Dimensional Structure

Crystallization was first achieved for chicken liver FPP synthase by Rilling and coworkers in 1975. However, no crystallographic characterization had been reported for a long time. After nearly 20 years, *B. stearothermophilus* FPP synthase overproduced in *E. coli* was crystallized, and preliminary X-ray diffraction data were determined to about 3Å resolution (Nakane *et al.*, 1993). An epoch-making achievement was made by Sacchettini, Poulter, and coworkers (Tarshis *et al.*, 1994). They determined the crystal structure of avian recombinant FPP synthase to 2.6 Å resolution, which is the first three-dimensional structure of any prenyltransferase. Strikingly, the synthase has a novel folding structure composed of all α-helices joined by connecting loops. According to their analysis, 10 core helices are arranged around a large central cavity whose wall carries highly conserved amino acid sequences, which are assumed to be essential for catalytic activity as shown later.

2.4. Comparative Primary Structures

Comparison of the amino acid sequences of prenyltransferases from various organisms reveals the presence of seven highly conserved regions as shown in Figure 1 (Koyama *et al.*, 1993). Regions II and VI, which contain DDxxxx

```
(6) N.c.GGPP   : MAVTSSSPGPAPLSLLSNNDDFIAPFNINTKFPSAIVPPRTSSNQPISVAIPSNRISSAGLAATQQAQTRKRKASVAQISLPSMLPTSFSPYT
(7) S.c. HPP   : MFQRSGAAHHIKLISSRRCRFKSSFAVALNAASKLVTPKILWNNPISLVSKEMNTLAKNIVALIGSGHPVLNKVTSYYFETEGKKVRPLLVLLLSRALSEIPMTERNH

                                                                    I                                                  II
(1) B.s.  FPP  : MA  QLSVEQFLNEQ   K QAVETALSRYIER LEGPAKLKKAM      AYSLEA   GGKRIRPLL LLSTVRALGKDPAVGLPVA          CAIEMIHTYSLIHDDLPSMDNDDLRRG KPTN
(2) E.c.  FPP  : MDFPQQLEACV    KQAN QAL    SRFIAP LPFQNTPVVETM     QYGALL   GGKRLRPFL VYATGHMFGVSTNTLDAPAAAVE      C I   HAYSLIHDDLPAMDDDDLRRG LPTC
(3) S.c.  FPP  : MASE KEIRRERFLNVFPKLVEELNA SL     LAYGMPKEACDWYAHS     LNYNTP GGKLNRG L SVVDTYAILSNKTVEQLGQEEYEKVAILGW C IELLQAYFLVADDM  MDKSITRRG QP C
(4) hum.  FPP  : MNGDQNSDVYAQE KQDFV QHF    SQIVRV LTEDEMGHPEIGDAIARLKEVLEYNAI GGKYNRG L TVVVAFRELVEPRKQDADSLQRAWTVGW  C VELLQAFFLVADD I MDSSLTRRG Q TC
(5) rat   FPP  :  MNGDQKLDVHNQE KQNFI QHF    SQIVKVLTEDELGHPEKGDAITRIKEVLEYNTV GGKYNRG L TVVQTFQELVEPRKQDAESLQRALTVGW  C VELLQAFFLVLDD I MDSSHTRRG QI C
(6) N.c.GGPP   : MAP QPPQPPPNPDRFATEDFF SP  SRRTWSEEKEKVLTGP      YDYLNGHP   GKDIRSQMV KAFDAWLDVPSESLEVIT         KVISMLHTASLLVDDVE  DNSVLRRG FPVA
(7) S.c.HexPP  : LKIDKSDVPEDPIYSKPSQNQLFQRPASSISP  LHILHGIKPLN  PLTKGPEPLPEETF D  KQRGIL PKQRRLA                EIVEMIHTASLLHDDVI DHSDTRRG RPSG

                              III                                IV                                                           V
(1) B.s.  FPP  : H KVFGEAMA  IL AGDG LLTYAFQ  LITEIDDERIPPSVRLRLIERLAKAAGPEGMVA GQAAD MEGEGKTLTLSELEYIHRH                              KT GKMLQYSVH  AGALIG GADA  RQTRELDEFAAHL
(2) E.c.  FPP  : HVKF GEANAI L AG DA LQTLAFS ILSDADMPEVSDRDRISMISELASASGIAGMCG GQALD LDAEGKHVPLDALERIHRH                              KT GALIRAAVRLGALSAGDKG  RRALPVLDKYAESI
(3) S.c.  FPP  : WYKVPEVGE  I AINDAFMLE AAI YKLLKSHFRNEKYYIDITELFHEVTFQTEL    GQLMD LITAPEDKVDLSKFSLKKHSFIVTF                         KT AYYSFYLPV ALAMYVAGITDEKDLKQARDVLIPL
(4) hum.  FPP  : WYQKPGVG   LDAINDANLLE ACI YRLLK    LYCREQPYYLNLIELFLQSSYQTEI GQTLD LLTAPQGNVDLVRFTEKRYKSIVKY                         KT AFYSFYLPIAA AMYMAGIDGEKEHANAKKILLEM
(5) rat   FPP  : WYQKPGIG   LDAINDALLLE AAI YRL    LKFYCREQPYYLNLLELFLQSSYQTEI GQTLD LITAPQGQVDLGRYTEKRYKSIVKY                         KT AFYSFYLPIAA AMYMAGIDGEKEHANALKILLEM
(6) N.c.GGPP   : HSIF GIPQT  INTSNYVYFYALQEL QK  LKNPKAVSIFSEE    L   L NLHR   GQGMD LFWRDTLTCPTEDDYLEMVSN                             KT GGL  FRLGIKL  MQAESRSPVDCVPLVNII
(7) S.c.HexPP  : NAAFTNKMAV  L AG DF LLGRATV SISRLHNPEVVELMSNSIANLVEGEFMQLKNTSI DADID TIENGHKLLPVPSKKLEVKEHDFRVPSRQQGLQLSHDQIIETAFEYYIHKTYL KT AALISKSCRCAAILSGASPAVIDECYDFGRNL

                        VI                                                                                                                      VII
(1) B.s.  FPP  : GLAFQIRDDILDIEGAEEKIGKPVGSD  QSNNKATYPA          LLSLAGAKEKLAFHIEAAQRHLRNADVDG                                  AA  L   AYICELVAA RDH*
(2) E.c.  FPP  : GLAFQVQDDILDVVGDTATLGKRQGAD Q QLGK STYPA          LLGLE   QARKKARDLIDDARQSLKQLAEQSLDTSA                              LE  ALADYII  QRNK*
(3) S.c.  FPP  : GEYFQIQDDYLDCFGTPEQIGKI GTD I QDNKCSWVINKAL          ELASAE  QRKTLDENYGKKDSVAEAKCKKIFNDLKIEQLYHEYEESIAKDLKAKISQVDESRGFKADV LT  AFLNKVY K RSK*
(4) hum.  FPP  : GEFFQIQDDYLDLFGDPSVTGKI GTD I QDNKCSWLVVQCLQRATPEQYQILKENYGQKEAEKVARVKALYEELDLPAVFLQYEEDSYSHIMALIEQYAAPLPPAVF      LGL A RK  IYK RRK*
(5) rat   FPP  : GEFFQIQDDYLDLFGDPSVTGKV GTD I QDNKCSWLVVQC         LLRATPQ QRQILEENYGQKDPEKVARVKALYEELDLRSVFFKYEEDSYNRLKSLIEQCSAPLPPSIF LEL AN K  IYK RRK*
(6) N.c.GGPP   : GLIFQIADDYHNLWNREYTANKGMCED LTEG KFSFPVIHSIRSNPSNMQ LLNILKQKTGDEEVKRYAVAYMESTGSFEYTRKVIKVLVDRARQMTEDIDDGRGKSGGIHKILDRIM  LHQEENVAQ  KNGKKE*
(7) S.c.HexPP  : GICFQLVDDMLDFTVSGKDLGKPSGAD LKLGIATAP VLFAWKEDPSLGP LISRNFSERGDVEKTIDSVRLHNGIAKTKILAEEYRDKALQNLRDSLPESDARSA       L EFLTNS  ILT RRK*
```

FIGURE 1. Comparison of conserved amino acid regions of prenyltransferases. FPP synthases from (1), *B. stearothermophilus* (Koyama *et al.*, 1993); (2), *E. coli* (Fujisaki *et al.*, 1990); (3) *S. cerevisiae* (Anderson *et al.*, 1989); (4), human (Wilkin *et al.*, 1990); (5), rat (Clarke *et al.*, 1987); GGPP synthase from (6), *Neurospora crrasa* (Carattoli *et al.*, 1991); HexPP synthase from (7), *S. cerevisiae* (Ashby and Edwards, 1990).

DxxxxRRG and GxxFQxxDDxxD motifs, respectively, are the most conservative and characteristic. Site-directed mutagenesis studies have shown that the conserved Asp and Arg residues in these two motifs are crucial for catalytic function (Marrero *et al.*, 1992; Joly and Edwards, 1993; Song and Poulter, 1994). Comparing these sequences with those of enzymes utilizing DMAPP as the only isoprenoid substrate, including tRNA: dimethylallyl transferase, Ashby *et al.* (1990) assumed that these motifs in Regions II and VI should be involved in the binding of the homoallylic and allylic substrates, respectively.

2.5. Site-Directed Mutagenesis

2.5.1. Mutations of Conserved Acidic and Basic Amino Acid Residues

By examining site-directed mutants in which the conserved Asp and Arg residues in these motifs were mutated to Glu and Lys, respectively, Joly and Edwards (1993) showed that the mutagenesis of Asp^{104}, Asp^{107}, Arg^{112}, and Arg^{113} in Region II and Asp^{243} in Region VI of rat FPP synthase resulted in approximately 1000-fold decreases of V_{max} as compared to the wild-type enzyme. However, no significant changes in K_m values for either IPP or GPP were observed in conservative mutations in which the charge of the residue remained unchanged. These results suggest that these amino acid residues are involved in catalytic efficiency rather than binding affinity. On the other hand, Asp^{103} (the first Asp in Region II) and Asp^{247} (the third Asp in Region VI) could be replaced with Glu without marked changes in kinetic properties. Using a recombinant form of yeast FPP synthase extended by a C-terminal EEF α-tublin epitope, Song and Poulter (1994) conducted mutagenesis studies in which a charged residue was replaced with an aliphatic residue. They showed that the replacement of the conserved Asp residues, including the first Asp in Region II (Asp^{100} for yeast FPP synthase), by Ala resulted in dramatic decreases of V_{max} (10^{-3}–10^{-5}) except for the third Asp in Region VI (Asp^{244} for the yeast enzyme). They also showed that Lys^{254}, which is conserved downstream in Region VII, can be replaced by Ala without marked changes in kinetic properties.

Coupled together, these results indicate the importance of the six Asp residues in the consensus $D^1D^2xxxxD^3xxxxRRG$ (Region II) and $GxxFQxxD^4D^5xxD^6$ motifs as follows: D^2, D^3, and D^4 are so critical that none of them can be replaced by even Glu without significant decrease in catalytic activity, whereas D^1 can be replaced by Glu but not by Ala. However, D^6 is so tolerant that even Ala can substitute for it. This is also supported by mutagenesis studies with *B. stearothermophilus* FPP synthase (Koyama *et al.*, 1996).

A number of mutant enzymes with markedly decreased k_{cat} values have been created, but only a few mutants showed significant changes in K_m values for either the homoallylic substrate (IPP) or the allylic substrates (GPP and DMAPP). One of the few examples is the replacement of the first Asp residue (Asp^{243}) in the

GxxFQxxDDxxD motif of rat FPP synthase by Glu, which resulted in a 26-fold increase in the K_m value for IPP (Marrero *et al.*, 1992).

Another remarkable finding is that the replacement of the conserved Lys residues (Lys^{47} and Lys^{183}) in Regions I and V of *B. stearothermophilus* FPP synthase with aliphatic amino acids resulted in a marked depression of the affinity of the enzyme for IPP, thereby causing a change in the distribution of products (Koyama *et al.*, 1996). Table I shows kinetic parameters obtained with purified mutants of *B. stearothermophilus* FPP synthase. Both K47I and K183A have k_{cat} values approximately 70-fold lower than that of the wild-type enzyme. Neither of them shows significantly changed K_m values for GPP, but they both show markedly increased K_m values for IPP. These facts indicate that both Lys residues conserved in Regions I and V contribute to the binding affinity for IPP. The K_m values of these mutants for IPP are the largest of all reported for the mutant enzymes derived from *B. stearothermophilus* FPP synthase (Koyama *et al.*, 1994a,b, 1995).

Blanchard and Karst (1993) isolated a mutant gene (erg20-2) from a yeast strain that is defective in FPP synthase but able to excrete geraniol. They showed that the unusual property of excreting geraniol is attributed to a one-point mutation resulting in a substitution of Glu for Lys^{197} of *Saccharomyces cerevisiae* FPP synthase, which is the conservative Lys in Region V and which corresponds to Lys^{183} in the *B. stearothermophilus* enzyme. In view of their finding that a FPP synthase assay with a crude extract of this mutant strain produced GPP in preference to FPP, they assumed that the $Lys^{197} \rightarrow$ Glu mutation might decrease the affinity of the enzyme for GPP, thereby causing a release of GPP. This is interesting in light of the above-mentioned observation by Koyama *et al.* (1996) that both

Table I
Kinetic Parameters of Wild-Type and Mutant FPP Synthases of *B. stearothermophilus* (I)

FPP synthase	K_m (IPP)[a] (μM)	K_m (GPP) (μM)	K_m (DMAPP) (μM)	k_{rel}[b]
Wild type	16.2	9.7	13.3	1
D224A	16.4	2.7	12.2	7.2×10^{-5}
D224E	17.7	3.3	22.3	1.0×10^{-4}
D225I	16.2	5.2	9.4	1.9×10^{-5}
D228A	160	8.6	77.9	8.9×10^{-2}
K238A	70.4	7.8	14.1	3.4×10^{-2}
K238R	57.9	8.9	11.4	3.9×10^{-2}
K47I	340	4.8	13.7	1.4×10^{-2}
K183A	303	11.8	ND[c]	1.5×10^{-2}

[a]For the reaction with GPP.
[b]Realative k_{cat} *value.*
[c] Not determined.

Table II
Relative Amounts of Products Formed by FPP Synthase Mutants[a]

	Percent product formation at			
	0.5 μM of IPP		25 μM of IPP	
FPP synthase	GPP (%)	FPP (%)	GPP (%)	FPP (%)
Wild type	11	89	2	98
K47I	61	39	6	94
K183A	55	45	15	85

[a]The concentration of DMAPP was 25 μM.

K47I and K183A showed markedly increased K_m values for IPP, since a decrease of the affinity for IPP might also cause an accumulation of GPP. In order to examine this possibility, experiments were carried out to measure the relative amounts of GPP and FPP produced from DMAPP and IPP in the reaction catalyzed by these mutant enzymes (Table II). When the concentration of IPP was the same as that of DMAPP, the accumulation of GPP in the reaction by these mutants was less than 20%. However, when the concentration of IPP was reduced to 1/50, the accumulation of GPP reached as much as approximately 60%, while the accumulation of GPP in the reaction catalyzed by the wild-type enzyme was 27% under the same conditions. Therefore, the increased accumulation of GPP in these mutants seems to reflect their lowered affinities for IPP, which causes decreases in the velocity of the condensation with GPP at reduced concentrations of IPP. This is similar to the observation made by Blanchard and Karst (1993) with the yeast mutant extract. It is noteworthy that K47I and K183A resemble each other with respect to both kinetic parameters and product distributions. These results raise the possibility that the release of GPP in the reaction catalyzed by the yeast mutant enzyme might also be due to depression of the binding affinity for IPP rather than for GPP. It is also conceivable that GPP synthase, which is found in higher plants but whose structure is unknown, might be a product of a simple mutation on FPP synthase.

2.5.2. Effects of Mutation of Cysteine Residues

It has been suggested that Cys residues play important roles in the catalytic function or the substrate binding of several prenyltransferases, including mammalian FPP synthase (Holloway and Popják, 1967; Barnard and Popják, 1981) and bacterial hexaprenyl diphosphate synthase (Yoshida *et al.*, 1989a), because these enzymes are inhibited by sulfhydryl-blocking reagents and activated by sulfhydryl compounds such as mercaptoethanol. The activation by sulfhydryl reagents is especially remarkable in human liver FPP synthase. Moreover, there are two chromatographically separable forms in FPP synthase from porcine liver, which are attributed to the oxidation–reduction behavior of Cys residues of the enzyme (Yeh and Rilling, 1977; Koyama *et al.*, 1977; Barnard *et al.*, 1978). Similar observations have also been made with geranylgeranyl diphosphate (GGPP) synthase from *Methanobacterium thermoautotrophicum* (Chen and Poulter, 1993).

B. stearothermophilus FPP synthase is thermostable and unique in that it possesses only two Cys residues, in contrast to FPP synthases from other sources that have four–six Cys residues. It is therefore advantageous to use this synthase for studying the role of Cys residues. As shown by the relative positions of the conserved regions and Cys residues (Figure 2), Cys^{73} appeared to be significant because it is well conserved and located in Region II, which is presumably involved in catalytic function. On the other hand, Cys^{289}, which is located in Region VII, is not conserved. In order to explore the significance of these Cys residues, site-directed mutagenesis studies were undertaken with this thermostable FPP synthase as follows (Koyama *et al.*, 1994a). The Cys residue at position 73 was changed to Phe or also to Ser to produce mutants (C73F and C73S) that have phenyl or hydroxyl groups in place of the sulfhydryl group, respectively. The other Cys residue at position 289 was also replaced with Phe or Ser to produce C239F and C239S, respectively. The mutant enzyme, C73S-C289S, in which both Cys^{73} and Cys^{289} are replaced with Ser, was made by connecting mutated gene fragments derived from the C73S and C289S mutants.

Overexpression and subsequent purification of the mutant enzymes were carried out as previously described for the wild-type enzyme, and these mutants were kinetically analyzed. As a result, all of the five mutants showed FPP synthase activity, indicating that neither of the Cys residues at positions 73 and 289 is essential for catalytic function. In view of the fact the wild-type enzyme is inactivated

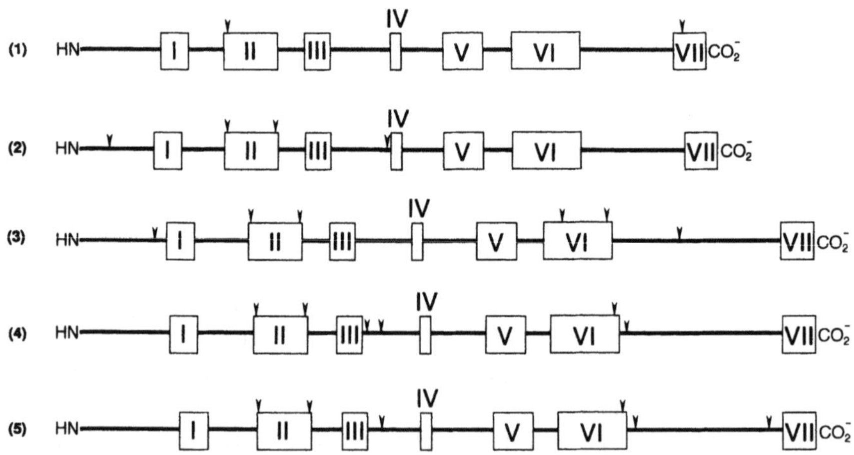

FIGURE 2. Relative positions of the conserved regions (I to VII) within each of the prenyltransferases. Notations of the prenyltransferases, (1) to (5), are the same as described in the legend to Figure 1. The relative positions of regions within the enzymes are drawn to scale. Arrow heads indicate the location of Cys residues.

Table III
Effect of PCMB on the Activities of Wild Type and C73S-C289S Synthase (I)

PCMB concentration (μM)	Percent enzyme activity	
	Wild type	C73S-C289S
0	100	100
50	63	104
100	51	90
150	17	89

by *p*-(chloromercuri)benzoic acid (PCMB), it is surprising that even the Cys[73] residue, which is fully conserved in Region II of all FPP synthases from various organisms, is not essential for catalytic activity. This will provide a warning when inhibitory effects by sulfhydryl-blocking reagents are interpreted. The sulfhydryl reagent may inactivate the enzyme by introducing groups that are bulky enough to cause a conformational change rather than by losing functional sulfhydryl groups. In fact, PCMB is inhibitory toward the wild-type enzyme but not at all toward the C73S-C289S mutant (Table III). Table IV shows kinetic parameters of the purified mutants, along with those of the wild-type enzyme. The mutant enzymes having replacements with Ser, C73S, C289S, and C73S-C289S showed k_{cat} values comparable to that of the wild-type enzyme. It is noteworthy that C73S and C289S show higher k_{cat} values than the wild-type enzyme. The replacement of Cys[73] with Phe resulted in a marked decrease in the heat stability of the enzyme, while the replacement of the same Cys with Ser had no such effect.

Table IV
Kinetic Parameters of the Wild Type and Mutant Enzymes (II)

Enzyme	K_m (GPP) μM	K_m (IPP)[a] μM	k_{cat}[b] (s^{-1})
Wild type	8.4	13.0	7.86
C289F	7.5	130	2.83
C73S	7.3	11.6	10.57
C289S	6.4	14.3	10.05
C73S-C289S	5.6	13.3	8.41

[a]For the reaction with GPP.
[b]k_{cat} was calculated from the *Vmax* value for the reaction with IPP and GPP.

Unexpectedly, the mutant C73S-C289S, as well as the wild-type enzyme, displayed two forms that were separable on Mono-Q chromatography. Therefore, the reported observation of chromatographic separation of the two forms of FPP synthase should not be ascribed only to a conformational change involving Cys residues. As far as the *B. stearothermophilus* enzyme is concerned, the resolution of the two forms is unrelated to modification of any Cys residues.

Although the K_m values of these mutant enzymes for GPP are similar to that of the wild-type enzyme, the K_m value of C289F for IPP is 10-fold greater than that of the wild-type enzyme. These facts suggest that the binding affinity for IPP is affected by the substitution of Cys^{289} with Phe. Similar tendencies have been observed in site-directed mutagenesis studies focused on the three C-terminal amino acids of this enzyme.

2.5.3. Effect of Modification in C-terminal Regions

Mammalian FPP synthases share a C-terminal region (Region VII) characterized as being crowded with basic amino acids as indicated by the KIYKRRK terminus (see Figure 1). However, only the Arg residue at the third position from the C-terminus is completely conservative if bacterial FPP synthases are taken into account. To examine the significance of this Arg in the *B. stearothermophilus* enzyme, it was mutated to Val (Koyama *et al.,* 1994b). This mutation, however, resulted in no marked decrease of catalytic activity, indicating that the conserved Arg residue in Region VII is not important for catalysis. Therefore, the Arg residues in the consensus RRG sequence in Region II, as already described, must be responsible for the inactivation of prenyltransferases by Arg-specific reagents such as 2,3-butanedione and phenylglyoxal (Barnard and Popják, 1980; Yoshida *et al.,* 1989a). This is more likely because the RRG motif is conserved not only in FPP synthases but also in all other prenyl diphosphate synthases. Experiments with hexaprenyl diphosphate synthase have indicated that the diphosphoryl groups of IPP and FPP in the presence of Mg^{2+} most efficiently protect the synthase against inactivation by 2,3-butadione and that, in contrast, the inactivation by *N*-ethylmaleimide is protected by the alkyl portion of FPP or its analog, farnesyl monophosphate (Yoshida *et al.,* 1989a). This also supports the idea that these Arg residues are directedly involved in the active site of enzyme.

The R295V mutant showed a three-fold increased K_m value for IPP but showed an unchanged K_m value for GPP compared with the wild type. Similarly, mutations of the penultimate Asp residue and the C-terminal His residue to Gly and Leu, respectively, resulted in considerable increases of K_m for IPP without affecting K_m for GPP (Koyama *et al.,* 1994b). A similar tendency has also been reported by Song and Poulter (1994), who showed that addition of a Glu-Glu-Phe epitope to the C-terminus of yeast FPP synthase caused a 14-fold increase in K_m for IPP but no significant change in K_m for GPP. Thus, it seems to be a general tendency that modification near the C-terminus causes a decrease in affinity for IPP.

2.5.4. Significance of Phenylalanine and Glutamine

As already described, some of the conserved Asp, Arg, and Lys residues are crucial for catalysis. These residues are assumed to be involved in the interaction with the diphosphoryl moieties of allylic and homoallylic substrates in association with divalent metal ions. However, in view of the structural requirement of the alkyl moieties of substrates, hydrophobic amino acid residues must also be involved in productive binding of the substrates. The Phe-Gln motif in Region VI appears interesting in this respect, because it is thoroughly conserved as shown in Figure 1.

To explore the significance of the conserved Phe-Gln motif, Koyama *et al.* (1995) conducted polymerase-chain reaction (PCR)-mediated site-directed mutagenesis of an expression plasmid prepared for *B. stearothermophilus* FPP synthase. The Phe^{220} residue was changed to Ala to analyze the difference between the aromatic and aliphatic hydrocarbon moieties at this position, and Gln^{221} was replaced with Glu to analyze the importance of the carbamoyl moiety. The resulting mutants, F220A and Q221E, were overproduced and purified to better than 95%.

As shown in Table V, both F220A and Q221E showed dramatically decreased catalytic activities, having only approximately 10^{-5} and 10^{-3} times of the wild-type activity, respectively. These facts indicate that both residues of the FQ motif are critical for enzymatic catalysis. Although the K_m values of the two mutants for IPP were comparable to that of the wild type, the K_m value of Q221E for DMAPP was 25-fold larger than that of the wild-type enzyme. On the other hand, substitution of Ala for Phe^{220} (F220A) caused a decrease in K_m for GPP but a slight increase in K_m for DMAPP. These facts indicate that Gln^{221} plays an important role in the binding for allylic substrates, especially for DMAPP. Preliminary experiments have shown that replacement of Gln^{221} by Asn also results in a marked loss of catalytic activity (Koyama *et al.*, unpublished results). Song and Poulter (1994) reported that the substitution of Ala for the third Asp in the

Table V
Kinetic Parameters of Wild Type and Mutant FPP Synthase (III)

FPP synthase	K_m (IPP)[a] μM	K_m (GPP) μM	K_m (DMAPP) μM	k_{cat}[b] (s^{-1})
Wild type	16.2	9.7	13.3	10.99
F220A	24.9	3.5	22.5	1.18×10^{-4}
Q221E	14.8	16.0	327.4	1.14×10^{-2}

[a]For the reaction with GPP.
[b]k_{cat} was calculated by the *V*max value for the reaction with IPP and GPP.

GxxFQxxDDxxD motif (Asp^{244} of yeast FPP synthase) resulted in a reduction in k_{cat} by a factor of 10, a five-fold increase in K_m for IPP, and a two-fold increase in K_m for GPP. A considerable number of FPP synthase mutants are known to show significantly increased K_m values for IPP (Marrero *et al.*, 1992; Joly and Edwards, 1993; Song and Poulter 1994; Koyama *et al.*, 1994a). However, Q221E is the first mutant to be found to have such a markedly increased K_m value for the allylic substrate, DMAPP.

In contrast, the K_m value for GPP was not significantly affected by this mutation. This may be explained by assuming that the affinity for GPP is dominated by a strong interaction between its long hydrocarbon chain and nearby hydrophobic amino acid residues, which probably form a cluster capable of accommodating various alkyl chains of proper size and shape, as suggested in previous studies on substrate specificity (Ogura *et al.*, 1970; Nishino *et al.*, 1971, 1972). It has also been found that *E*-2-heptenyl- and *E*-2-octenyl diphosphates are not active as substrates, whereas the corresponding 3-methyl derivatives are highly reactive substrates for FPP synthase (Ogura *et al.*, 1970). These findings indicate that the methyl group at the 3 position of 2-alkenyl diphosphates is an absolute requirement for being accepted as substrate. Taken together, these results raise the hypothesis that Phe^{220} might bind to the 3-methyl group so as to hold the DMAPP or GPP molecule in a proper direction necessary for the condensation with IPP.

Thus, Phe^{220} may be involved in catalysis by exerting an aromatic hydrophobicity on a methyl group or/and by exerting hydrophobic aromaticity on a prenyl cation, and Gln^{221} may exert its hydrogen-bonding capability on the prenyloxyl moiety to keep the allylic substrate aligned in a productive manner as well as to facilitate the cleavage of the P-O bond. This is also supported by the fact that

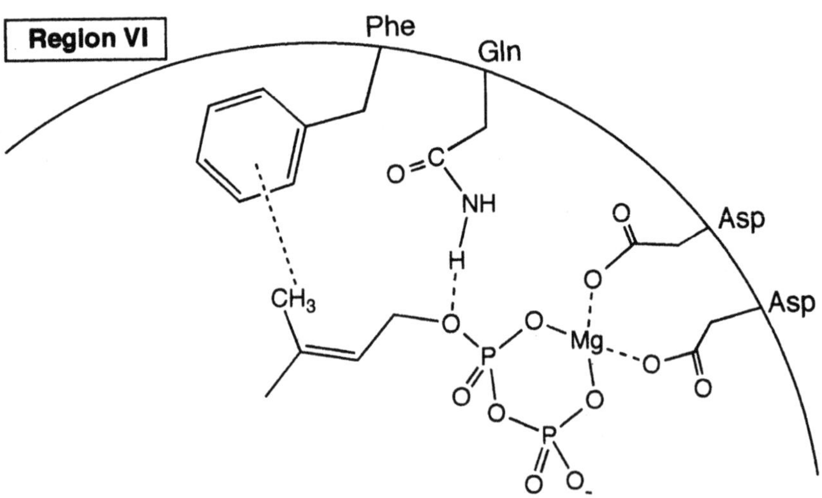

FIGURE 3. Hypothetical role of the two residues in the FQ motif in region VI.

the FQ motif is well conserved not only in all prenyl diphosphate synthases but also in tRNA-dimethylallyl transferase (Najarian *et al.,* 1987), which uses DMAPP as substrate. Figure 3 illustrates a hypothetical model of the roles of these two residues in the binding of an allylic diphosphate to region VI.

However, in view of the dramatic effect on k_{cat} of the replacement of the Phe group by Ala, Phe[220] might contribute to the acceleration of catalysis by stabilizing the prenyl cation, the formation of which is "rate limiting" through a π-cation interaction.

2.6. Mechanistic Considerations

Biosynthetic pathways often involve enzymes that catalyze reversible bond formations accompanied by concomitant liberation of inorganic pyrophosphate. Although the high energy involved in the P-O-P bond appears to be wasted in these reactions, it is ultimately utilized to displace the equilibrium of the reversible reaction in favor of the direction of synthesis, because the inorganic pyrophosphate liberated is irreversibly hydrolyzed to phosphate by the action of intracellular inorganic pyrophosphatase. However, the isoprenoid chain-elongation reaction catalyzed by prenyltransferase is unusual because this reaction is accompanied by liberation of inorganic pyrophosphate, despite the irreversibility of the reaction.

In order to explore the significance of the diphosphate moiety of prenyltransferase substrates, the effect of replacement of the diphosphate (POP) moiety (see Scheme 2) was studied (Gotoh *et al.,* 1987). The replacement of the POP moiety of DMAPP and GPP by a methylenediphosphonate (PCP) moiety (see Scheme 2) resulted in a marked (34 and 73 fold, respectively) decrease of V_{max},

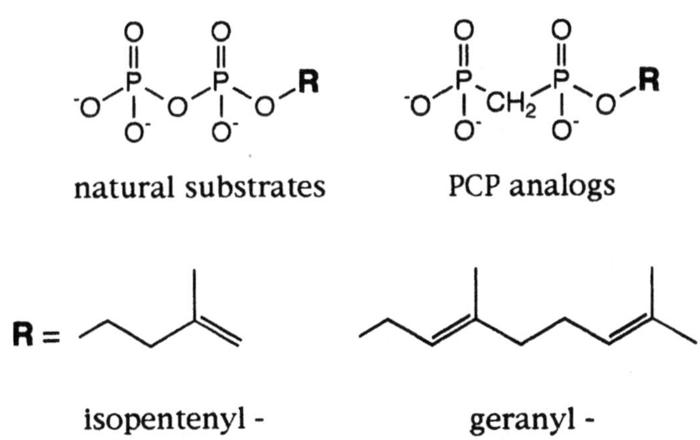

SCHEME 2. Structures of substrate analogs.

but the PCP analogues, dimethylallyl methylenediphosphonate (DMAPCP) and geranyl methylenediphosphonate (GPCP), were both accepted as substrates, with K_m values comparable to those of the corresponding diphosphate substrate. These results suggest that the high energy bond of the P-O-P linkage is not essential, but the diphosphate moiety is topologically and kinetically important. The decrease of V_{max} can be understood as a reflection of the poorer leaving nature of the PCP_i group compared with PP_i, and is consistent with the ionization–condensation–elimination mechanism proposed by Poulter *et al.* (1981). The velocity of the GPCP or DMAPCP reaction reached a maximum at a lower Mg^{2+} concentration than that of GPP or DMAPP. In addition, the optimal pH for GPCP was higher by 0.5 pH unit than for GPP. These phenomena can also be reflections of the stronger affinity for metal ions and the higher pK_a of PCP_i than of PP_i. These results suggest that the allylic diphosphate acts as a substrate in the form of the Mg^{2+} complex.

This suggests that the POP linkage of IPP could be replaced by a PCP linkage without marked loss of reactivity, because PP_i is released from allylic substrates but not from IPP. However, the V_{max} value of isopentenyl methylenediphosphonate (IPCP) in the reaction catalyzed by pig liver FPP synthase was also 18-fold lower that of IPP (Gotoh *et al.,* 1988, 1992). In addition, markedly different tendencies were observed between IPCP and GPCP with respect to the effects of pH and metal ion concentration. The optimal pH of the reaction of IPCP was around 6.6, whereas that of GPCP was 8.2. At pH 6.6, the IPCP reaction shows a hyperbolic curve as a function of $MgCl_2$ concentration, but at pH 8.5 it shows a maximum at about 0.5 mM $MgCl_2$ and then declines markedly as the Mg^{2+} concentration increases. Therefore, the behavior of IPCP is not consistent with the assumption that the IPCP-Mg^{2+} complex is the true substrate. The optimum pH of the reaction of IPCP was 6.6, which was smaller by 1.1 and 1.4 pH units than those for IPP and GPCP, respectively. The acidic pH should be unfavorable for substrate–metal complexation. In addition, the reactivity of IPCP at pH 8.2 was strongly inhibited when the concentration of Mg^{2+} was raised, whereas such inhibition was not observed at pH 6.6. These unexpected properties of IPCP as a function of pH and metal ion concentration can be interpreted by assuming that the homoallylic substrate, IPCP, reacts without binding to Mg^{2+}.

A possible mechanism of interaction between the enzyme and the substrates is as follows. The magnesium ion of the IPP-Mg complex is captured by the IPP binding site of the enzyme, and then transferred again to the newly formed allylic diphosphate, which is the product of condensation between the allylic diphosphate complexed with Mg^{2+} and the IPP molecule free of Mg^{2+}. During the condensation reaction, the enzyme may dynamically interact with the substrates and the products, including inorganic pyrophosphate, to maintain an efficient turnover of catalysis. For example, the PP_i liberated may bind to the enzyme, causing a con-

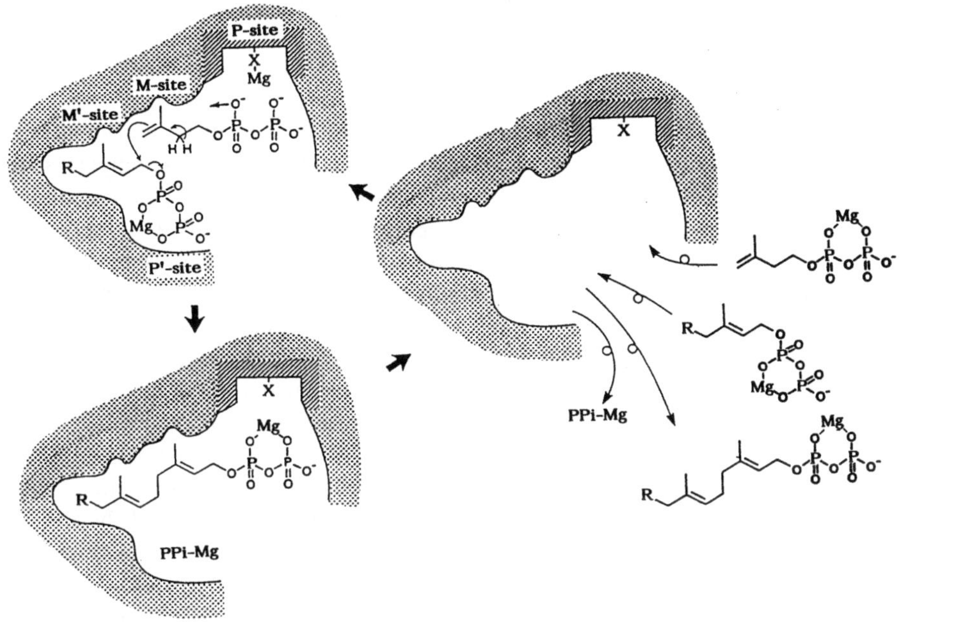

FIGURE 4. Hypothetical mechanism of the FPP synthase reaction.

formational change of the enzyme in such a way that the captured magnesium is released (Figure 4). The conformational change caused by the highly charged PP_i might also provide a driving force to move the intermediate allylic diphosphate into the proper position for the next condensation with IPP. The complex formation is favorable for the allylic substrate because it accelerates the release of PP_i to form the carbocation, but it seems unnecessary for IPP to be complexed with Mg^{2+} because PP_i is not released from IPP during condensation. IPP free of Mg^{2+} must be advantageous because the negatively-charged diphosphate moiety can act as a base to abstract the *pro-R* hydrogen at C-2 of IPP, thereby assisting the nucleophilic attack of the C-4 of IPP at the carbocation formed from the allylic substrate. The reason that nature selects the diphosphate moiety, and not monophosphate, in prenyltransferase reactions may be that this moiety can play dual functions by taking advantage of the ability to reversibly chelate with divalent cations under the control of the enzyme. Poulter and Rilling (1976) proposed a mechanism in which PP_i, when released from the allylic substrate, would be in a position for abstraction of the *pro-R* hydrogen at C-2 of IPP. Another possibility is that the proton would be abstracted by a basic residue in the enzyme active site. Crystallographic analysis of an enzyme-substrate complex is expected to show the conclusive mechanism that is operative.

3. GERANYLGERANYL DIPHOSPHATE SYNTHASE

3.1. History

After Lynen *et al.* (1958) discovered FPP synthase as an enzyme responsible for sterol biosynthesis in yeast, Kandutsch *et al.* (1964) found geranylgeranyl diphosphate (GGPP) synthase in a bacterium, *Micrococcus luteus,* which produces carotenoids but not steroids. This enzyme catalyzes the consecutive $C_5 \rightarrow C_{10} \rightarrow C_{15} \rightarrow C_{20}$ reactions to yield all-*E*-GGPP without accumulation of any intermediate.

The next study was focused on prenyltransferases in higher plants, which produce both sterols and carotenes, to see whether a single enzyme is responsible for the synthesis of both FPP and GGPP or whether two distinct enzymes are involved in the respective syntheses. In the latter case, it was also a major question as to whether the GGPP synthase is capable of catalyzing the $C_5 \rightarrow C_{10} \rightarrow C_{15} \rightarrow C_{20}$ reactions or responsible only for the $C_{15} \rightarrow C_{20}$ reaction. Using pumpkin fruit extracts as an enzyme source, Ogura *et al.* (1972) succeeded in the separation of FPP synthase and GGPP synthase from each other and demonstrated that these synthases catalyze the $C_5 \rightarrow C_{10} \rightarrow C_{15}$ and $C_5 \rightarrow C_{10} \rightarrow C_{15} \rightarrow C_{20}$ reactions, respectively. This fact indicates that the biosynthetic pathways to steroids and to carotenoids branch at the stage of C_5 so that the synthesis of GGPP can be independent of the production of FPP by FPP synthase.

3.2. Mammalian GGPP Synthase

During the search for prenyltransferases in various organisms, Sagami *et al.* (1981) found GGPP synthase with an unusual substrate specificity in pig liver. At that time, the occurrence of this enzyme in mammalian tissues was not generally accepted, since mammals lack the ability to synthesize diterpenes or carotenoids. Now, however, the significance of this enzyme has become undoubtedly clear because of the striking discovery of geranylgeranylated proteins as well as farnesylated proteins (Clarke, 1992). The purification of this GGPP synthase (Sagami *et al.,* 1992 and 1993b) was difficult to achieve because of its scanty occurrence in contrast to the abundant occurrence of FPP synthase, particularly in liver, which serves as a major site of cholesterol biosynthesis. However, chromatography on an affinity gel with a farnesylmethyl phosphonophosphonyl ligand was extremely efficient in the purification of GGPP synthase (Figures 5 and 6). Affinity chromatog-

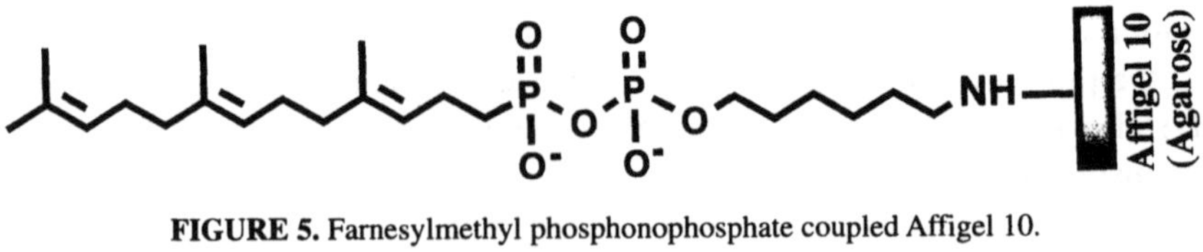

FIGURE 5. Farnesylmethyl phosphonophosphate coupled Affigel 10.

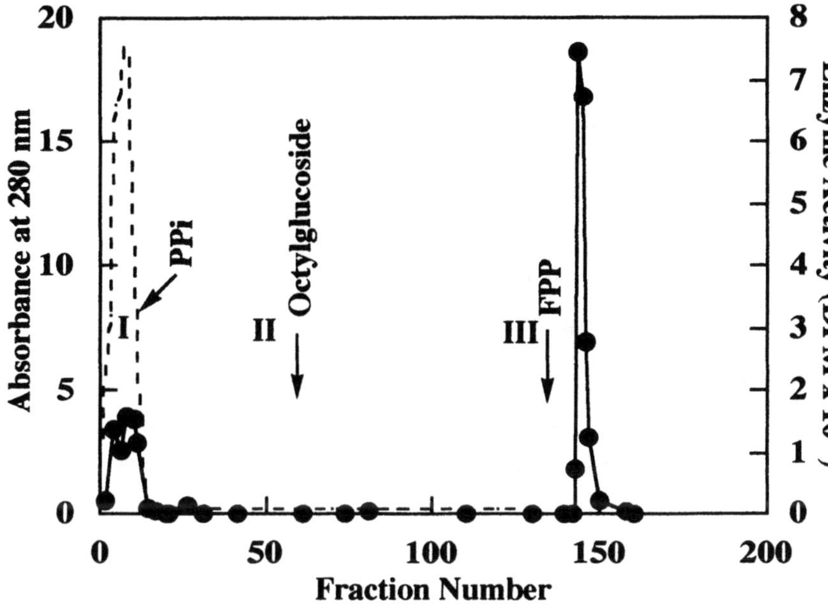

FIGURE 6. Affinity chromatography of GGPP synthase of bovine brain. A 40-60% ammonium sulfate fraction was dialyzed against buffer containing 0.1 M KCl and 1 mM PP_i (I) and applied to the affinity gel column. After the column was washed with buffer (I) and then with buffer containing 0.2% octyl glucoside (II), the proteins were eluted with a linear gradient of FPP (0-500 μM) (III). Dashed line, absorbance at 280 nm; ●, enzyme activity.

raphy followed by Mono-Q chromatography of bovine-brain extracts led to the first successful purification of mammalian GGPP synthase to homogeneity (Sagami *et al.*, 1994). The purified enzyme had a specific activity of 98 nmol/min/mg of protein as assayed with FPP and IPP. This synthase, unlike GGPP synthases from other organisms including plants and bacteria, does not catalyze the condensation of IPP with DMAPP, but catalyzes condensations of IPP with GPP and with FPP to ultimately yield all-*E*-GGPP. However, the cellular function of this enzyme must be to catalyze the condensation between IPP and FPP in view of the much better V_{max}/K_m value for FPP than for GPP. Therefore, the synthesis of GGPP in mammalian cells, unlike that in plants, is dependent on FPP synthase. The involvement of FPP synthase and GGPP synthase in isoprenoid biosynthesis is shown in Figure 7.

3.3. Subcellular Distribution and Regulation of Mammalian Prenyltransferases

In the earliest study, FPP synthase activity was found in a soluble fraction from rat liver homogenates (Goodman and Popják, 1960). The soluble FPP

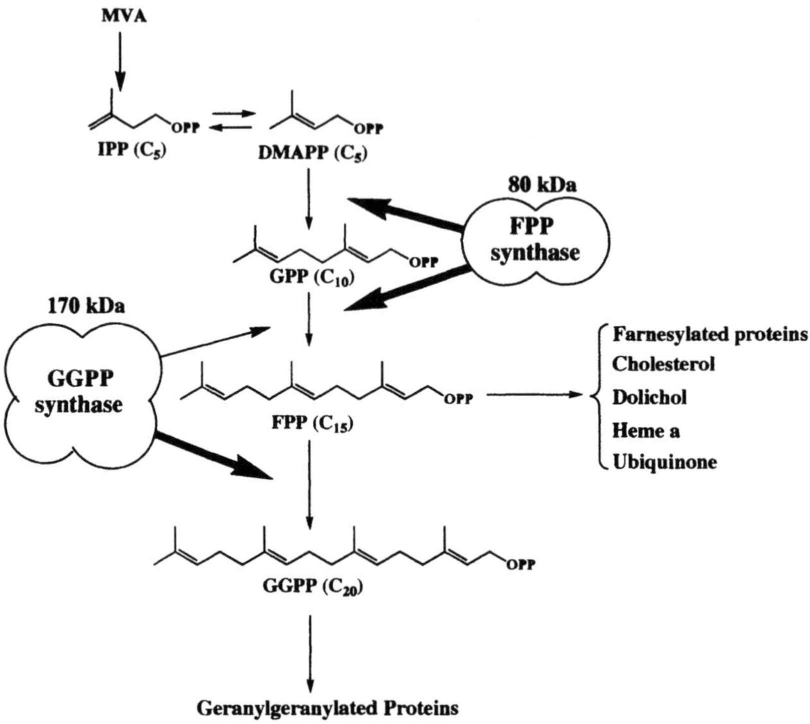

FIGURE 7. Isoprenoid pathways in mammalian cells.

synthase was then purified in homogeneous form from various mammalian sources and extensively studied as already described. GGPP synthase activity was first detected in a soluble fraction from pig liver extracts (Nandi and Porter, 1964), but it was reasonably attributed to a property of the FPP synthase since FPP synthase purified in homogeneous form from avian liver was shown to have a low capability to form GGPP from FPP and IPP (Reed and Rilling, 1975). Sagami *et al.* (1981), however, succeeded in chromatographically separating GGPP synthase and FPP synthase from pig liver soluble fractions. They also showed that the molecular weight of this GGPP synthase is much higher than that of FPP synthase, and that the GGPP synthase lacks dimethylallyl-transferring activity in contrast to FPP synthase. Therefore, it became evident that a GGPP synthase occurs in pig liver. At that time, however, the circumstances did not encourage them to study the GGPP synthase further, since its biological significance was unknown.

Recent findings of prenylated proteins, including farnesylated and geranylgeranylated proteins, have prompted many researchers to investigate the biosynthesis of shorter chain prenyl diphosphates. In particular, special attention has been paid to the GGPP synthase, since geranylgeranylated proteins were shown to be more abundant than farnesylated proteins (Epstein *et al.,* 1991). Runquist *et al.* (1992) found that soluble fractions of rat liver, brain, kidney, spleen, and testis homogenates were capable of converting IPP into GGPP as well as FPP. They demonstrated that the highest rate of GGPP synthesis was found in brain, testis, and spleen. Lutz *et al.* (1992) reported the incorporation of mevalonate into FPP and GGPP in rabbit reticulocyte cytosol. They demonstrated that these isoprenoids can downregulate their own synthesis *in vitro,* suggesting that this regulation may control the level of these polyisoprenoids *in vivo.* Sagami *et al.* (1993a) also investigated the synthesis of FPP and GGPP from IPP in soluble fractions of rat liver and brain. They observed that the formation of GGPP from either IPP or mevalonate was comparable to that of FPP, and that addition of DMAPP or GPP to the same enzyme systems in the presence of IPP caused a marked increase in the rate of formation of FPP without affecting the rate of formation of GGPP (Figure 8). A possible explanation for this observation is that GGPP synthase is saturated with FPP, even at low concentrations supplied from IPP by the action of IPP isomerase followed by FPP synthase, whereas FPP synthase is not saturated with DMAPP or GPP. The difference between the two synthases with respect to the degree of saturation seems to arise largely from their relative abundance rather than from their differential affinities for substrates. Therefore, the level of IPP isomerase must be important in regulating the flow of C_5 units to FPP, since GPP can only be supplied as an intermediate of the reaction of DMAPP and IPP catalyzed by FPP synthase. Ericsson *et al.* (1993) have also made similar observations with the soluble fraction of rat brain. These studies have demonstrated that the soluble fraction of mammalian cells contain FPP synthase and GGPP synthase that can be separated from each other. These soluble FPP and GGPP synthases seem to be responsible for supplying the lipid substrates for protein:farnesyltransferase and protein:geranylgeranyltransferase (Boguski and McCormick, 1993).

In addition to the soluble synthases, membrane-bound prenyl diphosphate synthases have been reported. In an earlier study (Ishii *et al.,* 1983a), microsomal FPP synthase was solubilized with Triton X-100 and characterized. The microsomal enzyme differs from the cytosolic enzyme in the pH optimum and susceptibility to metal ions. More markedly, their substrate specificities differ in that the dimethylallyl-transferring activity of the microsomal enzyme was much lower than the geranyl-transferring activity in contrast to the soluble enzyme, in which both activities are similar. The geranyl-transferring activity of the microsomal FPP synthase was activated about two-fold by Triton X-100, whereas the dimethylallyl-transferring activity was not affected. These results suggest that the allylic substrate for priming the chain elongation by the microsomal FPP synthase might be GPP rather than DMAPP, although more than one enzyme protein may be involved in the microsomal FPP synthase activity. It has been reported that rat liver microsomes contain an all-*E*-polyprenyl diphosphate synthase that utilizes GPP as the primer much more effectively than it does FPP (Teclebrhan *et al.,* 1993). FPP synthase is also found within peroxisomes and may thus supply

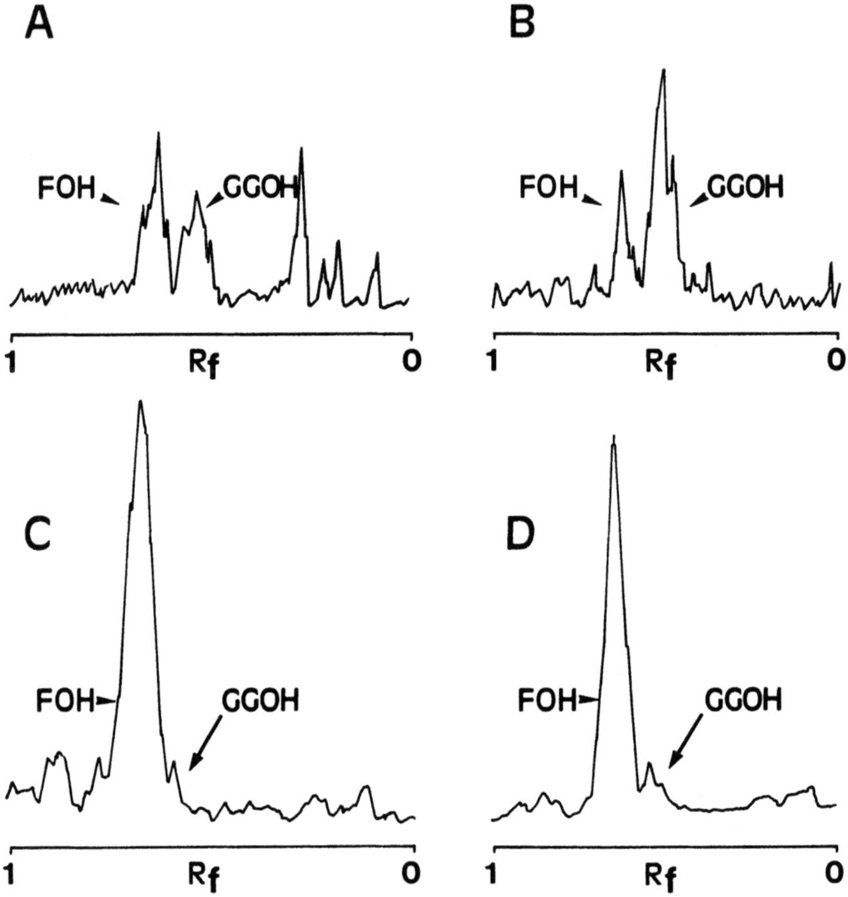

FIGURE 8. Radio-TLC of the prenols derived from the products of incubation of ^{14}C-IPP. ^{14}C-IPP (25 μM) was incubated with 100,000 x g supernatant of rat liver (A and C) or of brain (B and D) in the presence of 25 μM of GPP (C and D) or its absence (A and B). The products were hydrolyzed with alkaline phosphatase. FOH, farnesol; GGOH, geranylgeraniol.

substrates for squalene synthase and Z-prenyltransferase occurring in this organelle (Ericsson *et al.*, 1992). More recently, Krisans *et al.* (1994) analyzed FPP synthase activity and protein in fractionated rat liver together with immunofluorescent and immunoelectron microscopy studies, demonstrating that FPP synthase is largely localized in peroxisomes. Since dehydrodolichyl-PP synthase is located in microsomes (Wong and Lennarz, 1982; Adair and Keller, 1982; Sagami *et al.*, 1989; and Matsuoka *et al.*, 1991), it would be reasonable to assume that the product of microsomal FPP synthase is utilized as the allylic primer for dehydrodolichyl-PP synthase and as the substrate for squalene synthase. Sagami *et al.* (1991) found that rat liver microsomes catalyze the formation of *Z,E,E*-

GGPP as well as dehydrodolichyl-PP from FPP and IPP, suggesting the occurrence of *Z,E,E*-geranylgeranyl-PP synthase, which makes a connection between FPP synthase and dehydrodolichyl-PP synthase in microsomes. Similar observations were also made by Ericsson *et al.* (1993).

FPP synthase and GGPP synthase activities, as well as decaprenyl-PP synthase activity, have also been detected in pig liver mitochondria (Ishii *et al.*, 1983b). Although neither of the FPP and GGPP synthases has yet been characterized in detail, they seem to be implicated in the biosynthetic pathway to ubiquinone and a connection between IPP isomerase and decaprenyl-PP synthase, localized in the same organelle. Similar observations have been made with rat liver mitochondria, except that GGPP synthase activity was not detected (Runquist *et al.*, 1994).

In summary, compartmentation of these short- and long-chain prenyl-PP synthases into subcellular organelles such as microsomes, mitochondria, and peroxisomes would be required in mammalian cells for the syntheses of diverse end products such as all-*trans* long chain prenyl-PP for the side chain of ubiquinone, *cis-trans* mixed long-chain prenyl-PP for dolichyl-phosphate, squalene, and farnesylated and geranylgeranylated proteins. More detailed studies on organelle-bound short- and long-chain prenyltransferases will be necessary to understand the entire flow of isoprenoid biosynthesis. Detailed discussions on the branch-point reactions in the biosynthesis of cholesterol, dolichol, ubiquinone, and prenylated proteins are available in a review article (Grunler *et al.*, 1994).

The transcription of FPP synthase as well as 3-hydroxy-3-methylglutaryl (HMG)-CoA reductase is regulated by the sterol level of the cell. Edwards *et al.* showed that mutagenesis of the 6-base-pair sequence ATTGGC in the rat FPP synthase promoter impaired sterol-dependent transcription of this gene (Spear *et al.*, 1994). They also suggested that nuclear factor Y (NF-Y) plays a novel role in sterol-dependent regulation of the gene for both FPP synthase and HMG-CoA reductase, and that this role requires a specific interaction with the sterol regulatory element binding protein or related transcriptional factors (Jackson *et al.*, 1995).

4. BACTERIAL PRENYLTRANSFERASES

4.1. Medium-Chain Prenyl Diphosphate Synthases

Bacteria provide good sources to search for prenyltransferases that catalyze the synthesis of prenyl products with chain lengths longer than C_{20}, because they do not synthesize secondary metabolites but produce menaquinones or ubiquinones having polyprenyl substituents whose chain-lengths are varied in a species-specific manner (Jeffries *et al.*, 1967). Thus, it was of interest to see whether there would exist respective enzymes corresponding to the variety of the chain lengths of quinones. It was also important to explore bacterial enzymes, because *Z,E*-undecaprenyl phosphate is known as a lipid carrier of glycosyl transfer in the biosynthesis of cell wall polysaccharide components in bacteria.

Interestingly, studies with bacteria have led to unexpected results. The C_{30} chain elongation is catalyzed by a novel enzyme system. It catalyzes the synthesis of all-*E*-hexaprenyl diphosphate (HexPP) by adding three molecules of IPP to FPP, i.e., $C_{15} \rightarrow C_{20} \rightarrow C_{25} \rightarrow C_{30}$, but it cannot catalyze the synthesis of GPP or FPP from DMAPP and IPP, i. e., $C_5 \rightarrow C_{10} \rightarrow C_{15}$. This enzyme was found in *M. luteus* B-P 26, which produces menaquinone-C_{30} (Fujii *et al.*, 1982). Similarly, heptaprenyl diphosphate (HepPP) synthase has been found in *B. subtilis,* which produces menaquinone-C_{35} exclusively (Takahashi *et al.*, 1980). It catalyzes the $C_{15} \rightarrow C_{20} \rightarrow C_{25} \rightarrow C_{30} \rightarrow C_{35}$ reactions to give all-*E*-HepPP, but it is not able to catalyze the $C_5 \rightarrow C_{10} \rightarrow C_{15}$ reactions either. The inability of these synthases to catalyze the $C_5 \rightarrow C_{10} \rightarrow C_{15}$ reactions implies the occurrence of FPP synthase in bacteria, which was shown to be the case by isolating it from a variety of bacteria.

The purification of these enzymes was accompanied by unexpected difficulties. The difficulties were due not only to the unexpected substrate specificities as described above but also to the fact that the synthesis of HexPP requires two separable proteins, neither of which is catalytically active alone (Fujii *et al.*, 1982). Therefore, the HexPP synthase had been elusive before this fact was disclosed. Figure 9 shows the chromatographic separations of the two proteins responsible for the synthesis of HexPP.

Similarly, two different proteins responsible for HepPP synthesis were separated from *B. subtilis* (Fujii *et al.*, 1983). The components of HexPP synthase and HepPP synthase are so specific that neither of the two components of one enzyme is exchangeable with either of the two components of the other enzyme. They therefore appear to be novel heterodimeric enzymes, with the subunits being dissociated under physiological conditions.

Taking advantage of the thermostability of *B. stearothermophilus* enzyme, Koike-Takeshita *et al.* (1995) identified the genes encoding such two protein components that constitute a medium-chain polyprenyl diphosphate synthase. One of the two proteins constituting an enzyme system for the synthesis of HepPP synthase has 323 amino acid residues and it has a 32% sequence similarity to the FPP synthase of the same bacterium. In addition, it has highly conserved motifs that are assumed to be essentially involved in catalysis. The other protein, which is composed of 220 amino acids, has no such similarity, nor are similar protein entries found in databases. Therefore, it seems likely that the former carries substantial active sites for substrate binding and catalysis, while the latter plays an auxiliary, but essential, role in catalytic function. Several lines of indirect evidence have led to a proposal for the mechanism of this unusual enzyme system as shown in Figure 10. This might account for the ability of these soluble proteins to efficiently synthesize amphipathic molecules from water-soluble substrate without association with membrane components (Yoshida *et al.*, 1989b).

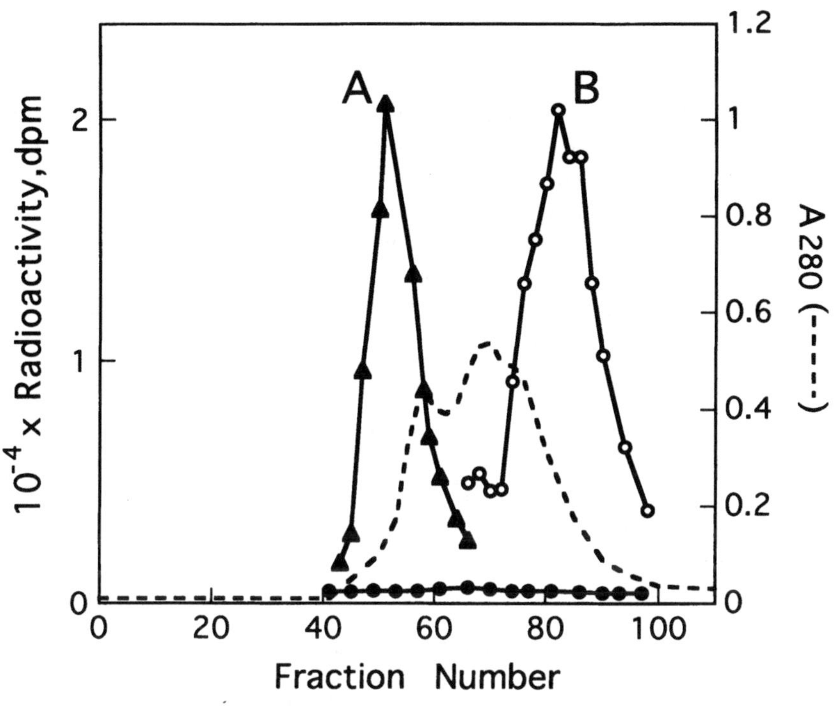

FIGURE 9. Chromatographic separation of the two protein components of HexPP synthase of *M. luteus* B-P 26. A 0–80% ammonium sulfate fraction was chromatographed on hydroxyapatite. Enzyme activity was assayed with ^{14}C-IPP in the presence of FPP (●), FPP plus 0.25 ml of fraction 54 (○), and FPP plus 0.25 ml of fraction 85 (▲).

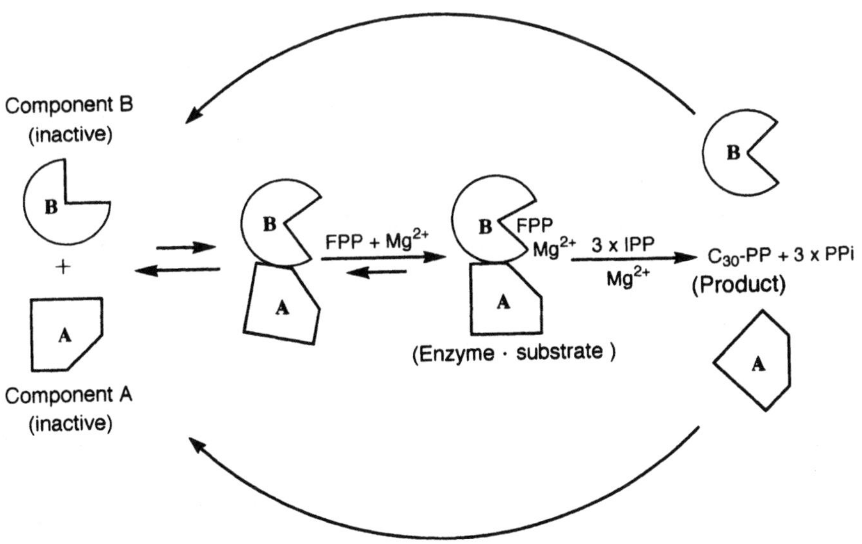

FIGURE 10. Possible mechanism of the two component system of HexPP synthase of *M. luteus* B-P 26.

In experiments with a yeast mutant in coenzyme Q biosynthesis, Ashby and Edwards (1990) isolated a gene from a plasmid containing a wild-type genomic DNA fragment that is able to complement the mutant and restore HexPP synthase activity. They also showed that this gene encodes a 473-amino acid protein having highly conserved domains characteristic of prenyl diphosphate synthases, as shown in Figure 11. Therefore, this protein seems to correspond to the larger protein component of *B. stearothermophilus* HepPP synthase (Koike-Takeshita *et al.*, 1995). However, it is not known whether the yeast 473 amino-acid protein acts as HexPP synthase by itself or in association with another gene product similar to the

FIGURE 11. Comparison of amino acid sequences of the larger component of *B. stearothermophilus* with several prenyltransferases. FPP synthases: (1), from *B. stearothermophilus*; (2), from *E. coli*; (3), from *S. cerevisiae*. GGPP synthases: (4), from *E. herbicola*; (5), from *E. uredovola*; (6), hexaprenyl PP synthase from *S. cerevisiae*; (7), the larger component of heptaprenyl PP synthase of *B. stearothermophilus*. Amino acids identical in at least 4 of 7 sequences in the conserved regions (I–VII) are shaded.

smaller protein. If the latter is the case, the mutant described above must be deficient in one of the two components of HexPP synthase. It would be interesting to learn whether dissociable heterodimeric systems are common to the medium-chain prenyl diphosphate synthase of both prokaryotic and eukaryotic cells.

All-*E*-pentaprenyl (all-*E*-geranylfarnesyl) diphosphate synthase has been purified from an archaebacterium (Tachibana, 1994). This synthase can catalyze the continuous $C_5 \rightarrow C_{10} \rightarrow C_{15} \rightarrow C_{20} \rightarrow C_{25}$ reactions, and it has no two-component system. Therefore, pentaprenyl diphosphate synthase should be grouped into the class of short-chain prenyl diphosphate synthases in terms of the mode of subunit composition.

4.2. Long-Chain *E*-Prenyl Diphosphate Synthases

By analogy with HexPP- and HepPP synthases, it was predicted that the enzymes for the synthesis of polyprenyl diphosphate with chain lengths longer than C_{35} would also have similar two-component systems. However, this prediction was incorrect. *E*-Nonaprenyl (solanesyl) diphosphate synthase was isolated from *M. luteus* as an enzyme catalyzing the chain elongation up to C_{45} from C_{10} (Sagami *et al.*, 1977), and it was found that this enzyme is a homodimeric protein that is functionally active by itself. In this respect, it resembles FPP synthase, but it differs in that it requires a protein factor to maintain an efficient catalytic turnover (Ohnuma *et al.*, 1991). This protein was separated as a high molecular weight component of a soluble fraction of the same bacterium. It dramatically stimulates the synthesis of solanesyl PP in a time-dependent manner, but it does not affect the stability of the enzyme. Bovine serum albumin or detergents such as Tween 80 can be a substitute for this protein, showing a similar manner of stimulation. These results indicate that the protein facilitates catalytic turnover by removing hydrophobic products that otherwise would inhibit the enzyme reaction from the active site of the enzyme (Figure 12). Therefore, this protein should be designated as polyprenyl

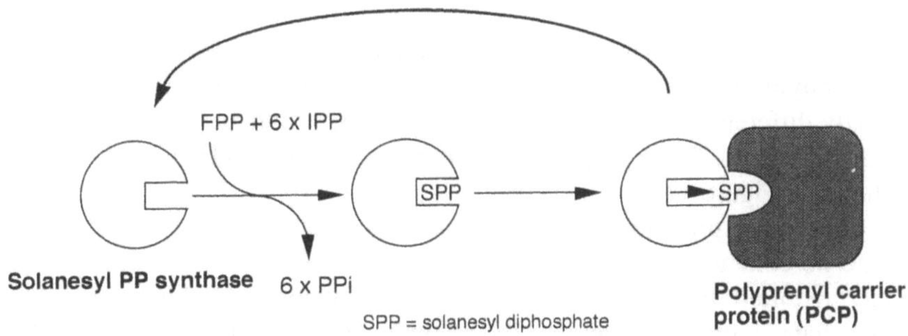

FIGURE 12. Turnover of polyprenyl PP synthases dependent on PCP. SPP, Solanesyl diphosphate.

carrier protein (PCP). All-*E*-octaprenyl PP synthase and all-*E*-decaprenyl PP synthase have been found and partially purified from *E. coli* (Fujisaki *et al.*, 1986) and *Paracoccus denitrificaus* (Ishii *et al.*, 1985), respectively. These polyprenyl PP synthases also have similar properties to those of the solanesyl PP synthase described above. The polyprenyl carrier protein (PCP) isolated from *M. luteus* acts effectively on these synthases as well. However, PCP shows no effect on undecaprenyl (all-*E*-farnesyl-all-*Z*-octaprenyl) PP synthase, which is also a homodimeric enzyme. Therefore, PCP acts specifically on all-*E*-polyprenyl diphosphate with chain lengths of C_{40}, C_{45}, and C_{50}.

4.3. *Z*-Prenyl Diphosphate Synthases

Undecaprenyl PP synthase occurs widely in bacteria and catalyzes the construction of a *Z*-prenyl chain onto all-*E*-FPP as a primer to yield a C_{55}-prenyl diphosphate with *E, Z*-mixed stereochemistry (Allen *et al.*, 1976; Baba and Allen, 1980; Takahashi *et al.*, 1982; Muth and Allen, 1984). This product is hydrolyzed to its monophosphate, which acts as the glycosyl carrier lipid in the biosynthesis of cell wall components. This synthase is so loosely associated with periplasmic membranes that it is easily solubilized, but it requires a phospholipid or detergents such as Triton X-100 for activity. *P. denitrificans,* which is assumed to be an ancestor of mitochondria, resembles higher animals in that it has ubiquinone-10 and all-*E*-decaprenyl PP synthase (Ishii *et al.*, 1985), which is responsible for the biosynthesis of the prenyl side chain of the quinone. More interestingly, however, this bacterium has all-*E*-farnesyl-all-*Z*-hexaprenyl (C_{45}) PP synthase (Ishii *et al.*, 1986) instead of the undecaprenyl PP synthase, which is common to general bacteria. Thus, a number of prenyltransferases corresponding to a variety of chain lengths and stereochemistry of prenyl diphosphates have been isolated from various organisms.

5. SUMMARY

It is noteworthy that in spite of the similarity of the reactions catalyzed by these prenyltransferases, the modes of expression of catalytic function are surprisingly different, varying according to the chain length and stereochemistry of reaction products. These enzymes are summarized and classified into four groups, as shown in Figure 13.

Short-chain prenyl diphosphates synthases such as FPP and GGPP synthases require no cofactor except divalent metal ions, Mg^{2+} or Mn^{2+}, which are commonly required by all prenyl diphosphate synthases. Medium-chain prenyl diphosphate synthases, including the enzymes for the synthesis of all-*E*-HexPP and all-*E*-HepPP, are unusual because they each consist of two dissociable dissimilar protein components, neither of which has catalytic activity. The enzymes

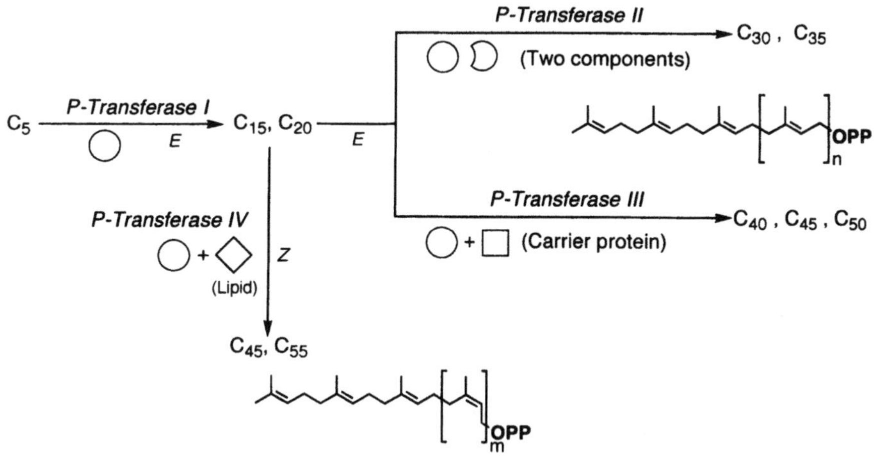

FIGURE 13. Classification of bacterial prenyltransferases.

for the synthesis of long-chain all-*E*-prenyl diphosphates, including octaprenyl (C_{40}), nonaprenyl-(C_{45}), and decaprenyl (C_{50}) diphosphates, require polyprenyl carrier proteins that remove polyprenyl products from the active sites of the enzymes to maintain efficient turnovers of catalysis. The enzymes responsible for *Z*-chain elongation include *Z,E*-nonaprenyl-(C_{45}) and *Z,E*-undecaprenyl (C_{55}) diphosphate synthases, which require a phospholipid.

The classification of mammalian synthases seems to be fundamentally similar to that of bacterial synthases except that no medium-chain prenyl diphosphate synthases are included. The *Z*-prenyl diphosphate synthase in mammalian cells is dehydrodolichyl PP synthase, which catalyzes much longer chain elongations than do bacterial enzymes. Dehydrodolichyl PP synthase will be a major target of future studies in this field in view of its involvement in glycoprotein biosynthesis.

6. REFERENCES

Adair, W. L., Jr., and Keller, R. K., 1982, Dolichol metabolism in rat liver: determination of the subcellular distribution of dolichyl phosphate and its site and rate of *de novo* biosynthesis, *J. Biol. Chem.* **257**:8990–8996.

Allen, M. C., Keenan, M. V., and Sack, J., 1976, *Lactobacillus plantarum* undecaprenyl pyrophosphate synthetase: purification and reaction requirements, *Arch. Biochem. Biophys.* **175**:236–248.

Anderson, M., S., Yarger J. G., Burck, C. L, and Poulter, C. D., 1989, Farnesyl diphosphate synthase: molecular cloning, sequence, and expression of an essential gene from *Saccharomyces cerevisiae*, *J. Biol. Chem.* **264**:19176–19184.

Ashby, M. N., and Edwards, P. A., 1990, Elucidation of the deficiency in two yeast coenzyme Q mutants: characterization of the structural gene encoding hexaprenyl pyrophosphate synthetase, *J. Biol. Chem.* **265**:13157–13164.

Ashby, M. N., Spear, D. H., and Edwards, P. A., 1990, Prenyltransferases: from yeast to man, in: *Molecular Biology of Atherosclerosis: Twentieth Steenbock Symposium*, pp. 27–34, Elsevier, New York.

Attucci, S., Aitken, S. M., Gulick, P. J., and Ibrahim, R. K., 1995, Farnesyl pyrophosphate synthetase from white lupin: molecular cloning, expression, and purification of the expressed protein, *Arch. Biochem. Biophys.* **321**:493–500.

Baba, T., and Allen, M. C., 1980, Prenyl transferases from *Micrococcus luteus*: characterization of undecaprenyl pyrophosphate synthetase, *Arch. Biochem. Biophys.* **200**: 474–484.

Barnard, G. F., Langton, B., and Popják, G., 1978, Pseudo-isoenzyme forms of liver prenyl transferase, *Biochem. Biophys. Res. Commun.* **85**:1097–1103.

Barnard, G. F., and Popják, G., 1980, Characterization of liver prenyl transferase and its inactivation by phenylglyoxal, *Biochim. Biophys. Acta* **617**:169–182.

Barnard, G. F., and Popják, G., 1981, Human liver prenyltransferase and its characterization, *Biochim. Biophys. Acta* **661**:87–99.

Bartlett, D. L., King, C.-H. R., and Poulter, C. D., 1985, Purification of farnesylpyrophosphate synthetase by affinity chromatography, *Methods Enzymol.* **110**:171–184.

Blanchard, L., and Karst, F., 1993, Characterization of a lysine-to-glutamic acid mutation in a conservative sequence of farnesyl diphosphate synthase from *Saccharomyces cerevisiae, Gene* **125:** 185–189.

Bloch, K., Chaykin, S., Phillips, A. H., and de Waard, A., 1959, Mevalonic acid pyrophosphate and isopentenylpyrophosphate, *J. Biol. Chem.* **234**:2595–2608.

Boguski, M. S., and McCormick, F., 1993, Proteins regulating Ras and its relatives, *Nature* **366:** 643–654.

Carattoli, A., Romano, N., Ballario, P., Morelli, G., and Macino, G., 1991, The *Neurospora crassa* carotenoid biosynthetic gene (albino 3) reveals highly conserved regions among prenyltransferases. *J. Biol. Chem.* **266**:5854–5859.

Clarke, C. F., Tanaka, R. D., Svenson, K., Wamsley, M., Fogelman, A. M., and Edwards, P. A., 1987, Molecular cloning and sequence of a cholesterol-repressible enzyme related to prenyltransferase in the isoprene biosynthetic pathway, *Mol. Cell. Biol.* **7**:3138–3146.

Clarke, S., 1992, Protein isoprenylation and methylation at carboxyl-terminal cysteine residues, *Annu. Rev. Biochem.* **61**:355–386.

Eberhardt, N. I., and Rilling, H. C., 1975, Prenyltransferase from *Saccharomyces cerevisiae*: purification to homogeneity and molecular properties, *J. Biol. Chem.* **250**:863–866.

Epstein, W. W., Lever, D., Leining, L. M., Bruenger, E., and Rilling, H. C., 1991, Quantitation of prenylcysteines by a selective cleavage reaction, *Proc. Natl. Acad. Sci. USA* **88**:9668–9670.

Ericsson, J., Appelkvist, E.-L., Thelin, A., Chojnacki, T., and Dallner, G., 1992, Isoprenoid biosynthesis in rat liver peroxisomes: characterization of *cis*-prenyltransferase and squalene synthetase, *J. Biol. Chem.* **267**:18708–18714.

Ericsson, J., Runquist, M., Thelin, A., Andersson, M., Chojnacki, T., and Dallner, G., 1993, Distribution of prenyltransferase in rat tissues: evidence for a cytosolic all-*trans*-geranylgeranyl diphosphate synthase, *J. Biol. Chem.* **268**:832–838.

Fujii, H., Koyama, T., and Ogura, K., 1982, Hexaprenyl pyrophosphate synthetase from *Micrococcus luteus* B-P 26, *J. Biol. Chem.* **257**:14610–14612.

Fujii, H., Koyama, T., and Ogura, K., 1983, Essential protein factors for polyprenyl pyrophosphate synthetase, *FEBS Lett.* **161**:257–260.

Fujisaki, S., Nishino, T., and Katsuki, H., 1986, Isoprenoid synthesis in *Escherichia coli*. Separation and partial purification of four enzymes involved in the synthesis, *J. Biochem.* **99**:1327–1337.

Fujisaki, S., Hara, H., Nishimura, Y., Horiuchi, K., and Nishino, T., 1990, Cloning and nucleotide sequence of the ispA gene responsible for farnesyl diphosphate synthase activity in *Escherichia coli, J. Biochem.* **108**:995–1000.

Goodman, D. W. S., and Popják, G., 1960, Studies on the biosynthesis of cholesterol: XII. Synthesis of allylic pyrophosphates from mevalonate and their conversion into squalene with liver enzymes, *J. Lipid Res.* **2:**286–300.

Gotoh, T., Koyama, T., and Ogura, K., 1987, The significance of the diphosphate linkage involved in the substrates for prenyltransferase. Geranyl- and dimethylallyl methylenediphosphonate as artificial substrates, *Chem. Lett.* 1627–1630.

Gotoh, T., Koyama, T., and Ogura, K., 1988, Different roles of the diphosphate moieties of allylic and homoallylic diphosphates in prenyltransferase reaction, *Biochem. Biophys. Res. Commun.* **156:** 396–402.

Gotoh, T., Koyama, T., and Ogura, K., 1992, Farnesyl diphosphate synthase and solanesyl diphosphate synthase reactions of diphosphate-modified allylic analogs: the significance of the diphosphate linkage involved in the allylic substrates for prenyltransferase, *J. Biochem.* **112:**20–27.

Grunler, J., Ericsson, J., and Dallner, G., 1994, Branch-point reactions in the biosynthesis of cholesterol, dolichol, ubiquinone and prenylated proteins, *Biochim. Biophys. Acta* **1212:** 259–277.

Holloway, P. W., and Popják, G., 1967, The purification of 3,3-dimethylallyl- and geranyl-transferase and of isopentenyl pyrophosphate isomerase from pig liver, *Biochem. J.* **104:** 57–70.

Ishii, K., Sagami, H., and Ogura, K., 1983a, Farnesyl pyrophosphate synthetase located in microsomes from pig liver, *J. Biochem.* **93:**1635–1639.

Ishii, K., Sagami, H., and Ogura, K., 1983b, Decaprenyl pyrophosphate synthetase from mitochondria of pig liver, *Biochem. Biophys. Res. Commun.* **116:**500–506.

Ishii, K., Sagami, H., and Ogura, K., 1985, Decaprenyl pyrophosphate synthetase from *Paracoccus denitrificans, Biochim. Biophys. Acta* **835:**291–297.

Ishii, K., Sagami, H., and Ogura, K., 1986, A novel prenyltransferase from *Paracoccus denitrificans, Biochem. J.* **233:**773–777.

Jackson, S. M., Ericsson, J., Osborne, T. F., and Edwards, P., 1995, NR-Y has a novel role in sterol-dependent transcription of two cholesterogenic genes, *J. Biol. Chem.* **270:**21445–21448.

Jeffries, L., Cawthorne, M. A., Harris, M., Diplock, A. T., Green, J., and Price, S. A., 1967, Distribution of menaquinones in aerobic Micrococcaceae, *Nature* **215:**257–259.

Joly, A., and Edwards, P. A., 1993, Effect of site-directed mutagenesis of conserved aspartate and arginine residues upon farnesyl diphosphate synthase activity, *J. Biol. Chem.* **268:** 26983–26989.

Kandutsch, A. A., Paulus, H., Levin, E., and Bloch, K., 1964, Purification of geranylgeranyl pyrophosphate synthetase from *Micrococcus lysodeikticus, J. Biol. Chem.* **239:**2507–2515.

Koike-Takeshita, A., Koyama, T., Obata, S., and Ogura, K., 1995, Molecular cloning and nucleotide sequences of the genes for two essential proteins constituting a novel enzyme system for heptaprenyl diphosphate synthesis, *J. Biol. Chem.* **270:**18396–18400.

Koyama, T., Saito, Y., Ogura, K., and Seto, S., 1977, Two forms of farnesyl pyrophosphate synthetase from hog liver, *J. Biochem.* **82:**1585–1590.

Koyama, T., Obata, S., Osabe, M., Takeshita, A., Yokoyama, K., Uchida, M., Nishino, T., and Ogura, K., 1993, Thermostable farnesyl diphosphate synthase of *Bacillus stearothermophilus*: molecular cloning, sequence determination, overproduction, and purification, *J. Biochem.* **113:**355–363.

Koyama, T., Obata, S., Saito, K., Takeshita-Koike, A., and Ogura, K., 1994a, Structural and functional roles of the cysteine residues of *Bacillus stearothermophilus* farnesyl diphosphate synthase, *Biochemistry* **33:**12644–12648.

Koyama, T., Saito, K., Ogura, K., Obata, S., and Takeshita, A., 1994b, Site-directed mutagenesis of farnesyl diphosphate synthase; effect of substitution on the three carboxyl-terminal amino acids, *Can. J. Chem.* **72:**75–79.

Koyama, T., Tajima, M., Nishino, T., and Ogura, K., 1995, Significance of Phe-220 and Gln-221 in the catalytic mechanism of farnesyl diphosphate synthase of *Bacillus stearothermophilus, Biochem. Biophys. Res. Commun.* **212:**681–686.

Koyama, T., Tajima, M., Sano, H., Doi, T., Koike-Takeshita, A., Obata, S., Nishino, T., and Ogura, K., 1996, Identification of Significant Residues in the Substrate Binding Site of *Bacillus stearothermophilus* Farnesyl Diphosphate Synthase, *Biochemistry,* **35:**9533–9538.

Krisans, S. K., Ericsson, J., Edwards, P. A., and Keller, G.-A., 1994, Farnesyl-diphosphate synthase is localized in peroxisomes, *J. Biol. Chem.* **269:**14165–14169.

Lutz, R. J., Mclain, T. M., and Sinensky, M., 1992, Feedback inhibition of polyisoprenyl pyrophosphate synthesis from mevalonate *in vitro, J. Biol. Chem.* **267:**7983–7986.

Lynen, F., Eggerer, H., Henning, U., and Kessel, I., 1958, Farnesyl-pyrophosphat und 3-Methyl-Δ3-butenyl-1-pyrophosphat, die Biologischen Vorstufen des Squalens. Zur Biosynthase der Terpene, III, *Angew. Chem.* **70:**738–742.

Lynen, F., Agranoff, B. W., Eggerer, H., Henning, U., and Möslein, E. M., 1959, γ.γ-Dimethylallyl-pyrophosphat und Geranyl-pyrophosphat, Biologische Vorstufen des Squalens. Zur Biosynthese der Terpene, VI, *Angew. Chem.* **71:**657–663.

Marrero, P. F., Poulter, C. D., and Edwards, P. A., 1992, Effects of site-directed mutagenesis of the highly conserved aspartate residues in domain II of farnesyl diphosphate synthase activity, *J. Biol.Chem.* **267:**21873–21878.

Matsuoka, S., Sagami, H., Kurisaki, A., and Ogura, K., 1991, Variable product specificity of microsomal dehydrodolichyl diphosphate synthase from rat liver, *J. Biol. Chem.* **266:** 3464–3468.

Muth, J. D., and Allen, M. C., 1984, Undecaprenyl pyrophosphate synthetase from *Lactobacillus plantrum*: a dimeric protein, *Arch. Biochem. Biophys.* **230:**49–60.

Nakane, H., Koyama, T., Obata, S., Osabe, M., Takeshita, A., Nishino, T., Ogura, K., and Miki, K., 1993, Crystallization and preliminary X-ray diffraction studies of *Bacillus stearothermophilus* farnesyl diphosphate synthase expressed in *Escherichia coli, J. Mol. Biol.* **233:**787–788.

Najarian, D., Dihanich, M. E., Martin, N. C., and Hopper, A. K., 1987, DNA sequence and transcript mapping of MOD5: features of the 5' region which suggest two translational starts, *Mol. Cell. Biol.* **7:**185–191.

Nandi, D. L., and Porter, J. W., 1964, The enzymatic synthesis of geranylgeranyl pyrophosphate by enzymes of carrot root and pig liver, *Arch. Biochem. Biophys.* **105:**7–19.

Nishino, T., Ogura, K., and Seto, S., 1971, Reactivity of artificial substrates for prenyltransferase, *Biochim. Biophys. Acta* **235:**322–325.

Nishino, T., Ogura, K., and Seto, S., 1972, Substrate specificity of farnesyl pyrophosphate synthetase, *J. Am. Chem. Soc.* **94:**6849–6853.

Ogura, K., Nishino, T., Koyama, T., and Seto, S., 1970, Enzymic condensation of 3-methyl-2-alkenyl pyrophosphates with isopentenyl pyrophosphate, *J. Am. Chem. Soc.* **92:**6036–6041.

Ogura, K., Shinka, T., and Seto, S., 1972, The purification and properties of geranylgeranyl pyrophosphate synthetase from pumpkin fruit, *J. Biochem.* **72:**1101–1108.

Ohnuma, S., Koyama, T., and Ogura, K., 1991, Purification of solanesyl-diphosphate synthase from *Micrococcus luteus*. A new class of prenyltransferase, *J. Biol. Chem.* **266:**23706–23713.

Poulter, C. D., and Rilling, H., 1976, Prenyltransferase: the mechanism of the reaction, *Biochemistry* **15:**1079–1083.

Poulter, C. D., Wiggins, P. I., and Le, A. T., 1981, Farnesylpyrophosphate synthetase. A stepwise mechanism for the 1′-4 condensation reaction, *J. Am. Chem. Soc.* **103:**3926–3927.

Reed, B. C., and Rilling, H. C., 1975, Crystallization and partial characterization of prenyltransferase from avian liver, *Biochemistry* **14:**50–54.

Runquist, M., Ericsson, J., Thelin, A., Chojnacki, T., and Dallner, G., 1992, Biosynthesis of *trans,trans,trans*-geranylgeranyl diphosphate by the cytosolic fraction from rat tissues, *Biochem. Biophys. Res. Commun.* **186:**157–165.

Runquist, M., Ericsson, J., Thelin, A., Chojnacki, T., and Dallner, G., 1994, Isoprenoid biosynthesis in rat liver mitochondria: studies on farnesyl pyrophosphate synthase and *trans*-prenyltransferase, *J. Biol. Chem.* **269:**5804–5809.

Sagami, H., Ogura, K., and Seto, S., 1977, Solanesyl pyrophosphate synthetase from *Micrococcus lysodeikticus, Biochemistry* **16:**4616–1622.

Sagami, H., Ishii, K., and Ogura K., 1981, Occurrence and unusual properties of geranylgeranyl pyrophosphate synthetase of pig liver, *Biochem. Int.* **3:**669–675.

Sagami, H., Lennarz, W. J., and Ogura, K., 1989, The biosynthesis of dehydrodolichyl phosphates by rat liver microsomes, *Biochim. Biophys. Acta* **1002:**218–224.

Sagami, H., Matsuoka, S., and Ogura, K., 1991, Formation of *Z,E,E*-geranylgeranyl diphosphate by rat liver microsomes, *J. Biol. Chem.* **266:**3458–3463.

Sagami, H., Korenaga, T., Ogura, K., Steiger, A., Pyun, H.-J., and Coates, R. M., 1992, Studies on geranylgeranyl diphosphate synthase from rat liver: specific inhibition by 3-azageranylgeranyl diphosphate, *Arch. Biochem. Biophys.* **297:**314–320.

Sagami, H., Korenaga, T., Kurisaki, A., and Ogura, K., 1993a, Biosynthesis of prenyl diphosphates by cell-free extracts from mammalian tissues, *J. Biochem.* **114:**112–117.

Sagami, H., Korenaga, T., and Ogura, K., 1993b, Geranylgeranyl diphosphate synthase catalyzing the single condensation between isopentenyl diphosphate and farnesyl diphosphate, *J. Biochem.* **114:**118–121.

Sagami, H., Morita, Y., and Ogura, K., 1994, Purification and properties of geranylgeranyl-diphosphate synthase from bovine brain, *J. Biol. Chem.* **269:**20561–20566.

Sheares, B. T., White, S. S., Molowa, D. T., Chan. K., Ding, V. D.-H., Kroon, P. A., Bostedor, R. G., and Karkas, J. D., 1989, Cloning, analysis, and bacterial expression of human farnesyl pyrophosphate synthetase and its regulation in Hep G2 cells, *Biochemistry* **28:**8129–8135.

Song, L., and Poulter, C. D., 1994, Yeast farnesyl-diphosphate synthase: site-directed mutagenesis of residues in highly conserved prenyltransferase domain I and II, *Proc. Natl. Acad. Sci. USA* **91:**3044–3048.

Spear. D. H., Ericsson, J., Jackson, S. M., and Edwards, P. A., 1994, Identification of a 6-Base pair element involved in the sterol-mediated transcriptional regulation of farnesyl diphosphate synthase, *J. Biol. Chem.* **269:**25212–25218.

Tachibana, A., 1994, A novel prenyltransferase, farnesylgeranyl diphosphate synthase, from the haloalkaliphilic archaeon, *Natronobacterium pharaonis, FEBS Lett.* **341:**291–294.

Takahashi, I., Ogura, K., and Seto, S., 1980, Heptaprenyl pyrophosphate synthetase from *Bacillus subtilis, J. Biol. Chem.* **255:**4539–4543.

Tarshis, L. C., Yan, M., Poulter, C. D., and Sacchettini, J. C., 1994, Crystal structure of recombinant farnesyl diphosphate synthase at 2.6-Å resolution, *Biochemistry* **33:**10871–10877.

Teclebrhan, H., Olsson, J., Swiezewska, E., and Dallner, G., 1993, Biosynthesis of the side chain of ubiquinone: *trans*-prenyltransferase in rat liver microsomes, *J. Biol. Chem.* **268:** 23081–23068.

Wilkin, D. J., Kutsunai, S. Y., and Edwards, P. A., 1990, Isolation and sequence of the human farnesyl pyrophosphate synthase cDNA. Coordinate regulation of the mRNAs for farnesyl pyrophosphate synthetase, 3-hydroxy-3-methylglutaryl coenzyme A reductase, and 3-hydroxy-3-methylglutaryl coenzyme A synthase by phorbol ester, *J. Biol. Chem.* **265:** 4607–4614.

Wong, T. K., and Lennarz, W. J., 1982, The site of biosynthesis and intracellular deposition of dolichol in rat liver, *J. Biol. Chem.* **257:**6619–6624.

Yeh, L.-S., and Rilling, H. C., 1977, Purification and properties of pig liver prenyltransferase. Interconvertible forms of the enzyme, *Arch. Biochem. Biophys.* **183:**718–725.

Yoshida, I., Koyama, T., and Ogura, K., 1989a, Protection of hexaprenyl-diphosphate synthase of *Micrococcus luteus* B-P 26 against inactivation by sulphydryl reagents and arginine-specific reagents, *Biochim. Biophys. Acta* **995:**138–143.

Yoshida, I., Koyama, T., and Ogura, K., 1989b, Formation of a stable and catalytically active complex of the two essential components of hexaprenyl diphosphate synthase from *Micrococcus luteus* B-P 26, *Biochem. Biophys. Res. Commun.* **160:**448–452.

Chapter 4

Antifungal Sterol Biosynthesis Inhibitors

De-an Guo, Anil T. Mangla, Wen Zhou, Monica Lopez, Zhonghua Jia, S. Derrek Nichols, and W. David Nes

1. INTRODUCTION

Herbal medicines and related drugs used to treat infective organisms were available in ancient cultures and detailed in the first pharmacopoeia reported in 1146 AD, titled "Antidotarium Niclai" (Grun, 1991). During the following millennium, there has been continued interest in drug discovery and application. However, only in the last fifty years have effective therapeutic agents for fungal disease been developed that exploit a fundamental difference between the biochemistry or physiology of the pathogens and their hosts. Sterol biosynthesis is one of the few areas of difference in primary metabolism between fungi and animals. Unlike animals that synthesize 24-desalkyl sterols, such as cholesterol (cholest-5-en-3β-ol), many pathogenic fungi synthesize 24-alkylsterols, such as ergosterol (24β-methyl cholesta-5,7,22-trien-3β-ol). The major difference in the two sterols stems structurally from the presence and nature of the 24-alkyl group.

Fungal sterols are a class of natural products of 27–30 carbon atoms. Figure 1 shows the sterol constitution and numbering system. In this review we follow the sterol nomenclature system proposed by Fieser and Fieser (1959) and subsequently modified by Nes and McKean (1977) and Popják *et al.* (1977). An

De-an Guo, Anil T. Mangla, Wen Zhou, Monica Lopez, Zhonghua Jia, S. Derrek Nichols, and W. David Nes Departments of Chemistry and Biochemistry, Texas Tech University, Lubbock, Texas 79409.

Subcellular Biochemistry Volume 28: Cholesterol: Its Functions and Metabolism in Biology and Medicine, edited by Robert Bittman. Plenum Press, New York, 1997.

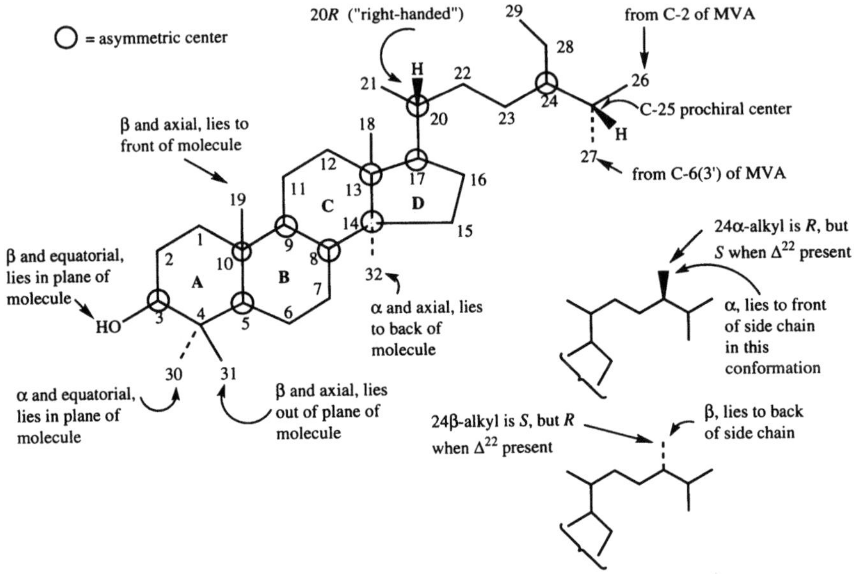

FIGURE 1. Sterol numbering system and stereochemistry of the side chain. The biosynthetic side chain rule designations are given in the figure. Note that C-26 originates from C-2 of mevalonic acid and C-27 originates from C-6 of mevalonic acid.

alternate system was recommended by the IUPAC commission in 1989. Parker and Nes (1992) examined the two systems. The biosynthetic side chain rule, introduced by Nes and McKean (1977) and Popják *et al.* (1977), is often used by biochemists to assign chemical shifts in ^{1}H-and ^{13}C-NMR spectroscopy and in distinguishing mechanisms related to the metabolism of the sterol 24,25-double bond (Parker and Nes, 1992).

Ergosterol is the major Δ^5-sterol produced by advanced-ascomycetous fungi. The structure has been characterized in stereochemical detail by chromatographic and spectroscopic methods (Adler *et al.,* 1977; Hull and Woolfsin, 1976; Nes *et al.,* 1985; Zhou *et al.,* 1996). Ergosterol is a sterol with three double bonds at positions 5,7,22 *trans,* a 24β-methyl group, 20α-hydrogen (i.e., "right-handed"), and 3β-hydroxyl group. No geometric or diastereoisomeric sterols co-occur with ergosterol in fungi, suggesting that biosynthesis leads to stereochemically pure end products. Some uncertainties remain as to whether all of the carbon atoms in the side chain structure are functionally necessary and whether the side chain orients into a preferred conformation in complexing with membrane lipids or with enzymes that act on the sterol molecule. Small changes in the length of the sterol side chain affect the interaction of sterols with phospholipids in bilayers and affect sterol-controlled growth support of fungi (Venkatramesh and Nes, 1995; Nes, 1987). On the other hand, the pseudocyclic sterol side chain conformation mediates the formation of some sterol–enzyme complexes (Nes *et al.*, 1991; Venkatramesh *et al.*, 1996).

Sterols are synthesized from precursor molecules that pass from the isopentenoid pathway (culminating in the production of squalene and 2,3-oxidosqualene) into the sterol transformation pathway. In fungi, the start of the sterol pathway has been operationally defined as beginning with Δ^8-sterol production (lanosterol) and ending with Δ^5-sterol production (e.g., ergosterol). The equilibrium of the pathway is directed towards Δ^5-sterol production during active cell proliferation, and there is relatively minor accumulation of intermediate sterol. Carbon flow is directed towards production of Δ^5-sterol end products that are inserted into the membrane first; subsequently, C-3 derivatization gives esters that form intracellular lipid droplets.

Cell growth is often used as a measure of the suitability of sterols to act as membrane inserts (Bloch, 1983; Nes *et al.*, 1993; Nes, W. R., 1987; Taylor and Parks, 1981). In the putative physiological conformation, the molecule is flat and the side chain orients to the right. These structural attributes, coupled with the amphipathic nature of sterols, make them suitable membrane inserts and promote their binding to enzymes. The evolution of sterol asymmetry appears to satisfy reproductive fitness, which may be different in primitive and advanced fungi, and is not necessarily the result of random mutations developed from Darwinian selection pressures on growth (Nes and Venkatramesh, 1994). These observations have lead to the tacit assumption that a mating of sterol biosynthesis and function exists, which provides the focus for study on inhibitors developed to interfere with sterol synthesis.

Antifungals that interfere with sterol biosynthesis may be natural products (antibiotics) produced by fungi or synthetic mimics. A class of antifungals that complex with sterols and prevent them from acting as membrane inserts are the macrolide polyene antibiotics (Hamilton-Miller, 1974). They are produced by the bacterium *Streptomycetes nodusus*. Each molecular type has a polyhydroxylic lactone ring of C-23–C-27 carbons and contain four–seven double bonds. The polyene antibiotics are currently used in therapy to overcome infection caused by topical and systemic fungi. Another group of sterol-complexing agents are the steroid saponins. Some steroid saponins are produced by marine organisms and marketed in Japan to control athlete's foot (Fusentani, 1988). Steroid and triterpenoid saponins produced by crop plants also function as antifungal compounds, but in this case they contribute uniquely to plant protection (Roddick, 1987). Polyene and saponin antibiotics possess a common detergent-like foaming property that effectively removes sterol from the lipid bilayer of the membrane through coprecipitation. (For a more detailed discussion see Chapter 16 by Severs.)

A second class of antibiotics is targeted for a selected metabolic enzyme, e.g., sinefungin, which interferes with sterol biomethylation (McCammon *et al.*, 1981).

This review focuses on the biochemistry of inhibitors that were designed to interfere with sterol biosynthesis. Major attention is given to the molecular basis of action of inhibitors that block the enzyme-catalyzed introduction of 24-alkyl groups in the sterol side chain. Some of these specific areas or related ones have been the subject of other reviews (Abe *et al.,* 1994; Berg and Plempel, 1988; Burden *et al.,* 1989; Janssen *et al.,* 1991; Mercer, 1993; Oehlschlager and Czyzewaska, 1992; St. Georgiev, 1988; Weete, 1987).

2. ANTIFUNGAL AGENTS AND STEROL BIOCHEMISTRY

The study of sterol biosynthesis and inhibition of sterol production and function involves five major approaches: (i) analyze the sterol composition from actively growing and stationary phase whole cells and subcellular fractions, qualitatively and quantitatively; (ii) incubate (e.g., pulse-chase type experiments) intact cells with isotopically-labeled compounds (these may be stable or radioactive isotopes); (iii) conduct *in vitro* enzyme assays; (iv) create mutants in the sterol pathway either by classic mutagenesis or through modern molecular biology techniques; and (v) examine cell growth, cellular sterol composition, and catalytic competence of sterol enzymes treated with rationally designed sterol biosynthesis inhibitors (SBIs). Using these approaches, investigators have recently reported differences in sterol distribution (Bansal and Knoche, 1981; Loefler *et al.,* 1992; Nes, W. D., 1987; Nes *et al.,* 1990; Nes *et al.,* 1988) and sensitivity of fungi to SBIs (Berg and Plempel, 1988; St. Georgiev, 1988). The ergosterol pathway in yeast is often regarded as the fungal paradigm to test SBI efficacy *in vivo* and *in vitro.* Figure 2 shows the hypothetical kinetically favored pathway from lanosterol to ergosterol and sites of interference by SBIs (for the structures of typical inhibitors, see Figure 3) (Berg and Plempel, 1988; St. Georgiev, 1988; Weete, 1987). Ergosterol is also a major sterol in other pathogenic and parasitic organisms, such as in the yeast-like algae (Janssen *et al.,* 1991; Nes *et al.,* 1990) and in protozoa. SBIs have been incubated with these organisms to inhibit their growth (Haughan *et al.,* 1995; Raederstorff and Rohmer, 1987; Rahman and Pascal, 1990).

2.1. Antibiotics

For more than three decades treatment of systemic fungal infections has rested with prescription of amphotericin B and nystatin. Nevertheless, there exist acute and chronic toxicities of amphotericin B and nystatin that have prompted efforts to develop more effective therapeutic agents for eradication of the disease. A renewed search began in the mid-1970s for new chemical classes of pharmaceuticals and to determine whether pharmaceuticals might also be useful as agrochemicals. From random biological screening of synthetic compounds, a novel group of dual action (i.e., used as pharmaceutical or agrochemical antifungal agents) fungicides was discovered—the azole-type compounds. The azoles are so-named because they contain an imidazole or triazole moiety as the active pharmacophore. Azole antifungals were found to inhibit fungal growth by inhibiting the activity of

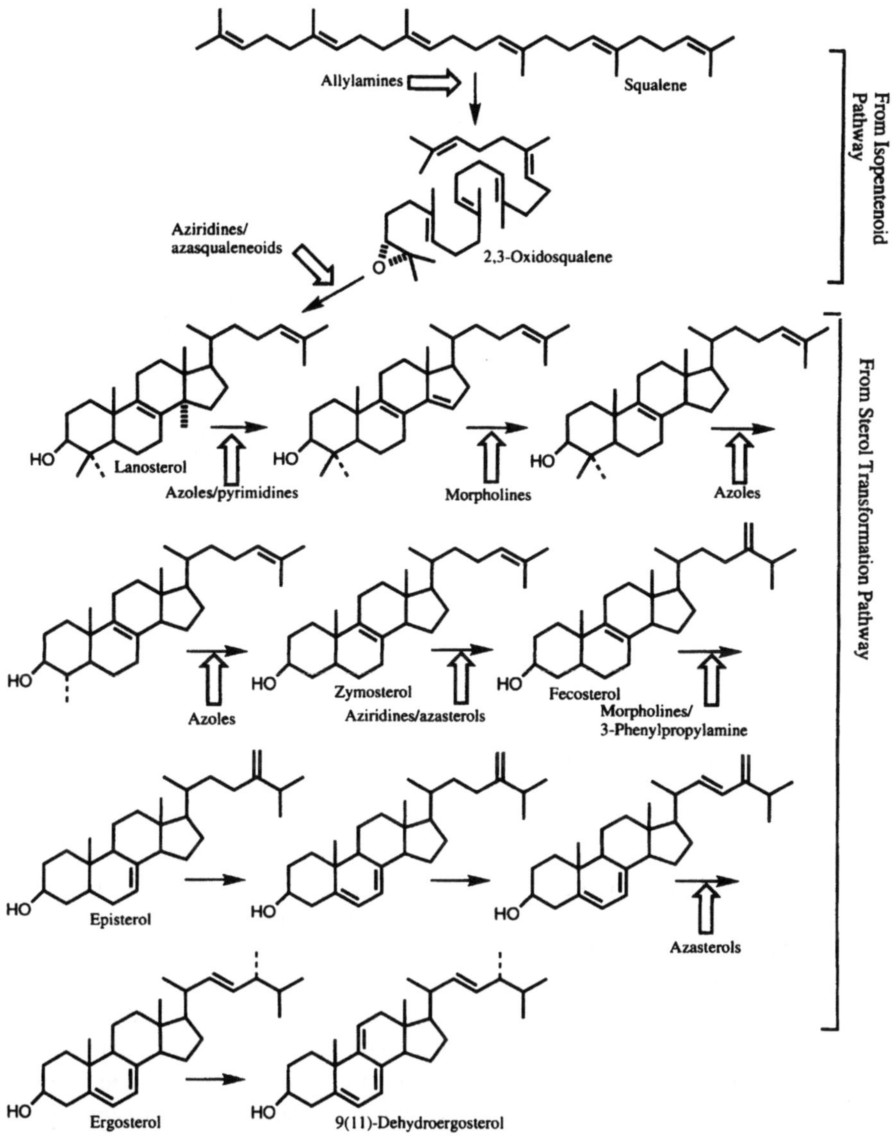

FIGURE 2. Hypothetical kinetically favored sterol pathway in *Saccharomyces cerevisiae*. The arrows show the sites of inhibition.

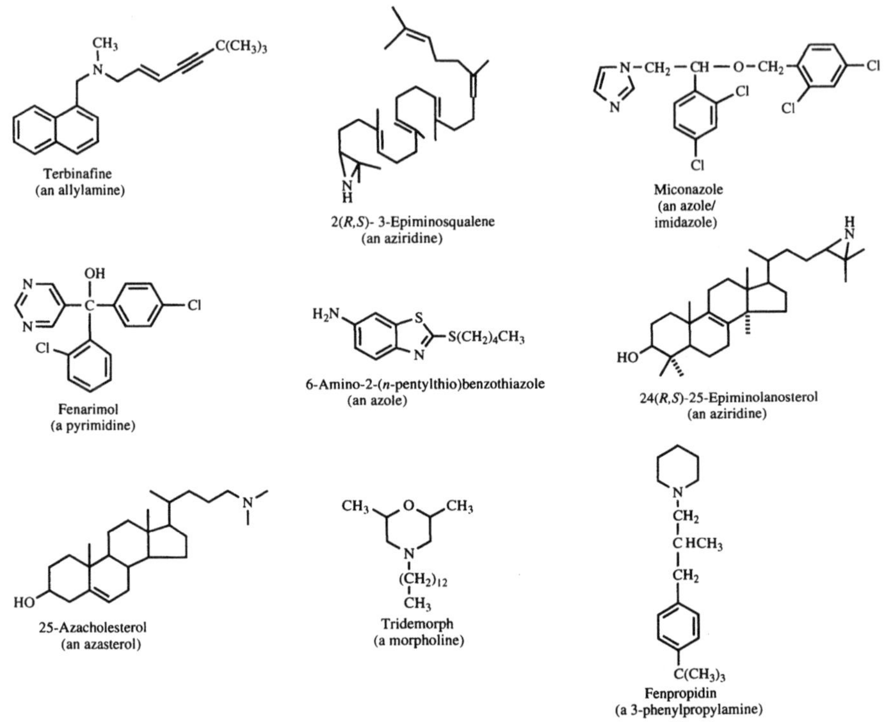

FIGURE 3. Structures of common sterol biosynthesis inhibitors (SBI) targeted for enzymes that act on sterol substrates, as discussed in the text. See Figure 2 for SBI sites of interference in the sterol pathway.

the 14-demethylase enzyme in sterol synthesis, which inhibits lanosterol turnover. This class of SBIs was marketed by Janssen Pharmaceutica in 1981. Azoles affect sterol turnover by decreasing the rate of ergosterol production and increasing the steady-state balance of intermediates, particularly 4,4,14-trimethylsterols. The change in ergosterol homeostasis leads to a modified sterol composition that apparently is harmful to cell physiology (Weete, 1987). Although the azoles have the advantage of being able to be administered orally, they are not fungicidal; clinical problems were discovered resulting in teratogenicity and enzyme specificity, as well as idiosyncratic hepatoxicity for the fungal 14-demethylase enzyme. These first generation azoles were further shown to possess limited selectivity relative to the mammalian enzyme (St. Georgiev, 1988). However, the selectivity of a second-generation azole, fluconazole, is markedly improved (Houston *et al.*, 1988).

2.2. Sterol Biosynthesis Inhibitors

SBIs prepared recently for commercial purposes are allylamine derivatives that were developed to block squalene conversion to squalene oxide, azole and pyrimidine derivatives that were developed to block C-4 and C-14 demethylations of lanosterol, and morpholines that were designed to block the rearrangement of fecosterol to episterol (Δ^8-to Δ^7-isomerase inhibitor); see Figures 2 and 3 (Mercer, 1993).

Cyclase inhibitors, such as 2(*R,S*),3-epiminosqualene (ISq) and the azasqualeneoids (Balliano *et al.*, 1988: Nes *et al.*, 1988; Nes and Parish, 1988; Pinto and Nes, 1983; Ryder, 1988; Viola *et al.*, 1995), have been used as research tools in elucidating fungal sterol biochemistry and function (Figure 4). Although many of the azasqualeneoids were found to be inefficiently absorbed by yeast (Balliano *et al.*, 1988), ISq was found to be actively absorbed by *Saccharomyces cerevisiae* cultured anaerobically (Pinto and Nes, 1983) and by branching mycelia of *Gibberella fujikuori* (Nes and Parish, 1988). ISq inhibited the conversion of squalene oxide to lanosterol, thereby inhibiting growth of *S. cerevisiae* and *G. fujikuroi.* We considered the possibility that ISq might initially block cyclase activity (Popják *et al.,* 1989), then undergo transformation to a safe nontoxic metabolite lanosterol, for which evidence was presented using the phytopathogenic fungus *G. fujikuori* (Nes and Parish, 1988; Nes *et al.,* 1989). Reported differences in the efficacy of SBIs as fungitoxic–fungistatic agents in different organisms may result from differential absorption and transport of the SBIs and from suspected differences in amino acid compositions of proteins that control sterol synthesis.

SBIs have been prepared for steps leading up to squalene production (Abe *et al.,* 1994). However, none of the commercial SBIs currently in use were designed to inhibit 3-hydroxy-3-methylglutaryl coenzyme A (HMG-CoA) reductase (HMGR) activity; instead, SBIs were designed to inhibit ergosterol production at steps from squalene epoxidation to ergosterol. At first glance, it is difficult to reconcile how some of these SBIs with structures that obviously fail to mimic sterol structures bind to the sterol enzymes and interfere mechanistically with catalysis. For demethylase inhibitors (DMIs), the fungitoxic action results from binding to a site on the cytochrome P-450 14-demethylase that is distinct from the sterol binding site. DMI inhibits enzyme activity by utilizing the nonbonded electron pair on the meta-*N* of the heterocycle to cause some interaction between the lipophilic moiety and the lipophilic binding site (Mercer, 1993). Rationally designed SBIs that may act as transition state analogs (TSAs) of the 14-demethylation reaction have been shown to be fungicidal (Mercer, 1993; Weete, 1987). The allylamines that might act as TSAs of the epoxidation reaction were found not to act as substrate analogs (Ryder, 1988). The inhibitor may bind to the squalene-binding site. Alternatively, other SBIs were rationally designed as TSAs (or high energy intermediate analogs). They block squalene-oxide cyclization to lanosterol, biomethylation of the 24,25-double bond, reduction of the 24(28)-double bond, and Δ^8-to-Δ^7-isomerization (Mercer, 1993). In each case, some structural resemblance of the inhibitor to the native substrate was demonstrated for biological activity.

3. STEROL BIOMETHYLATION AS A TARGET SITE FOR INHIBITION

Early studies on sterol biosynthesis in fungi indicated that the rate-limiting step in ergosterol production was at a stage in the pathway before lanosterol formation. The rate-limiting enzyme was thought to be HMGR, which regulates the conversion of HMG-CoA to mevalonic acid (Bard and Downing, 1981; Kawaguchi, 1970). However, HMGR may not be the rate-limiting enzyme controlling ergosterol genesis (Casey *et al.*, 1992; Parker and Nes, 1992), although HMGR activity may act as an important gate that controls carbon flow from the isopentenoid pathway into the sterol pathway (Pinto *et al.*, 1986). For some fungi, the critical slow step in the sterol pathway is biomethylation of lanosterol, whereas in other fungi, C-4 demethylation may be rate-limiting (Nes *et al.*, 1988). In *G. fujikuroi*, 24-methylation of lanosterol is the first structural modification of sterol (Nes and Le, 1990). Both fungi are Ascomycetes, suggesting phylogenetic differences in sterol sequencing to ergosterol.

Biomethylation of the sterol side chain is the characteristic feature that distinguishes sterol biosynthesis in fungi from animals. The importance of the biomethylation enzyme [(*S*)-adenosyl-L-methionine : Δ^{24}-sterol methyl transferase, SMT] in fungal physiology has been demonstrated (Nes *et al.*, 1978; Nes *et al.*, 1986; Nes, W. D., 1987). The activity of the SMT enzyme from yeast is downregulated after the enzyme complexes with ergosterol, but not cholesterol (Venkatramesh *et al.*, 1996b). Hence, it seems possible that when ergosterol sites are filled in the membrane and no C-3 esterification is possible, ergosterol may be retained on the endoplasmic reticulum, its site of synthesis (Zisner *et al.*, 1993), and complex with sterol-synthesizing enzymes. In cholesterol-containing yeast cells, SMT activity is downregulated by 50% compared with that from ergosterol-containing cells (Venkatramesh *et al.*, 1996b). The latter results indicate that SMT activity is sensitive to differences in the type and amount of cellular sterols. Nes *et al.* (1978) demonstrated that 24-methylation of the ergosterol side chain was necessary for yeast growth under anaerobic conditions. Several investigators subsequently found that 25-azasterols and related nitrogen-containing sterol side chains inhibited biomethylation (Figure 4) and fungal growth (Ator *et al.*, 1992; Oehlschlager *et al.*, 1984; Nes and Le, 1988) and reproduction (Nes *et al.*, 1986; Nes and Le, 1988). Harmful effects on growth and reproduction were associated with interference of the biomethylation step.

To test whether the azasterol was fungitoxic, 25-azacholesterol was incubated with a mutant *S. cerevisiae* strain GL7 auxotrophic for sterol. 25-Azasterol supplementation could not replace the native sterol ergosterol as a growth supplement in GL7 (Guo *et al.*, 1996). When GL7 cells were fed increasing levels of 25-azacholesterol and 10 ppm cholestanol in the growth medium (cholestanol served as a membrane insert), the inhibitor was shown to exhibit fungistatic behavior at a tenfold higher concentration than that found when inhibitor was fed to wild-type cells (Oehlschager *et al.*, 1984). Inhibition of SMT activity by 25-azacholesterol was

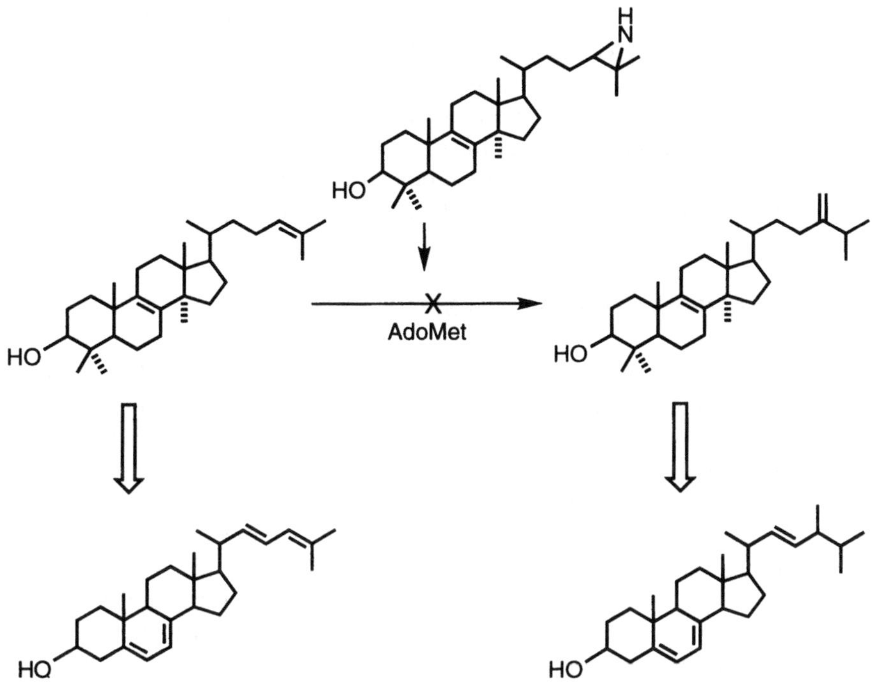

FIGURE 4. Inhibition of C-24 methylation by 24(*R,S*)-25-epiminolanosterol to lanosterol. Adapted from Nes and Parish (1988) and Nes *et al.* (1989).

confirmed using a microsome-bound SMT enzyme from GL7 (Venkatramesh *et al.,* 1996). The mutant has a leaky sterol pathway that produces trace levels of ergosterol (Nes and Dhanuka, 1988). This ergosterol, which apparently may also be produced in wild-type cells, is thought to possess a hormonal function that is distinct from the function of sterol as a membrane insert (Nes, W. R., 1987). The ability of 25-azacholesterol to inhibit growth of GL7 confirms that GL7 is leaky in ergosterol synthesis. When lanosterol was added to the medium of GL7, the cells converted all of the absorbed sterol to ergosterol. When 10 ppm lanosterol was added to the medium of GL7 with 10 ppm 25-azacholesterol, cell growth was partly inhibited (80% of ergosterol grown cells) and the resulting 24-desalkyl sterol composition was similar to that observed from 25-azacholesterol-treated wild-type cells (as percent total sterol: ergosterol, trace; cholesta-5,7,24-trienol, 17%; cholesta-5,7,22,24-tetraenol, 46%; cholesta-7,24-dienol, 1%; lanosterol, 36%) (Guo *et al.,* 1996). These results independently confirm that 25-azacholesterol is accumulated by GL7. The effect of sterol biomethylation inhibitors with modified side chains seems to be similar in fungi, but the type and amount of sterol produced in the treated cultures depends on the concentration of inhibitor in the medium and the growing time.

25-Azasterols and related biomethylation inhibitors (triparanol) were found to be harmful to animal physiology (Rosenblum *et al.*, 1979; Dukes, 1980). Therefore, they were never marketed as fungicides. Nevertheless, the initial results with the azasterols were promising; therefore, there has been continued interest in the commercial value of inhibitors that interfere with biomethylation activity (Ator *et al.*, 1992). The new focus of study on biomethylation inhibitors is based on a more exacting analysis of the kinetic and reaction mechanism, as reviewed next.

4. STEROL BIOMETHYLATION MECHANISMS

The involvement of (*S*)-adenosyl-L-methionine (SAM = AdoMet) as a biological methylating agent in a variety of biosynthetic processes, including sterol biomethylation, is well documented. According to the currently accepted hypothesis as originally formulated by Nes and coworkers (Castle *et al.*, 1963), the mechanism of the transfer reaction is visualized as a substitution on the methyl group of the alkylating agent by attack of an appropriate nucleophilic center of the substrate undergoing alkylation (Figure 5). Many studies on the biosynthesis of branched fatty acids and of sterols with modified side-chains have indicated that isolated C=C double bonds have sufficient nucleophilic character for participating in such alkylations. For a historical review of biomethylation biochemistry, see Nes and McKean, 1977).

4.1. Recent Findings in 24-Alkylsterol Phylogenesis

Recent studies on the distribution of 24-alkylsterols in fungi indicate that the sterol composition is much more varied than previously thought. The variation was found to be associated with the type (e. g., one or two methyl transfers) and

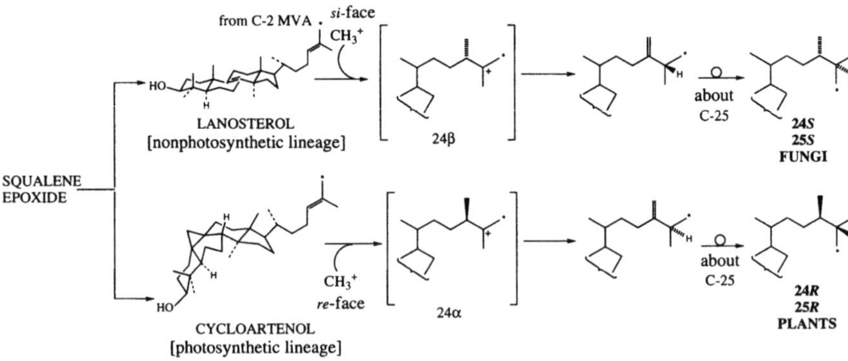

FIGURE 5. Two-pathway model of sterol biomethylation showing lanosterol–cycloartenol bifurcation giving rise to stereochemically different 24-alkyl sterols. Adapted from Bloch (1983) and Nes and McKean (1977).

nature (e. g., chirality) of biomethylation at C-24 of the sterol side chain. Biomimetic studies indicate that the sterol side chain may be alkylated at C-24 in various ways to give alternate 24-alkylsterol isomers (Julia and Marazana, 1985). In phylogenetic studies from this laboratory designed to establish the uniformity and differences in fungal sterol composition and biosynthesis, we identified several new 24-alkylsterols and confirmed their biosynthesis from incubations with ^{14}C-and ^{3}H-precursors. For instance, we found that *Saprolegnia ferax* synthesized 24(28)-methylene cholesterol and fucosterol [*E*-24(28)-ethylidine-cholesterol] (Nes and Le, 1988). *G. fujikuroi* was found to synthesize ergosterol and 24β-methylcholesterol by separate routes (Nes and Le, 1990). The non-photosynthetic yeastlike skin pathogen *Prototheca wickerhamii* was found to synthesize ergosterol from cycloartenol and cyclolaudenol (Nes *et al.*, 1990). Shimizu *et al.* (1992) reported the isolation of several new fungal 24-alkylsterols, including $\Delta^{25(27)}$-sterols and 24,25-cyclopropyl-containing sterols. Loefler *et al.* (1992) reported that several fungi produce 24(28)-methylene-cholesterol as the major cellular sterol, and Bansal and Knoche (1981) presented enzymatic evidence that *Uromyces phaseolis* could double bioalkylate the sterol side chain.

4.2. Alternate Biomethylation Pathways in Fungi

Mechanistically, the key to the methylation reaction is the positive charge on the sulfur atom of AdoMet. This renders the methyl group attached to the sulfur atom electrophilic, and facilitates a reaction with the 24,25-double bond that is nucleophilic because of the π-electrons. The reaction that ensues results in the cleavage of the S-CH_3 bond in AdoMet, and in the generation of an electron deficiency at C-25 (Figure 6). The C-25 carbenium ion may be eliminated by proton

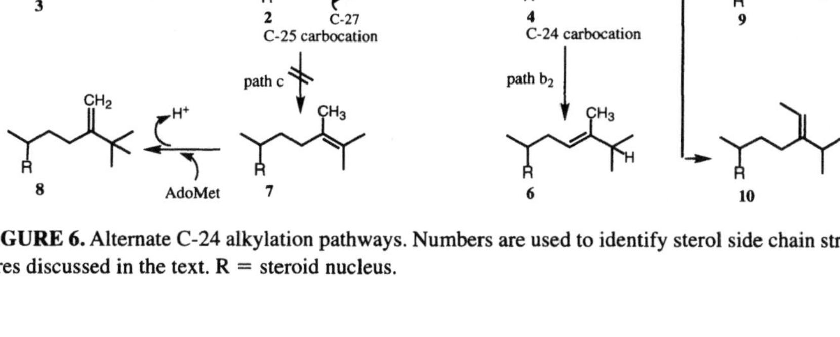

FIGURE 6. Alternate C-24 alkylation pathways. Numbers are used to identify sterol side chain structures discussed in the text. R = steroid nucleus.

loss from C-27 (path a) or from C-28 (path d); or rearrange, such that the C-24 hydrogen undergoes a 1,2-hydride shift to C-25, producing a carbenium ion with the electron deficiency at C-24 (path b). The latter charge may be eliminated by loss of a proton from the newly introduced methyl group (path b1) or from C-23 on the sterol side chain (path b2). As Arigoni recognized (1978), it is particularly striking that no *bona fide* biochemical evidence has been obtained for the operation of path c, which would result in the generation of Δ^{24}-unsaturated sterol directly from the biomethylation reaction. There is, however, support for such a mechanism from biomimetic studies on the biomethylation reaction (Julia and Marazano, 1985). The important difference between the biological-and chemical-induced 1,2-hydride shift following methylation is the juxtaposition of a base relative to the C=C bond to abstract the hydrogen from C-24 to produce an olefin. Clearly, no SMT enzyme has evolved this base, which must be disposed on the opposite face of the double bond that will be attacked by the methyl cation. There are cogent reasons to believe that the second biomethylation event is formed by the same SMT enzyme and by the same mechanism that gives rise to the first biomethylation product (Bansal and Knoche, 1981; Husselstein *et al.,* 1996; Parker and Nes, 1992).

The demonstration that several types of 24-alkylsterols are synthesized by fungi raises new concerns about sterol biosynthesis and function. In the consideration of biomethylation inhibitors, it is important to determine whether the different 24-alkene products produced by fungi are from mechanistically similar SMT enzymes. Moreover, information should be obtained about whether a family of SMT enzymes evolved in fungi, and whether the enzymes will be shown to be equally or differentially sensitive to the biomethy lation inhibitors. Studies of inhibitors of SMT activity provide information on the topology of the active site (Janssen and Nes, 1992; Venkatramesh *et al.*, 1996b).

The production of new active sites could have evolved in different organisms by alteration of the size and shape of the active site, or by introduction of specific amino acid residues that give rise to a catalytically more efficient enzyme or to an enzyme that produces novel 24-alkylsterol mixtures on the same enzyme surface. If SMT isoforms are synthesized by fungi and the expression of some isoforms are developmentally regulated, then we will have to determine which factors govern expression and whether gene expression is functionally significant to sterol fitness. An interesting possibility is that the mechanism of biomethylation is the same in fungi, but the substrate specificity of the SMT enzyme differs. We recently discovered that in corn two SMT enzymes exist with different sterol substrate requirements—one is the $\Delta^{24(28)}$-isoform and the other is the $\Delta^{23(24)}$-isoform; their activities were found to be developmentally regulated during seedling development (Guo *et al.,* 1996; Nes and Venkatramesh, 1996).

4.3. Sterol Specificity and the Importance of Flat Structures

Some investigators assume that plant sterols will not bind to the fungal SMT enzyme because they are bent structures (Bloch, 1983; Rahier *et al.*, 1984; Ourisson, 1994). The potential difference in sterol conformations in phytosterols and fungal sterols (Figure 6) may be useful in the design of biomethylation inhibitors. We considered another possibility, that is, squalene oxide may orient into a conformation on the squalene oxide to cycloartenol cyclase (synthase) that gives rise to a flat cycloartenol structure, similar to the conformation of lanosterol (Nes *et al.*, 1988; Nes *et al.*, 1991a). The solid-state conformations of lanosterol and cycloartenol were studied by x-ray crystallography (Griffin *et al.*, 1994; Nes *et al.*, 1991a; Yoshida *et al.*, 1989) and found to be similar. The conformation of the 4 rings in cycloartenol are chair/half-chair/twist-chair/envelope. Recently, we examined the solution conformation and solid-state conformation of a series of 9β,19-cyclopropyl sterols (CSs) and found they were uniformly flat (Griffin *et al.*, 1994). High-field (500-MHz) 1D-and 2D-nuclear magnetic resonance (NMR) spectroscopic studies of cycloartenol (Nes *et al.*, 1996) indicated that Milon *et al.* (1989) misassigned several important chemical shifts and erred in their molecular mechanic calculations. Earlier, we reported the 1D-NMR (nOe) of 24,25-dehydropollinastanol (Nes *et al.*, 1988), and concluded that CSs were flat in solution. The new spectroscopic results (Nes *et al.*, 1996b) confirm our hypothesis and show unambiguously that cycloartenol is flat in solution and no significant population of molecules interconvert from the flat to bent shape.

Further support for the lack of rapid equilibration of CS from the bent to the flat conformer was obtained with molecular modeling of cycloartenol (Griffin *et al.*, 1994; Nes *et al.*, 1996b). Venkatramesh and Nes (1995) isolated a bent Δ^7-14-desmethyl CS from GL7 as a metabolite of cycloartenol metabolism. The bent CS could not function as a membrane insert in GL7. Bent cycloartenol isomers occur naturally in plants, e.g., 10α-cucurbitacin (Nes *et al.*, 1991). However, they are turned-over (hydroxylated) to function as defense compounds in plant-insect interactions. Thus, the introduction of the Δ^7-bond in CS regulates the conformational stability of the molecule and may lock the C-ring into the chair conformation, thereby producing a CS that has A/B/C rings in the chair/pseudo-chair conformations. Alternatively, Δ^7-14-methyl-CSs are known in plants, and they orient into the flat structure (Kadota *et al.*, 1995).

The SMT enzyme from GL7 was found to biomethylate 24,25-dehydropollinastanol effectively in a cell-free microsome-bound enzyme preparation and *in vivo* (Venkatramesh *et al.*, 1996a; Venkatramesh and Nes, 1995). Neither lanosterol nor cycloartenol are biomethylated by the fungus. The failure of SMT to biomethylate cycloartenol probably arises from the presence of a geminal C-4 methyl group that prevents productive binding of the sterol to the SMT-enzyme, rather than because cycloartenol may be bent (Nes and Venkatramesh, 1995). Similar arguments were given for the inability of cycloartenol to replace ergosterol as a membrane insert in GL7 (Nes *et al.*, 1993). In *U. phaseolis,* Bansal and Knoche (1981) demonstrated that cycloartenol and lanosterol were efficiently biomethylated by SMT; zymosterol was the preferred sterol substrate for biomethylation.

4.4. The X-Group Mechanism

The main pathway for monoalkylation of the Δ^{24}-unsaturated steroidal side chain gives rise to a 24(28)-methylene structure (Arigoni, 1978). Cornforth (1968) considered a novel olefin alkylation biochemistry (X-group mechanism) that was applicable to sterol biomethylation, as critiqued by Mihailovic (1984). Figure 7 shows the addition of the methyl group to the *re*-face of the substrate double bond and concomitant antiperiplanar addition of an X-group (an unspecified residue on the enzyme) that would give a covalently bound intermediate. Rotation of C-24 by 120° (path a) would place the X-group and the migrating hydrogen in an antiperiplanar arrangement. Elimination of the X-group with concomitant hydride shift migration and deprotonation would give the 24-methylene intermediate. On the other hand, a rotation of C-25 by 120° (path b) would imply that methyl donation and protein elimination occur on the same face of the double bond. The X-group mechanism may allow for the C-25 pro-*E* methyl (*cis*) group to assume the pro-*R* or pro-*S* position, depending on the mechanism. For instance, if the methyl cation donor (attack from the bottom: *re*-face, also mechanistically equivalent to a direct biomethylation transfer to produce a 24α-methylsterol) and a donor group from the enzyme (attack from the top: *si*-face, also mechanistically equivalent to a direct

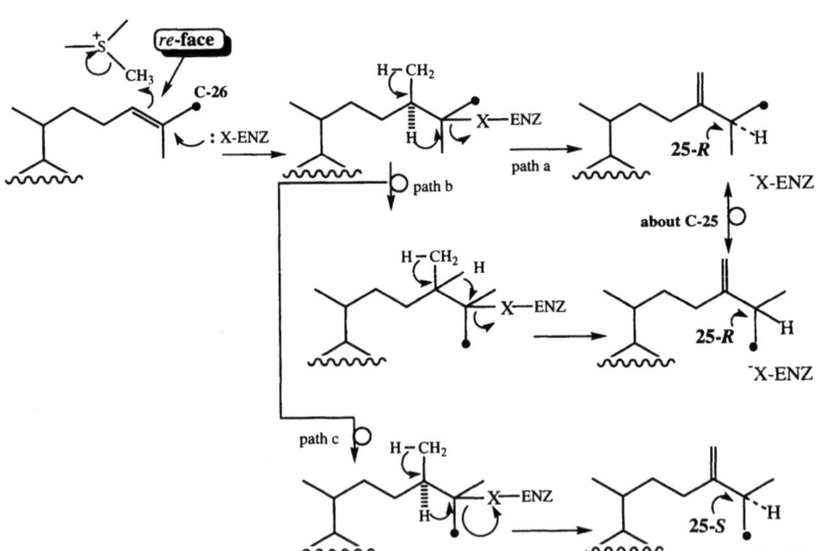

FIGURE 7. Hypothetical X-group mechanism of biomethylation.

biomethylation transfer to produce a 24β-methylsterol) proceed to give what is tantamount to a strongly stabilized C-25 carbenium ion that then undergoes H^- attack at C-25 from the opposite side, the final result of the latter process (inversion at C-25) would be to place CH_3^+ and H^- on the same side of the originally planar system (path c). Enzymological support of the Cornforth hypothesis was drawn from studies that indicated that the 1,2-hydride shift of H-24 to C-25 following methyl cation attack on the 24,25-double bond operated by the *re*-face mechanism in plants and that the activity of the plant enzyme was inhibited by 24*R*(α)-methyl-25-azacycloartenol to a greater degree than 24*S*(β)-methyl-25-azacycloartenol (Figure 5) and that sterol biomethylation in yeast operated by a ping-pong kinetic mechanism (consistent with covalent attachment of the enzyme to sterol or coenzyme during methylation) (Oehlschlager *et al.,* 1984). Studies on the sterol specificity on the biomethylation enzyme further indicated that the conformation of the sterol, i.e., bent versus flat, and stereochemistry of hydrogen migration from C-24 to C-25 during alkylation (from either the *re*-face or the *si*-face of the original 24,25-double bond) might be related mechanistically (Rahier *et al.,* 1984).

4.5. The Steric Course of Biomethylation

The mechanism of biomethylation catalyzed by SMT from *S. cerevisiae* has been characterized in stereochemical and kinetic detail (Arigoni, 1978; Akhtar *et al.,* 1967; Moore and Gaylor, 1970; Venkatramesh *et al.,* 1996; Zhou *et al.,* 1996). Early studies on sterol biomethylation in fungi involved incubation of ^{14}C-and ^{3}H-sterols with intact wild-type cells (reviewed by Nes and McKean, 1977; Parks, 1978). These studies were designed to show the type of methylation intermediate (viz., 24(28)-methylene versus 24(28)-cyclopropyl) and the nature of the 1,2-hydride shift of H-24 in ergosterol biosynthesis (whether C-24 hydrogen moves to C-25 or to C-23). The approach taken by several investigators was to incubate yeast cells with isotopically labeled lanosterol and isolate the labeled ergosterol from the cells. The ergosterol side chain was chemically degraded and the position of the labeled carbon was determined (Akhtar *et al.,* 1967; Nes and Le, 1990).

More recently, GL7 was used to prove the mechanism of sterol biomethylation. The mutant was fed several lanosterol and methionine samples that were isotopically labeled with stable isotopes ([24-^{2}H]lanosterol, [27-^{13}C]lanosterol, and [$^{2}H_3$-methyl]methionine). GL7 cells accumulated the lanosterol and methionine samples and converted them to isotopically labeled ergosterol, which then was examined by mass spectroscopy and ^{1}H-and ^{13}C-NMR spectroscopy (Zhou *et al.,* 1996). Zhou *et al.* (1996) found that the intensity of the proton signal at C-28 in the ^{1}H-NMR spectrum of [28-^{2}H-methyl]ergosterol decreased relative to the corresponding peak in the control ergosterol spectrum, following incubation with [$^{2}H_3$-methyl]AdoMet. This result confirms that AdoMet attacks C-24. In the same study, the investigators noted that there was an absence of a signal in the ^{13}C-NMR spectrum of [25-^{2}H]ergosterol following incubation with [24-^{2}H]lanosterol, which confirms the stereospecific 1,2-hydride shift of H-24 to C-25. Additionally, the position of the ^{13}C-signal in [27-^{13}C]ergosterol was determined (enhanced peak) after incubation with [27-^{13}C]lanosterol. This result indicated the C-25 hydrogen chirality and hence, face of H-24 movement to C-25.

The reaction progress may now be summarized with confidence as follows: (i) at the active site, a methyl group from AdoMet adds to C-24 of the sterol acceptor molecule. The methylation engages specifically the *si*-face of the 24,25-double bond and proceeds by an inversion mechanism at the methyl center (Arigoni,1978); (ii) following formation of the carbenium ion, a hydride ion migrates in a concerted manner from C-24 to C-25 and the positive charge at C-25 is eliminated by formation of the 24(28)-exocyclic methylene structure. The 1,2-hydride migration of H-24 to C-25 occurs in such a way that the pro-*E*-methyl group on the 24,25-double bond on the sterol acceptor molecule is transformed stereospecifically into the pro-*S*-methyl group of the product (Figure 8). This implies that methyl insertion and hydride migration occur on opposite faces of the original double bond. The chemical data coupled with the enzymic findings for the absence of a deuterium isotope effect (k_H/k_D) from [24-^{2}H]zymosterol/[24-H] zymosterol and [$^{3}H_3$-methyl]AdoMet incubated with the SMT enzyme, and the observation that the SMT enzyme operates by a random bi–bi kinetic mechanism (Venkatramesh *et al.*, 1996) establish conclusively that the X-group mechanism proposed by Cornforth (1968) is not operational in yeast.

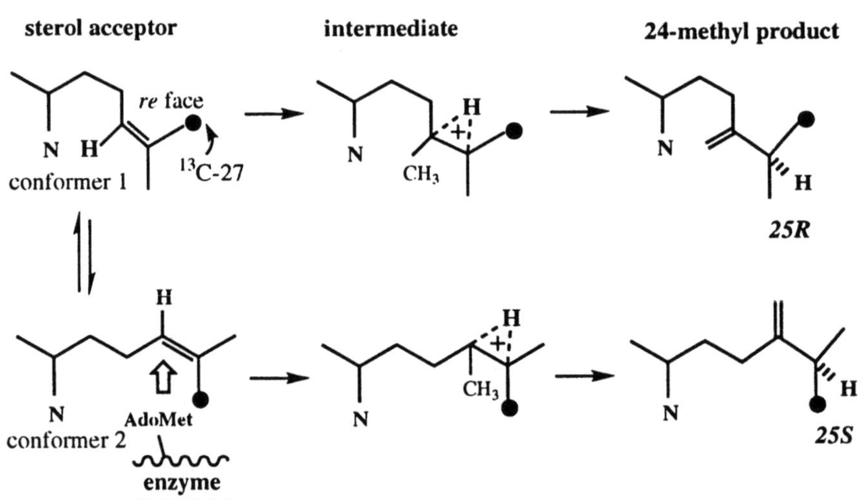

FIGURE 8. Mechanism of sterol biomethylation in ergosterol biosynthesis. (Adapted from Zhou *et al.*, 1996.) The biosynthetic side chain rule is adopted here, so that ^{13}C-27 of lanosterol is ^{13}C-27 of ergosterol, not ^{13}C-26 as IUPAC nomenclature would dictate by Conforth (1968) and critiqued by Mihailovic (1984).

Of the many speculative proposals that can be advanced to explain the topography of the sterol-AdoMet-SMT ternary enzyme complex, the proposal put forth by Arigoni (1978) and Akhtar and Jones (1978) seems the most probable as a guide for further elaboration. Their hypothesis is that a negatively charged counterion located next to the sulfonium center, so as to facilitate its fixation at the active site of the enzyme, is identical with the basic group ultimately responsible for abstraction of the proton (from the methyl on AdoMet). When the ternary complex includes a negative counterion (i.e., a nucleophile or base) at the active site, a conformational change or charge-relay system may take place to remove the negative charge from the sphere of the sulfonium cation. It is reasonable to assume that such a charge distribution will promote methyl cation transfer and provide an anion for the removal of a proton in the terminal step of the overall reaction.

On the assumption that the above mechanism operates generally in fungal biomethylations whereby the negative counterion plays two roles, then addition of the methyl group and removal of the proton from C-28 must occur from the same face of the plane of the original double bond of the substrate. A key point to this mechanism is that the formation of a $\Delta^{24(28)}$-unsaturated sterol from **2** in Figure 6 is now inhibited because the bridging proton lies on the wrong side of the molecule, whereas deprotonation of **2** and **4** (Figure 6) to the corresponding olefins is free from such a constraint (Arigoni, 1978). A two-base active site mechanism may be postulated in which the enzyme uses two different nucleophiles. One of the nucleophiles may interact with the C-25 carbenium ion of the 24-methylated sterol in the transition state of the reaction progress, that is, with the C-25 carbenium ion (Ator *et al.,* 1989; Oehlschlager *et al.,* 1984), and a second nucleophile may interact with the newly introduced methyl at C-24 in order to abstract a hydrogen from C-28. Based on recent results from this laboratory on the stereochemistry of hydrogen migration from C-24 to C-25 during ergosterol biosynthesis and from SMT enzyme kinetics and kinetic isotope effects using deuteriated sterols and coenzyme (Venkatramesh *et al.,* 1996a; Zhou *et al.,* 1996), we now hypothesize that a single base in the active site mediates double bond formation. The base may play two roles—one as a counterion and the other as the abstracting unit that removes the proton from C-28 following sterol biomethylation. An aromatic amino acid residue that is polar, yet hydrophobic, may be involved in this role. Dougherty (1996) has recently suggested that cation–pi interactions in AdoMet-mediated biochemistry may a play a role whereby the AdoMet lies on top of a tryptophan residue of the protein, with the methyl group projecting forward.

5. STERIC-ELECTRIC PLUG MODEL

We proposed a steric-electric plug model that shows the topological relationship of the sterol-AdoMet-enzyme complex (Parker and Nes, 1992). Historically, C-25 of the sterol was thought to be charged following methylation at C-24

(Nes and McKean, 1977). Thus many models, including models we had considered, illustrated sterol biomethylation with a two-base component with one of the bases on the face that was opposite to the incoming methyl cation. The base located opposite to the face of the double bond was supposed to stabilize the C-25 cation generated during catalysis. The plant (typified by sunflower) and fungal (typified by yeast) SMT enzymes were viewed as operating the same *si*-face methylation mechanism (Nes and Parker, 1992). The SMT enzymes were hypothesized to bind productively only to sterols that possess the following molecular features: (i) the nucleus must be flat, (ii) the side chain must be free to orient to the right into the pseudocyclic conformation, (iii) nucleophilic centers must be present at C-3 and C-24, and the C-3 hydroxyl group will act as a hydrogen acceptor in forming hydrogen bonds (its complementary amino acid residue in the yeast SMT enzyme may be a hydrogen donor, such as tyrosine), (iv) the sterol binds with its front face pointed toward the interior of the hydrophobic pocket, and (v) the 24,25-double bond is oriented in the enzyme cleft whereby the methyl cation may attack from the top of the pi-system, (from the *si*-face) and the hydrogen at C-24 migrates to C-25 from the *re*-face.

The sequence alignment of the SMT gene product reported by Hardwick and Pelham (1994) from *Saccharomyces cerevisiae,* a fungus, and that reported by Husselstein *et al.* (1996) (Table I) from *Arabidopsis Thalliana* and by Shi *et al.* (1996) from *Glycine max,* a plant, show several conserved motifs, of which one set of amino acid residues may function as a AdoMet-sterol binding site. Of the four apparent conserved regions shown in Table I (soybean sequence not shown), only region I, which corresponds to amino acids 80–92 (yeast sequence), has the necessary amino acid residue(s) to bind to sterol and AdoMet and catalyze the sterol biomethylation reaction in the manner described using the aromatic amino acid(s) as the negative point charge.

6. STEROL BIOMETHYLATION INHIBITORS

The study of the inhibition of SMT-catalyzed reactions has involved preparation of a series of steroids with a heteroatom in the side chain, usually nitrogen or sulfur and usually positively charged, replacing either C-24 or C-25 (Figure 9). These TSA inhibitors are expected to mimic the structural and electronic properties of the carbocation intermediates formed during the normal sterol biomethylation reaction and consequently bind tightly to the enzyme. As a first approximation, the kinetics of tight binding should be of the reversible-type pattern. The positive charge on the heteroatom arises by protonation of the neutral amine at physiological pH, such as occurs with 25-azacholesterol (Oehschlager *et al.,* 1984; Rahier *et al.,* 1984). The inhibitors that have been studied generally contain the cholesterol, lanosterol, and cycloartenol (and, sometimes, zymosterol) nucleus, and they possess the same flat nuclear conformation and charged side chain as might occur with the natural sterol substrates during

Table I
Comparison of the Predicted Mature Regions of Amino Acid Sequences Corresponding to the SMT Enzyme from *Saccharomyces* (top) and *Arabadopsis* (bottom)[a]

```
                 .         .         .         .         .
  1   MSETELRKRQA....QFTRELHGDDIGKKTGLSALMSKNNSAQKEAVQKY    46
      |.. .|   .|   .: : | . : :.:.| .|:  ...|   | ||.
  1   MDSLTLFFTGALVAVGIYWFLCVLGPAERKGKRAVDLSGGSISAEKVQDN    50
                 .         .                   I
 47   LRNWDGRTDKDAEERRLEDYNEATHSYYNVVTDFYEYGWGSSFHFSRFYK    96
      .:.: :   :. |    |. : ...:||:|||:||:|||  |||||. ..
 51   YKQYWSFFRRPKEIETAEKVPDFVDTFYNLVTDIYEWGWGQSFHFSPSIP   100
                 .         .       .II               .
 97   GESFAASIARHEHYLAYKAGIQRGDLVLDVGCGVGGPAREIARFTGCNVI   146
      |.|  ...  ||.  .   .:..|: :|||||||||| |.||. . .||:
101   GKSHKDATRLHEEMAVDLIQVKPGQKILDVGCGVGGPMRAIASHSRANVV   150
                 .         .         .         .         .
147   GLNNNDYQIAKAKYYAKKYNLSDQMDFVKGDFMKMDFEENTFDKVYAIEA   196
      |:. |:||:.:|:...|| .|..  :.| |:|:.|.|::|.|| .|.|||
151   GITINEYQVNRARLHNKKAGLDALCEVVCGNFLQMPFDDNSFDGAYSIEA   200
                 .       III               .         .
197   TCHAPKLEGVYSEIYKVLKPGGTFAVYEWVMTDKYDENNPEHRKIAYEIE   246
      ||||||||:||.|||:|||||: :. |||| |:|:..::.|| .:  :||
201   TCHAPKLEEVYAEIYRVLKPGSMYVSYEWVTTEKFKAEDDEHVEVIQGIE   250
                 .         .         .         .         .
247   LGDGIPKM.FHVDVARKALKNCGFEVLVSEDLADNDDEIPWYYPLTGEWK   295
       ||::| :  .||:| .| |..|||:: ..|||. ..| ||: .|
251   RGDALPGLRAYVDIAETA.KKVGFEIVKEKDLASPPAE.PWWTRL.....   293
                 .         .         .         .         .
296   YVQNLANLATFFRTSYLGRQFTTAMVTVMEKLGLAPEGSKEVTAALENAA   345
         .::.|| ::|. .:           | ::. :|:||.|. :| . | ..|
294   ...KMGRLA.YWRNHIV........VQILSAVGVAPKGTVDVHEMLFKTA   331
            IV .         .         .
346   VGLVAGGKSKLFTPMMLFVARKPENAETPSQTSQEATQ*   384
       .|. ||.. :|.|| :::.||||.:|..|
332   DCLTRGGETGIFSPMHMILCRKPESPEESS*.......   362
```

Bold regions: I = DFYEYGWGSSFH / DIYEWGWGQSFH; II = VLDVGCGVG / ILDVGCGVG; III = LKPGG / LKPGS; IV = GLVAGGKS / CLTRGGET.

[a]Percent similarity: 55.182 (connecting dots) and percent identity: 38.095 (connecting lines). Region **I:** putative AdoMet/sterol-binding site; Regions **II** and **III:** motifs common to other methyltransferases (from search of the Gene Bank); Region **IV:** putative ATP/GTP-binding site (match based on sequence alignments with other genes using computer program-GAP). Data were obtained from Venkatramesh *et al.* (1996a) and Husselstein *et al.* (1996).

the biomethylation reaction. In spite of the similarity of the reaction catalyzed by the fungal and plant SMT enzymes, the modes of interactions of the inhibitors with the enzymes may be different according to the nature of the charged species introduced into the sterol side chain. For instance, the aziridine-containing sterol may act as a TSA inhibitor or as a mechanism-based inhibitor by accepting a methyl group (Figure 10), depending on the detailed amino acid composition in the active site and on possible differences in protein structure that give the sterol access to different enzyme

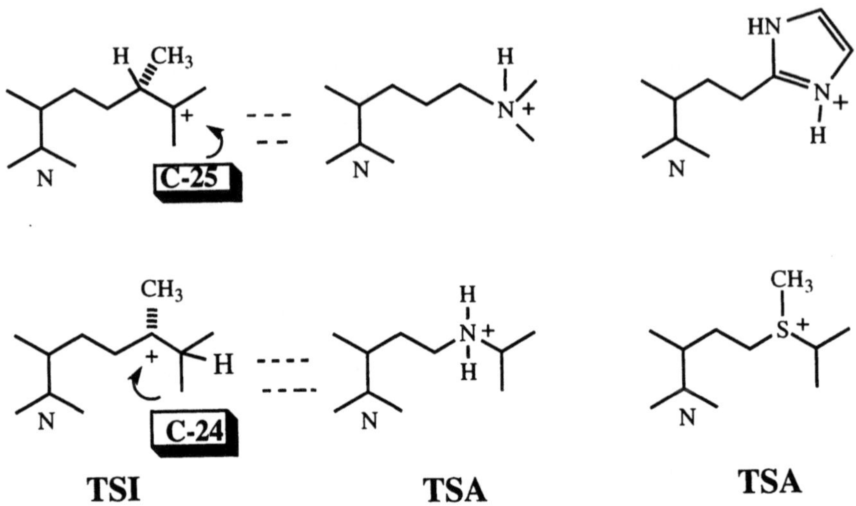

FIGURE 9. Typical rationally designed transition state analogs (TSAs) that mimic the C-24 and C-25 transition state (TSIs) in the biomethylation reaction progress. Four types of sterol nuclei can serve in the rational design of inhibitors (cholesterol, lanosterol, zymosterol, and cycloartenol). N = nucleus.

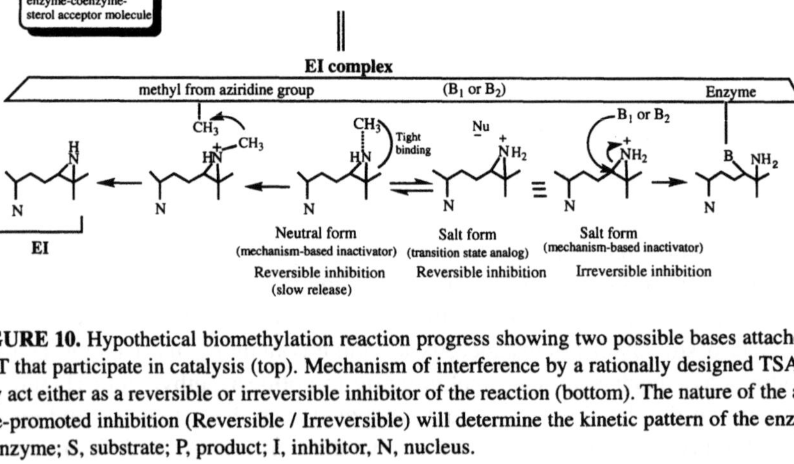

FIGURE 10. Hypothetical biomethylation reaction progress showing two possible bases attached to SMT that participate in catalysis (top). Mechanism of interference by a rationally designed TSA that may act either as a reversible or irreversible inhibitor of the reaction (bottom). The nature of the aziridine-promoted inhibition (Reversible / Irreversible) will determine the kinetic pattern of the enzyme. E, enzyme; S, substrate; P, product; I, inhibitor, N, nucleus.

conformations during turnover. The sulfur-containing side chain may also act as a mechanism-based inactivator by methylation of the sterol acceptor molecule from AdoMet followed by the sterol methylating a nucleophile on SMT (Ator *et al.,* 1989).

An alternative approach to the design of sterol biomethylation inhibitors is based on the predicted amino acid sequence of the SMT enzymes (from the steric-electric plug model, section 5) which indicates the presence of tryptophan and tyrosine residues in the conserved motif that may function in AdoMet/sterol binding. These electron-rich indole and phenol side chains from tryptophan and tyrosine residues are hypothesized to give rise to a hydrophobic pocket providing essential negative point charges that direct the methyl group on AdoMet to project toward the 24,25-double bond of the sterol acceptor molecule, and stabilize the AdoMet coenzyme through cation–pi interactions. Inhibitors that form favorable packing and charge-relay interactions with the negatively charged aromatics in the active site should inhibit SMT enzyme activity. Whereas positively charged 25-azasterols will fulfill this requirement, they do so as reversible inhibitors. The alternative approach is to prepare environmentally friendly inhibitors, such as the 24,25-aziridine sterol, which are designed to act as mechanism-based inactivators of SMT enzyme activity and will ultimately undergo metabolism by a microorganism to a safe byproduct (Table II) (Guo *et al.,* 1996b; Jia *et al.,* 1996).

The molecular features of the 25-azasterol involved in inhibition of the yeast SMT activity were shown to be related to the nuclear structure, e. g., 25-azazymosterol > 25-azacholesterol > 25-azalanosterol, and to correspond to the sterol features recognized in equivalent order by the SMT enzyme for sterol binding and catalysis (Oehlschlager *et al.,* 1984), i.e., zymosterol > desmosterol > lanosterol (Moore and Gaylor, 1970; Venkatramesh *et al.,* 1996). For this reason, 25-azasterols were thought to behave as sterol substrate mimics. In plants, 25-azasterols were found to exhibit a different set of molecular parameters than was found in the fungal system. The plant SMT enzyme recognized 25-azacycloartenol = 25-azalanosterol > 25-azacholesterol (Rahier *et al.,* 1984), and this order differed from the sterol binding specificity of the plant SMT enzymes: cycloartenol > lanosterol > desmosterol (cholesta-5,24-dienol) (Rahier *et al.,* 1986; Nes *et al.,* 1991). Rahier *et al.* (1984) interpreted the results with the 25-azasterols to imply that the molecular requirements for the tetracyclic moiety of the inhibitor are less strict than in the case of the substrates. They also concluded that the bent conformation of 25-azacycloartenol is not required in the inhibitor structure, since 25-azalanosterol, with a flat structure, was equally potent as a biomethylation inhibitor.

The effect of several different classes of sterol biomethylation inhibitors on SMT enzyme activity has been examined recently by Guo *et al.* (1996) and Venkatramesh *et al.* (1996a,b). As shown in Table II (Guo *et al.,* 1996; Nes *et al.,* 1996a), yeast SMT activity is inhibited by TSAs that possess different nuclear structures, but not necessarily in the same manner as suggested by Oehlschlager *et al.* (1984). Relative potency follows the order: 25-azacycloartenol = 25-azalanosterol >

Table II
Inhibition of SMT Activity by Substrate Analogs with Modified Side Chains[a]

Inhibitor	Nucleus	K_i	Kinetic pattern	Reversible (R)/ Irreversible (I)
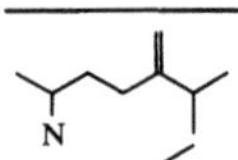	L	NA	NA	NA
	Z	54 μM	C	R
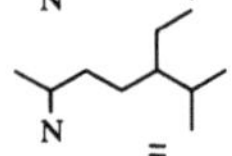	CH	NA	NA	NA
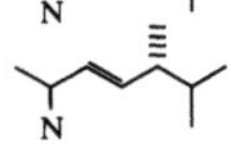	E	65 μM	C	R
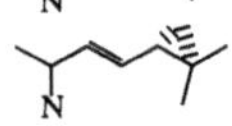	E	NA	NA	NA
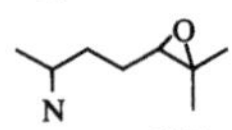	L	NA	NA	NA
	Z	NA	NA	NA
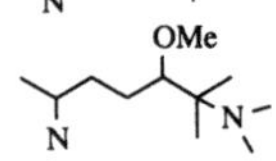	L	50μM	C	ND
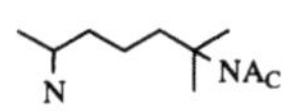	L	50 μM	C	ND
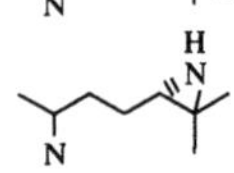	L	10 nM	NC	R
	Z	5 nM	NC	R
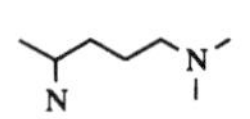	L	45 Nm	NC	R
	CH	10 nM	NC	R
	CY	50 nM	NC	R
	Z	15 nM	NC	R
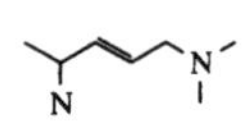	CH	250 nM	NC	R
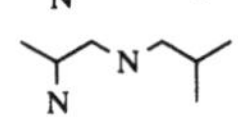	CH	75 nM	NC	R
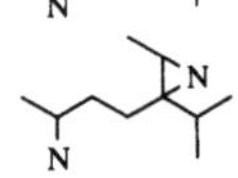	CH	3 μM	NC	R
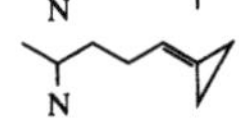	L	NA	NA	NA
	Z	30 μM	C	I

[a]NA = not active; C = competitive-type inhibition; NC = noncompetitive-type inhibition; ND = not determined. Enzyme kinetics were determined from Lineweaver–Burke and Dixon plots. CH = cholesterol nucleus; CY = cycloartenol nucleus; E = ergosterol nucleus; L = lanosterol nucleus; Z = zymosterol nucleus; N = nucleus. Data from Guo *et al.* (1996) and Venkatramesh *et al.* (1996a). k_{inact} = 1.5 min^{-1} for the cyclopropylidene-sterol.

25-azazymosterol = 25-azacholesterol, whereas there is no difference in inhibitory potency between 24,25-epiminozymosterol and 24,25-epiminolanosterol. These results suggest that the structure and charge in the inhibitor side chain are more important to inhibitory potency than the degree of methylation at C-4. In the sunflower SMT enzyme (Janssen and Nes, 1992), TSA inhibitors bind to the enzyme in the same sterol conformation as their corresponding non-functionalized sterol partner and adopt a sterol–enzyme conformation that fails to bind the native sterol substrate productively. Recent studies with 25-azasterols and related *N*-sterols are shown in Table II. The order of magnitude of diminished inhibitory effect of 24(28)-epiminofucosterol and 25-azacholesta-5,22-dienol compared with 24,25-epiminozymosterol and 25-azacholesterol, respectively, indicates that bulk and restricted rotation of the side chain are important to the inhibitor's potency and mode of action. Similar conclusions were made about the sterol binding site of the SMT enzyme (Venkatramesh *et al.,* 1996a). The ability of 25-azacycloartenol to inhibit SMT activity in a similar manner to 25-azalanosterol indicates that 25-azacycloartenol mimics the 3-dimensional shape of 25-azalanosterol in the inhibitor–enzyme complex. The positive charge of the ammonium group does not have to interact with a carbocation-stabilizing group in the active site to act as a reversible non-competitive inhibitor, but may interact with the nucleophile that is responsible for binding the positive charge on the sulfur from AdoMet (compare the steric-electric plug model).

The kinetics of interaction of the 24,25-aziridine-containing sterol assayed in sunflower and yeast have indicated distinctly different patterns of inhibition (Table II). In yeast, the aziridine-sterol is a reversible inhibitor, whereas in sunflower it is a irreversible inhibitor (Guo *et al.,* 1996; Janssen and Nes, 1992). The irreversible nature of the aziridine complexing with the SMT enzyme from *G. fujikuroi* may explain the accumulation of 24-desalkyl sterols in that fungus and the developmentally delayed metabolism of 24,25-epiminolanosterol to 25-aminolanosterol by intact mycelia (Nes *et al.,* 1989). The aziridine-containing sterols are weakly absorbed by wild-type yeast (Nes *et al.,* 1996a).

Mechanism-based inactivators that mimic the ground-state structure (i.e., 26,27-cyclopropylidenes, Figure 11) were also found to exhibit a structural resemblance to the native substrate in their inhibitory potency and to possess competitive-type inhibition kinetics. In contrast to the reversible inhibition on SMT activity by 24-methylated sterols, the mechanism-based substrate analogs show time-dependent irreversible inactivation of enzyme activity. The latter inhibitor may be used in active-site mapping of SMT through covalent modification (Abe and Prestwich, 1994; Croteau *et al.,* 1993). It is interesting that the TSAs inhibit SMT activity in the nanomolar range and show noncompetitive-type inhibition kinetics, whereas the 24(28)-methylene product inhibitors and mechanism-based inactivators inhibit SMT enzyme activity in the micromolar range (around the K_m of the substrate) and show competitive-type kinetics. Ergosterol was found to downregulate SMT enzyme activity in yeast specifically (Nes *et al.,* 1996*a*; Venkatramesh *et al.,* whereas sitosterol was found to down-regulate SMT activity in sunflower specifically (Janssen and Nes, 1992). Thus plant and fungal SMT enzymes respond differently to feedback regulation from 24-alkyl sterols and to sterol biomethylation inhibitors.

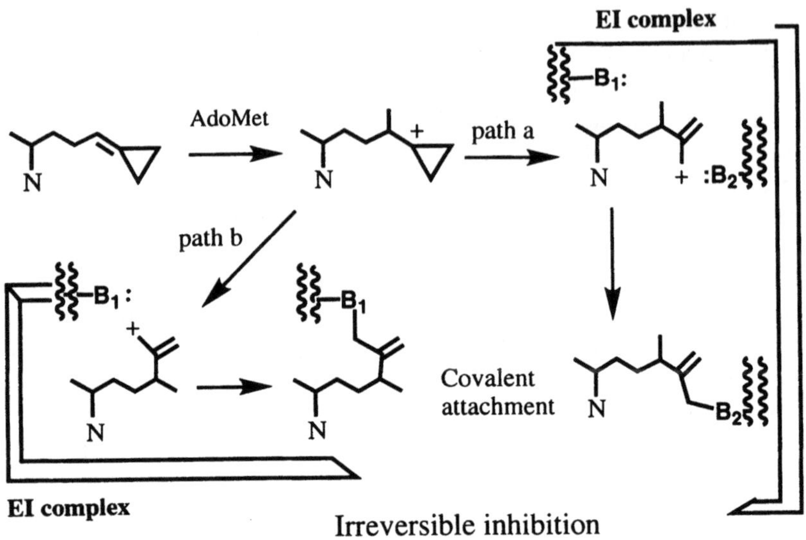

FIGURE 11. Hypothetical inactivation of SMT by a mechanism-based inactivator. The inhibitor may interact with either of the two presumptive bases B_1 or B_2 in the active site: B_1 abstracts a proton from C-28, whereas B_2 stabilizes the C-25 carbenium ion. Covalent modification of enzyme may result from SMT-promoted biomethylation of the sterol acceptor molecule. N = nucleus.

7. SUMMARY

During the course of the last decade, the development of SBIs, and particulary sterol biomethylation inhibitors, has been based on the rational design approach. Successful though this approach has been in elucidating sterol biomethylation enzymology, its limitations are becoming apparent from the findings that: (i) 24,25-double bond metabolism gives rise to cholesterol and ergosterol in a mechanistically similar manner, (ii) 25-azasterols are harmful to human physiology, and (iii) side-chain modified sterols designed to inhibit the SMT enzyme in *S. cerevisiae* may be ineffective or operate by another kinetic mechanism in a related organism, rendering it therapeutically nonuseful. Nevertheless, it may be possible to ultimately capitalize on the unique aspects of sterol biomethylation chemistry and enzymology to design taxa-specific inhibitors. With increased understanding of the structure and function of SMT enzymes in different fungi, it should be possible to prepare novel mechanism-based inactivators to control SMT activity uniquely and with high specific activity.

ACKNOWLEDGMENTS. We thank the Welch Foundation for support. WDN thanks his associates, Professors James Harman and Robert Shaw, for many stimulating discussions on molecular biology and mechanistic enzymology.

8. REFERENCES

Abe, I., and Prestwich, G. D., 1994, Active-site mapping of affinity-labeled rat oxidosqualene cyclase, *J. Biol. Chem.* **269:**802–804.

Abe, I., Tomesch, J. C., Wattanasin, S., and Prestwich, G. D., 1994, Inhibitors of squalene biosynthesis and metabolism, *Nat. Prod. Rep.* **11:**279–302.

Adler, J. H., Young, M., and Nes, W. R., 1977, Determination of the absolute configuration at C-20 and C-24 of ergosterol in Ascomycetes and Basdiomycetes by proton magnetic resonance spectroscopy, *Lipids* **12:**364–366.

Akhtar, M., and Jones, C., 1978, Some biological transformations involving unsaturated linkages: the importance of charge separation and neutralization in enzyme catalysis, *Tetrahedron* **34:**813–832.

Akhtar, M., Hunt, P. F., and Parvez, M. A., 1967, The transfer of hydrogen from C-24 to C-25 in ergosterol biosynthesis, *Biochem. J.* **103:**616–622.

Arigoni, D., 1978, Stereochemical studies of enzymic C-methylations, *Ciba Found. Symp.* **60:** 243–261.

Ator, M. A., Schmidt, S. J., Adams, J. L., and Dolle, R. E., 1989, Mechanism and inhibition of Δ^{24}-sterol methyl transferase from *Candida albicans* and *Candida tropicalis, Biochemistry* **28:** 9633–9640.

Ator, M. A., Schmidt, S. J., Adams, J. L., Dolle, R. E., Kruse, L. L., Frey, C. L., and Barone, J. M., 1992, Synthesis, specificity, and antifungal activity of inhibitors of the *Candida albicans* Δ^{24}-sterol methyl transferase, *J. Med. Chem.* **35:**100–106.

Balliano, G., Viola, F., Ceruti, M., and Cattel, L., 1988, Inhibition of sterol biosynthesis in *Saccharomyces cerevisae* by *N,N*-diethylazasqualene and derivatives, *Biochim. Biophys. Acta* **959:**9–19.

Bansal, S. R., and Knoche, H. W., 1981, Sterol methyl transferase from *Uromyces phaseoli:* an investigation of the first and second transmethylation reactions, *Phytochemistry* **20:**1269–1277.

Bard, M., and Downing, J. F., 1981, Genetic and biochemical aspects of yeast sterol regulation involving 3-hydroxy-3-methylglutaryl coenzyme A reductase, *J. Gen. Microbiol.* **125:**15–20.

Berg, D., and Plempel, M. (eds.), 1988, *Sterol Biosynthesis Inhibitors: Pharmaceutical and Agrochemical Aspects,* Ellis Horwood Ltd., Chichester, United Kingdom.

Bloch, K. E., 1983, Sterol structure and membrane function, *Crit. Rev. Biochem.* **14:**47–92.

Burden, R. S., Cooke, D. T., and Carter, G. F., 1989, Inhibitors of sterol biosynthesis and growth in plants and fungi, *Phytochemistry* **28:**1791–1804.

Casey, W. M., Keesler, G. A., and Parks, L. W., 1992, Regulation of partitioned sterol biosynthesis in *Saccharomyces cerevisiae, J. Bacteriol.* **174:**7283–7288.

Castle, M., Blondin, G., and Nes, W. R., 1963, Evidence for the origin of the ethyl group of β-sitosterol, *J. Am. Chem. Soc.* **85:**3306–3307.

Cornforth, J. W., 1968, Olefin alkylation in biosynthesis, *Angew. Chem. Int. Ed. Engl.* **7:**903–911.

Croteau, R., Alonso, W. R., Koepp, A. E., Shim, J. H., and Cane, D. E., 1993, Irreversible inactivation of monoterpene cyclase by a mechanism-based inhibitor, *Arch. Biochem. Biophys.* **307:**397–404.

Dougherty, D. A., 1996, Cation–pi interactions in chemistry and biology: a new view of benzene, Phe, Tyr, and Trp, *Science* **271:**163–168.

Dukes, M. N. G., 1980, Drugs affecting lipid metabolism, in: *Myler's Side Effects of Drugs,* 9th edition, (G. S. Smith, ed.) pp. 727–737, Elsevier, Amsterdam.

Fieser, L. F., and Fieser, M., 1959, *Steroids,* Reinhold, New York.

Fusentani, N., 1988, Antifungal substances from marine invertebrates, *Ann. New York Acad. Sci.* **44:**113–127.

Griffin, J. F., Nes, W. D., and Allinger, N. L., 1994, Evidence from crystallographic results and molecular mechanic calculations for the conformation of cycloartenol in membranes, INFORM **5:**509 (A).

Grun, B., 1991, *The Timetables of History,* 3rd revised edition, Simon & Schuster/Touchstone, New York.

Guo, D., Jia, Z., and Nes, W. D., 1996a, Phytosterol biosynthesis: Isotope effects associated with biomethylation formation to 24-alkene sterol isomers, Tetrahedron Letts., **37:**6823–6826.

Guo, D., Jia, Z., Zhou, W., and Nes, W. D., 1996b, Sterol biomethylation inhibitors of the (*S*)-adenosyl-L-methionine: $\Delta^{24(25)}$-sterol methyl transferase from *Saccharomyces cerevisiae,* (submitted).

Hamilton-Miller, J. M. T., 1974, Fungal sterols and the mode of action of the polyene antibiotics, *Adv. Appl. Microbiol.* **17:**109–125.

Hardwick, K. G., and Pelham, H. R. B., 1994, *SED6* is identical to *ERG6* and encodes putative methyl transferase required for ergosterol synthesis, *Yeast* **10:**265–269.

Haughan, P. A., Chance, M. L., and Goad, L. J., 1995, Effects of an azasterol inhibitor of sterol 24-transmethylation on sterol biosynthesis and growth of *Leishmania donovani* promastigotes, *Biochem. J.* **308:**31–38.

Houston, J. B., Humphrey, M. J., Mathew, D. E., and Tarbit, M. H., 1988, Comparison of two azole antifungal drugs, ketoconazole and fluconazole, as modifiers of rat hepatic monooxygenase activity, *Biochem. Pharmacol.* **37:**401–408.

Hull, S. E., and Woolfsin, M. M., 1976, The crystal structure of ergosterol monohydrate, *Acta Cryst.* **B32:**2370–2373.

Husselstein, D., Gachotte, D., Desprez, T., Bard, M., and Benveniste, P., 1996, Transformation of *Saccharomyces cerevisiae* with cDNA encoding a sterol methyl transferase from *Arabadopsis thaliania* results in synthesis of 24-ethyl sterols, *FEBS Lett.* **381:**87–92.

Janssen, G. G., and Nes, W. D., 1992, Structural requirements for transformation of substrates by the (*S*)-adenosyl-L-methionine: $\Delta^{24(25)}$-sterol methyl transferase. II. Inhibition by analogs of the transition state coordinate, *J. Biol. Chem.* **267:**25856–25863.

Janssen, G. G., Kalinowska, M., Norton, R. A., and Nes, W. D., 1991, (*S*)-Adenosyl-L-methionine: Δ^{24}-sterol methyl transferase: mechanism, enzymology, inhibitors, and physiological importance, in: *Physiology and Biochemistry of Sterols* (G. W. Patterson and W. D. Nes, eds.), pp. 83–117, Amer. Oil Chem. Soc. Press, Champaign, IL.

Jia, Z., Zhou, W., Guo, D., and Nes, W. D., 1996, Synthesis of rationally designed mechanism-based inactivators of the (*S*)-adenosyl-L-methionine: $\Delta^{24(25)}$-sterol methyl transferase, *Syn. Commun.* **26:**3841–3846.

Julia, M., and Marazano, C., 1985, Biomimetic methyl transfer to olefins, *Tetrahedron* **41:**3717–3724.

Kawaguchi, A., 1970, Control of ergosterol biosynthesis in yeast, *J. Biochem* **67:**219–227.

Loefler, R. S. T., Butters, J. A., and Hollomon, D. W., 1992, The sterol composition of powdery mildews, *Phytochemistry* **31:**1561–1563.

McCammon, M. T., and Parks, L. W., 1981, Inhibition of sterol transmethylation by *S*-adenosylhomocysteine analogs, *J. Bacteriol.* **145:**106–112.

Mercer, E. I., 1993, Inhibitors of sterol biosynthesis and their applications, *Prog. Lipid Res.* **32:**357–416.

Mihailovic, M. M, 1984, Biosynthesis of phytosterols in *Trebouxia* sp.: steric course of the C-alkylation step, pp. 1–140, ETH Dissertation No. 7535, Zürich.

Moore, T. S., and Gaylor, J. L., 1970, Investigation of an *S*-adenosylmethionine: Δ^{24}-sterol methyl transferase in ergosterol biosynthesis in yeast. II. Specificity of sterol substrates and inhibitors, *J. Biol. Chem.* **245:**4684–4688.

Nes, W. D., 1987, Biosynthesis and requirement for sterols in the growth and reproduction of Oomycetes, *Amer. Chem. Soc. Symp. Ser.* **325:**304–328.

Nes, W. D., Benson, M., Lundin, R. E., and Le, P. H., 1988a, Conformational analysis of 9β,19-cyclopropyl sterols: detection of the pseudoplanar conformer by nuclear Overhauser effects and its functional implications, *Proc. Natl. Acad. Sci. USA* **85:**5759–5763.

Nes, W. D., Guo, D., and Lopez, M., 1997, Studies on the regulation of fungal growth and sterol metabolism by plant sterols and sterol methylation inhibitors, 1996, (in preparation).

Nes, W. D., Hanners, P. K., and Parish, E. J., 1986, Control of fungal sterol C-24 transalkylation: importance to developmental regualtion, *Biochem. Biophys. Res. Commun.* **139:**410–415.

Nes, W. D., Heupel, R. C., and Le, P. H., 1985, Biosynthesis of ergosta-6(7),8(14),22(23)-trien-3β-ol by *Gibberella fujikuroi:* its importance to ergosterol's metabolic pathway, *J. Chem. Soc. Chem. Commun.* 1431–1433.

Nes, W. D., Janssen, G. G., and Bergenstrahle, A., 1991a, Structural requirements for transformation of substrates by the (*S*)-adenosyl-L-methionine:$\Delta^{24(25)}$-sterol methyl transferase, *J. Biol. Chem.* **266:**15202–15212.

Nes, W. D., Janssen, G. G., Crumley, F. G., Kalinowska, M., and Akihisa, T., 1993, The structural requirements of sterols for membrane function in *Saccharomyces cerevisiae, Arch. Biochem. Biophys.* **300:**724–733.

Nes, W. D., Jia, Z., Koike, K., Nikaido, T., Sakamoto, Y., Guo, D., Griffin, J. F., Allinger, A. L., 1997b, Proof that cycloartenol and biogenetically related 9β,19-cyclopropyl sterols are conformationally flat in solution and the solid state, (in preparation).

Nes, W. D., and Le, P. H., 1988, Regulation of sterol biosynthesis in *Saprolegnia ferax* by 25-azacholesterol, *Pest. Biochem. Physiol.* **30:**87–94.

Nes, W. D., and Le, P. H., 1990, Evidence for separate intermediates in the biosynthesis of 24β-methylsterol end products by *Gibberella fujikuroi, Biochim. Biophys. Acta* **1042:**119--125.

Nes, W. D., Norton, R. A., Crumley, F. G., Madigan, S. J., and Katz, E. R., 1990, Sterol phylogenesis and alga evolution, *Proc. Natl. Acad. Sci. USA* **87:**7565–7569.

Nes, W. D., and Parish, E. J., 1988, Metabolism of 2,3-epiminosqualene to 24,25-epiminolanosterol by *Gibberella fujikuroi, Lipids* **23:**375–376.

Nes, W. D., and Venkatramesh, M., 1994, Molecular asymmetry and sterol evolution, *Amer. Chem. Soc. Symp. Ser.* **562:**55–89.

Nes, W., and Venkatramesh, M., 1996, Enzymology of phytosterol transformations, in: *Biochemistry and Function of Sterols* (E. J. Parish and W. D. Nes, eds.), pp. 111–122, CRC Press, Boca Raton, FL.

Nes, W. D., Wong, R. Y., Benson, M., and Akihisa, T., 1991b, Conformational analysis of 10α-cucurbitacin, *J. Chem. Soc. Chem. Commun.* 1272–1274.

Nes, W. D., Xu, S., and Haddon, W. F., 1988b, Evidence for similarities and differences in the biosynthesis of fungal sterols, *Steroids* **53:**533–538.

Nes, W. D., Xu, S., Parish, E. J., 1989, Metabolism of 24(*R,S*),25-epiminolanosterol to 25-aminolanosterol and lanosterol by *Gibberella fujikuroi, Arch. Biochem. Biophys.* **272:**323–331.

Nes, W. R., 1987, Structure–function relationships for sterols in *Saccharomyces cerevisiae, Amer. Chem. Soc. Symp. Ser.* **325:**252–267.

Nes, W. R., and Dhanuka, I. C., 1988, Inhibition of sterol synthesis by Δ^5-sterols in a sterol auxotroph of yeast defective in oxidosqualene cyclase and cytochrome P-450, *J. Biol. Chem.* **263:** 11844–11850.

Nes, W. R., and McKean, M. L., 1977, *Biochemistry of Steroids and Other Isopentenoids,* University Park Press, Baltimore.

Nes, W. R., Sekula, B. C., Nes, W. D., and Adler, J. H., 1978, The functional importance of structural features of ergosterol in yeast, *J. Biol. Chem.* **253:**6218–6225.

Oehlschlager, A. C., and Czyzewaska, E., 1992, Rationally designed inhibitors of sterol biosynthesis, in: *Emerging Targets and Antifungal Therapy* (J. Sutcliffe and N. H. Georgopapadakou, eds.), pp. 437–475, Routledge, Chapman & Hall Press, New York.

Oehlschlager, A. C., Angus, R. H., Pierce, A. M., Pierce, H. D., Jr., and Srinivasan, R., 1984, Azasterol inhibition of Δ^{24}-sterol methyl transferase in *Saccharomyces cerevisiae, Biochemistry* **23:** 582–3589.

Ourisson, G., 1994, Pecularities of sterol biosynthesis in plants, *J. Plant Physiol.* **243:**434–439.

Parker, S. R., and Nes, W. D., 1992, Regulation of sterol biosynthesis and its phylogenetic implications, *Amer. Chem. Soc. Symp. Ser.* **497:**110–145.

Parks, L. W., 1978, Metabolism of sterols in yeast, *CRC Crit. Rev. Microbiol.* **6:**301–341.

Pinto, W. J., and Nes, W. R., 1983, Stereochemical specificity for sterols in *Saccharomyces cerevisiae, J. Biol. Chem.* **258:**4472–4476.

Pinto, W. J., Lozano, R., and Nes, W. R., 1986, Inhibition of sterol biosynthesis by ergosterol and cholesterol in *Saccharomyces cerevisiae, Biochim. Biophys. Acta* **826:**89–95.

Popják, G., Edmond, J., Anet, F. A. L., and Easton, N. R., Jr., 1977, Carbon-13 NMR studies on cholesterol biosynthesized from [^{13}C]mevalonates, *J. Am. Chem. Soc.* **99:**931–935.

Popják, G., Meenan, A., Parish, E. J., and Nes, W. D., 1989, Inhibition of cholesterol synthesis and cell growth by 24(*R,S*),25-iminolanosterol and triparanol in cultured rat hepatoma cells, *J. Biol. Chem.* **264:**6230–6238.

Raederstorff, D., and Rohmer, M., 1987, Sterol biosynthesis via cycloartenol and other biochemical features related to photosynthetic phyla in the amoeba *Naegleria lovaniensis* and *Naegleria gruberi, Eur. J. Biochem.* **164:**427–434.

Rahier, A., Génot, J-C., Schuber, F., Benveniste, P., and Narula, A. S., 1984, Inhibition of *S*-adenosyl-L-methionine sterol-C-24-methyltransferase by analogues of a carbocationic ion high-energy intermediate, *J. Biol. Chem.* **259:**15215–15223.

Rahier, A., Taton, M., Bouvier-Navé, P., Schmitt, P., Benveniste, P., Schuber, F., Narula, A.S., Cattel, L., Anding, C., and Place, P., 1986, Design of high energy intermediate analogues to study sterol biosynthesis in higher plants, *Lipids* **21:**52–62.

Rahman, M. D., and Pascal, R. A., Jr., 1990, Inhibitors of ergosterol biosynthesis and growth of the trypanosmatid protozoan *Crithida fasciculata, J. Biol. Chem.* **265:**4986–4996.

Roddick, J. G., 1987, Antifungal activity of plant steroids, *Amer. Chem. Soc. Symp. Ser.* **325:**286–303.

Rosenblum, E. R., Malloy, J. M., McManus, I. R., Naworal, J. D., and Campbell, I. M., 1979, Effect of 20, 25-diazacholesterol on viability and steroid synthesis capability of cultured chick embryo pectoral muscle cells, *Biochem. Biophys. Res. Commun.* **88:**1105–1110.

Ryder, N. S., 1988, Mechanism of action and biochemical selectivity of allylamine antimycotic agents, in: *Antifungal Drugs* (V. St. Georgiev, ed.), *Ann. New York Acad. Sci.* **544:**208–220.

Shi, J., Gonzales, R. A., and Bhattacharyya, M. K. Identification and characterization of an *S*-adenosyl-L-methionine: Δ^{24}-sterol-C-methyl transferase cDNA from soybean, 1996, *J. Biol. Chem.,* **271:**9384–9389.

Shimizu, S., Kawashima, H., Wada, M., and Yamada, H., 1992, Occurrence of a novel 24,25-methylenecholest-5-en-3β-ol in *Mortierela alpina* 1S-4, *Lipids* **27:**481–483.

St. Georgiev, V. (ed.), 1988, *Antifungal Drugs,* Volume 544, Annals of the New York Academy of Sciences, *Ann. New York Acad. Sci.,* New York.

Taylor, F. R., and Parks, L. W., 1981, An assessment of the specificity of sterol uptake and esterification in *Saccharomyces cerevisiae, J. Biol. Chem.* **256:**13048–13054.

Venkatramesh, M., and Nes, W. D., 1995, Novel sterol transformations promoted by *Saccharomyces cervevisiae* strain GL7: evidence for 9β,19-cyclopropyl to 9(11)-isomerization and for 14-demethylation to 8(14)-sterols, *Arch. Biochem. Biophys.* **324:**189–199.

Venkatramesh, M., Guo, D., Harman, J. G., and Nes, W. D., 1996a, Sterol specificity of the *Saccharomyces cerevisiae ERG6* gene product expressed in *Escherichia coli, Lipids* **31:**373–377.

Venkatramesh, M., Guo, D., Jia, Z., Nes, W. D., 1996b, Mechanism and structural requirements for transformation of substrates by the (*S*)-adenosyl-L-methionine: $\Delta^{24(25)}$-sterol methyltransferase from *Saccharomyces cerevisiae, Biochim. Biophys. Acta* **1299:**313–324.

Viola, F., Brusa, P., Balliano, G., Ceruti, M., Boutaud, O., Schuber, F., and Cattel, L., 1995, Inhibition of 2,3-oxidosqualene cyclase and sterol biosynthesis by 10-and 19-azasqualene derivatives, *Biochem. Pharmacol.* **50:**787–796.

Weete, J. D., 1987, Mechanism of fungal growth suppression by inhibitors of ergosterol biosynthesis, *Amer. Chem. Soc. Symp. Ser.* **325:**268–285.

Yoshida, K., Hirose, Y., Imai, Y., and Kondo, T., 1989, Conformational analysis of cycloartenol, 24-methylenecycloartanol and their derivatives, *Agric. Biol. Chem.* **53:**1901–1912.

Zhou, W., Guo, D., and Nes, W. D., 1996, Stereochemistry of hydrogen migration from C-24 to C-25 during biomethylation in ergosterol biosynthesis, *Tetrahedron Lett.* **37:**1339–1342.

Zisner, E., Paltauf, F., and Daum, G., 1993, Sterol composition of yeast organelle membranes and subcellular distribution of enzymes involved in sterol metabolism, *J. Bacteriol.* **175:**2853–2858.

Chapter 5

The Smith–Lemli–Opitz Syndrome: A Potentially Fatal Birth Defect Caused by a Block in the Last Enzymatic Step in Cholesterol Biosynthesis

G. S. Tint, Ashok K. Batta, Guorong Xu, Sarah Shefer, Akira Honda, Mira Irons, Ellen Roy Elias, and Gerald Salen

1. INTRODUCTION

The Smith-Lemli-Opitz or "RSH" syndrome, first described in 1964 (Smith *et al.*, 1964), is a severely debilitating and frequently fatal birth defect syndrome with an autosomal recessive mode of inheritance (Dellaire, 1969). Original estimates suggested an incidence of one in 20,000 births (Lowry and Yong, 1980) and a probable carrier frequency of about 1% (Chasalow, 1985). However, newer studies from

G.S. Tint and Gerald Salen Research Service and Medical Service, Department of Veterans Affairs Medical Center, East Orange, New Jersey 07019 and Department of Medicine and the Liver Center, University of Medicine and Dentistry—New Jersey Medical School, Newark, New Jersey 07103. **Ashok K. Batta, Guorong Xu, Sarah Shefer, and Akira Honda** Department of Medicine and the Liver Center, University of Medicine and Dentistry—New Jersey Medical School, Newark, New Jersey 07103. **Mira Irons and Ellen Roy Elias** Division of Clinical Genetics, Department of Pediatrics, New England Medical Center, Tufts University School of Medicine, Boston, Massachusetts 02111.

Subcellular Biochemistry, Volume 28: Cholesterol: Its Functions and Metabolism in Biology and Medicine, edited by Robert Bittman. Plenum Press, New York, 1997.

the Czech Republic may place its true prevalence closer to 1 in 9,000, with as many as one in 45–50 individuals carrying the defect (Opitz *et al.*, 1997). Although all of the above estimates were made in cases that were not biochemically confirmed, if the most recent observations prove to be correct, then the syndrome is as least as common as medium-chain acyl-CoA dehydrogenase (MCAD) deficiency (Zaideh *et al.*, 1995) and may well be the fourth most common recessively inherited disorder in Caucasians following cystic fibrosis, phenylketonuria, and hemochromatosis. Until the discovery in 1993 of the cholesterol biosynthetic defect in children with the syndrome (Irons *et al.*, 1993; Tint *et al.*, 1993; Tint *et al.*, 1994), its cause was entirely unknown (Gorlin *et al.*, 1990).

2. CLINICAL CHARACTERISTICS

The syndrome consists of a large and seemingly unrelated group of congenital abnormalities that can affect virtually every organ (Smith *et al.*, 1964; Blair and Martin, 1966; Dellaire, 1969; Hoefnaegel *et al.*, 1969; Deaton and Mendoza, 1973; Garcia *et al.*, 1973; Fierro *et al.*, 1977; Jeanty *et al.*, 1977; Lowry and Yong, 1980; Cherstvoy *et al.*, 1984; Chasalow, 1985; Donnai *et al.*, 1986; Lowry *et al.*, 1986; Thompson and Baraitser, 1986; Curry *et al.*, 1987; Jones, 1988; Gorlin *et al.*, 1990; McKeever and Young, 1990; Pober, 1990; Opitz and de la Cruz, 1994). The most diagnostically useful features are a set of dysmorphic craniofacial and limb abnormalities that include microcephaly, micrognathia, ptosis of eyelids, wide nasal bridge with anteverted nares, prominent epicanthal folds, low-set retroverted ears, cleft palate, syndactyly of the second and third toes, polydactyly, valgus foot deformities, and an unusually large number of digital whorls. Reports of functional and histological disorders of the central and peripheral nervous system are common. These include profound to severe mental retardation, abnormal electroencephalograms, abnormal muscle tone (hypotonia at birth followed sometime later by hypertonia), irregular gyri, small cerebellum, loss of nerve cells, astrocyte proliferation, absence of cerebellar vermis, hypomyelination of cerebellum, and optic nerve and Hirschprung's disease (i.e., congenital aganglosis of the large bowel).

Affected males often exhibit genital anomalies, including cryptorchidism, hypospadias, and microphalus, and the most severe cases generally have either ambiguous or female external genitalia (Dellaire, 1969; Jeanty *et al.*, 1977; Gorlin *et al.*, 1990). Functional and structural defects of the liver, endocrine system, kidneys and urinary system, heart, lungs, digestive organs, and skeleton are also frequent and can contribute to lack of survival in the most seriously affected individuals. Lipid abnormalities, except for a single report of unusually low plasma total, low-density, and high-density lipoprotein cholesterol in one infant (Curry *et al.*, 1987), had not appeared in the literature before 1993.

At birth, in addition to the dysmorphic appearance of the face and limbs, affected children usually present with severe failure to thrive, and all exhibit delayed development (Smith *et al.*, 1964; Dellaire, 1969; Cherstvoy *et al.*, 1984; Opitz *et al.*, 1994). There has been, however, the rare report of a subject with borderline normal intelligence (Lowry and Yong, 1980; Kelley, 1994). Early death is common but not inevitable (Blair and Martin, 1966; Hoefnaegel *et al.*, 1969; Deaton and Mendoza, 1973; Lowry and Yong, 1980; Thompson and Baraitser, 1986; Pierquin *et al.*, 1995).

Curry *et al.* (1987) suggested the existence of a more severe phenotype, which they termed Smith–Lemli–Opitz syndrome type II, characterized by multiple major organ abnormalities, a larger number of dysmorphic facial features, more frequent limb anomalies, a lack of masculinization in males, and death in the majority of cases before the age of six months. At the time there was considerable debate over whether type II represented a new genotype or if these patients just might represent the most seriously affected examples of a continuum of developmental abnormalities (Cherstvoy *et al.*, 1984; Bialer *et al.*, 1987; Curry *et al.*, 1987).

3. THE CHOLESTEROL BIOSYNTHETIC PATHWAY

To understand the biochemical defect in the Smith–Lemli–Opitz syndrome, it should be noted that cholesterol is synthesized in the endoplasmic reticulum of mammalian tissue (Figure 1) via a pathway in which mevalonate is condensed to squalene, and squalene is cyclized to lanosterol (4,4′,14-methyl-cholesta-8(9),24-diene-3β-ol) (Clayton, 1965; Frantz and Schroepfer, 1967; see Chapter 4 by Guo *et al.*). Lanosterol may then be demethylated and isomerized to form desmosterol (cholesta-5,24-dien-3β-ol) so that the final enzymatic reaction is the reduction of the desmosterol C-24 double bond to yield cholesterol (cholest-5-en-3β-ol) (left side of Figure 1). Alternatively (right side of Figure 1), cholesterol can be made by a branch of the same pathway in which the lanosterol C-24 double bond is immediately reduced and that proceeds by the intermediates 24,25-dihydrolanosterol (4,4′,14-methyl-cholest-8(9)-en-3β-ol), lathosterol (cholest-7-en-3β-ol), and 7-dehydrocholesterol (cholesta-5,7-dien-3β-ol) (Kandutsch and Russell, 1960; Schroepfer and Frantz, 1961; Clayton, 1965; Frantz and Schroepfer, 1967; Tint and Salen, 1974; Tint and Salen, 1977; Tint and Salen, 1982). It is generally agreed, however, that the above two pathways are not truly independent because they share common enzymes. That is, there exists only a single microsomal enzyme that reduces the steroid side-chain C-24 double bond, and can do so no matter what the state of the nucleus. There also exists another unique enzyme, 3β-hydroxysterol Δ^7-reductase, which saturates the C-7 double bond independently of the oxidation state of the side chain

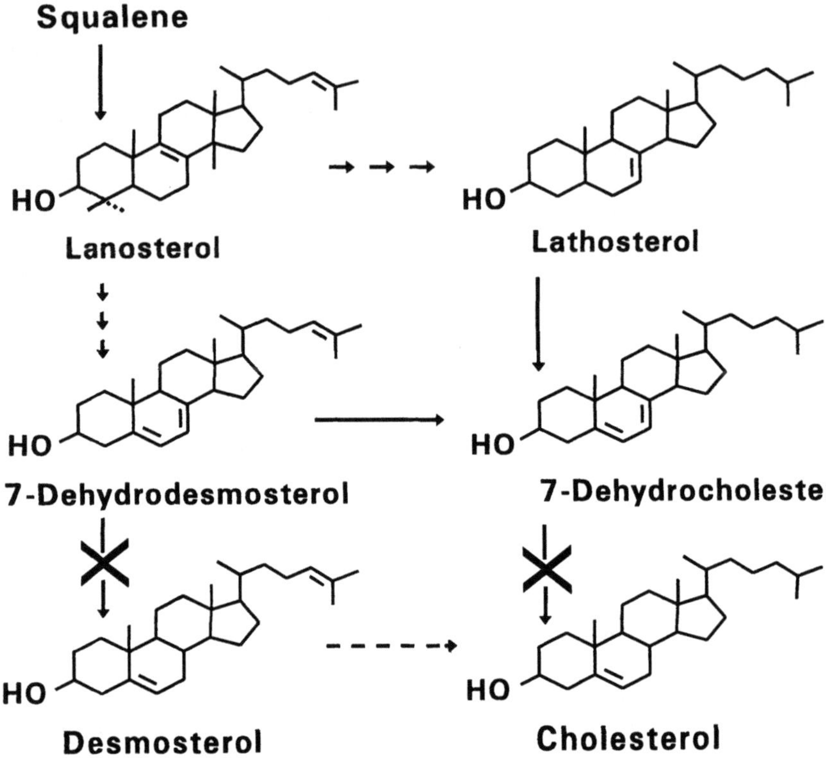

FIGURE 1. Biosynthesis of cholesterol from lanosterol. The left side depicts synthesis from C-24 unsaturated intermediates while the right side shows precursors with a saturated side chain. The large X indicates the enzyme block in the Smith–Lemli–Opitz syndrome.

(Steinberg and Avigan, 1960; Dvornik *et al.*, 1963; Fumagalli *et al.*, 1965; Koroly and Dempsey, 1981).

4. THE METABOLIC DEFECT IN THE SMITH–LEMLI–OPITZ SYNDROME

4.1. Plasma and Tissue Sterols

No laboratory or biochemical test for the syndrome existed until 1993–1994 when a block in cholesterol biosynthesis in several affected children was reported (Irons *et al.*, 1993; Tint *et al.*, 1993; Tint *et al.*, 1994). By autumn of 1995, plasma sterol compositions in nearly 100 additional patients had been described (Kelley, 1994; Seedorf *et al.*, 1994; Kelley, 1995; Tint *et al.*, 1995a; Honda *et al.*, 1995; Seedorf *et al.*, 1995). It is now abundantly clear that 95% of individuals with clinically defined Smith–Lemli–Opitz syndrome will present with abnormally reduced plasma cholesterol levels and that all of them will exhibit markedly elevated plasma concentrations of the cholesterol precursor 7-dehydrocholesterol and its epimer cholesta-5,8-dien-3β-ol (8-dehydrocholesterol) (Batta *et al.*, 1995b). It is also clear that patients with a phenotype not inconsistent with the Smith–Lemli–Opitz syndrome but without the biochemical abnormalities also exist (Tint *et al.*, 1995b; Kelley, 1995).

Our own study (Table I) now includes 50 well-documented subjects, aged 1 day to 35 years, with plasma cholesterol, 7-dehydrocholesterol, and 8-dehydrocholesterol levels (mean ± S.D.) of 49 ± 41 mg/dl, 16 ± 10 mg/dl and 12 ± 5 mg/dl, respectively. Calculations based on published data (Berenson *et al.*, 1979; Parker *et al.*, 1983; Hayes *et al.*, 1992; Wong *et al.*, 1993) yield estimates of lower 5th percentile plasma cholesterol levels for normal newborn, 6-month, and 12-month-old infants of about 40, 85, and 90 mg/dl, respectively, and 120 mg/dl for children over the age of 2–3 years (Lipid Research Clinics, 1980). In contrast, in young children from our study, plasma cholesterol levels averaged 53 ± 18 for 4–8-month-old children ($n = 4$) and 56 ± 22 mg/dl for those 10–14 months of age ($n = 5$). Cholesterol concentrations in 47 of the 50 subjects (94%) were below 110 mg/dl. Mean 7-dehydrocholesterol plasma concentrations for healthy children and adults have been reported to be in the range of 0.005 to 0.01 mg/dl (Axelson, 1991; Kelley, 1995) while control levels of 8-dehydrocholesterol appear to be of the same magnitude (Axelson, 1991).

Table I

Plasma Cholesterol, 7-Dehydrocholesterol (7-DHC), 8-Dehydrocholesterol (8-DHC), and Total Sterols in 50 Subjects with the Smith–Lemli–Opitz Syndrome

Sterol	All subjects ($n = 50$)	Type I ($n = 36$)	Type II ($n = 14$)	Youngest type I ($n = 11$)
	mg/dl (mean ± S.D. [Range])			
Cholesterol	49 ± 41	65 ± 38[a]	9.4 ± 6.5	65 ± 52[a]
	[3.7–190]	[18–190]	[3.7–29]	[18–190]
7-DHC	16 ± 10	14 ± 7[b]	20 ± 14	13 ± 7
	[0.15–61]	[0.15–30]	[6–61]	[5–24]
8–DHC	12 ± 5	11 ± 4[b]	15 ± 7	11 ± 4
	[1–26]	[1–19]	[7–26]	[6–18]
Total sterols[c]	79 ± 37	91 ± 34[a]	48 ± 25	90 ± 46[d]
	[22–200]	[32–200]	[22–104]	[32–200]
Age (years)[e]	7.2 ± 9.9	9.9 ± 11	0.26 ± 0.66	0.44 ± 0.40
	[1d–35y]	[1d–35y]	[1d–2.5y]	[1d–1.4y]

[a] $p < 0.001$ vs. type II.
[b] $p < 0.05$ vs. type II.
[c] Includes 19-nor-5,7,9(10)-cholestatrien-3β-ol.
[d] $p < 0.01$ vs. type II.
[e] d = day, y = year.

Although at present, nothing is known about the kinetics of the dehydrocholesterols in humans, Kaiser and Stocker (1978) demonstrated that in the minipig 7-dehydrocholesterol obeys two-pool kinetics with kinetic parameters similar to those of cholesterol.

On occasion, either the clinical presentation of the syndrome or plasma sterol concentrations in subjects in whom the syndrome is suspected is questionable. Elevated 7-dehydrocholesterol tissue levels have recently been used to diagnose an atypical case of a fetus with a number of features consistent with the Smith–Lemli–Opitz syndrome but with a missing fifth digit on the hands (oligodactyly) rather than an extra digit (polydactyly) (Seller *et al.*, 1995). Nevertheless, analysis of a number of tissues *post mortem* unequivocally demonstrated the biochemical defect (Table II, subject F). Also, in a few somewhat less severely affected older subjects, plasma 7-dehydrocholesterol may be only moderately elevated above control values, i.e., to 0.2–0.3 mg/dl, while cholesterol concentrations can be normal (Honda *et al.*, 1995; Kelley, 1995). For example, a 26-year-old moderately affected patient was found to have plasma cholesterol and 7-dehydrocholesterol concentrations of 154 and 0.2 mg/dl, respectively (Honda *et al.*, 1995). However, in spite of the near normal plasma sterol levels that now are present (his values as a child are unknown), the phenotype was evident and the diagnosis was easily made from the clinical presentation when the patient was five years old (Johnson, 1975). In contrast to plasma sterol levels, the enzyme defect was obvious in cultured fibroblasts from this subject (see below). It should also be

Table II
Plasma and Tissue Cholesterol, 7-Dehydrocholesterol (7-DHC), and 8-Dehydrocholesterol (8-DHC) in a Five Month Old (#1) and a One Day Old Type II Female (#3) and in a 20 Week Male Fetus (F) with Clinical Features of the Severe Form of the Smith-Lemli-Opitz Syndrome

Subject	Tissue	Total[a] (mg/g)	Cholesterol	7-DHC (% of total sterols)[a]	8-DHC
#1	Plasma	64 mg/dl	13	50	37
	RBC[b]	nd[c]	10	64	26
	Lens	nd	28	36	36
#3	Plasma	32 mg/dl	20	40	40
	Non-brain[d]	4.2 ± 1.1	20 ± 2	37 ± 5	35 ± 5
	Brain	5.1	4.1	61	31
	Bile	34 mg/dl	3.5	56	30
F	Non-brain[e]	1.9 ± 0.9	17 ± 11	44 ± 8	29 ± 8
	Brain	3.5	2.6	60	29

[a]Includes 19-nor-5,7,9(10)-cholestatrien-3β-ol.
[b]Erythrocytes.
[c]Not determined.
[d]Muscle, adrenal, kidney, liver, adipose, lungs, thymus.
[e]Muscle, adrenal, kidney, testis.

noted that plasma concentrations of 7-dehydrocholesterol in non-affected subjects may be elevated as much as fourfold above baseline (i.e., to levels of nearly 0.04 mg/dl) when cholesterol biosynthesis is high due to therapy with bile acid sequestrants or following ileal resection (Axelson, 1991).

Plasma sterol composition and phenotype are usually similar among siblings and close relatives (Table III). Because these individuals are expected to inherit the same gene defect from the parents this result suggests that plasma sterol concentrations are determined, to a great extent, by the cholesterol biosynthetic rate.

In addition to 7- and 8-dehydrocholesterol, a fourth compound, previously denoted as sterol III (Tint *et al.*, 1994), and which is seen in plasma from most affected subjects in concentrations of 0.3–9.6 mg/dl, has recently been identified as the 26-carbon sterol 19-nor-5,7,9(10)-cholestatrien-3β-ol (Batta *et al.*, 1995a). Its significance has not been established. Furthermore, at least one triunsaturated C_{27} sterol (Kelley, 1995; Batta *et al.*, 1994) can also be detected in plasma in somewhat lower concentrations than the C_{26}-triene, but its structure is not yet certain.

The two aberrant C_{27} diunsaturated sterols (dehydrocholesterols) are well esterified in plasma with 65% of 7-dehydrocholesterol and 83% of 8-dehydrocholesterol transported as the ester, while both compounds accumulate to a considerable degree in erythrocytes, lens, and other tissues (Tint *et al.*, 1994; Tint *et al.*, 1995). Another striking finding was that in the lens from a 5-month-old type II girl, in seven non-brain tissues, and plasma from a type II female who died at 27 hours and in muscle, adrenal gland, kidney, and testes from a severely affected male fetus aborted at week 20, cholesterol constituted only 16–20% of total sterols (Table II). The major constituents were, instead, 7-dehydrocholesterol and 8-dehydrocholesterol in a ratio that ranged from 1:1 to 3:2. However, total sterol concentrations (i.e., the sum of cholesterol, 7-dehydrocholesterol, and 8-dehydrocholesterol) were generally in the low normal range for each organ (Tint

Table III
Plasma Cholesterol, 7-Dehydrocholesterol (7-DHC), and 8-Dehydrocholesterol (8-DHC) in Affected Siblings

Subjects (age)	Relationship	Cholesterol	7-DHC (mg/dl)	8-DHC (mg/dl)
#6 (1.5 y)	Brother	33	9.7	10
#7 (2 mo)	Sister	29	19	18
#21 (25 y)	Brother	36	10	8.5
#22 (18 y)	Sister	50	9.6	8.5
#23 (16 y)	1st Cousin	36	6.2	8.2
#28 (31 y)	Sister	132	11	9.5
#29 (25 y)	Brother	154	0.2	1.0

et al., 1995). Furthermore, the cholesterol:dehydrocholesterol ratio in erythrocytes closely paralleled the ratio in plasma (Tint *et al.*, 1994). These results suggest that membrane sterol composition probably reflects the relative levels of these compounds in the plasma.

It is noteworthy that the effects of the cholesterol biosynthetic defect were most pronounced in the brain, with cholesterol representing only 3–4% of total sterols but, as in the other organs, total sterol concentrations were only moderately reduced. It should also be noted that desmosterol, an intermediate whose synthesis requires reduction of the sterol C-7 double bond, constituted 7–8% of control fetal brain sterols but was not detected in the brain of the affected fetus (Tint *et al.*, 1995). The latter observation further confirmed the hypothesis that the enzyme 3β-hydroxysterol Δ^7-reductase (7-dehydrocholesterol Δ^7-reductase) is defective in individuals with the syndrome.

The biochemical defect is the same in the type I and the more severe type II phenotypes, but the block is more complete in the latter subjects (Table I). This result had been more or less predicted by Bialer *et al.* (1987) following their analysis of the spectrum of reported abnormalities for these children, and it reinforces their suggestion that the type II classification is somewhat artificial when it is based only on a clinical description of the patient. Nevertheless, in spite of numerous objections (Opitz and de la Cruz, 1994), the designation remains a useful clinical tool in describing the most severely affected infants with the poorest prognosis and probably will remain with us for a while. The results shown in Table I confirm our earlier report in 33 subjects (Tint *et al.*, 1995a) that cholesterol concentrations are abnormally low while levels of 7-dehydrocholesterol and 8-dehydrocholesterol are markedly elevated in the plasma of both phenotypes. However, mean cholesterol levels were found to be 7-fold higher and dehydrocholesterols were significantly lower (by 50%) in plasma from type I as compared to type II (Table I), suggesting that the enzymatic defect is more profound in the latter subjects. Although the type II subjects constituted a much younger group of children than those classified as type I, the differences in plasma sterols cannot be attributed to age alone. Analysis of plasma sterols in all 11 type I subjects in our 50 patient study group who were younger than 1.5 years (Table I) revealed the same differences between types I and II. Also, a plot (Figure 2) of plasma cholesterol levels vs. age in type I demonstrated no strong correlations (correlation coefficient $r^2 = 0.22$ and Spearmen rank-correlation coefficient $r = 0.24$). A similar lack of affect of age has been reported previously (Kelley, 1994; Kelley, 1995).

Reduced survival is predicted by low plasma cholesterol concentrations in type II (15 ± 9 mg/dl in five surviving children vs. 6.5 ± 2.6 mg/dl in eight deceased children, $p < 0.02$). In our study subjects, no child with a plasma cholesterol level ≤7 mg/dl has lived beyond 30 weeks. In contrast, plasma dehydrocholesterol and total sterol concentrations tended to be considerably higher in the survivors (Table IV). Thus, although the most important predictor of early death in the severely affected child is an extremely low plasma cholesterol level, greatly

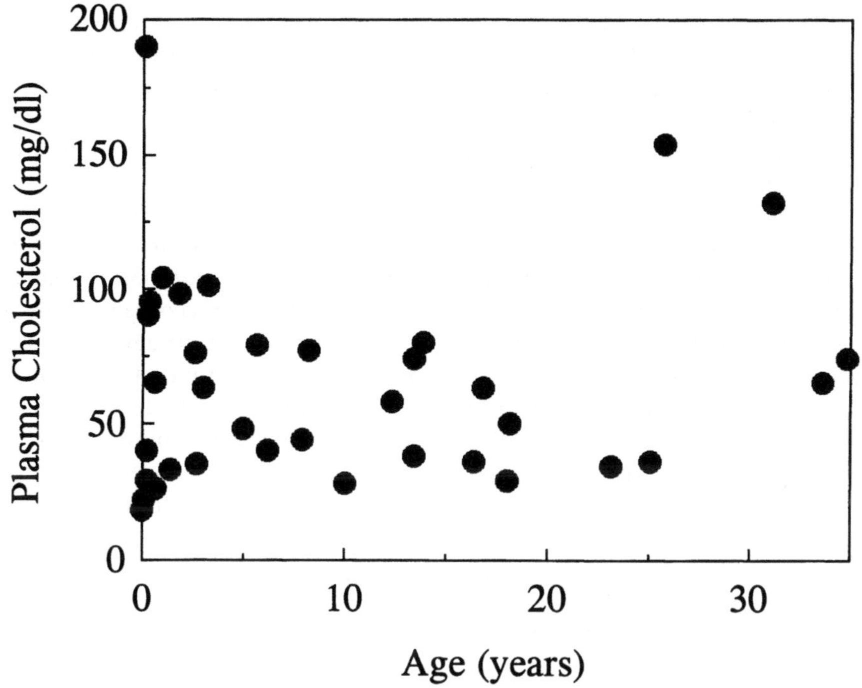

FIGURE 2. Lack of correlation between plasma cholesterol concentration and age in 36 subjects with the less severe type I phenotype of the Smith–Lemli–Opitz syndrome (correlation coefficient $r^2 = 0.22$).

Table IV
Plasma Cholesterol, 7-Dehydrocholesterol (7-DHC), and 8-Dehydrocholesterol (8-DHC) Concentrations in Five Severely Affected (Type II) Patients Who Survive, Compared to Levels in the Nine Type II Subjects Who Did Not

Sterol	Alive (n = 5)	Dead (n = 9)[a]	p
	mg/dl (mean ± S.D. [Range])		
Cholesterol	15 ± 9 [8.1–29]	6.5 ± 2.6[3.7–12]	<0.02
7-DHC	24 ± 8 [12–32]	18 ± 17 [6–61]	NS[b]
8-DHC	20 ± 6 [10–26]	16 ± 6 [7–26]	0.059
Total sterols	61 ± 19 [34–87]	41 ± 26 [22–104]	NS

[a]Age of death, 8 ± 9 [0.14–30] weeks.
[b]NS, not significant.

elevated levels of the dehydrocholesterols, in place of cholesterol, do appear to aid in survival. These results also suggest that the types I and II phenotypes are allelic forms of the same genotype, but that the particular gene mutation(s) in type II result in a 7-dehydrocholesterol Δ^7-reductase enzyme with considerably less net activity than that seen in type I. More recent studies with cultured fibroblasts have confirmed these hypotheses (see below).

It cannot be, however, that abnormally low plasma cholesterol concentrations alone are responsible for the developmental anomalies. In the codominantly inherited disorder hypobetalipoproteinemia, a defect in cholesterol transport caused by truncated apolipoprotein B, patients can have plasma cholesterol concentrations as low as 15 mg/dl (Linton *et al.*, 1993) but do not exhibit any of the developmental abnormalities associated with the Smith–Lemli–Opitz syndrome. Instead, they demonstrate symptoms traceable to deficiencies in various fat-soluble vitamins. In contrast to the Smith–Lemli–Opitz syndrome, cholesterol biosynthesis is markedly elevated in hypobetalipoproteinemia (Illingworth *et al.*, 1989), so that tissues can compensate for the reduced availability of cholesterol from plasma.

4.2. Fecal and Biliary Neutral Sterols and Bile Acids

Fecal and biliary neutral sterols (Tint *et al.*, 1994; Batta *et al.*, 1995a; Batta *et al.*, 1995b) and bile acids (Tint *et al.*, 1994), as well as urinary bile acids (Batta *et al.*, 1994; Natowicz and Evans, 1994), have been described in several affected subjects, with 7-dehydrocholesterol, 8-dehydrocholesterol, and cholesterol being excreted by all subjects. Comparable cholesterol:dehydrocholesterol ratios in plasma (1:7 and 1:1.8) and feces (1:6 and 1:1.2, respectively) from two subjects known to be consuming low cholesterol diets (Tint *et al.* 1994; Irons *et al.*, 1994; Batta *et al.*, 1995b) suggest that the liver secretes all three sterols more or less equally. Cholic acid, chenodeoxycholic acid, or their bacterial metabolites were detected in feces from all type I phenotypes, while cholic, chenodeoxycholic, and deoxycholic acids at 0.3, 0.9, and 0.1 mg/ml, respectively, were noted postmortem in bile from a 1-day-old child (Tint *et al.*, 1995b). Few bile acids have been identified as yet in the colon contents from this latter child, but we did note 3β-hydroxy-cholest-5-enoic acid as well as two compounds, in high concentrations, with mass spectra consistent with a monohydroxy-diunsaturated bile acid structure (G. S. Tint, unpublished observations).

It has also been reported (Tint *et al.*, 1994) that in a young type II phenotype no fecal bile acids were detected when the subject was first examined at age 8 months, but that chenodeoxycholic acid and cholic acid were both found in bile and feces several weeks after the subject was placed on a high cholesterol diet (Irons *et al.*, 1994).

Preliminary analyses of urinary bile acids have proved to be extremely interesting and may have diagnostic significance. Natowicz and Evans (1994) analyzed urinary bile acids by means of direct injection fast atom bombardment (FAB) mass spectrometry and reported spectra consistent with mono-, di-, and tri-oxygenated monounsaturated bile acid sulfates. Batta *et al.* (1994) detected the presence of a monounsaturated dihydroxy bile acid by gas chromatography–electron impact (EI) mass spectrometry. Structural confirmation will have to await the synthesis of appropriate reference compounds.

4.3. Amniotic Fluid Sterols

It is possible to diagnose an affected fetus by assaying amniotic fluid sterols by gas chromatography or gas chromatography–mass spectrometry (McGaughran *et al.*, 1994; Abuelo *et al.*, 1995; Gelman-Kohan and Chemke, 1995; Kelley, 1995; McGaughran *et al.*, 1995; Rossiter *et al.*, 1995; Stewart *et al.*, 1995). In control amniotic fluid, 7-dehydrocholesterol and 8-dehydrocholesterol levels are about 0.005 µg/ml (McGaughran *et al.*, 1994; Kelley, 1995; McGaughran *et al.*, 1995), whereas in 14–22 week amniotic fluids from Smith–Lemli–Opitz syndrome pregnancies, they are increased a thousand-fold. In contrast, cholesterol levels in both control and affected pregnancies are similar. The sterol assay method appears to be far more sensitive and specific than ultrasonography, especially with respect to detecting the less severely affected type I phenotypes (Abuelo *et al.*, 1995; Stewart *et al.*, 1995) but no measurements of amniotic fluid sterols have yet been reported for pregnancies in which both parents are carriers but in which the fetus is unaffected.

5. CHOLESTEROL METABOLISM IN CULTURED CELLS

The coupling of reduced plasma cholesterol levels combined with massively elevated concentrations of the cholesterol precursor 7-dehydrocholesterol led to the natural postulate that the enzyme 3β-hydroxysteroid Δ^7-reductase (7-dehydrocholesterol C-7 reductase) must be defective in subjects with the Smith–Lemli–Opitz syndrome (Irons *et al.*, 1993; Tint *et al.*, 1993; Tint *et al.*, 1994). The hypothesis has recently been confirmed.

5.1. Conversion of Sterol Precursors to Cholesterol

A series of experiments measuring the transformation of radiolabeled lathosterol to cholesterol has demonstrated conclusively that the defect in the Smith–Lemli–Opitz syndrome is markedly reduced conversion of 7-dehydrocholesterol to cholesterol (Honda *et al.*, 1995). Fibroblasts from type I and type II subjects, parents, controls, and amniocytes from a 14 week type II fetus were incubated with [^{3}H]lathosterol (Dempsey *et al.*, 1964), the immediate precursor of 7-dehydrocholesterol (Figure 1), and cholesterol, 7-dehydrocholesterol, and lathosterol were extracted, isolated, and purified. The results are compelling (Table V); in control and carrier cells most label was recovered as cholesterol while

Table V
Conversion of Radiolabeled Lathosterol to 7-Dehydrocholesterol and Cholesterol Following a 24-Hour Incubation with Cultured Fibroblasts from Smith–Lemli–Opitz (SLO) Syndrome Type I and Type II Patients, Parents (Obligate Carriers), and Controls and with Amniocytes from an Affected Fetus

Cells	[^{3}H] in sterol fraction compared to total label extracted from medium 7-Dehydrocholesterol	 Cholesterol
	% of total (mean ± S.E.M.)	
Fibroblasts:		
SLO type I (n = 7)	50 ± 1[a]	1.1 ± 0.2[a]
SLO type II (n = 8)	45 ± 4[a]	0.03 ± 0.03[a,b]
Parents (n = 14)	7.9 ± 1.2[c]	59 ± 2
Controls (n = 8)	3.2 ± 0.7	64 ± 3
Amniocytes:		
SLO (n = 1)	34	0.0
Controls (n = 3)	3.0 ± 1.3	39 ± 19

[a] $p < 0.0001$ vs. parents and controls.
[b] $p < 0.0005$ vs. type I SLO.
[c] $p < 0.01$ vs. controls.

little label remained as 7-dehydrocholesterol. In contrast, the [^{3}H]cholesterol fraction obtained after incubating cells from types I and II Smith–Lemli–Opitz patients averaged only 1.1 ± 0.2% and 0.03 ± 0.0% of recovered label, respectively ($p < 0.001$ vs. controls, $p < 0.005$, I vs. II). Instead, most label was found as 7-dehydrocholesterol. Thus, lathosterol can be converted to 7-dehydrocholesterol in affected cells by the enzyme lathosterol C-5 desaturase, but the resulting 7-dehydrocholesterol is very poorly transformed to cholesterol. Also, significantly more label was found in the 7-dehydrocholesterol fraction from parents' cells than was seen in control fibroblasts, which suggests that a single copy of an abnormal gene for 7-dehydrocholesterol Δ^7-reductase will lead to measurably reduced net enzyme activity. Although there was too much overlap to use the method in its present form as an assay for unaffected heterozygotes, these results may, in the future, serve as a basis for a viable carrier test.

The entire cholesterol biosynthetic pathway was assayed by incubating cells from two patients and one parent for 24 hours with radiolabeled mevalonolactone (Honda *et al.*, 1995). Recoveries in the lathosterol, 7-dehydrocholesterol, and cholesterol fractions from the carrier fibroblasts were 24%, 2%, and 65%, respectively, compared to an average of 31%, 27%, and 0.5%, respectively, in cells from the affected children, demonstrating that the pathway from mevalonate to cholesterol is probably intact in the Smith–Lemli–Opitz syndrome except for the final step.

5.2. Sterol Composition in Cultured Fibroblasts

Cholesterol concentrations in fibroblasts from affected children, their parents (obligate carriers), and controls maintained in culture medium with the addition of 10% fetal bovine serum are similar (60–70 μg/mg cellular protein). In contrast, 7-dehydrocholesterol levels are 50–100 times greater in fibroblasts from affected individuals (1.5–2.5 μg/mg protein) compared to levels in carrier or control cells (Kelley, 1995; Honda *et al.*, 1997b; Pierquin *et al.*, 1995). 7-Dehydrocholesterol concentrations in fibroblasts from parents were found to be higher than that in control cells when cultured either in a cholesterol rich or a delipidated medium. But again, there was considerable overlap so that the differences did not reach statistical significance (Honda *et al.*, 1997b). When fibroblasts from patients were grown for 2–4 weeks in a medium containing delipidated fetal bovine serum to up-regulate cholesterol biosynthesis (Brown and Goldstein, 1973), 7-dehydrocholesterol increased to become 30–35% (Honda *et al.*, 1997b; Kelley, 1995) of total sterols. Similarly, when control cells were grown in delipidated medium for 4 weeks with 50 μM BM 15.766 (Boehringer–Mannheim), an efficient inhibitor of 7-dehydrocholesterol Δ^7-reductase (Pill *et al.*, 1990; Pill *et al.*, 1985), the concentration of 7-dehydrocholesterol increased to become 50% of total sterols.

7-Dehydrocholesterol was well esterified ($40 \pm 5\%$ vs. $36 \pm 6\%$ of cholesterol) in cells from affected individuals grown in delipidated fetal bovine serum (Honda *et al.*, 1997b). The observation of 7-dehydrocholesterol esters in both plasma and cultured cells suggests that not only is lecithin cholesterol acyltransferase (LCAT) active toward 7-dehydrocholesterol but 7-dehydrocholesterol can also be efficiently esterified by acyl-coenzyme A cholesterol acyltransferase (ACAT).

It should be noted that in no fibroblast culture was the concentration of 8-dehydrocholesterol greater than 2% of total sterols. This latter result suggests that the synthesis of this unusual sterol, thought to be made from 7-dehydrocholesterol by steroid-8-ene isomerase acting in the reverse direction (Wilton *et al.*, 1969; Paik *et al.*, 1986), is likely to be tissue specific.

5.3. Cholesterol Biosynthetic Enzymes in Cell-Free Preparations

Activities of lathosterol C-5-desaturase and 7-dehydrocholesterol Δ^7-reductase (Figure 1) were measured in isolated liver microsomes from four type II phenotypes and seven unaffected young children (Shefer *et al.*, 1995) by incubating [^{3}H]lathosterol and [^{3}H]7-dehydrocholesterol, respectively, with liver microsomes (Shefer *et al.*, 1986; Shefer *et al.*, 1992) for periods ranging from 5–60 min. Lathosterol C-5 desaturase activities were found to be the same in microsomes from controls and from patients: 120 ± 8 and 100 ± 7 pmol/mg protein/min, respectively, mean ± S.D. In contrast, the activity of 7-dehydrocholesterol C-7 reductase in controls was nearly 10-fold greater than that observed in the four type II subjects: 365 ± 4 vs. 40 ± 4 pmol/mg/min, respectively ($p < 0.0001$). In addition, when labeled 7-dehydrocholesterol was incubated with microsomes from affected children considerable [^{3}H]8-dehydrocholesterol was detected, suggesting that 7-dehydrocholesterol is the source of 8-dehydrocholesterol and that the transformation can occur efficiently in liver.

This latter result further supports the idea that in humans production of 8-dehydrocholesterol is likely to be tissue specific. A similar effect was predicted in the rat by studies that described elevations in liver, but not brain, concentrations of 8-dehydrocholesterol when 7-dehydrocholesterol Δ^7-reductase was chemically inhibited (Fumagalli *et al.*, 1980).

It is important to note that the measurements of hepatic 7-dehydrocholesterol Δ^7-reductase described by Shefer *et al.* (1995) were carried out under saturating conditions (a large excess of substrate) and, as such, are good approximations to V_{max}. In contrast, when the corresponding assays were performed in whole fibroblasts (Honda *et al.*, 1995) no unlabeled 7-dehydrocholesterol was added, so that the reaction depended on the availability of endogenous 7-dehydrocholesterol. Although both methods demonstrated conclusively that the enzyme is indeed defective in the Smith–Lemli–Opitz syndrome, because of the different techniques and tissues employed, it is not a surprise that the magnitude of the differences in enzyme activity between controls and affected individuals were dissimilar.

6. ANIMAL MODELS

6.1. Teratogenic Effects of 7-Dehydrocholesterol Δ^7-Reductase Inhibitors

The link between defective reduction of the C-7 double bond of 7-dehydrocholesterol and the kinds of birth defects seen in the Smith–Lemli–Opitz syndrome is provided by a considerable literature describing the effects of pharmacologic inhibitors of this reaction on fetal and embryonic rats. Both AY-9944 [*trans*-1,4-bis(2-chlorobenzylaminomethyl)cyclohexane dihydrochloride] (Ayerst) and BM 15.766 [4-(2-[1-(4-chlorocinamyl)piperazin-4-yl]ethyl)-benzoic acid] (Boehringer-Mannheim) administered to adult rats for 3–14 d induce as much as a threefold reduction in plasma cholesterol levels and an increase in 7-dehydrocholesterol concentrations from undetectable levels before treatment to nearly 20 mg/dl (Dvornik *et al.*, 1963; Dvornik *et al.*, 1966; Dvornik and Hill, 1968; Kaiser and Stocker, 1978; Pill *et al.*, 1985; Pill *et al.*, 1990). These values model plasma sterols in the Smith–Lemli–Opitz syndrome (Table I).

AY-9944 given to pregnant rats or added to cultured rat embryos generate a large number of abnormalities, which have their counterpart in children with the Smith–Lemli–Opitz syndrome. The central and peripheral nervous system is adversely affected with reports of abnormal muscle tone (Suzuki and De Paul, 1971) and degeneration of neurological tissue with reduced myelination (Fumagalli *et al.*, 1969; Smith *et al.*, 1970; Suzuki and De Paul, 1971; Suzuki *et al.*, 1973;

Igarashi *et al.*, 1975; Rawlins and Lópes-Jiménez, 1979; Repetto *et al.*, 1990; see Chapter 7 by Snipes and Suter). These observations of neurological abnormalities in treated rats are not entirely unexpected since 25–30% of mammalian myelin consists of cholesterol (Norton and Cammer, 1984) and virtually all of the cholesterol that is incorporated into myelin in either the central or peripheral nervous system must be synthesized *in situ* (Edmond *et al.*, 1991; Jurevics and Morell, 1994; Jurevics and Morell, 1995). Other studies have reported increased fetal mortality (Roux *et al.*, 1966; Roux *et al.*, 1979), lower body weight (Roux and Aubry, 1966; Igarashi *et al.*, 1975; Repetto *et al.*, 1990), reduced brain weight (Igarashi *et al.*, 1975), pituitary agenesis (Barbu *et al.*, 1988; Repetto *et al.*, 1990), nephrotic kidney disorders and cryptorchidism (Roux and Aubry, 1966; Roux *et al.*, 1979), club foot and hair lip (Roux and Aubry, 1966), maxillary and mandibular hypoplasia, and reduced head size (Roux *et al.*, 1980). In a similar manner, the Δ^7-reductase inhibitor YM9429 (*cis*-1-[4-(*p*-monothane-8-yloxy)phenyl]piperidine) (Yamanouchi Pharmaceutical Co.) (Ohata *et al.*, 1983) given to pregnant rats between gestational days seven and 17 generated cleft palate and skeletal abnormalities in the offspring (Shibata, 1993).

Malformations in mouse embryos that resemble some of those noted in the Smith–Lemli–Opitz syndrome also follow the destruction of specific cell populations by the acute administration of metabolic poisons such as ethanol (Kotch *et al.*, 1990; Gage and Sulik, 1991; Kotch *et al.*, 1991; Cohen and Sulik, 1992; Kotch and Sulik, 1992; Sulik and Sadler, 1993). It has been suggested (Goldstein and Chin, 1981; West and Goodlett, 1990; West *et al.*, 1990) that the observed craniofacial anomalies in this model might result from cellular vulnerability in the developing central nervous system because of neuronal maturation, which requires the deposition of large quantities of lipids. Thus, the relative paucity of cholesterol in the syndrome may prevent the formation of mature neurons.

HMG-CoA reductase and squalene synthase inhibitors, agents that interfere with earlier steps in cholesterol biosynthesis, are reported to cause mainly skeletal defects (Minsker *et al.*, 1983) and demyelination of peripheral nerves (Toews *et al.*, 1991; Wagner-Recio *et al.*, 1991). The latter effects, however, appear to be specific for tellurium salts. Other studies, in contrast, have demonstrated that atherogenic diets can help prevent birth defects associated with AY-9944, even though 7-dehydrocholesterol persists as 10–12% of total plasma sterols (Roux *et al.*, 1979; Roux *et al.*, 1980; Barbu *et al.*, 1988). These results suggest that it is the lack of cholesterol, rather than the accumulation of 7-dehydrocholesterol, that is more detrimental to the embryo and fetus and it has been demonstrated that 7-dehydrocholesterol is not in itself a mutagen (Ansari *et al.*, 1982).

Additional support for the hypothesis that an adequate supply of cholesterol is critical early in development is provided by the finding of severe cranio-facial abnormalities (i.e., exencephaly and hydrocephaly) in embryonic mice in which the apolipoprotein B gene has been altered (Homanics *et al.*, 1993; Homanics *et al.*, 1995). Moreover, complete knockout of the normal murine apolipoprotein B gene is lethal to mouse embryos, but viable offspring can be produced by these same animals when they were made homozygous for the human apolipoprotein B transgene (Farese *et al.*, 1995). These results suggest that any interference with cholesterol transport from mother to offspring early in murine gestation will seriously hinder development.

However, the substitution of 7-dehydrocholesterol for cholesterol in erythrocytes does significantly alter membrane properties. Osmotic fragility and glycerol permeability are both reduced (Bruckdorfer *et al.*, 1969), gramicidin A inhibition of cation membrane permeability is abolished (Schagina *et al.*, 1992), and orientational order in phospholipid bilayer membranes is greatly enhanced (Korstanje *et al.*, 1990). 7-Dehydrocholesterol is also known to be a strong antioxidant and it inhibits iron-dependent lipid peroxidation (Wiseman, 1993).

It has recently been noted that holoprosencephaly can be generated in embryonic rats by administering BM 15.766 on gestational days 8–12 by a mechanism that seems to involve the loss of cellular adhesion in the group of rapidly growing cells at the rim of the cranial neural folds (Dehart *et al.*, 1995). This observation might be explained by variations in either the width or hydrophobicity of the plasma membrane arising from changes in sterol content or composition. Such alterations can affect the activity of a number of transmembrane enzymes, perhaps by modifying the placement of the hydrophilic amino acid domain within the membrane (Chen *et al.*, 1995). An attractive candidate compound that might link the above observation to changes in membrane sterol composition is the transmembrane glycoprotein C-CAM. This is a cellular adhesion molecule that has recently been identified in the developing rat brain and which may be involved in adhesion between endothelial cells, pericytes, and astrocytes (Sawa *et al.*, 1994). Changes in plasma membrane structure in patients with the Smith–Lemli–Opitz syndrome are also likely in light of our observations of moderately low total sterol concentrations in fetal and newborn tissue and the replacement of up to 96% and 80% of brain and peripheral tissue cholesterol, respectively, with 7- and 8-dehydrocholesterol (Tint *et al.*, 1995b).

Thus it is not at all certain whether it is a paucity of cholesterol that might be essential at certain key times during gestation, or perhaps the accumulation of unusual sterols in membranes, or some combination of both that leads to the developmental abnormalities in children with the syndrome.

6.2. Replicating the Smith–Lemli–Opitz Syndrome in Animals: Using the Rat in a Search for Therapies

Because 7-dehydrocholesterol Δ^7-reductase inhibitors administered to rats reproduce both the biochemical and morphological abnormalities seen in the Smith–Lemli–Opitz syndrome, this animal is an attractive model for investigating the effects of possible therapies. We have treated rats by combining the inhibitor BM 15.766 with cholesterol (2% in diet), cholic acid (1% in diet), and lovastatin (0.04% in diet), alone or in combination. Plasma cholesterol and 7-dehydrocholesterol levels, biliary bile acid flux, and hepatic HMG-CoA reductase and cholesterol 7α-hydroxylase activities and mRNA levels were determined (Xu *et al.*, 1995a; Xu *et al.*, 1995b) (Table VI, Figure 3). After two weeks of administration, the inhibitor mimicked the biochemical defect in the Smith–Lemli–Opitz syndrome by reducing plasma cholesterol levels to one-third of pre-treatment levels and raising 7-dehydrocholesterol concentrations to equal those of cholesterol (Figure 3). Because of the reduced cholesterol levels, HMG-CoA reductase mRNA levels doubled during treatment with inhibitor alone. Cholesterol feeding, however, returned the activities to control values. Cholic acid therapy lowered mRNA levels while the addition of cholesterol to the diet reduced them still further to one-half of control levels. Although plasma cholesterol levels were highest with this treatment, so were 7-dehydrocholesterol concentrations (Figure 3). As one might expect, lovastatin increased HMG-CoA reductase mRNA relative concentrations 10-fold but adding cholesterol to the diet along with the cholic acid returned them to 70% of control levels. mRNA values for cholesterol 7α-hydroxylase increased 50% during treatment with the inhibitor alone, increased to twice control levels when cholesterol was added, and were suppressed markedly by cholic acid feeding. In the only other studies of this kind, hepatic cholesterol 7α-hydroxylase activity was reported to be increased twofold after 4 days of treatment (Van Cantfort, 1971) or inhibited 40% after a single bolus of the 3β-hydroxysteroid C-7 reductase inhibitor AY 9944 (Pandak *et al.*, 1991).

Table VI
Effect of Various Therapies on Hepatic HMG-CoA Reductase and Cholesterol 7α-Hydroxylase Activities and Relative mRNA Levels in Rats Treated with the 7-Dehydrocholesterol Δ^7-Reductase Inhibitor BM 15.766

	HMG-CoA reductase		Cholesterol 7α-hydroxylase	
Treatment[a]	Activity (pmol/min/mg)	mRNA (relative units)	Activity (pmol/min/mg)	mRNA (relative units)
		mean ± S.D. (*n*)		
Control	53 ± 9 (6)	24 ± 7 (6)	20 ± 6 (7)	22 ± 8 (7)
I14	92 ± 34 (9)[b]	49 ± 15 (6)[b]	19 ± 5 (9)	39 ± 16 (6)
I + C	24 ± 9 (5)[c]	25 ± 10 (4)[c]	23 ± 3 (5)	48 ± 24 (5)[b]
I + CA	47 ± 16 (5)[c]	19 ± (3)[c]	7 ± 3 (5)[b,c]	5 ± 4 (5)[b,c]
I + CA + C	18 ± 4 (5)[c]	13 ± 5 (5)[c]	15 ± 3 (4)	15 ± 9 (5)
L	5010 ± 680 (7)[c]	270 ± 140 (4)[c]	19 ± 4 (7)	66 ± 7 (4)[b]
L + C	86 ± 16 (7)	17 ± 3 (5)[c]	16 ± 6 (7)	54 ± 8 (5)[b]

[a] I14 = Inhibitor only for 14 days; I + C = inhibitor 14 days + cholesterol, 2% in diet, for last 7 days; I + CA = inhibitor for 14 days + cholic acid, 1% in diet, for last 7 days; I + CA + C = inhibitor 14 days plus cholic acid and cholesterol for final 7 days; L = inhibitor 14 days + lovastatin, 0.04% in diet, for last 7 days; L + C = inhibitor 14 days + lovastatin and cholesterol for last 7 days.
[b] $p < 0.05$ vs. controls.
[c] $p < 0.05$ vs. inhibitor alone.

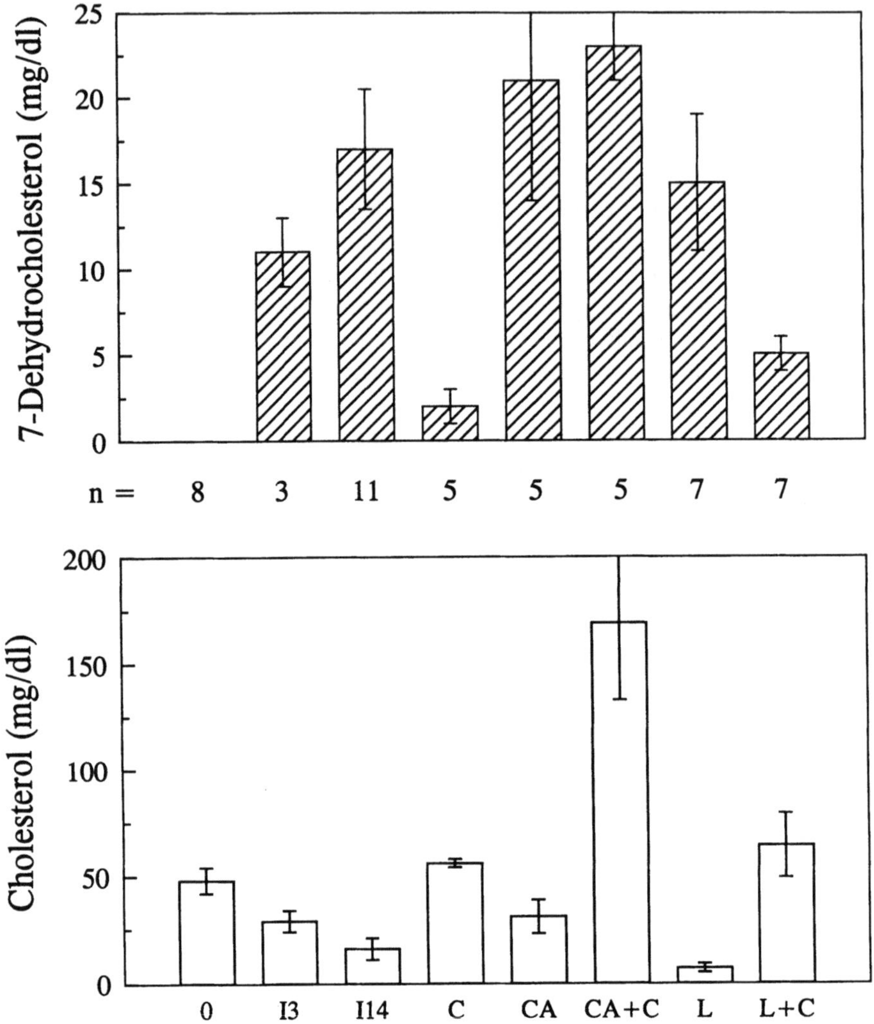

FIGURE 3. Plasma cholesterol (lower plot) and 7-dehydrocholesterol (upper plot) levels in rats treated with 30 mg/kg/day of the 7-dehydrocholesterol Δ^7-reductase inhibitor BM 15.766 and subjected to various experimental therapies (n = number of animals in each group). 0 = Controls; I3 = inhibitor only for 3 days; I14 = inhibitor only for 14 days; I + C = inhibitor for 14 days + cholesterol, 2% in diet, for last 7 days; I + CA = inhibitor for 14 days + cholic acid, 1% in diet, for last 7 days; I + CA + C = inhibitor for 14 days plus cholic acid and cholesterol for final 7 days; L = inhibitor for 14 days + lovastatin, 0.04% in diet, for last 7 days; L + C = inhibitor for 14 days + lovastatin and cholesterol for last 7 days.

Although chronic (i. e., two week) administration of BM 15.766 inhibitor produced a 73% increase ($p < 0.05$) in hepatic HMG-CoA reductase activities (Xu *et al.*, 1995a), the short-term effects are reversed. After 3 days of administration of 30 mg/kg/day BM 15.766, HMG-CoA reductase activities were reduced from 53 ± 9 pmol/mg protein/min to 34 ± 8 pmol/mg/min ($p < 0.05$). Pandak *et al.* (1990) demonstrated a statistically significant 60% reduction in a similar model four hours after the infusion of a bolus of 1 mg/kg AY 9944. It is possible that the change noted in enzyme activity after 2 weeks of inhibitor therapy may represent the well-known ability of the rat to adjust hepatic enzyme levels greatly in an attempt to maintain cholesterol homeostasis in the face of a challenge (Dietschy *et al.*, 1993; Xu *et al.*, 1995a, Xu *et al.*, 1995b). The response seen at 4 hours or 3 days, however, may arise from the immediate suppression of enzyme activity because of a rapid rise in the microsomal 7-dehydrocholesterol concentration.

The above data indicate most forcefully that the single most important dietary agent for treating the plasma abnormalities in the Smith–Lemli–Opitz syndrome is cholesterol. They also suggest that the addition of bile acids, especially in the more severe type II phenotype, may be useful. Although the combination of lovastatin plus cholesterol seems to be quite effective, the rats sometimes became lethargic and stopped eating. Thus, in agreement with the conclusions drawn by Hoffmann *et al.* (1993) from their experience in infants with mevalonic aciduria, HMG-CoA reductase inhibitors should be used with great care in children with low plasma cholesterol levels.

7. GENE DEFECT

Although the location of the gene has not, as yet, been proven, there have been two well-documented reports of the association of the syndrome with translocations involving the long arm of chromosome 7 (Curry *et al.*, 1987; Berry *et al.*, 1989). For this reason, it is particularly noteworthy that Muenke *et al.* (1994) reported a child with both the Smith–Lemli–Opitz syndrome and holoprosencephaly (failure of the embryonic brain to properly divide into quadrants). Holoprosencephaly is frequently associated with defects in chromosome 7 in the vicinity of q32-q34 (Cohen and Sulik, 1992) and autosomal dominant holoprosencephaly has recently been linked to a locus at q36 (Muenke *et al.*, 1994). Recently, Alley *et al.*. (1995) reported the creation of a yeast artificial chromosome (YAC) that spans the translocation breakpoint at 7q32.1 in a biochemically confirmed infant with "type II" Smith–Lemli–Opitz syndrome (Wallace *et al.*, 1994).

8. THERAPY

While one cannot expect to extrapolate completely from animals to humans, the consistent finding that one can suppress many of the teratogenic effects of the

3β-hydroxysterol Δ^7-reductase inhibitor AY-9944 by intervening early in pregnancy with a high cholesterol diet suggested a possible therapy (Roux *et al.*, 1979; Barbu *et al.*, 1988; Tint *et al.*, 1994). Following a pilot study in one subject (Irons *et al.*, 1994), a dietary therapy was initiated in five affected children which used 40 mg/kg/day cholesterol plus 15 mg/kg/day ursodeoxycholic and 7.5 mg/kg/day chenodeoxycholic acids to aid absorption. After a mean of 36 weeks of therapy, cholesterol increased from 42 ± 29 to 66 ± 31 mg/dl ($p < 0.01$) while cholesterol as percent of total sterols increased from 53 ± 27 to $67 \pm 17\%$ ($p = 0.056$). Bile acids were undetectable in feces from our first subject (Tint *et al.*, 1994) before treatment, but after one month of therapy, cholic, chenodeoxycholic, and lithocholic acids were found at concentrations of 0.35, 1.20, and 0.41 mg/g, respectively.

The study has since been expanded to include an additional seven patients (Irons *et al.*, 1997). Seven female and five males (age newborn to 16 years) have been treated for 6–15 months, with three patients receiving cholesterol alone and nine getting cholesterol plus bile acids (chenodeoxycholic and ursodeoxycholic acids either alone or in combination). Plasma cholesterol concentrations increased 143% and 150% while the cholesterol:7-dehydrocholesterol ratio increased by 40% and 70% in the two groups, respectively. The greatest relative increases occurred in patients with baseline concentrations below 40 mg/dl, with plasma cholesterol levels rising approximately 200%.

9. METHODS OF DETECTION

7-Dehydrocholesterol is unstable in air and is easily oxidized, so that care must be taken in its processing. Because it has strong absorption maxima at 271, 282, and 293 nm (*Rodd's Chemistry of Carbon Compounds,* 1970), Kandutsch, more than 30 years ago, used spectrophotometric methods to monitor 7-dehydrocholesterol concentrations in studies demonstrating cholesterol biosynthesis from Δ^{24} saturated intermediates (Kandutsch, 1962). Similar techniques, which depend on light absorption, are used by most clinical laboratories to measure plasma cholesterol levels. It has been our experience that cholesterol levels reported by clinical laboratories in the United States from plasma of individuals with the Smith–Lemli–Opitz syndrome are a reasonably good approximation to the sum of plasma cholesterol, 7-dehydrocholesterol, and 8-dehydrocholesterol. However, this may not always be so because some common colorimetric assays previously in use are known to yield very different optical densities from equal concentrations of 7-dehydrocholesterol and cholesterol (Sarkar and Cenedella, 1982).

The first measurements of elevated 7-dehydrocholesterol in plasma and tissues of children with the Smith–Lemli–Opitz syndrome (Tint *et al.*, 1993; Irons *et al.*, 1993; Tint *et al.*, 1994; Tint *et al.*, 1995b) and in the amniotic fluid of affected pregnancies (Abuelo *et al.*, 1995; Rossiter *et al.*, 1995), however, were made by analyzing hydrolyzed lipid extracts using standard capillary gas-liquid-chromatographic methods with both polar (Carbowax) and apolar (dimethylpolysiloxane) columns. Soon after the first reports of the cholesterol biosynthetic defect in the syndrome, 7-dehydrocholesterol levels in plasma (Batta *et al.*, 1994; Honda *et al.*, 1997a) and amniotic fluid (Seedorf *et al.*, 1994) were successfully measured by optical absorption methods. Almost at the same time as the latter report, diagnosis of the syndrome in a fetus by means of gas-liquid chromatography selected ion monitoring mass spectrometry (GC-MS) of amniotic fluid was described (McGaughran *et al.*, 1994; McGaughran *et al.*, 1995). Recently Kelley (1995) provided a detailed description of several of these methods and suggested that use of the peak at m/z 325 (M-A ring) is optimum when employing GC-MS selected ion monitoring. Two groups of investigators have also described the detection of 7-dehydrocholesterol in plasma (Seedorf *et al.*, 1995) and blood card spots (Zimmerman *et al.*, 1996) by time-of-flight–secondary-ion (TOF–SIM) mass spectrometry. The latter report is extremely exciting as it will permit, for the first time, an accurate estimate of the frequency of the syndrome. Functional assays for the 3β-hydroxysteroid Δ^7-reductase are also useful for the diagnosis of the syndrome (Shefer *et al.*, 1995; Honda *et al.*, 1995). However, both methods rely on the conversion of easily oxidized radiolabeled 7-dehydrocholesterol to cholesterol. For this reason, a new assay was developed (Honda *et al.*, 1996) which uses selected ion monitoring GC-MS to measure the saturation of the C-7 double bond of the fungal sterol ergosterol (ergosta-5,7,22-diene-3β-ol) added to cultured cells. Ergosterol and its Δ^7-reduced product are not normally detected in human tissue, are far more stable than radiolabeled 7-dehydrocholesterol and are inexpensive to buy.

ACKNOWLEDGMENTS. Work from the authors' laboratories reported in this manuscript was supported in part by grants from the Department of Veterans Affairs Research Service, by NIH grants HD-31932, DK-18707, and HL-17818, and by a grant from the Herman Goldman Foundation.

10. REFERENCES

Abuelo, D., Tint, G. S., Kelley, R., Batta, A. K., Shefer, S., and Salen, G., 1995, Prenatal diagnosis of the cholesterol biosynthetic defect in Smith-Lemli-Opitz syndrome by the analysis of amniotic fluid, *Am. J. Med. Genet.* **56**:281–285.

Alley, T. L., Gray, B. A., Lee, S.-H., Scherer, S., Tsui, L.-C., Tint, G. S., Williams, C. A., Zori, R., and Wallace, M. R., 1995, Identification of a yeast artificial chromosome clone spanning a translocation breakpoint at 7q32.1 in a Smith-Lemli-Opitz syndrome patient, *Am. J. Hum. Genet.* **56**:1411–1416.

Ansari, G. A. S., Walker, R. D., Smart, V. B., and Smith, L. L., 1982, Further investigations of mutagenic cholesterol preparations, *Chem. Toxic.* **20**:35–41.

Axelson, M., 1991, Occurrence of isomeric dehydrocholesterols in human plasma, *J. Lipid Res.* **32**:1441–1448.

Barbu, V., Roux, C., Lambert, D., Dupuis, R., Gardette, J., and Maziere, J.-C., 1988, Cholesterol prevents the teratogenic action of AY9944: importance of the timing of cholesterol supplementation to rats, *J. Nutr.* **118**:774–779.

Batta, A. K., Salen, G., Tint, G. S., and Shefer, S., 1995a, Identification of 19-nor-5,7,9(10)-cholestratrien-3β-ol in plasma and stool of patients with the Smith-Lemli-Opitz syndrome, *J. Lipid Res.* **36**:2413–2418.

Batta, A. K., Tint, G. S., Abuelo, D., Shefer, S., and Salen, G., 1995b, Identification of 8-dehydrocholesterol (cholesta-5,8-dien-3β-ol) in patients with Smith-Lemli-Opitz syndrome, *J. Lipid Res.* **36**:705–713.

Batta, A. K., Tint, G. S., Salen, G., Shefer, S., Irons, M., and Elias, E. R., 1994, Identification of 7-dehydrocholesterol and related sterols in patients with Smith-Lemli-Opitz syndrome, *Am. J. Med. Genet.* **50**:334 (abstract).

Berenson, G. S., Blonde, C. V., Farris, R. P., Foster, T. A., Frank, G. C., Srinivasan, S. R., Voors, A. W., and Webber, L. S., 1979, Cardiovascular disease risk factor variables during the first year of life, *Am. J. Dis. Child.* **133**:1049–1057.

Berry, R., Wilson, H., Robinson, J., Sandlin, C., Tyson, W., Campbell, J., Porreco, R., and Manchester, D., 1989, Apparent Smith-Lemli-Opitz syndrome and Miller-Dieker syndrome in a family with segregating translocation t(7;17)(q34;p13.1), *Am. J. Med. Genet.* **34**:358–365.

Bialer, M. G., Penchezade, V. B., Kahn, E., Libes, R., Krigsman, G., and Lesser, M. L., 1987, Female external genitalia and Müllerian duct derivatives in a 46, XY infant with the Smith-Lemli-Opitz syndrome, *Am. J. Med. Genet.* **28**:723–731.

Blair, H. R., and Martin, J. K., 1966, A syndrome characterized by mental retardation, short stature, craniofacial dysplasia, and genital anomalies occurring in siblings, *J. Pediatr.* **69**:457–459.

Brown, M. S., and Goldstein, J. L., 1973, Regulation of 3-hydroxy-3-methylglutaryl coenzyme A reductase activity in human fibroblasts by lipoproteins, *Proc. Natl. Acad. Sci. USA* **70**:2162–2166.

Bruckdorfer, K. R., Demel, R. A., De Gier, J., and Van Deenen, L. L. M., 1969, The effect of partial replacements of membrane cholesterol by other steroids on the osmotic fragility and glycerol permeability of erythrocytes, *Biochim. Biophys. Acta* **183**:334–345.

Chasalow, F. I., 1985, Possible abnormalities of steroid secretion in children with Smith-Lemli-Opitz Syndrome and their parents, *Steroids* **46**:827–843.

Chen, M., Mason, R. P., and Tulenko, T. N., 1995, Atherosclerosis alters the composition, structure and function of arterial smooth muscle plasma membranes, *Biochim. Biophys. Acta* **1272**:101–112.

Cherstvoy, E. D., Lazjuk, G. I., Ostrovskaya, T. I., Shved, I. A., Kravtzova, G. I., Lurie, I. W., and Gerasimovich, A. I., 1984, The smith-lemli-opitz syndrome. A detailed pathological study as a clue to a etiological heterogeneity, *Virchows Arch. [Pathol. Anat.]* **404**:413–425.

Clayton, R. B., 1965, Biosynthesis of sterols, steroids and terpenoids. Part 1. Biogenesis of cholesterol and the fundamental steps in terpenoid biosynthesis, *Quart. Rev. Chem. Soc.* **19**:168–200.

Cohen, M. M., Jr., and Sulik, K. K., 1992, Perspectives on holoprosencephaly: Part II. Central nervous system, craniofacial anatomy, syndrome commentary, diagnostic approach, and experimental studies, *J. Craniofac. Genet. Dev. Biol.* **12**:196–244.

Curry, C. J. R., Carey, J. C., Holland, J. S., Chopra, D., Fineman, R., Golabi, M., Sherman, S., Pagon, R. A., Allanson, J., Shulman, S., Barr, M., McGravey, V., Dabiri, C., Schmike, N., Ives, E., and Hall, B. D., 1987, Smith-Lemli-Opitz syndrome-Type II: Multiple congenital anomalies with male pseudohermaphroditism and frequent early lethality, *Am. J. Med. Genet.* **26**:45–57.

Deaton, J. G., and Mendoza, L. O., 1973, Smith-Lemli-Opitz syndrome in a 23-year-old man, *Arch. Intern. Med.* **132**:422–423.

Dehart, D. B., Lanoue, L., Tint, G. S., and Sulik, K. K., 1995, Altered cholesterol biosynthesis in rats: a model for Smith-Lemli-Opitz syndrome, *Teratology* **51**:165 (abstract).

Dellaire, L., 1969, Syndrome of retardation with urogenital and skeletal anomalies (Smith-Lemli-Opitz syndrome): clinical features and mode of inheritance, *J. Med. Genet.* **6**:113–120.

Dempsey, M. E., Seaton, J. D., Schroepfer, G. J., Jr., and Trockman, R. W., 1964, The intermediary role of $\Delta^{5,7}$-cholestadien-3β-ol in cholesterol biosynthesis, *J. Biol. Chem.* **239**:1381–1387.

Dietschy, J. M., Turley, S. D., and Spady, D. K., 1993, Role of liver in the maintenance of cholesterol and low density lipoprotein homeostasis in different animal species, including humans, *J. Lipid Res.* **34**:1637–1659.

Donnai, D., Young, I. D., Owen, W. G., Clark, S. A., Miller, P. F. W., and Knox, W. F., 1986, The lethal multiple congenital anomaly syndrome of polydactyly, sex reversal, renal hypoplasia, and unilobular lungs, *J. Med. Genet.* **23**:64–71.

Dvornik, D., and Hill, P., 1968, Effect of long-term administration of AY 9944, an inhibitor of 7-dehydrocholesterol Δ^7-reductase, on serum and tissue lipids in the rat, *J. Lipid Res.* **9**:587–595.

Dvornik, D., Kraml, M., and Bagli, J. F., 1966, Agents affecting lipid metabolism. XVIII. A 7-dehydrocholesterol Δ^7-reductase inhibitor (AY-9944) as a tool in studies of Δ^7-sterol metabolism, *Biochemistry* **5**:1060–1064.

Dvornik, D., Kraml, M., Dubuc, J., Givner, M., and Gaudry, R., 1963, A novel mode of inhibition of cholesterol biosynthesis, *J. Am. Chem. Soc.* **85**:3309.

Edmond, J., Korsak, R. A., Morrow, J. W., Torok-Both, G., and Catlin, D. H., 1991, Dietary cholesterol and the origin of cholesterol in the brain of developing rats, *J. Nutr.* **121**:1323–1330.

Farese, R. V., Jr., Ruland, S. L., Flynn, L. M., Stokowski, R. P., and Young, S. G., 1995, Knockout of the mouse apolipoprotein B genes results in embryonic lethality in homozygotes and protection against diet-induced hypercholesterolemia in heterozygotes, *Proc. Natl. Acad. Sci. USA* **92**:1774–1778.

Fierro, M., Martinez, A. J., Harbison, J. W., and Hay, S. H., 1977, Smith-Lemli-Opitz syndrome: neuropathological and ophthalmological observations, *Dev. Med. Child. Neurol.* **19**:57–62.

Frantz, I. D., Jr., and Schroepfer, G. J., Jr., 1967, Sterol biosynthesis, *Annu Rev. Biochem.* **36**:691–726.

Fumagalli, R., Bernini, F., Galli, G., Anastasia, M., and Feicchi, A., 1980, The identification of a novel C_{27} diene, cholesta-5,8-dien-3β-ol, in tissues of rats given AY-9944 (*trans*-1,4-bis(2-dichlorobenzylaminoethyl)cyclohexane) in pregnancy, *Steroids* **35**:665–672.

Fumagalli, R., Niemiro, R., and Paoletti, R., 1965, Investigation of the biogenetic reaction sequence of cholesterol in rat tissues, through inhibition with AY-9944, *J. Am. Oil Chem. Soc.* **42**: 1018–1023.

Fumagalli, R., Smith, M. E., Urna, G., and Paoletti, R., 1969, The effect of hypocholesteremic agents on myelinogenesis, *J. Neurochem.* **16**:1329–1339.

Gage, J. C., and Sulik, K. K., 1991, The pathogenesis of ethanol-induced hydronephrosis and hydroureter is demonstrated following in vivo exposure of mouse embryos, *Teratology* **44**:299–312.

Garcia, C. A., McGarry, P. A., Voirol, M., and Duncan, C., 1973, Neurological involvement in the Smith-Lemli-Opitz syndrome: clinical and neuropathological findings, *Develop. Med. Child Neurol.* **15**:48–55.

Gelman-Kohan, Z., and Chemke, J., 1995, Letter to the editor: Prenatal diagnosis of SLO syndrome, *Am. J. Med. Genet.* **56**:288.

Goldstein, D. B., and Chin, J. H., 1981, Interaction of ethanol with biological membranes, *Fed. Proc.* **40**:2073–2076.

Gorlin, R. J., Cohen, M. M., and Levin, L. S., 1990, Well-known miscellaneous syndromes, in: *Syndromes of the Head and Neck (3rd edition)*, pp. 890–895, Oxford University Press, New York.

Hayes, K. C., Pronczuk, A., Wood, R. A., and Guy, D. G., 1992, Modulation of infant formula fat profile alters the low-density lipoprotein/high-density lipoprotein ratio and plasma fatty acid distribution relative to those with breast-feeding, *J. Pediatr.* **120**:S109–S116.

Hoefnaegel, D., Wurster, D., Pomeroy, J., and Benz, R., 1969, The Smith-Lemli-Opitz syndrome in an adult male, *J. Ment. Defic. Res.* **13**:249–257.

Hoffmann, G. F., Charpentier, C., Mayatepek, E., Mancini, J., Leichsenring, M., Gibson, K. M., Divry, P., Hrebicek, M., Lehnert, W., Sartor, K., Trefz, F. K., Rating, D., Bremer, H. J., and Nyhan, W. L., 1993, Clinical and biochemical phenotype in 11 patients with mevalonic aciduria, *Pediatr.* **91**:915–921.

Homanics, G. E., Maeda, N., Traber, M. G., Kayden, H. J., Dehart, D. B., and Sulik, K. K., 1995, Exencephaly and hydrocephaly in mice with targeted modification of the apolipoprotein B (Apob) gene, *Teratology* **51**:1–10.

Homanics, G. E., Smith, T. J., Zhang, S. H., Lee, D., Young, S. G., and Maeda, N., 1993, Targeted modification of the apolipoprotein B gene results in hypobetalipoproteinemia and develpmenta abnormalites in mice, *Proc. Natl. Acad. Sci. USA* **90**:2389–2393.

Honda, A., Tint, G. S., Salen, G., Batta, A. K., Chen, T. S., and Shefer, S., 1995, Defective conversion of 7-dehydrocholesterol to cholesterol in cultured skin fibroblasts from Smith-Lemli-Opitz syndrome homozygotes, *J. Lipid Res.* **36**:1595–1601.

Honda, A., Batta, A. K., Salen, G., Tint, G. S., Chen, T. S., and Shefer, S., 1997a, Screening for abnormal cholesterol biosynthesis in the Smith-Lemli-Opitz syndrome: rapid determination of plasma 7-dehydrocholesterol by ultraviolet spectrometry, *Am. J. Med. Genet.*, in press.

Honda, A., Tint, G. S., Batta, A. K., Kelley, R. I., Honda, M., Chen, T., Shefer, S., and Salen, G., 1997b, 7-Dehydrocholesterol concentrations are markedly increased in cultured skin fibroblasts from Smith-Lemli-Opitz syndrome homozygotes, *Am. J. Med. Genet.*, in press.

Honda, M., Tint, G. S., Honda, A., Batta, A. K., Chen, T. S., Shefer, S., Salen, G., 1996, Measurement of 3β-hydroxysteroid Δ^7-reductase activity in cultured skin fibroblasts utilizing ergosterol as a substrate: a new method for the diagnosis of the Smith-Lemli-Opitz syndrome, *J. Lipid. Res.* **37**:2433–2438.

Igarashi, M., Suzuki, K., Chen, S. M., and Suzuki, K., 1975, Changes in brain hydrolytic enzyme activities in rats treated with cholesterol biosynthesis inhibitor, AY-9944, *Brain Res.* **90**:97–114.

Illingworth, D. R., Papu, A. S., and Gregg, R. E., 1989, Increased urinary mevalonic acid excretion in patients with abetalipoproteinemia and homozygous hypobetaproteinemia, *Arteriosclerosis* **76**:21–27.

Irons, M., Elias, E. R., Bull, M. J., Greene, C. L., Johnson, V. P., Keppen, L., Schanen, C., Tint, G. S., and Salen, G., 1997, Treatment of Smith-Lemli-Opitz syndrome (SLOS): results of a multicenter trial, *Am. J. Med. Genet.*, in press.

Irons, M., Elias, E. R., Salen, G., Tint, G. S., and Batta, A. K., 1993, Defective cholesterol synthesis in Smith-Lemli-Opitz syndrome, *Lancet* **341**:1414.

Irons, M., Elias, E. R., Tint, G. S., Salen, G., Frieden, R., Buie, T. M., and Ampola, M., 1994, Abnormal cholesterol metabolism in the Smith-Lemli-Opitz syndrome: report of clinical and biochemical findings in 4 patients and treatment in 1 patient, *Am. J. Med. Genet.* **50**:347–352.

Jeanty, P., Delbake, D., Lemli, L., and Dorchy, H., 1977, Smith-Lemli-Opitz syndrome without failure to thrive, *Acta Paediatr. Belg.* **30**:175–178.

Johnson, V. P., 1975, Smith-Lemli-Opitz syndrome: review and report of two affected siblings, *Z. Kinderheilk.* **119**:221–234.

Jones, K. L., 1988, Smith-Lemli-Opitz Syndrome, in: *Smith's Recognizable Patterns of Human Malformation,* 4th ed., pp. 104–105, Saunders, Philadelphia.

Jurevics, H. E., and Morell, P., 1994, Sources of cholesterol for kidney and nerve during development, *J. Lipid. Res.* **35**:112–120.

Jurevics, H., and Morell, P., 1995, Cholesterol for synthesis of myelin is made locally, not imported into brain, *J. Neurochem.* **64:**895–901.

Kaiser, W., and Stocker, K., 1978, Evaluation of a nonisotopic technique for studies of in vivo cholesterol metabolism in mini-pigs using inhibition of 7-dehydrocholesterol reductase by AY 9944, *Res. Exp. Med. (Berlin)* **174**:79–108.

Kandutsch, A. A., 1962, Enzymatic reduction of the Δ^7 bond of 7-dehydrocholesterol, *J. Biol. Chem.* **237**:358–362.

Kandutsch, A. A., and Russell, A. E., 1960, Preputial gland sterols. III. A metabolic pathway from lanosterol to cholesterol, *J. Biol. Chem.* **235**:2256–2261.

Kelley, R. I., 1994, Quantification of neutral sterols by gas liquid chromatography/mass fragmentography for diagnosis of inborn errors of sterol metabolism, *Am. J. Med. Genet.* **50**:335 (abstract).

Kelley, R. I., 1995, Diagnosis of Smith-Lemli-Opitz syndrome by gas chromatography/mass spectrometry of 7-dehydrocholesterol in plasma, amniotic fluid and cultured skin fibroblasts, *Clin. Chim. Acta* **236**:45–58.

Koroly, M. J., and Dempsey, M. E., 1981, Synthesis of $\Delta^{5,22}$-cholestadien-3β-ol from $\Delta^{5,7,22}$-cholestatrien-3β-ol by a liver enzyme, *Lipids* **16**:755–758.

Korstanje, L. J., van Ginkel, G., and Levine, Y. K., 1990, Effects of steroid molecules on the dynamical structure of dioleoylphosphatidylcholine and digalactosyldiacylglycerol bilayers, *Biochim. Biophys. Acta* **1022**:155–162.

Kotch, L. E., and Sulik, K. K., 1992, Experimental fetal alcohol syndrome: proposed pathogenic basis for a variety of associated facial and brain anomalies, *Am. J. Med. Genet.* **44**:168–176.

Kotch, L. E., Webster, W. S., Dehart, D. B., and Sulik, K. K., 1991, Pathogenic basis of ethanol-induced exencephaly and its association with midline facial clefts in early mouse embryos, *Teratology* **43**:439 (abstract).

Kotch, L. E., Webster, W. S., and Sulik, K. K., 1990, Patterns of excessive cell death in mouse embryos and their relationnship to alcohol-related birth defects, *Teratology* **41**:572 (abstract).

Linton, M. F., Farnese, R. V., Jr., and Young, S. G., 1993, Familial hypobetalipoproteinemia, *J. Lipid Res.* **34**:521–541.

The Lipid Research Clinics population studies data book, *Vol. I, The prevalence study,* 1980, p. 29, U.S. Department of Health and Human Services, Bethesda, MD.

Lowry, R. B., Miller, J. R., and MacLean, J. R., 1986, Micrognathia, polydactyly, and cleft palate, *J. Pediatr.* **72**:959–961.

Lowry, R. B., and Yong, S.-L., 1980, Borderline intelligence in the Smith-Lemli-Opitz syndrome, *Am. J. Med. Genet.* **5**:137–143.

McGaughran, J. M., Clayton, P. T., Mills, S., Rimmer, S., Moore, L., and Donnai, D., 1995, Prenatal diagnosis of Smith-Lemli-Opitz syndrome, *Am. J. Med. Genet.* **56**:269–271.

McGaughran, J., Donnai, D., Clayton, P., and Mills, K., 1994, Diagnosis of Smith-Lemli-Opitz syndrome. Letter to the editor, *N. Eng. J. Med.* **330**:1685–1686.

McKeever, P. A., and Young, I. D., 1990, Smith-Lemli-Opitz syndrome II: a disorder of the fetal adrenals? *J. Med. Genet.* **27**:465–466.

Minsker, D. H., Macdonald, J. S., Robertson, R. T., and Bokelman, D. L., 1983, Mevalonate supplementation in pregnant rats suppresses the teratogenicity of mevinolic acid, an inhibitor of 3-hydroxy-3-methylglutaryl-coenzyme A reductase, *Teratology* **28**:449–456.

Muenke, M., Gurrieri, F., Bay, C., Yi, D. H., Collins, A. L., Johnson, V. P., Hennekam, R. C. M., Schaefer, G. B., Weik, L., Lubinsky, M. S., Daak-Hirsch, S., Moore, C. A., Dobyns, W. B., Murray, J. C., and Price, R. A., 1994a, Linkage of a human brain malformation, familial holoprosencephaly, to chromosome 7 and evidence for genetic heterogeneity, *Proc. Natl. Acad. Sci. USA* **91**:8102–8106.

Muenke, M., Hennekam, R. C. M., and Kelley, R. I., 1994b, Holoprosencephaly as a manifestation in Smith-Lemli-Opitz syndrome, *Am. J. Hum. Genet.* **55**:A35.

Natowicz, M. R., and Evans, J. E., 1994, Abnormal bile acids in the Smith-Lemli-Opitz syndrome, *Am. J. Med. Genet.* **50**:364–367.

Norton, W. T., and Cammer, W., 1984, Isolation and characterization of myelin, in *Myelin* (P. Morell, ed.), pp. 147–95, Plenum, New York.

Ohata, I., Sakamoto, N., Nitatori, R., and Maeno, H., 1983, Aminophenyl ether compound (YM-95831), a new hypolipidemic agent with a high density lipoprotein elevating activity in rats, *Arzneim.-Forsch./Drug Res.* **33**:237–241.

Opitz, J. M., and de la Cruz, F., 1994, Cholesterol metabolism in the RSH/Smith-Lemli-Opitz syndrome: summary of an NICHD conference, *Am. J. Med. Genet.* **50**:326–338.

Opitz, J. M., Penchaszadeh, V. B., Holt, M. C., Spano, L. M., and Smith, V. L., 1994, Smith-Lemli-Opitz (RSH) syndrome bibliography: 1964–1993, *Am. J. Med. Genet.* **50**:339–343.

Opitz, J. M., Seemanová, E., and Benesková, D., 1997, RSH ('Smith-Lemli-Opitz') syndrome as a common, paradigmatic metabolic malformation syndrome, *Am. J. Med. Genet.*, in press.

Pandak, W. M., Vlahcevic, Z. R., Heuman, D. M., and Hylemon, P. B., 1990, Regulation of bile acid synthesis. V. Inhibition of conversion of 7-dehydrocholesterol to cholesterol is associated with down-regulation of cholesterol 7α-hydroxylase activity and inhibition of bile acid synthesis, *J. Lipid Res.* **31**:2149–2158.

Parker, C. R., Jr., Carr, B. R., and Simpson, E. R., 1983, Decline in the concentration of low-density lipoprotein-cholesterol in human fetal plasma near term, *Metabolism* **32**:919–923.

Pierquin, G., Peeters, P., Roels, F., Vamos, E., Brucher, J. M., Tint, G. S., Honda, A., and Van Regemorter, N., 1995, Brief clinical report: severe Smith-Lemli-Opitz syndrome with prolonged survival and lipid abnormalities, *Am. J. Med. Genet.* **56**:276–280.

Pill, J., Schmidt, F. H., Stegmeier, K., and Witte, E.-C., 1985, Effects of BM 15.766 on serum lipids in rats, *Horm. Metabol. Res.* **17**:543–544.

Pill, J., Stegmeier, F. H., and Schmidt, K., 1990, Effects of drugs affecting cholesterol biosynthesis pathway on BM 15.766-induced 7-dehydrocholesterol accumulation in rats. An animal model for testing compounds reducing cholesterol synthesis, *Meth. Find. Exp. Pharmacol.* **12**:167–174.

Pober, B., 1990, Smith-Lemli-Opitz syndrome, in: *Birth Defects Encyclopedia* (M. L. Buyse, ed.), pp. 1570–1572, Blackwell Scientific Publications, Cambridge, MA.

Rawlins, F. A., and Lópes-Jiménez, C., 1979, Estudio comparativo de la mielina aislada de cerebros de ratas jovenes y adultas tratadas con un inhibidor de la biosintesis del colesterol, *Acta Cient. Venezolana* **30**:551–558.

Repetto, M., Maziere, J. C., Citadelle, D., Dupuis, R., Meier, M., and Biade, S., 1990, Teratogenic effect of the cholesterol synthesis inhibitor AY 9944 on rat embryos in vitro, *Teratology* **42**:611–618.

Rodd's Chemistry of Carbon Compounds, 2nd edition, Vol. IID: Steroids, 1970, (S. Coffey, ed.), p. 106, Elsevier, Amsterdam.

Rossiter, J. P., Hofman, K. J., and Kelley, R. I., 1995, Smith-Lemli-Opitz syndrome: prenatal diagnosis by quzntification of cholesterol precursors, *Am. J. Med. Genet.* **56:** 272–275.

Roux, C., and Aubry, M., 1966, Action tératogène chez le rat d'un inhibiteur de la synthèse du cholestérol, le AY 9944, *C. R. Soc. Biol.* **160**:1353–1357.

Roux, C., Dupuis, R., Horvath, C., and Talbot, J.-N., 1980, Teratogenic effect of an inhibitor of cholesterol synthesis (AY 9944) in rats: correlation with maternal cholesterolemia, *J. Nutr.* **110**:2310–2312.

Roux, C., Horvath, C., and Dupuis, R., 1979, Teratogenic action and embryo lethality of AY 9944: prevention by a hypercholesterolemia-provoking diet, *Teratology* **19**:35–38.

Sarkar, C. P., and Cenedella, R. J., 1982, Reactivity of key metabolic sterols in standard colorimetric assays for cholesterol, *Lipids* **17**:46–49.

Sawa, H., Kamada, K., Sato, H., Sendo, S., Kondo, A., Saito, I., Edlund, M., and Obrink, B., 1994, C-CAM expression in the developing rat central nervous system, *Brain Res. Dev.* **78**:35–43.

Schagina, L. V., Korchev, Y. E., Grinfeldt, A. E., Lev, A. A., and Blaskò, K., 1992, Sterol specific inactivation of gramicidin A induced membrane cation permeability, *Biochim. Biophys. Acta* **1109**:91–96.

Schroepfer, G. J., Jr., and Frantz, I. D., Jr., 1961, Conversion of Δ^7-cholestenol-4-^{14}C and 7-dehydrocholesterol-4-^{14}C to cholesterol, *J. Biol. Chem.* **236**:3137–3140.

Seedorf, U., Fobker, M., Voss, R., Meyer, K., Kannenberg, F., Meschede, D., Ulrich, K., Horst, J., Benninghoven, A., and Assmann, G., 1995, Smith-Lemli-Opitz syndrome diagnosed by using time-of-flight secondary-ion mass spectrometry, *Clin. Chem.* **41**:548–552.

Seedorf, U., Walter, M., and Assman, G., 1994, Diagnosis of Smith-Lemli-Opitz syndrome. Letter to the editor, *N. Eng. J. Med.* **330**:1686–1687.

Seller, M. J., Russell, J., and Tint, G. S., 1995, An unusual case of Smith-Lemli-Opitz syndrome type II, *Am. J. Med. Genet.* **56**:265–268.

Shefer, S., Nguyen, L. B., Salen, G., Ness, G. C., Chowdharry, I. R., Lerner, S., Batta, A. K., and Tint, G. S., 1992, Differing effects of cholesterol and taurocholate on steady state hepatic HMG-CoA reductase and cholesterol 7α-hydroxylase and mRNA levels in the rat, *J. Lipid Res.* **33**: 1193–1200.

Shefer, S., Salen, G., and Batta, A. K., 1986, Methods of assay, in: *Cholesterol 7α-Hydroxylase (7α-Monooxygenase)* (R. Fears and J. R. Sabine, eds.), pp. 43–49, CRC Press, Boca Raton, FL.

Shefer, S., Salen, G., Batta, A. K., Honda, A., Tint, G. S., Irons, M., Elias, E., Holick, M. F., and Chen, T. -C., 1995, Markedly inhibited 7-dehydrocholesterol-Δ^7-reductase activity in liver microsomes from Smith-Lemli-Opitz homozygotes, *J. Clin. Invest.* **96**:1779–1785.

Shibata, M., 1993, A new potent teratogen in CD rats inducing cleft palate, *J. Toxicol. Sci.* **18**:171–178.

Smith, D. W., Lemli, L., and Opitz, J. M., 1964, A newly recognized syndrome of multiple congenital anomalies, *J. Pediatr.* **64**:210–217.

Smith, M. E., Hasinoff, C. M., and Fumagalli, R., 1970, Inhibitors of cholesterol synthesis and myelin formation, *Lipids* **5**:665–671.

Steinberg, D., and Avigan, J., 1960, Studies of cholesterol biosynthesis. II. The role of desmosterol in the biosynthesis of cholesterol, *J. Biol. Chem.* **235**:3127–3129.

Stewart, F. J., Nevin, N. C., and Dornan, J. C., 1995, Letter to the editor: prenatal diagnosis of Smith-Lemli-Opitz syndrome, *Am. J. Med. Genet.* **56**:286–287.

Sulik, K. K., and Sadler, T. W., 1993, Postulated mechanisms underlying the development of neural tube defects. Insights from in vitro and in vivo studies, *Ann. N.Y. Acad. Sci.* **678**:8–21.

Suzuki, K., and De Paul, L. D., 1971, Cellular degeneration in developing central nervous system of rats produced by hypocholesteremic drug AY9944, *Lab. Invest.* **25**:546–555.

Suzuki, K., Zagoren, J. C., Gonates, J., and Suzuki, K., 1973, Ultrastructural study of neuronal cytoplasmic inclusions produced by hypercholesterolemic drug, AY9944, *Acta Neuropath.* **26**: 185–197.

Thompson, E., and Baraitser, M., 1986, An autosomal recessive mental retardation syndrome with hepatic fibrosis and renal cysts, *Am. J. Med. Gen.* **24**:151–158.

Tint, G. S., Irons, M., Elias, E. R., Batta, A. K., Frieden, R., Chen, T. S., and Salen, G., 1994, Defective cholesterol biosynthesis associated with the Smith-Lemli-Opitz syndrome, *N. Eng J. Med.* **330**:107–113.

Tint, G. S., and Salen, G., 1974, Transformation of 5α-cholest-7-en-3β-ol to cholesterol and cholestanol in cerebrotendinous xanthomatosis, *J. Lipid Res.* **15**:256–262.

Tint, G. S., and Salen, G., 1977, Evidence for the early reduction of the 24,25 double bond in the conversion of lanosterol to cholesterol in cerebrotendinous xanthomatosis, *Metabolism* **26**:721–729.

Tint, G. S., and Salen, G., 1982, Biosynthesis of cholesterol, lanosterol, and Δ^7-cholestenol, but not cholestanol, in cultured fibroblasts from normal individuals and patients with cerebrotendinous xanthomatosis, *J. Lipid Res.* **23**:597–603.

Tint, G. S., Salen, G., Batta, A. K., Shefer, S., Irons, M., Ampola, M., and Frieden, R., 1993, Abnormal cholesterol and bile acid biosynthesis in an infant with a defect in 7-dehydrocholesterol (7-DHC)-Δ^7 reductase, *Gastroenterology* **104**:A1008.

Tint, G. S., Salen, G., Batta, A. K., Shefer, S., Irons, M., Elias, E., Abuelo, D. N., Johnson, V. P., Lambert, M., Lutz, R., Schanen, C., Morris, C. A., Hoganson, G., and Hughes-Benzie, R., 1995a, Severity and outcome correlate with plasma sterol levels in type I and type II variants of the Smith-Lemli-Opitz syndrome, *J. Pediatr.* **127**:82–87.

Tint, G. S., Seller, M., Hughes-Benzie, R., Batta, A. K., Shefer, S., Genest, D., Irons, M., Elias, E., and Salen, G., 1995b, Markedly increased concentrations of 7-dehydrocholesterol combined with low levels of cholesterol are characteristic of the Smith-Lemli-Opitz syndrome, *J. Lipid Res.* **36**:89–95.

Toews, A. D., Goodrum, J. F., Lee, S. Y., Eckermann, C., and Morell, P., 1991, Tellurium-induced alterations in 3-hydroxy-3-methylglutaryl-CoA reductase gene expression and enzyme activity: differential effects in sciatic nerve and liver suggest tissue-specific regulation of cholesterol synthesis, *J. Neurochem.* **57**:1902–1906.

Van Cantfort, J., 1971, Action de l'AY 9944 sur l'activité de la cholestérol-7α-hydroxylase chez rat, *C. R. Acad. Sci. Paris, Series D* **273**:1438–1440.

Wagner-Recio, M., Toews, A. D., and Morell, P., 1991, Tellurium blocks cholesterol synthesis by inhibiting squalene metabolism: preferential vulnerability to this metabolic block leads to peripheral nervous system demyelination, *J. Neurochem.* **57**:1891–1901.

Wallace, M., Zori, R. T., Alley, T., Whidden, E., Gray, B. A., and Williams, C. A., 1994, Smith-Lemli-Opitz syndrome in a female with a de novo, balanced translocation involving 7q32: probable disruption of an SLOS gene, *Am. J. Med. Genet.* **50**:368–374.

West, J. R., and Goodlett, C. R., 1990, Teratogenic effects of alcohol on brain development, *Ann. Med.* **22**:319–325.

West, J. R., Goodlett, C. R., Bonthius, D. J., Hamre, K. M., and Marcussen, B. L., 1990, Cell population depletion associated with fetal alcohol brain damage: mechanisms of BAC-dependent cell loss, *Alcoholism: Clin. Exp. Res.* **14**:813–818.

Wiseman, H., 1993, Vitamin D is a membrane antioxidant. Ability to inhibit iron-dependent lipid peroxidation in liposomes compared to cholesterol, ergosterol and tamoxifen and relevance to anticancer action, *FEBS Lett.* **326**:285–288.

Wong, W. W., Hachey, D. L., Insull, W., Opekun, A. R., and Klein, P. D., 1993, Effect of dietary cholesterol on cholesterol synthesis in breast-fed and formula-fed infants, *J. Lipid Res.* **34**: 1403–1411.

Xu, G., Salen, G., Shefer, S., Ness, G. C., Chen, T. C., Zhao, Z., Salen, L., and Tint, G. S., 1995a, Treatment of the cholesterol biosynthetic defect as seen in the Smith-Lemli-Opitz syndrome reproduced in rats by BM 15.766, *Gastroenterology* **109**:1301–1307.

Xu, G., Salen, G., Shefer, S., Ness, G. C., Chen, T. S., Zhao, Z., and Tint, G. S., 1995b, Reproducing abnormal cholesterol biosynthesis as seen in the Smith-Lemli-Opitz syndrome by inhibiting the conversion of 7-dehydrocholesterol to cholesterol in rats, *J. Clin. Invest.* **95**:76–81.

Zaideh, R., Hoffman, E. P., Finegold, D. N., Hoop, R. C., Brackett, J. C., Strauss, A. W., and Naylor, E. W., 1995, Medium chain acyl-CoA dehydrogenase deficiency in Pennsylvania: neonatal screening shows high incidence and unexpected mutation frequencies, *Pediatr. Res.* **37**:675–678.

Zimmerman, P. A., Hercules, D. M., and Naylor, E. W., 1996, Direct analysis of filter paper blood specimens for identification of the Smith-Lemli-Opitz syndrome using time-of-flight secondary-ion mass spectrometry, *Am. J. Med. Genet.*, in press.

Chapter 6

Has Nature Designed the Cholesterol Side Chain for Optimal Interaction with Phospholipids?

Robert Bittman

1. SCOPE

In view of the many roles played by cholesterol in membranes, it is not surprising that the interactions of cholesterol with phospholipids, and their consequences with respect to membrane structure and function, have been studied intensively for more than three decades. This chapter reviews recent studies on the importance of the isooctyl side chain of cholesterol on ordering of the acyl chains, as well as how variations in the alkyl side chain length affect the conformational order of fatty acyl chains of phospholipids in bilayer membranes and the molecular packing of phospholipids in monolayers. The discussion focuses on recent biophysical studies of liposomes and monolayers in which the alkyl chain length of the sterol and/or the acyl chain length of the phospholipid were varied. The results are interpreted according to the hydrophobic mismatch effect. Also reviewed in this chapter are some recent kinetic studies of intermembrane movement and intracellular transport of sterols having a side-chain structure different from that of cholesterol. For a recent review of other aspects of sterol–phospholipid interactions,

Robert Bittman Department of Chemistry and Biochemistry, Queens College of The City University of New York, Flushing, New York 11367–1597.

Subcellular Biochemistry, Volume 28: Cholesterol: Its Functions and Metabolism in Biology and Medicine, edited by Robert Bittman. Plenum Press, New York, 1997.

including a more detailed discussion of the use of NMR spectroscopy to establish phase diagrams than are presented here, and an overview of differential interactions of cholesterol with subclasses of phospholipids, the reader is referred to Keough *et al.* (1996) and references cited therein.

2. THE FUNCTIONS OF CHOLESTEROL IN MEMBRANES

Cholesterol, or a closely related sterol, is an abundant component of plasma membranes of eukaryotic cells, and may be as high as 50 mol% with respect to total phospholipid. As discussed in several other chapters in this volume, cholesterol is essential to cell growth. Many biophysical investigations of the effects of cholesterol on the structure of phospholipid bilayers have provided support for the widely accepted notion that cholesterol plays a variety of important roles in biological membranes via its effects on the physical state of membrane lipids. Among its effects on the properties of phospholipid bilayers are modulation of the orientational order of the phospholipid hydrocarbon chains (increasing order in the liquid–crystalline phase and decreasing order in the gel phase), disruption of polar head-group interactions involving adjacent phospholipids and lowering of the surface charge density, removal of the cooperative gel-to-liquid–crystalline lipid phase transition, increasing bilayer thickness and compressibility, and regulation of membrane passive permeability properties (decreasing permeability in the liquid–crystalline phase and increasing it in the gel phase) to cations, anions, and small, neutral solutes. Intracellular membranes generally contain small amounts of cholesterol; a small amount of cholesterol may stabilize particular conformations of lipid-requiring enzymes and influence the activity of membrane-bound enzymes and transporters (for reviews see Presti, 1985; Bittman, 1988; Yeagle, 1988; Dahl, 1993). The structure and function of caveolae depend on the presence of cholesterol. Binding of cholesterol to VIP21/caveolin, a protein that plays a key role in the biogenesis of caveolae, appears to induce a conformational change in the protein (Murata *et al.*, 1995).

As a consequence of the roles outlined above, cholesterol affects the biochemical and physiological properties of the cell. Cholesterol modifies the properties of both the gel and liquid–crystalline phases of phospholipid bilayers. By decreasing the fluidity of phospholipid hydrocarbon chains in the liquid–crystalline phase and increasing the hydrocarbon chain fluidity in the gel phase, cholesterol produces an intermediate or buffered state of fluidity in the lipid bilayer, i.e., a state that appears somewhat closer to the gel phase than to the liquid–crystalline phase. Thus, cholesterol regulates the permeability of the bilayer and the lateral diffusion of molecules present in the bilayer. In cholesterol-rich membranes, there is a universality of the buffered environment that is independent of the large variation in lipid composition that exists in cell membranes. At low to moderate cholesterol content (up to about 25 mol %), a "liquid-disordered" (l_d)

phase exists, whereas at higher cholesterol content a "liquid-ordered" (l_o) phase exists (see section 5).

3. STRUCTURAL MODEL OF CHOLESTEROL–PHOSPHOLIPID CONTACTS

3.1. Localization of Cholesterol in Bilayer Membranes

The hydroxyl group and hydrocarbon-like nucleus create the amphipathic nature of cholesterol, and the tetracyclic nucleus and isooctyl side chain create the bulky wedge-type shape of the molecule. X-ray and neutron diffraction data showed that cholesterol is aligned in bilayers with its 3β-hydroxyl group in the vicinity of the ester bonds of diacyl glycerophospholipids and its tetracyclic ring system immersed in the bilayer interior, in close contact with a portion of the phospholipid fatty acyl chains (Franks, 1976; Worcester and Franks, 1976). Neutron diffraction measurements with C_3-deuterated cholesterol indicate that the distance between the C_3 atom of cholesterol and the egg PC (phosphatidylcholine) bilayer center is 18 Å, which places the hydroxyl group at 19.5 Å from the center of the bilayer (Worcester and Franks, 1976). Since the length of the cholesterol molecule, including the isooctyl tail in an all-*trans* energy-minimum conformation, is approximately 20 Å, it appears that the length of the cholesterol molecule is matched closely to the thickness of each leaflet of the egg PC bilayer. However, many naturally occurring phospholipids have average acyl chains that are appreciably shorter than 20 Å. For example, 20 Å corresponds to the length of the acyl chains of di-C19:0-PC in the gel phase, and a typical membrane contains many phospholipids with appreciably shorter chains than 19 carbons; moreover, in the liquid–crystalline phase, where *gauche* conformers exist (especially in the absence of high cholesterol content), the hydrocarbon chain length decreases by as much as one-third.

On the other hand, some conformational freedom may exist in the alkyl side chain of cholesterol (although part of the side chain is probably rigid because of restrictions imposed by the C18- and C21-methyl groups; for the sterol numbering system see Chapter 4 by Guo *et al.* and Chapter 15 by Salen *et al.*), so that the average total hydrophobic length of cholesterol would also be less than 20 Å. NMR data suggest about 17.5 Å as the effective hydrophobic length of cholesterol in phospholipid bilayers at the main phase transition temperature, which is the gel-phase length anticipated for a 17-carbon chain (Wu and Chi, 1991).

Figure 1 shows the preferred conformers of di-C16:0-PC and cholesterol, based on minimum energies of both lipids calculated using a molecular modeling program. To show the spatial relationship between the two lipids, they were aligned parallel to each other, with the cholesterol hydroxyl group aligned at the level of the *sn*-2 ester group. The latter alignment is somewhat arbitrary, since the hydrogen bond at this site is speculative (see below).

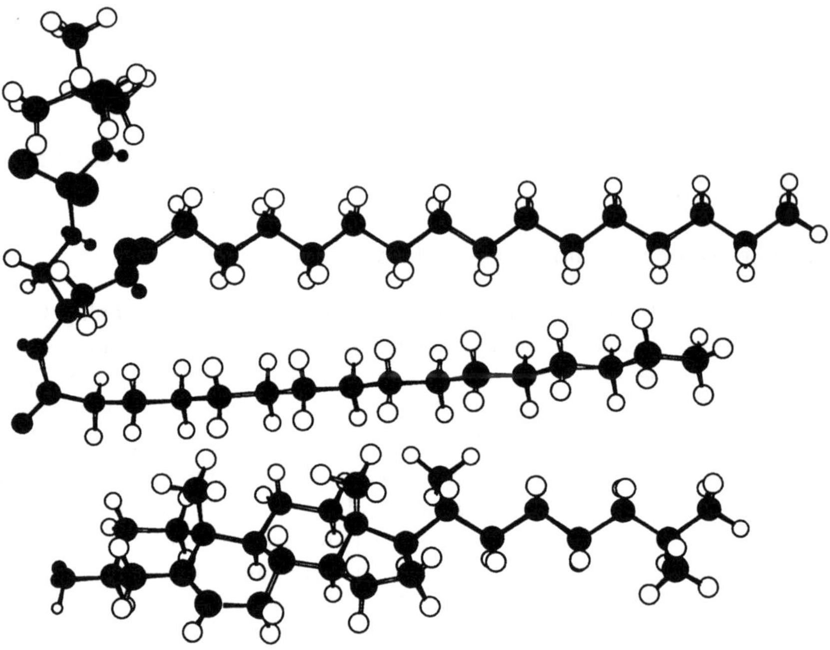

FIGURE 1. Spatial relationships between cholesterol and di-C16:0-PC, calculated by using the ChemDraw 3D+ program, which uses a modified Allinger molecular mechanics appproach. The preferred conformations shown are based on minimum energies. The total conformational energy includes van der Waals, hydrogen bonding, charge-dipole, and rotational interaction terms.

If cholesterol has as its nearest neighbor a phospholipid with a chain length shorter than about 17 carbon atoms, a void volume would be expected, unless it could adopt a less extended conformation or unless part of the cholesterol side chain could be partially interdigitated into the opposing monolayer of the bilayer (Sankaram and Thompson, 1990); the driving force for an interdigitated type of bilayer would increase at temperatures above the T_m of the pure phospholipid, where the chain length decreases. [Alternatively, at physiological temperatures or above, cholesterol may move from the center of the bilayer toward the polar head group of the phospholipid (Reinl *et al.*, 1992).] On the other hand, in bilayers that are formed from very long chain phospholipids, part of the fatty acyl chains of the phospholipid would penetrate into the opposing leaflet. Figure 2 shows models of the transbilayer distribution of cholesterol and phospholipids in the l_o (Figure 2A) and l_d (Figure 2B) phases of bilayers in which the hydrophobic ends of each molecule are not aligned with each other in the bilayer center.

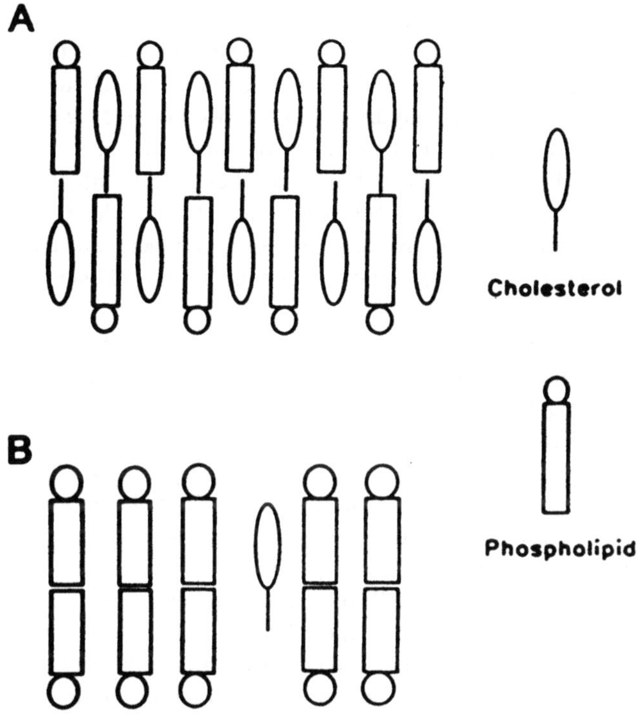

FIGURE 2. Proposed models of the time-averaged locations of cholesterol in PC bilayers in the liquid-ordered (l_o) and liquid-disordered (l_d) phases. A, cholesterol-rich fluid l_o phase at high cholesterol content. Cholesterol molecules are present in both leaflets of the bilayer. The alkyl tail is proposed to interdigitate partially into the opposing leaflet. The terminal methyl groups of the phospholipid and cholesterol alkyl chain do not meet at the center of the bilayer. B, the cholesterol-poor, l_d fluid phase at low cholesterol content. Cholesterol is proposed to span the bilayer, and is thus present simultaneously in both monolayers. These models are based on ^{13}C-NMR studies of di-C16:0-PC bilayers at 50°C in the presence and absence of spin labels in the PC acyl chains. (Reprinted with permission from Sankaram and Thompson, 1991.) A recent magic-angle-sample spinning ^{13}C-NMR spin-lattice relaxation study of egg PC-cholesterol (1:1) bilayers in the presence and absence of a paramagnetic ion was consistent with proposed packing model A (Villalaín, 1996).

3.2. Localization of the Hydroxy Group of Cholesterol in the Interface

3.2.1. Postulated Hydrogen Bonding to an Oxygen at the *sn*-1 or *sn*-2 Position

A hydrogen bond serves to anchor the cholesterol hydroxyl group near the lipid–water interface. Unlike the model shown in Figure 2, in which cholesterol is free to bob up and down within the bilayer, space-filling models have been proposed that include a hydrogen bond between the cholesterol hydroxy group and the oxygen atom in the ester linkage to the *sn*-2 glycerol carbon atom (Presti *et al.*, 1982; Finean, 1990). However, no experimental support for this predicited hydrogen bond has yet been obtained, and an energy minimization study of di-C14:0-PC-cholesterol dimers in the anhydrous state showed no hydrogen bond between cholesterol and the glycerol ester oxygen at either the *sn*-1 or *sn*-2 position (Vanderkooi, 1994). Furthermore, cholesterol was able to affect the permeability and thermotropic properties of bilayers formed from a phospholipid in which a hydrocarbon chain, without any oxygen atom, is attached to the carbon atom at the 2 position of the three-carbon backbone of the phospholipid (Bittman *et al.*, 1981).

A number of investigators have proposed a hydrogen bond between the hydrogen of the cholesterol hydroxy group and the *sn*-2 carbonyl oxygen of diacyl-PCs. However, experimental support for this type of putative hydrogen bond formation has failed to materialize (see, for example, Higinbotham *et al.*, 1993; Guo and Hamilton, 1995), perhaps because the 3β-hydroxy group of cholesterol in PC bilayers projects toward the water, and thus would be accessible for hydrogen bonding with interfacial water molecules. The latter would effectively compete with phospholipid oxygens for hydrogen bond formation (Dufourc *et al.*, 1984; Murari *et al.*, 1986). ^{13}C-NMR relaxation studies using cholesterol enriched with ^{13}C at the 4 position indicate that cholesterol is located in a similar position with respect to the surface in ether-linked, as in ester-linked, phospholipid vesicles (Bittman *et al.*, 1984), again casting doubt on the importance of a hydrogen bond between cholesterol and an oxygen atom in the glycerol ester linkage.

When a hydrogen bond was forced to form between cholesterol and the *sn*-2 carbonyl oxygen, an energy-minimized structure was found that showed the C_{26}/C_{27} carbon atoms of cholesterol extending to a position intermediate between the terminal methyl groups of the two chains of di-C14:0-PC (Vanderkooi, 1994); this is consistent with the cholesterol-di-C16:0-PC model shown in Figure 1. It should be noted that the *sn*-1 and *sn*-2 chains are of unequal length in a symmetric-chain diacyl-PC, owing to the sharp bend in the *sn*-2 chain at the lipid–water interface.

3.2.2. Postulated Hydrogen Bonding to an Oxygen at the *sn*-3 Position

Molecular dynamics simulations of hydrated bilayers of di-C14:0-PC with 11 mol % cholesterol indicate that the cholesterol hydroxyl group can be solvated by water or by the oxygens of the *sn*-3 phosphate head group, but when cholesterol resides more deeply into the bilayer it can be hydrogen bonded to both of the carbonyl oxygens (Robinson *et al.*, 1995). Experimentally, however, there appears to be no evidence for a direct interaction between the hydroxyl group of cholesterol and the polar head groups of zwitterionic glycerophospholipids in hydrated bilayers. Cholesterol increases the degree of hydration of phospholipid head groups (Lundberg *et al.*, 1978), suggesting that it does not bind to the phospholipid head group. ^{13}C-NMR studies show that cholesterol affects the polar head group region of di-C14:0-PC bilayers at 30°C to a lesser extent than it decreases the rate or amplitude of the fatty acyl chains and glycerol backbone region (Le Guernevé and Auger, 1995). However, if a net negative charge is present in the polar head group, as in phosphatidylglycerol and phosphatidylserine, stronger interactions with cholesterol may occur, leading to phase separation (Borochov *et al.*, 1995). Nevertheless, it is clear that with all of the classes of phospholipids investigated, nonspecific van der Waals attractive forces provide the dominant contribution to the stabilization of cholesterol–phospholipid interactions in membranes.

3.3. Nonuniform Effects of Cholesterol at Different Sites along the Phospholipid Acyl Chain

The motions that take place in different regions of a phospholipid molecule, in the presence and absence of cholesterol, have been examined by a variety of spectroscopic methods. Since there is a nearly parallel alignment between the sterol nucleus and the phospholipid acyl chains, it is expected that the segments of the phospholipid chain that make close contact with the sterol nucleus would have their motion restricted and would thus be the most ordered part of the phospholipid chain; this is indeed the case.

3.3.1. Magnetic Resonance Studies of the Effects of Cholesterol on Phospholipid Acyl Chain Order

The structural model for cholesterol–PC interactions discussed above predicts a depth dependence for the effects of cholesterol on the extent of phospholipid acyl chain conformational order/disorder. ^{2}H-NMR studies of bilayers with phospholipids labeled with deuterium at different sites along the acyl chain have provided an estimate of the orientational order at different positions of the chain. In the presence of cholesterol, the acyl chain region of phospholipids from C_2 (the carbon next to the ester moiety) to about the seventh to tenth carbon into the bilayer (the top half of the chain, i.e., the portion nearest to the glycerol backbone) exists in a relatively ordered state because of van der Waals attractive interactions between the rigid sterol ring system and the polymethylene groups of the phospholipid. From the middle of the chain to the bilayer center (the bottom half of the chain), the conformational freedom of the phospholipid acyl chain segments increases because these CH_2 groups come into contact with the relatively flexible isooctyl side chain attached to C_{17} of cholesterol.

The interactions of the sterol rings with the phospholipid acyl chain segments in the top half region of each leaflet of the bilayer produce a fairly constant ordering of the top half of the chain, and only *trans* conformers are considered to exist there at high cholesterol levels because close contact with the sterol nucleus inhibits bond rotation to form *gauche* conformers. ^{2}H-NMR estimates of the distance of the phospholipid chain from C_2 to C_{14} in di-C14:0-PC bilayers indicate that cholesterol at 30 mol % increases the length of the *sn*-2 chain at the transition temperature by 2.3 Å (Oldfield *et al.*, 1978).

3.3.2. Estimation of Phospholipid Chain Order Parameters by Molecular Dynamics Simulations

Chain order parameters for hydrated bilayers have also been estimated by molecular dynamics simulations (Robinson *et al.*, 1995). Calculations of order parameters from 400 ps of molecular dynamics show that the two chains of a symmetric-chain PC (such as di-C14:0-PC), which are known to be located at slightly different depths in the bilayer, are affected differently by cholesterol (11 mol %). For the *sn*-1 chain of di-C14:0-PC, the degree of ordering is higher in the upper part (C_2–C_6) because of contacts with cholesterol, whereas the order of the lower half of the myristoyl chain (C_7–C_{14}) is unaffected by cholesterol. In contrast, for the *sn*-2 chain the situation is reversed, with the lower half of the chain undergoing an increase in order parameter in the presence of cholesterol but the upper half of the chain remaining unaffected.

3.3.3. Infrared Studies of Conformational Order in Phospholipid Acyl Chains

Using specifically deuterated di-C16:0-PC in bilayers, Mendelsohn *et al.* (1989) analyzed CD_2 rocking modes by Fourier-transform infrared spectroscopy in order to quantitatively estimate the degree of conformational disorder in phospholipid chains at 50°C, which is above the gel-to-liquid–crystalline phase transition of di-C16:0-PC. Thus, the CD_2 group was inserted into each chain and used as a probe to determine what conformational state would be experienced by a CH_2 group in the same position of the chain. Spectral subtraction of bilayers prepared with nondeuterated phospholipids was necesssary to calculate the number of *gauche* bonds. As expected, the percentage of *gauche* rotamers was lower in the presence of 33 mol % cholesterol than in the absence of cholesterol. Fewer *gauche* bonds were found within the upper half of the palmitoyl chain (at positions 4 and 6) by factors of 6–9 when cholesterol was incorporated to 33 mol % in the bilayers (Mendelsohn *et al.*, 1991). In the lower half of the palmitoyl chain (at positions 10, 12, and 13), which is in contact with the sterol aliphatic side chain, *gauche* bond formation was reduced by a factor of only 1.5–2 on cholesterol incorporation. Thus the ability of cholesterol to reduce the disorder in the acyl chains of PC displays a pronounced depth dependence.

Monolayers of di-C16:0-PC in which each palmitoyl chain was half-deuterated in the lower half of the chain (between C_9 and C_{16}) had higher conformational order closer to the head group, and more disorder near the tail end (Gericke *et al.*, 1996).

3.3.4. Effects of a Spin Label in PC on Carbon Nuclei in Cholesterol

The observation that ^{13}C-NMR spin-lattice relaxation rates are enhanced when they are in the vicinity of an unpaired electron, as in a spin label, led to a study of the relaxation rates of different carbon atoms of cholesterol in equimolar di-C16:0-PC/cholesterol bilayers containing 1 mol % of a C5-spin-labeled phospholipid (Sankaram and Thompson, 1991). The enhanced relaxation rates of the carbon nuclei in rings A and B of cholesterol indicated that these rings are aligned near C_5 of the *sn*-2 chain of di-C16:0-PC (see Figure 2A) in the cholesterol-rich domain of the bilayer in the liquid–crystalline phase (see section 8.1 for a discussion of domain formation). In the cholesterol-poor liquid-disordered domain of di-C16:0-PC, in which the membrane is thinner than in the cholesterol-rich domain because of the depletion of cholesterol, the spin-lattice relaxation rates of carbon nuclei at the end of the isooctyl chain (C_{25}, C_{26}, and C_{27}) were found to be increased by the C_5-spin-labeled phospholipid (although to a lesser extent than carbon nuclei in the A ring of cholesterol). This was an unexpected observation, since the conformational flexibility of this region of the cholesterol molecule is expected to result in more motional freedom, rather than less. The model (Figure 2B) proposed to account for this observation shows the tail end of cholesterol partially interdigitated into the opposing monolayer of the bilayer, reaching to at least the C_5 position of the phospholipid acyl chain.

4. CHOLESTEROL ANALOGS: ABILITY TO FUNCTION AS CHOLESTEROL SUBSTITUTES IN NATURAL MEMBRANES

4.1. Sterols with Substituents in the α or β Face

Sterols that have a planar fused tetracyclic ring system, a free 3β-hydroxyl group, and a flexible aliphatic side chain generally produce cholesterol-like characteristic effects on the physical properties of bilayers. Although a broad range of sterol structures can substitute for cholesterol in modulating lipid order, sterols that have one or more methyl groups protruding from the α face of the molecule, such as lanosterol, are generally less effective than cholesterol in ordering the lipid bilayer above the phase transition temperature (Dahl *et al.*, 1980a; Yeagle, 1988) and in supporting mycoplasma growth (Dahl *et al.*, 1980a). In cholesterol, the α face is studded with only axial hydrogen atoms (see Figure 3A). The presence of a bulky group in the α face of the molecule, as with the 14α-methyl group of lanosterol (see Figure 3B), would block a close contact between the sterol nucleus and phospholipid chains (Yeagle *et al.*, 1977; Bloch, 1983).

The plant sterol cycloartenol, which has a cyclopropyl group between C9, C10, and C19, is isomeric with lanosterol and has a 14α-methyl group, yet has

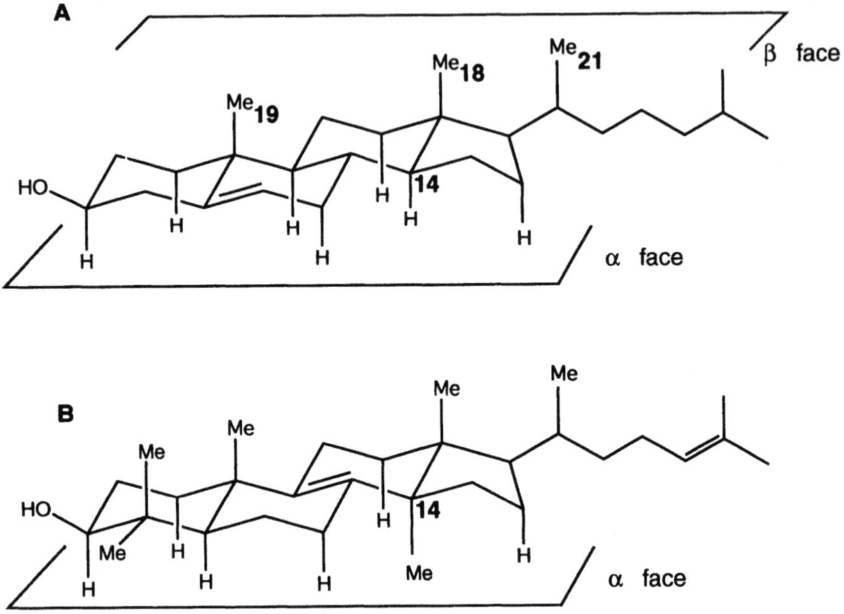

FIGURE 3. A, Chair conformation of cholesterol, showing its α and β faces. The angular methyl groups (carbons 18 and 19) project into the β face of the steroid nucleus, whereas only axial hydrogen atoms project into the flat α face. The C21 methyl group is also in the β face. B, Chair conformation of lanosterol, showing the presence of the axial methyl group at C_{14} projecting into the α face.

the capacity to order phospholipid acyl chains and support mycoplasma growth (Dahl *et al.*, 1980b). Its conformation was postulated to differ from that of lanosterol; it was proposed that the cyclopropyl group in the β face of the molecule causes ring B to assume a boat conformation and the molecule to have a bow-shape, raising the C_{14}-methyl group into alignment with some of the axial hydrogen atoms and restoring some of the planar character to the α face (Bloch, 1983). However, x-ray diffraction of cycloartenol crystals shows the B ring in a half-chair and the C ring in an unusual twist-boat conformation, giving a very flat molecule, similar to cholesterol, with the 14-methyl group in a similar position to that in lanosterol (Duax *et al.*, 1988; see section 4.3 of Chapter 4 by Guo *et al.*). Moreover, recent NMR studies showed that the rings of cycloartenol are flat in solution (see section 4.3 of Chapter 4 by Guo *et al.*). Therefore, it is unlikely that the orientation of the 14-methyl group in the α face can account for the different effects of lanosterol and cycloartenol on membranes, unless the depth of insertion into the α face differs.

The angular methyl groups (C18 and C19, see numbering system of cholesterol, Figure 2 of Chapter 15 by Salen *et al.*) and the isooctyl chain of cholesterol may be neatly accommodated into packing with phospholipids if there is a kink (gtg′) conformation in the phospholipid acyl chain (Huang, 1977).

4.2. Biohopanoids

Many pentacyclic terpenes in the hopanoid family also have the ability to decrease the fluidity of liquid–crystalline phase bilayers (Kannenberg *et al.*, 1983). They have a widespread occurrence in bacterial membranes, and may perform sterol-like functions in bacterial membranes (Ourisson and Rohmer, 1992). Although these "ancestors" of cholesterol resemble cholesterol in rigidity, general cylindrical shape, and overall amphilicity, they have different molecular dimensions than cholesterol. The increase in the order of phospholipid acyl chains induced by polyterpenoid incorporation into bilayers falls short of that induced by cholesterol (Krajewski-Bertrand *et al.*, 1992). Thus the precise molecular dimensions of cholesterol may have evolved in order to influence the packing of membrane phospholipids in a highly efficient manner.

4.3. Marine Sterols

A low-sensitivity differential-scanning calorimetry study indicated that desmosterol, fucosterol, and sutinasterol (for structures, see Table I) affect the thermotropic properties of 1-stearoyl-2-oleoyl-PC liposomes; the ability of desmosterol to decrease the overall main transition temperature (T_m) and ΔH appears to be somewhat less than that of cholesterol, fucosterol, or sutinasterol (Dai *et al.*, 1991). High-sensitivity differential-scanning calorimetry studies of the effects of various sterols on di-C16:0-PC and 1-stearoyl-2-oleoyl-PC liposomes are discussed in section 7.2.

Table I
Structural Features of the Δ^5 Sterols Discussed in This Chapter[a]

Sterol	C_{24} substituent	Additional features
Cholesterol	H	
Campesterol	α-methyl	
Sitosterol	α-ethyl	
Ergosterol	β-methyl	Δ^7, Δ^{22} (*trans*)
Stigmasterol	α-ethyl	Δ^{22} (*trans*)
Desmosterol		Δ^{24} (*trans*)
Fucosterol	ethyl	24-*cis*-ethylidene
Androstenol	none	no side chain at C_{17}

[a]Sutinasterol is a Δ^7 sterol, with an exomethylene group at C_{27}, a gem-dimethyl group at C_{26}, and an ethyl group at C_{24}.

4.4. Effects of Modification in the Alkyl Side Chain of the Sterol on Membrane Function

4.4.1. Examples of Recognition of Sterol Side-Chain Structure; Failure of Side-Chain Analogs of Cholesterol to Serve as Cholesterol Mimics

Cholesterol analogs such as the phytosterols sitosterol (which differs from cholesterol only at C_{24}, where the hydrogen of cholesterol is replaced by an ethyl group), stigmasterol, and ergosterol (for sterol structures, see Table I, and Chapter 15 by Salen *et al.*) do not substitute for cholesterol as a growth factor in a mouse fibroblast sterol auxotroph cell culture (Rujanavech and Silbert, 1986), perhaps because cholesterol has a higher capacity to order the fluid bilayer at the growth temperature in this system or because cholesterol participates in other cellular functions to regulate specific metabolic processes. Growth rates of mycoplasma cells appear to be correlated with an increase in hydrocarbon chain order of bilayer lipids induced by cholesterol uptake (Huang *et al.*, 1991).

Sterols that bear alkyl substitution or a double bond in the side chain are taken up by mycoplasma cells from the growth medium. These sterols were found to be predominantly incorporated into and accumulated in the outer leaflet of the membrane of mycoplasma cells, whereas sterols that have the isooctyl side chain, as in cholesterol, but bear a modification in the ring nucleus (removal of the Δ^5 bond or introduction of additional unsaturation in the rings), have the same transbilayer distribution as cholesterol (Bittman, 1993). Alkylation at C_{24} and/or unsaturation at Δ^{22} of the sterol side chain, as in sitosterol, ergosterol, and *cis*- and *trans*-22-dehydrocholesterol, may impede the translocation of the sterol into the inner leaflet of the bilayer because the increased bulk of the side chain leads to steric interference. Accumulation of sterols with modified side-chain structure in the outer leaflet of the mycoplasma bilayer may influence the asymmetry of membrane proteins and phospholipids. However, the physiological significance of transbilayer sterol distribution in the two halves of the mycoplasma membrane bilayer is not yet known, since sitosterol, ergosterol, and stigmasterol serve as growth-supporting sterols for *M. capricolum* almost as well as cholesterol, but differ in their localization within the bilayer (Clejan *et al.*, 1981; Clejan and Bittman, 1984).

A comparison was made of the uptake of sterols from negatively charged liposomes into macrophages and their subsequent intracellular transport (Sato *et al.*, 1995). Liposomes that contained campesterol, sitosterol, stigmasterol, or fucosterol (see Table I for structural features) were taken up by macrophages as efficiently as liposomes that contained cholesterol. However, these analogs of cholesterol remained in phagocytotic vacuoles, instead of being further transported to other subcellular organelles, as is cholesterol. Thus modifications in the side chain (alkylation at C_{24} as in campesterol and sitosterol, and both alkylation and unsaturation as in stigmasterol and fucosterol) affect the endocytotic routing of sterols.

Cholesterol is required for the process of liver uptake of chylomicron-like remants from plasma. Lower rates of plasma clearance of cholesteryl oleate were found when the emulsions were prepared with androst-5-en-3β-ol (referred to below as androstenol), a sterol that lacks the side chain at C_{17} but is otherwise identical to cholesterol (Mortimer *et al.*, 1995). However, the rates of clearance of triolein and cholesteryl oleate from artificial emulsion particles prepared with campesterol or sitosterol were similar to those of cholesterol-containing emulsions. Thus, there is a requirement for an alkyl chain at C_{17} for a cholesterol congener to function in chylomicron metabolism. Synthetic short-chain analogs of cholesterol mimic the behavior of androstenol, whereas a sterol with seven carbon atoms in the side chain mimics the behavior of cholesterol (Redgrave *et al.*, 1997).

Campesterol and desmosterol were ineffective in reconstitution of the activity of the γ-aminobutyric acid transporter in liposomes, a system in which cholesterol is effective (Shouffani and Kanner, 1990). Similarly, sitosterol, ergosterol, and stigmasterol were ineffective in replacing cholesterol for support of Na^+–Ca^{2+} exchange and Na^+, K^+-ATPase activity in reconstituted vesicles (Vermuri and Philipson, 1989). These studies suggest that sterols bearing unsaturation or alkyl substituents within the sterol side chain fail to mimic the bilayer environment created when cholesterol is present.

A specific type of alkyl side chain at C_{17} is also required for a sterol to serve as a substrate for acyl-CoA: cholesterol acyltransferase (ACAT) in microsomes; analogs of cholesterol with minor changes in side-chain structure were esterified poorly by ACAT (Tavani *et al.*, 1982).

Cholesterol undergoes spontaneous exchange between the surfaces of membranes, as monitored by analyzing the movement of radiolabeled cholesterol from a donor membrane to an acceptor membrane prepared with unlabeled cholesterol (see Chapter 9 by Johnson *et al.*). The process of desorption of cholesterol from the outer leaflet of donor vesicles into the aqueous medium is the rate-limiting step during the exchange or transfer to an excess of acceptor vesicles. The rate of sterol exchange is sensitive to the interactions that take place between the sterols and their neighboring phospholipids in the host bilayer, as well as to the extent of sterol aqueous solubility; the former appear to dominant over the latter (Kan *et al.*, 1992). The rate of spontaneous desorption of sitosterol is significantly slower than that of cholesterol, but at equilibrium the extent of sterol exchange is the same (Kan and Bittman, 1990, 1991). The rate ratio of sitosterol/cholesterol exchange is unaffected by variation in the membrane phospholipid composition and extent of saturation in the phospholipid acyl chains. The additional van der Waals forces made possible by substitution of the ethyl group at C_{24}, together with the lower aqueous solubility of sitosterol, appear to contribute to the slower desorption of sitosterol from lipid–water interfaces.

The rates of association of the polyene antibiotics filipin and amphotericin B with sterols in PC vesicles is also dependent on sterol side-chain structure (Clejan and Bittman, 1985).

4.4.2. Examples of Lack of Recognition of Sterol Side-Chain Structure

Some of the mechanisms involved in metabolic feedback inhibition appear to not recognize differences in sterol side-chain structure. In the presence of sterols, the activity of HMG-CoA reductase is downregulated; side-chain analogs of cholesterol are as effective as cholesterol in depressing the conversion of acetate into sterol (Erickson and Nes, 1982).

As noted above, the precise structure of the alkyl side chain is not critical for the physiological clearance of chylomicron-like emulsions (Mortimer *et al.*, 1995); however, for high clearance rates it is essential that the length of the side chain be similar to the isooctyl chain of cholesterol (Redgrave *et al.*, 1997).

5. PHASE DIAGRAM OF CHOLESTEROL-PC MIXTURES

The composition of bilayers is easily controlled and manipulated, making possible the study of phospholipid–cholesterol interactions over a wide range of sterol concentrations. Phase diagrams, which are plots of temperature vs. composition, have been determined for cholesterol/phospholipid mixtures from detailed biophysical studies in bilayers. The cholesterol concentrations at which the phase properties change are calculated to give the phospholipid–cholesterol phase diagram. Properties that are sensitive to the thermodynamic phase as a function of temperature, such as the lineshape in NMR or ESR spectroscopy and excess specific heat in differential-scanning calorimetry, are monitored in order to compute the high and low temperature limits of coexisting phases for a given bilayer composition.

Three distinct phases have been identified in the PC-cholesterol system: (1) the solid-ordered (s_o) phase (also known as the gel phase and L_β phase), in which the hydrocarbon chains are highly ordered; (2) a fluid phase known as the liquid-disordered (l_d) phase, in which the hydrocarbon chains are disordered and undergo rapid axially symmetric reorientation on the NMR timescale; and (3) another fluid phase known as the liquid-ordered phase (l_o) (called the β phase by Vist and Davis, 1990), which is immiscible in the l_d phase. The phase boundaries of s_o, l_d, and l_o have been constructed from data obtained on cholesterol-PC binary mixtures (for recent reviews see Davis, 1993; Keough *et al.*, 1996).

At low cholesterol content (less than ~22 mol% cholesterol), pure phospholipid and phospholipid–cholesterol domains coexist, but above 22 mol% cholesterol the domains of pure phospholipid are not detected. A phase diagram was proposed recently with two pure phospholipid phases from 0–22 mol % cholesterol, coexisting with two liquid-ordered or β phases, which coexist from about 1 to about 50 mol % cholesterol (McMullen and McElhaney, 1995, 1996). Models that show a stoichiometry of about 3.5 phospholipid molecules per cholesterol molecule (i.e., 22 mol % sterol) have been proposed to account for the behavior of mixtures of cholesterol and various PCs (see references in Presti, 1985).

Detailed thermodynamic and spectroscopic studies are required in order to fully understand the interactions of sterol analogs with phospholipids. At present, there are insufficient data available to construct a phase diagram for any bilayers composed of cholesterol analogs and phospholipids.

6. THE HYDROPHOBIC MISMATCH EFFECT: VARIATIONS IN STEROL SIDE-CHAIN STRUCTURE AND PHOSPHOLIPID ACYL-CHAIN STRUCTURE

The effect of a sterol on the phase behavior of PC bilayers depends on how closely the hydrophobic lengths of the sterol and the PC in the host bilayer are matched with each other. The hydrophobic length of the sterol molecule is defined as the length from the edge of the A ring to the end of its alkyl chain. Since the ring system of cholesterol or related sterols is highly rigid, the hydrophobic length of a sterol is not considered to become appreciably shorter in the temperature range at which the phospholipid undergoes the main melting transition from the gel to liquid–crystalline phase. The hydrophobic length of cholesterol incorporated into PC bilayers appears to be ~17.5 Å, which was estimated as outlined below (McMullen *et al.*, 1993). Androstenol, which retains the same ring structure and 3β-hydroxyl group as cholesterol but lacks the aliphatic side chain entirely, is about 5 Å shorter than cholesterol (as estimated using molecular modeling), and thus has a hydrophobic length of ~12.5 Å (McMullen *et al.*, 1994). To analyze the role of the cholesterol side chain on interactions between cholesterol and PCs, a variety of side-chain analogs of cholesterol have been synthesized, with the aliphatic chain ranging from being shorter to longer than that of cholesterol, and the effects of these sterols on the phase properties and organization of bilayers and monolayers of PCs having different acyl-chain structure have been tested (Chia *et al.*, 1993; Slotte *et al.*, 1994; McMullen *et al.*, 1995; Vilchèze *et al.*, 1996), as described below.

7. RECENT STUDIES OF THERMOTROPIC PHASE BEHAVIOR OF PCs IN BILAYERS

7.1. Conformational Order in Model Membranes as Evaluated from Infrared CH_2 Wagging Progression Intensities

In section 3.3.3, the use of FT–IR spectroscopy for analysis of phospholipid chain order was discussed, using phospholipids deuterated at specific methylene groups as probes. The CD_2 rocking modes are extremely weak vibrations. Further study of the conformational order made use of the intensity of another spectral parameter, i.e., the CH_2 wagging progression between 1150 and 1300 cm^{-1}, after subtraction of the overlapping phosphate symmetric stretching band (Chia *et al.*,

1993). The all-*trans* conformation of the acyl chains is considered to be responsible for the intensity of the CH_2 wagging progression. At temperatures higher than T_m, the all-*trans* conformation is lost because *gauche* rotamers are introduced, as is the CH_2 wagging progression.

7.1.1. Variation in Sterol Side-Chain Length

The effect of different sterols on the temperature-dependent intensity of CH_2 wagging was analyzed. Figure 4 (Chia *et al.*, 1993) shows the decay of the intensity of one of the components of the di-C16:0-PC wagging progression in 2:1 PC:sterol (mol/mol) bilayers as the temperature is increased. Sterols with a short side chain (such as 3 carbons) or no side chain (androstenol) induced disorder into the di-C16:0-PC acyl chains. However, with sterols that have a longer side chain, such as cholesterol or a sterol with a 10-carbon aliphatic chain, a higher temperature was needed to introduce *gauche* conformers into the phospholipid acyl chains. For di-C16:0-PC/androstenol bilayers, the IR progression intensity approached zero at a temperature very close to the T_m for pure di-C16:0-PC, indicative of a poor degree of interaction between these two lipids, which have a large hydrophobic mismatch. Androstenol-containing bilayers were calculated to contain an additional 1 to 1.5 *gauche* bonds per chain of di-C16:0-PC compared with cholesterol-containing bilayers (Senak *et al.*, 1992). However, as the sterol side-chain length was increased from zero to three, four, five, and six carbons, the IR progression intensity persisted to higher temperatures, indicative of increasing order in the hydrocarbon chains, but still less than the order induced by the same content of cholesterol at a given temperature. The IR intensity profiles as a function of temperature were very similar for cholesterol and a sterol with a seven-carbon side chain, whereas sterols with a side chain of more than eight carbons induced higher conformational order into the palmitoyl chains than did cholesterol.

7.1.2. Variation in Phospholipid Acyl-Chain Length

IR studies were made of the ability of cholesterol to reduce the number of *gauche* conformers in a series of symmetric-chain diacyl-PC bilayers (2:1 molar ratio of PC:cholesterol) as a function of temperature (Chia *et al.*, 1993). Figure 5 shows a comparison of the decay of intensity of the wagging progression with temperature, reflecting the introduction of *gauche* conformers into the acyl chains as temperature increases. In order to compare the effect of temperature on conformational disorder in different PCs, a "reduced temperature" scale was used, which is defined as T-T_m, i.e., the difference between the temperature of the IR measurement and the T_m of bilayers prepared from the pure PC. Cholesterol was more effective in ordering the acyl chains of di-C14:0-PC than of di-C16:0-PC and di-C20:0-PC (Figure 5). The plot of the decay of the progression intensity of pure

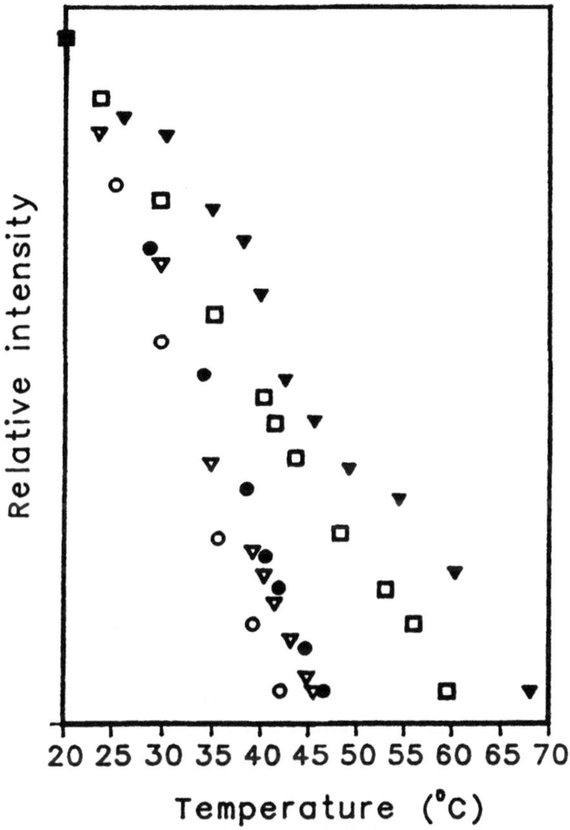

FIGURE 4. Effect of variation in sterol side-chain structure on the relative intensities of one of the components of the IR CH_2 wagging progression of di-C16:0-PC in 2:1 PC:sterol mixtures. The decay in intensity of the wagging progression with increasing temperature reflects the introduction of *gauche* conformers into the phospholipid acyl chains. Intensity data were normalized at 20°C. Short side-chain sterols [(o), androstenol; (∇), *iso*-C3-sterol, and (●) *n*-C4-sterol] induce disorder into the PC acyl chains. With cholesterol (□) and other long side-chain sterols [(▼), *iso*-C10-sterol], the number of *trans-gauche* isomerizations in the phospholipid acyl chains is decreased, so introduction of *gauche* conformers requires higher temperatures. For di-C16:0-PC bilayers containing *iso*-C6-, *n*-C7-, and *iso*-C7-sterols, the decrease in relative intensity with increasing temperature was similar to that found with cholesterol. (Reprinted with permission from Chia *et al.*, 1993.)

di-C20:0-PC bilayers vs. temperature was very similar to that found for the di-C20:0-PC/cholesterol mixture, and the IR progression intensity reached zero near a reduced temperature of only 5°C. Therefore, cholesterol was ineffective in stabilizing the all-*trans* conformation of this PC when the temperature was increased. In contrast, for di-C14:0-PC bilayers, the progression intensity persisted to about 40°C above the T_m of pure di-C14:0-PC. The experimental data are

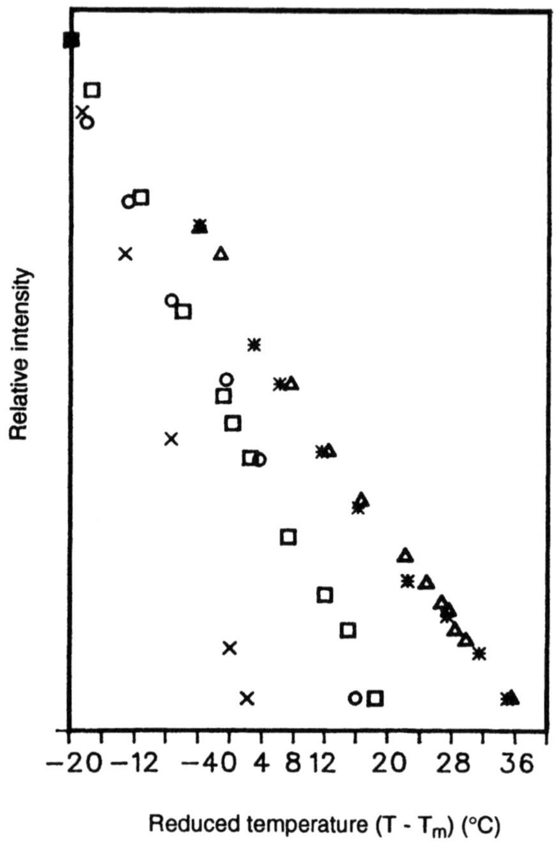

FIGURE 5. Relative ability of cholesterol to induce conformational order in the acyl chains of bilayers formed from various symmetric-chain disaturated PCs as a function of increasing temperature. The molar ratio of PC/cholesterol was 2:1. The persistence of the IR wagging progression to temperatures significantly higher than the chain melting temperature of the pure PC in bilayers of di-C14:0-PC (△) and in di-C12:0-PC (*) is indicative of the ordering ability of cholesterol. The T_m values of the pure PCs are: di-C12:0-PC,-1°C; di-C14:0-PC, 24.0°C; (□) di-C16:0-PC, 41.5°C; (o) di-C20:0-PC, 64.1°C, and (x) di-C22:0-PC, 75.0°C. The reduced temperature scale (T-T_m) allows a comparison of the effects of sterols on conformational disorder in PCs having different T_m values. (Reprinted with permission from Chia *et al.*, 1993.)

interpreted according to the "hydrophobic mismatch hypothesis." Most of the CH_2 groups of the myristoyl chains of di-C14:0-PC are in van der Waals contacts with cholesterol, since the two lipids have similar hydrophobic lengths, permitting the all-*trans* conformation to persist to a high temperature. The acyl chains of di-C20:0-PC are much longer than the hydrophobic length of cholesterol, so a significant fraction of the CH_2 groups in the C20 chain are not in contact with any part of the cholesterol molecule. Therefore, cholesterol cannot inhibit the formation of *gauche* conformers in the acyl chains of di-C20:0-PC bilayers when the temperature is increased. It is likely that part of the very long acyl chains interdigitate into the opposite leaflet of the bilayer in order to prevent the formation of a void volume.

7.2. Differential-Scanning Calorimetry Studies of Sterol-Induced Changes in PC Phase Behavior

7.2.1. Effect of Variation in PC Acyl-Chain Length

The overall endotherm of a sterol/PC mixture observed by calorimetry is a composite of two overlapping components, which are decomposed or deconvoluted into a sharp (sterol-poor) and a broad (sterol-rich) component. The T_m of the cholesterol-rich domain of bilayers formed from di-C17:0-PC and cholesterol was not altered on increasing incorporation of cholesterol; however, introduction of increasing cholesterol concentration into the same cholesterol-rich domains of PC bilayers with chains shorter than C17:0 increased the T_m, and decreased the T_m of PCs with acyl chains longer than C17:0-PC (McMullen *et al.*, 1993).

Why does a molecule (such as cholesterol) that does not undergo an appreciable change in length with increasing temperature have a dramatic effect on the relative stabilities of the gel and liquid–crystalline phases (as reflected in the temperature of the chain melting transition) of the host PC bilayer? The working hypothesis is that cholesterol would stabilize the gel phase of the host PC bilayer when the hydrophobic length of cholesterol is longer than the mean hydrophobic length of the PC molecule, which is defined as the phospholipid acyl-chain length midway between the hydrophobic lengths of the PC in the gel and liquid–crystalline phases. In this situation, the best match with respect to hydrophobic length would be obtained between cholesterol and the gel-state PC, and a large mismatch in hydrophobic length would occur on PC chain melting, which shortens the acyl chains. Conversely, cholesterol would destabilize the gel phase of the host PC bilayer and stabilize the liquid–crystalline phase when the mean hydrophobic length of the PC exceeds that of cholesterol. In this situation, the T_m of the broad (cholesterol-rich) transition would be decreased with cholesterol incorporation. The hydrophobic mismatch effect explains why the T_m values of the broad component of di-C13:0-to di-C16:0-PC bilayers are increased (via gel-phase stabilization) on increasing uptake of cholesterol, whereas the T_m of the broad component of di-C18:0-to di-C21:0-PC bilayers is decreased (via destabilization of the gel phase) (McMullen *et al.*, 1993). The best match with respect to hydrophobic lengths is predicted between cholesterol and di-C17:0-PC ($\sim$17.5 Å), and between andostenol and di-C12:0-PC ($\sim$12.5 Å). In confirmation of the hypothesis, the T_m of the broad component does not change in bilayers containing cholesterol/di-C17:0-PC and androstenol/di-C12:0-PC mixtures (McMullen *et al.*, 1993, 1994).

The IR data for CH_2 wagging progression intensities discussed above are also consistent with this hypothesis, since at a given content of cholesterol more conformational disorder was found in the gel phase of bilayers of long-chain PCs than of bilayers of shorter-chain PCs (Chia *et al.*, 1993; Figure 5). Furthermore, the CH_2 wagging progression of complexes of a given PC with sterols having different side-chain structure diminished sharply as the temperature was increased when the hydrophobic mismatch was large (Figure 4).

Measurements of the influence of cholesterol on the thickness of PC bilayers by x-ray studies show a dependence on phospholipid acyl-chain length; incorporation of cholesterol at temperatures below the main phase transition temperature increases the thickness of di-C12:0-and di-C16:0-PC bilayers, but decreases the thickness of di-C18:0-PC bilayers (McIntosh, 1978).

In summary, the relative stabilization of the gel and liquid–crystalline phases of PC bilayers seems to depend on the extent of the sterol/phospholipid hydrophobic mismatch, and even small changes in PC length result in substantial differences in thermotropic phase behavior. Recent DSC studies show that the hydrophobic mismatch effect is not limited to PC bilayers, since the thermotropic phase properties (T_m and ΔH of sharp and broad components, cooperativity) of phosphatidylethanolamines of different acyl-chain lengths (di-C14:0 to di-C18:0) respond differently to incorporation of cholesterol (McMullen and McElhaney, 1997).

7.2.2. Variation in Sterol Side-Chain Structure

Studies with synthetic sterols also have shown that small changes in the length of the aliphatic chain alter the thermotropic phase behavior of PC bilayers (Chia *et al.*, 1993; McMullen *et al.*, 1995; Vilchèze *et al.*, 1996) and monolayers (Slotte *et al.*, 1994). Sterols with short side chains appear to disrupt the gel-state packing and increase conformational disorder in 1-stearoyl-2-oleoyl-PC bilayers, since these sterols cause the broad-component T_m and the cooperativity of the acyl-chain melting to decrease (Vilchèze *et al.*, 1996). However, sterols with saturated side chains of moderate length, as well as those with longer length than the isooctyl chain, have less effect on the transition temperature and cooperativity of the broad component. In fact, sterols having 9 and 10 carbons in the side chain cause the broad T_m to increase slightly, indicative of increased conformational order in the PC chains. Similarly, with di-C16:0-PC/sterol mixtures, short-chain sterols decrease the T_m of the broad component melting, and long-chain sterols increase the broad-component T_m (McMullen *et al.*, 1995). These results are consistent with the hydrophobic mismatch effect, as noted above in other PC/sterol mixtures. Replacement of the terminally branched alkyl chain of the sterol with a straight chain of the same number of carbon atoms does not affect the ability of the sterol to modify the phospholipid thermotropic properties.

8. RECENT STUDIES OF THE EFFECTS OF STEROL SIDE-CHAIN STRUCTURE ON STEROL–PC INTERACTIONS IN MONOLAYERS

8.1. Effect of Sterol Side-Chain Length on Lateral Domain Formation in Monolayers

The side-chain structure of the sterol has a marked effect on the ability of sterol/phospholipid mixed monolayers to form phase-separated domains (Mattjus *et al.*, 1995). Domains can be visualized by using 22-(*N*-(7-nitrobenz-2-oxa-1,3-diazol-4-yl)amino)-23,24-bisnor-5-cholen-3β-ol (NBD-cholesterol), which partitions preferentially into liquid-expanded (phospholipid-rich) domains and is excluded from sterol-rich domains. Sterols that have shorter side chains than five carbon atoms do not induce domain formation in di-C16:0-PC and *N*-C16:0-sphingomyelin monolayers, presumably because they interact less favorably with these phospholipids than do sterols with side chains of 5–10 carbons.

8.2. Rate of Sterol Oxidation with Cholesterol Oxidase

The rates of oxidation of cholesterol catalyzed by cholesterol oxidase at 22°C are slower in di-C14:0-to di-C17:0-PC monolayers than in either short-chain (di-C10:0-and di-C12:0-PC) or long chain (di-C18:0-and di-C20:0-PC) monolayers (Mattjus *et al.*, 1994). The faster rates of oxidation indicate a greater accessibility of cholesterol to the enzyme (Slotte, 1992). When the hydrophobic length of PC is either much shorter or longer than that of cholesterol, laterally heterogeneous domains may exist in the monolayer in which there are cholesterol-rich and cholesterol-poor domains, which can be detected by using fluorescent lipid probes (Slotte and Mattjus, 1995).

In vesicles of 1-stearoyl-2-oleoyl-PC, a faster rate of sterol oxidation catalyzed by cholesterol oxidase was found with sterols having side chains with four, five, and six carbons than with a C7-side-chain sterol and with cholesterol (Slotte *et al.*, 1994). The difference in oxidation susceptibility suggests that the molecular packing density is looser in vesicles with the shorter chain sterols than with a sterol having a side-chain similar to that of cholesterol.

8.3. Cyclodextrin-Mediated Removal of Cholesterol from Model Membranes

The cyclic heptamer of α-D-glucose, β-cyclodextrin (CD), can bind to a variety of lipids, including fatty acids, fatty acyl-CoAs, glycerides, bile acids, and cholesterol. A molecular inclusion complex or "guest-host" complex is formed, in which part of the lipid (guest) molecule is encapsulated into the relatively apolar interior (~7Å internal diameter) of the β-CD host (Bloch, 1977; Ueno *et al.*, 1990; Claudy *et al.*, 1991; Laurent *et al.*, 1994). CDs are relatively soluble in water since their exterior is hydrophilic; for use of CD as a lipid-lowering agent, the formation of CD–lipid complexes that retain appreciable water solubility is particularly desirable. Indeed, chemically modified CDs with enhanced water solubility, such as derivatives in which one or more hydroxy groups are replaced with *O*-methyl or *O*-2-hydroxypropyl ether groups or with amino groups, are new pharmacological agents with the potential for decreasing plasma cholesterol and altering the lipid composition of membranes (Frijlink *et al.*, 1991; Carpenter *et al.*, 1995; Favier *et al.*, 1995). β-CD appears to act as an efficient receptor of cholesterol and bile acids, and extracts cholesterol from erythrocytes and cells in culture (Kilsdonk *et al.*, 1995; Klein *et al.*, 1995). Therefore, CDs represent a new research tool for examining the role of cholesterol and other lipids in membranes (Gimpl *et al.*, 1995; Rawyler and Siegenthaler, 1996).

When injected into the subphase of pure cholesterol or mixed phospholipid–cholesterol monolayers, β-CD extracts cholesterol from the monolayer and transports it into the aqueous phase (Ohtani *et al.*, 1989). Ohvo and Slotte (1996) have shown that the CD-mediated rate of sterol desorption from phospholipid–sterol monolayers may complement the use of cholesterol oxidase kinetics as a probe of the interaction between cholesterol and phospholipids.

9. CONCLUSIONS

The degree of hydrophobic mismatch between the effective lengths of the sterol and host PC molecules can account for many of the observed effects of sterols on the thermotropic phase behavior of PCs and the miscibility of sterols in PC monolayers and bilayers. Thus, the answer to the question presented in the title of this chapter is affirmative, provided that the acyl chain of the phospholipid is not abnormally short. Di-C12:0-PC would have a mean hydrophobic thickness in close register to the hydrophobic length of androstenol, so that optimal interactions can occur with this sterol and this phospholipid even though no aliphatic side chain is present in the sterol. The degree of hydrophobic mismatch between androstenol and PCs increases when the phospholipid chain length is longer than 12 carbons, resulting in increased immiscibility of the sterol in the host bilayer as the phospholipid acyl chain length increases. A systematic evaluation of the effects of sterols on PCs in model membranes, with variations in both the sterol side-chain length and phospholipid acyl-chain length and degree of unsaturation, has shown a clear trend: a small change in chain length of either lipid results in a significant difference in stabilization of either the gel or liquid–crystalline phase.

This chapter has shown that sophisticated spectroscopic methods, such as ^{2}H- and ^{13}C-NMR, spin label, Fourier-transform IR, and fluorescence studies, together with diffraction, high-sensitivity differential scanning calorimetry, kinetic, computational chemistry, and molecular modeling methods have been applied to analyze the effects of cholesterol on phospholipids in model membranes. As a result, our knowledge about cholesterol–phospholipid, and in particular cholesterol–PC, interactions has reached a rather advanced state. Further work is required on the *functional* role of the hydrophobic mismatch of sterols and constituent phospholipids using organisms that can incorporate sterols with modified side chains and cannot synthesize different membrane fatty acids in response to the absence of the normal cholesterol structure.

10. REFERENCES

Bittman, R., 1988, Sterol exchange between mycoplasma membranes and vesicles, in: *Biology of Cholesterol* (P. L. Yeagle, ed.), pp. 173–195, CRC Press, Boca Raton, FL.

Bittman, R., 1993, A review of the the kinetics of cholesterol movement between donor and acceptor bilayer membranes, in: *Cholesterol in Membrane Models* (L. Finegold, ed.), pp. 45–65, CRC Press, Boca Raton, FL.

Bittman, R., Clejan, S., Jain, M. K., Deroo, P. W., and Rosenthal, A. F., 1981, Effects of sterols on permeability and phase transitions of bilayers from phosphatidylcholines lacking acyl groups, *Biochemistry* **20:**2790–2795.

Bittman, R., Clejan, S., Lund-Katz, S., and Phillips, M. C., 1984, Influence of cholesterol on bilayers of ester- and ether-linked phospholipids. Permeability and ^{13}C-nuclear magnetic resonance measurements, *Biochim. Biophys. Acta* **772:**117–124.

Bloch, K. E., 1977, Control mechanisms for fatty acid synthesis in *Mycobacterium smegmatis*, *Adv. Enzymol.* **45:**1–84.

Bloch, K. E., 1983, Sterol structure and membrane function, *Crit. Rev. Biochem.* **14:**47–92.

Borochov, N., Wachtel, E. J., and Bach, D., 1995, Phase behavior of mixtures of cholesterol and saturated phosphatidylglycerols, *Chem. Phys. Lipids* **76:**85–92.

Carpenter, T. O., Gerloczy, A., and Pitha, J., 1995, Safety of parenteral hydroxypropyl β-cyclodextrin, *J. Pharm. Sci.* **84:**222–225.

Chia, N.-C., Vilchèze, C., Bittman, R., and Mendelsohn, R., 1993, Interactions of cholesterol and synthetic sterols with phospholipids as deduced from infrared CH_2 wagging progression intensities, *J. Am. Chem. Soc.* **115:**12050–12055.

Claudy, P., Létoffé, J. M., Germain, P., Bastide, J. P., Bayol, A., Blasquez, S., Rao, R. C., and Gonzalez, B., 1991, Physicochemical characterization of cholesterol-beta cyclodextrin inclusion complexes, *J. Thermal Anal.* **37:**2497–2506.

Clejan, S., and Bittman, R., 1984, Distribution and movement of sterols with different side-chain structures between the two leaflets of the membrane bilayer of mycoplasma cells, *J. Biol. Chem.* **259:** 449–455.

Clejan, S., and Bittman, R., 1985, Rates of amphotericin B and filipin association with sterols. A study of changes in sterol structure and phospholipid composition of vesicles, *J. Biol. Chem.* **260:** 2884–2889.

Clejan, S., Bittman, R., and Rottem, S., 1981, Effects of sterol structure and exogenous lipids on the transbilayer distribution of sterols in the membrane of *Mycoplasma capricolum*, *Biochemistry* **20:** 2200–2204.

Dahl, C. E., Dahl, J. S., and Bloch, K., 1980a, Effect of alkyl-substituted precursors of cholesterol on artificial and natural membranes and on the viability of *Mycoplasma capricolum*, *Biochemistry* **19:**1462–1467.

Dahl, C. E., Dahl, J. S., and Bloch, K., 1980b, Effects of cycloartenol and lanosterol in model membranes, *Biochim. Biophys. Acta* **733:**111–116.

Dahl, J., 1993, The role of cholesterol in mycoplasma membranes, in: ***Subcellular Biochemistry, Volume 20: Mycoplasma Cell Membranes*** (S. Rottem and I. Kahane, eds.), pp. 167–188, Plenum Press, New York.

Dai, M. C., Chiche, H. B., Düzgünes, N., Ayanoglu, E., and Djerassi, C., 1991, Phospholipid studies of marine organisms. 26. Interactions of some marine sterols with 1-stearoyl-2-oleoyl-phosphatidylcholine (SOPC) in model membranes, *Chem. Phys. Lipids* **59:**245–253.

Davis, J. H., 1993, The molecular dynamics, orientational order, and thermodynamic phase equilibria of cholesterol/phosphatidylcholine mixtures: ^{2}H nuclear magnetic resonance, in: *Cholesterol in Membrane Models* (L. Finegold, ed.), pp. 67–135, CRC Press, Boca Raton, FL.

Duax, W. L., Wawrzak, Z., Griffin, J. F., and Cheer, C., 1988, Sterol conformation and molecular properties, in: *Biology of Cholesterol* (P. L. Yeagle, ed.), pp. 1–18, CRC Press, Boca Raton, FL.

Dufourc, E. J., Parish, E. J., Chitrakorn, S., and Smith, I. C. P., 1984, Structural and dynamical details of cholesterol–lipid interaction as revealed by deuterium NMR, *Biochemistry* **23:**6062–6071.

Erickson, K. A., and Nes, W. R., 1982, Inhibition of hepatic cholesterol synthesis in mice by sterols with shortened and stereochemically varied side chains, *Proc. Natl. Acad. Sci. USA* **79:**4873–4877.

Favier, M.-L., Rémésy, C., Moundras, C., and Demigné, C., 1995, Effect of cyclodextrin on plasma lipids and cholesterol metabolism in the rat, *Metab., Clin. Exp.* **44:**200–206.

Finean, J. B., 1990, Interaction between cholesterol and phospholipid in hydrated bilayers, *Chem. Phys. Lipids* **54:**147–156.

Franks, N. P., 1976, Structural analysis of hydrated egg lecithin and cholesterol bilayers. I. X-ray diffraction, *J. Mol. Biol.* **100:** 345–358.

Frijlink, H. W., Eissens, A. C., Hefting, N. R., Poelstra, K., Lerk, C. F., and Meijer, D. K. F., 1991, The effect of parenterally administered cyclodextrins on cholesterol levels in the rat. *Pharm. Res.* **8:**9–16.

Gericke, A., Brauner, J. W., Erukulla, R. K., Bittman, R., and Mendelsohn, R., 1996, In-situ investigation of partially deuterated fatty acid and phospholipid monolayers at the air/water interface by IR reflection-absorption spectroscopy, *Thin Solid Films*, **284–285:**428–431.

Gimpl, G., Klein, U., Reiländer, H., and Fahrenholz, F., 1995, Expression of the human oxytocin receptor in baculovirus-infected insect cells: high-affinity binding is induced by a cholesterol-cyclodextrin complex, *Biochemistry* **34:** 13794–13801.

Guo, W., and Hamilton, J. A., 1995, A multinuclear solid-state NMR study of phospholipid-cholesterol interactions. Dipalmitoylphosphatidylcholine-cholesterol binary system. *Biochemistry* **34:** 14174–14184.

Higinbotham, J., Beswick, P. H., Malcolmson, R. J., Reed, D., Parkinson, J. A., and Sadler, I. H., 1993, ^{13}C-NMR determination of the molecular dynamics of cholesterol in DMPC vesicles, *Chem. Phys. Lipids* **66:**1–11.

Huang, C.-H., 1977, A structural model for the cholesterol-phosphatidylcholine complexes in bilayer membranes, *Lipids* **12:**348–356.

Huang, T.-H., DeSiervo, A. J., and Yang, Q.-X., 1991, Effect of cholesterol and lanosterol on the structure and dynamics of the cell membrane of *Mycoplasma capricolum*. Deuterium nuclear magnetic resonance study, *Biophys. J.* **59:**691–702.

Kan, C-C., and Bittman, R., 1990, Constraint of the spontaneous intermembrane movement of sitosterol by its 24α-ethyl group, *J. Am. Chem. Soc.* **112:**884–886.

Kan, C-C., and Bittman, R., 1991, Spontaneous rates of sitosterol and cholesterol exchange between phospholipid vesicles and between lysophosholipid dispersions: evidence that desorption rate is impeded by the 24α-ethyl group, *J. Am. Chem. Soc.* **113:**6650–6656.

Kan, C-C., Yan, J., and Bittman, R., 1992, Rates of spontaneous exchange of sterols between lipid vesicles, *Biochemistry* **31:**1866–1874.

Kannenberg, E., Blume, A., McElhaney, R. N., and Poralla, K., 1983, Sterols in membranes: growth characteristics and membrane properties of *Mycoplasma capricolum* cultured on cholesterol and lanosterol, *Biochemistry* **19:**1467–1472.

Keough, K. M. W., Morrow, M. R., and Bittman, R., 1996, Phospholipid–cholesterol bilayers, in: *Encyclopedia of Nuclear Magnetic Resonance* (D. M. Grant and R. K. Harris, eds.), pp. 3601–3608, Wiley, Chichester, U. K.

Kilsdonk, E. C., Yancey, G., Stoudt, G. W., Bangerter, F. W., Johnson, W. J., Phillips, M. C., and Rothblat, G. H., 1995, Cellular cholesterol efflux mediated by cyclodextrins, *J. Biol. Chem.* **270:**17250–17256.

Klein, U., Gimpl, G., and Fahrenholz, F., 1995, Alteration of the myometrial plasma membrane cholesterol content with β-cyclodextrin modulates the binding affinity of the oxytocin receptor, *Biochemistry* **34:**13784–13793.

Krajewski-Bertrand, M.-A., Milon, A., Nakatani, Y., and Ourisson, G., 1992, The interaction of various cholesterol "ancestors" with lipid membranes: a ^{2}H-NMR study on oriented bilayers, *Biochim. Biophys. Acta* **1105:**213–220.

Laurent, S., Ivanova, M. G., Pioch, D., Graille, J., and Verger, R., 1994, Interactions between β-cyclodextrin and insoluble glyceride monomolecular films at the argon/water interface: application to lipase kinetics, *Chem. Phys. Lipids* **70:**35–42.

Le Guernevé, C., and Auger, M., 1995, New approach to study fast and slow motions in lipid bilayers: Application to dimyristoylphosphatidylcholine–cholesterol interactions, *Biophys. J.* **68:** 1952–1959.

Lundberg, B., Svens, E., and Ekman, S., 1978, The hydration of phospholipids and phospholipid–cholesterol complexes, *Chem. Phys. Lipids* **22:**285–292.

Mattjus, P., Bittman, R., Vilchèze, C., and Slotte, J. P., 1995, Lateral domain formation in cholesterol/phopsholipid monolayers as affected by the sterol side chain conformation, *Biochim. Biophys. Acta* **1240:**237–247.

Mattjus, P., Hedström, G., and Slotte, J. P., 1994, Monolayer interaction of cholesterol with phospholipids: effects of acyl chain length, *Chem. Phys. Lipids* **74:**195–203.

McIntosh, T. J., 1978, The effect of cholesterol on the structure of phosphatidylcholine bilayers, *Biochim. Biophys. Acta* **513:**43–58.

McMullen, T. P. W., and McElhaney, R. N., 1995, New aspects of the interaction of cholesterol with dipalmitoylphosphatidylcholine bilayers as revealed by high-sensitivity differential scanning calorimetry, *Biochim. Biophys. Acta* **1234:**90–98.

McMullen, T. P. W., and McElhaney, R. N., 1996, Physical studies of cholesterol/phospholipid interactions, *Curr. Opin. Colloid Interf. Sci.* **1:**83–90.

McMullen, T. P. W., and McElhaney, R. N., 1997, Differential scanning calorimetric studies of the interaction of cholesterol with phosphatidylethanolamines, *Biophys. J.*, submitted.

McMullen, T. P. W., Lewis, R. N. A. H., and McElhaney, R. N., 1993, Differential scanning calorimetry study of the effect of cholesterol on the thermotropic phase behavior of a homologous series of linear saturated phosphatidylcholines, *Biochemistry* **32:**516–522.

McMullen, T. P. W., Lewis, R. N. A. H., and McElhaney, R. N., 1994, Comparative differential scanning calorimetric and FTIR and ^{31}P-NMR spectroscopic studies of the effects of cholesterol and androstenol on the thermotropic phase behavior and organization of phospholipid bilayers, *Biophys. J.* **66:**741–752.

McMullen, T. P. W., Vilchèze, C., McElhaney, R. N., and Bittman, R., 1995, Differential scanning calorimetric study of the effect of sterol side chain length on dipalmitoylphosphatidylcholine thermotropic phase behavior, *Biophys. J.* **69:**169–176.

Mendelsohn, R., Davies, M. A., Brauner, J. W., Schuster, H. F., and Dluhy, R. A., 1989, Quantitative determination of conformational disorder in the acyl chains of phospholipid bilayers by infrared spectroscopy, *Biochemistry* **28:**8934–8939.

Mendelsohn, R., Davies, M. A., Schuster, H. F., Xu, Z., and Bittman, R., 1991, CD_2 rocking modes as quantitative infrared probes of one-, two-, and three-bond conformational disorder in dipalmitoylphosphatidylcholine and dipalmitoylphosphatidylcholine/cholesterol mixtures, *Biochemistry* **30:**8558–8563.

Mortimer, B.-C., Tso, P., Phan, C. T., Beveridge, D. J., Wen, J., and Redgrave, T. G., 1995, Features of cholesterol struucture that regulate the clearance of chylomicron-like lipid emulsions, *J. Lipid Res.* **36:**2038–2053.

Murari, R., Murari, M. P., and Baumann, W. J., 1986, Sterol orientations in phosphatidylcholine liposomes as determined by deuterium NMR, *Biochemistry* **25:**1062–1067.

Murata, M., Peränen, J., Schreiner, R., Wieland, F., Kurzchalia, T. V., and Simons, K., 1995, VIP21/caveolin is a cholesterol-binding protein, *Proc. Natl. Acad. Sci. USA* **92:**10339–10343.

Ohtani, Y., Irie, T., Uekama, K., Fukunaga, K., and Pitha, J., 1989, Differential effects of the α-, β-, and γ-cyclodextrins on human erythrocytes, *Eur. J. Biochem.* **186:**17–22.

Ohvo, H., and Slotte, J. P., 1996, Cyclodextrin-mediated removal of sterols from monolayers: effects of sterol structure and phospholipids on desorption rate, *Biochemistry* **35:**8018–8024.

Oldfield, E., Meadows, M., Rice, D., and Jacobs, R., 1978, Spectroscopic studies of specifically deuterium labeled membrane systems. Nuclear magnetic resonance investigation of the effects of cholesterol in model systems, *Biochemistry* **17:**2727–2740.

Ourisson, G., and Rohmer, M., 1992, Hopanoids. 2. Biohopanoids a novel class of bacterial lipids, *Acc. Chem. Res.* **25:**403–408.

Presti, F. T., 1985, The role of cholesterol in regulating membrane fluidity, in: *Membrane Fluidity in Biology, Vol. 4, Cellular Aspects* (R. C. Aloia and J. M. Boggs, eds.), pp. 97–146, Academic Press, New York.

Presti, F. T., Pace, R. J., and Chan, S. I., 1982, Cholesterol–phospholipid interaction in membranes. 2. Stoichiometry and molecular packing of cholesterol-rich domains, *Biochemistry* **21:**3831–3835.

Rawyler, A., and Siegenthaler, P. A., 1996, Cyclodextrins: a new tool for the controlled lipid depletion of thylakoid membranes, *Biochim. Biophys. Acta* **1278:**89–97.

Redgrave, T. G., Martins, I., Mortimer, B.-C., Vilchèze, C., and Bittman, R., 1997, in preparation.

Reinl, H., Brumm, T., and Bayerl, T. M., 1992, Changes of the physical properties of the liquid-ordered phase with temperature in binary mixtures of DPPC with cholesterol. A ^{2}H-NMR, FT–IR, DSC, and neutron scattering study, *Biophys. J.* **61:**1025–1035.

Robinson, A. J., Richards, W. G., Thomas, P. J., and Hann, M. M., 1995, Behavior of cholesterol and its effect on head group and chain conformations in lipid bilayers: a molecular dynamics study, *Biophys. J.* **68:**164–170.

Rujanavech, C., and Silbert, D. F., 1986, LM cell growth and membrane lipid adaptation to sterol structure, *J. Biol. Chem.* **261:**7196–7203.

Sankaram, M. B., and Thompson, T. E., 1990, Modulation of phospholipid acyl chain order by cholesterol. A solid-state ^{2}H nuclear magnetic resonance study, *Biochemistry* **29:**10676–10684.

Sankaram, M. B., and Thompson, T. E., 1991, Cholesterol-induced fluid-phase immiscibility in membranes, *Proc. Natl. Acad. Sci. USA* **88:**8686–8690.

Sato, Y., Nishikawa, K., Aikawa, K., Mimura, K., Murakami-Murofushi, K., Arai, H., and Inoue, K., 1995, Side-chain structure is critical for the transport of sterols from lysosomes to cytoplasm, *Biochim. Biophys. Acta* **1257:**38–46.

Senak, L., Moore, D., and Mendelsohn, R., 1992, CH_2 wagging progressions as IR probes of slightly disordered phospholipid acyl chain states, *J. Phys. Chem.* **96:**2749–2754.

Shouffani, A., and Kanner, B. I., 1990, Cholesterol is required for the reconstitution of the sodium-and chloride-coupled, γ-aminobutyric acid transporter from rat brain, *J. Biol. Chem.* **265:**6002–6008.

Slotte, J. P., and Mattjus, P., 1995, Visualization of lateral phases in cholesterol and phosphatidylcholine monolayers at the air/water interface—a comparative study with different reporter molecules, *Biochim. Biophys. Acta* **1254:**22–29.

Slotte, J. P., Junger, M., Vilchèze, C., and Bittman, R., 1994, Effect of sterol side-chain structure on sterol-phosphatidylcholine interactions in monolayers and small unilamellar vesicles, *Biochim. Biophys. Acta* **1190:**435–443.

Tavani, D. M., Nes, W. R., and Billheimer, J. T., 1982, The sterol substrate specificity of acyl CoA cholesterol acyltransferase from rat liver, *J. Lipid Res.* **23**:774–781.

Ueno, A., Minato, S., Suzuki, I., Fukushima, M., Ohkubo, M., Osa, T., Hamada, F., and Murai, K., 1990, Host-guest sensory system of dansyl-modified β-cyclodextrin for detecting steroidal compounds by dansyl fluorescence, *Chem. Lett.* 605–608.

Vanderkooi, G., 1994, Computation of mixed phosphatidylcholine–cholesterol bilayer structures by energy minimization, *Biophys. J.* **66**: 1457–1468.

Vermuri, R., and Philipson, K. D., 1989, Influence of sterols and phospholipids on sarcolemmal and sarcoplasmic reticular cation transporters, *J. Biol. Chem.* **264**:8680–8685.

Vilchèze, C., McMullen, T. P. W., McElhaney, R. N., and Bittman, R., 1996, The effect of side-chain analogues of cholesterol on the thermotropic phase behavior of 1-stearoyl-2-oleoyl-phosphatidylcholine bilayers: a differential scanning calorimetry study, *Biochim. Biophys. Acta* **1279**:235–242.

Villalaín, J., 1996, Location of cholesterol in model membranes by magic-angle-sample spinning NMR, *Eur. J. Biochem.* **241**:585–593.

Vist, M. R., and Davis, J. H., 1990, Phase equilibria of cholesterol/DPPC mixtures: ^{2}H nuclear magnetic resonance and differential scanning calorimetry, *Biochemistry* **29**:451–464.

Worcester, D. L., and Franks, N., 1976, Structural analysis of hydrated egg lecithin and cholesterol bilayers. II. Neutron diffraction, *J. Mol. Biol.* **100**:359–378.

Wu, W.-G., and Chi, L.-M., 1991, Conformational change of cholesterol side chain in lipid bilayers, *J. Am. Chem. Soc.* **113**:4683–4685.

Yeagle, P. L., 1988, Cholesterol and the cell membrane, in: *Biology of Cholesterol* (P. L. Yeagle, ed.), pp. 121–145, CRC Press, Boca Raton, FL.

Yeagle, P. L., Martin, R. B., Lala, A. K., Lin, H.-K., and Bloch, K., 1977, Differential effects of cholesterol and lanolesterol on artificial membranes, *Proc. Natl. Acad. Sci. USA* **74**:4924–4926.

Chapter 7

Cholesterol and Myelin

G. Jackson Snipes and Ueli Suter

1. INTRODUCTION

Although it is well known that the central nervous system (CNS) and the peripheral nervous system (PNS), and in particular their myelin, contain large quantities of cholesterol, the specific role of this lipid in the structure and function of the nervous system remains elusive. In this review, we will summarize progress in our understanding of the role of cholesterol in the CNS and PNS with a particular emphasis on myelin. In the first part, we will provide an overview of the role of cholesterol in the biochemical composition and the structure of myelin, followed by a summary of the current knowledge of the biosynthesis and metabolism of cholesterol within neural tissues. Finally, we will review the association of cholesterol with various pathologies affecting the nervous system, including Alzheimer's disease, tellurium toxicity, and the Smith–Lemli–Opitz syndrome.

2. MYELIN BIOCHEMISTRY

2.1. Biochemical Composition of the Brain

The CNS is comprised of gray and white matter that can be readily differentiated by visual inspection of postmortem specimens and by magnetic

G. Jackson Snipes Departments of Neuropathology, Neurology, and Neurosurgery, Montreal Neurological Institute, McGill University, Montreal, Quebec, H3A 2B4, Canada **Ueli Suter** Department of Cell Biology, Swiss Federal Institute of Technology, ETH-Hönggerberg, 8093 Zürich, Switzerland.

Subcellular Biochemistry, Volume 28: Cholesterol: Its Functions and Metabolism in Biology and Medicine, edited by Robert Bittman. Plenum Press, New York, 1997.

resonance imaging (MRI) techniques *in vivo*. White matter consists mainly of myelin, which surrounds individual axons. In addition to myelinated axons, white matter contains glial cells (oligodendrocytes, astrocytes, and microglia) and blood vessels. Gray matter is characterized by the presence of neuronal cell bodies and their dendritic extensions, relatively little myelin, as well as glial cells and blood vessels. The different appearances of these two brain regions reflect their unique biochemical compositions. The white matter, so named for its visual appearance on gross inspection, belies the greater amount of lipids in myelin as compared to the cerebral cortex and underlying nuclei. Myelin cholesterol has been suggested to be the major determinant of the characteristic MRI appearance of white matter (Koenig, 1991), again reflecting the unique lipid composition of myelin. Biochemically, the normal adult human brain is composed of approximately 80% water, 10–15% lipids, and a minor but functionally significant protein fraction (for review on the biochemical composition of myelin, see Morell *et al.,* 1994 and Table I). The chemical and physical properties of white matter are mainly determined by myelin that, as its largest nonwater constituent, comprises approximately half of its dry weight. As discussed below, the biochemical composition appears to reflect the highly specialized function of myelin as an electrical insulator. By dry weight, purified myelin consists of 70% lipid and 30% protein (myelin has a water content of approximately 40% by weight). By comparison, most other biological membranes contain proportionally less lipid.

Table I
Composition of Human Brain and CNS Myelin[a]

Substance[b]	Gray matter	White matter	Myelin
Protein	55.3	39.0	30.0
Lipid	32.7	54.9	70.0
Cholesterol	22.0	27.5	27.7
Cerebroside	5.4	19.8	22.7
Sulfatide	1.7	5.4	3.8
Total galactolipid	7.3	26.4	27.5
Ethanolamine phosphatides	22.7	14.9	15.6
Phosphatidylcholine	26.7	12.8	11.2
Sphingomyelin	6.9	7.7	7.9
Phosphatidylserine	8.7	7.9	4.8
Phosphatidylinositol	2.7	0.9	0.6
Plasmalogens	8.8	11.2	12.3
Total Phospholipids	69.5	45.9	43.1

[a]Modified from Morell *et al.*, 1994.
[b]Protein and lipid expressed as percentage of dry weight; all others expressed as percentage of total lipids.

2.1.1. Myelin Lipids

The lipid composition of myelin has been studied intensively. Interestingly, no myelin-specific lipids have been described, although myelin is enriched in cholesterol and cerebrosides (see Table I). A variety of minor lipids can also be found, including sulfatides, ethanolamine phosphatides, phosphatidylcholine, sphingomyelin, phosphatidylserine, and phosphatidylinositol. PNS myelin contains roughly the same lipids as the CNS with some quantitative differences. Specifically, PNS myelin contains decreased levels of cerebroside and sulfatide but an increased proportion of sphingomyelin (Norton and Cammer, 1984). A comparative phylogenetic analysis of myelin similarly reveals little difference in the composition of myelin lipids, except for an approximately twofold increase of cerebrosides in rodents as compared to cartilaginous fishes (Burgisser *et al.*, 1986).

2.1.2. Myelin Proteins

Despite the fact that proteins are minor myelin components, studies of spontaneous mouse mutants and various human diseases affecting myelination have demonstrated that most of the known myelin proteins are required for the correct development and maintenance of the nervous system. Interestingly, the protein compositions of CNS and PNS myelin are distinct, a finding that is likely to be of major functional significance with respect to the unique ability of PNS nerves to regenerate (in sharp contrast to the CNS). While some proteins are restricted either to the CNS (e.g., proteolipid protein, PLP) or the PNS (e.g., protein zero, P0), other proteins are found in both tissues (e.g., myelin basic protein; MBP).

In the mammalian CNS, PLP is the major myelin protein component (approximately 50% of total protein; reviewed by Lemke, 1993). PLP has an approximate molecular weight of 30 kDa and biochemical evidence suggests that it is an integral membrane protein containing four transmembrane domains (Popot *et al.*, 1991). PLP contains cysteine-bound long chain fatty acids that are responsible for the demonstrated solubility of PLP in organic, but not aqueous, solvents (Weimbs and Stoffel, 1992). Naturally occurring PLP mutations resulting in severe CNS myelin deficiencies have been identified in various species, including humans (Pelizaeus–Merzbacher disease), dog, rabbit, rat, and mice (Nave, 1994). The combined genetic evidence from natural, as well as artificially generated, mouse mutants indicates that PLP is crucially involved in the formation and stabilization of the myelin structure (Boison and Stoffel, 1994). However, the exact biological function of PLP in the normal organism has remained largely elusive. It is anticipated that more insights into role of PLP will be gained from the analysis of its variant DM-20, which originates from the PLP gene by an alternative splicing mechanism (Macklin *et al.*, 1987; Nave *et al.*, 1987). Like PLP, DM-20 is also found in minor quantities in CNS myelin and some studies suggest that PLP/DM-20 might have an important role in early development of the oligodendrocyte lineage (Schneider *et al.,* 1992). Recently, several novel member of the PLP/DM-20 gene family have been identified (Yan *et al.,* 1993). It is likely that these novel PLP/DM-20 related proteins will be valuable tools for the elucidation of the function of this group of proteins.

The major glycoprotein of compact PNS myelin, P0 is a single transmembrane glycoprotein with a molecular weight of 28 kDa (for review, see Snipes and Suter, 1995a). P0 expression is restricted to Schwann cells and accounts for approximately 50% of total PNS myelin protein. The extracellular domain of P0 contains an immunoglobulin (Ig)-like domain (Lemke and Axel, 1985), and convincing evidence from *in vitro* (D'Urso *et al.,* 1990; Filbin *et al.,* 1990) and *in vivo* studies (Giese *et al.,* 1992) has shown that P0 is involved in the maintenance of myelin structure. In particular, the adhesive extracellular domain of P0 is responsible for maintaining the intraperiod line, and the highly basic cytoplasmic domain appears to be important for the integrity of the major dense line. In support of this hypothesis, transgenic mice that have been genetically engineered to lack P0 expression were shown to have severe PNS myelin deficiencies (Giese *et al.,* 1992). Similarly, mutations affecting the P0 gene have been found in patients suffering from the rare hereditary demyelinating disorder Charcot–Marie–Tooth disease type 1B (reviewed in Snipes and Suter, 1995b).

MBPs are cytoplasmic proteins found in both the CNS (approximately 30% of total protein) and in the PNS (5–15% of total protein) (Lees and Brostoff, 1984). This group of proteins has been studied extensively because injection into animals can elicit the CNS autoimmune disease experimental allergic encephalomyelitis (EAE), which is characterized by focal inflammation and demyelination (Raine, 1985). Though controversial, MBP-induced EAE is widely used as an animal model for the study of multiple sclerosis (Macklin *et al.,* 1987). The individual MBPs vary in molecular weight from 12–22 kDa, with major qualitative and quantitative differences between species (Greenfield *et al.,* 1982; Barbarese *et al.,* 1977). The heterogeneity of MBPs is generated via differential splicing from a single gene and by various postranslational modifications (reviewed in Kamholz and Wrabetz, 1992). Evidence from the naturally occurring mouse mutant *shiverer* (partial deletion in the MBP gene) suggests that MBPs are crucially involved in maintaining the stability of CNS myelin (Inoue *et al.,* 1981) and myelin thickness (Privat *et al.* 1979). A similar but redundant function of MBPs in the PNS has been proposed based on the phenotype of animals with both reduced P0 and MBP levels (Martini *et al.* 1995).

Minor CNS and PNS myelin protein components have also been identified. Myelin-associated protein (MAG) is a transmembrane glycoprotein found throughout the nervous system, primarily at the margin of the mature myelin sheath immediately adjacent to the axon (Sternberger *et al.,* 1979; reviewed by Martini, 1994). MAG belongs to the Ig-protein superfamily and has been suggested to play an important role in axon–glia interactions, particularly during the initiation of myelin formation (Owens *et al.*, 1990). Furthermore, *in vitro* studies suggest that MAG may act as an inhibitor of axonal outgrowth (McKerracher *et al.*, 1994; Mukhopadhyay *et al.*, 1994). Additional proteins found in both CNS and PNS myelin include the 15 kDa cytosolic lipid-binding P2 protein, whose expression is restricted to myelinating cells (Martenson and Uyemura, 1992; Trapp *et al.*, 1984), and the 18-kDa protein plasmolipin, a four-transmembrane domain proteolipid protein that is also found in the kidney as a putative cation-specific channel (Fischer *et al.* 1994). Examples of other proteins enriched in CNS myelin are myelin-oligodendrocyte protein (MOG), another member of the Ig superfamily with two putative transmembrane domains and a suggested function in cellular recognition processes (Gardinier *et al.*, 1992), and the enzyme 2′,3′-cyclic nucleotide 3′-phosphodiesterase (CNPase) whose function in myelin remains to determined (Sprinkle, 1989).

Prominent proteins found in PNS but not in CNS myelin include, besides P0, the peripheral myelin protein 22 (PMP22) and periaxin (Gillespie *et al.*, 1994). PMP22 is a putative four-transmembrane domain protein in compact myelin that recently received much attention since mutations affecting the PMP22 gene are associated with peripheral nerve myelin deficits in *Trembler* and *Trembler-J* mice and with the most common forms of human hereditary motor and sensory neuropathies, Charcot–Marie–Tooth disease type 1A, hereditary neuropathy with liability to pressure palsies, and Dejerine–Sottas Syndrome (reviewed in Snipes and Suter, 1995b; Suter and Snipes, 1995). Functionally, PMP22 has been suggested to be required for the correct development of peripheral nerves, the maintenance of axons, and the determination of myelin thickness and stability. Besides its high expression in PNS myelin, PMP22 is also expressed in the adult gastrointestinal tract, lung, and heart (Spreyer *et al.*, 1991; Welcher *et al.*, 1991) as well as in various tissues throughout mouse development (Baecher *et al.*, 1995). Furthermore, PMP22 has been directly implicated in the regulation of Schwann cell growth in *in vitro* experiments (Zoidl *et al.*, 1995). Periaxin has been characterized as a large repetitive PNS-specific protein that partially colocalizes with MAG (Gillespie *et al.*, 1994). Based on this observation, periaxin has been suggested to play a role in the ensheathment of axons by Schwann cells. Finally, it should be noted that the list of myelin-associated proteins has increased considerably over the last few years and biochemical evidence suggests that more myelin proteins are yet to be discovered.

3. CHOLESTEROL AND MYELIN

In considering the role of cholesterol in myelin, it is necessary to consider the general role of myelin in the nervous system. And, as structure determines function, a consideration of cholesterol in myelin structure is also warranted.

Functionally, myelin acts as a discontinuous electrical insulator along the length of large caliber axons. This insulation allows for the saltatory conduction of electrical impulses necessary for the rapid propagation of action potentials along nerve fibers. Structurally, this requires the repeated wrapping of the oligodendrocyte (CNS) or Schwann cell (PNS) plasma membrane around the axon, presumably for the purpose of increasing the electrical resistance across multiple membranes. As discussed below, the composition of myelin lipids is well suited for minimal ion permeability. The wrapping process brings both the extracellular leaflets of the plasma membrane together, and the intracellular leaflets into apposition, which are observed in cross-section as alternating spirals termed the intraperiod line and the major dense line, respectively (Peters *et al.*, 1991). Part of the process of myelin formation involves the elimination of most of the cytoplasm between the apposing myelin layers at the major dense line, a process termed compaction. Two specialized regions of myelin remain uncompacted, the Schmidt–Lantermann clefts and the paranodal myelin loops (see Figure 1). These regions provide a conduit for active transport and diffusion of cytoplasmic components through the entire thickness of the myelin sheath. These specialized re-

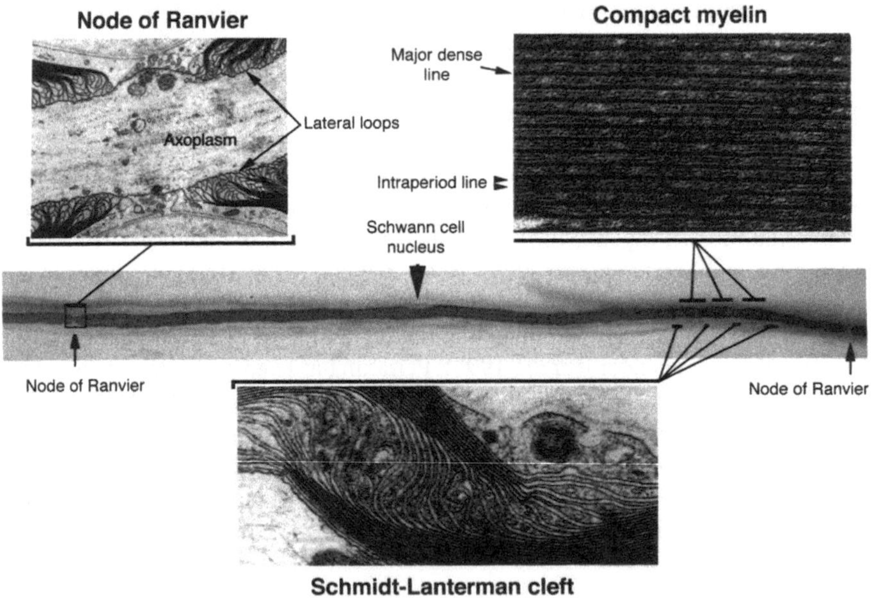

FIGURE 1. Composite showing the organization of the compacted and uncompacted (lateral loops of myelin at the node of Ranvier and the Schmidt–Lanterman clefts) portions of myelin in a single internodal segment produced by a single Schwann cell. The center figure is an isolated myelinated peripheral nerve axon (stained with osmium tetroxide); the others are electron micrographs. The length of the myelin internodal segment shown is approximately 0.7 millimeters. (Adapted from Morell *et al.*, 1994 and Thomas and Ochoa, 1984. We thank Drs. R.H.M. King, P.K. Thomas, and C.S. Raine for granting permission to use their electron micrographs.)

gions of myelin are believed to be important during myelin assembly and maintenance, and probably during myelin disassembly. Their role in mature myelin is uncertain, but since they provide a direct connection between the myelin-forming cell and the axon, they may be involved in axon–glial interactions. Given that myelin has various subdomains and an ongoing need for remodeling, the organization of compact myelin is deceptively simple. As we will see, the function of myelin as an electrical insulator is probably served by its constituent lipids, including cholesterol. The structure and assembly of myelin probably depends on the correct interaction of these lipids with the structural proteins of myelin.

3.1. Cholesterol May Regulate Ion Permeability in Myelin

The lipid composition of myelin membranes appears to reflect the unique function of this structure as an electrical insulator. Using liposomes prepared from myelin lipids, Diaz and Monreal (1994) have demonstrated that myelin lipids, free of protein, exhibit extremely poor permeability to H^+ ions. These authors note that membranes containing cholesterol (Zhou *et al.*, 1991), and particularly cholesterol in association with sphingomyelin, demonstrate poor permeability to ions, perhaps by reducing the water accessibility to the hydrophobic core (Perochon *et al.*, 1992) or by decreasing packing defects of the lipid bilayer. Lipids enriched in myelin, such as the sulfatides, reduce ion permeability of lipid vesicles (Chen *et al.*, 1992), while other membrane components, such as free fatty acids that are associated with increased cation/proton permeability (Macri *et al.*, 1991; Cooper *et al.*, 1990), are reduced in myelin.

3.2. Myelin Cholesterol May Affect the Function of Myelin Proteins

As described above, a number of mutations resulting in dysmyelination have been identified in the structural genes of myelin, specifically PLP, MBP, PMP22, and P0 protein. Only P0 has a known function that readily explains the phenotype associated with its mutations. Of the remaining proteins, PLP and PMP22 have a number of similarities in that both are relatively small molecules that have four putative transmembrane domains. These proteins are quite different in that they have no significant sequence homology, and while PLP comprises up to half of the protein content of CNS myelin, PMP22 comprises only about 2–5% of PNS myelin protein. Though the functions of these proteins are unknown, they have been hypothesized to be involved in intracellular adhesion or in the formation of channels or pores through the myelin membranes. Since many of the disease-causing mutations affect transmembrane domains, it appears possible that abnormal associations in the plane of the myelin membranes are responsible for the abnormal phenotypes. Given the relative abundance of cholesterol and lipids in myelin membranes, it is interesting to consider the possibility that some of the critical associations that are disrupted are not exclusively protein–protein, but also protein-lipid interactions. Important protein–lipid interactions have been hypothesized for the MBPs and the P2 protein of myelin. The MBPs, which are soluble highly positively charged proteins, have been suggested to neutralize the repulsive forces of the phospholipid head groups at the major dense line. The P2 protein has significant homology with the retinol-binding proteins, which also bind lipids.

The lipid and cholesterol composition of membranes are known to affect the activity of integral membrane proteins. For example, Shouffani and Kanner (1990) have shown that the γ-aminobutyric acid (GABA) transporter from rat brain has a specific requirement for cholesterol. Other channel and transport proteins, such as the Na^{+}–Ca^{2+} exchanger and the Na^{+}, K^{+}-ATPase, require cholesterol for activity in reconstitution experiments (Vemuri and Philipson, 1989). The mechanism by which cholesterol exerts its effect on transmembrane proteins is uncertain, but there are a number of possible explanations. Cholesterol could directly bind to, and activate, proteins. Alternatively, cholesterol could modulate the physical properties of membranes, such as fluidity (North and Fleischer, 1983). At the extreme, cholesterol content could cause phase separations in the plane of the membrane or alter the phase behavior of other membrane lipids, thereby facilitating or stabilizing protein–protein interactions between membrane proteins (Shouffani and Kanner, 1990). By analogy, the high levels of cholesterol in myelin could serve to regulate either the biological activities or the intermolecular associations of myelin proteins. This is a particularly attractive hypothesis for PLP and PMP22, which are largely embedded in membranes. In addition, many mutations in the putative transmembrane domains of these proteins cause myelin deficiencies. Finally, dysmyelination may also be caused by underexpression or overexpression of the wild type genes for these proteins, suggesting that the correct stoichiometry of these proteins in an oligomeric complex is crucial for normal myelin function. Interestingly, the PLP protein appears to increase the permeability of lipid membranes to protons (Lin and Lees, 1982). Whether this represents a function of the PLP protein as a channel or pore protein, or, alternatively, its ability to nonspecifically disrupt artificial membranes, is not clear.

3.3. Cholesterol Is an Important Component of CNS Myelin

As outlined previously, cholesterol is a major component of the brain (reviewed in Davison, 1965), comprising approximately 10% of the dry weight of the whole brain. For comparison, although cholesterol is a component of many cell membranes, cholesterol represents only about 1% of the dry weight of most other organs. Within the brain, cholesterol is found predominantly in white matter (ratio of white matter/gray matter cholesterol is approximately 4:1) (Davison and Wajda, 1962). Indeed, it has been estimated that 70% of the cholesterol in the whole brain is incorporated into myelin (Laatsch *et al.*, 1962). Thus, the brain contains a disproportionate amount of cholesterol relative to other organs and the majority of cholesterol that is present in the brain is incorporated into myelin.

4. CHOLESTEROL BIOSYNTHESIS, METABOLISM, AND REGULATION IN THE NERVOUS SYSTEM

4.1. Cholesterol Is Synthesized Locally in the Brain

There are two major mechanisms that can serve to provide the brain with sterols; endogenous synthesis or delivery of sterols via circulating lipoproteins. The liver is well known as a major site of synthesis for sterols, though many peripheral organs are independently capable of cholesterol synthesis (Spady and Dietschy, 1983). Accordingly, cholesterol is synthesized at a high rate in the liver. The high cholesterol content of the adult brain, at least in rodents, is not associated with a correspondingly high rate of cholesterol synthesis, which is approximately 20–80-fold less than that in the liver (Spady and Dietschy, 1983). Nevertheless, the ability of the brain to synthesize cholesterol has been amply demonstrated *in vivo* (Dietschy and Spady, 1984), in tissue slices (Andersson *et al.,* 1990), and *in vitro* (Ramsey, 1973). There has been considerable controversy in the literature regarding the contribution of lipoprotein-derived cholesterol to the total pool of brain cholesterol. Early studies established that the rate of cholesterol synthesis in the adult brain, unlike other organs, including adrenal, kidney, adipose tissue, and gut, was independent of circulating lipoprotein concentration (Anderson and Dietschy, 1977), indicating the brain's independence of systemic feedback mechanisms for regulating cholesterol synthesis observed in many other organs. Part of this feedback insensitivity by the brain could have been due either to inaccessibility of the adult brain to circulating lipoproteins or a more general downregulation of brain cholesterol synthesis in older animals. Nonetheless, most, but not all (Haque and Mozaffor, 1992), recent studies have favored the view that the majority of brain cholesterol is synthesized locally. The strength of this conclusion derives in part from studies of Edmond *et al.* (1991), who fed rat pups deuterated cholesterol, which was significantly incorporated into liver and lung, but not brain, tissue.

This conclusion is also strongly supported by the recent work of Jurevics and Morell (1995) who measured the rate of incorporation of tritiated water into newly synthesized cholesterol, coupled with the rate of accumulation of total sterols in the brain, and demonstrated that local synthesis is sufficient to account for approximately 95% of total brain sterol (desmosterol and cholesterol). In agreement with earlier studies that seemed to contradict the local synthesis hypothesis by demonstrating cholesterol influx into the brains of young animals (Serougne *et al.,* 1976; Serourgne and Chevellier, 1974; Dobbing, 1963), Jurevics and Morell (1995) concluded that approximately 3–5% of cholesterol was probably derived from systemic sources during the first 8 postnatal days in the rat.

It is important to clarify several unique aspects of the nervous system to fully appreciate some of the controversy regarding the biosynthesis of brain cholesterol. First, there is a mechanical barrier to diffusion of large molecules at the blood–brain barrier that may not be fully developed at birth, but certainly is present in adults. Tight junctional attachments between adjacent capillary endothelial cells provide the major structural basis for the diffusion barrier to entry of large molecules into the brain (Reese and Karnovsky, 1967). Thus, the blood–brain barrier could serve to isolate the brain from circulating lipoproteins, necessitating local synthesis. However, it appears that brain endothelial cells do express luminal LDL receptors that can be upregulated by astrocytes, possibly via a soluble factor (Dehouck *et al.,* 1994). Secondly, there is probably very little turnover of myelin proteins or cholesterol in the CNS of uninjured adult animals (Davison, 1965). Thus, compared to developing brains that need to synthesize significant amounts of sterols during the process of myelination, mature brains appear to have no extraordinary requirement for ongoing cholesterol synthesis. The critical period of myelination varies significantly among different species. In rodents, the period of myelination in CNS occurs postnatally, predominantly during the first 4 weeks of life (Wells and Dittmer, 1967). In humans, myelination begins during the second trimester and extends through the second year of life (Kinney *et al.,* 1988). At least in rodents, there is a dramatic increase in cholesterol biosynthesis that correlates well with the myelination process (Kinney *et al.,* 1994; Wells and Dittmer, 1967). Interestingly, myelination in the CNS is immediately preceded by an increase in cholesterol esters (Kinney *et al.,* 1994), desmosterol (24-dehydrocholesterol), and 7-dehydrodesmosterol (Bourre *et al.,* 1990). Although these intermediates and metabolites can be detected at low levels in the adult brain, cholesterol is overwhelmingly the major sterol in the adult CNS (Macala *et al.,* 1983; Wells and Dittmer, 1967). The high levels of cholesterol esters and desmosterol preceding myelination has not been fully explained, but may represent accumulation of cholesterol and its precursors following the initiation of myelination by axon–oligodendrocyte interactions prior to coordinated myelin synthesis (Hinse and Shah, 1971). It has been estimated that an actively myelinating oligodendrocyte synthesizes three times its weight as myelin membranes each day (Norton, 1977). Comparisons between astrocytes and oligodendrocytes *in vitro* indicate that both can synthesize lipids and cholesterol from acetoacetate and glucose, and that oligodendrocytes produce more cholesterol than astrocytes (Koper *et al.,* 1984). This finding is entirely consistent with the unique role of the oligodendrocyte in myelination.

4.2. The Regulation of Cholesterol Synthesis in the CNS Is Not Unique to the Brain

While the regulation of cholesterol biosynthesis has been explored extensively in many cell types, relatively little is known about the specifics of cholesterol biosynthesis in myelin-forming glia. The rat C6 glioma cell line, with features of both astrocytes and oligodendrocytes, has been used to study the regulation of sterol synthesis. In this cell, as in non-neural cells, 3-hydroxy-3-methylglutaryl coenzyme A reductase (HMG-CoA reductase, EC 1.1.1.34) appears to be the rate-limiting enzyme for cholesterol biosynthesis (Langan *et al.*, 1987). In addition, HMG-CoA synthase (EC 2.3.1.9) appears to be the rate-limiting enzyme when the HMG-CoA reductase activity in C6 cells is stimulated by the addition of exogenous lipids such as linoleate (Volpe and Obert, 1981). This is consistent with the major known sites of sterol regulation in many cells involving the two sequential enzymes in mevalonate synthesis, HMG-CoA synthase and HMG-CoA reductase (Goldstein and Brown, 1990).

A third common site for the cellular regulation of cholesterol is via regulation of low density lipoprotein (LDL) receptors. *In situ* mRNA hybridization studies for LDL receptors and HMG-CoA synthase in nervous tissue shows similar distributions for both mRNAs in glia and in selected neuronal populations (Hofmann *et al.*, 1987). The regulation of HMG-CoA reductase by exogenous lipoproteins (LDLs) in primary glial cell cultures (predominantly astrocytes) and in C6 cells provides further evidence that lipoproteins and their receptors function coordinately in the nervous system (Langan *et al.*, 1987). As expected, HMG-CoA reductase activity and sterol synthesis were inversely related to the presence of LDLs in the incubation medium, suggesting that astrocytes express LDL receptors. Interestingly, the addition of HDLs stimulated sterol synthesis and HMG-CoA reductase activity, perhaps by acting as acceptors for cholesterol egress from the cultured cells.

4.3. The Sterol Composition of the PNS Is Similar, but Not Identical, to the CNS

We have already outlined differences in the structure of myelin, the constituent myelin proteins, and lipid composition between the CNS and PNS. There are three additional functional differences between the CNS and PNS that are important for understanding their comparative sterol metabolism. First, axons in the PNS can regenerate after injury, while, with rare exceptions, those in the CNS of mammals cannot. Second, the degeneration of axons and myelin following injury (Wallerian degeneration) occurs much more rapidly in the PNS than in the CNS. For example, it may take up to 6 months for a severed myelinated axon in the CNS to reach the same stage of degeneration that can be achieved by a severed myelinated PNS axon in 2 weeks. Third, in rodents, PNS myelination commences before CNS myelination (Rawlins and Smith, 1971).

Classically, two major pathways have been described for the conversion of lanosterol to cholesterol. The major route in most mammalian cells includes 7-dehydrocholesterol as an intermediate, while the alternative route utilizes desmosterol (7-dehydrodesmosterol) as an intermediate (see Figure 2). It has been implied

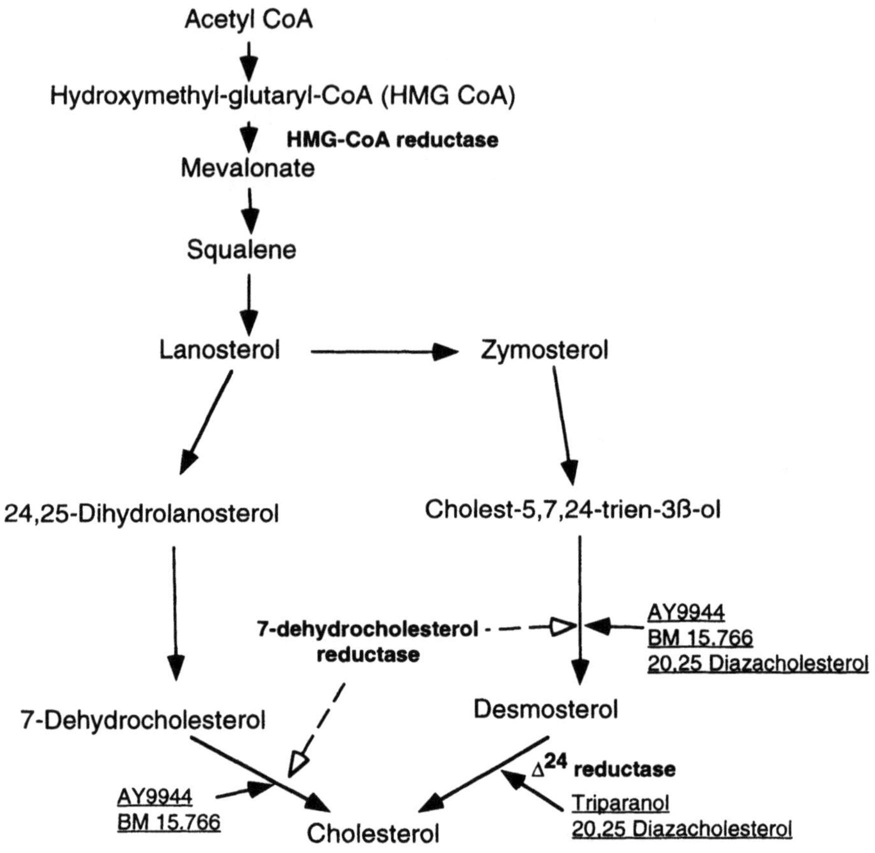

FIGURE 2. Schematic showing the two major pathways for cholesterol biosynthesis referred to in the text (many intermediate steps are missing). Enzyme names are in boldface; cholesterol synthesis inhibitors and their major sites of action are underlined.

that Schwann cells in the PNS might preferentially utilize the 7-dehydroxycholesterol pathway for sterol synthesis, while oligodendrocytes in the CNS may preferentially utilize the 7-dehydrodesmosterol pathway (Bourre *et al.*, 1990). This suggestion is based, in part, on the observation that systemic administration of diazacholesterol, a drug that blocks the conversion of 7-dehydrodesmosterol (cholest-5,7,24-trien-3β-ol) to desmosterol and desmosterol to cholesterol (see Figure 2), causes accumulation of desmosterol in both the CNS and PNS, but accumulation of 7-dehydrodesmosterol in the PNS only (Ramsey, 1977). In addition, the rate of conversion of desmosterol to cholesterol (reduction of the Δ^{24} double bond) has been shown to correlate with CNS myelination and has been suggested as rate limiting for CNS cholesterol biosynthesis (Hinse and Shah, 1971). In the PNS, on the other hand, 7-dehydrocholesterol was 40-fold more concentrated in nerve tissue extracts at 8 days postpartum compared with brain tissue extracts (Bourre *et al.,* 1990). This hypothesis is also supported by the finding that dysmyelinating mice carrying mutations in myelin genes (e.g., the MBP gene in *shiverer* and the PMP22 gene in *Trembler*) have normal levels of 7-dehydrocholesterol in CNS, but markedly abnormal levels in the PNS (Bourre *et al.,* 1989). The enzymatic basis for these apparent differences in PNS and CNS sterol synthesis pathways remains to be determined.

4.4. Cholesterol Metabolism in the PNS: A Role for Apolipoprotein E

The PNS injury model provides an excellent experimental system for studying myelin formation and metabolism. In this paradigm, a peripheral nerve (generally the rodent sciatic nerve) is injured by a focal crush lesion. Following the lesion, the axons lose their connection to their neuronal cell bodies and degenerate, and the axonal degeneration induces the ensheathing Schwann cells to undergo demyelination. The period of demyelination is followed by axon regeneration and remyelination. In the rat, significant axon regeneration and remyelination associated with functionally significant motor recovery is apparent by 40 days following a focal crush lesion to the sciatic nerve at the mid-thigh level (e.g., see Snipes *et al.,* 1992).

During the first 3 days following nerve injury, the mRNAs for the major myelin structural proteins, P0 and PMP22, are dramatically downregulated (Snipes *et al.,* 1992; Trapp *et al.,* 1988). This change is paralleled by a similar downregulation of HMG-CoA reductase activity (Goodrum, 1990). However, during the next few weeks as axons regenerate and remyelination ensues, the levels of myelin mRNA and protein rise, while HMG-CoA reductase activity remains largely unchanged (Goodrum, 1990). The down-regulation of HMG-CoA reductase is paralleled by a decrease in the rate of ^{14}C acetate incorporation into cholesterol, which remains below the cholesterol synthesis rate of uninjured control nerves.

Analysis of the total lipid composition during this period reveals a marked increase in the quantity of cholesterol esters peaking at 2 weeks following nerve injury. The cholesterol esters and other lipids can be observed by microscopy as lipid droplets in Schwann cells and macrophages during the degenerative phase (Goodrum, 1990). The peak of cholesterol ester accumulation at 2 weeks coincides with a marked induction of apolipoprotein E (apoE) synthesis and secretion (Ignatius *et al.,* 1986; Snipes *et al.,* 1986) by Schwann cells and macrophages (Boyles *et al.,* 1989). The newly synthesized apoE is physically associated with cholesterol, cholesterol esters, and polar lipids (Goodrum, 1991). Together, this complex exhibited a buoyant density in KBr gradients consistent with a lipoprotein (Boyles *et al.,* 1989).

Additional studies have suggested that apolipoproteins A-I and A-IV (apoA-I and apoA-IV) derived from serum may be associated with the apoE and apoD lipoprotein complexes from Schwann cells and macrophages (Boyles *et al.,* 1990).

The lipoprotein complexes are probably reutilized for new myelin formation and as a source of membrane lipids for growing axons. In support of this hypothesis, functional LDL receptors have been observed on the growing tips of regenerating neurites and on Schwann cells following denervation (Boyles *et al.*, 1989). Autoradiographic studies of peripheral nerves that have incorporated [^{3}H]cholesterol into their myelin during development further support the notion that most of the cholesterol is reutilized locally for myelin formation during the regeneration process (Rawlins *et al.*, 1972; Rawlins *et al.*, 1970).

4.5. Cholesterol Metabolism in the CNS: A Role for Lipoprotein Receptors

Similar but less dramatic increases in the synthesis and secretion of apoE have been observed during development (Snipes *et al.*, 1987) and following lesion of CNS myelinated fiber pathways (Snipes *et al.*, 1986; Mueller *et al.*, 1985), suggesting a role for apoE in brain cholesterol metabolism. In the brain, the major source of apoE production appears to be astrocytes (Boyles *et al.*, 1985), though expression of apoE by oligodendrocytes following nerve injury has been described (Stoll *et al.*, 1989). Lesser amounts of apoE appear to be present in the cytoplasm of neurons (Han *et al.*, 1994) possibly reflecting uptake of exogenous lipoproteins. Lipoproteins appear to be capable of circulating through the brain since apoE-containing lipoproteins have been isolated from human cerebrospinal fluid (Pitas *et al.*, 1987). This expanded role for lipoproteins and their receptors in the nervous system has prompted further analysis of lipoprotein receptors in the brain. Northern blot analysis for the LDL receptor mRNA revealed a widespread distribution throughout the brain, particularly in regions such as the brainstem and spinal cord, which are rich in myelin (Hofmann *et al.*, 1987). *In situ* mRNA studies additionally revealed the presence of significant amounts of LDL receptor mRNA in neurons in sensory ganglia and in specialized sensory and motor nuclei (Swanson *et al.*, 1988).

More recent studies have identified a family of proteins structurally and functionally related to the LDL receptor that mediate endocytotic targeting and degradation of their respective ligands (Krieger and Herz, 1994). This family of receptors includes the low density lipoprotein receptor-related protein (LRP), the very low density lipoprotein (VLDL) receptor, and glycoprotein 330 (gp330). LRP is a multifunctional receptor that binds not only apoE, but also proteinases such as tissue-type plasminogen activator and urokinase-type plasminogen activator, proteinase inhibitor complexes such as α2-macroglobulin (α2M)-proteinase complexes, serpin-proteinase complexes, matrix proteins such as thrombospondin, hepatic lipase, lipoprotein lipase (see Bu *et al.*, 1994a, and references therein), and secreted amyloid precursor protein (APP) (Kounnas *et al.*, 1995, see below). LRP is expressed in many tissues, but is enriched in liver and brain (Moestrup *et al.*, 1992). In the CNS, LRP mRNA is predominantly synthesized by neurons where the receptor protein can be identified in the plasma membrane (Bu *et al.*, 1994b).

Glial cell lines and reactive glia also synthesize LRP (Bu *et al.*, 1994a). The VLDL receptor and gp330 have also been identified in the brain (Kounnas *et al.*, 1994; Sakai *et al.*, 1994; Zheng *et al.*, 1994), the latter receptor in association with the apical surfaces of ependyma and choroid plexus epithelial cells. While it is true that most of the cholesterol in the brain is associated with myelin, and that apoE expression is associated with myelin cholesterol metabolism, it appears that lipoproteins and their receptors have additional specialized functions in the brain that are unrelated to myelin.

4.6. Cholesterol as a Precursor of Neurosteroids

The term "neurosteroid" has been applied to steroids synthesized in the brain either directly from cholesterol or from steroid hormone precursors (for reviews see Robel and Baulieu, 1994; Majewska, 1992). A series of recent studies has established that the brain can synthesize steroid hormones *de novo* that are derived from the side chain cleavage of cholesterol by the mitochondrial cytochrome P450scc (cholesterol side chain cleavage enzyme) (for a review of steroid hormone synthesis see Miller, 1988), which has recently been identified in oligodendrocytes (Hu *et al.*, 1987) and mixed glial cell cultures (Mellon and Deschepper, 1993; Jung-Testas *et al.*, 1989). Synthetic pathways for pregnenolone, dehydroepiandrosterone (DHEA), progesterone, and other neurosteroids, as well as the steroids themselves, have been identified in nervous tissue (Jung-Testas *et al.*, 1989). Intracellular receptors for steroid hormones are widespread in the brain but each class of receptors shows a unique expression pattern (McEven, 1991a; McEven, 1991b; Robel and Baulieu, 1994). Although the physiological relevance of neurosteroids still requires confirmation and the molecular mechanisms of action are poorly understood, the specific interplay between steroid hormones and their receptors are thought to be involved in the determination of behavior, aging, neurotransmitter regulation, and organization of neural circuits during development and in adulthood.

4.6.1. Role of Neurosteroids in the Regulation of Myelination

In a parallel line of studies, a number of steroids have been tested for their effects on myelination or the expression of myelin components. Jung-Testas *et al.* (1992) showed that treatment of mixed glial cultures with progesterone resulted in decreased cell growth, differentiation of oligodendrocytes and astrocytes, and increased synthesis of MBP in oligodendrocytes and glial fibrillary acidic protein (GFAP) in astrocytes. Kumar *et al.* (1989) showed that hydrocortisone increased MBP and PLP gene expression at the posttranscriptional level in primary oligodendrocytes and, interestingly, the serum levels of hydrocortisone parallel the time course of myelination in the brain (Hof *et al.*, 1990). Indeed, a nine base-pair region has been identified in the 5′-untranslated region of the MBP mRNA that confers glucocorticoid responsiveness to the rate of translation of the MBP protein (Verdi and Campagnoni, 1990). Furthermore, the translation efficiencies of the PLP and CNPase mRNAs are also modulated by glucocorticoids. Recently, it has been demonstrated that Schwann cells *in vitro* can convert pregnenolone to progesterone and that progesterone markedly stimulates myelin production by Schwann cells *in vivo* (Koenig *et al.,* 1995). Interestingly, the levels of progesterone were markedly reduced in the PNS of the hypomyelinating *Trembler* mice, possibly reflecting a general downregulation of myelin formation.

4.7. Cholesterol Regulation and Axon–Glial Interactions

There is abundant evidence suggesting the importance of axon–glial interactions in regulating the expression of myelin protein genes. These mechanisms appear to be necessary because of the extraordinary requirement for the rapid, but transient, synthesis of myelin during development. Since the myelin cholesterol requirement parallels the need for myelin proteins, there should be mechanisms for the axon-dependent regulation of cholesterol synthesis by oligodendrocytes and Schwann cells. In nonneural cells, the major sites for the regulation of cholesterol biosynthesis are those regulating HMG-CoA reductase and synthetase, as well as the LDL receptors (Goldstein and Brown, 1990). Of these, HMG-CoA reductase has been shown to be rapidly downregulated following nerve injury in the PNS in a time course that parallels the rapid downregulation of the major PNS myelin proteins (Goodrum, 1993; Snipes *et al.,* 1992). Since only the axons, but not the Schwann cells themselves, are injured following a focal nerve injury, at least HMG-CoA reductase (and perhaps other components of the cholesterol biosynthetic machinery) levels are responsive to axon injury. So far, the molecular mechanisms underlying axon–glial interactions have remained elusive, but a great deal is known about the regulation of HMG-CoA reductase in nonglial cells. HMG-CoA reductase can be regulated by modifying its level of transcription through the action of specific activators on sterol response elements in its promoter, by changing the stability of its mRNA, or by altering the rate of degradation of the enzyme itself (Goldstein and Brown, 1990). It would be very interesting to determine which, if any, of these mechanisms are operative in myelinating glial cells. It would also be of particular interest to determine if any of these responses are specifically regulated by axon–glial interactions.

5. CHOLESTEROL IN DISEASES OF THE NERVOUS SYSTEM

5.1. Cholesterol Metabolism and Alzheimer's Disease

The most surprising finding to emerge from the study of brain cholesterol metabolism is the strong association with mechanistic pathways that have been directly implicated in the pathogenesis of Alzheimer's disease (AD). As is well known, Alzheimer's disease is a relentlessly progressive dementing illness associated with synaptic and neuronal loss and the pathologic hallmarks of neurofibrillary tangles and neuritic plaques. Most cases of AD occur sporadically in the general population, but familial forms have been described and linked to at least three genetic loci (Levy-Lahad *et al.*, 1995a; Levy-Lahad *et al.*, 1995b). Familial AD has been associated with two genes that are of particular relevance to cholesterol homeostasis in the brain; namely, the genes for apoE (APOE) and the amyloid precursor protein (APP).

The human APOE gene has three major allelic variants, designated APOE ε2-ε4. Independently, the human APOE gene was linked to chromosome 19 (Das *et al.*, 1985) at 19q13.2, and parallel studies revealed that chromosomal region 19q13.1-q13.3 is associated with late-onset familial AD (Pericak-Vance *et al.*, 1991). Subsequently, an increased frequency and dosage of the APOE e4 allele was associated with familial late-onset AD (Corder *et al.*, 1993; Strittmatter *et al.*, 1993) and with sporadic AD (Roses *et al.*, 1994; Rebeck *et al.*, 1993). ApoE is found in amyloid deposits in senile plaques, neuritic tangles, and blood vessels displaying congophilic angiopathy in the brains of AD patients (Namba *et al.*, 1991). *In vitro*, apoE binds purified β amyloid (see below) with high avidity (Strittmatter *et al.*, 1993). Among the interesting biological differences between the apoE isoforms is the ability of the ε3 isoform (when incorporated into β-VLDLs) to increase neurite outgrowth, while the ε4 isoform decreased neurite outgrowth from dorsal root ganglion neurons (Nathan *et al.*, 1994). In addition, apoE3 and apoE4 are taken up in a similar fashion by neuronal cells (Neuro-2a), but apoE4 has a different intracellular distribution and is specifically associated with the depolymerization of microtubules (Nathan *et al.*, 1995). Thus, the major apolipoprotein associated with cholesterol metabolism in the brain, apoE, is a major determinant for AD.

Biochemical and genetic studies have identified mutations in the amyloid precursor protein (APP) on human chromosome 21 as a cause of familial AD (Goate *et al.*, 1991). The amyloid precursor protein is so named because it encodes the 4 kDa β amyloid (A4) peptide that is apparently generated by aberrant proteolysis of the mature APP protein (Tanzi *et al.*, 1987). Like apoE, the β amyloid polypeptide accumulates in senile plaques and abnormal blood vessels with congophilic change (Rewcastle, 1991). While the structure and possible functions for the various APP isoforms are beyond the scope of this review, the significance of the APP protein in cholesterol metabolism derives from the fact that it binds apoE with high avidity (Strittmatter *et al.*, 1993) and that a soluble form of the APP protein [generated by cleavage of the extracellular domain of the transmembrane APP770 isoform that contains a Kunitz protease inhibitor (KPI) domain] shares affinity for the same LRP receptor that is important for apoE (and presumably cholesterol) metabolism in the CNS (Kounnas *et al.*, 1995). In agreement with these findings, complexes of LRP and apoE, as well as additional LRP ligands, have been found to decorate senile plaques, suggesting a possible link between AD, LRP, and apoE (Rebeck *et al.*, 1993).

Recently, Kounnas *et al.* (1995) have integrated these data by demonstrating that LRP binds secreted APP and is responsible for its endocytosis and subsequent lysosomal degradation. Given the fact that LRP also binds apoE-enriched lipoproteins and that the inheritance of the ε4 allele of the apoE gene is a risk factor for AD, these findings provide a link via a shared metabolic pathway of two molecules implicated in AD. Interestingly, Sparks *et al.* (1994) reported that rabbits fed a high cholesterol diet (2% cholesterol) developed abnormal β amyloid immunoreactivity in hippocampal neurons. It remains to be determined whether the observed connection between LRP, apoE, and APP turns out to be the crucial common pathway in AD. Furthermore, it remains to be seen whether VLDL receptors and gp330 are similarly involved in APP clearance.

5.2. Tellurium Toxicity

Tellurium toxicity to the peripheral nervous system has become a model for the study of demyelination/remyelination because of the rapidity and the extent to which these processes occur after administration of the toxin. As opposed to Wallerian degeneration, the majority of the axons remain intact. Thus, the toxin appears to selectively damage the nerve sheath cells. Tellurium appears to induce a systemic block in cholesterol biosynthesis and, thus, is also a model for the role of cholesterol in myelin. That tellurium specifically blocks cholesterol biosynthesis is supported by the finding that overall cholesterol synthesis in sciatic nerve is reduced and that the cholesterol precursor squalene accumulates (Harry *et al.*, 1989) when animals are fed tellurium. It was later shown, by direct *in vitro* enzyme inhibition experiments using the probable active metabolite tellurite (TeO_3^-), that the effect occurs specifically at the level of squalene epoxidase (Wagner-Recio *et al.*, 1991). Additional indirect evidence was provided by the observation that squalene epoxidase activity was strongly upregulated in liver and sciatic nerve after tellurium feeding of rats, presumably due to a compensation effect (Wagner *et al.*, 1995).

If young rats are put on a diet containing approximately 1% tellurium just after weaning, they develop a neurotoxic response manifested as a primary demyelination of the PNS (Lampert and Garrett, 1971; Lampert *et al.*, 1970), in particular in the sciatic nerve. As pointed out previously, cholesterol is enriched in myelin with a putatively important role in maintaining the structural integrity of myelin. Thus, it is thought that the block of cholesterol biosynthesis in the developmental time period just after weaning (when accumulation of myelin is at its peak) results in demyelination. Furthermore, it has been suggested that the more myelin has to be synthesized and maintained in a given myelinating unit, the more cholesterol, other lipids, and myelin proteins are needed for appropriate structural support (Morell *et al.*, 1994). This hypothesis is supported by the finding that the sciatic nerve is particularly affected by tellurium toxicity, presumably since this nerve contains many large caliber axons surrounded by prominent myelin sheaths.

In agreement with these observations, larger internodal segments are preferentially affected (Bouldin *et al.,* 1988). Interestingly, CNS myelin is not or is only slightly (for a different experimental paradigm see Jackson *et al.,* 1989) affected by tellurium toxicity. This finding can be explained by the fact that the required rate of myelin synthesis in the CNS is much lower compared to the PNS (Rawlins and Smith, 1971), and thus the need for cholesterol in the sciatic nerve is much higher at peak myelination. Therefore, this tissue is particularly susceptible to tellurium insults. As discussed earlier, the nervous system has only very limited, if any, access to circulating cholesterol for possible compensation; the vast majority of the necessary cholesterol is synthesized within the tissue itself.

Tellurium toxicity is characterized by segmental demyelination of the sciatic nerve without detectable signs of axonal abnormalities (Lampert and Garrett, 1971). Demyelination starts 2 days after the start of tellurium feeding and reaches its maximum at day 5 (Toews *et al.,* 1990). At this time, the animals show pronounced weakness of the hind limbs, a diagnostic criterion for peripheral neuropathy. After 7 days, first signs of remyelination are observed, even if the animals continue to live on a tellurium diet. However, recovery is hastened if tellurium treatment is discontinued.

Demyelination and remyelination are accompanied by profound changes in gene expression. First, the mRNAs for the major myelin proteins P0 and MBP are profoundly downregulated during the demyelination period, but they recover rapidly with the beginning of remyelination (Toews *et al.,* 1990). In addition, several markers of non-myelinating Schwann cells, including the low affinity nerve growth factor receptor and glial fibrillary acidic protein, are expressed at higher levels during tellurium-induced demyelination (Toews *et al.,* 1992).

The effects of tellurium neurotoxicity on Schwann cell gene expression closely resemble the demyelination and remyelination events in the distal nerve stump after peripheral nerve crush injury. In contrast to nerve lesions, however, tellurium-induced neuropathy appears not to involve damage to the axon. Thus, the observed gene regulation in the tellurium toxicity model suggests that the conversion of myelinating Schwann cells to a precursor cell type is independent of axonal degeneration (Toews *et al.,* 1992). These findings have been interpreted as suggesting that demyelination and remyelination are regulated by a coordinate gene regulation pathway (Toews *et al.,* 1992).

At first glance, it seems that the liver should also be affected by tellurium toxicity since this organ is of critical importance in cholesterol biosynthesis and metabolism. Interestingly, however, although there is accumulation of squalene in the livers of tellurium-fed rats, cholesterol biosynthesis is still normal (Toews *et al.,* 1991; Wagner-Recio *et al.,* 1991) Based on the finding of increased levels of HMG-CoA reductase in the liver of tellurium-treated rats (Toews *et al.,* 1991), it has been hypothesized that squalene biosynthesis is rapidly upregulated, allowing for a substrate-driven compensation of the tellurium-induced inhibition of squalene epoxidase.

5.3. Smith–Lemli–Opitz (RSH) Syndrome

The probable biochemical defect underlying the pathogenesis of the Smith–Lemli–Opitz syndrome has recently been described as a defect in 7-dehydrocholesterol (cholestest-5,7-dien-3β-ol) reductase (Tint *et al.,* 1994; also see chapter 5 by Tint *et al.*). This enzyme appears to be involved in the conversion of lanosterol to desmosterol, as well as the conversion of 7-dehydrocholesterol to cholesterol. Thus, both major branches of the cholesterol biosynthetic pathway described above are blocked (see Figure 2). Accordingly, serum cholesterol levels in SLO patients are very low and there is a significant accumulation (greater than 2000-fold) of the precursor 7-dehydrocholesterol. This syndrome is of interest to our discussion of cholesterol in myelin for two reasons. First, it represents a unique opportunity to study the effects of greatly reduced cholesterol levels systemically. More importantly for this discussion, however, is the nervous system lesions that have been described in this disorder.

A myriad of nonspecific clinical features have been associated with the Smith–Lemli–Opitz syndrome (Opitz, 1994; see Opitz *et al.,* 1994, for a recent bibliography on SLO), including microcephaly; dysmorphic facial features; limb abnormalities; genital disorders; endocrine, cardiac, and kidney anomalies; and mental retardation. Several detailed neuropathological autopsy studies have been performed on patients with SLO (Cherstvoy *et al.,* 1984; Fierro *et al.,* 1977; Cherstvoy *et al.,* 1975; Garcia *et al.,* 1973). Like the systemic manifestations, the CNS involvement is nonspecific and somewhat variable. These include features of holoprosencephaly, neuronal loss, gliosis, attenuation of white matter (loss of myelin), partial aplasia of the cerebellar vermis and various regional hypoplasias and dysplasias. Taken together, these changes probably result from general developmental stress originating during the first trimester *in utero.* To date, there are no clear descriptions of the biochemical findings in the brains of SLO patients with regard to their cholesterol and dehydrocholesterol compositions. Thus, it is not clear from human studies whether these defects result from lack of cholesterol or from accumulation of the abnormal precursor. Also, since this biochemical defect in cholesterol biosynthesis has only recently been described, there has been no effort to specifically examine the effects of a block in cholesterol synthesis with its effects on myelin. Fortunately, animal studies have provided some insight into these issues.

5.4. The Effects of Cholesterol Synthesis Inhibitors on Myelin

Several pharmacological agents, including AY9944 (*trans*-1,4-bis-(2-chlorobenzylaminomethyl)cyclohexane dihydrochloride), Triparanol (1-*p*-(β-diethylaminoethoxy)-phenyl-1-(*p*-tolyl)-2-(*p*-chlorophenyl)ethanol), 20,25-diazacholesterol, and BM 15.766 (4-2-[1-(4-chlorocinnamyl)piperazin-4-yl]ethyl)-benzoic acid) have been synthesized and shown to block cholesterol biosynthesis (see Figure 2). Specifically, AY9944 (Kraml *et al.,* 1964) and BM 15.766 (Pill *et al.,* 1985) appear to inhibit 7-dehydrocholesterol-Δ^7-reductase and, as such, provide useful models for the animal studies of the Smith–Lemli–Opitz syndrome. Triparanol inhibits desmosterol-24-reductase (Steinberg and Avigan, 1960), resulting in overall lowering of cholesterol levels and accumulation of desmosterol. 20,25-Diazacholesterol appears mainly to block Δ^{24} reductase (Fumagalli *et al.,* 1969, Thompson *et al.,* 1963), though there is some evidence that it also blocks 7-dehydrocholesterol reductase (Suzuki *et al.,* 1974). With the caveat that these drugs may be shown to have additional effects, we will base the following discussion on the assumption that their major effect is on cholesterol biosynthesis. The pharmacologic and teratogenic effects of these drugs in developing rodents have been investigated and, recognizing the close association between cholesterol biosynthesis and myelin formation, several studies have specifically examined the effects of these drugs on oligodendrocytes and myelin.

Dvornik and Hill (1968) demonstrated that long-term administration of AY9944 to adult rats (over 12 months of age) resulted in lowered brain cholesterol levels that were largely compensated for by the increased levels of 7-dehydroxycholesterol. Interestingly, when AY9944-treated animals were provided a diet rich in cholesterol, there was no change in the CNS levels of cholesterol or 7-dehydrocholesterol, whereas the liver and serum levels of cholesterol were greatly elevated. This finding again argues for the autonomy of local brain sterol synthesis in adult animals. When AY9944 is administered to young postnatal rats (from 5–45 days post-partum–the period of most active CNS myelination), treated animals were found to have significantly lower body weights (35% of controls) and somewhat lower brain weights (70% of controls) (Igarashi *et al.,* 1975). Morphological studies revealed oligodendroglial cell degeneration, demyelination, and astrocytosis in the experimental animals that was associated with both abnormal cytoplasmic inclusions, probably lysosomal in nature, in most brain cell types, including oligodendrocytes (Igarashi *et al.* 1975; Suzuki and De Paul, 1971), and a generalized increase in lysosomal enzyme activities (Igarashi *et al.,* 1975). 7-Dehydrocholesterol comprised up to 90% of brain sterols in 45-day-old animals that had been treated with AY9944 since postnatal day 5. It is unclear from these studies whether the lack of cholesterol or the increase in the precursor 7-dehydrocholesterol is responsible for the pathologic and biochemical changes.

It seems likely that 7-dehydrocholesterol can substitute for cholesterol in myelin. This assertion is supported by animal studies in which the cholesterol biosynthesis inhibitors AY9944 (and 20,25-diazacholesterol) were administered to postnatal rats. It was demonstrated that myelination in the CNS (Fumagalli *et al.,* 1969) and PNS (Rawlins and Uzman, 1970) of these animals was retarded following drug treatment. Ultimately, the treated animals had reduced amounts of morphologically normal myelin (Smith *et al.,* 1970), but purified myelin preparations from the CNS had apparently normal ratios of phospholipid-galactolipid-sterol-protein, although the sterols were enriched in the precursor forms (Fumagalli *et al.*, 1969). Nonspecific lowering of plasma cholesterol levels by treatment with clofibrate had a small effect on brain sterol levels, which were qualitatively normal in composition.

In summary, it appears that cholesterol synthesis inhibitors affect myelin during postnatal development by delaying the onset, and limiting the extent of, myelination. This could be explained by a limiting supply of sterol. In addition, there is evidence for a toxic effect of AY9944, which may either reflect direct toxicity of 7-dehydrocholesterol or an effect of lower cholesterol levels. The lack of a similar toxic effect of cholesterol lowering by 20,25-diazacholesterol treatment (Suzuki *et al.*, 1974) argues for a specific toxicity of 7-dehydrocholesterol (or AY9944), but it is difficult to control for the bioavailability of the two drugs to allow for a direct comparison. In either case, it appears that the myelin that is formed is morphologically normal and that the precursors desmosterol and 7-dehydrocholesterol are able to functionally substitute for cholesterol in myelin.

The effects of cholesterol biosynthesis inhibitors have also been investigated during gestation in rodents. The major CNS manifestation of *in utero* exposure to AY9944 beginning at the fourth day of gestation appears to be holoprosencephaly, most often manifested as agenesis of the pituitary gland (Roux *et al.*, 1979a). The major teratogenic effects of AY9944 can be counteracted by feeding the pregnant rats a high cholesterol diet (Barbu *et al.*, 1988; Roux *et al.*, 1979b). From these studies, it appears that the early CNS developmental defects due to a block in 7-dehydrocholesterol reductase can be counteracted by maternally derived systemic cholesterol. This is in contrast to adult animals where the accumulation of abnormal sterols following treatment with AY9944 is independent of dietary and systemic cholesterol levels (Dvornik and Hill, 1968). It remains to be seen whether the favorable effect of cholesterol supplementation can be applied to the treatment of SLO patients (see Acosta, 1994; Barness, 1994; and Irons *et al.*, 1994, for a more thorough discussion), where the major malformations arise during early gestation. As a reference for comparison, myelination in the human CNS begins as early as 12–14 weeks gestation in the spinal cord, but the majority of CNS myelination occurs between midgestation and the second year of life (Kinney *et al.*, 1994; Kinney *et al.*, 1988).

In summary, SLO appears to result from a defect in cholesterol biosynthesis at the level of 7-dehydrocholesterol reductase. Patients with SLO have been reported to have hypomyelination associated with other CNS abnormalities. It is not clear from human studies whether the myelin defects are due primarily to direct effects on the oligodendrocyte or whether the hypomyelination is secondary to neuronal and axon loss. Results of administration of AY9944 to rodents suggest that the major teratogenic effects of 7-dehydrocholesterol reductase inhibition can be uncoupled from its effect on myelination according to the time of administration of the drug; early gestational administration leads to developmental malformations such as holoprosencephaly, while postnatal administration leads to total brain hypoplasia and oligodendrocyte death. Thus, animal studies would suggest that SLO patients have both primary and secondary myelin loss that may be reversible, in part, by a maternal diet rich in cholesterol.

6. CONCLUDING REMARKS

In this review, we have established that cholesterol is an important constituent of myelin, which is a major site of cholesterol utilization. The exact role of cholesterol in myelin has not been firmly established, but probably relates to the role of myelin as an electrical insulator. It appears that the majority of cholesterol is synthesized locally in the nervous system. Developing and regenerating nerves need to rapidly mobilize cholesterol into myelin membranes. Experimental (tellurium toxicity, cholesterol synthesis inhibitors) or naturally occurring blocks in these pathways (SLO) may give rise to developmental abnormalities or demyelination depending on the timing of administration and the enzyme affected by the block. We have also described the remarkable association between pathways of cholesterol metabolism and Alzheimer's disease. The realization of the importance of cholesterol in diseases affecting myelin and the nervous system is paralleled by important findings implicating myelin proteins in the etiology of hereditary dysmyelinating disorders such as Charcot–Marie–Tooth disease and Pelizaeus–Merzbacher disease. It can be anticipated that the synthesis of these parallel endeavors will result in a better understanding of the role of cholesterol in myelin, the diseases of myelin, and a few, but important, diseases of the nervous system.

7. REFERENCES

Acosta, P. B., 1994, RSH/SLO (Smith–Lemli–Opitz) syndrome: designing a high cholesterol diet for the SLO syndrome, *Am. J. Med.Genet.* **50**:358–363.

Anderson, J. M., and Dietschy, J. M., 1977, Regulation of sterol synthesis in 15 tissues of rat. II. Role of rat and human high and low density plasma lipoproteins and of rat chylomicron remnants, *J. Biol. Chem.* **252**:3652–3659.

Andersson, M., Elmberger, P. G., Edlund, C., Kristensson, K., and Dallner, G., 1990, Rates of cholesterol, ubiquinone, dolichol and dolichyl-P biosynthesis in rat brain slices, *FEBS Lett.* **269**:15–18.

Baecher, D., Lier, T., Hameister, H., Altenberger, H., Grehl, H., Suter, U., and Rautenstrauss, B., 1995, Widespread expression of the peripheral myelin protein-22 gene (pmp22) in neural and non-neural tissues during murine development, *J. Neurosci. Res.* **42**:733–741.

Barbarese, E., Braun, P. E., and Carson, J. H., 1977, Identification of prelarge and presmall basic proteins in mouse myelin and their structural relationship to large and small basic proteins, *Proc. Natl. Acad. Sci. USA* **74**:3360–3364.

Barbu, V., Roux, C., Lambert, D., Dupuis, R., Gardette, J., Maziere, J.-C., Maziere, C., Elefant, E., and Polonovski, J., 1988, Cholesterol prevents the teratogenic action of AY 9944: Importance of the timing of cholesterol supplementation to rats, *J. Nutr.* **118**:774–779.

Barness, L. A., 1994, Nutritional requirements of infants and children with respect to cholesterol and related compounds, *Am J. Med. Genet.* **50**:353–354.

Boison, D., and Stoffel, W., 1994, Disruption of the compacted myelin sheath of axons of the central nervous system in proteolipid protein-deficient mice, *Proc. Natl. Acad. Sci. USA* **91:** 11709–11713.

Bouldin, T. W., Samsa, G., Earnhardt, T., and Krigman, M. R., 1988, Schwann-cell vulnerability to demyelination is associated with internodal length in tellurium neuropathy, *J. Neuropathol. Exp. Neurol.* **47:**41–47.

Bourre, J. M., Clement, M., Gerard, D., and Chaudiere, J., 1989, Alterations of cholesterol synthesis precursors (7-dehydrocholesterol, 7-dehydrodesmosterol, desmosterol) in dysmyelinating neurological mutant mouse (quaking, shiverer, and trembler) in the PNS and CNS, *Biochim. Biophys. Acta* **1004:**387–390.

Bourre, J. M., Clement, M., Gerard, D., Legrand, R., and Chaudiere, J., 1990, Precursors for cholesterol synthesis (7-dehydrocholesterol, 7-dehydrodesmosterol, and desmosterol): cholesterol/7-dehydrocholesterol ratio as an index of development and aging in PNS but not in CNS, *J. Neurochem.* **54:**1196–1199.

Boyles, J. K., Notterpek, L. M., and Anderson, L. J., 1990, Accumulation of apolipoproteins in the regenerating and remyelinating mammalian peripheral nerve, *J. Biol. Chem.* **265:**17805–17815.

Boyles, J. K., Zoellner, C. W., Anderson, L. J., Kosik, L. M., Pitas, R. E., Weisgraber, K. H., Hui, D. Y., Mahley, R. W., Gebicke-Haerter, P. J., Ignatius, M. J., and Shooter, E. M., 1989, A role for apolipoprotein E, apolipoprotein A-I, and low density lipoprotein receptors in cholesterol transport during regeneration and remyelination of the rat sciatic nerve, *J. Clin. Invest.* **83:**1015–1031.

Boyles, J. K., Pitas, R. E., Wilson, E., Mahley, R. W., and Taylor, J. M., 1985, Apolipoprotein E associated with astrocytic glia of the central nervous system and with nonmyelinating glia of the peripheral nervous system, *J. Clin. Invest.* **76:**1501–1513.

Bu, G., Maksymovitch, E. A., Geuze, H., and Schwartz, A. L., 1994a, Subcellular localization and endocytic function of low density lipoprotein receptor-related protein in human glioblastoma cells, *J. Biol. Chem.* **269:**29874–29882.

Bu, G., Maksymovitch, E. A., Nerbonne, J. M., and Schwartz, A. L., 1994b, Expression and function of the low density lipoprotein receptor-related protein (LRP) in mammalian central neurons, *J. Biol. Chem.* **269:**18521–18528.

Burgisser, P., Matthieu, J. M., Jeserich, G., and Waehneldt, T. V., 1986, Myelin lipids: a phylogenetic study, *Neurochem. Res.* **11:**1261–1272.

Chen, D., Li, Q.-T., and Lee, K. H., 1992, Studies on the entrapment and stability of liposomes containing sulfatide from human brain, *Biochem. Int.* **28:**297–304.

Cherstvoy, E. D., Laxjuk, G. I., Lurie, I. W., Nedzved, M. K., and Usoev, S. S., 1975, The pathological anatomy of the Smith–Lemli–Opitz syndrome, *Clin. Genet.* **7:**382–387.

Cherstvoy, E. D., Lazjuk, G. I., Ostrovskaya, T. I., Shved, I. A., Kravtzova, G. I., Lurie, I. W., and Gerasimovich, A. I., 1984, The Smith–Lemli–Opitz syndrome. A detailed pathological study as a clue to a etiological heterogeneity, *Virchows Arch. [Pathol. Anat.]* **404:**413–425.

Cooper, C. E., Wrigglesworth, J. M., and Nicholls, P., 1990, The mechanism of potassium movement across the liposomal membrane, *Biochem. Biophys. Res. Commun.* **173:**1008–1012.

Corder, E. H., Saunders, A. M., Strittmatter, W. J., Schmechel, D. E., Gaskell, P. C., Small, G. W., Roses, A. D., Haines, J. L., and Pericak-Vance, M. A., 1993, Gene dose of apolipoprotein E type 4 allele and the risk of Alzheimer's disease in late onset families, *Science* **261:**921–923.

D'Urso, D., Brophy, P. J., Staugaitus, S. M., Gillespie, C. S., Frey, A. B., Stempack, J. G., and Colman, D. R., 1990, Protein zero of peripheral nerve myelin: biosynthesis, membrane insertion and evidence for homotypic interaction, *Neuron* **2:**449–460.

Das, H. K., McPherson, J., Bruns, G. A. P., Karathanasis, S. K., and Breslow, J. L., 1985, Isolation, characterization, and mapping to chromosome 19 of the human apolipoprotein E gene, *J. Biol. Chem.* **260:**6240–6247.

Davison, A. N., 1965, Brain sterol metabolism, *Adv. Lipid Res.* **3:**171–196.

Davison, A. N., and Wajda, M., 1962, Cerebral lipids in multiple sclerosis, *J. Neurochem.* **9:**427–432.

Dehouck, B., Dehouck, M. P., Fruchart, J. C., and Cecchelli, R., 1994, Upregulation of the low density lipoprotein receptor at the blood-brain barrier: intercommunications between brain capillary endothelial cells and astrocytes, *J. Cell Biol.* **126:**465–473.

Diaz, R. S., and Monreal, J., 1994, Unusual low proton permeability of liposomes prepared from the endogenous myelin lipids, *J. Neurochem.* **62:**2022–2029.

Dietschy, J. M., and Spady, D. K., 1984, Measurement of rates of cholesterol synthesis using tritiated water, *J. Lipid Res.* **25:**1469–1476.

Dobbing, J., 1963, The entry of cholesterol into rat brain during development, *J. Neurochem.* **10:** 739–742.

Dvornik, D., and Hill, P., 1968, Effect of long-term administration of AY-9944, an inhibitor of 7-dehydrocholesterol Δ^7-reductase, on serum and tissue lipids in the rat, *J. Lipid Res.* **9:**587–595.

Edmond, J., Korsak, R. A., Morrow, J. W., Torok, B. G., and Catlin, D. H., 1991, Dietary cholesterol and the origin of cholesterol in the brain of developing rats, *J. Nutr.* **121:**1323–1330.

Fierro, M., Martinez, A. J., Harbison, J. W., and Hay, S. H., 1977, Smith-Lemli-Opitz syndrome: Neuropathological and ophthalmological observations, *Devel. Med. Child Neurol.* **19:**57–62.

Filbin, M. T., Wash, F. S., Trapp, B. D., Pizzey, J. A., and Tennekoon, G. I., 1990, Role of myelin Po-protein as a homophilic adhesion molecule, *Nature* **344:** 871–872.

Fischer, I., and Sapirstein, V. S., 1994, Molecular cloning of plasmolipin, *J. Biol. Chem.* **269:** 24912–24919.

Fischer, I., Durrie, R., and Sapirstein, V. S., 1994, Plasmolipin: the other myelin proteolipid. A review of studies on its structure, expression, and function, *Neurochem. Res.* **19:**959–966.

Fumagalli, R., Smith, M. E., Urna, G., and Paoletti, R., 1969, The effect of hypocholesterolemic agents on myelinogenesis, *J. Neurochem.* **16:**1329–1339.

Garcia, C. A., McGarry, P. A., Voirol, M., and Duncan, C., 1973, Neurological involvement in the Smith–Lemli–Opitz syndrome: clinical and neuropathological findings, *Develop. Med. Child Neurol.* **15:**48–55.

Gardinier, M. V., Amiguet, P., Linington, C., and Matthieu, J. M., 1992, Myelin/oligodendrocyte glycoprotein is a unique member of the immunoglobulin superfamily, *J. Neurosci. Res.* **33:**177–187.

Giese, K. P., Martini, R., Lemke, G., Soriano, P., and Schachner, M., 1992, Mouse P0 gene disruption leads to hypomyelination, abnormal expression of recognition molecules, and degeneration of myelin and axons, *Cell* **71:**565–576.

Gillespie, C. S., Sherman, D. L., Blair, G. E., and Brophy, P. J., 1994, Periaxin, a novel protein of myelinating Schwann cells with a possible role in axonal ensheathment, *Neuron* **12:**497–508.

Goate, A., Chartier-Harlin, M.-C., and Mullan, M., 1991, Segregation of missense mutation in the amyloid precursor protein gene with familial Alzheimer's disease, *Nature* **349:**704–707.

Goldstein, J. L., and Brown, M. S., 1990, Regulation of the mevalonate pathway, *Science* **343:** 425–430.

Goodrum, J. F., 1990, Cholesterol synthesis is down-regulated during regeneration of peripheral nerve, *J. Neurochem.* **54:**1709–1715.

Goodrum, J. F., 1991, Cholesterol from degenerating nerve myelin becomes associated with lipoproteins containing apolipoprotein E. *J. Neurochem.* **56:**2082–2086.

Goodrum, J. F., 1993, Cholesterol synthesis in regenerating peripheral nerve is not influenced by serum cholesterol levels, *J. Neurochem.* **60:**1564–1566.

Greenfield, S., Weise, M. J., Gantt, G., Hogan, E. L., and Brostoff, S. W., 1982, Basic proteins of rodent peripheral nerve myelin: immunochemical identification of the 21.5K, 18.5K, 17K, 14K and P2 proteins, *J. Neurochem.* **39:** 1278–1282.

Han, S.-H., Einstein, G., Weisgraber, K. H., Strittmatter, W. J., Saunders, A. M., Pericak-Vance, M., Roses, A. D., and Schmechel, D. E., 1994, Apolipoprotein E is localized to the cytoplasm of human cortical neurons: A light and electron microscopic study, *J. Neuropathol. Exp. Neurol.* **53:**535–544.

Haque, Z. U., and Mozaffor, Z., 1992, Importance of dietary cholesterol for the maturation of mouse brain myelin, *Biosci. Biotech. Biochem.* **56:**1351–1354.

Harry, G. J., Goodrum, J. F., Bouldin, T. W., Wagner-Recio, M., Toews, A. D., and Morell, P., 1989, Tellurium-induced neuropathy: metabolic alterations associated with demyelination and remyelination in rat sciatic nerve, *J. Neurochem.* **52:**938–945.

Hinse, C. H., and Shah, S. N., 1971, The desmosterol reductase activity of rat brain during development, *J. Neurochem.* **18:**1989–1998.

Hof, L., Hayman, S. R., and Csiza, C. K., 1990, Glucocorticoid levels and effects of glial enzymes in myelin deficient rats, *Trans. Am. Soc. Neurochem.* **21:**234.

Hofmann, S. L., Russell, D. W., Goldstein, J. L., and Brown, M. S., 1987, mRNA for low density lipoprotein receptor in brain and spinal cord of immature and mature rabbits, *Proc. Natl. Acad. Sci. USA* **84:**6312–6316.

Hu, Z. Y., Borreau, E., Jung-Testas, I., Robel, P., and Baulieu, E. E., 1987, Neurosteroids: oligodendrocyte mitochondria convert cholesterol to pregnenolone, *Proc. Natl. Acad. Sci. USA* **84:** 8215–8219.

Igarashi, M., Suzuki, K., Chen, S. M., and Suzuki, K., 1975, Changes in brain hydrolytic enzyme activities in rats treated with cholesterol biosynthesis inhibitor, AY9944, *Brain Res.* **90:**97–114.

Ignatius, M. J., Gebicke-Harter, P. J., Hangen, D. H., Skene, J. H. P., Schilling, J. W., Weisgraber, K. H., Mahley, R. W., and Shooter, E. M., 1986, Expression of apolipoprotein E during nerve degeneration and regeneration, *Proc. Natl. Acad. Sci. USA* **86:**1125–1129.

Inoue, Y., Nakamura, R., Mikoshiba, K., and Tsukada, Y., 1981, Fine structure of the central myelin sheath in the myelin deficient mutant shiverer mouse with special reference to the pattern of myelin formation by oligodendroglia, *Brain Res.* **219:**85–94.

Irons, M., Elias, E. R., Tint, G. S., Salen, G., Frieden, R., Buie, T. M., and Ampola, M., 1994, Abnormal cholesterol metabolism in the Smith–Lemli–Opitz syndrome: report of clinical and biochemical findings in four patients and treatment in one patient, *Am. J. Med. Genet.* **50:**347–352.

Jackson, K. F., Hammang, J. P., Worth, S. F., and Duncan, I. D., 1989, Hypomyelination in neonatal rat central and peripheral nervous systems following tellurium intoxification, *Acta Neuropathol.* **78:**301–309.

Jung-Testas, I., Hu, Z. Y., Baulieu, E. E., and Robel, P., 1989, Steroid synthesis in rat brain cell cultures, *J. Steroid Biochem.* **34:**511–519.

Jung-Testas, I., Renoir, J. M., Greene, G. L., and Baulieu, E. E., 1992, Demonstration of steroid hormone receptors and steroid action in primary cultures of glial cells, *J. Steroid Biochem. Mol. Biol.* **41:**621–631.

Jurevics, H., and Morell, P., 1995, Cholesterol for synthesis of myelin is made locally, not imported into brain, *J. Neurochem.* **64:**895–901.

Kamholz, J., and Wrabetz, L., 1992, Molecular genetics of myelin basic protein, in: *Myelin: Biology and Chemistry* (R. E. Martenson, ed.), pp. 367–385, CRC Press, Boca Raton, FL.

Kinney, H. C., Brody, B. A., Kloman, A. S., and Gilles, F. H., 1988, Sequence of central nervous system myelination in human infancy. II. Patterns of myelination in autopsied infants, *J. Neuropathol. Exp. Neurol.* **47:**217–234.

Kinney, H. C., Karthigasan, J., Borenshteyn, N. I., Flax, J. D., and Kirschner, D. A., 1994, Myelination in the developing human brain: biochemical correlates, *Neurochem. Res.* **19:**983–996.

Koenig, H. L., Schumacher, M., Ferzaz, B., Do Thi, A. N., Ressouches, A., Guennoun, R., Jung-Testas, I., Robel, P., Akwa, Y., and Baulieu, E.-E., 1995, Progesterone synthesis and myelin formation by Schwann cells, *Science* **268:**1500–1503.

Koenig, S. H., 1991, Cholesterol of myelin is the determinant of gray-white contrast in MRI of brain, *Magn. Reson. Med.* **20:**285–291.

Koper, J. W., Zeinstra, E. C., Lopez-Cardozo, M., and Van Golde, L. M. G., 1984, Acetoacetate and glucose as substrates for lipid synthesis by rat brain oligodendrocytes and astrocytes in serum-free culture, *Biochim. Biophys. Acta* **796:**20–26.

Kounnas, M. Z., Moir, R. D., Rebeck, G. W., Bush, A. I., Argraves, W. S., Tanzi, R. E., Hyman, B. T., and Strickland, D. K., 1995, LDL receptor-related protein, a multifunctional apoE receptor, binds secreted β-amyloid precursor protein and mediates its degradation, *Cell* **82:**331–340.

Kounnas, M. Z., Haudenschild, C. C., Strickland, D. K., and Argraves, W. S., 1994, Immunological localization of glycoprotein 330, low density lipoprotein receptor related protein and 39 kDa receptor associated protein in embryonic mouse tissue, *In Vivo* **8:**342–352.

Kraml, M., Bagli, J. F., and Dvornik, D., 1964, Inhibition of the conversion of 7-dehydrocholesterol to cholesterol by AY 9944, *Biochem. Biophys. Res. Commun.* **15:**455–457.

Krieger, M., and Herz, J., 1994, Structures and functions of multiligand lipoprotein receptors: macrophage scavenger receptors and LDL receptor-related proteins (LRP), *Annu. Rev. Biochem.* **63:**601–637.

Kumar, S., Cole, R., Chiapelli, F., and de Vellis, J., 1989, Differential regulation of oligodendrocyte markers by glucocorticoids: post-transcriptional regulation of both proteolipid protein and myelin basic protein and transcriptional regulation of glycerol phosphate dehydrogenase, *Proc. Natl. Acad. Sci. USA* **86:**6807–6811.

Laatsch, R. H., Kies, M. W., Gordon, S., and Alvord, E. C., 1962, The encephalitogenic activity of myelin isolated by ultracentrifugation, *J. Exp. Med.* **115:**77–78.

Lampert, P. W., and Garrett, R. S., 1971, Mechanism of demyelination in tellurium neuropathy. Electron microscopic observations, *Lab. Invest.* **25:**380–388.

Lampert, P., Garro, F., and Pentshew, A., 1970, Tellurium neuropathy, *Acta Neuropathol.* **15:**308–317.

Langan, T. J., Iimori, Y., White, G., and Volpe, J. J., 1987, Regulation of sterol synthesis and of 3-hydroxy-3-methylglutaryl coenzyme A reductase by lipoproteins in glial cells in primary culture, *J. Neurosci. Res.* **17:**361–366.

Lees, M. B., and Brostoff, S. W., 1984, Proteins of myelin in: *Myelin* (P. Morell, ed.), pp. 197–224, Plenum Press, New York.

Lemke, G., 1993, The molecular genetics of myelination—an update, *Glia* **7:**263–271.

Lemke, G., and Axel, R., 1985, Isolation and sequence of a cDNA encoding the major structural protein of peripheral myelin, *Cell* **40:**501–508.

Levy-Lahad, E., Wasco, W., Poorkaj, P., Romano, D. M., Oshima, J., Yu, C. E., Jondro, P. D., Schmidt, S. D., Wang, K., Crowley, A. C., Fu, Y.-H., Guenette, S. Y., Galas, D., Nemens, E., Wijsman, E. M., Bird, T. D., Schellenberg, T. D., and Tanzi, R. E., 1995a, Candidate gene for the chromosome 1 familial Alzheimer's disease locus, *Science* **269:**973–977.

Levy-Lahad, E., Wijsman, E. M., Nemens, E., Anderson, L., Goddard, K. A., Weber, J. L., Bird, T. D., and Schellenberg, G. D., 1995b, A familial Alzheimer's disease locus on chromosome 1, *Science* **269:**970–973.

Lin, L.-F. H., and Lees, M. B., 1982, Interactions of dicyclohexylcarbodiimide with myelin proteolipid, *Proc. Natl. Acad. Sci. USA* **79:**941–945.

Macala, L. J., Yu, R. K., and Ando, S., 1983, Analysis of brain lipids by high performance thin-layer chromatography and densitometry, *J. Lipid Res.* **24:**1243–1250.

Macklin, W. B., Campagnoni, C. W., Deininger, P. L., and Gardinier, M. V., 1987, Structure and expression of the mouse myelin proteolipid protein gene, *J. Neurosci. Res.* **18:**383–394.

Macri, F., Vianello, A., Braidot, E., and Zancani, M., 1991, Free fatty acids dissipate proton electrochemical gradients in pea stem microsomes and submitochondrial particles, *Biochim. Biophys. Acta* **1058:**249–255.

Majewska, M. D., 1992, Neurosteroids: endogenous bimodal modulators of the GABA-A receptor. Mechanism of action and physiological significance, *Prog. Neurobiol.* **38:**379–395.

Martenson, R. E., and Uyemura, K., 1992, Myelin P2, a neuritogenic member of the family of cytoplasmic lipid-binding proteins, in: *Myelin: Biology and Chemistry* (R. E. Martenson, ed.), pp. 509–528, CRC Press, Boca Raton, FL.

Martini, R., 1994, Expression and functional roles of neural cell surface molecules and extracellular matrix components during development and regeneration of peripheral nerves, *J. Neurocytol.* **23:**1–28.

Martini, R., Mojahajeri, M. H., Kasper, M. H., Giese, K. P., and Schachner, M., 1995, Mice doubly deficient in the genes for P0 and myelin basic protein show that both proteins contribute to the formation of the major dense line in peripheral nerve myelin, *J. Neurosci.* **15:**4488–4495.

McEven, B., 1991a, Non-genomic and genomic effects of steroids on neural activity, *Trends Pharmacol. Sci.* **12:**141–147.

McEven, B., 1991b, Steroid hormones are multifunctional messengers in the brain, *Trends Endocrin. Metab.* **2:**62–67.

McKerracher, L., David, S., Jackson, D. L., Kottis, V., Dunn, R. J., and Braun, P. E., 1994, Identification of myelin-associated glycoprotein as a major myelin-derived inhibitor of neurite growth, *Neuron* **13:**805–811.

Mellon, S. H., and Deschepper, C. F., 1993, Neurosteroid biosynthesis: genes for adrenal steroidogenic enzymes are expressed in the brain, *Brain Res.* **629:**283–292.

Miller, W. L., 1988, Molecular biology of steroid hormone synthesis, *Endocr. Rev.* **9:**295–318.

Moestrup, S. K., Gliemann, J., and Pallesen, G., 1992, Distribution of the a2-macroglobulin receptor/lipoprotein receptor-related protein in human tissues, *Cell Tissue Res.* **269:**375–382.

Morell, P., Quarles, R. H., and Norton, W. T., 1994, Myelin formation, structure, and biochemistry, in: *Basic Neurochemistry: Molecular, Cellular and Medical Aspects* (G. J. Siegel, B. W. Agranoff, R. W. Albers, and P. B. Molinoff, eds.), pp. 117–143, Raven Press, New York.

Mueller, H. W., Gebicke-Haerter, P. J., Hangen, D. H., and Shooter, E. M., 1985, A specific 37,000-Dalton protein that accumulates in regenerating but not in nonregenerating mammalian nerves, *Science* **228:**499–501.

Mukhopadhyay, G., Doherty, P., Walsh, F. S., Crocker, P. R., and Filbin, M. T., 1994, A novel role for myelin-associated glycoprotein as an inhibitor of axonal regeneration, *Neuron* **13:**757–767.

Namba, Y., Tomonaga, M., Kawasaki, H., Otomo, E., and Ikeda, K., 1991, Apolipoprotein E immunoreactivity in cerebral amyloid deposits and neurofibrillary tangles in Alzheimer's disease and kuru plaque amyloid in Creutzfeldt-Jacob disease, *Brain Res.* **541:**163–166.

Nathan, B. P., Bellosta, S., Sanan, D. A., Weisgraber, K. H., Mahley, R. W., and Pitas, R. E., 1994, Differential effects of apolipoproteins E3 and E4 on neuronal growth in vitro, *Science* **264:**850–852.

Nathan, B. P., Chang, K.-C., Bellosta, S., Brisch, E., Ge, N., Mahley, R. W., and Pitas, R. E., 1995, The inhibitory effect of apolipoprotein E4 on neurite outgrowth is associated with microtubule depolymerization, *J. Biol. Chem.* **270:**19791–19799.

Nave, K.-A., 1994, Neurological mouse mutants and the genes of myelin, *J. Neurosci. Res.* **38:** 607–612.

Nave, K.-A., Lai, C., Bloom, F. E., and Milner, R. J., 1987, Splice site selection in the proteolipid protein (PLP) gene transcript and primary structure of the DM20 protein of central nervous system, *Proc. Natl. Acad. Sci. USA* **84:**5665–5669.

North, P., and Fleischer, S., 1983, Alteration of synaptic membrane cholesterol/phospholipid ratio using a lipid transfer protein, *J. Biol. Chem.* **258:**1242–1253.

Norton, W. T., 1977, Isolation and characterization of myelin, in: *Myelin* (P. Morell, ed.), pp. 161–199, Plenum Press, New York.

Norton, W. T., and Cammer, W., 1984, Isolation and characterization of myelin in: *Myelin* (P. Morell, ed.), pp. 147–195, Plenum Press, New York.

Opitz, J. M., 1994, RSH/SLO ("Smith–Lemli–Opitz") syndrome: historical, genetic, and developmental considerations, *Am. J. Med. Genet.* **50:**344–346.

Opitz, J. M., Penchaszadeh, V. B., Holt, M. C., Spano, L. M., and Smith, V. L., 1994, Smith–Lemli–Opitz (RSH) syndrome bibliography: 1964–1993, *Am. J. Med. Genet.* **50:**339–343.

Owens, G. C., Boyd, C. J., Bunge, R. P., and Salzer, J., 1990, Expression of recombinant myelin-associated glycoprotein in primary Schwann cells promotes the initial investment of axons by myelinating Schwann cells, *J. Cell Biol.* **111:**1171–1182.

Pericak-Vance, M. A., Bebout, J. L., Gaskell, P. C. J., Yamaoka, L. H., Hung, W. Y., Alberts, M. J., Walker, A. P., Bartlett, R. J., Haynes, C. A., Welsh, K. A., Earl, N. A., Heyman, A., Clark, C. M., and Roses, A. D., 1991, Linkage studies in familial Alzheimer disease: evidence for chromosome 19 linkage, *Am. J. Hum. Genet.* **48:**1034–1050.

Perochon, E., Lopez, A., and Tocanne, J. F., 1992, Polarity of lipid bilayers. A fluorescence investigation, *Biochemistry* **31:**7672–7682.

Peters, A., Palay, S. L., and Webster, H. D., 1991, *The Fine Structure of the Nervous System,* Oxford University Press, New York.

Pill, J., Schmidt, F. H., Stegmeier, K., and Witte, E.-C., 1985, Effects of BM 15.766 on serum lipids in rats, *Horm. Metabol. Res.* **17:**543–544.

Pitas, R. E., Boyles, J. K., Lee, S. H., Hui, D., and Weisgraber, K. H., 1987, Lipoproteins and their receptors in the central nervous system: characterization of the lipoproteins in cerebrospinal fluid and identification of apolipoprotein B, E (low density lipoprotein) receptors in the brain, *J. Biol. Chem.* **262:**14352–14360.

Popot, J.-L., Pham Dinh, D., and Dautigny, A., 1991, Major myelin proteolipid: the 4-a-helix topology, *J. Membr. Biol.* **120:**233–246.

Privat, A. C., Jacque, C., Bourre, J. M., Dupouey, P., and Baumann, N., 1979, Absence of the major dense line in myelin of the mutant mouse 'shiverer', *Neurosci. Lett.* **12:**107–112.

Raine, C. S., 1985, Experimental allergic encephalomyelitis and experimental allergic neuritis, in: *Handbook of Clinical Neurology,* Vol. 3 (P. J. Vinken, G. W. Bruyn, H. L. Klawans, and J. C. Koetsier, eds.), pp. 429–466, North Holland, Amsterdam, (revised series).

Ramsey, R. B., 1973, New concepts in brain cholesterol metabolism, *Biochem. Soc. Trans.* **1:**341–348.

Ramsey, R. B., 1977, Effect of extended hypocholesterolemic drug treatment on peripheral and central nervous system sterol content of the rat, *Lipids* **12:**841–846.

Rawlins, F. A., and Smith, M. E., 1971, Myelin synthesis in vitro: a comparative study of central and peripheral nervous tissue, *J. Neurochem.* **18:**1861–1870.

Rawlins, F. A., and Uzman, B. G., 1970, Retardation of peripheral nerve myelination in mice treated with inhibitors of cholesterol biosynthesis. A quantitative electron microscopic study, *J. Cell Biol.* **46:**505–517.

Rawlins, F. A., Hedley, W. E., Villegas, G., and Uzman, B. G., 1970, Reutilization of cholesterol-1,2-H^3 in the regeneration of peripheral nerve. An autoradiographic study, *Lab. Invest.* **22:** 237–240.

Rawlins, F. A., Villegas, G. M., Hedley, W. E., and Uzman, B. G., 1972, Fine structural localization of cholesterol-1,2-^3H in degenerating and regenerating mouse sciatic nerve, *J. Cell Biol.* **52:** 615–625.

Rebeck, G. W., Reiter, J. S., Strickland, D. K., and Hyman, B. T., 1993, Apolipoprotein E in sporadic Alzheimer's disease: Allelic variation and receptor interactions, *Neuron* **11:**575–580.

Reese, T. S., and Karnovsky, M. J., 1967, Fine structural localization of a blood-brain barrier to exogenous peroxidase, *J. Cell Biol.* **34:**207–217.

Rewcastle, N. B., 1991, Degenerative diseases of the central nervous system, in: *Textbook of Neuropathology* (R. L.Davis and D. M. Robertson, eds.), pp. 904–961, Williams & Wilkins, Baltimore, MD.

Robel, P., and Baulieu, E.-E., 1994, Neurosteroids, biosynthesis and function, *Trends Endocrin. Metab.* **5:**1–8.

Roses, A. D., Strittmatter, W. J., Pericak-Vance, M. A., Corder, E. H., Saunders, A. M., and Schmechel, D. E., 1994, Clinical application of apolipoprotein E genotyping to Alzheimer's disease, *Lancet* **343:**1564–1565.

Roux, C., Horvath, C., and Dupuis, R., 1979a, Interpretation of the isolated agenesis of the hypophysis, *Teratol.* **19:**39–41.

Roux, C., Horvath, C., and Dupuis, R., 1979b, Teratogenic action and embryo lethality of AY 9944: Prevention by a hypercholesterolemia-provoking diet, *Teratol.* **19:**35–38.

Sakai, J., Hoshino, A., Takahashi, S., Miura, Y., Ishii, H., Suzuki, H., Kawarabayasi, Y., and Yamamoto, T., 1994, Structure, chromosome location, and expression of the human very low density lipoprotein receptor gene, *J. Biol. Chem.* **269:**2173–2182.

Schneider, A., Montague, P., Griffiths, I., Fanarraga, M., Kennedy, P., Brophy, P., and Nave, K.-A., 1992, Uncoupling of hypomyelination and glial cell death by a mutation in the proteolipid protein gene, *Nature* **358:** 758–761.

Serourgne, C., and Chevellier, F., 1974, Microscopic radioautography of adult rat brain cholesterol. Problem of the blood-brain barrier, *Exp. Neurol.* **44:**1–9.

Serougne, C., Lefevre, C., and Chevallier, F., 1976, Cholesterol transfer between brain and plasma in the rat: a model for the turnover of cerebral cholesterol, *Exp. Neurol.* **51:**229–240.

Shouffani, A., and Kanner, B. I., 1990, Cholesterol is required for the reconstruction of the sodium-and chloride-coupled, gamma-aminobutyric acid transporter from rat brain, *J. Biol. Chem.* **265:** 6002–6008.

Smith, M. E., Hasinoff, C. M., and Fumagalli, R., 1970, Inhibitor of cholesterol synthesis and myelin formation, *Lipids* **5:**665–671.

Snipes, G. J., and Suter, U., 1995a, Molecular anatomy and genetics of myelin proteins in the peripheral nervous system, *J. Anat.* **186:**483–494.

Snipes, G. J., and Suter, U., 1995b, Molecular basis of common hereditary motor and sensory neuropathies in humans and in mouse models, *Brain Pathol.* **5:**233–247.

Snipes, G. J., McGuire, C. B., Norden, J. J., and Freeman, J. A., 1986, Nerve injury stimulates the secretion of apolipoprotein E by nonneuronal cells, *Proc. Natl. Acad. Sci. USA* **83:**1130–1134.

Snipes, G. J., Costello, B., McGuire, C. B., Mayes, B. N., Bock, S. S., Norden, J. J., and Freeman, J. A., 1987, Regulation of specific neuronal and non-neuronal proteins during development and following injury in the rat CNS, *Prog. Brain Res.* **71:**155–175.

Snipes, G. J., Suter, U., Welcher, A. A., and Shooter, E. M., 1992, Characterization of a novel peripheral myelin protein (PMP-22/SR13), *J. Cell Biol.* **117:**225–238.

Spady, D. K., and Dietschy, J. M., 1983, Sterol synthesis in vivo in 18 tissues of the squirrel monkey, guinea pig, rabbit, hamster, and rat, *J. Lipid Res.* **24:**303–315.

Sparks, D. L., Scheff, S. W., Hunsaker, J. 3., Liu, H., Landers, T., and Gross, D. R., 1994, Induction of Alzheimer-like β-amyloid immunoreactivity in the brains of rabbits with dietary cholesterol, *Exp. Neurol.* **126:**88–94.

Spreyer, P., Kuhn, g., Hanemann, C. O., Gillen, C., Schaal, H., Kuhn, R., Lemke, G., and Muller, H. W., 1991, Axon-regulated expression of a Schwann cell transcript that is homologous to a 'growth-arrest-specific' gene, *EMBO J.* **10:**3661–3668.

Sprinkle, T. J., 1989, 2′,3′-Cyclic nucleotide 3′-phosphodiesterase, an oligodendrocyte-Schwann cell and myelin-associated enzyme of the nervous system, *Crit. Rev. Neurobiol.* **4:**235–301.

Steinberg, D., and Avigan, J., 1960, Studies of cholesterol biosynthesis. II. The role of desmosterol in the biosynthesis of cholesterol, *J. Biol. Chem.* **235:**3127–3129.

Sternberger, N. H., Quarles, R. H., Itoyama, Y., and Webster, H. D. F., 1979, Myelin associated glycoprotein demonstrated immunochemically in myelin and myelin-forming cells of developing rat, *Proc. Natl. Acad. Sci. USA* **76:**1510–1514.

Stoll, G., Mueller, H. W., Trapp, B. D., and Griffin, J. W., 1989, Oligodendrocytes but not astrocytes express apolipoprotein E after injury of rat optic nerve, *Glia* **2:**170–176.

Strittmatter, W. J., Saunders, A. M., Schmechel, D., Pericak-Vance, M., Enghild, J., Salvesen, G. S., and Roses, A. D., 1993, Apolipoprotein E: High-avidity binding to β-amyloid and increased frequency of type 4 allele in late-onset familial Alzheimer disease, *Proc. Natl. Acad. Sci. USA* **90:**1977–1981.

Suter, U., and Snipes, G. J., 1995, Biology and genetics of hereditary motor and sensory neuropathies, *Annu. Rev. Neurosci.* **18:**45–75.

Suzuki, K., and De Paul, L. D., 1971, Cellular degeneration in developing central nervous system of rats produced by hypocholesterolemic drug AY9944, *Lab. Invest.* **25:**546–555.

Suzuki, K., Zagoren, J. C., Chen, S. M., and Suzuki, K., 1974, Effects of triparanol and 20,25-diazacholesterol in CNS of rat: Morphological and biochemical studies, *Acta Neuropathol.* **29:**141–156.

Swanson, L. W., Simmons, D. M., Hofmann, S. L., Goldstein, J. L., and Brown, M. S., 1988, Localization of mRNA for low density lipoprotein receptor and a cholesterol synthetic enzyme in rabbit nervous system by *in situ* hybridization, *Proc. Natl. Acad. Sci. USA* **85:**9821–9825.

Tanzi, R. E., Gusela, J. F., Watkins, P. C., Bruns, G. A. P., St. George-Hyslop, P. H., Van Keuren, M. L., Patterson, D., Pagan, S., Kurnit, D. M., and Neve, R. L., 1987, The amyloid β protein gene: cDNA cloning, mRNA distribution, and genetic linkage near the Alzheimer locus, *Science* **235:**880–894.

Thomas, P. K., and Ochoa, J., 1984, Microscopic anatomy of peripheral nerve fibers, in: *Peripheral Neuropathy, 2nd edition* (P. J. Dyck, P. K. Thomas, E. H. Lambert, and R. Bunge, eds.), pp. 39–96, W. B. Saunders Co., Philadelphia, PA.

Thompson, M. J., Dupont, J., and Robbins, W. E., 1963, The sterols of liver and carcass of 20,25-diaza-cholesterol fed rats, *Steroids* **2:**99–104.

Tint, G. S., Irons, M., Elias, E. R., Batta, A. K., Frieden, R., Chen, T. S., and Salen, G., 1994, Defective cholesterol biosynthesis associated with the Smith–Lemli–Opitz syndrome, *N. Eng. J. Med.* **330:**107–113.

Toews, A. D., Lee, S. Y., Popko, B., and Morell, P., 1990, Tellurium-induced neuropathy: a model for reversible reductions in myelin protein gene expression, *J. Neurosci. Res.* **26:**501–507.

Toews, A. D., Goodrum, J. F., Lee, S. Y., Eckermann, C., and Morell, P., 1991, Tellurium-induced alterations in 3-hydroxy-3-methylglutaryl-CoA reductase gene expression and enzyme activity: differential effects in sciatic nerve and liver suggest tissue-specific regulation of cholesterol synthesis, *J. Neurochem.* **57:**1902–1906.

Toews, A. D., Griffiths, I. R., Kyriakides, E., Goodrum, J. F., Eckermann, C. E., Morell, P., and Thomson, C. E., 1992, Primary demyelination induced by exposure to tellurium alters Schwann cell gene expression: a model for intracellular targeting of NGF receptor, *J. Neurosci.* **12:** 3676–3687.

Trapp, B. D., Dubois-Dalcq, M., and Quarles, R. H., 1984, Ultrastructural localization of the P2 protein in actively myelinating rat Schwann cells, *J. Neurochem.* **43:**944–948.

Trapp, B. D., Hauer, P., and Lemke, G., 1988, Axonal regulation of myelin protein mRNA levels in actively myelinating Schwann cells, *J. Neurosci.* **8:**3515–3521.

Vemuri, R., and Philipson, K. D., 1989, Influence of sterols and phospholipids on sarcolemmal and sarcoplasmic reticular cation transporters, *J. Biol. Chem.* **264:**8680–8685.

Verdi, J. M., and Campagnoni, A. T., 1990, Translational regulation by steroids, *J. Biol. Chem.* **265:**20314–20320.

Volpe, J. J., and Obert, K. A., 1981, Coordinate regulation of cholesterol synthesis and 3-hydroxy-3-methylglutaryl coenzyme A synthase but not 3-hydroxy-3-methylglutaryl coenzyme A reductase in C-6 glia, *Arch. Biochem. Biophys.* **212:**88–97.

Wagner, M., Toews, A. D., and Morell, P., 1995, Tellurite specifically affects squalene epoxidase: investigations examining the mechanism of tellurium-induced neuropathy, *J. Neurochem.* **64:** 2169–2176.

Wagner-Recio, M., Toews, A. D., and Morell, P., 1991, Tellurium blocks cholesterol synthesis by inhibiting squalene metabolism: preferential vulnerability to this metabolic block leads to peripheral nervous system demyelination, *J. Neurochem.* **57:**1891–1901.

Weimbs, T., and Stoffel, W., 1992, Proteolipid protein (PLP) of CNS myelin: Positions of free, disulfide-bonded, and fatty acid thioester-linked cysteine residues and implications for the membrane topology of PLP, *Biochemistry* **31:**12289–12296.

Welcher, A. A., Suter, U., De Leon, M., Snipes, G. J., and Shooter, E. M., 1991, A myelin protein is encoded by the homologue of a growth arrest-specific gene, *Proc. Natl. Acad. Sci. USA* **88:** 7195–7199.

Wells, M. A., and Dittmer, J. C., 1967, A comprehensive study of the postnatal changes in the concentration of the lipids of developing rat brain, *Biochemistry* **6:**3169–3175.

Yan, Y., Lagenaur, C., and Narayanan, V., 1993, Molecular cloning of M6: Identification of a PLP/DM20 gene family, *Neuron* **11:**423–431.

Zheng, G., Bachinsky, D. R., Stamenkovic, I., Strickland, D. K., Brown, D., Andres, G., and McCluskey, R. T., 1994, Organ distribution in rats of two members of the low-density lipoprotein receptor gene family, gp330 and LRP/a2-macroglobulin receptor, and the receptor-associated protein (RAP), *J. Histochem. Cytochem.* **42:**531–542.

Zhou, Q., Jimi, S., Smith, T. L., and Kummerow, F. A., 1991, The effect of cholesterol on the accumulation of intracellular calcium, *Biochim. Biophys. Acta* **1085:**1–6.

Zoidl, G., Blass-Kampmann, S., D'Urso, D., Schmalenbach, C., and Muller, H. W., 1995, Retroviral-mediated gene transfer of the peripheral myelin protein PMP22 in Schwann cells: modulation of cell growth, *EMBO J.* **14:**1122–1128.

Chapter 8

Regulation of Mitochondrial Cholesterol Metabolism

Federico Martínez and Jerome F. Strauss III

1. INTRODUCTION

Cholesterol is required for the growth and replication of mammalian cells. As an integral component of membranes, particularly the plasmalemma, cholesterol facilitates membrane curvature and increases resistance to physical damage. The insertion of cholesterol into membranes increases fluidity (Chapman, 1975), a property closely correlated with the cholesterol/phospholipid (C/PL) ratio. Cholesterol is also the substrate for the production of bile acids and steroid hormones. Because of cholesterol's importance to cellular function, intracellular cholesterol concentrations are highly regulated (Yeagle, 1985; Lange and Ramos, 1983). Indeed, increases or decreases in this sterol can produce major alterations in cell physiology.

Most cells can synthesize sufficient amounts of cholesterol *de novo* to meet their metabolic requirements. However, extracellular cholesterol can be incorporated into nucleated cells, mostly through the uptake of low density lipoprotein particles (LDLs) mediated by lipoprotein receptors localized to the plasma membrane. After LDLs are internalized by endocytosis and hydrolyzed in lysosomes,

Federico Martínez Department of Biochemistry, Faculty of Medicine, UNAM, Mexico. **Jerome F. Strauss III** Department of Obstetrics and Gynecology, University of Pennsylvania School of Medicine, Philadelphia, Pennsylvania 19104.

Subcellular Biochemistry, Volume 28: Cholesterol: Its Functions and Metabolism in Biology and Medicine, edited by Robert Bittman. Plenum Press, New York, 1997.

cholesterol is transferred to various cellular membranes (Faust *et al.*, 1977; Goldstein and Brown, 1984). Other receptors, highly expressed on certain steroidogenic cells, mediate the selective transfer of sterol esters from high density lipoproteins (HDL) through a process that bypasses the normal endocytic pathway (Nestler *et al.*, 1990; Acton *et al.*, 1996). Whether cholesterol is derived from extracellular or intracellular sources, its distribution is tightly controlled.

Subcellular membranes contain different amounts of cholesterol (DePierre and Ernster, 1977; Daum, 1985); the plasma membrane contains 30 μg of cholesterol per mg of protein, whereas mitochondria possess between 3–5 μg of sterol per mg protein (Daum, 1985). Biological membranes also have microenvironments with different lipid and protein concentrations. To achieve the specific distribution of cholesterol within these membranes and microdomains, cells must have specialized mechanisms to deliver cholesterol to particular membranes and partition the sterol within the membrane. However, these mechanisms have not been fully elucidated.

Membrane composition, specifically, protein composition, could be a factor regulating the cholesterol content of membranes (Kleemann and McConnell, 1976; Cherry *et al.*, 1980; Sackmann *et al.*, 1984). It has been suggested that outer mitochondrial membranes possess a high cholesterol content (40 μg/mg protein) because their protein content is low (5%), whereas cholesterol concentrations in the inner membrane are low (3 μg/mg protein), because the relative amount of protein is high (55%) (Crémel *et al.*, 1990; Parsons and Yano, 1967).

Certain lipids may influence membrane cholesterol concentrations. Vesicles or particles containing specific phospholipids bind more cholesterol than vesicles made up of other phospholipids (Demel *et al.*, 1977; Bloj and Zilversmit, 1977; Lange *et al.*, 1979). Cellular membranes also exhibit variations in their phospholipid content, which appear to be correlated with membrane cholesterol content (Wattenberg and Silbert, 1983). Patton (1970) reported that the cholesterol content of several biological membranes was correlated with the concentration of sphingomyelin. Yet, despite the correlation between sphingomyelin and cholesterol in some membranes, this relationship is not invariably found.

2. MITOCHONDRIAL CHOLESTEROL DYNAMICS

2.1. Cholesterol Content of Mitochondria

Although mitochondria are relatively independent organelles, containing their own DNA and RNA, most of their membrane lipids are synthesized in the endoplasmic reticulum and then inserted into the mitochondrial membranes. The mechanism of insertion of specific lipids into mitochondria and other subcellular membranes remains unknown (Houslay and Stanley, 1990).

In a comprehensive review, Daum (1985) described the role of several lipids in mitochondria from animals, plants, and microorganisms. The content of cholesterol in mitochondria is lower than in other cell membranes, although there are some exceptions, such as the mitochondria of the gastric mucosa (Sen and Ray, 1980), rabbit heart muscle (Nagamoto *et al.,* 1980), and rat heart (Palmer *et al.,* 1981). Most of the cholesterol in mitochondria is located in the outer membranes (Table I).

Other lipids are uniquely distributed in mitrochondira. Cardiolipin is a characteristic phospholipid of mitochondria; its concentration in mitochondrial membranes is 4–18 times higher than in other cellular membranes. It has been suggested that cardiolipin is located exclusively in the inner membrane, which is the site of its synthesis (Colbeau *et al.,* 1971; Comte *et al.,* 1976).

The lipid composition of mitochondria is modified by diet as well as by disease. Feo *et al.* (1975) demonstrated that the cholesterol content of rat liver mitochondria is increased on a high cholesterol diet. Mitochondria from hepatoma cells also have an increased cholesterol content (Brossa *et al.,* 1980).

Rouslin *et al.* (1982) found that cholesterol increases in mitochondria isolated from ischemic pig heart. The increase in cholesterol content was twice that of controls (62 vs. 33 μg of cholesterol/mg), but the concentration of mitochondrial phospholipids was not altered. Thus, the C/PL ratio changed and, consequently, the microviscocity of the membranes was modified. The mitochondrial cholesterol content returned to the control level upon recovery from ischemia.

2.2. Enrichment of Mitochondrial Membranes with Cholesterol

The cholesterol content of mitochondria can be experimentally manipulated. Several methods have been described to increase the cholesterol content of mitochondria *in vitro.* Cholesterol presented in liposomes was used to enrich mitochondria with sterol by Toaff *et al.* (1979), Schneider *et al.* (1982), and Madden *et al.* (1980). Graham and Green (1970) used serum lipoproteins to deliver cholesterol to rat liver mitochondria. However, the amount of cholesterol incorporated was relatively low using this method. Coleman *et al.* (1978) developed a procedure based on the adsorption of cholesterol to Sephadex resin. This methodology could not be reproduced by Dietzen and Davis (1994).

Bovine serum albumin–cholesterol complexes have been used to increase cholesterol in isolated rat liver mitochondria (Martínez *et al.,* 1988; Echegoyen *et al.,* 1993). With this system, it was possible to increase cholesterol content in a controlled fashion. This method produces an increase in small vesicles between the outer and inner membranes (Echegoyen *et al.,* 1993), suggesting that cholesterol transport from outer to inner membrane could follow the model of vesiculation. However, the incorporated cholesterol remained in the outer membranes.

Other proteins have been employed to modify the cholesterol content of biological membranes, including nonspecific lipid transfer protein, also known as sterol carrier protein$_2$ (SCP2) (North and Fleischer, 1983). Modifications of the cholesterol content of red blood cells and mouse spleen cells have been achieved

Table I
Lipid Composition of the Inner and Outer Mitochondrial Membranes[a]

		P-lipids (μg/mg protein)	Sterols	CL Percentage of total phospholipids	PE	PC	PI	Lyso	SPM	PC-Plas	PE-Plas
Mammalian cells											
Liver											
Rat	IM	155–310	1–6	14–23	32–39	38–45	2–7	1–3	1–2		
	OM	280–600	21–66	3–10	20–35	44–59	5–20	2–3	2–4		
Guinea pig	IM	301	5	22	28	45	4			2	
	OM	878	30	3	25	55	14			3	
Heart											
Pig	IM	310	2	25	38	27	3	2	1	7	20
	OM	510	10	<1	28	56	9	1	5	23	12

[a]Abbreviations: IM, inner mitochondrial membrane; OM, outer mitochrondrial membrane; P-lipids, total phospholipids; CL, cardiolipin; PE, phosphatidylethanolamine; PC, phosphatidylcholine; PI, phosphatidylinositol; Lyso, lysophosphoglycerides; SPM, sphingomyelin; PC-Plas, phosphatidylcholine plasmalogens; PE-Plas, phosphatidylethanolamine plasmalogens. From Daum (1985). Reprinted with permission.

using human heat-inactivated plasma enriched with dispersed cholesterol (Shinitzky, 1978). These methods could be applied to the study of mitochondria.

Diets rich in cholesterol have been used by Dietzen and Davis (1994) and Feo *et al.* (1975) to alter mitochondrial sterol contents *in vivo*.

2.3. Metabolic Consequences of Altering Mitochondrial Cholesterol Content

Mitochondria enriched with cholesterol have been reported to display a variety of alterations. However, the literature is conflicting regarding the nature and the significance of these changes. Schneider *et al.* (1982) showed that cholesterol incorporation via liposomes into mitoplasts prepared from rat liver mitochondria caused lateral displacement of proteins, resulting in the formation of aggregates in the membrane and the physical separation of the proteins responsible for electron transport.

In mitoplasts and submitochondrial particles, incorporation of cholesterol stimulates the adenine ATP–ADP translocator (Krämer, 1982), whereas incorporation of cholesterol into intact organelles *in vitro* decreased ATPase activity (Pitotti *et al.*, 1980; Bruni *et al.*, 1975). In contrast, Rindina *et al.* (1986) reported that an increase in mitochondrial cholesterol raises ATPase activity in parallel to the effect of cholesterol on the diffusion coefficient of ubiquinone (Chazzote *et al.*, 1991). In this setting, cholesterol is thought to induce a lateral phase separation that progressively excludes ubiquinone from cholesterol-containing domains. A linear relationship between the content of cholesterol in the membranes and the passive proton permeability has also been observed (Baggetto *et al.*, 1992).

Graham and Green (1970) did not find changes in respiratory activity when mitochondria were enriched with cholesterol, although the succinate-cytochrome c reductase activity was higher than in control mitochondria. ATPase and succinate dehydrogenase activities were not changed when rat liver mitochondria contained seven times more cholesterol than controls (Echegoyen *et al.*, 1993). Coleman *et al.* (1978) and Parlo and Coleman (1984) demonstrated that increases in cholesterol in rat liver mitochondria *in vitro* produce changes similar to those observed in mitochondria isolated from hepatomas, including a low respiratory control ratio and an increase in H^+-ATPase activity. Dietzen and Davis (1994) observed that in rats fed a high cholesterol diet, the level of sterol in isolated liver mitochondria increased to the same degree as observed in hepatoma mitochondria. However, these investigators did not find metabolic alterations in the organelles.

Several investigators have suggested that increasing the cholesterol content of mitochondrial membranes raises their resistance to physical damage. Dietzen and Davis (1994) found that the sterol-rich mitochondria from tumor cells remained intact after incubation in water. Jain (1975) suggested that increased cholesterol protects yeast cells against the effects of high temperature. Feo *et al.* (1975) observed that increasing the cholesterol content of mitochondria from rat liver reduces hypoosmotic and phosphate-induced swelling.

2.4. Mechanisms of Intramitochondrial Sterol Movement

Several mechanisms have been proposed by which cholesterol could be transferred among mitochondrial membranes. One mechanism is desorption, in which excess cholesterol from a membrane is released into the aqueous phase and then incorporated into an acceptor membrane (McLean and Phillips, 1981; Clejan and Bittman, 1984; Bellini *et al.,* 1984; Bruckdorfer and Sherry, 1984; Fugler *et al.,* 1985).

Physical contact between membranes could also be a mechanism of cholesterol transfer. In this model, a membrane rich in cholesterol transfers sterol to another with a low sterol content by fusion or by a vesiculation mechanism (Steck *et al.,* 1988; Lange and Matthies, 1984; Billheimer and Reinhart, 1990). In this regard, the attachment points between outer and inner membranes of mitochondria could be important, since in these membranous structures the outer membrane is rich in cholesterol and the inner membrane is sterol-poor (Levrat and Louisot, 1992; Ardail *et al.,* 1990). Studies examining the lipid content and asymmetry in mitochondrial membranes support this idea (DePierre and Ernster, 1977). Attachment sites between the outer and inner membranes are located at sites of protein import into mitochondria. Hence, cholesterol could move from the outer to the inner membrane in association with the transport of proteins into the mitochondrial matrix as a consequence of the formation of an organized pore (Werner and Neupert, 1972; Stuart *et al.,* 1994; Pfanner *et al.,* 1994).

Using a rapid freezing technique, Ohlendieck *et al.* (1986) isolated a fraction from rat liver mitochondria containing boundary membrane contact sites. In cross fractures, larger vesicles containing smaller ones were observed; occasionally the membranes of these inner and outer vesicles appeared to be attached. These contact sites may have a hetergeneous lipid composition (Dorbani *et al.,* 1987). Freeze-fracture electron microscopy demonstrated that the attachment points have a hexagonal lipid phase (Van Venetie and Verkleij, 1982; Kachar and Reese, 1982).

There is evidence suggesting that excess cholesterol in membranes facilitates the transition of the bilayer to a hexagonal phase that facilitates membrane fusion (Van Venetie and Verkleij, 1982; Kachar and Reese, 1982; Cornelius, 1990: Kimura, 1986). Cholesterol in association with Ca^{2+} and phosphate could induce fusion of outer and inner membranes (Ohki and Leonards, 1984) and transform the bilayer into a hexagonal phase (Van Venetie and Verkleij, 1982; Kachar and Reese, 1982).

3. METABOLISM OF CHOLESTEROL IN STEROIDOGENIC TISSUES

Steroidogenic tissues (adrenals, gonads, placenta) transform cholesterol into hormones. Steroid hormones are also produced in the brain and possibly fetal intestine and uterine decidua, but in much smaller quantities than the above noted organs.

The mechanism of cholesterol transport to and within mitochondria of steroidogenic tissues is of considerable interest, since this sterol transport process is regulated by trophic hormones [e. g., adrenocorticotropin (ACTH) and gonadotropins] in the adrenal cortex and gonads (Hall, 1985; Rennert *et al.,* 1993; Roff *et al.,* 1994). Here we will focus on the aspects of adrenal steroidogenesis, almost all of which are shared by the steroidogenic cells of the gonads.

3.1. Intracellular Pools of Steroid Hormone Precursors

Adrenal cortex cells contain cytoplasmic lipid droplets in which cholesterol is stored. About 90% of the cholesterol in these lipid droplets is esterified. When the cells receive a hormonal stimulus, a sequence of steps leading to mobilization of free cholesterol from lipid droplets and its transport to mitochondria is initiated (Figure 1).

The amount of cholesterol in mitochondria isolated from the adrenal gland is relatively high, and closely correlated with their metabolic function (della-Cioppa *et al.,* 1986) (Table II). Cheng *et al.* (1985) showed that the cholesterol

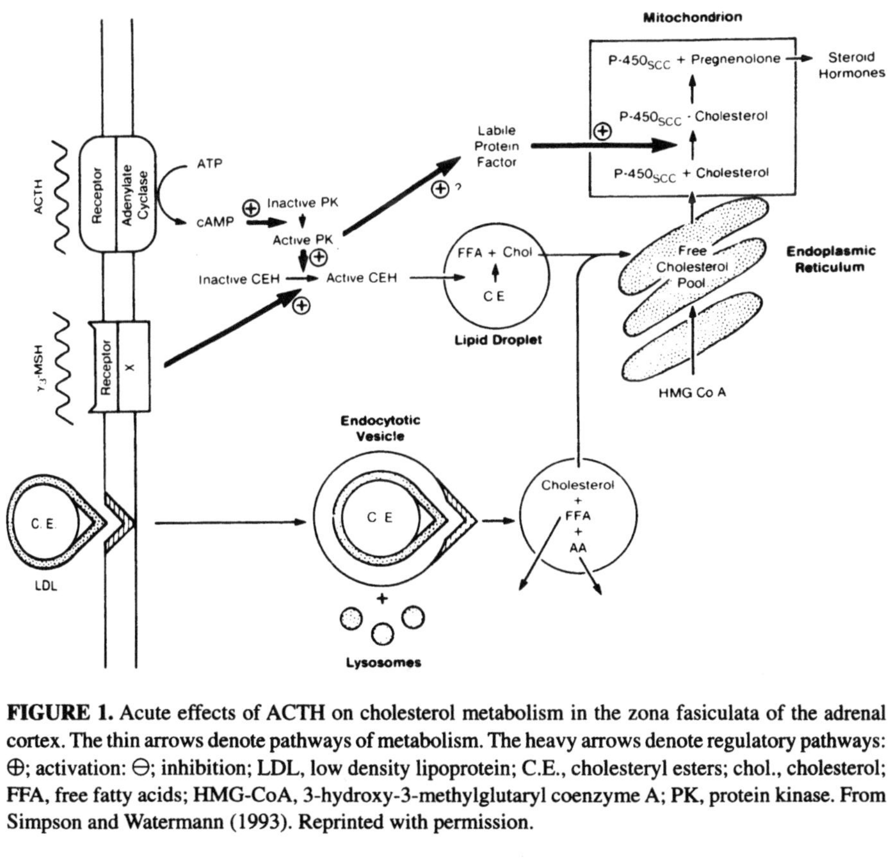

FIGURE 1. Acute effects of ACTH on cholesterol metabolism in the zona fasiculata of the adrenal cortex. The thin arrows denote pathways of metabolism. The heavy arrows denote regulatory pathways: ⊕; activation: ⊖; inhibition; LDL, low density lipoprotein; C.E., cholesteryl esters; chol., cholesterol; FFA, free fatty acids; HMG-CoA, 3-hydroxy-3-methylglutaryl coenzyme A; PK, protein kinase. From Simpson and Watermann (1993). Reprinted with permission.

Table II
Composition of Mitochondrial Membranes from Y-1 Cells[a]

	Cholesterol		Cholesterol/ phospholipid	Protein	
	(μg/mg protein)	(mg/mg phospholipid)	molar ratio	(mg/mg phospholipid)	Total (μg)
Whole mitochondria	29.7 ± 5.1	91.3 ± 2.5	4.7 ± 1.3	3.1 ± 0.7	275
Outer membrane	193.0 ± 24.0	137.0 ± 13.0	7.1 ± 0.7	0.73 ± 0.07	15
Inner membrane	18.0 ± 6.7	31.5 ± 6.1	1.6 ± 0.3	2.48 ± 0.13	184
Outer/inner (ratio)	10.7	4.3	4.4	0.29	0.08

[a]Values are means ± S.E. For outer membrane $n = 4$; for inner membrane $n = 3$. From della-Cioppa *et al.* (1986). Reprinted with permission.

from the inner mitochondrial membrane was quickly and efficiently used for steroid hormone synthesis during hormonal stimulation of bovine adrenal mitochondria.

Cholesterol is transformed into pregnenolone, the precursor of all other steroid hormones, by cytochrome P450scc, a mixed function oxidase that is located on the inner mitochondrial membranes. Steroidogenic mitochondria possess two electron transport chains. One is the respiratory chain involved in the synthesis of ATP; the other is the electron transport chain coupled to cytochrome P450scc. This electron transport chain consists of adrenodoxin reductase, adrenodoxin, and the side-chain cleavage cytochrome P450 (P450scc). Adrenodoxin reductase is a flavoprotein containing FMN as an hydrogen acceptor; its substrate is NADPH + H^+. Adrenodoxin is a shuttle protein that carries electrons between adrenodoxin reductase and cytochrome P450scc.

The pathway for cholesterol transformation by cytochrome P450scc has been elucidated (Orme-Johnson, 1990) (Figure 2). The first step is 22-hydroxylation, in which 1 mole of NADPH is reduced by 1 mole of cholesterol in the presence of 1 mole of O_2, forming 1 mole of 22*R*-hydroxycholesterol. In the second step, another hydroxylation occurs at C-20 to yield 20α,22*R*-dihydroxycholesterol; in this step 1 mole of NADPH and 1 mole of O_2 are required for each mole of substrate (Simpson and Waterman, 1983). After these two hydroxylations, P450scc cleaves the side chain of 20α,22*R*-dihydroxycholesterol to release pregnenolone and isocapraldehyde. The latter reaction consumes 1 mole of NADPH and 1 mole of O_2 per mole of substrate. Therefore, the stoichiometry of the total reaction for NADPH, oxygen, and cholesterol is 3:3:1. Once bound to P450scc, cholesterol undergoes the three successive cycles of oxidation and reduction without dissociating from the enzyme as a consequence of increased affinity of the hydroxylated intermediates for P450scc. Pregnenolone inhibits the metabolism of cholesterol by cytochrome P450scc. Hence, a regulatory function has been ascribed to intramitochondrial pregnenolone.

In adrenal cells, ACTH stimulates the synthesis of pregnenolone via cAMP, which exerts several effects on cholesterol metabolism. First, cholesterol must be moved from the cytoplasmic lipid droplets stores to mitochondria. Adrenal cells maintain about 90% of their cholesterol in the esterified form. Polyunsaturated fatty acids like arachidonic acid are usually esterified to cholesterol (Pedersen and Brownie, 1987). ACTH causes a rapid reduction in the total amount of cholesterol in adrenal cortex by mobilizing cholesterol esters from lipid droplets and inhibiting sterol ester synthesis. The mobilization of sterol esters is mediated by the cAMP-induced phosphorylation of a cytoplasmic neutral sterol esterase by protein kinase A.

Phosphorylation of the enzyme increases its catalytic activity and may also affect its subcellular distribution. A phosphorylation cascade may inhibit cholesterol acyl-CoA acyl transferase (ACAT). In addition, progestins inhibit ACAT, preventing the re-esterification of free cholesterol (Rennert *et al.*, 1993).

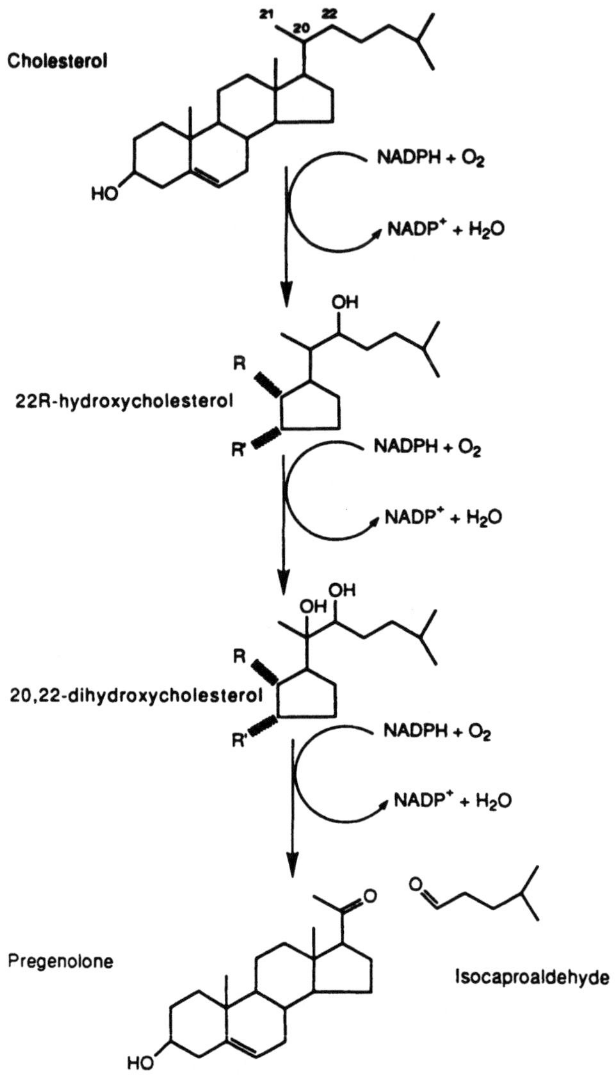

FIGURE 2. Reaction catalyzed by cytochrome P-450$_{scc}$. The probable steroid intermediates and the other reactants needed for each conversion are indicated. This reaction is the initial and the rate-determining reaction in glucocorticoid synthesis. From Orme-Johnson (1990). Reprinted with permission.

Freeman (1989) proposed an alternative pathway by which the steroidogenic substrate reaches the mitochondria. He suggested that cholesterol used in steroidogenesis by MA-10 Leydig cells is derived primarily from the plasma membrane. Although a relatively high concentration of cholesterol was found in mitochondria, this was largely inert as substrate for steroidogenesis; plasma membrane cholesterol was incorporated into steroid hormones without ever equilibrating with the mitochondrial cholesterol pool. Freeman suggested that cholesterol must enter the mitochondria via some specialized membrane contact areas.

Pedersen and Brownie (1983) determined the distribution of cholesterol in rat adrenal cells, outer and inner mitochondrial membranes. These authors observed that in the presence of aminoglutethimide, an inhibitor of P450scc, ether anesthesia stress, which acutely raises ACTH, caused accumulation of cholesterol in the inner mitochondrial membranes. This cholesterol was available to cytochrome P450scc, as demonstrated by its conversion into pregnenolone *in vitro* after removal of the inhibitor. The cholesterol increase in the inner membrane was blocked by simultaneous administration of the protein synthesis inhibitor, cycloheximide (CHX). In the presence of CHX, sterol settled in the outer membranes in response to stress. These findings led to the conclusion that CHX blocks the transfer of cholesterol to inner mitochondrial membranes, an idea consistent with earlier observations of Ferguson and co-workers (Ferguson and Morita, 1964) that the acute steroidogenic response of adrenals to trophic stimulation is blocked by protein synthesis inhibitors. These and other results indicated that a short-lived or labile protein factor is involved in the transfer of cholesterol from the outer membranes to the inner membranes of adrenal cortical mitochondria.

The artificial enrichment of the plasma membrane of Y-1 cells with cholesterol (higher C/PL ratio) is associated with higher levels of cholesterol in the cytoplasm and higher concentration of cholesterol in the inner membranes of mitochondria. Pregnenolone synthesis stimulated by dibutyryl-cAMP was higher in cells with a high C/PL ratio than in those cells with a low C/PL ratio (Iida *et al.*, 1989), but no differences were observed in mitoplasts. With these results, the authors suggested that cholesterol transport to the inner mitochondrial membrane of adrenal mitochondria could be regulated by at least two processes, namely, the supply of substrate cholesterol, which can saturate a transport mechanism, and some mechanism(s) stimulated by cAMP capable of enhancing transport above the maximal levels produced by high concentrations of cholesterol alone. The latter mechanism requires the synthesis of new protein(s).

3.2. Factors that Regulate Cholesterol Transport to Mitochondria

3.2.1. The Cytoskeleton

Participation of the cytoskeleton actin filaments, intermediate filaments, and microtubules, in cholesterol transport from various pools to mitochondria has been proposed. Cyto-chalasin B, an antimicrofilament agent, and vinblastine, an anti microtubule drug, block steroid synthesis by adrenal cells, but not by isolated mito chondria (Crivello and Jefcoate, 1980).

The steroidogenic response elicited by ACTH is inhibited by anti-actin antibodies, indicating that the cellular reaction to ACTH requires a pool of monomeric

(G-) actin that can be polymerized to F-actin. ACTH reorganizes the cytoskeleton and changes the shape of adrenal cells, suggesting a mechanism for transport of cholesterol that involves reorganization and contraction of actin microfilaments that may, in turn, cause movement of lipid droplets and mitochondria towards common points of attachment to intermediate filaments (Hall and Almahbobi, 1992). Almahbobi *et al.* (1992) reported that both lipid droplets and mitochondria are directly anchored to intermediate filaments in adrenal cells, as well as in rat Leydig cells (Almahbobi *et al.*, 1993).

3.2.2. Sterol Carrier Protein$_2$ (SCP2)

SCP2, a protein with a molecular weight of 13.5 kDa (Scallen *et al.*, 1985; Kesav *et al.*, 1992), may be involved in the transport of cholesterol to mitochondria. SCP2 stimulates the enzymatic conversion of sterol intermediates between lanosterol and cholesterol (Noland *et al.*, 1980), and is thought to participate in key steps in the utilization of cholesterol and intracellular transfer of sterol among cellular organelles (Chanderbhan *et al.*, 1982; Vahouny *et al.*, 1984). Addition of homogeneous rat liver SCP2 or of an adrenal cytoplasm fraction enhanced pregnenolone production by adrenal mitochondria, probably by improving the access of cholesterol to cytochrome P450scc (Vahouny *et al.*, 1983). In addition, overexpression of SCP2 in COS-7 cells engineered to produce progestins stimulated steroid synthesis (Yamamoto *et al.*, 1991).

Although there are reports of sterol binding to SCP2, Kesav *et al.* (1992) suggested that a possible mechanism of action of SCP2 could be the formation of a binary complex between SCP2 and the membrane, or the formation of a ternary complex involving SCP2 and two membranes. Despite the fact that existing data strongly implicate SCP2 in intracellular sterol movement, definitive evidence for a critical role for this protein in the steroidogenic process is still lacking (Pfeifer *et al.*, 1993).

3.2.3. Formation of Mitochondrial Membrane Contact Points

It has been observed that ACTH stimulates a very rapid net synthesis of phospholipids, apparently secondary to *de novo* synthesis of phosphatidic acid (Cozza *et al.*, 1990; Igarashi and Kimura, 1984). This increase in lipid synthesis results in an elevation of phosphatidylinositol (PI) and phosphatidylethanolamine (PE) in the mitochondrial outer membranes (Igarashi and Kimura, 1984). These findings, in conjunction with the enrichment of cardiolipin at the membrane contact sites (Ardail *et al.*, 1990), raise the possibility of alterations in mitochondrial membranes, including formation of hexagonal phases, which promote membrane fusion.

In a comprehensive review, Kimura (1986) described several factors that induce hexagonal phase (H_{II}) formation in membranes, through which cholesterol could be transported from outer to inner mitochondrial membranes (Figure 3). Kimura summarized the factors that stimulate H_{II} as follows: (i) an increase in temperature, (ii) a decrease in water content, (iii) an increase in cholesterol content, (iv) a decrease in phospholipid head group size, (v) a decrease in head group charge, (vi) an increase in fatty acyl chain length, and (vii) an increase in fatty acyl chain unsaturation. Kimura also proposed that Ca^{2+} and polylysine could increase the H_{II} phase by canceling the phospholipid charge and favoring the junction between mitochondrial membranes. In this model, the presence of cholesterol by itself is able is to transform the characteristics of the membrane, so that it could be a regulator of its own translocation (see Clejan and Bittman, 1984).

McNamara and colleagues described heterogeneous pools of cholesterol whose use depends on the source of reducing equivalents (McNamara *et al.*, 1990a, 1990b; Yamazaki *et al.*, 1993). They suggested a model for cholesterol transfer (Figure 4) that may occur following a labile fusion of inner and outer membranes in hexagonal phase membrane domains (Jefcoate *et al.*, 1992).

Changes in mitochondrial volumes could be related to hormonal stimulation. These changes in volume could produce physical and morphological changes, increasing the area of contact between the outer and inner membranes. Lambeth and Stevens (1985) suggested that the contact points in mitochondria could be favored by the action of ACTH, either by the synthesis of a protein that connects both membranes, or by changes in the matrix volume. Whatever the mechanism of action of ACTH, an increase in the mitochondrial matrix volume will increase the size and number of contact sites, which in turn could facilitate the passive transfer of cholesterol from the outer cholesterol-rich membrane to the inner cholesterol-poor membrane. When the mitochondrial matrix volumes of stressed rats were compared in the presence or absence of CHX, an increase in the matrix volume of rats exposed to stress was observed in the absence of the protein synthesis inhibitor, whereas in the rats exposed to stress in the presence of CHX, no differences were observed as compared to control animals (Stevens *et al.*, 1985). Under all condi-

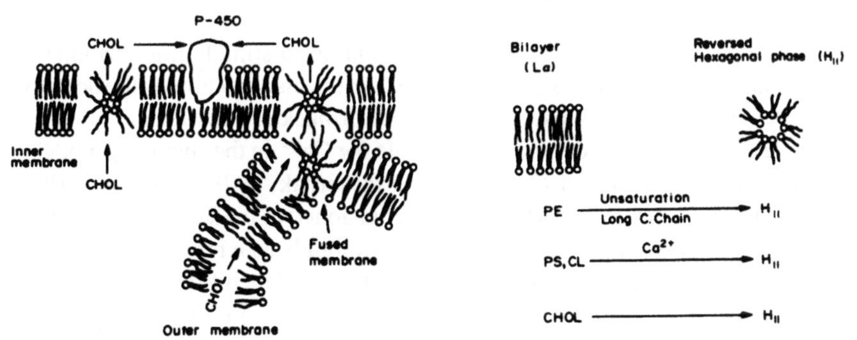

FIGURE 3. Factors stimulating reversed hexagonal phase (left) and permeation of cholesterol across membrane (right). From Kimura (1985). Reprinted with permission.

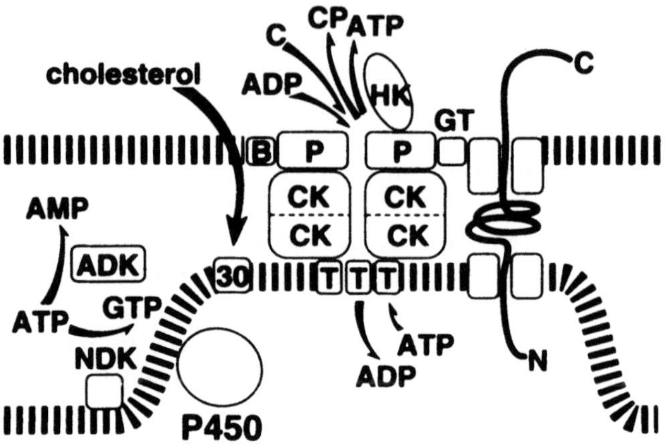

FIGURE 4. A model for cholesterol transfer at contact sites. Contact sites comprise ADP/ATP translocator (T), octameric creatine kinase (CK), porin (P), hexokinase (HK), benzodiazepine receptor (B), glutathione-transferase (GT), and protein translocation complex. The inner membrane contains a 30-kDa phosphoprotein (Jefcoate *et al.*, 1992) that is associated with active steroidogenesis and that may facilitate cholesterol transfer. Adenylate kinase (ADK) and nucleotide diphosphate kinase (NDK) are kinases located in the intermembrane space. From Jefcoate *et al.* (1992). Reprinted with permission.

tions, the total volumes of mitochondria were the same, which indicated that the intermembrane space decreased in response to stress (ACTH) in the absence of CHX.

3.2.4. Calcium and Nucleotides

Ca^{2+} can induce membrane fusion in addition to stimulating hormone synthesis in steroidogenic tissues (Hall, 1985). Calcium also modulates the activity of certain dehydrogenases (Hansford, 1985; McCormack *et al.*, 1990).

Several studies have demonstrated a role for Ca^{2+} in the fusion of model and natural membranes (Lucy, 1978; Poste and Pasternak, 1978; Poste and Allison, 1973; Kulka and Loyter, 1979). Calcium is envisioned to bind to membrane lipids, producing rigid, crystalline domains of phospholipids as a result of lateral phase separation of acidic phospholipids within a mixed lipid membrane. The formation of such domains is proposed to create instability and a highly permeable membrane susceptible to fusion with other similarly perturbed membranes (Papahadjopoulos *et al.*, 1977a, 1977b).

In adrenal cells, it has been suggested that Ca^{2+} has an effect on malic enzyme activity (Pfeiffer and Tchen, 1973), as well as on the mitochondrial enzymes involved in the transformation of cholesterol to pregnenolone, since the presence of ruthenium red, an inhibitor of Ca^{2+} transport in the mitochondria, inhibits the Ca^{2+}-induced cortisol secretion from permeabilized bovine adrenocortical cells (Yoshida *et al.,* 1993). Increases in cytoplasmic free Ca^{2+} resulting from hormonal stimulation could produce several effects, including activation of phospholipase A_2, which would increase the release of lysophospholipids, which in turn could produce membrane fusion (Siliprandi, 1980; Beatrice *et al.,* 1984). The possibility that Ca^{2+} participates in the assembly of the attachment points between outer and inner membranes in adrenal mitochondria, providing a portal of entry of outer membrane sterol to P450scc was suggested by Kimura (1986).

Kowluru *et al.* (1995) studied the effect of Ca^{2+} and GTP on the synthesis of pregnenolone by mitochondria isolated from rat adrenals. They observed that mitochondrial metabolism of exogenous cholesterol could be substantially stimulated by physiological levels of free Ca^{2+}. These authors also suggested that Ca^{2+} triggers the formation of contact sites between the outer and inner membranes, but proposed that the effect of Ca^{2+} could be at two levels: the metabolism of exogenous cholesterol, which is blocked by ruthenium red, and at a level which is not sensitive to ruthenium red.

Xu *et al.* (1989) demonstrated that in isolated adrenal mitochondria, GTP increases the synthesis of pregnenolone, an effect that disappears when mitochondria are disrupted by sonication. The effect was specific for GTP, since other nucleotides were unable to stimulate the transformation of cholesterol into pregnenolone. It was also suggested that the GTP-activated state is reversed after GTP hydrolysis achieved by a mitochondrial GTPase. The possibility was considered that the action of GTP could be associated with the activity of certain proteins thought to be involved in the transfer of cholesterol to mitochondria, such as steroidogenic activator peptide (SAP) and SCP2 (Xu *et al.,* 1991). They found that SAP was unable to stimulate steroidogenesis, whereas the combination of SAP plus GTP increased the synthesis of pregnenolone. SCP2 also stimulated the transformation of cholesterol, an effect that was increased when GTP was added. Xu *et al.* suggested that SCP2 acts preferentially by mobilizing cholesterol from specific extramitochondrial sites, leaving the substrate ready to be translocated within the mitochondria. It was concluded that GTP and SAP act synergistically, probably on the cholesterol translocation machinery, whereas SCP2 acts through a different mechanism.

These results were reproduced by Kowluru *et al.* (1995), who observed that in the presence of cholesterol the addition of GTP stimulates the synthesis of pregnenolone, and the metabolism of exogenous cholesterol increased 2–3 times. The effect of GTP was blocked by the presence GTP_{γ} s, a non-hydrolyzable GTP analog, suggesting that hydrolysis of GTP is a necessary step for the GTP effect on steroidogenesis.

The presence of GTP-binding proteins in mitochondria of bovine adrenal cortex was recently described. By using photoaffinity labeling with [^{32}P]GTP, a 45-kDa GTP-binding protein was identified in the outer mitochondrial membranes and a 19-kDa protein at the contact points between outer and inner membranes (Thomson *et al.*, 1995). Further studies are needed to clarify the role of these GTP-binding proteins in steroidogenesis.

3.3. Proteins Regulating Intramitochondrial Cholesterol Transport

Steroidogenesis stimulated by ACTH depends on protein synthesis, but not on mRNA synthesis (Ferguson and Morita, 1964). The cholesterol pool in the inner mitochondrial membranes is utilized in the initial phase of steroidogenesis, and hormonal regulation of steroidogenesis requires mobilization of cholesterol from other cellular sites to increase the pool of cholesterol localized in the inner mitochondrial membrane (Pedersen, 1985; Lambeth and Stevens, 1985). The kinetics of CHX's inhibitory action on steroidogenesis suggest that the "labile protein" has a half-life of several minutes. This conclusion has been reached by several groups (Lowry and McMartin, 1974; Pedersen, 1985).

3.3.1. Basic Polypeptides

Kido and Kimura (1981) suggested that the association of cholesterol with cytochrome P450scc could be facilitated by a basic peptide of low molecular weight (less than 10 kDa). This prediction was made based on their studies with reconstituted cholesterol-P450scc complex from bovine adrenal, liposome-associated cholesterol, and synthetic polylysine and polyarginine peptides. There have been several studies that show that polylysine stimulates mitochondrial cytochrome P450 side-chain cleavage activity (Mason *et al.*, 1978), presumably because the positive charges of polylysine bind to the negative charges of phospholipids, modifying the membrane structure and disturbing the proteins localized in it (Inouye *et al.*, 1982).

3.3.2. Steroidogenic Activator Polypeptide (SAP)

Several attempts have been made to identify the labile factor. Pedersen and Brownie (1983) purified a protein of about 2.2 kDa obtained from the 105,000 x g supernatant from ACTH-stimulated rat adrenal gland. The protein, named steroidogenic acute polypeptide (SAP), had the following characteristics: it was sensitive to pronase but not to heat; its synthesis was inhibited by cycloheximide; and it favored the binding of cholesterol to cytochrome P450scc. Pregnenolone synthesis by mitochondria in the presence of 1 μM of SAP increased about five-fold at 5 min of incubation. However, further studies suggested that SAP was a product of proteolysis of the minor heat shock protein, glucose-regulated protein-78 (GRP78), since their amino acid sequences were highly homologous (Pedersen and Brownie, 1987; Mertz and Pedersen, 1989). Xu *et al.* (1991) observed that SAP did not stimulate the transformation of cholesterol into pregnenolone by adrenal mitochondria unless GTP was present. Subsequent studies suggested that SAP is not directly related to the transport of cholesterol into mitochondria (Fristaci *et al.,* 1989).

3.3.3. Endozepine and the Peripheral-Type Benzodiazepine Receptor (PBR)

Yanagibashi *et al.* (1988) isolated a 8.2-kDa protein with the ability to stimulate pregnenolone synthesis from bovine adrenals. The protein had the following characteristics: (i) It produced a concentration-dependent increase in the synthesis of pregnenolone by adrenocortical mitochondria if cholesterol was added to the medium; (ii) it increased the concentration of cholesterol in both the outer and inner mitochondrial membranes when added to an incubation medium containing mitochondria and cholesterol; (iii) it accelerated the transport of cholesterol from the outer to inner membranes; (iv) it increased the capacity of side chain cleavage by P450scc; (v) it was weakly acidic (isoelectric point 6.5); (vi) its concentration in unstimulated adrenal cells was low; and (vii) it stimulated the synthesis of pregnenolone in both mitochondria and mitoplasts. The authors suggested that this protein increases cholesterol transport to mitochondria, which in turn stimulates cytochrome P450 activity. It was also postulated that this protein is not the only one responsible for cholesterol transport to mitochondria, and that the activity of the 8.2-kDa protein could be complemented by other proteins described in the literature. Further studies led to the identification of the 8.2-kDa protein as des-(Gly-Ile)-endozepine, which was postulated to act on the peripheral benzodiazepine receptor (Besman *et al.,* 1989). The same research group showed that addition of des-endozepine to mitochondria increased the substrate-bound fraction of the side-chain cleavage enzyme without addition of exogenous cholesterol (Yanagibashi *et al.,* 1988). This suggested that des-endozepine not only affects transport of cholesterol from the outer to inner mitochondrial membrane but may also promote the mobilization of inner membrane cholesterol (Besman *et al.,* 1989). Later studies of the regulation of endozepine expression in steroidogenic tissues revealed that levels of the protein are not acutely changed in response to tropic stimulation.

The discovery of the effects of endozepine on steroidogenesis coincided with the recognition of a novel class of binding sites for diazepam and certain isoquinoline compounds that are mainly located in the outer membrane of mitochondria of endocrine tissues. These receptors were natural candidates for binding endozepine and mediating endozepine effects on steroidogenesis. Braestrup and Squires (1977) observed a high density of diazepam binding sites in the kidney. Since then, many papers have reported that this class of benzodiazepine binding sites apparently is present in all tissues examined, including the CNS. However, because of their abundance in peripheral tissues, and in order to distinguish them from the $GABA_A$ or central benzodiazepine receptors, this class of benzodiazepine binding sites was named "peripheral-type benzodiazepine binding site" or "peripheral-type benzodiazepine receptors" (PBR).

The possibility that PBRs have an endocrine role was first raised by Anholt *et al.* (1985), who showed that hypophysectomy caused a significant decrease in PBR density in both the adrenal gland and testis. The first direct evidence that benzodiazepines could affect steroidogenesis was obtained in *in vitro* studies on decapsulated testes, in which diazepam and the inhibitor of diazepam binding, Ro5-4864, were shown to stimulate androgen production (Ritta *et al.*, 1987). PBR ligands have been shown to stimulate steroid production by adrenal mitochondria (Besman *et al.*, 1989; Yanagibashi *et al.*, 1989), human term placental tissue (Barnea *et al.*, 1989), and ovarian granulosa cells (Amsterdam and Suh, 1991), further supporting the hypothesis that PBRs are involved in steroidogenesis.

Studies to determine the effects of benzodiazepines on the synthesis of steroids, as well as the exact site of their action, demonstrated that PBR ligands stimulated pregnenolone synthesis by isolated mitochondria supplied with exogenous cholesterol in a dose-dependent manner. This effect was not observed in mitoplasts, suggesting that PBRs are localized in the outer mitochondrial membrane (Mukhin *et al.*, 1989). These studies also implicated PBRs in the acute stimulation of steroidogenesis, possibly by mediating the entry, distribution, and/or availability of cholesterol within mitochondria. Later research on the possible steps involved in PBR regulation revealed that the PBR ligands cause a translocation of cholesterol from the outer to inner mitochondrial membranes (Krueger and Papadopoulos, 1990). The mechanism by which this translocation is effected may involve the association of the 18-kDa isoquinoline binding site (PBR) with a 32-kDa voltage dependent anion channel (VDAC) (McEnery *et al.*, 1992). The latter protein is known to bind cholesterol, and a ligand-induced alteration in the PBR-VDAC association could be envisioned to result in sterol movement to the inner mitochondrial membrane. Because endozepine levels in steroidogenic tissues are not rapidly changed when they are stimulated with tropic hormones (Brown *et al.*, 1992), acute changes in PBR affinity for endozepine have been postulated to account for the fast translocation of mitochondrial cholesterol to P450scc (Figure 5) (Papadopoulos, 1993).

3.3.4. Steroidogenic Acute Regulatory Protein (StAR)

3.3.4.1. Identification and Molecular Cloning. Stocco and colleagues identified a group of phosphoproteins of 30 kDa in the mitochondrial fraction of stimulated MA-10 mouse Leydig tumor cells, as well as in mitochondria from unstimulated R2C rat Leydig tumor cells, which constitutively produce steroids at a high rate (Stocco and Kilgore, 1988; Stocco and Chaudhary, 1990; Stocco and Chen, 1991). Similar observations were made by Orme-Johnson and associates

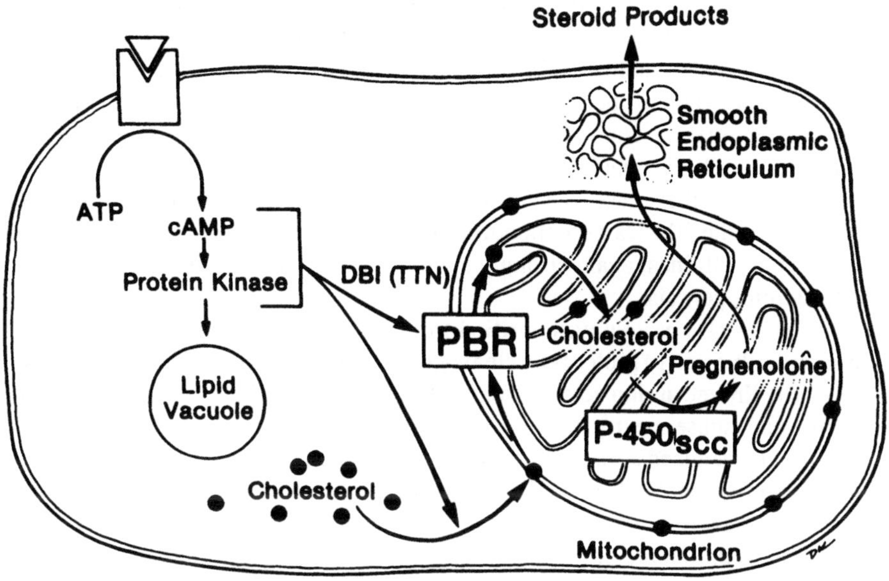

FIGURE 5. Schematic representation of the cellular mechanisms underlying steroid biosynthesis. Polypeptide hormones initially interact with a cell surface receptor stimulating cAMP production. Activation of the cAMP-dependent protein kinase results in the liberation of free cholesterol from different intracellular stores. This cholesterol is then transported, under the influence of cAMP, to the mitochondrion and is incorporated into the outer mitochondrial membrane. PBR is shown as mediating the rate-limiting step in steroid biosynthesis, which is transport of cholesterol from outer to inner mitochondrial membranes. Once associated with the inner mitochondrial membrane, cholesterol is accessible to the cytochrome P450 side-chain cleavage enzyme ($P450_{scc}$) where it is converted to pregnenolone, the first intermediate in the synthesis of all steroid hormones. The endogenous ligand for PBR, DBI, and its biologically active naturally occurring processing product, TTN, are also shown. From Papadopoulos (1993). Reprinted with permission.

(Krueger and Orme-Johnson, 1983; Pon and Orme-Johnson, 1988; Epstein *et al.*, 1989; Alberta *et al.*, 1989; Epstein and Orme-Johnson, 1991) examining adrenal and ovarian steroidogenic cells. These proteins appeared rapidly and were induced in a dose-dependent manner by tropic hormones and AMP analogs. The 30-kDa proteins were shown to arise from precursor proteins with molecular masses of 37-and 32-kDa, which were also associated with mitochondria (Stocco and Sodeman, 1991). It was suggested that 30-kDa proteins are transferred to the inner mitochondrial membrane by a process requiring both proteolytic removal of the mitochondrial targeting sequences and an electrical potential across the inner mitochondrial membrane, since 1,10-phenanthroline, an inhibitor of the mitochondrial metalloproteinase that cleaves pre-sequences, and cyanide *m*-chlorophenylhydrazone, which collapses the electrical potential across the mitochondrial membranes, inhibited their appearance (Krueger and Orme-Johnson, 1983). Orme-Johnson and colleagues suggested that the mitochondrial 30-kDa protein that is most closely correlated with activation of steroidogenesis is cotranslationally phosphorylated.

Stocco and Ascoli (1993) transfected MA-10 mouse Leydig tumor cells with a mutant gene of the type 1 regulatory subunit of the cAMP-dependent protein kinase under the control of a metallothionein promoter MA-10 (K3). Another transfected cell, designed MA-10 (P + 29), is a constitutive overproducer of a cAMP-phosphodiesterase. The authors showed that production of steroids and synthesis of the 30-kDa mitochondrial proteins were highly correlated in the two transfected cell lines, again strongly suggesting the relationship of these proteins with steroidogenesis.

The 30-kDa protein was purified from MA-10 Leydig tumor cell mitochondria, and subjected to proteolytic cleavage and amino acid analysis of isolated peptides. The limited amino acid sequence allowed cloning of the cDNA encoding the murine and, subsequently, the human protein (Clark *et al.*, 1994; Sugawara *et al.*, 1995a). The deduced amino acid sequence revealed a protein with a calculated mol wt. of approximately 32-kDa. The amino terminus consisted of an amino acid sequence characteristic of proteins imported into mitochondria, comprised of mostly basic and hydrophobic amino acids. A number of consensus sites for phosphorylation by protein kinase A were identified. Remarkably, the protein showed no obvious homologies to protein sequences deposited in existing data bases. The protein was named steroidogenic acute regulatory protein (StAR).

To demonstrate that StAR has biological activity, the cDNAs were introduced into MA-10 cells and COS-1 cells cotransfected with the cholesterol side-chain cleavage enzyme system; in all cases StAR produced a pronounced stimulation of steroidogenesis. Western blot analyses documented that a pre-protein running in SDS-PAGE at a molecular weight of about 37 kDa was converted into a 30-kDa protein, representing the mature form of StAR.

StAR mRNA was found in adrenals, ovaries, and testis in rodents and humans. Notably, StAR mRNA was absent from human placenta and choriocarcinoma cells, which secrete substantial amounts of progesterone, and the liver, which converts large amounts of cholesterol into bile acids through pathways involving hydroxylation at C_{27}, which is catalyzed by a mitochondrial P450.

The human and murine StAR genes were subsequently isolated. They both span approximately 8 kb and consist of seven exons (Clark *et al.*, 1995; Sugawara *et al.*, 1995b). The human StAR gene was mapped to chromosome 8p11.2. A StAR pseudogene was characterized on human chromosome 13 (Sugawara *et al.*, 1995a,b).

3.3.4.2. Mutations in the StAR Gene Cause Congenital Lipoid Adrenal Hyperplasia. The congenital adrenal hyperplasias (CAH) are a group of inborn errors of steroid hormone synthesis. The most severe form of CAH is congenital lipoid adrenal hyperplasia (lipoid CAH), which is characterized by a marked deficiency of all circulating or urinary steroid hormones and high basal concentrations of ACTH and plasma renin activity. Moreover, no steroid response can be elicited to long-term treatment with high doses of ACTH or hCG. The adrenal glands are grossly enlarged and laden with cholesterol and cholesterol esters (Sandison, 1955; Kirkland *et al.*, 1973; Hauffa *et al.*, 1985).

Lin *et al.* (1991) found no gross deletion, duplication, or rearrangements in the P450scc genes of patients with lipoid CAH nor alterations in the synthesis of cytochrome P450scc, adrenedoxin reductase, and adrenedoxin mRNAs. Since no alterations were found in the P450scc gene sequence and RNA levels, this group suggested that the defect responsible for lipoid CAH does not lie within the cholesterol side-chain cleavage system. They also suggested that differences in known proteins related to the transport of cholesterol, including SCP2, endozepine, and SAP, were not the cause of lipoid CAH. The mRNAs for these proteins were present in normal size in the patient's tissue. Using reverse transcriptase-PCR analysis of testicular RNA from a patient with lipoid CAH, Lin *et al.* (1993) also showed that the sequences of the mRNA and the genomic DNA for the PBR component of mitochondrial benzodiazepine receptors were not altered in the disease.

Lin *et al.* (1995) subsequently examined the possibility that defects in the StAR gene cause lipoid CAH. They used a primer based on a sequence, termed S1, in the 5′ untranslated region that distinguishes the authentic StAR gene from its pseudogene, and they amplified the expected 974-base pair StAR cDNA from normal controls and the two patients. All patient clones from independent RT-PCR reactions were identical to the wild-type sequence except for a C→T transition in codon 193 in one patient, and a C→T transition in codon 258 in the second patient. These mutations generated premature stop codons, resulting in encoded proteins that lacked 93 and 28 COOH-terminal amino acid residues, respectively. The authors investigated the functional consequences of these premature stop codons in the StAR gene by analyzing the wild-type and mutant proteins for their ability to enhance steroidogenesis. Coexpression of wild-type StAR with the cholesterol side chain cleavage system in COS-1 cells resulted in an approximately eight-fold increase in pregnenolone production, with endogenous cholesterol as a substrate. Both mutant StAR proteins were inactive, suggesting that each of two nonsense mutations caused lipoid CAH. Unlike cholesterol, 20α-hydroxycholesterol readily diffuses into mitochondria and thereby bypasses the mitochondrial cholesterol transport system. With 20α-hydroxycholesterol as a substrate, there were no significant differences in pregnenolone production between wild-type and mutant proteins (Table III). The differential effects of StAR on pregnenolone production with cholesterol and 20α-hydroxycholesterol as substrates suggest that StAR mediates the transport of cholesterol into mitochondria. Other mutations in the StAR gene causing lipoid CAH have recently been reported (Tee *et al.*, 1995).

Lipoid CAH is the only inborn disorder of steroid hormone synthesis known that is not caused by a defective steroidogenic enzyme. Identification of mutant StAR proteins as the cause of lipoid CAH now permits the prenatal molecular diagnosis of this devastating disease.

Table III
Effect of Nonsense Mutations on StAR Activity[a]

	Pregnenolone (nanograms per dish)			
	scc-Adx		F2	
Vector	Cholesterol	20α	Cholesterol	20α
pSPORT	20 ± 1	158 ± 21	17 ± 3	60 ± 7
pSTAR	175 ± 19	138 ± 15	131 ± 23	60 ± 11
Patient 1	19 ± 2	99 ± 23	18 ± 5	56 ± 7
Patient 2 or 3	25 ± 4	168 ± 35	22 ± 4	75 ± 7

[a]With the use of lipofectamine, nonsteroidogenic COS-1 monkey kidney cells were transfected with the vector pSPORT or with pSPORT containing normal human StAR cDNA (pStAR) or mutant StAR cDNA from patients 1 or 2 (or 3) (Lin *et al.*, 1995). The cells were also transfected with either vectors encoding bovine P450scc and bovine adrenodoxin (scc-Adx) or a pECE vector encoding a fusion protein, termed F2, that consists of the human cholesterol side chain cleavage system (H_2N-P450scc-adrenodoxin reductase-adrenodoxin-COOH) (see Lin *et al.*, 1995). The substrate was either cellular and serum cholesterol or added 20α-hydroxycholesterol (5 μg/ml) (20α). After incubation of cells for 48 h, the medium was collected and assayed for pregnenolone by immunoassay. Values are means ± S. D. from four separate transfections. [From Lin *et al.* (1995). Reprinted with permission].

Lipoid CAH due to nonfunctional StAR is comparable to the effect of a StAR gene knockout, showing that StAR is indispensable for adrenal and gonadal steroidogenesis. Thus, StAR is the first protein identified that plays an essential role for cholesterol access to P450scc. The sparing of placental steroidogenesis in fetuses with lipoid CAH and the absence of StAR expression in placenta and other steroidogenic tissues (Sugawara *et al.*, 1995a) indicates that different mechanisms may exist to facilitate cholesterol transport in mitochondria in these tissues. The importance of this finding was stressed by Waterman (1995), who proposed new alternatives and questions that have not yet been answered.

3.3.4.3. Is StAR the "Labile Protein" and How Does It Act? The compelling evidence for a critical role of StAR in the control of steroidogenesis, provided by the demonstration that mutations in the StAR gene cause lipoid CAH, makes StAR the leading candidate for the "labile protein." The earlier kinetic studies on the formation of labeled mitochondrial phosphoproteins, now known to be StAR, further strengthens the argument that StAR is the "labile protein." None of the other putative regulators, including SAP and the endozepine-PBR system, easily fit all of the requisite kinetic and biochemical criteria.

How are StAR levels regulated? If StAR mediates the acute, cAMP-triggered translocation of cholesterol within mitochondria, its levels of activity must be regulated. StAR mRNA and protein are increased by stimulation of cells with tropic hormones or cAMP analogs (Clark *et al.*, 1995; Sugawara *et al.*, 1995a). However, these changes take hours, while the acute steroidogenic response occurs within minutes. Thus, it is likely that cAMP causes a co- or post-translational change, such as phosphorylation, in pre-existing StAR, and this leads to enhanced biological activity. Unfortunately, at this writing this notion remains only a speculation.

The mechanism by which StAR affects the translocation of cholesterol within mitochondria is also obscure. One hypothesis is that during the mitochondrial import of StAR, the outer and inner membranes are brought together, allowing flow of sterol to P450scc. There is, however, no direct evidence as of yet for this model. It is evident that StAR's action is not restricted to the supply of cholesterol to P450scc since StAR can enhance cholesterol 27-hydroxylase activity in the same engineered COS-1 cell system employed to demonstrate its steroidogenesis enhancing properties (Sugawara *et al.,* 1995b). Thus, StAR can stimulate cholesterol metabolism by other mitochondrial P450 enzymes. Continuing exploration of the structure/function relationships will undoubtedly provide greater insight into its mechanism of action and the ways in which its biological activities are controlled.

4. REFERENCES

Acton, S., Rigotti, A., Landschulz, K. T., Xu, S., Hobbs, H. H., and Krieger, M., 1996, Identification of scavenger receptor SR-BI as a high density lipoprotein receptor, *Science* **271:**518–520.

Alberta, J. A., Epstein, L. F., Pon, L. A., and Orme-Johnson, N. R., 1989, Mitochondrial localization of a phosphoprotein that rapidly accumulates in adrenal cortex cells, *J. Biol. Chem.* **264:** 2368–2372.

Almahbobi, G., Williams, L. J., and Hall, P. F., 1992a, Attachment of mitochondria to intermediate filaments in adrenal cells: relevance to the regulation of steroid synthesis, *Exp. Cell. Res.* **200:**361–369.

Almahbobi, G., Williams, L. J., and Hall, P. F., 1992b, Attachment of steroidogenic lipid droplets to intermediate filaments in adrenal cells, *J. Cell. Sci.* **101:**383–393.

Almahbobi, G., Williams, L. J., Han, X-G., and Hall, P. F., 1993, Binding of lipid droplets and mitochondria to intermediate filaments in rat Leydig cells, *J. Reprod. Fertil.* **98:**209–217.

Amsterdam, A., and Suh, B. S., 1991, An inducible functional peripheral benzodiazepine receptor in mitochondria of steroidogenic granulose cells, *Endocrinology* **128:**503–510.

Anholt, R. R. H., DeSouza, E. B., Kuhar, M. J., and Snyder, S. H., 1985, Depletion of peripheral-type benzodiazepine receptors after hypophysectomy in rat adrenal gland and testis, *Eur. J. Pharmacol.* **110:**41–46.

Ardail, D., Privat, J-P., Egret-Charlier, M., Levrat, C., Lerme, F., and Louisot, P., 1990, Mitochondrial contact sites. Lipid composition and dynamics, *J. Biol. Chem.* **265:**18797–18802.

Baggetto, L. G., Clottes, E., and Vial, C., 1992, Low mitochondrial proton leak due to high membrane cholesterol content and cytosolic creatine kinase as two features of the deviant bioenergetics of Ehrlich and AS30-D tumor cells, *Cancer Res.* **52:**4935–4941.

Barnea, E. R., Fares, F., and Gavish, M., 1989, Modulatory action of benzodiazepines on human term placental steroidogenesis in vitro, *Mol. Cell. Endocrinol.* **64:**155–159.

Beatrice, M. C., Stiers, D. L., and Pfeiffer, D. R., 1984, The role of glutathione in the retention of Ca^{2+} in liver mitochondria, *J. Biol. Chem.* **259:**1279–1287.

Bellini, F., Phillips, M. C., Pickell, C., and Rothblat, G. H., 1984, Role of the plasma membrane in the mechanism of cholesterol efflux from cells, *Biochim. Biophys. Acta* **777:**209–215.

Besman, M. J., Yanagibashi, K., Lee, T. D., Kawamura, M., Hall, P. F., and Shively, J. E., 1989, Identification of des-(Gly-Ile)-endozepine as an effector of corticotropin-dependent adrenal steroidogenesis: Stimulation of cholesterol delivery is mediated by the peripheral benzodiazepine receptor, *Proc. Natl. Acad. Sci. USA* **86**:4897–4901.

Billheimer, J. T., and Reinhart, M. P., 1990, Intracellular trafficking of sterols, in: *Subcellular Biochemistry,* Volume 16 (H. J. Hildersson, ed.), pp. 301–331, Plenum Press, New York.

Bloj, B., and Zilversmit, D. B., 1977, Complete exchangeability of cholesterol in phosphatidylcholine/cholesterol vesicles of different degrees of unsaturation, *Biochemistry* **16**:3943–3948.

Braestrup, C., and Squires, R. F., 1977, Specific benzodiazepine receptors in rat brain characterized by high-affinity [^{3}H]diazepam binding, *Proc. Natl. Acad. Sci. USA* **74**:3805–3809.

Brossa, O., Garcea, R., Canuto, R. A., Gautero, B., and Feo, F., 1980, Cholesterol and phospholipid contents of the inner and outer membranes of highly purified mitochondria from the Yoshida hepatoma AH-130, *IRCS Med. Sci.* **8**:73–74.

Brown, A. S., Hall, P. F., Shoyab, M., and Papadopoulos, V., 1992, Endozepine/diazepam binding inhibitor in adrenocortical and Leydig cell lines: absence of hormonal regulation, *Mol. Cell Endocrinol.* **83**:1–9.

Bruckdorfer, K. R., and Sherry, M. K., 1984, The solubility of cholesterol and its exchange between membranes, *Biochim. Biophys. Acta* **769**:187–196.

Bruni, A., VanDijck, P. W. M., and DeGier, J., 1975, The role of phospholipid acyl chains in the activation of mitochondrial ATPase complex, *Biochim. Biophys. Acta* **406**:315–328.

Chanderbhan, B. J., Noland, B. J., Scallen, T. J., and Vahouny, G., 1982, Sterol carrier protein$_2$. Delivery of cholesterol from adrenal lipid droplets to mitochondria for pregnenolone synthesis, *J. Biol. Chem.* **257**:8928–8934.

Chapman, D., 1975, Lipid dynamics in cell membranes, in: *Cell Membranes. Biochemistry, Cell Biology & Pathology* (G. Weissmann and R. Clairborne, eds.), pp. 13–22, HP Publishing Co., New York.

Chazotte, B., Wu, E. S., and Hackenbrock, C. R., 1991, The mobility of a fluorescent ubiquinone in model lipid membranes. Relevance to mitochondrial electron transport, *Biochim. Biophys. Acta* **1058**:400–409.

Cheng, B., Hsu, D. K., and Kimura, T., 1985, Utilization of mitochondrial membrane cholesterol by cytochrome P-450-dependent cholesterol side-chain cleavage reaction in bovine adrenocortical mitochondria: steroidogenic and non-steroidogenic pools of cholesterol in the mitochondrial inner membranes, *Mol. Cell. Endocrinol.* **40**:233–243.

Cherry, R. J., Müller, U., Holenstein, C., and Heyn, M. P., 1980, Lateral segregation of proteins induced by cholesterol in bacteriorhodopsin-phospholipid vesicles, *Biochim. Biophys. Acta* **596**:145–151.

Clark, B. J., Wells, J., King, S. R., and Stocco, D. M., 1994, The purification, cloning, and expression of a novel luteinizing hormone-induced mitochondrial protein in MA-10 mouse Leydig tumor cells, *J. Biol. Chem.* **269**:2831–2834.

Clark, B. J., Soo, S.-C., Caron, K. M., Ikeda, Y., Parker, K. L., and Stocco, D. M., 1995, Hormonal and developmental regulation of the steroidogenic acute regulatory protein, *Mol. Endocrinol.* **9**: 1346–1355.

Clejan, S., and Bittman, R., 1984, Kinetics of cholesterol and phospholipid exchange between *Mycoplasma gallisepticum* cells and lipid vesicles. Alterations in membrane cholesterol and protein content, *J. Biol. Chem.* **259**:441–448.

Colbeau, A., Nachbaur, J., and Vignais, P. M., 1971, Enzymatic characterization and lipid composition of rat liver subcellular membranes, *Biochim. Biophys. Acta* **249**:462–492.

Coleman, P. S., 1986, Membrane cholesterol and tumor bioenergetics, *Ann. N.Y. Acad. Sci.* **488**: 451–467.

Coleman, P. S., and Parlo, R. A., 1984, Cholesterol-rich tumor mitochondria encourage altered metabolic patterns and cell proliferation, *Ann. N.Y. Acad. Sci.* **435**:129–132.

Coleman, P. S., Lavietes, B., Born, R., and Weg, A., 1978, Cholesterol enrichment of normal mitochondria in vitro: a model system with properties of hepatoma mitochondria, *Biochem. Biophys. Res. Commun.* **84**:202–207.

Comte, J., Maisterrena, B., and Gautheron, D. C., 1976, Lipid composition and protein profiles of outer and inner membranes from pig heart mitochondria. Comparison with microsomes, *Biochim. Biophys. Acta* **419:**271–284.

Cornelius, F., 1990, Functional reconstitution of the sodium-pump. Kinetics of exchange reactions performed by reconstituted Na/K-ATPase, *Biochim. Biophys. Acta* **1071:**19–66.

Cozza, E. N., del Carmen Vila, M., Acevedo-Duncan, M., Gomez-Sanchez, C. E., and Farese, R. V., 1990, ACTH increases de novo synthesis of diacylglycerol and translocates protein kinase C in primary cultures of calf adrenal glomerulose cells, *J. Steroid Biochem.* **35:**343–351.

Crémel, G., Filliol, D., Jancsik, V., and Rendon, A., 1990, Cholesterol distribution in rat liver and brain mitochondria as determined by stopped-flow kinetics with filipin, *Arch. Biochem. Biophys.* **278:**142–147.

Crivello, J. F., and Jefcoate, C. R., 1980, Intracellular movement of cholesterol in rat adrenal cells. Kinetic and effects of inhibitors, *J. Biol. Chem.* **255:**8144–8151.

Daum, G., 1985, Lipids of mitochondria, *Biochim. Biophys. Acta* **822:**1–42.

della-Cioppa, G., Muffly, K. E., Yanagibashi, K., and Hall, P. F., 1986, Preparation and characterization of submitochondrial fractions from adrenal cells, *Mol. Cell. Endocrinol.* **48:**111–120.

Demel, R. A., Jansen, J. W. C. M., Van Dijk, P. W. M., and Van Deenen, L. L. M., 1977, The preferential interaction of cholesterol with different classes of phospholipids, *Biochim. Biophys. Acta* **465:**1–10.

DePierre, J. W., and Ernster, L., 1977, Enzyme topology of intracellular membranes, *Annu. Rev. Biochem.* **46:**201–262.

Dietzen, D. J., and Davis, E. J., 1994, Excess membrane cholesterol is not responsible for metabolic and bioenergetic changes in AS-30D hepatoma mitochondria, *Arch. Biochem. Biophys.* **309:** 341–347.

Dorbani, L., Jancsik, V., Linden, M., Leterrier, J. F., Nelson, B. D., and Rendon, A., 1987, Subfractionation of the outer membrane of rat brain mitochondria: evidence for the existence of a domain containing the porin-hexokinase complex, *Arch. Bichem. Biophys.* **252:**188–196.

Echegoyen, S., Oliva, E. B., Sepúlveda, J., Díaz-Zagoya, J. C., Espinosa-García, M. T., Pardo, J. P., and Martínez, M., 1993, Cholesterol increase in mitochondria: its effect on inner-membrane functions, submitochondrial localization and ultrastructutal morphology, *Biochem. J.* **289:** 702–708.

Epstein, L. F., Alberta, J. A., Pon, L. A., and Orme-Johnson, N. R., 1989, Subcellular localization of a protein produced in adrenal cortex cells in response to ACTH, *Endocr. Res.* **15:**177–127.

Epstein, L. F., and Orme-Johnson, N. R., 1991, Regulation of steroid hormone biosynthesis. Identification of precursors of a phosphoprotein targeted to the mitochondrion in stimulated rat adrenal cortex cells, *J. Biol. Chem.* **266:**19739–19745.

Faust, J. R., Goldstein, J. L., and Brown, M. S., 1977, Receptor-mediated uptake of low density lipoprotein and utilization of its cholesterol for steroid synthesis in culture mouse adrenal cells, *J. Biol. Chem.* **252:**4861–4871.

Feo, F., Canuto, R. A., Garcea, R., and Gabriel, L., 1975, Effect of cholesterol content on some physical and functional properties of mitochondria isolated from adult rat liver, fetal liver, cholesterol enriched liver and hepatomas AH-130, 3924A and 5123, *Biochim. Biophys. Acta* **413:**116–134.

Ferguson, J. J., Jr., and Morita, Y., 1964, RNA synthesis and adrenocorticotropin responsiveness, *Biochim. Biophys. Acta* **87:**348–350.

Freeman, D. A., 1989, Plasma membrane cholesterol: removal and insertion into the membrane and utilization as substrate for steroidogenesis, *Endocrinology* **124:**2527–2534.

Fristaci, J., Mertz, L. M., and Pedersen, R. C., 1989, Steroidogenesis activator peptide (SAP) in the guinea pig adrenal cortex, *Mol. Cell. Endrocr.* **64:**137–143.

Fugler, L., Clejan, S., and Bittman, R., 1985, Movement of cholesterol between vesicles prepared with different phospholipid or sizes, *J. Biol. Chem.* **260:**4098–4102.

Goldstein, J. L., and Brown, M. S., 1984, Progress in understanding the LDL receptor and HMG-CoA reductase, two membrane proteins that regulate the plasma cholesterol, *J. Lipid Res.* **25:** 1450–1461.

Graham, J. M., and Green, C., 1970, The properties of mitochondria enriched in vitro with cholesterol, *Eur. J. Biochem.* **12:**58–66.

Hall, P. F., 1985, Trophic stimulation of steroidogenesis: in search of the elusive trigger, in: *Recent Progress in Hormone Research,* Volume 41, Academic Press, New York.

Hall, P. F., and Almahbobi, G., 1992, The role of the cytoskeleton in the regulation of steroidogenesis, *J. Steroid Biochem. Molec. Biol.* **43:**769–777.

Hauffa, B., Miller, W. L., Grumbach, M. M., Conte, F. A., and Kaplan, S. L., 1985, Congenital adrenal hyperplasia due to deficient cholesterol side-chain cleavage activity (20,22 desmolase) in a patient treated for 18 years, *Clin. Endocrinol.* **23:**481–493.

Houslay, M. D., and Stanley, K. K. (eds.), 1990, *Dynamics of Biological Membranes, Influence on Synthesis, Structure and Function,* Wiley, New York.

Iida, S., Papadopoulos, V., and Hall, P. F., 1989, The influence of exogenous cholesterol on steroid synthesis in cultured adrenal cells, *Endocrinology* **124:**2619–2624.

Igarashi, Y., and Kimura, T., 1984, Adrenocorticotropic hormone-mediated changes in rat adrenal mitochondrial phospholipids, *J. Biol. Chem.* **259:**10745–10753.

Inouye, S., Soberon, X., Franceschini, T., Nakamura, K., Itakura, K., and Inouye, M., 1982, Role of positive charge on the amino-terminal region of the signal peptide in protein secretion across the membrane, *Proc. Natl. Acad. Sci. USA* **79:**3438–3441.

Jain, M. K., 1975, Role of cholesterol in biomembranes and related systems, in: *Current Topics in Membranes and Transport,* Volume 6 (F. Bronner and A. Kleinzeller, eds.), pp. 1–57, Academic Press, New York.

Jefcoate, C. R., McNamara, B. C., Artemenko, I., and Yamazaki, T., 1992, Regulation of cholesterol movement to mitochondrial cytochrome P450scc in steroid hormone synthesis, *J. Steroid Biochem. Mol. Biol.* **43:**751–767.

Kachar, B., and Reese, T. S., 1982, Evidence for the lipidic nature of tight juntion strands, *Nature* **296:**464–466.

Kesav, S., McLaughlin, J., and Scallen, T. J., 1992, Participation of sterol carrier protein-2 in cholesterol metabolism, *Biochem. Soc. Trans.* **20:**818–824.

Kido, T., and Kimura, T., 1981, Stimulation of cholesterol binding to steroid-free cytochrome P450scc by poly-(L-lysine). The implication in functions of labile protein factor for adrenocortical steroidogenesis, *J. Biol. Chem.* **256:**8561–8566.

Kimura, T., 1986, Transduction of ACTH signal from plasma membrane to mitochondria in adrenocortical steroidogenesis. Effects of peptide, phospholipid, and calcium, *J. Steroid Biochem.* **25:** 711–716.

Kirkland, R. T., Kirkland, J. L., Johnson, C. M., Horning, M. G., Librik, L., and Clayton, G. W., 1973, Congenital lipoid adrenal hyperplasia in an eight-year-old phenotypic female, *J. Clin. Endocrinol. Metab.* **36:**488–496.

Kleemann, W., and McConnell, H. M., 1976, Interactions of proteins and cholesterol with lipids in bilayer membranes, *Biochim. Biophys. Acta* **419:**206–222.

Kowluru, R., Yamazaki, T., McNamara, B. C., and Jefcoate, C. R., 1995, Metabolism of exogenous cholesterol by rat adrenal mitochondria is stimulated equally by physiological level of free Ca^{2+} and GTP, *Mol. Cell. Endocrinol.* **107:**181–188.

Krämer, R., 1982, Cholesterol as activator of ADP–ATP exchange in reconstituted liposomes and mitochondria, *Biochim. Biophys. Acta* **693:**296–304.

Krueger, K. E., and Papadopoulos, V., 1990, Peripheral-type benzodiazepine receptors mediate translocation of cholesterol from outer to inner mitochondrial membranes in adrenocortical cells, *J. Biol. Chem.* **265:**15015–15022.

Krueger, R. J., and Orme-Johnson, N. R., 1983, Acute adrenocorticotropic hormone stimulation of adrenal corticosteroidogenesis. Discovery of a rapidly induced protein, *J. Biol. Chem.* **258:** 10159–10167.

Kulka, R. G., and Loyter, A., 1979, The use of fusion methods for the microinjection of animal cells, in: *Current Topics in Membranes and Transport,* Volume 12, pp. 365–430, Academic Press, New York.

Lambeth, J. D., and Stevens, V. L., 1985, Cytochrome P-450scc: enzymology, and the regulation of intramitochondrial cholesterol delivery of the enzyme, *Endocrine Res.* **10:**283–309.

Lange, Y., and Matthies, H. J. G., 1984, Transfer of cholesterol from its site of synthesis to the plasma membrane, *J. Biol. Chem.* **59:**14624–14630.

Lange, Y., and Ramos, B. V., 1983, Analysis of the distribution of cholesterol in the intact cell, *J. Biol. Chem.* **258:**15130–15134.

Lange, Y., O'Alessandro, J. S., and Small, D. M., 1979, The affinity of cholesterol for phosphatidylcholine and sphingomyelin, *Biochim. Biophys. Acta* **556:**388–398.

Levrat, C., and Louisot, P., 1992, Dual localization of the mitochondrial phospholipase A_2: outer membrane contact sites and inner membrane, *Biochem. Biophys. Res. Commun.* **183:**719–724.

Lin, D., Gitelman, S. E., Saenger, P., and Miller, W. L., 1991, Normal genes for the cholesterol side chain cleavage enzyme, P450scc, in congenital lipoid adrenal hyperplasia, *J. Clin. Invest.* **88:** 1955–1962.

Lin, D., Chang, Y. J., Strauss, J. F. III, and Miller, W. L., 1993, The human peripheral benzodiazepine receptor gene: cloning and characterization of alternative splicing in normal tissues and in a patient with congenital lipoid adrenal hyperplasia, *Genomics* **18:**643–650.

Lin, D., Sugawara, T., Strauss, J. F. III, Clark, B. J., Stocco, D. M., Seanger, P., Rogol, A., and Miller, W. L., 1995, Role of steroidogenic acute regulatory protein in adrenal and gonadal steroidogenesis, *Science* **267:**1828–1831.

Lowry, P. J., and McMartin, C., 1974, Measurement of the dynamics of stimulation and inhibition of steroidogenesis in isolated rat adrenal cells by using column perfusion, *Biochem. J.* **142:**287–294.

Lucy, J. A., 1978, Mechanism of chemically induced cell fusion, in: *Membrane Fusion,* Volume I (G. Poste and G. L. Nicolson, eds.), pp. 267–304, Elsevier/North-Holland Biomedical Press, New York.

Madden, T. D., Vigo, C., Bruckdorfer, K. R., and Chapman, D., 1980, The incorporation of cholesterol into inner mitochondrial membranes and its effect on lipid phase transition, *Biochim. Biophys. Acta* **599:**528–537.

Martínez, F., Echegoyen, S., Briones, R., and Cuellar, A., 1988, Cholesterol increase in mitochondria: a new method of cholesterol incorporation, *J. Lipid Res.* **29:**1005–1011.

Mason, J. I., Arthur, J. R., and Boyd, G. S., 1978, Regulation of cholesterol metabolism in rat adrenal mitochondria, *Mol. Cell. Endocrinol.* **10:**209–223.

McLean, L. R., and Phillips, M. C., 1981, Mechanism of cholesterol and phosphatidylcholine exchange or transfer between unilamellar vesicles, *Biochemistry* **20:**2893–2900.

McNamara, B. C., and Jefcoate, C. R., 1990a, Heterogeneous pools of cholesterol side-chain cleavage activity in adrenal mitochondria from ACTH-treated rats: differential responses to different reducing precursors, *Mol. Cell. Endocrinol.* **73:**123–134.

McNamara, B. C., and Jefcoate, C. R., 1990b, Heterogeneous pools of cholesterol side-chain cleavage activity in adrenal mitochondria from adrenocorticotropic hormone-treated rats: reconstitution of the isocitrate response with succinate and low concentrations of isocitrate, *Arch. Biochem. Biophys.* **283:**464–471.

Mertz, L. M., and Pedersen, R. C., 1989, Steroidogenesis activator polypeptide may be a product of glucose regulated protein 78 (GRP78), *Endocrine Res.* **15:**101–115.

Mukhin, A. G., Papadopoulos, V., Costa, E., and Krueger, K. E., 1989, Mitochondrial benzodiazepine receptors regulate steroid biosynthesis, *Proc. Natl. Acad. Sci. USA* **86:**9813–9816.

Nagatomo, T., Hattori, K., Ikeda, M., and Shimada, K., 1980, Lipid composition of sarcolemma, mitochondria and sarcoplasmic reticulum from newborn and adult rabbit cardiac muscle, *Biomed. Med.* **23:**108–118.

Nestler, J. E., Takagi, K., and Strauss, J. F. III, 1990, Lipoprotein and cholesterol metabolism in cells that synthesize steroid hormones, in: *Advances in Cholesterol Research* (M. Esfahani and J. B. Swaney, eds.), pp. 133–170, Telford Press, Caldwell, NJ.

North, P., and Fleischer, S., 1983, Use of a nonspecific lipid transfer protein to modify the cholesterol content of synaptic membranes, *Meth. Enzymol.* **98:**599–613.

Ohki, S., and Leonards, K. S., 1984, A possible role of cholesterol in membrane adhesion, *Biochemistry* **23:**5578–5581.

Ohlendieck, K., Riesinger, I., Adams, V., Krause, J., and Brdiczka, D., 1986, Enrichment and biochemical characterization of boundary membrane contact sites from rat-liver mitochondria, *Biochim. Biophys. Acta* **860:**672–689.

Orme-Johnson, N. R., 1990, Distinctive properties of adrenal cortex mitochondria, *Biochim. Biophys. Acta* **1020:**213–231.

Palmer, J. W., Schmid, P. C., Pfeiffer, D. R., and Schmid, H. H. O., 1981, Lipids and lipolytic enzyme activities of rat heart mitochondria, *Arch. Biochem. Biophys.* **211:**674–682.

Papadopoulos, V., 1993, Peripheral-type benzodiazepine/diazepam binding inhibitor receptor: biological role in steroidogenic cell function, *Endocrine Rev.* **14:**222–240.

Papahadjopoulos, D., 1977, Effect of bivalent cations and proteins on the thermotropic properties of phospholipid membranes: implications for the molecular mechanism of fusion and endocytosis, *J. Colloid Interface Sci.* **58:**459–469.

Papahadjopoulos, D., Vail, W. J., Newton, C., Nir, S., Jacobson, K., Poste, G., and Lazo, R., 1977, Studies on membrane fusion. III. The role of calcium-induced phase changes, *Biochim. Biophys. Acta* **465:**579–598.

Parlo, R. A., and Coleman, P. S., 1984, Enhanced rate of citrate export from cholesterol-rich hepatoma mitochondria, *J. Biol. Chem.* **259:**9997–10003.

Parsons, D. F., and Yano, Y., 1967, The cholesterol content of the outer and inner membranes of guinea-pig liver mitochondria, *Biochim. Biophys. Acta* **135:**362–364.

Patton, S., 1970, Correlative relationship of cholesterol and sphingomyelin in cell membranes, *J. Theor. Biol.* **29:**489–491.

Pedersen, R. C., 1985, Polypeptide activators of cholesterol side-chain cleavage, *Endocrine Res.* **10:**533–561.

Pedersen, R. C., and Brownie, A. C., 1983, Cholesterol side-chain cleavage in the rat adrenal cortex: Isolation of a cycloheximide-sensitive activator peptide, *Proc. Natl. Acad. Sci. USA* **80:** 1882–1886.

Pedersen, R. C., and Brownie, A. C., 1987, Steroidogenesis-activator polypeptide isolated from a rat Leydig cell tumor, *Science* **236:**188–190.

Pfanner, N., Craig, E. A., and Meijer, M., 1994, The protein import machinery of the mitochondrial inner membrane, *Trends Biochem. Sci.* **19:**368–372.

Pfeifer, S. M., Furth, E. E., Ohba, T., Chang, Y. J., Rennert, H., Sakuragi, N., Billheimer, J. T., and Strauss, J. F. III, 1993, Sterol carrier protein 2: a role in steroid hormone synthesis? *J. Steroid Biochem. Molec. Biol.* **47:**167–172.

Pfeiffer, D. R., and Tchen, T. T., 1973, The role of Ca^{2+} in the control of malic enzyme activity in bovine adrenal cortex mitochondria, *Biochem. Biophys. Res. Commun.* **50:**807–813.

Pitotti, A., Dobbeni-Sala, F., and Bruni, G., 1980, Phospholipid-dependent assembly of mitochondrial ATPase complex, *Biochim. Biophys. Acta* **600:**79–90.

Pon, L. A., and Orme-Johnson, N. R., 1988, Acute stimulation of corpus luteum cells by gonadotropin or adenosine 3′,5′-monophosphate causes accumulation of a phosphoprotein concurrent with acceleration of steroid synthesis, *Endocrinology* **123:**1942–1948.

Poste, G., and Allison, A. C., 1973, Membrane fusion, *Biochim. Biophys. Acta* **300:**421–465.

Poste, G., and Pasternak, C. A., 1978, Virus-induced cell fusion, in: *Membrane Fusion,* Volume I (G. Poste and G. L. Nicolson, eds.), pp. 305–367, Elsevier/North-Holland Biomedical Press, New York.

Rennert, H., Chang, Y. J., and Strauss, III, J. F., 1993, Intracellular cholesterol dynamics related to steroid hormone synthesis: a contemporary view, in: *The Ovary* (E. Y. Adashi and P. C. K. Leung, eds.), pp. 147–164, Plenum Press, New York.

Rindina, F. C., Baracca, A., Solaini, G., Rabbi, A., and Castelli, G. P., 1986, Effects of cholesterol on the kinetics of mitochondrial ATPase, *FEBS Lett.* **198:**353–356.

Ritta, M. N., Campos, M. B., and Calandra, R. A., 1987, Effect of GABA and benzodiazepines on testicular androgen production, *Life Sci.* **40:**791–798.

Roff, C. F., Strauss, III, J. F., Brady, R. O., and Pentchev, P. G., 1994, A contemporary view of intracellular cholesterol movement, *Trends in Glycoscience and Glycotechnology* **6:**75–91.

Rouslin, W., MacGee, J., Gupte, S., Wesselman, A., and Epps, D. E., 1982, Mitochondrial cholesterol content and membrane properties in porcine myocardial ischemia, *Am. J. Physiol.* **242:**H254–H259.

Sackmann, E., Kotulla, R., and Heiszler, F-J., 1984, On the role of lipid bilayer elasticity for the lipid—protein interactions and the indirect protein–protein coupling, *Can. J. Biochem. Cell. Biol.* **62:**778–788.

Sandison, A. T., 1955, A form of lipoidosis of the adrenal cortex in an infant, *Arch. Dis. Child.* **30:** 538–541.

Scallen, T. J., Pastuszyn, A., Noland, B. J., Chanderbhan, R., Kharroubi, A., and Vahouny, G. V., 1985, Sterol carrier and lipid transfer proteins, *Chem. Phys. Lipids* **38:**239–261.

Schneider, H., Höchli, M., and Hackenbrock, C. R., 1982, Relationship between the density distribution of intramembrane particles and electron transfer in the mitochondrial inner membrane as revealed by cholesterol incorporation, *J. Cell Biol.* **94:**387–392.

Sen, P. C., and Ray, T. K., 1980, Characterization of gastric mucosal membranes. II. Lipid compositions of purified gastric mitochondria from pig, rabbit and frog, *Biochim. Biophys. Acta* **618:** 300–307.

Shinitzky, M., 1978, An efficient method for modulation of cholesterol level in cell membranes, *FEBS Lett.* **85:**317–320.

Siliprandi, N., 1980, Factors affecting calcium transport in mitochondria, in: *New Horizons in Biological Chemistry* (M. Koike, T. Nagatsu, J. Okuda, and T. Ozawa, eds.), pp. 143–155, Japanese Sciences Press, Tokyo.

Simpson, E. R., and Waterman, M. R., 1983, Regulation by ACTH of steroid hormone biosynthesis in the adrenal cortex, *Can. J. Biochem. Cell Biol.* **61:**692–707.

Steck, T. L., Kezdy, F. J., and Lange, Y., 1988, An activation-collision mechanism for cholesterol transfer between membranes, *J. Biol. Chem.* **263:**13023–13031.

Stevens, V. L., Tribble, D. L., and Lambeth, J. D., 1985, Regulation of mitochondrial compartment volumes in rat adrenal cortex by ether stress, *Arch. Biochem. Biophys.* **242:**324–327.

Stocco, D. M., and Ascoli, M., 1993, The use of genetic manipulation of MA-10 Leydig tumor cells to demonstrate the role of mitochondrial proteins in the acute regulation of steroidogenesis, *Endocrinology* **132:**959–967.

Stocco, D. M., and Chaudhary, L. R., 1990, Evidence for the functional coupling of cyclic AMP in MA-10 mouse Leydig tumor cells, *Cell. Signal.* **2:**161–170.

Stocco, D. M., and Chen, W., 1991, Presence of identical mitochondrial proteins in unstimulated constitutive steroid-producing R2C rat Leydig tumor and stimulated nonconstitutive steroid-producing MA-10 mouse Leydig tumor cells, *Endocrinology* **128:**1918–1926.

Stocco, D. M., and Kilgore, M. W., 1988, Induction of mitochondrial proteins in MA-10 Leydig tumor cells with human chorionicgonadotropin, *Biochem. J.* **249:**95–103.

Stocco, D. M., and Sodeman, T. C., 1991, The 30-kDa mitochondrial proteins induced by hormone stiumulation in Ma-10 mouse Leydig tumor cells are processed from larger precursors, *J. Biol. Chem.* **266:**19731–19738.

Stuart, R. A., Cyr, D. M., Craig, E. A., and Neupert, W., 1994, Mitochondrial molecular chaperones: their role in protein translocation, *Trends Biochem. Sci.* **19:**87–92.

Sugawara, T., Holt, J. A., Driscoll, D., Strauss, J. F. III, Lin, D., Miller, W. L., Patterson, D., Clancy, K. P., Hart, I. M., Clark, B. J., Stocco, D. M., 1995a, Human steroidogenic acute regulatory protein: funtional activity in COS-1 cells, tissue-specific expression, and mapping of the structural gene to 8p11.2 and a pseudogene to chromosome 13, *Proc. Natl. Acad. Sci. USA* **92:**4778–4782.

Sugawara, T., Lin, D., Holt, J. A., Martin, K. O., Javitt, N. B., Miller, W. L., and Strauss, III, J. F., 1995b, Structure of the human steroidogenic acute regulatory protein (StAR) gene: StAR stimulates mitochondrial cholesterol 27-hydroxylase activity, *Biochemistry* **34:**12506–12512.

Tee, M.-K., Lin, D., Sugawara, T., Holt, J. A., Guigen, Y., Buckingham, B., Strauss, III, J. F., and Miller, W. L., 1995, T-A transversion 11 bp from a splice acceptor site in the human gene for steroidogenic acute regulatory protein causes congenital lipoid adrenal hyperplasia, *Human Mol. Genetics* **4:**2299–2305.

Thomson, M., Korn, M., and Hall, P. F., 1995, GTP-binding proteins in adrenocortical mitochondria, *Biochim. Biophys. Acta* **1248:**159–169.

Toaff, M. E., Flickinger, G. L., and Strauss III, J. F., 1979, Relationship of cholesterol supply to lateal mitochondrial steroid synthesis, *J. Biol. Chem.* **254:**3977–3982.

Vahouny, G. V., Dennis, R., Chanderbhan, G., Fiskum, G., Nolan, B. J., and Scallen, T. J., 1984, Sterol carrier protein 2 (SCP2)-mediated transfer of cholesterol to mitochondrial membranes, *Biochem. Biophys. Res. Commun.* **122:**509–515.

Vahouny, G. V., Chanderbhan, R., Nolan, B. J., Irwin, D., Dennis, P., Lambeth, J. D., and Scallen, T. J., 1983, Sterol carrier protein2. Identification of adrenal sterol carrier protein2 and site of action for mitochondrial cholesterol utilization, *J. Biol. Chem.* **258:**11731–11737.

Van Venetie, R., and Verkleij, A. J., 1982, Possible role of non-bilayer lipids in the structure of mitochondria, a freeze-fracture electron microscopy study, *Biochim. Biophys. Acta* **692:**397–405.

Waterman, M. R., 1995, A rising StAR: an essential role in cholesterol transport, *Science* **267:** 1780–1781.

Wattenberg, B. W., and Silbert, D. F., 1983, Sterol partitioning among intracellular membranes, *J. Biol. Chem.* **258:**2284–2292.

Werner, S., and Neupert, W., 1972, Functional and biogenetical heterogeneity of the inner membrane of rat-liver mitochondria, *Eur. J. Biochem.* **25:**379–396.

Xu, X., Xu, T., Robertson, D. G., and Lambeth, J. D., 1989, GTP stimulates pregnenolone generation in isolated rat adrenal mitochondria, *J. Biol. Chem.* **264:**17674–17680.

Xu, T., Bowman, E. P., Glass, D. B., and Lambeth, J. D., 1991, Stimulation of adrenal mitochondrial cholesterol side-chain cleavage by GTP, steroidogenesis activator peptide(SAP), and sterol carrier protein2. GTP and SAP act synergistically, *J. Biol. Chem.* **266:**6801–6807.

Yamazaki, T., McNamara, B. C., and Jefcoate, C. R., 1993, Competition for electron transfer between cytochromes P450scc and P45011 in rat adrenal mitochondria, *Mol. Cell. Endocrinol.* **95:**1–11.

Yanagibashi, K., Ohno, Y., Kawamura, M., and Hall, P. F., 1988, The regulation of intracellular transport of cholesterol in bovine adrenal cells: purification of a novel protein, *Endocrinology* **123:** 2075–2082.

Yanagibashi, K., Ohno, Y., Nakamichi, N., Matsui, T., Hayashida, K., Takamura, M., Yamada, K., Tou, S., and Kawamura, M., 1989, Peripheral-type benzodiazepine receptors are involved in the regulation of cholesterol side chain cleavage in adrenal mitochondria, *J. Biochem.* **106:**1026–1029.

Yeagle, P. L., 1985, Cholesterol and the cell membrane, *Biochim. Biophys. Acta* **822:**267–287.

Yoshida, T., Mio, M., and Tasaka, K., 1993, Ca^{2+}-induced cortisol secretion from permeabilized bovine adrenocortical cells: the roles of calmodulin, protein kinase C and cyclic AMP, *Pharmacology* **46:**181–192.

Chapter 9

Lipoproteins and Cellular Cholesterol Homeostasis

William J. Johnson, Michael C. Phillips, and
George H. Rothblat

1. OVERVIEW

Appropriate membrane structure and function in eukaryotic cells requires the presence of cholesterol (for reviews, see Yeagle, 1985, and Chapters 6 and 7 in this book). This cell cholesterol originates from endogenous and exogenous sources, with the latter being provided by plasma lipoproteins. In light of this link, it is necessary to know the essential features of both lipoprotein and cell cholesterol metabolism to understand how cholesterol homeostasis is achieved.

1.1. Lipoprotein Metabolism

Lipoprotein particles carry unesterified (free) cholesterol, cholesteryl esters (CE), and triglyceride (TG) molecules through the entero-hepatic and hepato-peripheral cycles of cholesterol transport (for a review, see Havel and Kane, 1995). Exogenous cholesterol in the diet is introduced into the entero-hepatic circulation as chylomicrons after dietary fat and cholesterol are absorbed by enterocytes and packaged into these particles. The core of nascent chylomicrons consists largely

William J. Johnson, Michael C. Phillips, and George H. Rothblat Department of Biochemistry, Allegheny University of the Health Sciences, Philadelphia, Pennsylvania 19129.

Subcellular Biochemistry, Volume 28: Cholesterol: Its Functions and Metabolism in Biology and Medicine, edited by Robert Bittman. Plenum Press, New York, 1997.

of TG, and the particles secreted by enterocytes into the interstitium contain apolipoproteins (apo) A-I, II, and IV, and apo B-48. These chylomicrons enter the circulation via the thoracic duct and acquire C apolipoproteins and apo E by transfer from high density lipoproteins (HDL). The chylomicrons bind to lipoprotein lipase (LPL) on the surface of capillary endothelial cells where most of the TG and some of the phospholipids (PL) are hydrolyzed. As the particles decrease in size due to the shrinkage of the core, they lose affinity for apo C. The remnant particle is released into the circulation and is taken up by hepatocytes. This endocytic process involves apo E-mediated binding to the low density lipoprotein (LDL) receptor (Goldstein and Brown, 1977) and the LDL receptor-related protein (LRP) (Strickland *et al.,* 1995), The remnant particle is hydrolyzed in lysosomes and the dietary cholesterol becomes available to the hepatocyte. Cholesterol in hepatocytes can enter pathways leading to formation of bile acids, be secreted into bile directly, be incorporated into nascent HDL and very low density lipoprotein (VLDL) particles, or be converted into CE (Figure 1).

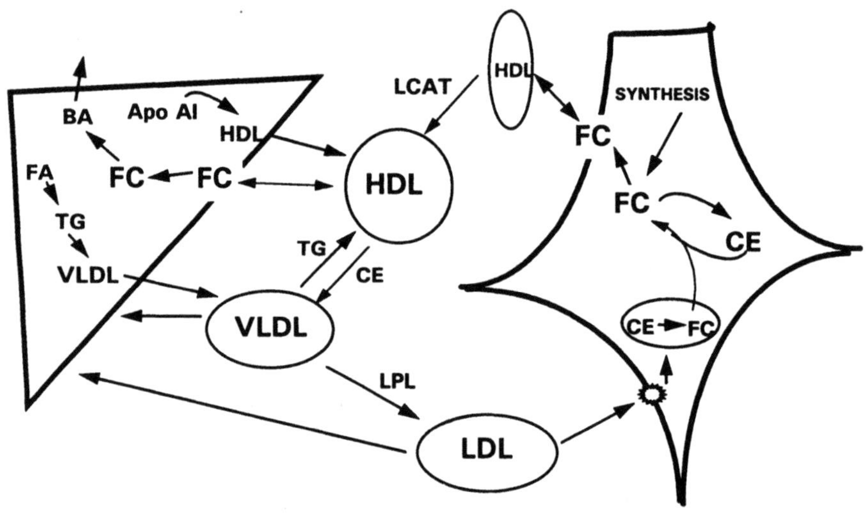

FIGURE 1. Overview of the pathways involved in lipoprotein-mediated cholesterol transport between liver and peripheral tissue in human. The delivery of cholesterol to the periphery is initiated by the assembly of VLDL in liver. The secreted VLDL, which contains apolipoproteins B, C, and E, is converted to LDL in the plasma compartment and the LDL is then taken into peripheral cells by LDL receptor-mediated endocytosis. The pathways involved in cholesterol homeostasis in peripheral cells are shown in more detail in Figure 2. Reverse transport of cholesterol from periphery to the liver is mediated by HDL; see text for discussion of the role of HDL subspecies. In addition to direct delivery of HDL FC to the liver, CE from HDL is transferred to VLDL (and LDL) by CETP, where it can be taken up by hepatocytes by LDL receptor-mediated endocytosis. Some of the FC that is delivered to hepatocytes is catabolized to bile acids. See text for more detail. Abbreviations: BA, bile acid; FA, fatty acid; CE, cholesteryl ester; FC, free (unesterified) cholesterol; TG, triacylglycerol; LPL, lipoprotein lipase; LCAT, lecithin-cholesterol-acyltransferase.

The VLDL particles that are secreted into the hepato-peripheral circulation are 30–80 nm in diameter and contain a CE/TG core and apo B-100, C, and E in the surface. The VLDL in the circulation enters a lipolysis cascade as a substrate for LPL so that TG is removed progressively. This ultimately leads to the formation of 20- to 25-nm diameter LDL particles that contain apo B-100 as their sole protein constituent. The excess surface components (PL and apoprotein) join the HDL pool. LDL particles are the primary carriers of cholesterol in the plasma and they deliver cholesterol to both peripheral cells and hepatocytes (Figure 1). This delivery involves recognition by the LDL receptor, which binds either apo B-100 or apo E. Cholesterol delivered to peripheral cells in this fashion redistributes to various membranes in the cell as well as entering an intracellular cycle of metabolic steps designed to maintain appropriate free cholesterol levels (see sections 1.2 and 3.1). Since peripheral cells (apart from steroidogenic cells) are not able to degrade cholesterol, a pathway by which intact cholesterol molecules can be removed from the cell is essential for cholesterol homeostasis. The process by which cholesterol from peripheral tissues is returned to the liver for catabolism is called "reverse cholesterol transport" (Fielding and Fielding, 1995), and HDL is critical in this pathway (Figure 1) (Johnson *et al.,* 1991b; Pieters *et al.,* 1994).

The major apolipoproteins of HDL are A-I and A-II; the former is synthesized in liver and intestine while the latter is liver-derived. Most HDL measured in plasma is not secreted as such by liver or intestine but is assembled in plasma as a consequence of the lipolysis of chylomicrons and VLDL described above. In the plasma compartment, cholesterol in the various precursor HDL particles (some of which originates from peripheral cells) is converted to CE by the action of lecithin-cholesterol acyltransferase (LCAT), for which apo A-I is a cofactor (Jonas, 1987). This formation of CE is associated with a maturation of the precursor HDL to spherical plasma HDL particles (Eisenberg, 1984). Interconversion of HDL particles between the HDL_2 and HDL_3 classes occurs by the combined actions of LCAT, hepatic lipase, and cholesteryl ester transfer protein (CETP), which exchanges CE and TG (Tall, 1993). The smaller, denser HDL_3 particle is the preferred substrate for LCAT; the formation of CE by this enzyme leads to the formation of larger HDL_2 particles. CETP exchanges CE in HDL_2 with TG in LDL and VLDL. The CE is transferred to LDL and then delivered to the liver by whole particle catabolism as part of reverse cholesterol transport. The TG exchanged into the HDL_2 particle becomes a substrate for hepatic lipase so that the HDL_2 is reconverted to an HDL_3 particle where it can again participate in the cycle of delivery of cholesterol to LCAT. The enrichment of HDL particles with CE can lead to the formation of larger particles that bind apo E; these particles are then taken up by the liver via the LDL receptor. The major sites of catabolism of apo A-I are the kidney and liver, while the HDL cholesterol is degraded in either liver or steroidogenic cells. Mutations in the genes of the proteins such as LCAT and CETP involved in lipoprotein transport affect plasma HDL-cholesterol levels (Breslow, 1995).

Reflecting the diverse pathways for formation and interconversion of HDL particles, circulating plasma HDL mostly comprises a heterogeneous population of spherical lipid-protein particles with diameters in the range 7–12 nm. Spherical HDL particles contain roughly equal masses of lipid and proteins: they comprise a core of neutral CE and TG molecules encapsulated by a surface monolayer of PL and apolipoprotein (Havel and Kane, 1995). Free cholesterol is able to partition between the core and surface with most of it being in the surface (Lund-Katz and Phillips, 1984); the cholesterol/PL molar ratio in the surface of human HDL is about 0.2. The apo A-I and apo A-II molecules stabilize the particles and their interactions with lipid are mediated by amphipathic helical segments (Brouillette and Anantharamaiah, 1995). The structure is dynamic in the sense that both the lipid and protein components of HDL particles can exchange with other lipoproteins in the plasma compartment.

HDL subclasses (von Eckardstein *et al.,* 1994; Fielding and Fielding, 1995; Forte and McColl, 1994; Skinner, 1994) have been demonstrated by various techniques. For instance, centrifugal studies have demonstrated heterogeneity in density and size; several classes of HDL, such as HDL_2 and HDL_3, have been defined by this technique. Non-denaturing gradient gel electrophoresis has resolved HDL on the basis of particle size into even more subspecies, and immunoaffinity chromatography and gradient gel electrophoresis have demonstrated that there are apoprotein-specific populations of plasma HDL within the different size and density classes. One population contains apo A-I only and another contains both apo A-I and apo A-II. Electrophoresis of plasma on agarose gels shows the presence of a small population of HDL particles containing only apo A-I that exhibit pre-β mobility (Fielding and Fielding, 1995). Clearly, a complete understanding of reverse cholesterol transport requires definition of the roles of the various HDL subclasses in cholesterol transport.

It is important to note that the HDL particles that are involved in the initial transport of cholesterol away from peripheral cells are not simply those present in plasma, because these cells are bathed by interstitial fluid that is separated from the plasma compartment by the capillary endothelium. Interstitial HDL particles arise both by filtration of plasma HDL across this barrier after which they are remodelled, and by direct secretion of apolipoproteins by some peripheral cells (Reichl, 1994). The lymph HDL particles are relatively enriched in free cholesterol relative to CE, and the HDL concentration in human peripheral lymph is about 10% of that in plasma.

Abnormalities in the lipoprotein-mediated transport of cholesterol between the liver and the periphery (Figure 1) can lead to pathological situations. Thus, premature coronary artery disease is associated with elevated plasma LDL levels; an early stage in the formation of the atherosclerotic lesion in the artery wall is accumulation of free and esterified cholesterol to give fatty streaks. Since macrophages contain receptors for modified LDL and VLDL (the latter is present at higher concentrations in hypercholesterolemic states) and these receptors are poorly downregulated in response to increased cell cholesterol, this scavenger pathway in macrophages is probably involved in formation of the CE-loaded or foam cells observed in atherosclerotic plaques (Brown and Goldstein, 1983). While enhanced delivery of cholesterol to cells can stimulate atherogenesis, it is probable that reduced reverse cholesterol transport can lead to a similar result. Epidemiological studies have shown that low plasma HDL levels are also associated with premature coronary artery disease. Reduced levels of HDL and apo A-I presumably lead to inefficient reverse cholesterol transport since HDL is the primary cholesterol carrier in this pathway. The concept of HDL playing a key role in reverse cholesterol transport is supported by cell culture experiments showing that HDL can clear excess cholesterol from some cells (see section 2).

In the last few years, studies with transgenic mice (for reviews, see Breslow, 1995; Schultz and Rubin, 1994) have provided insights into the metabolism of HDL and have clarified the role of HDL in atherogenesis. To address the question of whether HDL is directly antiatherogenic, mice transgenic for human apo A-I were fed atherogenic diets, and atherosclerotic lesions in the aorta were measured (Rubin *et al.,* 1991). The finding that overexpression of the human apo A-I gene protected against fatty streak development provided convincing *in vivo* evidence that HDL directly prevents atherogenesis. The protein composition of HDL affects its role in atherogenesis because apo A-I-HDL is more antiatherogenic than apo A-I/A-II-HDL (Schultz and Rubin, 1994).

1.2. Cellular Cholesterol Homeostasis

Generally, the homeostasis of cholesterol in a cell involves a balance of the fluxes of cholesterol into and out of the cell and the rate of *de novo* synthesis of cholesterol (for reviews, see Brown and Goldstein, 1983, 1986; Johnson *et al.,* 1991b; Rothblat and Phillips, 1991). The pathways involved in cellular cholesterol metabolism are summarized in Figure 2. In macrophages, exogenous cholesterol can be delivered by the three influx pathways indicated in Figure 2. The endocytic pathway for catabolism of the LDL particle after it binds to the LDL (apo B/E) receptor has been elucidated in detail (Brown and Goldstein, 1986). Macrophages are unusual among peripheral cells because, in addition to the LDL receptor, they have at least two other classes of receptors which recognize modified LDL. Macrophages possess a scavenger receptor (Kodama *et al.,* 1990) that recognizes both acetyl LDL and oxidized LDL, and an oxidized LDL receptor that recognizes only oxidatively modified LDL (Ottnad *et al.,* 1995). The receptor-mediated endocytosis of normal or modified LDL leads to the ultimate delivery of the lipoprotein particle via a series of endocytic vesicles to mature lysosomes. However, a vital difference is that expression of the modified LDL receptors is not downregulated by cell sterol levels. Consequently, macrophages exposed to the modified LDL can overaccumulate cholesterol and eventually become foam cells (Brown and Goldstein, 1983). An additional pathway by which macrophages and

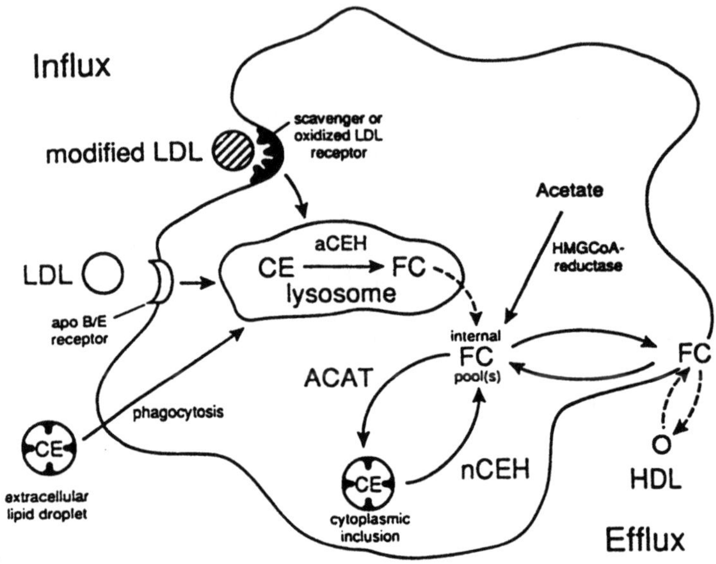

FIGURE 2. Cholesterol metabolism in mammalian cells. Cells acquire cholesterol by endocytic uptake of lipoproteins and extracellular lipid droplets (influx), and by biosynthesis (conversion of acetate carbon to cholesterol). Phagocytosis of extracellular lipid droplets (which arise from the lysis of dead foam cells and aggregation of lipoproteins) may be an important process in atherosclerotic lesions. Biosynthesis and the lysosomal degradation of exogenous CE contribute to internal (non-plasma membrane) pools of free cholesterol (FC). There is continual exchange of cholesterol between the internal pools and cytoplasmic CE droplets via a cycle of reactions catalyzed by acyl CoA: cholesterol acyltransferase (ACAT) and neutral cholesteryl ester hydrolase (nCEH). Cellular cholesterol homeostasis is maintained by the efflux of cholesterol from the plasma membrane to particles in the HDL fraction of plasma and interstitial fluid. The removal of cholesterol from internal pools requires preliminary transport to the plasma membrane. See text for further details.

smooth muscle cells can become foam cells involves the phagocytosis of CE-rich lipid droplets (Wolfbauer *et al.*, 1986) (see section 3.1).

As far as endogenous cholesterol is concerned, the rate-limiting step in the biosynthetic pathway for cholesterol is the reduction of β-hydroxy-β-methylglutaryl coenzyme A (HMG-CoA) to mevalonic acid (Brown and Goldstein, 1986). The transcription of the gene for HMG-CoA reductase is repressed when cells are loaded with sterols. The promoter for this gene involves a sterol regulatory element that is common to the promoter for the LDL receptor gene. This leads to coordinate regulation of the expression of both proteins, with both being subject to endproduct feedback regulation by cholesterol, so that as endocytosis of LDL leads to a rise in cell cholesterol levels, the expression of both proteins is suppressed.

As depicted in Figure 2, CE delivered to lysosomes by the various influx pathways is hydrolyzed to free cholesterol by acid CE hydrolase (aCEH). aCEH may be regulated by second messengers (Glick, 1990). If the rate of hydrolysis of CE is greater than the rate of removal of free cholesterol from the lysosome, accumulation of free cholesterol ensues, with subsequent formation of cholesterol crystals (see section 3.1). Free cholesterol leaves lysosomes by ill-defined pathways and becomes part of the intracellular free cholesterol pool(s). Some of this free cholesterol comprises a pool that activates acyl-CoA:cholesterol acyltransferase (ACAT), which is located in the rough endoplasmic reticulum. The activity of this enzyme appears to be regulated primarily by substrate availability, so that free cholesterol which is in excess of cellular requirements is esterified by ACAT and the CE product is stored in cytoplasmic inclusions. This CE is a substrate for a neutral CE hydrolase (nCEH), and there is a cycle involving the hydrolase and ACAT by which the stored CE is continuously turned over with expenditure of metabolic energy. In macrophages, at least some of the nCEH activity is due to hormone-sensitive lipase (Khoo *et al.*, 1993). The rate of CE hydrolysis is related to the physical state of the CE in the inclusions (Adelman *et al.*, 1984). Free cholesterol molecules can move between the intracellular pool(s) and the plasma membrane. In some cells, these translocation steps may be regulated by second messenger systems (see section 3.3). Cholesterol molecules in the plasma membrane become available for efflux from the cell to extracellular acceptor particles (see section 2). This process occurs through the desorption of free cholesterol molecules, possibly from different domains in the membrane (Rothblat *et al.*, 1992). Free and esterified cholesterol molecules can also move from HDL into the cell and lead to nonreceptor-mediated cellular cholesterol deposition.

It is apparent from the above information that the mechanisms by which cholesterol accumulates in cells are relatively well understood. In contrast, many of the details about the pathways of cellular cholesterol efflux are not known. With regard to reverse cholesterol transport and the flux of cholesterol between cells and HDL, the transport from the various metabolic and physical pools of cholesterol within cells to HDL is an area of active investigation. The remainder of this chapter focuses on the mechanisms involved in the intracellular trafficking and efflux from cells of cholesterol.

2. EFFLUX OF CELLULAR CHOLESTEROL

2.1. Efflux to Serum

Although several types of cholesterol acceptors have been used extensively in experiments designed to probe various aspects of cell cholesterol efflux, the use of whole serum or plasma for such studies have been relatively infrequent. Serum and plasma represent physiologically relevant acceptors for studies of reverse cholesterol transport since all of the potential cholesterol acceptors are present, together with other associated factors such as LCAT and CETP. On the other hand this very complexity limits the use of whole serum to studies that: i) identify the most efficient acceptor particles in serum, ii) study the interaction of serum components that participate in efflux and the subsequent movement of cell-derived cholesterol among serum components, and iii) determine physiologically relevant factors that modulate the movement of cholesterol from cells to acceptors. Two general protocols have been used for whole serum/plasma investigation of cell cholesterol efflux. In the first method, either serum or plasma at 100% concentration is added to cell monolayers, and incubations are generally in the range of 1–10 min. This experimental approach, if coupled with a rapid separation of lipoprotein particles, is particularly suitable for identifying the lipoproteins that serve as initial acceptors of cell cholesterol, and for the subsequent tracking of the released cell cholesterol as it exchanges with other plasma particles and undergoes esterification by LCAT. The second general approach is to use diluted serum (usually in the range 5–20%) as the acceptor. Although this approach has generated considerable kinetic data and can be employed to establish the correlation between efflux and the various components in serum that participate in reverse cholesterol transport, it has not been employed to identify the initial acceptor particles in serum because of their low concentrations in serum.

The protocol for the use of 100% plasma for the study of cell cholesterol efflux was first developed by Fielding and colleagues (Castro and Fielding, 1988; Miida *et al.,* 1992; Kawano *et al.,* 1993). In this assay, freshly collected human plasma, to which protease inhibitors have been added, is incubated for short periods with human skin fibroblasts that have been previously radiolabeled with [^{3}H]cholesterol. The subsequent distribution of the released cell cholesterol among serum lipoproteins is monitored by a two-dimensional gel analysis (Miida *et al.,* 1992; Kawano *et al.,* 1993). This approach shows that a particularly efficient acceptor of cholesterol is a small, PL-rich apo A-I-containing lipoprotein, termed preβ1-HDL (Castro and Fielding, 1988). This particle can be found in both plasma and lymph and represents the smallest of a series of preβ-migrating particles termed, in order of increasing size, preβ1-, preβ2-, and preβ3-HDL (Castro and Fielding, 1988; Fielding and Fielding, 1995). Preβ1-HDL has an apparent molecular weight of 60–70 kDa, a diameter of approximately 5 nm, and is comprised of apo A-I together with equal proportions of phosphatidylcholine (PC) and sphingomyelin (SM) (Fielding and Fielding, 1995). Its concentration in serum is reduced through the action of LCAT (Fielding and Moser, 1982) and is increased through the transfer of lipid by CETP (Kunitake *et al.,* 1992).

The hypothesis underlying the concept that preβ-HDL particles play a primary role in cell cholesterol efflux is that the small, lipid-poor, preβ1-HDL is the primary acceptor of cholesterol, which is then transferred to the larger preβ2-HDL. On this latter lipoprotein the cell-derived cholesterol is esterified by LCAT, which results in the formation of the spherical preβ3 particle. (For a more detailed discussion of this hypothesis and the supporting data see the review by Fielding and Fielding, 1995). Using efflux protocols and two-dimensional gel separation techniques similar to those used by other investigators (Lefevre *et al.*, 1987), von Eckardstein and colleagues (Huang *et al.*, 1994; von Eckardstein *et al.*, 1995) demonstrated that a number of different particles in plasma can act as early acceptors of cell-derived cholesterol. These particles include lipoproteins containing apo E, designated gamma-HDL, and other particles composed of apo A-IV with no apo A-I (von Eckardstein *et al.*, 1995). Thus, it is becoming obvious that a number of different particles can rapidly acquire cell cholesterol when whole plasma is incubated with radiolabeled cells.

Among the first detailed studies on cholesterol efflux using diluted human sera were those of Fielding *et al.* (1982, 1983). Initial studies using fibroblasts as the cholesterol donor cells established that sera from individuals differ in their ability to produce changes in net cellular cholesterol efflux. These differences were correlated to a number of clinical conditions that are known to influence serum lipoprotein patterns, such as hypertriglyceridemia (Fielding *et al.*, 1983) and diabetes (Fielding *et al.*, 1982). Clearly, factors in both the d = 1.062-1.21 and in the $d > 1.21$ g/ml fractions of serum can stimulate cholesterol efflux and net cholesterol removal from many cells (Oram, 1983). More recently, an experimental system was developed to test large numbers of serum specimens with the goal of establishing the extent to which human sera differ in ability to promote cell cholesterol efflux, and to establish correlations between cell cholesterol efflux and the different apolipoproteins and lipoproteins in the sera (de la Llera Moya *et al.*, 1994). The system quantitates the fractional release of labelled cholesterol from rat Fu5AH hepatoma cells to the individual serum samples diluted to 5%. The results of a study of more than 100 serum samples obtained from men indicate that there is no correlation between the extent of efflux and serum parameters associated with apoB-containing lipoproteins (Figure 3). The highest correlation to cholesterol efflux is with total HDL cholesterol, followed closely by a number of other HDL-related parameters. These correlations suggest that a single HDL subclass is not responsible for the efflux but rather a number of particles contribute to the process of removal of cell cholesterol.

HDL subclasses are usually separated and classified according to either their density or their apoprotein content. None of the techniques used for separation has clearly identified the particles responsible for modulating cholesterol efflux. In the HDL_2/HDL_3 combination, the general perception from studies with isolated lipoproteins is that HDL_3 is a more efficient acceptor (Oram *et al.*, 1981). Immunoaffinity techniques yield isolated HDL subclasses containing either apo A-I alone (LpAI) or both apo A-I and apo A-II (LpAI/AII) (Cheung and Albers, 1984; Fruchart *et al.*, 1992) but the roles played by these particles in cholesterol efflux remain controversial. Thus, measurements of the decrease in cell cholesterol content of the rat adipose cell line OB17 have shown that LpAI promotes net cholesterol efflux whereas incubation with LpAI/AII does not reduce cell cholesterol

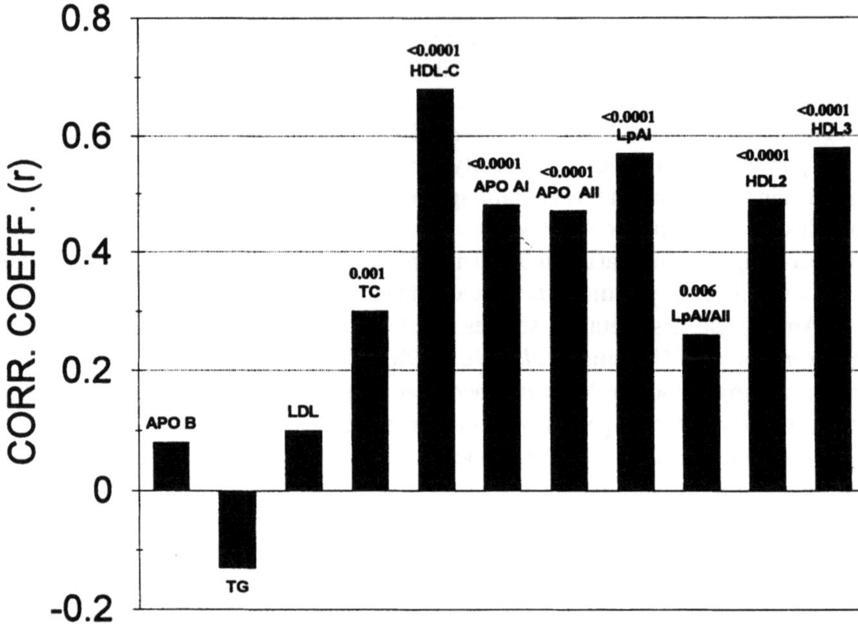

FIGURE 3. Bar graph showing correlation coefficients (corr. coeff., *r*) between lipid, apoprotein, and lipoprotein parameters of human serum samples, and the extent of efflux of cholesterol from rat Fu5AH hepatoma cells. The serum samples were obtained from 113 men, the majority of whom were individuals exhibiting varying degrees of hyperlipemia. Apo B indicates apolipoprotein (apo) B; TG, triglycerides; LDL, low density lipoprotein; TC, total cholesterol; HDL, high density lipoprotein; HDL-C, HDL cholesterol; and Lp, lipoprotein. The P value is shown above each bar. See de la Llera Moya *et al.* (1994) for more details.

content (Barbaras *et al.*, 1987; Barkia *et al.*, 1991). On the other hand, no differences in efflux were observed when immunopurified preparations of LpAI or LpAI/AII were added to other types of cells in culture (Johnson *et al.*, 1991; Hodenberg *et al.*, 1991; Okawa *et al.*, 1993). Among the factors that could explain the divergent results is the extent of contamination of the particles with LCAT. Indeed, it has been demonstrated that the reduction of macrophage cholesterol content by LpAI is, in part, linked to the association of LCAT with the LpAI particles (Ohta *et al.*, 1992). This observation supports the view that a combination of factors present in whole serum may participate in the regulation of cell cholesterol efflux so that purified subfractions of lipoproteins may not always yield the same data as whole serum. The correlation coefficients obtained between efflux and either LpAI or LpAI/AII concentration in whole serum studies suggest that there is a greater association between the concentration of LpAI and cell cholesterol efflux (de la Llera Moya *et al.*, 1994). The above results are consistent with the conclusion that LpAI and HDL_3 have the strongest influence on cell cholesterol efflux.

The recent development of a large variety of transgenic strains of mice (Hayek *et al.,* 1992; Schultz *et al.,* 1993; Nakashima *et al.,* 1994; Melchior *et al.,* 1994) has provided the opportunity to determine how the expression of specific serum factors influences the ability of serum to promote the efflux of cellular cholesterol. Using the cell cholesterol assay discussed above, the efflux potential was determined for sera obtained from transgenic mice expressing either human apo A-I, human CETP, or both apo AI and CETP (Atger *et al.,* 1996). In addition, studies were performed using serum obtained from mice in which the apo A-I gene was inactivated. The fractional efflux values for cholesterol released from the donor cells to the sera from the different groups of mice were in the order: background = apo A-I transgenic > CETP transgenic > apo A-I/CETP transgenic > apo A-I knockout. The markedly reduced efflux to serum lacking the mouse apo A-I demonstrates that apo A-I-containing particles are primarily responsible for serum-stimulated efflux of cell cholesterol. This is consistent with studies demonstrating reduced efflux when apo A-I-containing lipoproteins are removed by immunoaffinity chromatography (Fielding and Moser, 1982). The absence of apo A-I does not completely eliminate the efflux potential of serum, and it could be estimated that approximately 25% of the efflux potential of mouse serum is stimulated by lipoproteins lacking apo A-I. The nature of these lipoproteins has not been established, but they probably comprise apo A-IV-and apo E-containing particles. Analysis of correlations between fractional efflux and serum components helps to clarify the differences in efflux potential observed between the different groups of transgenic mice. In every group, efflux correlates best to the concentration of HDL-cholesterol in the test sera. The similar efflux efficiency of the serum from human apo A-I transgenic mice compared to that of control animals is probably due to the production of increased amounts of apo A-I without a proportionate increase in the availability of PL. If this is the case, an overproduction of apo A-I in apo A-I transgenic animals, without a concomitant increase in available PL, could result in the formation of lipid-free, or poorly lipidated, apo A-I that would have preβ mobility but a low capacity for cholesterol efflux (Atger *et al.,* 1995).

2.2. Mechanism of Cell Cholesterol Efflux

The studies summarized above show that HDL concentration correlates best with the relative efflux potential of serum. The mechanisms by which HDL removes cholesterol from cells have been investigated by simplifying the system and using purified HDL preparations.

When HDL is incubated with a cell monolayer, a bidirectional flux of cholesterol molecules occurs. The mechanism of this process has been investigated extensively. Efflux is defined as the rate of movement of free cholesterol from the cells to the medium, and influx is the rate of movement in the opposite direction: Each process has a characteristic rate constant, designated here as k_e and k_i for efflux and influx, respectively. The methods for obtaining k_e and k_i have been described elsewhere (Johnson *et al.*, 1986; Rothblat *et al.*, 1986). As summarized in Figure 4, the bidirectional flux of cholesterol between human HDL and rat Fu5AH hepatoma cells containing normal levels of cholesterol can be described quantitatively by assuming homogeneous cell and HDL free cholesterol pools (Johnson *et al.*, 1986). It follows that both translocation of cholesterol molecules in the plasma membrane and movement of cholesterol from internal sites to the plasma membrane are rapid relative to the rate of transfer of cholesterol molecules to the HDL particles. As discussed in detail elsewhere (Phillips *et al.*, 1987), the influx and efflux of free cholesterol mass between cells and HDL can be expressed in terms of k_i and k_e, and the concentrations of free cholesterol in the HDL and cell pools. Free cholesterol surface transfer processes drive the cell free cholesterol content towards a steady state level determined by the external concentration of free cholesterol and the ratio of k_i and k_e. The net movement of free cholesterol between cells and HDL seems to involve a diffusion of cholesterol molecules down their free energy gradient from regions with a high free cholesterol/PL ratio to regions with a low ratio. Enrichment of the plasma membrane with cholesterol does not alter k_e very much (Johnson *et al.*, 1988); the flux of cholesterol out of the cell is increased because of the enlarged pool of plasma membrane cholesterol. The concept of the partition of cholesterol between HDL and cell being primarily governed by the free cholesterol/PL ratio is supported by experiments showing that the free cholesterol/PL ratio of red blood cells can be increased or decreased in direct proportion to the cholesterol/PL ratio of extracellular liposomes (Cooper *et al.*, 1978). HDL seems to participate in simple exchange (i.e., efflux = influx) with cells containing normal amounts of cholesterol, whereas HDL can cause the net release of cholesterol from cells enriched in cholesterol. The direction of net cholesterol flux between cells and HDL is also affected by HDL cholesterol and PL composition; HDL particles with molar ratios of cholesterol/PL > 0.2 (cholesterol/protein $> 50–60$ μg/mg) stimulate cellular cholesterol accumulation, whereas particles with molar ratios less than 0.2 generally deplete cells of cholesterol (Johnson *et al.*, 1988). The variations observed in the net movement of cholesterol between cells and HDL preparations having different cholesterol contents are reflections of the size of the lipoprotein-free cholesterol pool and are not a result of changes in either k_i and k_e. Thus, the rate constants for flux remain unchanged when fibroblasts are incubated with HDL preparations that are either enriched or depleted in cholesterol (Johnson *et al.*, 1988).

Two enzymes that are thought to play an important role in the reverse cholesterol transport process by modulating HDL lipid compositions are LCAT and hepatic lipase. LCAT provides a driving force in reverse cholesterol transport by esterifying the cholesterol cellular cholesterol removed by HDL, thereby maintaining a cholesterol gradient favoring net efflux. Cell culture studies have documented the ability of LCAT to modulate the movement of free cholesterol between HDL and cells. Thus, exposure of cells to control HDL results in cholesterol exchange without a change in content, whereas incubation of cells with HDL that has

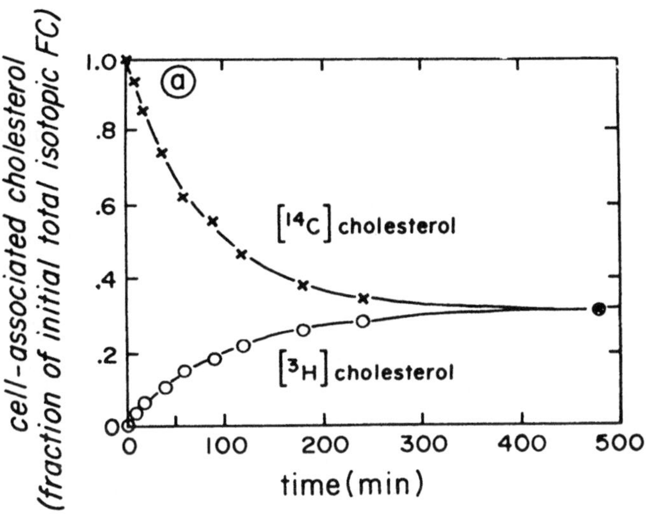

FIGURE 4. The bidirectional flux of cholesterol between human HDL and rat Fu5AH hepatoma cells. The HDL and cells were prelabeled, respectively, with 7-^{3}H-labeled free cholesterol and 4-^{14}C-labeled free cholesterol under conditions which prevented esterification of the tracers. The HDL was diluted into tissue culture medium and applied to cells. At intervals, the cells and medium were separated and analyzed for isotopic composition by liquid scintillation counting. The fractional retention of ^{14}C-labeled free cholesterol (X) and the fractional uptake of ^{3}H-labeled free cholesterol (O) by the cells are plotted against incubation time in min. The smooth curves were obtained by the computerized fitting of each set of kinetic measurements to a model for free cholesterol (FC) equilibration between two pools. This model predicts that the time course for ^{14}C-labeled free cholesterol retention by cells should conform to the relationship

$$\frac{^{14}\text{C cells}}{^{14}\text{C cells}, t = 0} = \frac{k_e}{k_i + k_e} e^{-(k_i + k_e)t} + \frac{k_i}{k_i + k_e},$$

and that for ^{3}H-labeled free cholesterol uptake by the cells to:

$$\frac{^{3}\text{H cells}}{^{3}\text{H med.}\ t = 0} = -\frac{k_i}{k_i + k_e} e^{-(k_i + k_e)t} + \frac{k_i}{k_i + k_e},$$

where

$$\frac{^{14}\text{C cells}}{^{14}\text{C cells}, t = 0} \quad \text{and} \quad \frac{^{3}\text{H cells}}{^{3}\text{H med.}\ t = 0}$$

are, respectively, the fractional retention of ^{14}C-labeled free cholesterol and the fractional uptake of ^{3}H-labeled free cholesterol, and k_e and k_i are the rate constants for free cholesterol efflux and influx, respectively. (For complete details, see Johnson *et al.,* 1986.)

previously been modified by exposure to LCAT produces a decrease in cell cholesterol content (Ray *et al.*, 1980). Just as LCAT functions to produce a cholesterol gradient that stimulates cholesterol efflux from peripheral cells, so might hepatic lipase function to establish a cholesterol gradient within the liver that favors the net movement of free cholesterol from HDL to hepatocytes. The hypothesis that hepatic lipase, through its phospholipase activity, depletes HDL of PL and enhances cholesterol uptake was first proposed by Jansen and Hulsmann (1980).

With respect to the molecular mechanism, it is well established that the exchange of free cholesterol molecules is a surface transfer process in the sense that internalization of the lipoprotein particles by the cell is not required. The kinetics of cholesterol flux in the surface transfer process are consistent with a so-called "aqueous diffusion" mechanism (for reviews, see Phillips *et al.*, 1987; Johnson *et al.*, 1991b; Brown, 1992). In the aqueous diffusion process (Figure 5), cholesterol molecules desorb from the donor lipid–water interface and diffuse through the intervening aqueous layer until they collide with, and are absorbed by, an acceptor particle; when excess acceptor particles are present, the rate-limiting step is the desorption of cholesterol molecules from the donor particle. The following kinetic criteria are characteristic of the process: i) The rate of free cholesterol exchange is first order with respect to the concentration of free cholesterol in the given kinetic pool in the donor particles; ii) the free cholesterol exchange process is strongly temperature dependent and the activation energy is about 70 kJ/mol; and iii) at high acceptor/donor particle ratios, the rate of exchange is zero order with respect to the concentration of acceptor particles. Transfer of cholesterol molecules to the aqueous phase proceeds through a transition-state complex in which the cholesterol molecule is attached to the donor lipid–water interface by the tip of its hydrophobic tail. The height of the activation energy barrier for formation of this transition state is affected by the interaction energy of the cholesterol molecule with neighboring molecules at the surface of the donor particle. Thus, if the van der Waals attraction between cholesterol and the host PL is relatively low, the rate of desorption of cholesterol molecules is enhanced. For this reason, decreases in the degree of saturation of the host PL molecules and the SM content of the lipid–water interface are accompanied by increases in the cholesterol exchange rate.

As an alternative to the aqueous diffusion mechanism to explain HDL-mediated efflux of cellular cholesterol, it has been postulated that a cell-surface receptor for HDL is involved (Oram *et al.*, 1983; Oram, 1990). As summarized in Figure 5, the original hypothesis suggested that binding of HDL to the receptor facilitates movement of cholesterol molecules between the plasma membrane and the extracellular HDL particle. However, several lines of evidence show that the efflux of plasma membrane cholesterol is not coupled to the high affinity binding of HDL to the cell surface (for reviews, see Phillips *et al.*, 1987; Johnson *et al.*, 1991b). There is now general agreement that binding of HDL to the cell surface does not affect the efflux of plasma membrane cholesterol (Phillips *et al.*, 1987;

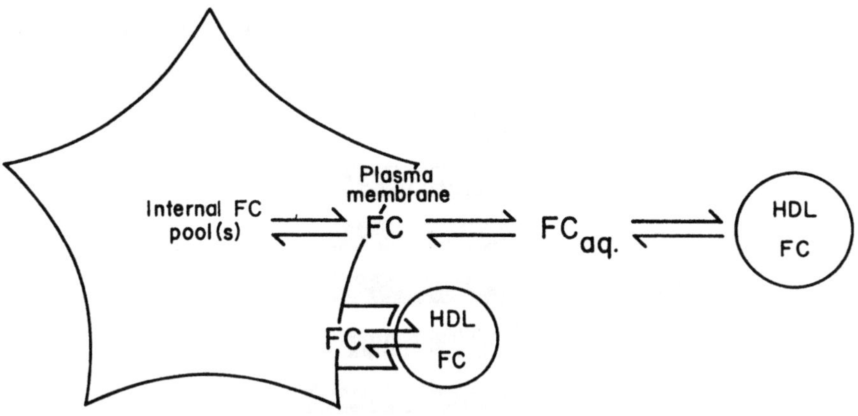

FIGURE 5. Schematic representation of the "aqueous diffusion" and receptor-mediated mechanisms of unesterified (free) cholesterol (FC) flux between the cell plasma membrane and extracellular HDL particles. FC_{aq} in the aqueous diffusion mechanism represents cholesterol molecules that have desorbed into the aqueous phase and are diffusing between the cell surface and HDL. In the receptor-mediated process, the high affinity binding of HDL apoprotein to the cell surface receptor is postulated to facilitate the movement of cholesterol from the plasma membrane to the HDL particles. See text for more detail.

Oram, 1990; Johnson *et al.,* 1991b). However, as summarized in section 3.2, interaction of HDL with the plasma membrane may stimulate efflux of intracellular cholesterol.

Since the packing of cholesterol molecules in a cell plasma membrane influences the rate of transfer to an acceptor HDL particle by the aqueous diffusion mechanism, it is to be expected that the rate of cholesterol release to a common acceptor particle is dependent upon the cell type. Figure 6 shows that the halftimes ($t_{1/2}$) for cell cholesterol efflux are typically in the range 1–24 h. The fact that intact cells and isolated plasma membranes release cholesterol at the same rate (Bellini *et al.,* 1984) confirms that variations in plasma membrane structure can account for differences in kinetics of cell cholesterol release. An example of this effect is the influence of the SM content (for reviews, see Bittman, 1988, 1993). Comparison of the kinetics of cholesterol efflux from different species of red blood cells with SM/PC weight ratios ranging from 5/1 to 0.3/1 shows that increasing the membrane SM content raises the $t_{1/2}$ for exchange (Gold and Phillips, 1990). Interestingly, the influence of SM is less in the erythrocyte membrane than in model SM/PC membranes, perhaps because of modulating factors such as membrane proteins and the presence of a more complex PL mixture. Manipulation of the red blood cell to immobilize or remove the cytoskeleton has no affect on $t_{1/2}$, suggesting that the cholesterol/PL interactions in the outer leaflet of the plasma membrane are not affected by the PL bilayer–protein interactions involved in the attachment of the cytoskeleton (Gold and Phillips, 1992). However, in more

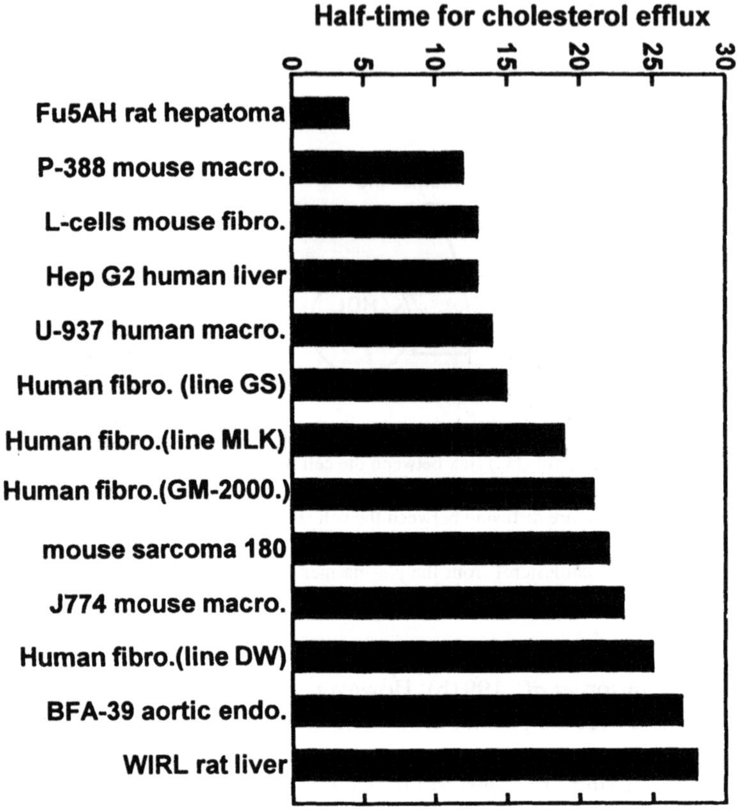

FIGURE 6. Comparison of the half-times for cholesterol efflux (in h) for a variety of cells. Cells were prelabeled with [^{3}H]cholesterol and incubated with apoHDL/egg PC acceptor particles at a concentration of 1 mg PC/ml in minimal essential medium for 6 h. (1) Fu5AH rat hepatoma; (2) mouse P-388 macrophage; (3) mouse L-cells; (4) human HepG2 liver; (5) human U-937 macrophage; (6) human preputial fibroblast (line GS); (7) human preputial fibroblast (line MLK); (8) human skin fibroblast (GM-2000, LDL-receptor negative); (9) mouse sarcoma 180; (10) mouse J774 macrophage; (11) human preputial fibroblast (line DW); (12) bovine fetal aortic endothelial BFA-39; (13) WIRL rat liver.

complex cells, membrane proteins are likely to modulate the cholesterol packing in the membrane and affect the kinetics of the cholesterol exchange. Such proteins can cause formation of domains in the membrane; these influence cholesterol efflux (Bittman, 1988; Schroeder *et al.,* 1991, 1995; Rothblat *et al.,* 1992; Fielding and Fielding, 1995). Transbilayer and lateral domains of cholesterol can exist in the plasma membrane, and an area of active investigation is the relative availability of cholesterol molecules in these pools for efflux to different acceptor particles, such as the various HDL subspecies.

2.3. Effects of HDL Structure

The cell cholesterol efflux experiments using whole serum summarized in section 2.1 show that different types of HDL particles exhibit varying capabilities to remove cell cholesterol. The presence of apo A-I in an appropriately lipidated state seems to give rise to the most efficient efflux. To understand the mechanistic basis for this effect, many studies have been conducted using isolated HDL subfractions as well as reconstituted HDL particles (for reviews, see Phillips *et al.*, 1987; Johnson *et al.*, 1991b; Fielding and Fielding, 1995). Experiments with reconstituted HDL particles have provided insights into the contributions of the lipid and protein components to the functioning of HDL in cell cholesterol efflux.

The data in Figure 7 show that a given concentration of PL in the extracellular medium gives much greater efflux of cell cholesterol when it is present as a discoidal complex with apo A-I instead of a small unilamellar vesicle. The discoidal apo A-I/PC complex and serum HDL_3 particles behave similarly in this assay when present at the same PL concentration. Apo A-I alone gives very little

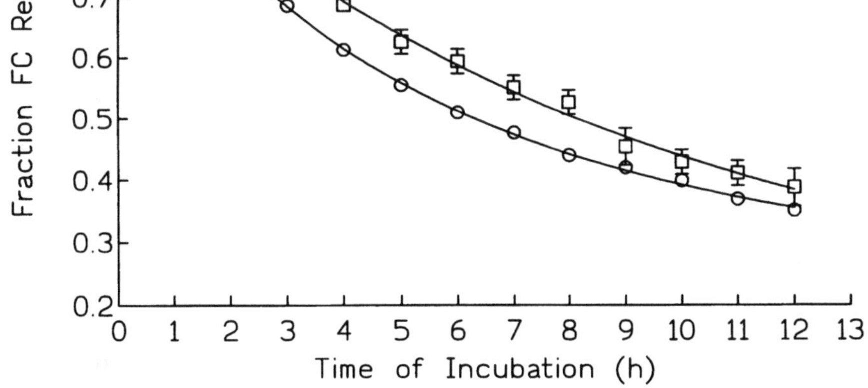

FIGURE 7. Time courses showing relative efflux of unesterified (free) cholesterol (FC) from mouse L-cell fibroblasts to palmitoyl-oleoyl PC (POPC) small unilamellar vesicles (△), reconstituted discoidal HDL particles made with POPC and human apo A-I (o) and human HDL_3 particles (□). The cells were labeled with [^{3}H]cholesterol and incubated with the acceptor particles present at 400 μg PL/ml. The POPC particles were prepared as described in Davidson *et al.* (1995b).

efflux from mouse L-cell fibroblasts under the conditions of this experiment (compare section 2.4). In large part, apo A-I promotes cellular cholesterol efflux because it solubilizes PL so that the PL is present in the extracellular medium as many small particles (DeLamatre *et al.*, 1986; Davidson *et al.*, 1995a). The rate of cholesterol efflux from the cell is commonly influenced by the rate constant describing the propensity of cholesterol molecules to collide with, and incorporate into, the acceptor particle. This term can be influenced by the diffusion coefficient of the acceptor particle and the fraction of the cholesterol–acceptor collisions that result in successful incorporation into the acceptor particle. The number of PL molecules in the small unilamellar vesicle is about ten times that in the discoidal HDL particle, so that a given concentration of PL gives more discoidal particles. The presence of many more small discoidal particles that can diffuse relatively rapidly causes more rapid efflux of cell cholesterol to the discoidal apo A-I/PC particles (Figure 7). It should be noted that at a given concentration of acceptor particles, larger particles are more efficient than smaller ones (Davidson *et al.*, 1995a).

The ability of apo A-I/PL discoidal particles to remove cholesterol from cells is dependent upon the acyl chain composition of the PL present (Davidson *et al.*, 1995b). Generally, the ability of a given HDL particle to accept cellular cholesterol is related to PL acyl chain length and degree of unsaturation; decreases in PL acyl chain length and increases in unsaturation tend to result in more efficient cholesterol acceptor particles. It seems that HDL particles that contain highly fluid surfaces sequester cholesterol molecules that have diffused from the cell plasma membrane at a significantly faster rate than those containing highly organized lipid surfaces with restricted PL acyl chain mobility. The influence of apolipoprotein structure on the ability of HDL to remove cell cholesterol is being investigated actively. Synthetic, amphipathic, α-helical peptides can form discoidal HDL particles that remove cholesterol from cells with similar or greater efficiency than apo A-I-containing particles (Davidson *et al.*, 1994). Since the peptides have no amino acid sequence homology to apo A-I, it follows that efflux does not require a specific apolipoprotein conformation, but simply the presence of amphipathic α-helices to solubilize the PL. Consistent with this conclusion, discoidal HDL particles reconstituted with exchangeable apoproteins of the A, C, and E classes function as acceptors of cell cholesterol (Mahlberg *et al.*, 1991; Stein *et al.*, 1995). Differences in the effectiveness of various HDL particles of the same size are probably due to differences in protein conformation. This effect is probably the reason that apo A-I-containing particles are the most efficient acceptors (Mahlberg *et al.*, 1991; Lagrost *et al.*, 1995). Monoclonal antibodies have been used to determine the conformation or domain of apo A-I that is critical for mediating cell cholesterol efflux. Antibodies that bind to epitopes in the central region of the apo A-I molecule have been shown in two studies to inhibit cholesterol efflux from monocytic cells and adipocytes, although the precise regions of the apo A-I molecule involved are different in the two studies (Banka *et al.*, 1994; Luchoomun *et al.*,

1994). In summary, the conformation and organization of apo A-I in an efficient acceptor particle such as preβ1-HDL still requires elucidation.

2.4. Efflux to Lipid-Free Apolipoproteins

In addition to studies of apo A-I present in reconstituted HDL particles, the ability of the lipid-free molecule to mediate cell cholesterol efflux has been investigated. The relevance to reverse cholesterol transport is that some portions of apo A-I and apo A-IV in serum and interstitial fluid are not associated with lipoprotein. It has been estimated that about 3% of the apo A-I in human serum is in this form (Neary and Gowland, 1987). The first studies on cholesterol efflux using free apolipoproteins indicated that the efficiency of these proteins was relatively low when compared to HDL, but efflux was increased with the addition of PL together with apoproteins (Jackson *et al.,* 1975; Stein *et al.,* 1986). However, it has been demonstrated that lipid-free apo A-I, apo A-II, and apo E are all able to stimulate significant cholesterol efflux and reduce the level of cholesterol in cells (Hara and Yokoyama, 1991; Bielicki *et al.,* 1992; Kritharides *et al.,* 1995). Although a number of lipid-free apoproteins can stimulate cell cholesterol efflux, they differ in their apparent efficiencies, with apo A-I, apo A-II, and apo A-IV being equally efficient on a molar basis and more efficient than either apo E or apo C. The enrichment of cells with cholesterol results in enhanced efflux of the cholesterol to lipid-free apoproteins, and the fractional release of cholesterol differs considerably between cell types, with smooth muscle cells being relatively poor cholesterol donors and macrophages being highly efficient donors (Bielicki *et al.,* 1992; Komaba *et al.,* 1992; Akira *et al.,* 1993; Yancey *et al.,* 1995).

Cellular PL also undergo efflux to extracellular acceptors such as HDL or lipid-free apolipoproteins (Bielicki *et al.,* 1991; Hara and Yokoyama, 1991; Bielicki *et al.,* 1992; Akira *et al.,* 1993; Forte *et al.,* 1995; Yancey *et al.,* 1995). Just as in the case of cholesterol release, the movement of cell PL to lipid-free apoproteins differs among cell types and is enhanced if the cells are enriched with cholesterol. The predominant PL recovered in the culture medium are SM and PC, and the PL composition of the nascent particle differs depending on cell type, incubation conditions, and apoprotein. It appears that there is a common mechanism because the concentrations of lipid-free apo A-I and apo A-IV that promote half-maximum efflux of cell cholesterol are similar to those for PL release. The kinetics of release of the two cell lipids differ, with PL appearing in the medium more rapidly than cholesterol.

To obtain more detailed information on the structural requirements for a lipid-free apoprotein to promote lipid efflux from cells, a series of investigations was performed using synthetic peptides of defined structure (Yancey *et al.,* 1995). Apo A-I and the other exchangeable apoproteins are composed of a varying number of repeating, 22-residue, α-helical domains that mediate interaction of the protein with lipid (Segrest *et al.,* 1992). Peptide 18A is an 18 amino acid peptide that forms a single amphipathic α-helix; there is no sequence homology between this synthetic peptide and naturally occurring apolipoproteins. Peptide 37pA contains two 18A peptides linked by a proline residue (Anantharamaiah *et al.*, 1985). Thus, this peptide consists of two amphipathic helical segments and has higher lipid binding affinity than 18A. Blocked-18A (Ac-18A-NH_2), which is the 18A molecule modified to contain an acetyl group at the N-terminal and an amide group at the *C*-terminal, has a higher lipid-binding affinity than 18A. The ability of the above peptides to remove cell cholesterol and PL demonstrates that it is critical for the proteins to contain α-helices that are amphipathic and have the lipid-binding characteristics of the natural apoproteins. It is evident that the efficiency with which lipid is released from cells to proteins is influenced by the number of amphipathic helical segments per molecule, since the efficiency with which the peptides stimulate lipid efflux is 37pA $>$ 18A. Increasing the length of the amphipathic helix within a molecule also increases the efficiency with which the peptide stimulates lipid efflux (Ac-18A-NH_2 $>$ 18A).

The mechanism by which lipid-free apoproteins or amphipathic peptides stimulate the efflux of cell cholesterol and PL remains to be elucidated. The lack of specificity in the acceptor is demonstrated by the finding that peptides with no amino acid sequence homology to apolipoproteins stimulate efflux. Therefore, it seems unlikely that there is a specific interaction between the acceptor protein/peptide and receptors on the cell membrane. The most likely mechanism is one in which there is a reversible interaction between the lipid-free acceptor and the membrane, at which time the apoprotein or peptide acquires cell membrane lipid. This non-specific interaction could be a reflection of the lipid affinity of the protein/peptide, and the acceptor could interact more readily with specific lipid domains in the membrane (Rothblat *et al.*, 1992). The availability of such lipid domains may differ among cell types and may be increased upon the enrichment of the membrane with cholesterol. Based on the kinetics of release of PL and cholesterol, it is probable that there is a sequential release of lipids, with the acceptor protein first acquiring membrane PL and subsequently stimulating the efflux of membrane cholesterol. The latter event could occur either while the apoprotein is associated with the membrane, or the cholesterol could be incorporated into the nascent apoprotein/PL complexes by an aqueous diffusion mechanism.

The extent to which lipid-free or lipid-poor apolipoproteins participate in the efflux of peripheral cell cholesterol and reverse cholesterol transport remains to be established. Unquestionably, such acceptors can promote efflux of cholesterol from cells in culture but, since most studies have been conducted exposing cells to only the apoprotein, it has not been established what effect the presence of alternative acceptor particles, as are present in serum, would have on efflux. Studies using serum from transgenic rodents overexpressing human apo A-I, or experiments in which apo A-I was added to serum, indicate that the presence of high concentrations of apo A-I without associated PL does not stimulate efflux (see section 2.1 and Atger *et al.*, 1996). The failure of lipid-free apo A-I to stimulate efflux when added to serum could have several explanations and more work is required to resolve the issue.

3. TRANSPORT OF DIFFERENT SUBCELLULAR POOLS OF CHOLESTEROL

3.1. Lysosomal Cholesterol

Lysosomal cholesterol is derived from a wide range of sources (Figure 2), including the receptor-mediated endocytosis of lipoproteins (Brown and Goldstein, 1986; Yla-Herttuala *et al.,* 1989; Runquist and Havel, 1991), the phagocytosis of extracellular debris (Sato *et al.,* 1995; Tangirala *et al.,* 1994), and the autophagic turnover of membranes and organelles (Holtzman, 1989; Koval and Pagano, 1990). The relative importance of these sources is not clearly established, but under "normal" healthy circumstances, the predominant source for most cells is likely to be endocytosed lipoproteins. In atherosclerosis the unregulated endocytosis of oxidized lipoproteins and the phagocytosis of cholesterol-rich debris from necrotic cells may become of major importance in cells of the arterial wall (Tangirala *et al.,* 1994). Lysosomal cholesterol is used ubiquitously as a structural component of the plasma membrane (Bretscher and Munro, 1993) and as a substrate for ACAT-mediated esterification in the rough endoplasmic reticulum (RER) (Suckling and Stange, 1985). In specialized tissues, it is also utilized for steroid hormone synthesis (Freeman, 1987) and bile acid synthesis (Esnault-Dupuy *et al.,* 1987). These metabolic processes are initiated in mitochondria (Lin *et al.,* 1995) and the smooth ER (Chiang *et al.,* 1990), respectively. Several recent reports describe the synthesis and subsequent efflux of 27-hydroxycholesterol in a variety of extrahepatic cells (Javitt, 1994; Bjorkhem *et al.,* 1994). The substrate for this synthesis is at least partially lysosomal in origin, whereas the sterol 27-hydroxylase is mitochondrial (Axelson and Larsson, 1995). These patterns of utilization suggest a complex system for the trafficking of lysosomal cholesterol, involving eventual delivery to the plasma membrane, mitochondria, and subcompartments of the endoplasmic reticulum.

The transport of cholesterol from lysosomes to the plasma membrane appears to be a major route of sterol trafficking in mammalian cells. This is evidenced most clearly by the rapid availability of LDL-derived cholesterol for efflux to extracellular acceptors that function by promoting sterol desorption from the cell surface (Johnson *et al.,* 1990). This pathway has been examined in a variety of cell types in culture, including macrophages, fibroblasts, hepatoma cells, and Chinese hamster ovary (CHO) cells (Brasaemle and Attie, 1990; Johnson *et al.,* 1990, 1991; Mahlberg *et al.,* 1991; Sato *et al.,* 1995). The usual methods for studying this process involve the preloading of lysosomes with labeled cholesterol by incubation with labeled lipoproteins or negatively charged liposomes, followed by assessment of the delivery of labeled sterol from lysosomes to the plasma membrane by such methods as measurement of efflux, treatment of cells with exogenous cholesterol oxidase, and subcellular fractionation. Experiments using these methods largely agree that this transport proceeds constitutively, and is accomplished quite rapidly, with a half-time of 40 min or less. Other important features include the ability to disrupt the transport by genetic mutation and with a variety of drugs. These findings and their implications regarding the transport mechanism are considered in more detail below.

Several pharmacological agents have been reported to inhibit the transport of lysosomal cholesterol to the plasma membrane (see Chapter 14 by Pentchev *et al.*). The most prominent of these agents are progesterone (Butler *et al.*, 1992), the Upjohn androstene derivative U18666A (Liscum and Faust, 1989) and other amphipathic amines (Rodriguez-Lafrasse *et al.*, 1990; Lange and Steck, 1994), as well as bafilomycin A_1, a specific inhibitor of the ATP-dependent proton pump (Furuchi *et al.*, 1993). The effects of progesterone and U18666A occur without inhibition of lysosomal hydrolytic function and thus appear to be distinct from the effects of bafilomycin. It has been noted that progesterone and the cationic inhibitors share the property of being able to interact with the P-glycoprotein drug-efflux pumps that are responsible for some forms of drug-resistant cancer (Lange and Steck, 1994). Although this relationship is intriguing, there is no direct evidence that lysosomal sterol trafficking depends on this type of transport protein. Liscum and Collins (1991) reported the isolation of U18666A-resistant CHO cells. The ability to isolate such mutant cells suggests that the effect of this compound may be due to its interaction with a specific transport protein. Whether these cells are resistant to progesterone or other amphipathic amines was not described. Agents that do not affect lysosomal sterol trafficking include energy-metabolism poisons, monensin (a cation-selective ionophore with a variety of effects on mammalian cells, but often viewed as an inhibitor of protein secretion at the level of the trans-Golgi apparatus), and brefeldin A (which causes resorption of the Golgi apparatus into the ER preventing vesicle-mediated protein secretion) (Liscum and Underwood, 1995). Other types of perturbations that have been reported not to affect this transport are lack of sterol carrier protein-2 (SCP_2, also known as nonspecific lipid transfer protein), the stimulation of ACAT-mediated cholesterol esterification in the RER, and variation in the type of extracellular sterol acceptor provided in the incubation medium (Johnson *et al.*, 1990, 1991; Johnson and Reinhart, 1994).

Whether cholesterol moves directly from lysosomes to other intracellular sites of utilization is not clear. Regarding delivery to the RER, Xu and Tabas (1991) suggested that lysosomal cholesterol may first transit through the plasma membrane, based on experiments with macrophages showing that prior to esterification, lipoprotein-derived cholesterol mixes with the pool of cellular cholesterol available for efflux. Similarly, in hepatoma cells, lipoprotein-derived cholesterol appears to be accessible to exogenous cholesterol oxidase either before or concomitantly with its availability for esterification (Lange *et al.*, 1993). In addition, several reports have documented that plasma membrane cholesterol can be transferred readily to the ACAT substrate pool (Lange *et al.*, 1993; Slotte and Bierman, 1988). Thus, the plasma membrane or an organelle in rapid communication with the plasma membrane (such as the endosome-trans-Golgi membrane system) may be an obligatory intermediate site for lysosomal cholesterol that is destined for the RER. A more complex scheme may operate for the cholesterol that is destined for steroidogenesis. For instance, in MA-10 Leydig tumor cells, LDL cholesterol is delivered to the plasma membrane and from there can contribute to the steroid substrate pool; however, during stimulation of the cells with cAMP, it also appears that LDL cholesterol can be diverted directly to mitochondria for steroid production without transit through the plasma membrane (Freeman, 1987). Thus, the plasma membrane appears to serve as a reservoir for lysosomal cholesterol that ultimately will be used in steroidogenesis, but transit of cholesterol through this reservoir is not essential. Details regarding the delivery of cholesterol to mitochondria in steroidogenic cells are reviewed in Liscum and Dahl (1992). Recent reports suggest an important rate-limiting role for steroidogenic acute regulatory protein (StAR) in the transport of cholesterol between the outer and inner mitochondrial membranes in steroidogenic cells (Sugawara *et al.*, 1995; see Chapter 8 by Martínez and Strauss).

The transport of lysosomal cholesterol to the RER is usually assessed by availability of the sterol for ACAT-mediated esterification, with appropriate controls to show that agents intended to perturb sterol transport do not have a direct effect on the function of ACAT. The results of various studies that have used this approach indicate that the transport of lysosomal sterol to the RER is sensitive to many of the same perturbations that reduce sterol delivery to the plasma membrane. For instance, esterification is reduced by the same mutations that reduce efflux of lysosomal cholesterol (described below). Esterification is also reduced by progesterone, bafilomycin, compound U18666A, and other amphipathic amines (Liscum and Underwood, 1995). In addition, delivery from the plasma membrane to ACAT is reduced in mutant cells lacking cytoplasmic vimentin filaments, suggesting a role for this component of the cytoskeleton (Sarria *et al.*, 1992). The effects of progesterone and U18666A may be twofold, since they reduce both efflux of lysosomal cholesterol (suggesting slower transport to the cell surface) and the esterification of cholesterol that originates in the plasma membrane (suggesting an effect on the movement of sterol from the plasma membrane to the RER) (Lange, 1994; Harmala *et al.*, 1994). Using macrophages that had phagocytosed negatively charged cholesterol–phospholipid liposomes, the steroid structural requirements necessary for the inhibition of cholesterol delivery to the RER were examined by Aikawa *et al.* (1994). These studies demonstrated the primary importance of an oxo functional group at carbon 17 or 20 of the steroid. In addition, during incubation with these steroids, exogenous cholesterol accumulated in phagolysosomes, suggesting direct inhibition of cholesterol egress from lysosomes. In other studies with this system, Sato *et al.* (1995) showed that lysosomal phytosterols, such as campesterol and sitosterol, were esterified less efficiently than lysosomal cholesterol, and again showed that the sterols accumulated in phagolysosomes, suggesting that the major factor slowing esterification was reduced transport, rather than reduced affinity of the phytosterols for ACAT.

The disruption of lysosomal cholesterol transport by genetic mutation has attracted considerable attention in recent years. Disruption of this transport appears to be the main defect in the Niemann–Pick type C (NPC) lipid storage disease (reviewed in Pentchev *et al.,* 1995, and Chapter 14 by Pentchev *et al.*). Similar spontaneously arising metabolic diseases have been described in mice, cats, and dogs (Pentchev *et al.,* 1995). In addition, defective transport has been produced by mutagenesis of CHO cells (Cadigan *et al.,* 1990; Dahl *et al.,* 1992). In each case, the defect results in the accumulation of exogenous cholesterol in morphologically identifiable lysosomes and in reduced delivery to the plasma membrane and the RER. In human NPC fibroblasts, cholesterol also appears to accumulate in the trans cisternae of the Golgi apparatus, suggesting this location as a potential distribution center for lysosomal cholesterol, and that the NPC defect may involve egress from this organelle (Coxey *et al.,* 1993). These results have raised the question of whether the primary defect in NPC cells is a transport blockage at the Golgi apparatus with secondary "back-up" of cholesterol into lysosomes, or reduced cholesterol egress from both sites due to dependence on the same gene product. This issue has not been resolved. The transport defect in NPC cells can be corrected by incubation with 2% dimethyl sulfoxide (Blanchette-Mackie *et al.,* 1989). However, this agent did not produce consistent therapeutic benefit in human patients (Pentchev *et al.,* 1995). In humans the NPC mutation is inherited in an autosomal recessive fashion and has been narrowed to a small region of chromosome 18 (Pentchev *et al.,* 1995). As yet there are no reports of the cloning of the gene or its cDNA. The mutations in CHO cells fall into two complementation groups, suggesting the existence of at least two gene products involved in lysososomal cholesterol trafficking (Dahl *et al.,* 1994).

Important insights into the role of lysosomes in atherosclerosis have been provided by studies on the degradation of phagocytosed CE droplets by macrophages and other cells (Tangirala *et al.,* 1994; Chapter 12 by Kruth). Such extracellular droplets are thought to be generated by cell death and lipoprotein aggregation in atherosclerotic lesions. In studies with macrophages in culture, it has been found that these droplets can be phagocytosed and delivered to lysosomes efficiently, leading to the production of a foam-cell phenotype after only a few hours of incubation (compare section 3.3). The lysosomal hydrolysis of the esters causes the rapid production of large amounts of unesterified cholesterol. This rapid production of cholesterol in the limited space of the lysosome apparently overwhelms the machinery for cholesterol egress, leading to the formation of lysosomal cholesterol crystals (Figure 8). Similar processes occurring in the atherosclerotic lesion would contribute to cholesterol retention and the progression of this disease. An explanation for the formation of cholesterol crystals in foam cells is important since cholesterol in this physical state is particularly

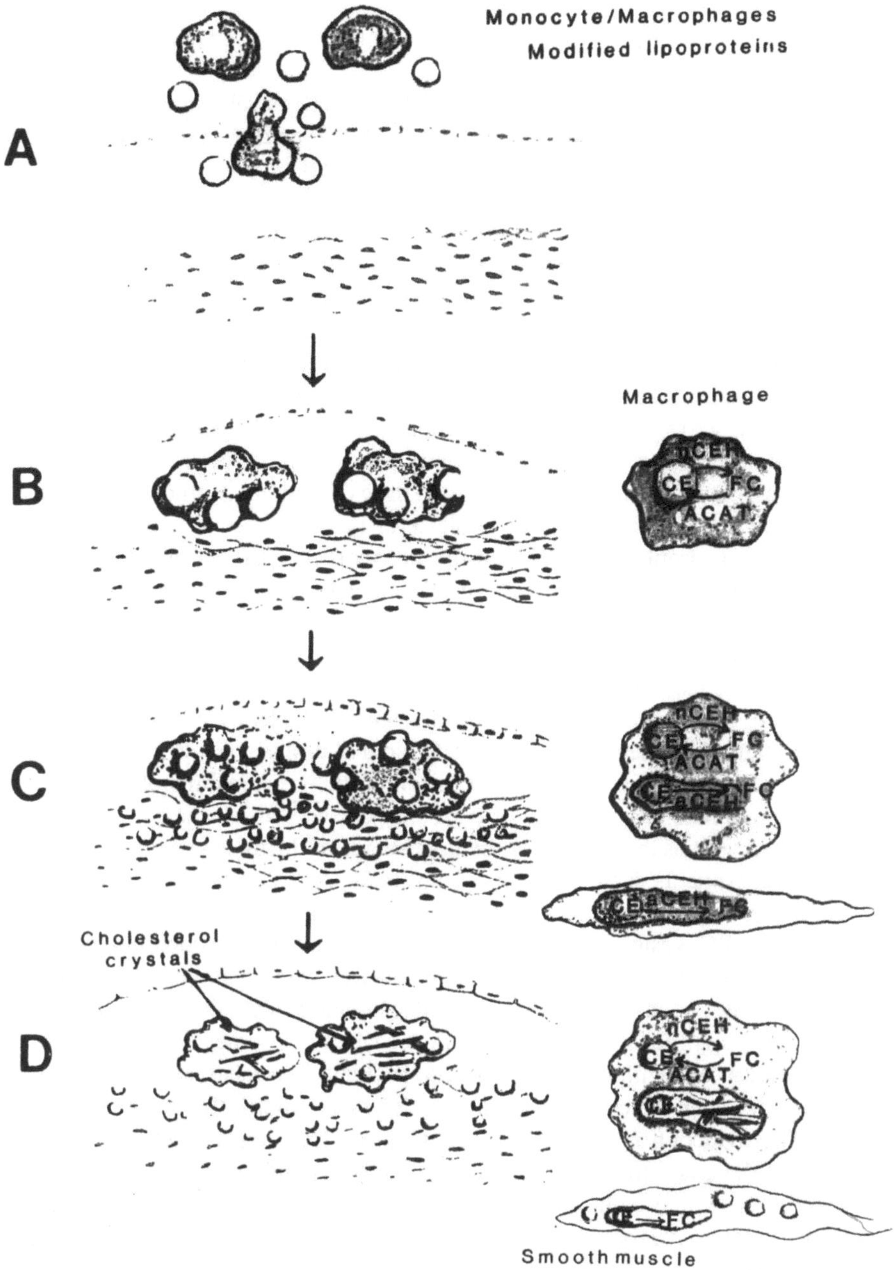

FIGURE 8. A model for the recycling of CE droplets and cholesterol accumulation in foam cells during progression of atherosclerosis. A) Infiltration of circulating monocytes into the arterial wall followed by the incorporation of excess cholesterol from modified lipoproteins by the monocyte/macrophages could lead to the formation of earliest macrophage-derived foam cells. B) In the early lesions, the foam cells are primarily macrophage-derived foam cells with the cholesterol deposited primarily as cytoplasmic inclusions. The phagocytic incorporation of the liberated inclusions by macrophages leads to lysosomal CE hydrolysis by the action of acid cholesteryl ester hydrolase (aCEH) with the generation of FC. Rapid CE hydrolysis and FC accumulation in the macrophage-foam cells results in FC deposition in the lipid-filled lysosomes. D) The accumulation of FC in the lipid-filled lysosomes of macrophage foam cells leads to the formation of cholesterol monohydrate crystals. The formation of FC crystals could eventually result in the lysis and death of the cells. (From Tangirala *et al.*, 1994.)

difficult to solubilize and remove from the lesion. The development of methods to prevent and reverse the formation of these crystals is an important long-term goal, and can be addressed with these cell-culture systems.

As indicated above, considerable information has been generated on the pathways, kinetics, and potential perturbants of lysosomal sterol trafficking. However, the mechanism of the transport remains poorly defined. Hypothetically, three distinct but nonexclusive mechanisms may be involved: unmediated diffusion of cholesterol in the aqueous cytosol, protein-mediated diffusion, and intermembrane vesicle trafficking. The best studied pathway, and that which seems most likely to involve egress directly from lysosomes, is delivery to the plasma membrane. Present evidence regarding the rate-limiting step in this transport favors some form of mediated diffusion. Arguing for specific mediation are the abilities to mutate the transport and inhibit it selectively with drugs. In addition, studies showing accumulation and crystallization of cholesterol in lysosomes of cells loaded with exogenous CE droplets suggest that the transport mechanism may be saturable. Several observations argue against vesicle-mediated transport. These include the lysosomal accumulation of phagocytosed phytosterols (which ought to partition as efficiently as cholesterol into transport vesicles derived from the lysosomal membrane) and the lack of effect of energy poisons and drugs, such as monensin and brefeldin A. In addition, it can be inferred from the existence of lysosomal storage diseases caused by the absence of specific acid hydrolases (Holtzman, 1989) that in most cells there is very little direct fusion of the lysosomal membrane with the plasma membrane or trafficking of lysosome-derived vesicles to the cell surface. These observations thus argue for the importance of diffusional mechanisms.

In experiments designed to characterize the contribution of unmediated diffusion, the transfer of cholesterol from CHO cell-derived lysosomes to small unilamellar PC vesicles was examined during cell-free incubation (Johnson, 1996). The results demonstrated efficient cell-free degradation of LDL within intact lysosomes, a maximal rate of cholesterol release of about 3% per hour ($t_{1/2} \sim$ 23 h), and a significant reduction in transfer when lysosomes were isolated from mutant CHO cells with defective sterol trafficking. The relatively slow release of cholesterol under these conditions argues for mediation of the transport in intact cells. The reduction in transfer using lysosomes from mutant cells supports the view that there is a transport blockage at the lysosomal membrane, and that the transport in normal cells may involve diffusional release of sterol, rather than vesicle budding. A prominent candidate for a protein mediator of sterol transport is SCP_2. However, as noted above, the trafficking of lysosomal cholesterol in cells does not appear to be dependent on this protein (van Heusden *et al.*, 1992; Johnson and Reinhart, 1994). Thus, present understanding suggests that lysosomal sterol transport in cells is likely to depend on protein-mediated transport between membranes that maintain their individual identities. However, there are no clear candidates for this mediator. Continued progress in this area probably will depend on further application of cell-free incubation methods and on the isolation and characterization of the gene products defective in NPC disease and related animal and cell models.

3.2. Biosynthetic Cholesterol

Sterol synthesis is initiated in the cytosol and then continued in the membranes of the smooth ER (Faust *et al.,* 1988; Chesterton, 1968). In liver it may be completed within the ER (Chesterton, 1968). In extrahepatic cells, sterol intermediates (e.g., lanosterol, zymosterol, and desmosterol) leave the ER before final conversion to cholesterol, and accumulate in the plasma membrane, as demonstrated by a variety of measures (e.g., subcellular fractionation, cholesterol oxidase, efflux) (Echevarria *et al.,* 1990; Johnson *et al.,* 1995; Tabacik *et al.,* 1991). The completion of cholesterol synthesis in extrahepatic cells appears to occur in non-ER membranes, but the precise location of this metabolism is not clear (Lange and Muraski, 1988). Return flux of at least one intermediate, zymosterol, from the plasma membrane to internal membranes and conversion to cholesterol has been demonstrated (Lange *et al.,* 1991). Cholesterol also is synthesized in peroxisomes (Thompson *et al.,* 1987); the eventual fate and specific functions of this biosynthetic sterol pool are unknown and are intriguing questions. In addition to deposition in the plasma membrane, synthesized cholesterol ultimately reaches the mitochondria (Lange *et al.,* 1988) and the rough ER (Tabas *et al.,* 1988), where it is metabolized to steroid hormones and CE, respectively. These observations suggest a complex system for the trafficking of biosynthetic sterols that involves the movement of both cholesterol and sterol intermediates among several subcellular compartments.

The movement of biosynthetic sterol from the ER to the plasma membrane has been studied directly by the groups of Lange (Lange and Mattheis, 1984) and Simoni (Kaplan and Simoni, 1985). Typical methods for these studies have involved pulsing mammalian cells in culture with water-soluble precursor (e.g., [^{3}H]acetate) at 15°C, followed by chase incubations at 37°C, during which delivery to the plasma membrane was monitored by subcellular fractionation or access to exogenous cholesterol oxidase. The major findings have been that delivery to the plasma membrane is prevented by incubation at 15°C, but at 37°C this transport proceeds rapidly ($t_{½} < 60$ min), is energy dependent, and appears to be mediated by a class of sterol-rich vesicles with a density similar to, or somewhat less than, that of the plasma membrane. These vesicles are distinguishable from the plasma membrane, Golgi apparatus, and ER by appropriate combinations of enzyme and density-shift treatments, suggesting that they may constitute a unique system for cellular sterol transport (Lange and Steck, 1985). The transport of synthesized sterol to the plasma membrane is not prevented by brefeldin A, suggesting that it bypasses the Golgi apparatus and is distinct from the constitutive protein secretory pathway (Urbani and Simoni, 1990). It is normal in cells from NPC patients (Liscum *et al.,* 1989), indicating that the transport of lysosomal and biosynthetic sterol occur by different processes. It is impaired partially in fibroblasts from

Zellweger syndrome patients (who lack the mature form of SCP_2) or in normal fibroblasts treated with SCP_2 antisense oligonucleotide, suggesting some dependence on this intracellular transport protein (Puglielli *et al.*, 1995).

The transport of biosynthetic sterol to the plasma membrane was studied under cell-free conditions by Waits *et al.* (1990), using subcellular fractions derived from guinea pig testis. The plasma membrane acceptor in this system was pre-immobilized on nitrocellulose strips, providing for convenient separation of acceptor from donor when incubations were ended. Transport in this system was reduced at low temperature and in the absence of ATP, but was not dependent on cytosol. In studies using the same methodology with fractions derived from rat liver, this group has also examined lipid transport from the ER to the Golgi apparatus and from the Golgi apparatus to the plasma membrane. The ER-to-Golgi transport was found to be stimulated by ATP and cytosol, and was inhibitable with *N*-ethylmaleimide and GTP-gamma-*S*, consistent with a vesicle-mediated mechanism (Moreau *et al.*, 1991). When plasma membrane was substituted for Golgi-derived membranes as the acceptor, no ATP-dependent transport was apparent (Moreau *et al.*, 1991). Thus, the liver-derived system may not reproduce the Golgi-independent transport of sterol that has been reported by the groups of Lange and Simoni, as described above. In experiments addressing Golgi-to-plasma membrane lipid transport, ATP was not stimulatory. In contrast, transport was stimulated by a combination of NADH and cytosolic proteins, and was inhibited by *N*-ethylmaleimide or GTP-gamma-*S* (Rodriguez *et al.*, 1992). The NADH stimulation could be enhanced by adding ascorbate radical to the incubations. The stimulation by NADH and ascorbate was interpreted to indicate that vesicle budding from the Golgi apparatus may be energized by an electron transport system, which is thought to exist in the membranes of the *trans*-Golgi cisternae (Rodriguez *et al.*, 1992).

Oram and colleagues (Oram, 1990) have emphasized biosynthetic sterols as markers for the general intracellular pool of cholesterol that contributes to cytoplasmic CE deposition in foam cells (see section 3.3 for a description of CE deposition in foam cells). They have found that when cells are enriched with unesterified cholesterol, the small amount of sterol that is synthesized seems to be sequestered internally unless the medium contains specific acceptors, such as HDL, able to interact with the putative HDL receptor (compare section 2.2). When an appropriate acceptor is present in the medium, the biosynthetic sterols move readily to the plasma membrane and become available for efflux or oxidation by exogenous cholesterol oxidase (Slotte *et al.*, 1987). The intracellular signal for the regulation of this transport appears to be diacylglycerol, suggesting a role for protein kinase C (Mendez *et al.*, 1991). Chromatographic analysis indicated that both cholesterol and desmosterol-like sterols participated in this process (Aviram *et al.*, 1989; Hokland *et al.*, 1993). It was unclear whether there was any discrimination among sterols. The regulated translocation to the plasma membrane was reported to be sensitive to brefeldin A and monensin, suggesting a route that involves transit through the Golgi apparatus (Mendez, 1994). Thus, it may differ from the pathway followed by biosynthetic sterols in unenriched cells, as described above. In studies that relied on reverse phase HPLC to resolve cholesterol from biosynthetic sterol intermediates, a major portion of the sterols that were released from extrahepatic cells consisted of desmosterol and sterols of similar polarity (Johnson *et al.*, 1995) (Figure 9). This was true for both HDL and non-HDL acceptors, although it was also found that HDL selectively enhanced the efflux of biosynthetic cholesterol in comparison to the polar sterol precursors and to the lysosomal and plasma membrane pools of cholesterol (Johnson *et al.*, 1995). The mechanism of this differentiation of sterols has not been established.

The nature of the HDL receptor and its role in sterol trafficking remain uncertain (see section 2.2). Receptors for this lipoprotein and an apparent role in cellular sterol efflux have been reported in a variety of cell types (Oram, 1990; Barbaras *et al.*, 1987; Tozuka and Fidge, 1989). A cDNA for a human HDL receptor has been cloned and sequenced (McKnight *et al.*, 1992). The cDNA sequence indicated a protein gene product of 1268 amino acids, somewhat larger than the 110-and 130-kDa products observed in HDL ligand-blotting studies. In transfection experiments, the primary translation product was processed to lower molecular weight proteins, providing an explanation for this discrepancy. Secondary structural analysis did not indicate the presence of transmembrane domains, but rather several amphipathic helical segments that would be expected to embed only partially in a bilayer membrane (McKnight *et al.*, 1992). Thus, the structure of this protein differs from that of other receptors involved in transmembrane signalling. Whether overexpression of this gene provides for increased sterol delivery to the plasma membrane has not been reported.

3.3. Cytoplasmic Cholesteryl Esters

Under conditions in which free cholesterol accumulates in cells, the excess cholesterol is converted to CE by ACAT and the CE is stored in lipid inclusions (Figure 2). The presence of such CE inclusions in macrophages in the vessel wall gives these cells the appearance of "foam cells," which is a hallmark of the atherosclerotic plaque. The esterified cholesterol in inclusions is in a dynamic state and undergoes hydrolysis by a cytoplasmic neutral CE hydrolase (nCEH) (Brown and Goldstein, 1983; Brown *et al.*, 1980; Glick *et al.*, 1987), the reaction of which results in the production of free cholesterol. Thus, a CE cycle is established resulting in the movement of cholesterol between free and esterified pools. This cycle can be interrupted by the diversion of free cholesterol out of the cells, which removes the free cholesterol substrate of ACAT. Even though ACAT is inhibited, the nCEH continues to function, resulting in the clearance of stored esterified cholesterol. Since CE cannot directly undergo efflux and must first be hydrolyzed to free cholesterol, the presence of efficient extracellular cholesterol acceptors results in the reduction in the synthesis of CE, and promotes the net hydrolysis of CE.

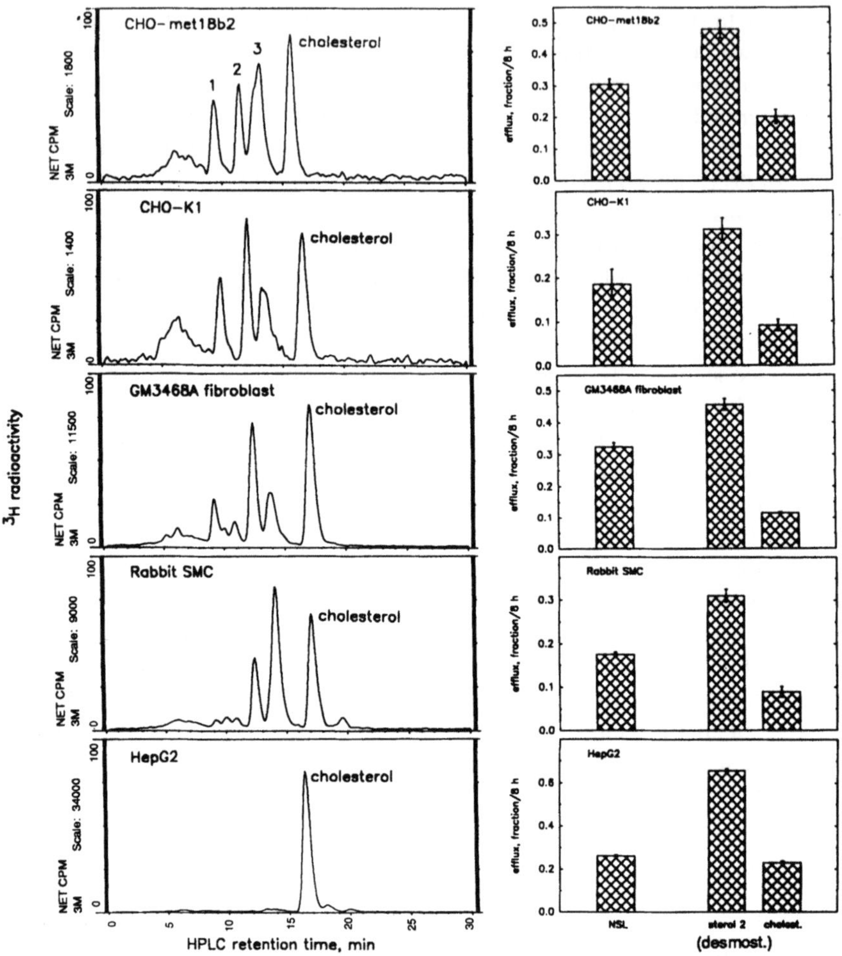

FIGURE 9. Production and efflux of newly synthesized sterols in a variety of cell types. The five indicated cell types were compared in a single experiment. The base medium for CHO cells was Ham's F-12, and for the other cells minimum essential medium. Otherwise, the conditions for all cells were identical. Cells were incubated for 8 h in media containing [^{3}H]acetate (50 μCi/ml) and either BSA (0.2%) or BSA plus HDL_3 (1 mg of protein/ml), and then non-saponifiable lipid (NSL) were prepared from cells and media and analyzed by HPLC. The panels on the left show HPLC profiles of NSL from cells incubated with BSA in the absence of HDL_3. The panels on the right show fractional efflux of [^{3}H]-labeled NSL, cholesterol and sterol 2 to media containing HDL_3. In CHO cells, sterol 2 was identified as desmosterol. In addition, components 1 and 3 were partially precipitable with digitonin, indicating that they contain some 3-β-hydroxy sterols. (From Johnson *et al.*, 1995.)

The ability to obtain net removal of stored esterified cholesterol in tissue culture models of foam cells depends on both the cell type and the acceptor particle. Since the final step in the clearance of esterified cholesterol from cells is the movement of free cholesterol from the plasma membrane to the extracellular acceptor, the clearance of esterified cholesterol is modulated, in large part, by the same factors that influence the efflux of free cholesterol. Thus, small HDL-like particles, with a high PL/cholesterol ratio, will be the most efficient in reducing intracellular cholesterol pools (compare section 2.2). Since the net removal of CE is dependent on the reduction of free cholesterol, which serves as a substrate for ACAT, an acceptor that contains free cholesterol and promotes bidirectional flux of cholesterol may not efficiently reduce esterified cholesterol mass, even though it may promote the reduction of intracellular radiolabeled CE.

In addition to factors related to acceptor composition, cells exhibit marked differences in their ability to reduce CE stores. For example, it was clearly demonstrated by the early studies of Brown and Goldstein (1983) that exposure of mouse peritoneal cells to HDL produced a reduction in esterified cholesterol. However, although all cells will clear esters if a sufficiently large cell to acceptor cholesterol gradient is provided, the impressive esterified cholesterol clearance observed with mouse macrophages exposed to HDL was not as readily apparent with other cell types. For example, in Fu5AH rat hepatoma cells, which accumulate large pools of CE that can be depleted by growth in apoprotein/PL complexes, addition of native HDL to the culture medium results in an additional accumulation of esterified cholesterol, rather than clearance (Yau-Young *et al.*, 1982). In comparative studies using mouse peritoneal macrophages and the J774 cell line, Bernard and colleagues (1990) demonstrated that the same preparations of HDL that stimulated clearance from the peritoneal cells were unable to produce significant reductions in the CE content of the J774 cells.

The reason for differences between cell types in their ability to clear CE to acceptors such as HDL probably lies in the extent to which a sufficient cholesterol gradient can be generated to favor net movement of cholesterol from the plasma membrane to the extracellular acceptor. Three factors that are important in determining this gradient are: i) the content and distribution of cholesterol in the plasma membrane, ii) the rate of turnover of the CE cycle, and iii) the transport of the free cholesterol from the site of CE hydrolysis to the plasma membrane. Although there have been numerous studies on the distribution and physical state of cholesterol in artificial and biological membranes, little data are available on the content and organization of cholesterol in the plasma membrane of macrophages, particularly the cholesterol-enriched macrophage foam cell. Considerably more information is available on the relationship between the rate of turnover of the CE cycle and the clearance of esterified cholesterol. For example, a recent study comparing the rate of turnover of CE in peritoneal macrophages from rats, mice, and rabbits (Hakamata *et al.*, 1994) demonstrated differences in the turnover rate of esterified cholesterol among the cell types in the order mouse > rat > rabbit (Table I). This

Table I
Estimated Rates of CE Hydrolysis in Various Types of Macrophages[a]

Macrophage	Half-time (h)	Reference
J774	38	Bernard *et al.*, 1990
Mouse peritoneal	9	Bernard *et al.*, 1990; Hakamata *et al.*, 1994
Rat peritoneal	19	Hakamata *et al.*, 1994
Rabbit peritoneal	>200	Hakamata *et al.*, 1994
White Carneau peritoneal	89	Yancey and St. Clair, 1994
Show Racer peritoneal	37	Yancey and St. Clair, 1994

[a]Half-times are estimated from data on CE clearance obtained either in the presence of an ACAT inhibitor or high concentrations of apoprotein/phospholipid complexes. The hydrolysis of CE is assumed to follow first-order kinetics.

same relationship was maintained when the extent of HDL-induced clearance of CE was determined (Hakamata *et al.*, 1994). It has also been demonstrated that the ability to clear CE from cultured peritoneal macrophages derived from White Carneau pigeons is slower than that from macrophages from Show Racer pigeons (Yancey and St. Clair, 1992), and a comparative study of the activity of ACAT and nCEH in macrophages from these pigeons demonstrated that the defect in clearance from the White Carneau is associated with a reduced rate of CE hydrolysis (Yancey and St. Clair, 1994), relative to Show Racer macrophages (Table I). Since the rate of clearance of intracellular CE cannot be any faster than its rate of hydrolysis, this will be the rate-limiting step under conditions that produce the maximum rate of efflux of unesterified cholesterol from the cell.

Even if there is efficient hydrolysis of intracellular esterified cholesterol and net accumulation of free cholesterol, the cholesterol still must be transported to the plasma membrane to undergo efflux to HDL. The mechanism of this intracellular cholesterol transport process, and the factors that regulate it, remain unresolved. In the presence of a cholesterol-free acceptor a sufficient gradient is achieved to promote net movement of cholesterol from all cell types. However, if the acceptor is HDL, which does not establish a steep cholesterol gradient, then a more active intracellular transport process is needed to promote net efflux. This transport appears to be linked to intracellular signaling, particularly through the action of cAMP. Thus, J774 macrophages, which normally are resistant to HDL-induced CE clearance, will reduce their CE stores if treated with cAMP (Bernard *et al.*, 1991). cAMP has been shown to increase nCEH activity (for a review of CE hydrolase and its role in cells, see Glick, 1990), and there is evidence that a secondary effect of cAMP is at the level of intracellular transport of free cholesterol (Picardo *et al.*, 1986; Hokland *et al.*, 1993). It has been suggested that the interaction of HDL with cell membrane binding sites triggers a signal transduction pathway that results in the mobilization of newly synthesized cholesterol (section 3.2 and Mendez *et al.*, 1991), and this model has been applied to the mobilization of free cholesterol generated within the CE cycle. However, this does not seem to be a highly specific effect because studies with synthetic amphipathic peptides have demonstrated that a net reduction of stored esterified cholesterol can be achieved with these lipid-free acceptors even though they have no amino acid homology with native apolipoproteins (Yancey *et al.,* 1995). The CE cycle and cholesterol efflux pathway are two homeostatic mechanisms that function together to maintain the necessary cholesterol concentrations compatible with proper membrane function. This is exemplified by studies demonstrating that, in CE-loaded macrophages, the disruption of the CE cycle by inhibiting ACAT results in the accumulation of free cholesterol and an associated destabilization of cell membranes; however, if HDL is present, the excess free cholesterol is effectively removed, thereby maintaining cell integrity (Warner *et al.,* 1995).

4. SUMMARY AND CONCLUSIONS

Cholesterol homeostasis in peripheral cells involves a balance between the influx and efflux processes. The acquisition of cholesterol by such cells is mediated by a variety of receptor and non-receptor processes involving both normal and modified lipoproteins. The offsetting efflux process is mediated by HDL and especially particles containing only apo A-I. An efficient reverse cholesterol transport by HDL of cholesterol from peripheral cells to the liver protects against the development of atherosclerosis. In cells that do not contain excess cholesterol, the cholesterol is distributed as unesterified cholesterol molecules between the plasma membrane and the membranes of the intracellular organelles. In cholesterol-loaded cells such as macrophage foam cells, the membranes became enriched in unesterified cholesterol and, in addition, cytoplasmic CE droplets and lysosomal cholesterol crystals can form. The ways in which cholesterol molecules move between intracellular sites and the plasma membrane to become available for efflux to extracellular acceptor particles are becoming known. Cholesterol molecules in the plasma membrane can desorb and diffuse through the aqueous phase and be sequestered by HDL particles. The cell cholesterol available for efflux can exist in different kinetic pools, and these pools, such as those in various domains in the plasma membrane, require further definition. The cholesterol molecules present in intracellular pools also efflux with different kinetics and by different pathways. Thus, newly synthesized cholesterol is actively transported by a vesicle system from the ER to the plasma membrane, whereas lysosomal cholesterol seems to be transported to the plasma membrane by a protein-mediated, diffusional process. Clearance of cytoplasmic CE is dependent upon the rate of turnover of the CE cycle and the magnitude of the cholesterol gradient between the plasma membrane and the extracellular acceptor particle. It can be expected that the interdependence of the pathways and the molecular mechanisms underlying the intracellular trafficking of cholesterol will be elucidated in the near future.

ACKNOWLEDGMENTS. We are indebted to our colleagues who have contributed to various aspects of the work from this laboratory described here. This research was supported by NIH Program Project Grant HL22633 and Institutional Training Grant HL07443.

5. REFERENCES

Adelman, S. J., Glick, J. M., Phillips, M. C., and Rothblat, G. H. 1984, Lipid composition and physical state effects on cellular cholesteryl ester clearance, *J. Biol. Chem.* **259:**13844–13850.

Aikawa, K., Furuchi, T., Fujimoto, Y., Arai, H., and Inoue, K., 1994, Structure-specific inhibition of lysosomal cholesterol transport in macrophages by various steroids, *Biochim. Biophys. Acta* **1213:**127–134.

Akira, Q. L., Komaba, A., and Yokoyama, S., 1993, Cholesterol is pooly available for free apolipoprotein-mediated cellular lipid efflux from smooth muscle cells, *Biochemistry* **32:** 4597–4603.

Anantharamaiah, G. M., Jones, J. L., Brouillette, C. G., Schmidt, C. F., Chung, B. H., Hughes, T. A., Bhown, A. S., and Segrest, J. P., 1985, Studies of synthetic peptide analogs of the amphipathic helix, *J. Biol. Chem.* **260:**10248–10255.

Atger, V., de la Llera Moya, M., Bamberger, M., Francone, O., Casgrove, P., Tall, A., Walsh, A., Moatti, N., and Rothblat, G., 1995, Cholesterol efflux potential of sera from mice expressing human CETP and/or human apolipoprotein AI, *J. Clin. Invest.* **96:**2613–2622.

Aviram, M., Bierman, E. L., and Oram, J. F., 1989, High density lipoprotein stimulates sterol translocation between intracellular and plasma membrane pools in human monocyte-derived macrophages, *J. Lipid Res.* **30:**65–76.

Axelson, M., and Larsson, O., 1995, Low density lipoprotein (LDL) cholesterol is converted to 27-hydroxycholesterol in human fibroblasts, *J. Biol. Chem.* **270:**15102–15110.

Banka, C. L., Black, A. S., and Curtiss, L. K., 1994, Localization of an apolipoprotein A-I epitope critical for lipoprotein-mediated cholesterol efflux from monocytic cells, *J. Biol. Chem.* **269:**10288–10297.

Barbaras, R., Puchois, P., Fruchart, J.-C., and Ailhaud, G., 1987, Cholesterol efflux from cultured adipose cells is mediated by LpAI particles but not by LpAI:AII particles, *Biochem. Biophys. Res. Commum.* **142:**63–69.

Barkia, A., Puchois, P., Ghalim, N., Torpier, G., Barbaras, R., Ailhaud, G., and Fruchart, J. C., 1991, Differential role of apolipoprotein AI-containing particles in cholesterol efflux from adipose cells, *Atherosclerosis* **87:**135–146.

Bellini, F., Phillips, M. C., Pickell, C., and Rothblat, G. H., 1984, Role of the plasma membrane in the mechanism of cholesterol efflux from cells, *Biochim. Biophys. Acta* **777:**209–215.

Bernard, D. W., Rodriguez, A., Rothblat G. H., and Glick, J. M., 1990, Influence of high density lipoprotein on esterified cholesterol stores in macrophages and hepatoma cells, *Arteriosclerosis* **10:**135–144.

Bernard, D. W., Rodriguez, A., Rothblat G. H., and Glick, J. M., 1991, cAMP stimulates cholesteryl ester clearance to high density lipoproteins in J774 macrophages, *J. Biol. Chem.* **266:**710–716.

Bielicki, J. K., Johnson, W. J., Weinberg, R. B., Glick J. M., and Rothblat, G. H., 1991, Efflux of phospholipids from fibroblasts with normal and elevated levels of cholesterol, *Biochim. Biophys. Acta* **1084:**7–14.

Bielicki, J. K., Johnson, W. J., Weinberg, R. B., Glick, J. M., and Rothblat, G. H., 1992, Efflux of lipid from fibroblasts to apolipoproteins: dependence on elevated levels of cellular unesterified cholesterol, *J. Lipid Res.* **33:**1699–1710.

Bittman, R., 1988, Sterol exchange between mycoplasma membranes and vesicles, *in*: *Biology of Cholesterol* (P. L. Yeagle, ed.), pp. 173–195, CRC Press, Boca Raton, FL.

Bittman, R., 1993, Mycoplasma membrane lipids: chemical composition and transbilayer distribution, *in*: *Subcellular Biochemistry, Vol. 20* (S. Rottem and I. Kahane, eds.), pp. 29–52, Plenum Press, New York.

Bjorkhem, I., Andersson, O., Diczfalusy, U., Sevastik, B., Xiu,R. J., Duan,C., and Lund, E., 1994, Atherosclerosis and sterol 27-hydroxylase: evidence for a role of this enzyme in elimination of cholesterol from human macrophages, *Proc. Natl. Acad. Sci. USA* **91**:8592–8596.

Blanchette-Mackie, E. J., Dwyer, N. K., Vanier, M. T., Sokol, J., Merrick, H. F., Comly, M. E., Argoff, C. E., and Pentchev, P. G., 1989, Type C Niemann–Pick disease: dimethyl sulfoxide moderates abnormal LDL-cholesterol processing in mutant fibroblasts, *Biochim. Biophys. Acta* **1006**:219–226.

Brasaemle, D. L., and Attie, A. D., 1990, Rapid intracellular transport of LDL-derived cholesterol to the plasma membrane in cultured fibroblasts, *J. Lipid Res.* **31**:103–111.

Breslow, J. L., 1995, Familial disorders of high-density lipoprotein metabolism, in: *The Metabolic and Molecular Basis of Inherited Disease,* (C. R. Scriver, A. L. Beaudet, W. S. Sly, and D. Valle, eds.), pp. 2031–2052, McGraw-Hill, New York.

Bretscher, M. S., and Munro, S., 1993, Cholesterol and the Golgi apparatus, *Science* **261**:1280–1281.

Brouillette, C., and Anantharamaiah, G. M., 1995, Structural models of human apolipoprotein A-I, *Biochim. Biophys. Acta* **1256**:103–129.

Brown, M. S., and Goldstein, J. L., 1983, Lipoprotein metabolism in the macrophage: implications for cholesterol deposition in atherosclerosis, *Annu. Rev. Biochem.* **52**:223–261.

Brown, M. S., and Goldstein, J. L., 1986, A receptor-mediated pathway for cholesterol homeostasis, *Science* **232**:34–47.

Brown, M. S., Ho, Y. K., and Goldstein, J. L., 1980, The cholesteryl ester cycle in macrophage foam cells. Continual hydrolysis and re-esterification of cytoplasmic cholesteryl esters, *J. Biol. Chem.* **255**:9344–9352.

Brown, R. E., 1992, Spontaneous lipid transfer between organized lipid assemblies, *Biochim. Biophys. Acta* **1113**:375–389.

Butler, J. D., Blanchette-Mackie, J., Goldin, E., O'Neill, R. R., Carstea, G., Roff, C. F., Patterson, M. C., Patel, S., Comly, M. E., Cooney, A., Vanier, M. T., Brady, R. O., and Pentchev, P. G., 1992, Progesterone blocks cholesterol translocation from lysosomes, *J. Biol. Chem.* **267**:23797–23805.

Cadigan, K. M., Spillane, D. M., and Chang, T.-Y., 1990, Isolation and characterization of Chinese hamster ovary cell mutants defective in intracellular low density lipoprotein-cholesterol trafficking, *J. Cell Biol.* **110**:295–308.

Castro, G. R., and Fielding, C. J., 1988, Early incorporation of cell-derived cholesterol into pre-β-migrating high-density lipoprotein, *Biochemistry* **27**:25–29.

Chesterton, C. J., 1968, Distribution of cholesterol precursors and other lipids among rat liver intracellular structures, *J. Biol. Chem.* **243**:1147–1151.

Cheung, M. C., and Albers, J. J., 1984, Characterization of lipoprotein particles isolated by immunoaffinity chromatography, *J. Biol. Chem.* **259**:12201–12209.

Chiang, J. Y. L., Miller, W. F., and Lin, G. M., 1990, Regulation of cholesterol 7 α-hydroxylase in the liver, *J. Biol. Chem.* **265**:3889–3897.

Cooper, R. A., Leslie, M. H., Fischkoff, S., Shinitzky, M., and Shattil, S. J., 1978, Factors influencing the lipid composition and fluidity of red cell membranes in vitro: production of red cells possessing more than two cholesterols per phospholipid, *Biochemistry* **17**:327–331.

Coxey, R. A., Pentchev, P. G., Campbell, G., and Blanchette-Mackie, E. J., 1993, Differential accumulation of cholesterol in Golgi compartments of normal and Niemann–Pick type C fibroblasts incubated with LDL: a cytochemical freeze-fracture study, *J. Lipid Res.* **34**:1165–1176.

Dahl, N. K., Reed, K. L., Daunais, M. A., Faust, J. R., and Liscum, L., 1992, Isolation and characterization of Chinese hamster ovary cells defective in the intracellular metabolism of low density lipoprotein-derived cholesterol, *J. Biol. Chem.* **267**:4889–4896.

Dahl, N. K., Daunais, M. A., and Liscum, L., 1994, A second complementation class of cholesterol transport mutants with a variant Niemann–Pick type C phenotype, *J. Lipid Res.* **35:**1839–1849.

Davidson, W. S., Lund-Katz, S., Johnson, W. J., Anantharamaiah, G. M., Palgunachari, N., Segrest, J. P., Rothblat, G. H., and Phillips, M. C., 1994, The influence of apolipoprotein structure on the efflux of cellular free cholesterol to high density lipoprotein, *J. Biol. Chem.* **269:** 22975–22982.

Davidson, W. S., Rodrigueza, W. V., Lund-Katz, S., Johnson, W. J., Rothblat, G. H., and Phillips, M. C., 1995a, Effects of acceptor particle size on the efflux of cellular free cholesterol, *J. Biol. Chem.* **270:**17106–17113.

Davidson, W. S., Gillotte, K. L., Lund-Katz, S., Johnson, W. J., Rothblat, G. H., and Phillips, M. C., 1995b, The effect of high density lipoprotein phospholipid acyl chain composition on the efflux of cellular free cholesterol, *J. Biol. Chem.* **270:**5882–5890.

de la Llera Moya, M., Atger, V., Paul, J. L., Fournier, N., Moatti, N., Giral, P., Friday, K. E., and Rothblat, G. H., 1994, A cell culture system for screening human serum for ability to promote cellular cholesterol efflux: relationships between serum components and efflux, esterification and transfer, *Arterioscler. Thromb.* **14:**1056–1065.

DeLamatre, J., Wolfbauer, G., Phillips, M. C., and Rothblat, G. H., 1986, Role of apolipoproteins in cellular cholesterol efflux, *Biochim. Biophys. Acta* **875:**419–428.

Echevarria, F., Norton, R. A., Nes, W. D., and Lange, Y., 1990, Zymosterol is located in the plasma membrane of cultured human fibroblasts, *J. Biol. Chem.* **265:**8484–8489.

Eisenberg, S., 1984, High density lipoprotein metabolism, *J. Lipid Res.* **25:**1017–1058.

Esnault-Dupuy, C., Chanussot, F., Lafont, H., Chabert, C., and Hauton, J., 1987, The relationship between HDL-, LDL-, liposome-free cholesterol, biliary cholesterol and bile salts in the rat, *Biochim. Biophys. Acta* **69:**45–52.

Faust, J. R., Trzaskos, J. M., and Gaylor, J. L., 1988, Cholesterol biosynthesis, *in*: *Biology of Cholesterol* (P. L. Yeagle, ed.), pp. 19–38, CRC Press, Boca Raton, FL.

Fielding, C. J., and Fielding, P. E., 1995, Molecular physiology of reverse cholesterol transport, *J. Lipid Res.* **36:**211–228.

Fielding, C. J., and Moser, K., 1982, Evidence for the separation of albumin-and apo A-I-dependent mechanisms of cholesterol efflux from cultured fibroblasts into human plasma, *J. Biol. Chem.* **257:**10955–10960.

Fielding, C. J., Reaven, G. M., and Fielding, P. E., 1982, Human noninsulin-dependent diabetes: Identification of a defect in plasma cholesterol transport normalized in vivo by insulin and in vitro by selective immunoadsorption of apolipoprotein E, *Proc. Natl. Acad. Sci. USA* **79:**6365–6369.

Fielding, P. E., Fielding, C. J., Havel, R. J., Kane, J. P., and Tun, P., 1983, Cholesterol net transport, esterification, and transfer in human hyperlipidemic plasma, *J. Clin. Invest.* **71:**449–460.

Forte, T. M., and McCall, M. R., 1994, The role of apolipoprotein A-I-containing lipoproteins in athero-
sclerosis, *Curr. Opin. Lipidol.* **5:**354–364.

Forte, T. M., Bielicki, J. K., Goth-Goldstein, R., Selmek, J., and McCall, M. R., 1995, Recruitment of cell phospholipids and cholesterol by apolipoproteins A-II and A-I: Formation of nascent apolipoprotein-specific HDL that differ in size, phospholipid composition, and reactivity with LCAT, *J. Lipid Res.* **36:**148–157.

Freeman, D. A., 1987, Cyclic AMP mediated modification of cholesterol traffic in Leydig tumor cells, *J. Biol. Chem.* **262:**13061–13068.

Fruchart, J. C., and Ailhaud, G., 1992, Apolipoprotein A-containing lipoprotein particles: physiological role, quantification, and clinical significance, *Clin. Chem.* **38:**793–797.

Furuchi, T., Aikawa, K., Arai, H., and Inoue, K., 1993, Bafilomycin A1, a specific inhibitor of vacuolar-type H+-ATPase, blocks lysosomal cholesterol trafficking in macrophages, *J. Biol. Chem.* **268:** 27345–27348.

Glick, J. M., 1990, Intracellular cholesteryl ester hydrolysis and clearance, *in*: *Advances in Cholesterol Research* (M. Esfahani, and J. B. Swaney, eds.), pp. 167–197, Telford Press, Caldwell, NJ.

Glick, J. M., Adelman, S. A., and Rothblat, G. H., 1987, Cholesteryl ester cycle in cultured hepatoma cells, *Atherosclerosis* **64:**223–230.

Gold, J. C., and Phillips, M. C., 1990, Effects of membrane lipid composition on the kinetics of cholesterol exchange between lipoproteins and different species of red blood cells, *Biochim. Biophys. Acta* **1027:**85–92.

Gold, J. C., and Phillips, M. C., 1992, Effects of membrane lipids and-proteins and cytoskeletal proteins on the kinetics of cholesterol exchange between high density lipoprotein and human red blood cells, ghosts and microvesicles, *Biochim. Biophys. Acta* **1111:**103–110.

Goldstein, J. L., and Brown, M. S., 1977, The low density lipoprotein pathway and its relation to atherosclerosis, *Annu. Rev. Biochem.* **46:**897–930.

Hakamata, H., Miyazaki, A., Sakai, M., Suginohara, Y., Sakamoto, Y.-I., and Horiuchi, S., 1994, Species difference in cholesteryl ester cycle and HDL-induced cholesterol efflux from macrophage foam cells, *Arterioscler. Thromb.* **14:**1860–1865.

Hara, H., and Yokoyama, S., 1991, Interaction of free apolipoproteins with macrophages, *J. Biol. Chem.* **266:**3080–3086.

Harmala, A., Porn, M. I., Mattjus, P., and Slotte, J. P., 1994, Cholesterol transport from plasma membranes to intracellular membranes is inhibited by 3β-[2-(diethylamino)ethoxy]androst-5-en-17-one, *Biochim. Biophys. Acta* **1211:**317–325.

Havel, R. J., and Kane, J. P., 1995, Introduction: Structure and metabolism of plasma lipoproteins, *in*: *The Metabolic and Molecular Basis of Inherited Disease* (C. R. Scriver, A. L. Beaudet, W. S. Sly, and D. Valle, eds.), pp. 1841–1851, McGraw-Hill, New York.

Hayek, T., Chajek-Shaul, T., Walsh, A., Agellon, L. B., Moulin, P., Tall, A. R., and Breslow, J. L., 1992, An interaction between the human cholesteryl ester transfer protein (CETP) and apolipoprotein A-I genes in transgenic mice results in a profound CETP-mediated depression of high density lipoprotein cholesterol levels, *J. Clin. Invest.* **90:**505–510.

Hodenberg, E., Heinen, S., Howell, K. E., Luley, C., Kubler, W., and Bond, H. M., 1991, Cholesterol efflux from macrophages mediated by high-density lipoprotein subfractions,which differ principally in apolipoprotein A-I and apolipoprotein A-II ratios, *Biochem. Biophys. Res. Commun.* **1086:**173–184.

Hokland, B. M., Slotte, J. P., Bierman, E. L., and Oram, J. F., 1993, Cyclic AMP stimulates efflux of intracellular sterol from cholesterol-loaded cells, *J. Biol. Chem.* **268:**25343–25349.

Holtzman, E., 1989, *Lysosomes,* Plenum Press, New York.

Huang, Y., von Eckardstein, A., Wu, S., Maeda, N., and Assmann, G., 1994, A plasma lipoprotein containing only apolipoprotein E and with gamma mobility on electrophoresis releases cholesterol from cells, *Proc. Natl. Acad. Sci. USA* **91:**1834–1838.

Jackson, R. L., Gotto, A. M., Stein, O., and Stein, Y., 1975, A comparative study on the removal of cellular lipids from landschutz ascites cells by human plasma apolipoproteins, *J. Biol. Chem.* **250:**7204–7209.

Jansen, H., and Hulsmann, W. C., 1980, Heparin-releasable (liver) lipase(s) may play a role in the uptake of cholesterol by steroid-secreting tissues, *Trends Biochem. Sci.* **5:**265–268.

Javitt, N. B., 1994, Bile acid synthesis from cholesterol: regulatory and auxiliary pathways, *FASEB J.* **8:**1308–1311.

Johnson, W. J., 1996, Cell-free transfer of cholesterol from lysosomes to phospholipid vesicles, *J. Lipid Res.* **37:**54–66.

Johnson, W. J., and Reinhart, M. P., 1994, Lack of requirement for sterol carrier protein-2 in the intracellular trafficking of lysosomal cholesterol, *J. Lipid Res.* **35:**563–573.

Johnson, W. J., Bamberger, M. J., Latta, R. A., Rapp, P. E., Phillips, M. C., and Rothblat, G. H., 1986, The bidirectional flux of cholesterol between cells and lipoproteins, *J. Biol. Chem.* **261:** 5766–5776.

Johnson, W. J., Mahlberg, F. H., Chacko, G. K., Phillips, M. C., and Rothblat, G. H., 1988, The influence of cellular and lipoprotein cholesterol contents on the flux of cholesterol between fibroblasts and high density lipoprotein, *J. Biol. Chem.* **263:**14099–14106.

Johnson, W. J., Chacko, G. K., Phillips, M. C., and Rothblat, G. H., 1990, The efflux of lysosomal cholesterol from cells, *J. Biol. Chem.* **265:**5546–5553.

Johnson, W. J., Kilsdonk, E. P. C., Van Tol, A., Phillips, M. C., and Rothblat, G. H., 1991a, Cholesterol efflux from cells to immunopurified subfractions of human high density lipoproteins: LP-AI and LP-AI/AII, *J. Lipid Res.* **32:**1993–2000.

Johnson, W. J., Mahlberg, F. H., Rothblat, G. H., and Phillips, M. C., 1991b, Cholesterol transport between cells and high density lipoproteins, *Biochim. Biophys. Acta* **1085:**273–298.

Johnson, W. J., Fischer, R. T., Phillips, M. C., and Rothblat, G. H., 1995, Efflux of newly synthesized cholesterol and biosynthetic sterol intermediates from cells: dependence on acceptor type and on enrichment of cells with cholesterol, *J. Biol. Chem.* **270:**25037–25046.

Jonas, A., 1987, Lecithin cholesterol acyltransferase, *in*: *Plasma Lipoproteins* (A. M. Gotto, Jr., ed.), pp. 299–333, Elsevier, Amsterdam.

Kaplan, M. R., and Simoni, R. D., 1985, Transport of cholesterol from the endoplasmic reticulum to the plasma membrane, *J. Cell Biol.* **101:**446–453.

Kawano, M., Miida, T., Fielding, C. J., and Fielding, P. E., 1993, Quantitation of preβ-HDL-dependent and nonspecific components of the total efflux of cellular cholesterol and phospholipid, *Biochemistry* **32:**5025–5028.

Khoo, J. C., Reue, K., Steinberg, D., and Schotz, M. C., 1993, Expression of hormone-sensitive lipase mRNA in macrophages, *J. Lipid Res.* **34:**1969–1974.

Kodama, T., Freeman, M., Rohrer, L., Zabrecky, J., Matsudaira, P., and Krieger, M., 1990, Type I macrophage scavenger receptor containing α-helical and collagen-like coiled coils, *Nature* **343:** 531–535.

Komaba, A., Li, Q., Hara, H., and Yokoyama, S., 1992, Resistance of smooth muscle cells to assembly of high density lipoproteins with extracellular free apolipoproteins and to reduction of intracellularly accumulated cholesterol, *J. Biol. Chem.* **267:**17560–17566.

Koval, M., and Pagano, R. E., 1990, Sorting of an internalized plasma membrane lipid between recycling and degradative pathways in normal and Niemann–Pick, type A fibroblasts, *J. Biol. Chem.* **111:**429–442.

Kritharides, L., Jessup, W., Mander, E. L., and Dean, R. T., 1995, Apolipoprotein A-I-mediated efflux of sterols from oxidized LDL-loaded macrophages, *Arterioscler. Thromb.* **15:**276–289.

Kunitake, S. T., Mendel, C. M., and Hennessy, L. K., 1992, Interconversion between apolipoprotein A-I-containing lipoproteins of pre-beta and alpha electrophoretic mobilities, *J. Lipid Res.* **33:** 1807–1816.

Lagrost, L., Dengremont, C., Athias, A., De Geitere, C., Fruchart, J. C., Lallemant, C., Gambert, P., and Castro, G., 1995, Modulation of cholesterol efflux from Fu5AH hepatoma cells by the apolipoprotein content of high density lipoprotein particles, *J. Biol. Chem.* **270:**13004–13009.

Lange, Y., 1994, Cholesterol movement from plasma membrane to rough endoplasmic reticulum, *J. Biol. Chem.* **269:**3411–3414.

Lange, Y., and Matthies, H. J. G., 1984, Transfer of cholesterol from its site of synthesis to the plasma membrane, *J. Biol. Chem.* **259:**14624–14630.

Lange, Y., and Muraski, M. F., 1988, Topographic heterogeneity in cholesterol biosynthesis, *J. Biol. Chem.* **263:**9366–9373.

Lange, Y., and Steck, T. L., 1985, Cholesterol-rich intracellular membranes: a precursor to the plasma membrane, *J. Biol. Chem.* **260:**15592–15597.

Lange, Y., and Steck, T. L., 1994, Cholesterol homeostasis. Modulation by amphiphiles, *J. Biol. Chem.* **269:**29371–29374.

Lange, Y., Schmit, V. M., and Schreiber, J. R., 1988, Localization and movement of newly synthesized cholesterol in rat ovarian granulosa cells, *Endocrinol.* **123:**81–86.

Lange, Y., Echevarria, F., and Steck, T. L., 1991, Movement of zymosterol, a precursor of cholesterol, among three membranes in human fibroblasts, *J. Biol. Chem.* **266:**21439–21443.

Lange, Y., Strebel, F., and Steck, T. L., 1993, Role of the plasma membrane in cholesterol esterification in rat hepatoma cells, *J. Biol. Chem.* **268:**13838–13843.

Lefevre, M., Goudey-Lefevre, J. C., and Roheim, P. S., 1987, Gradient gel electrophoresis-immunoblot analysis (GGEI): a sensitive method for apolipoprotein profile determinations, *J. Lipid Res.* **28:**1495–1507.

Lin, D., Sugawara, T., Strauss, J. F. I., Clark, B. J., Stocco, D. M., Saenger, P., Rogol. A., and Miller, W. L., 1995, Role of steroidogenic acute regulatory protein in adrenal and gonadal steroidogenesis, *Science* **267:**1828–1831.

Liscum, L., and Collins, G. J., 1991, Characterization of Chinese hamster ovary cells that are resistant to 3-β-[2-(diethylamino)ethoxy]androst-5-en-17-one inhibition of low density lipoprotein-derived cholesterol metabolism, *J. Biol. Chem.* **266:**16599–16606.

Liscum, L., and Dahl, N. K., 1992, Intracellular cholesterol transport, *J. Lipid Res.* **33:**1239–1254.

Liscum, L., and Faust, J. R., 1989, The intracellular transport of low density lipoprotein-derived cholesterol is inhibited in Chinese hamster ovary cells cultured with 3-β-[2-(diethylamino) ethoxy] androst-5-en-17-one, *J. Biol. Chem.* **264:**11796–11806.

Liscum, L., and Underwood, K. W., 1995, Intracellular cholesterol transport and compartmentation, *J. Biol. Chem.* **270:**15443–15446.

Luchoomun, J., Theret, N., Clavey, V., Duchateau, P., Rosseneu, M., Brasseur, R., Denefle, P., Fruchart, J. C., and Castro, G. R., 1994, Structural domain of apolipoprotein A-I involved in its interaction with cells, *Biochim. Biophys. Acta* **1212:**319–326.

Lund-Katz, S., and Phillips, M. C., 1984, Packing of cholesterol molecules in human high-density lipoproteins, *Biochemistry* **23:**1130–1138.

Mahlberg, F. H., Glick, J. M., Lund-Katz, S., and Rothblat, G. H., 1991, Influence of apolipoproteins AI, AII and C on the metabolism of membrane and lysosomal cholesterol in macrophages, *J. Biol. Chem.* **266:**19930–19937.

McKnight, G. L., Reasoner, J., Gilbert, T., Sundquist, K. O., Hokland, B., McKernan, P. A., Champagne, J., Johnson, C. J., Bailey, M. C., Holly, R., O'Hara, P. J., and Oram, J. F., 1992, Cloning and expression of a cellular high density lipoprotein-binding protein that is up-regulated by cholesterol loading of cells, *J. Biol. Chem.* **267:**12131–12141.

Melchior, G. W., Castle, C. K., Murray, R. W., Blake, W. L., Dinh, D. M., and Marotti, K. R., 1994, Apolipoprotein A-I metabolism in cholesteryl ester transfer protein transgenic mice. Insights into the mechanisms responsible for low plasma high density lipoprotein levels, *J. Biol. Chem.* **269:**8044–8051.

Mendez, A. J., 1995, Monensin and brefeldin A inhibit high density lipoprotein-mediated cholesterol efflux from cholesterol-enriched cells, *J. Biol. Chem.* **270:**5891–5990.

Mendez, A. J., Oram, J. F., and Bierman, E. L., 1991, Protein kinase C as a mediator of high density lipoprotein receptor-dependent efflux of intracellular cholesterol, *J. Biol. Chem.* **266:** 10104–10111.

Miida, T., Kawano, M., Fielding, C. J., and Fielding, P. E., 1992, Regulation of the concentration of pre-β high-density lipoprotein in normal plasma by cell membranes and lecithin-cholesterol acyltransferase activity, *Biochemistry* **31:**11112–11117.

Moreau, P., Rodriguez, M., Cassagne, C., Morre, D. M., and Morre, J. D., 1991, Trafficking of lipids from the endoplasmic reticulum to the Golgi apparatus in cell-free system from rat liver, *J. Biol. Chem.* **266:**4322–4328.

Nakashima, Y., Plump, A. S., Raines, E. W., Breslow, J. L., and Ross, R., 1994, ApoE-deficient mice develop lesions of all phases of atherosclerosis throughout the arterial tree, *Arterioscler. Thromb.* **14:**133–140.

Neary, R. H., and Gowland, E., 1987, Stability of free apolipoprotein A-I concentration in serum, and its measurement in normal and hyperlipidemic subjects, *Clin. Chem.* **33:**1163–1169.

Ohta, T., Nakamura, R., Ikeda, Y., Shinohara, M., Miyazaki, A., Horiuchi, S., and Matsuda, I., 1992, Differential effect of subspecies of lipoprotein containing apolipoprotein A-I on cholesterol efflux from cholesterol-loaded macrophages: functional correlation with lecithin: cholesterol acyltransferase, *Biochim. Biophys. Acta* **1165:**119–128.

Okawa, S., Mendez, A. J., Oram, J. F., Bierman, E. L., and Cheung, M. C., 1993, Effects of high-density lipoprotein particles containing apo A-I, with or without A-II, on intracellular cholesterol efflux, *Biochim. Biophys. Acta* **1165:**327–334.

Oram, J. F., 1983, Effects of high density lipoprotein subfractions on cholesterol homeostasis in human fibroblasts and arterial smooth muscle cells, *Arteriosclerosis* **3:**420–432.

Oram, J. F., 1990, Cholesterol trafficking in cells, *Curr. Opin. Lipidology* **1:**416–421.

Oram, J. F., Albers, J. J., Cheung, M. C., and Bierman, E. L., 1981, The effects of subfractions of high density lipoprotein on cholesterol efflux from cultured fibroblasts, *J. Biol. Chem.* **256:**8348–8356.

Oram, J. F., Brinton, E. A., and Bierman, E. L., 1983, Regulation of high density lipoprotein receptor activity in cultured human skin fibroblasts and human arterial smooth muscle cells, *J. Clin. Invest.* **72:**1611–1621.

Oram, J. F., Mendez, A. J., Slotte, J. P., and Johnson, T. F., 1991, High density lipoprotein apolipoproteins mediate removal of sterol from intracellular pools but not from plasma membranes of cholesterol-loaded fibroblasts, *Arterioscler. Thromb.* **11:**403–414.

Ottnad, E., Parthasarathy, S., Sambrano, G. R., Ramprasad, M. P., Quehenberger, O., Kondratenko, N., Green, S., and Steinberg, D., 1995, A macrophage receptor for oxidized low density lipoprotein distinct from the receptor for acetyl low density lipoprotein: partial purification and role in recognition of oxidatively damaged cells, *Proc. Natl. Acad. Sci. USA* **92:**1391–1395.

Pentchev, P. G., Vanier, M. T., Suzuki, K. and Patterson, M. C., 1995, Niemann–Pick disease type C: a cellular cholesterol lipidosis, *in*: *The Metabolic and Molecular Bases of Inherited Disease* (C. R. Scriver, A. L. Beaudet, W. S. Sly, and D. Valle, eds.), pp. 2625–2639, McGraw-Hill, New York.

Phillips, M. C., Johnson, W. J., and Rothblat, G. H., 1987, Mechanisms and consequences of cellular cholesterol exchange and transfer, *Biochim. Biophys. Acta* **906:**223–276.

Picardo, M., Massey, J. B., Kuhn, D. E., Gotto, J., Gianturco, S. H., and Pownall, H. J., 1986, Partially reassembled high density lipoproteins. Effects on cholesterol flux, synthesis, and esterification in normal human skin fibroblasts, *Arteriosclerosis* **6:**434–441.

Pieters, M. N., Schouten, D., and Van Berkel, T. J. C., 1994, In vitro and in vivo evidence for the role of HDL in reverse cholesterol transport, *Biochim. Biophys. Acta* **1225:**125–134.

Puglielli, L., Rigotti, A., Greco, A. V., Santos, M. J., and Nervi, F., 1995, Sterol carrier protein-2 is involved in cholesterol transfer from the endoplasmic reticulum to the plasma membrane in human fibroblasts, *J. Biol. Chem.* **270:**18723–18726.

Ray, E., Bellini, F., Stoudt, G., Hemperly, S., and Rothblat, G. H., 1980, Influence of lecithin:cholesterol acyltransferase on cholesterol metabolism in hepatoma cells and hepatocytes, *Biochim. Biophys. Acta* **617:**318–334.

Reichl, D., 1994, Extravascular circulation of lipoproteins: their role in reverse transport of cholesterol, *Atherosclerosis* **105:**117–129.

Rodriguez, M., Moreau, P., Paulik, M., Lawrence, J., Morre, D. J., and Morre, D. M., 1992, NADH-activated cell-free transfer between Golgi apparatus and plasma membranes of rat liver, *Biochim. Biophys. Acta* **1107:**131–138.

Rodriguez-Lafrasse, C., Rousson, R., Bonnet, J., Pentchev, P. G., Louisot, P., and Vanier, M. T., 1990, Abnormal cholesterol metabolism in imipramine-treated fibroblast cultures. Similarities with Niemann–Pick/type C disease, *Biochim. Biophys. Acta* **1043:**123–128.

Rothblat, G. H., and Phillips, M. C., 1991, Cholesterol efflux from arterial wall cells, *Curr. Opin. Lipidol.* **2:**288–294.

Rothblat, G. H., Bamberger, M., and Phillips, M. C., 1986, Reverse cholesterol transport, *in*: *Methods in Enzymology, Vol. 129* (J. Albers, and J. Segrest, eds.) pp. 628–644, Academic Press, New York.

Rothblat, G. H., Mahlberg, F. H., Johnson, W. J., and Phillips, M. C., 1992, Apolipoprotein, membrane cholesterol domains, and the regulation of cholesterol efflux, *J. Lipid Res.* **33:**1091–1098.

Rubin, E. M., Ishida, B. Y., Clift, S. M., and Krauss, R. M., 1991, Expression of human apolipoprote in A-I in transgenic mice results in reduced plasma levels of murine apolipoprotein A-I and the appearance of two new high denisty lipoprotein size subclasses, *Proc. Natl. Acad. Sci. USA* **88:**434–438.

Runquist, E. A., and Havel, R. J., 1991, Acid hydrolases in early and late endosome fractions from rat liver, *J. Biol. Chem.* **266:**22557–22563.

Sarria, A. J., Panini, S. R., and Evans, R. M., 1992, A functional role for vimentin intermediate filaments in the metabolism of lipoprotein-derived cholesterol in human SW-13 cells, *J. Biol. Chem.* **267:**19455–19463.

Sato, Y., Nishikawa, K., Aikawa, K., Mimura, K., Murakami-Murofushi, K., Arai, H., and Inoue, K., 1995, Side-chain structure is critical for the transport of sterols from lysosomes to cytoplasm, *Biochim. Biophys. Acta* **1257:**38–46.

Schroeder, F., Jefferson, J. R., Kier, A. B., Knittel, J., Scallen, T. J., Wood, W. G., and Hapala, I., 1991, Membrane cholesterol dynamics: cholesterol domains and kinetic pools, *Proc. Soc. Exp. Biol. Med.* **196:**235–252.

Schroeder, F., Woodford, J. K., Kavecansky, J., Wood, W. G., and Joiner, C., 1995, Cholesterol domains in biological membranes, *Mol. Membrane Biol.* **12:**113–119.

Schultz, J. R., and Rubin, E. M., 1994, The properties of HDL in genetically engineered mice, *Curr. Opin. Lipidol.* **5:**126–137.

Schultz, J. R., Verstuyft, J. G., Gong, E. L., Nichols, A. V., and Rubin, E. M., 1993, Protein composition determines the anti-atherogenic properties of HDL in transgenic mice, *Nature* **365:**762–764.

Segrest, J. P., Jones, M. K., De Loof, H., Brouillette, C. G., Venkatachalapathi, Y. V., and Anantharamaiah, G. M., 1992, The amphipathic helix in the exchangeable apolipoproteins: a review of secondary structure and function, *J. Lipid Res.* **33:**141–166.

Skinner, E. R., 1994, High-density lipoprotein subclasses, *Curr. Opin. Lipidol.* **5:**241–247.

Slotte, J. P., and Bierman, E. L., 1988, Depletion of plasma-membrane sphingomyelin rapidly alters the distribution of cholesterol between plasma membranes and intracellular cholesterol pools in cultured fibroblasts, *Biochem. J.* **250:**653–658.

Slotte, J. P., Oram, J. F., and Bierman, E. L., 1987, Binding of high density lipoproteins to cell receptors promotes translocation of cholesterol from intracellular membranes to the cell surface, *J. Biol. Chem.* **262:**12904–12907.

Stein, O., Stein, Y., Lefevre, M., and Roheim, P. S., 1986, The role of apolipoprotein A-IV in reverse cholesterol transport studied with cultured cells and liposomes derived from an ether analog of phosphatidylcholine, *Biochim. Biophys. Acta* **878:**7–13.

Stein, O., Dabach, Y., Hollander, G., Ben-Naim, M., Oette, K., and Stein, Y., 1995, Effects of interactions of apolipoprotein A-II with apolipoproteins A-I or A-IV on [^{3}H]cholesterol efflux and uptake in cell culture, *Biochim. Biophys. Acta* **1257:**174–180.

Strickland, D. K., Kounnas, M. Z., and Argraves, W. S., 1995, LDL receptor-related protein: a multiligand receptor for lipoprotein and proteinase catabolism, *FASEB J.* **9:**890–898.

Suckling, K. E., and Stange, E. F., 1985, Role of acyl-CoA:metabolism acyltransferase in cellular cholesterol metabolism, *J. Lipid Res.* **26:**647–671.

Sugawara, T., Holt, J. A., Driscoll, D., Strauss, J. F. I., Lin, D., Miller, W. L., Patterson, D., Clancy, K. P., Hart, I. M., Clark, B. J., and Stocco, D. M., 1995, Human steroidogenic acute regulatory protein: Functional activity in COS-1 cells, tissue-specific expression, and mapping of the structural gene to 8p11.2 and a pseudogene to chromosome 13, *Proc. Natl. Acad. Sci. USA* **92:** 4778–4782.

Tabacik, C., Valentin, J.-P., Aliau, S., and Descomps, B., 1991, Active cholesterol biosynthesis in cultured aortic smooth muscle cells: evolution during the lifespan of the culture, *Atherosclerosis* **86:**123–137.

Tabas, I., Rosoff, W. J., and Boykow, G. C., 1988, Acyl coenzyme A:cholesterol acyl transferase in macrophages utilizes a cellular pool of cholesterol oxidase-accessible cholesterol as substrate, *J. Biol. Chem.* **263:**1266–1272.

Tall, A. R., 1993, Plasma cholesteryl ester transfer protein, *J. Lipid Res.* **34:**1255–1274.

Tangirala, R. K., Jerome, W. G., Jones, N. L., Small, D. M., Johnson, W. J., Glick, J. M., Mahlberg, F. H., and Rothblat, G., 1994, Formation of cholesterol monohydrate crystals in macrophage-derived foam cells, *J. Lipid Res.* **35:**93–104.

Thompson, S. L., Burrows, R., Laub, R. J., and Krisans, S. K., 1987, Cholesterol synthesis in rat liver peroxisomes. Conversion of mevalonic acid to cholesterol, *J. Biol. Chem.* **262:**17420–17425.

Tozuka, M., and Fidge, N., 1989, Purification and characterization of two high-density-lipoprotein-binding proteins from rat and human liver, *Biochem. J.* **261:**239–244.

Urbani, L., and Simoni, R. D., 1990, Cholesterol and vesicular stomatitis virus G protein take separate routes from the endoplasmic reticulum to the plasma membrane, *J. Biol. Chem.* **265:**1919–1923.

van Heusden, G. P., van Beckhover, J. R. C. M., Thieringer, R., Raetz, C. R. H., and Wirtz, K. W. A., 1992, Increased cholesterol synthesis in Chinese hamster ovary cells deficient in peroxisomes, *Biochim. Biophys. Acta* **1126:**81–87.

von Eckardstein, A., Huang, Y., and Assmann, G., 1994, Physiological role and clinical relevance of high-density lipoprotein subclasses, *Curr. Opin. Lipidol.* **5:**404–416.

von Eckardstein, A., Huang, Y., Wu, S., Funke, H., Noseda, G., and Assmann, G., 1995, Reverse cholesterol transport in plasma of patients with different forms of familial HDL deficiency, *Arterioscler. Thromb.* **15:**691–703.

Waits, L., Dunkle, S., Wilkinson, F. E., Moreau, P., Safranski, K., Reust, T., Morre, D. M., and Morre, D. J., 1990, Cell-free transfer of sterols from dictyosome-like structures to plasma membrane vesicles of guinea pig testes, *Protoplasma* **154:**8–15.

Warner, G. J., Stoudt, G., Bamberger, M., Johnson, W. J., and Rothblat, G. H., 1995, Cell toxicity induced by inhibition of acyl coenzyme A cholesterol acyltransferase and accumulation of unesterified cholesterol, *J. Biol. Chem.* **270:**5772–5778.

Wolfbauer, G., Glick, J. M., Minor, L. K., and Rothblat, G. H., 1986, Development of the smooth muscle foam cell:uptake of macrophage lipid inclusions, *Proc. Natl. Acad. Sci. USA***83:**7760–7764.

Xu, X., and Tabas, I., 1991, Lipoprotein active acyl-coenzyme A:cholesterol acyltransferase in macrophages only after cellular cholesterol pools are expanded to a critical threshold level, *J. Biol. Chem.* **266:**17040–17048.

Yancey, P. G., and St. Clair, R. W., 1992, Cholesterol efflux is defective in macrophages from atherosclerosis-susceptible White Carneau pigeons ralative to resistant Show Racer pigeons, *Arterioscler. Thromb.* **12:**1291–1304.

Yancey, P. G., and St. Clair, R. W., 1994, Mechanism of the defect in cholesteryl ester clearance from macrophages of atherosclerosis-susceptible White Carneau pigeons, *J. Lipid Res.* **35:**2114–2129.

Yancey, P. G., Bielicki, J. K., Johnson, W. J., Lund-Katz, S., Palgunachari, M. N., Anantharamaiah, G. M., Segrest, J. P., Phillips, M. C., and Rothblat, G. H., 1995, The efflux of cellular cholesterol and phospholipid to lipid-free apolipoproteins and class A amphipathic peptides, *Biochemistry* **34:**7955–7965.

Yau-Young, A. O., Rothblat, G. H., and Small, D. M., 1982, Mobilization of cholesterol from cholesterol ester enriched tissue culture cells by phospholipid dispersions, *Biochim. Biophys. Acta* **710:**181–187.

Yeagle, P. L., 1985, Cholesterol and the cell membrane, *Biochim. Biophys. Acta* **822:**267–287.

Yla-Herttuala, S., Palinski, W., Rosenfeld, M. E., Parathasarathy, S., Carew, T. E., Butler, S., Witztum, J. L., and Steinberg, D., 1989, Evidence for the presence of oxidatively modified low density lipoprotein in atherosclerotic lesions of rabbit and man, *J. Clin. Invest.* **84:**1086–1095.

Chapter 10

Cholesterol–Sphingomyelin Interactions in Cells—Effects on Lipid Metabolism

J. Peter Slotte

1. INTRODUCTION

Both cholesterol and sphingomyelin are important constituents of cellular plasma membranes. The molecules are chemically and functionally very different, yet they appear to be attracted to each other in the membrane compartment. It is the aim of this review to discuss how alterations in their membrane interactions may affect lipid homeostasis in cells, and to suggest a molecular explanation for their mutual affinity in membranes. Recent reviews dealing with the subcellular distribution and transport of cholesterol (Liscum and Dahl, 1992; Liscum and Faust, 1994; Liscum and Underwood, 1995), with cellular lipid traffic (van Meer, 1989; Pagano, 1990; Voelker, 1991; Allan and Kallen, 1993), with transport and metabolism of sphingomyelin (Koval and Pagano, 1991), and with the role of sphingolipids in cell signaling (Kolesnick, 1991, 1994; Hannun and Bell, 1993; Hannun, 1994), may also be of interest to the reader.

2. CELLULAR DISTRIBUTION OF CHOLESTEROL

2.1. Cell Cholesterol Homeostasis

Cholesterol is an essential component of cellular plasma membranes in higher organisms. Cholesterol interacts with membrane phospholipids and

J. Peter Slotte Department of Biochemistry and Pharmacy, Åbo Akademi University, FIN 20520 Turku, Finland.

Subcellular Biochemistry, Volume 28: Cholesterol: Its Functions and Metabolism in Biology and Medicine, edited by Robert Bittman. Plenum Press, New York, 1997.

influences their physico-chemical properties. The important membrane properties that are directly or indirectly influenced by membrane levels of cholesterol include solute permeability (for a review, see Yeagle, 1985), phospholipid acyl chain mobility (Gally *et al.*, 1976; Stockton and Smith, 1976, Yeagle, 1985) and lateral packing (Chapman *et al.*, 1969; Lund-Katz *et al.*, 1988; Smaby *et al.*, 1994). Cholesterol also has a marked influence on lateral phase separations (Smutzer and Yeagle, 1985), and on the effective free volume of membranes (Straume and Litman, 1987), two parameters that are directly related to the flexibility of membrane proteins (e.g., ion channels, enzymes) and hence to their function in membranes.

Since cholesterol is such an essential and ubiquitous molecule in cells, most eucaryotic cells have retained the capacity to synthesize cholesterol from two-carbon precursor molecules. An important and early rate-limiting step in the biosynthesis of cholesterol is the conversion of β-hydroxy-β-methylglutaryl-CoA to mevalonate, catalyzed by β-hydroxy-β-methylglutaryl-CoA reductase (HMG-CoA reductase; Brown *et al.*, 1973). The regulation of HMG-CoA reductase includes (long-term) transcriptional regulation, probably by oxygenated sterols, as well as inactivation/activation of the enzyme by an AMP-dependent protein kinase/phosphatase cascade (Beg *et al.*, 1978; Ingebritsen *et al.*, 1981; Brown and Goldstein, 1986; Mehrabian *et al.*, 1986; Gibson and Parker, 1987).

An alternative source of cellular cholesterol is provided by low density lipoproteins (LDL), which bind to plasma membrane apo B/E receptors (Goldstein and Brown, 1977). The receptor/LDL complex is internalized by a process termed receptor-mediated endocytosis (Brown and Goldstein, 1986), in which the receptor ligand is transported to the lysosomal compartment for partial or complete degradation (Brown *et al.*, 1975). Unesterified cholesterol in the lysosomes (released from the LDL particles or derived from the hydrolysis of LDL-derived cholesteryl esters) is transported directly to the plasma membrane compartment. Eventually, any excess of plasma membrane unesterified cholesterol may be transported to the endoplasmic reticulum for conversion to cholesteryl esters (Tabas *et al.*, 1988). Rates of LDL internalization and cellular cholesterol esterification (by acyl-CoA:cholesterol acyl transferase; ACAT) are sensitive to cellular levels of unesterified cholesterol (Goldstein and Brown, 1977). An unregulated pathway for cellular uptake of lipoprotein-derived unesterified cholesterol involves the direct (surface) transfer of lipoprotein surface cholesterol to the cellular plasma membrane compartment. This pathway may contribute significantly to cell cholesterol accumulation when the extracellular level of LDL is high (Slotte *et al.*, 1988).

Since extrahepatic and nonsteroidogenic cells cannot metabolize cholesterol, save for the conversion of cholesterol to cholesteryl esters, routes must exist whereby cells can rid themselves of excess cholesterol. High-density lipoproteins have been shown to act as cholesterol acceptors, and initiate the reverse-cholesterol transport system first proposed by Glomset in 1968 (Glomset, 1968; Rothblat and Phillips, 1991). It is at present unclear which of the HDL subtypes are the best cholesterol acceptor particles, although their capacities to initiate cholesterol efflux from cells clearly differ (Rothblat *et al.*, 1992; Mahlberg and Rothblat, 1992; Davidson *et al.*, 1994). Whereas the actual transfer of cholesterol from cellular plasma membranes to extracellular acceptor particles apparently do not require binding proteins (Johnson *et al.*, 1991), the interaction of HDL-like particles with specific cell surface binding proteins may lead to a regulated flow (translocation) of intracellular cholesterol to the cell surface (Slotte *et al.*, 1987; Mendez *et al.*, 1991; Pörn *et al.*, 1991a).

2.2. Subcellular Distribution of Cholesterol

Most of the cellular unesterified cholesterol can be found in the plasma membrane compartment. Depending on the cell type examined and the assay method used, plasma membranes have been reported to contain from about 50% to as much as 90% of the total cellular unesterified cholesterol (Lange, 1991; Warnock *et al.*, 1993). It appears that cells with an extensive recycling of plasma membranes through the endocytic system have a larger fraction of intracellular cholesterol (Allan and Kallen, 1993). It is not surprising that cellular plasma membranes contain such a high concentration of cholesterol, since several of the cellular cholesterol transport pathways are directed toward the cell surface. Newly synthesized cholesterol is transported vectorially from the intracellular site of synthesis to the cell surface, where it appears with a transfer half-time of 10–18 min (DeGrella and Simoni, 1982; Lange *et al.*, 1991; Urbani and Simoni, 1990; Kaplan and Simoni, 1985). Exogenously derived cholesterol, entering the cell with LDL particles, is also quantitatively transported from lysosomes to the cell surface, where it accumulates (Tabas *et al.*, 1988). Only when the solubilizing capacity of the plasma membrane compartment is saturated does the excess cholesterol move to the endoplasmic reticulum for esterification by ACAT (Xu and Tabas, 1991). It is now established that the capacity of plasma membranes to solubilize cholesterol is largely a function of its sphingomyelin content (see section 5).

3. SYNTHESIS AND DISTRIBUTION OF SPHINGOMYELIN

3.1. Site of Sphingomyelin Synthesis in Cells

Sphingomyelin appears to be synthesized by the enzymatic transfer of the phosphocholine moiety of phosphatidylcholine to ceramide (catalyzed by sphingomyelin synthase), thus yielding sphingomyelin and diacylglycerol (Voelker and Kennedy, 1982; Spence *et al.*, 1983; Marggraf and Kanfer, 1984; Kishimoto, 1993). The direct transfer of CDP-choline to ceramide is apparently a pathway of minor significance in most cells (Voelker and Kennedy, 1982). There is currently a dispute regarding the major site of sphingomyelin synthesis, since synthetic activities have been demonstrated (directly or indirectly) in several different cellular compartments (Futerman *et al.*, 1990; van Echten *et al.*, 1990; Jeckel *et al.*, 1990; Kallen *et al.*, 1993, 1994). In studies using subcellular fractionation techniques (rat liver), it was observed that *cis* Golgi fractions accounted for the largest activity of sphingomyelin synthesis, whereas only a small fraction of the synthetic activity could be attributed to the plasma membrane fraction (Futerman *et al.*, 1990; Jeckel *et al.*, 1990). In baby hamster kidney cells, however, it was reported that a substantial synthesis of sphingomyelin occurred in a compartment distinct from the *cis/medial* Golgi. This compartment was presumed to include the endocytic recycling pathway (Kallen *et al.*, 1993, 1994). There is always the possibility that the predominant site of synthesis may differ from one cell type to another, as a different synthetic topology may be needed for functionally specialized cells. This is indeed discussed in a recent report, where it was shown that the plasma and myelin membrane compartment accounted for at least half of the sphingomyelin synthase activity in oligodendrocytes (Vos *et al.*, 1995).

3.2. Subcellular Distribution of Sphingomyelin

The distribution of sphingomyelin mass among intracellular organelles has been examined using cell fractionation studies and sphingomyelinase degradation experiments. Results with both techniques suggest that more than half of the cellular sphingomyelin mass is confined to the plasma membranes (with most of it being in the exoleaflet; for a review, see Koval and Pagano, 1991). With fibroblasts it was reported that as much as 90% of the total cell sphingomyelin mass was in the plasma membrane compartment (Lange, 1991). Again, as was the case with cholesterol distribution, it appears that cells with an extensive recycling of plasma membranes through the endocytic system also have a larger fraction of sphingomyelin in intracellular compartments (inaccessible to exogenously applied sphingomyelinase) (Allan and Kallen, 1993). The predominant exoleaflet orientation of plasma membrane sphingomyelin apparently results from the topological orientation of sphingomyelin synthase, since sphingomyelin flip-flop from one leaflet to the other is very limited, and since a membrane translocase acting on sphingomyelin has not yet been observed.

4. EFFECTS OF SPHINGOMYELIN ON CELL CHOLESTEROL METABOLISM

4.1. Increased Sphingomyelin Levels versus Cholesterol Homeostasis

The older observations that there appeared to exist a positive correlation between the contents of cholesterol and sphingomyelin mass in the membranes of rat liver hepatocytes (Patton, 1970), and the more recent data indicating colocalization of these lipid classes in the plasma membrane compartment (as well as in the exocytic cycling pathway), may suggest that one of these lipid classes affect the distribution of the other (directly or indirectly). Early evidence indicating that plasma membrane levels of sphingomyelin might affect the content of cholesterol in the membrane was obtained from sphingomyelin loading experiments in cells. The incorporation of sphingomyelin mass from liposomes into fibroblast membranes was observed to result in a reduced esterification of cell cholesterol, and to an increased formation of newly synthesized cholesterol (Gatt and Bierman, 1980; Kudchodkar *et al.,* 1983). Since both the ACAT-catalyzed esterification reaction and the biosynthesis of cholesterol are acutely regulated by the cellular level of unesterified cholesterol, the sphingomyelin loading experiments suggested that sphingomyelin incorporation directly led to a net flow of cholesterol from intracellular sites to the plasma membrane compartment. Plausibly, the incorporation of sphingomyelin mass in the cellular membranes increased the capacity of cells to solubilize cholesterol, which in turn resulted in the observed metabolic responses.

4.2. Decreased Sphingomyelin Levels versus Cholesterol Homeostasis

4.2.1. Effects on Cholesterol Esterification

Since an increased level of sphingomyelin mass in cell membranes led to a marked flow of cell cholesterol toward the cell surface, it seemed likely that a redistribution in the opposite direction would occur in cells deprived of cell surface sphingomyelin. Experiments along these lines were first performed by Slotte and Bierman (1988), who showed that the degradation of sphingomyelin mass (with the aid of exogenous sphingomyelinase) in cultured fibroblasts led to a dramatic activation of the endogenous esterification of cholesterol. Since the ACAT-reaction in the endoplasmic reticulum is sensitive to the flow of substrate cholesterol, its activation strongly suggested that sphingomyelin degradation resulted in a flow of cholesterol from the cell surface into the substrate pool of ACAT. This pioneering study was subsequently followed by related studies with different cell types, in which similar effects of sphingomyelin degradation on the activation of ACAT consistently have been observed (Gupta and Rudney, 1991; Stein *et al.,* 1992). The sphingomyelinase-induced activation of ACAT was not seen when cells were exposed to degradation products of sphingomyelin such as phosphorylcholine, ceramide, or sphingosine (Slotte and Bierman, 1988; Gupta and Rudney, 1991).

4.2.2. Effects on Cholesterol Biosynthesis

In addition to affecting the rate of endogenous cholesterol esterification, the sphingomyelinase-induced cholesterol translocation also affected the rate of cholesterol biosynthesis. Slotte and Bierman (1988) originally demonstrated that the incorporation of sodium [^{14}C]acetate into *de novo* sterols was reduced markedly in cells treated with sphingomyelinase as compared to control fibroblasts. Later, Gupta and Rudney (1991) extended these experiments to include rat intestinal epithelial cells, human skin fibroblasts, and HepG2 cells, and demonstrated that sphingomyelin degradation, in conjunction with the resulting cholesterol flow, directly downregulated the activity of HMG-CoA reductase, the key regulatory enzyme in cholesterol biosynthesis.

4.2.3. Effects on Steroidogenesis

In steroidogenic cells, cholesterol is converted to various steroid hormones by the action of cytochrome $P450_{scc}$ in the mitochondrial inner membrane. For this conversion to take place, cholesterol must first be transported from cellular storage sites (cholesteryl ester droplets, from which free cholesterol is released by hydrolysis; plasma membrane cholesterol pools) to the mitochondria (Bisgaier *et al.*, 1985; Freeman, 1989). Transfer of cholesterol from the outer mitchondrial membrane to the inner appears to involve a steroidogenic acute regulatory protein (Lin *et al.*, 1995).

In our experiments with Leydig tumor cells, the degradation of cell surface sphingomyelin was observed to increase the formation of steroid hormones (Pörn *et al.*, 1991b). These results were interpreted to demonstrate that a significant fraction of cell surface cholesterol, which was induced to translocate by the action of sphingomyelinase, was transferred to the mitochondrial compartment to serve as a substrate for cytochrome P450scc, to yield newly synthesized steroid hormones.

4.2.4. Cellular Recovery from Sphingomyelin Degradation

The recovery of cells from the effects of sphingomyelin degradation were tested with several different cell types (Pörn and Slotte, 1990; Slotte *et al.*, 1990). When cells that initially were exposed to sphingomyelinase (to yield sphingomyelin degradation) were allowed to recover in the absence of exogenous sphingomyelinase, a restoration of sphingomyelin mass was observed to take place. The rate of sphingomyelin resynthesis was, however, very different in different cell types. A very slow restoration of sphingomyelin levels was observed in human fibroblasts and SH-SY5Y neuroblastoma cells, whereas a much faster restoration was seen with BHK-21 cells. However, with all cell types tested, the restoration of sphingomyelin mass also restored cholesterol enrichment at the cell surface, and decreased the activation of the ACAT-reaction (Pörn and Slotte, 1990; Slotte *et al.*, 1990).

4.2.5. Determination of Cellular Cholesterol Distribution in Sphingomyelin-Depleted Cells

In an attempt to quantitate cell surface cholesterol in whole cells, using a modified cholesterol oxidase assay (Slotte *et al.*, 1989), it was observed that the complete degradation of sphingomyelin mass from the external leaflet of the cellular plasma membrane invariably led to a rapid and substantial decrease of the fraction of cell cholesterol susceptible to oxidation by cholesterol oxidase (Slotte *et al.*, 1989; Pörn and Slotte, 1990; Slotte *et al.*, 1990). This decrease in cell cholesterol oxidizability has been interpreted to indicate a substantial translocation of cell surface cholesterol to intracellular membranes. The extent of cholesterol translocation appear to differ from one cell type to another (between 20 and 60% of the cellular free cholesterol), and also within different lines of a similar cell type (i.e., human fibroblasts; J. P. Slotte, M. I. Pörn, and A. S. Härmälä, unpublished observations).

A recent study, in which the cellular distribution of cholesterol was determined using filipin staining and fluorescence microscopy, showed that the filipin-staining pattern was not significantly different in control and sphingomyelin-depleted cells (Pörn and Slotte, 1995). These results do not invalidate previous findings about the sphingomyelinase-induced cholesterol translocation in cells (Slotte *et al.*, 1989; Pörn and Slotte, 1990; Slotte *et al.*, 1990), but do suggest, however, that the extent of cholesterol translocation is not as dramatic as suggested by the decreased cholesterol oxidation susceptibility. In light of both the filipin-staining results and the cholesterol oxidation results, it appears that sphingomyelin depletion results in a massive redistribution of cholesterol within the plasma membrane structure, but that only a smaller fraction is actually transferred to intracellular sites (e.g., mitochondria and the endoplasmic reticulum).

4.2.6. Phosphatidylcholine Degradation Does Not Affect Cholesterol Homeostasis

To examine whether a degradation of other plasma membrane phospholipids would also lead to a redistribution of cell surface cholesterol, cell surface phosphatidylcholine was selectively degraded in fibroblasts, using a phosphatidylcholine-specific phospholipase C from *Bacillus cereus*. Surprisingly, the selective degradation of phosphatidylcholine did not cause a detectable cholesterol translocation in cells (Pörn *et al.*, 1993). Under the conditions used, the enzyme degraded about 15% of the total cellular phosphatidylcholine, corresponding to 50 nmol phosphatidylcholine per mg cell protein. The partial degradation of phosphatidylcholine left the cells morphologically intact, as evidenced by their exclusion of Trypan blue. Treatment of the same cells with sphingomyelinase, which resulted in a marked cholesterol redistribution and activation of cholesterol esterification, led to a degradation of about 22 nmol sphingomyelin per mg cell protein. It appears that sphingomyelin depletion alone affects cellular cholesterol distribution, since a selective degradation of plasma membrane phosphatidylcholine was shown not to have measurable effects on cell cholesterol homeostasis.

5. EFFECTS OF SPHINGOMYELIN ON METABOLISM OF EXOGENOUSLY DERIVED CHOLESTEROL

A substantial fraction of the uptake of exogenously derived cholesterol into cells is mediated by receptor-coupled internalization of LDL particles. This cholesterol is eventually transported from lysosomes (the site of LDL degradation) to the plasma membrane compartment. Subsequently, if there is excess free cholesterol in the cell, cholesterol may be transported to the endoplasmic reticulum for esterification by ACAT. Tabas and colleagues have demonstrated that lipoprotein-derived cholesterol mixes rapidly with cellular cholesterol in macrophages, and only move to the substrate pool of ACAT when the cellular cholesterol mass has reached a critical threshold level (Xu and Tabas, 1991). Interestingly, but maybe not surprisingly, it has been shown that this threshold level is regulated by the sphingomyelin content of the cells (Okwu *et al.*, 1994). These results are entirely compatible with the notion that the sphingomyelin content of cell membranes (mainly plasma membranes) is the major determinant of the solubility of cholesterol in the membranes.

Another interesting effect of plasma membrane sphingomyelin on cell cholesterol metabolism was observed in the human intestinal CaCo-2 cell line, in which the content of sphingomyelin in the apical membrane was found to regulate cholesterol uptake from bile salt micelles (Chen *et al.*, 1992). A 60% degradation of cell sphingomyelin resulted in a 50% decrease of cholesterol absorption, as well as in a downregulation of endogenous cholesterol biosynthesis. In addition, the basolateral secretion of cholesterol was likewise reduced in sphingomyelin-depleted CaCo-2 cells (Chen *et al.*, 1992). These authors suggested that the sphingomyelin content of the intestinal cells directly affects cholesterol flux, and that the overall process may be regulated by endogenous sphingomyelinases present in pancreatic juice.

6. CHOLESTEROL–SPHINGOMYELIN INTERACTIONS IN MODEL MEMBRANES

Based on evidence from cell culture systems, it is clear that the cellular homeostasis of cholesterol is directly influenced by sphingomyelin. This fact suggests that these two lipids associate directly in the membrane compartment. What is not clear is why cholesterol prefers to interact with sphingomyelin. An answer to this question is not likely to be gained from experiments at the cellular level, but rather from model membrane studies, in which the level of complexity is markedly reduced and the possibility for accurate interpretation of results is higher.

6.1. Affinity of Cholesterol–Sphingomyelin Interaction Compared to Other Phospholipid Systems

Many lines of evidence suggest that the strength or mode of interaction of cholesterol with phosphatidylcholines and sphingomyelins differ. First, the molecular packing properties in cholesterol/sphingomyelin membranes have been suggested to differ from the corresponding cholesterol/phosphatidylcholine model membranes (Lund-Katz *et al.*, 1988), although contradictory results have recently been reported (Smaby *et al.*, 1994). It has also been shown that water permeability is lower in cholesterol/sphingomyelin bilayer membranes than it is in cholesterol/phosphatidylcholine systems (Barenholz, 1986), a finding that would be consistent with a more dense lateral packing in cholesterol/sphingomyelin membranes. In addition, it has been established that cholesterol desorption from the membrane into the aqueous phase is markedly reduced in sphingomyelin-containing membranes as compared with phosphatidylcholine membranes of similar acyl chain composition (Lund-Katz *et al.*, 1988; Bittman, 1993). The slower desorption of cholesterol from sphingomyelin membranes is probably a result of the formation of more attractive van der Waals forces between cholesterol and sphingomyelin. The compressibility modulus has also been reported to be larger in cholesterol/sphingomyelin membranes than in cholesterol/phosphatidylcholine membranes (Needham and Nunn, 1990). Finally, using cholesterol oxidase as a probe of the strength of interaction between cholesterol and sphingomyelin or phosphatidylcholine, it was observed that the oxidation susceptibility of cholesterol was much reduced in a sphingomyelin-rich membrane, suggesting a much stronger cholesterol interaction with sphingomyelin than phosphatidylcholine (Slotte, 1992; Mattjus and Slotte, 1994). Based on all of these results, one can safely postulate that marked differences exist in how cholesterol interacts with sphingomyelin on one hand, and with phosphatidylcholine on the other.

6.2. Effects of Alterations in the Sphingomyelin Molecule on Cholesterol–Sphingomyelin Interactions

In an effort to examine the molecular explanation(s) for the high affinity of association between cholesterol and sphingomyelin, synthetic analogues of sphingomyelin were prepared in which different functional groups were altered systematically (Figure 1). These analogues were incorporated into model membrane systems, together with cholesterol, in order to examine how the functional alterations in the sphingomyelin molecule affected the strength of interaction with cholesterol, as probed with cholesterol oxidase. When the hydroxy group at carbon 3 of the sphingosine backbone was replaced by a hydrogen atom (to yield 3-deoxy-sphingomyelin), the oxidation susceptibility of

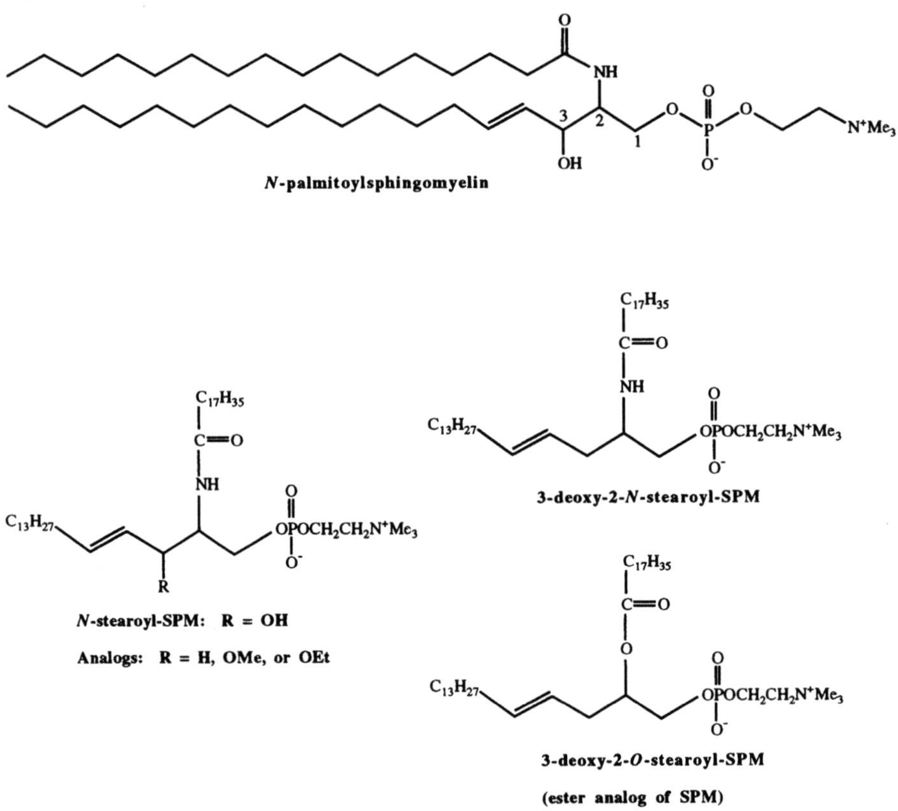

FIGURE 1. Schematic molecular structure of *N*-palmitoylsphingomyelin. See the text for a discussion of the effects of structural alterations in the sphingomyelin molecule on cholesterol–sphingomyelin interactions in model membranes.

cholesterol was slightly reduced (Grönberg *et al.*, 1991). However, when other functional groups were attached to the 3-OH group (i.e., a methoxy or ethoxy function) no change was observed in the rate of cholesterol oxidation, suggesting that alterations of the 3-OH group has little or no effect on cholesterol–sphingomyelin interaction in model membranes. The length and degree of unsaturation (saturated vs mono-unsaturated) of the amide-linked fatty acid was expectedly shown to affect the strength of cholesterol–sphingomyelin interaction (Slotte, 1992).

When the amide-linkage fatty acid was replaced by an ester-linked acyl chain, the strength of interaction between the sphingomyelin and cholesterol decreased markedly (Bittman *et al.*, 1994). This result implies that the amide-linked fatty acid function in sphingomyelin has a profound effect on the sterol–sphingomyelin interaction. It is possible that the amide-linked fatty acid influences the lateral interaction between sphingomyelin and cholesterol, especially in the polar interface region of the membrane, thereby allowing for a more favorable packing geometry.

6.3. Stoichiometry of the Cholesterol–Sphingomyelin Interaction

The interaction between cholesterol and phosphatidylcholine in model membranes appears to be of an equimolar nature (Lecuyer and Dervichian, 1969; Phillips and Finer, 1974; Collins and Phillips, 1982). When the cholesterol/phospholipid molar ratio is below 1, clusters of free phospholipid coexist with equimolar cholesterol/phospholipid "complexes" (McLean and Phillips, 1982). On the other hand, if the cholesterol/phospholipid ratio exceeds 1, free cholesterol clusters coexist with cholesterol/phospholipid complexes (Lecuyer and Dervichian, 1969; Collins and Phillips, 1982). When the equimolar stoichiometry is exceeded with respect to cholesterol, free sterol-rich clusters apparently form in the membranes (probably in laterally segregated domains). The observation that phosphatidylcholines can only solubilize an equimolar amount of cholesterol has been demonstrated using several different experimental techniques, both in bilayer and in monolayer membranes. However, when the maximal solubilizing capacity of sphingomyelin was determined in monolayer membranes using cholesterol oxidase as a probe, it was found that one mole of sphingomyelin could solubilize two moles of cholesterol (Slotte, 1992). This 2:1 cholesterol to sphingomyelin stoichiometry appears to hold, irrespective of the nature of the amide-linked fatty acid function in the sphingomyelin molecule (Slotte, 1992; Slotte, unpublished observations). Even replacement of the amide-linked fatty acid chain by an ester acyl function did not alter the 2:1 stoichiometry (Bittman *et al.*, 1994). Interestingly, phosphatidylcholines with one or two alkyl-linked chains (either in *sn*-1 or *sn*-2 positions, or both) also have the capacity to solubilize two moles of cholesterol per mole of the alkyl phosphatidylcholine (Mattjus *et al.*, 1996).

7. EFFECTS OF CHOLESTEROL ON PHOSPHOLIPID HOMEOSTASIS

Since the membrane level of sphingomyelin, by affecting the subcellular distribution of cholesterol, can indirectly regulate the rate of cholesterol biosynthesis, one can ask whether changes in plasma membrane cholesterol levels, or changes in cell cholesterol homeostasis in general, also may affect rates of phospholipid biosynthesis. In a preliminary study in which the incorporation of radiolabeled acetate into phosphatidylcholine was examined as a function of cholesterol surface transfer from extracellular vesicles, it was observed that a net flow of cholesterol mass into cultured rat aortic smooth muscle cells led to an increased formation of radiolabeled phosphatidylcholine (Slotte and Lundberg, 1983). In a later study, it was demonstrated that cholesterol loading into macrophages stimulated the biosynthesis of phosphatidylcholine (Shiratori *et al.*, 1994). In another study with intestinal CaCo-2 cells, it was demonstrated that the regulation of HMG-CoA reductase (cholesterol biosynthesis) and serine palmitoyltransferase (sphingosine synthesis) were independent of each other (Chen *et al.*, 1993). However, it was observed that when the membrane mass of either cholesterol or sphingomyelin were altered, parallel changes occurred in the rate of synthesis of these two lipids. In a recent study with Chinese hamster ovary cells, it was noted that 25-hydroxycholesterol stimulated sphingomyelin synthesis (Ridgway, 1995). This may be of relevance during cholesterol loading of cells, since cells apparently can oxidize cholesterol to 25-and 27-hydroxycholesterol (Björkhem *et al.*, 1994), which in turn may regulate the formation of sphingomyelin. A recent review on phospholipid biosynthesis in eucaryotic cells may also be of interest to the reader (Kent, 1995).

8. CONCLUDING REMARKS

Studies performed so far suggest strongly that plasma membrane sphingomyelin is the important determinant of cholesterol distribution in the cell. This suggestion does not exclude the possibility that other additional properties of plasma membranes also affect the cellular cholesterol distribution. Although it has not been directly proved that sphingomyelin and cholesterol actually interact with each other in the membrane compartment—such an interaction is suggested based on results obtained from work with both biological and model membranes. Another question which has yet to be addressed adequately is which species comes first to the plasma membrane—cholesterol or sphingomyelin? And, if sphingomyelin is first, is this enough of a reason for cholesterol to also accumulate in this same compartment? Current efforts in many laboratories to elucidate both intracellular cholesterol transfer routes and their regulation, and sphingomyelin transfer routes and possible sorting events, is expected to help clarify this picture and allow us to better understand the underlying reasons.

9. REFERENCES

Allan, D., and Kallen, K.-J., 1993, Transport of lipids to the plasma membrane in animal cells, *Prog. Lipid Res.* **32:**195–219.

Barenholz, Y., 1984, Sphingomyelin-lecithin balance in membranes: composition, structure, and function relationships, in: *Physiology of Membrane Fluidity,* Volume 1 (M. Schinitsky, ed.), pp. 131–174, CRC Press, Boca Raton, FL.

Beg, Z. H., Stonic, J. A., and Brewer, H. B., 1978, 3-Hydroxy-3-methylglutaryl coenzyme A reductase: regulation of enzymatic activity by phosphorylation and dephosphorylation, *Proc. Natl. Acad.Sci. USA* **75:**3678–3681.

Bisgaier, C. L., Chanderbhan, R., Hinds, R. W., Vahouny, G. V., 1985, Adrenal cholesterol esters as substrate source for steroidogenesis, *J. Steroid Biochem.* **23:**967–974.

Bittman, R., 1993, A review on the kinetics of cholesterol movement between donor and acceptor bilayer membranes, in: *Cholesterol in Model Membranes* (L. X. Finegold, ed.), pp. 45–65, CRC Press, Boca Raton, FL.

Bittman, R., Kasireddy, C. R., Mattjus, P., and Slotte, J. P., 1994, Interaction of cholesterol with sphingomyelin in monolayers and vesicles, *Biochemistry* **33:**11776–11781.

Björkhem, I., Andersson, O., Diczfalusy, U., Sevastik, B., Xiu, R. J., Duan, C., Lund, E., 1994, *Proc. Natl. Acad. Sci. USA* **91:**8592–8596.

Brown, M. S., and Goldstein, J. L., 1986, A receptor-mediated pathway for cholesterol homeostasis, *Science* **232:**34–47.

Brown, M. S., Dana, S. E., and Goldstein, J. L., 1975, Receptor-dependent hydrolysis of cholesteryl esters contained in plasma low density lipoproteins, *Proc. Natl. Acad. Sci. USA* **72:**2925–2929.

Brown, M. S., Dana, S. E., and Goldstein, J. L., 1973, Regulation of 3-hydroxy-3-methylglutaryl-coenzyme A reductase activity in human fibroblasts by lipoproteins, *Proc. Natl. Acad. Sci. USA* **70:**2162–2166.

Chapman, D., Owens, N. F., Phillips, M. C., and Walker, D. A., 1969, Mixed monolayers of phospholipids and cholesterol, *Biochim. Biophys. Acta* **183:**458–465.

Chen, H., Born, E., Mathur, S. N., Johlin, F. C., and Field, F. J., 1992, Sphingomyelin content of intestinal cells membranes regulate cholesterol absorption. Evidence for pancreatic and intestinal cell sphingomyelinase activity, *Biochem. J.* **286:**771–777.

Chen, H., Born, E., Mathur, S. N., and Field, F. J., 1993, Cholesterol and sphingomyelin syntheses are regulated independently in cultured human intestinal cells, CaCo-2: role of membrane cholesterol and sphingomyelin content, *J. Lipid Res.* **34:**2159–2167.

Collins, J. J., and Phillips, M. C., 1982, Stability and structure of cholesterol-rich codispersions of cholesterol and phosphatidylcholine, *J. Lipid Res.* **23:**291–298.

Davidson, W. S., Lund-Katz, S., Johnson, W. J., Anantharamaiah, G. M., Palgunachari, M. N., Segrest, J. P., Rothblat, G. H., and Phillips, M. C., 1994, The influence of apolipoprotein structure on the efflux of cellular free cholesterol to high density lipoprotein, *J. Biol. Chem.* **269:**22975–22982.

DeGrella, R. F., and Simoni, R. D., 1982, Intracellular transport of cholesterol to the plasma membrane, *J. Biol. Chem.* **257:**14256–14262.

Demel, R. A., Bruckdorfer, K. R., and van Deenen, L. L. M., 1972, The effect of sterol structure on the permeability of liposomes to glucose, glycerol, and Rb^+, *Biochim. Biophys. Acta* **255:**321–330.

Freeman, D. A., 1989, Plasma membrane cholesterol: removal and insertion into the membrane and utilization as substrate for steroidogenesis, *Endocrinol.* **124:**2527–2534.

Futerman, A. H., Stieger, B., Hubbard, A. L., and Pagano, R. E., 1990, Sphingomyelin synthesis in rat liver occurs predominantly at the cis and medial cisternae of the Golgi apparatus, *J. Biol. Chem.* **265:**8650–8657.

Gally, H. U., Seelig, A., and Seelig, J., 1976, Cholesterol induced rod-like motion of fatty acyl chains in lipid bilayers, a deuterium magnetic resonance study, *Hoppe-Zeyler's Z. Physiol. Chem.* **357:**1447–1450.

Gatt, S., and Bierman, E. L., 1980, Sphingomyelin suppresses the binding and utilization of LDL by skin fibroblasts, *J. Biol. Chem.* **255:**3371–3376.

Gibson, D. M., and Parker, R. A., 1987, Hydroxymethylglutaryl coenzyme A, in: *The Enzymes,* Volume XVIII (P. D. Boyer, and E. G. Krebs, eds.), pp. 179–215, Academic Press, NY.

Glomset, J. A., 1968, The plasma lecithin:cholesterol acyl transferase reaction, *J. Lipid Res.* **9:**155–167.

Goldstein, J. L., and Brown, M. S., 1977, The low-density lipoprotein pathway and its relation to atherosclerosis, *Annu. Rev. Biochem.* **46:**897–930.

Grönberg, L., Ruan, Z.-s., Bittman, R., and Slotte, J. P., 1991, Interaction of cholesterol with synthetic sphingomyelin derivatives in mixed monolayers, *Biochemistry* **30:**10746–10754.

Gupta, A. K., and Rudney, H., 1991, Plasma membrane sphingomyelin and the regulation of HMG-CoA reductase activity and cholesterol biosynthesis in cell cultures, *J. Lipid Res.* **32:**125–136.

Hannun, Y. A., 1994, The sphingomyelin cycle and second messenger function of ceramide, *J. Biol. Chem.* **269:**3125–3128.

Hannun, Y. A., and Bell, R. M., 1993, The sphingomyelin cycle: a prototypic sphingolipid signaling pathway, *Adv. Lipid Res.* **25:**27–41.

Ingebritsen, T. S., Parker, R. A., and Gibson, D. M., 1981, Regulation of HMG-CoA reductase by a bicyclic phosphorylation system, *J. Biol. Chem.* **256:**1138–1144.

Jeckel, D., Karrenbauer, A., Birle, R., Schmidt, R. R., and Wieland, F., 1990, Sphingomyelin is synthesized in the cis Golgi, *FEBS Lett.* **261:**155–157.

Johnson, W. J., Mahlberg, F. H., Rothblat, G. H., and Phillips, M. C., 1991, Cholesterol transport between cells and high density lipoproteins, *Biochim. Biophys. Acta* **1085:**273–298.

Kallen, K.-J., Quinn, P., and Allan, D., 1993, Monensin inhibits synthesis of sphingomyelin by blocking transport of ceramide through the Golgi: evidence for two sites of sphingomyelin synthesis in BHK cells, *Biochim. Biophys. Acta* **1166:**305–308.

Kallen, K.-J., Allan, D., Whatmore, J. L., and Quinn, P., 1994, Synthesis of surface sphingomyelin in the plasma membrane recycling pathway of BHK cells, *Biochim. Biophys. Acta* **1191:**52–58.

Kaplan, M. R., and Simoni, R. D., 1985, Transport of cholesterol from the endoplasmic reticulum to the plasma membrane, *J. Cell Biol.* **101:**446–453.

Kent, C., 1995, Eucaryotic phospholipid biosynthesis, *Annu. Rev. Biochem.* **64:**315–343.

Kishimoto, Y., 1993, Sphingolipid formation, in: *The Enzymes,* Volume XVI (P. D. Boyer, ed.), pp. 358–407, Academic Press, NY.

Kolesnick, R. N., 1991, Sphingomyelin and derivatives as cellular signals, *Prog. Lipid Res.* **30:**1–38.

Kolesnick, R., 1994, Signal transduction through the sphingomyelin pathway, *Mol. Chem.-Neuropathol.* **21:**287–297.

Koval, M., and Pagano, R. E., 1991, Intracellular transport and metabolism of sphingomyelin, *Biochim. Biophys. Acta* **1082:**113–125.

Kudchodkar, B. J., Albers, J. J., and Bierman, E. L., 1983, Effects of positively charged sphingomyelin liposomes on cholesterol metabolism of cells in culture, *Atherosclerosis* **46:**353–367.

Lange, Y., 1991, Disposition of intracellular cholesterol in human fibroblasts, *J. Lipid Res.* **32:** 329–339.

Lange, Y., Echevarria, F., and Steck, T., 1991, Movement of zymosterol, a precursor of cholesterol, among three membranes in human fibroblasts, *J. Biol. Chem.* **266:**21439–21443.

Lecuyer, H., and Dervichian, D. G., 1969, Structure of aqueous mixtures of lecithin and cholesterol, *J. Mol. Biol.* **45:**39–57.

Lin, D., Sugawara, T., Strauss III., J. F., Clark, B. J., Stocco, D. M., Saenger, P., Rogol, A., Miller, W. L., 1995, Role of steroidogenic acute regulatory protein in adrenal and gonadal steroidogenesis, *Science* **267:**1828–1831.

Liscum, L., and Dahl, N. K., 1992, Intracellular cholesterol transport, *J. Lipid Res.* **33:**1239–1254.

Liscum, L., and Faust, J. R., 1994, Compartmentation of cholesterol within the cell, *Curr. Opin. Lipidol.* **5**:221–226.

Liscum, L., and Underwood, K. W., 1995, Intracellular cholesterol transport and compartmentation, *J. Biol. Chem.* **270**:15443–15446.

Lund-Katz, S., Laboda, H. M., McLean, L. R., and Phillips, M. C., 1988, Influence of molecular packing and phospholipid type on cholesterol exchange, *Biochemistry* **27**:3416–3423.

Mahlberg, F. H., and Rothblat, G. H., 1992, Cellular cholesterol efflux—role of cell membrane kinetic pools and interactions with apoproteins A-I, A-II, and Cs, *J. Biol. Chem.* **267**:4541–4550.

Marggraf, W.-D., and Kanfer, J. N., 1984, *Biochim. Biophys. Acta* **793**:346–353.

Mattjus, P., and Slotte J. P., 1994, Availability for enzyme-catalyzed oxidation of cholesterol in mixed monolayers containing both phosphatidylcholine and sphingomyelin, *Chem. Phys. Lipids* **71**: 73–81.

Mattjus, P., Bittman, R., and Slotte, J. P., 1996, Molecular interaction and lateral domain formation in monolayers containing cholesterol and phosphatidylcholines with acyl- or alkyl-linked C16 chains, *Langmuir* **12**:1284–1290.

McLean, L. R., and Phillips, M. C., 1982, Cholesterol desorption from clusters of phosphatidylcholine and cholesterol in unilamellar vesicle bilayers during lipid transfer or exchange, *Biochemistry* **21**:4053–4059.

Mehrabian, M., Callaway, K. A., Clarke, C. F., Tanaka, R. D., Greenspan, M., Lusis, A. J., Sparkes, R. S., Mohandas, T., Edmond, J., and Edwards, P. A., 1986, *J. Biol. Chem.* **261**:16249–16255.

Mendez, A. J., Oram, J. F., and Bierman, E. L., 1991, Protein kinase C as a mediator of HDL-receptor dependent efflux of intracellular cholesterol, *J. Biol. Chem.* **266**:10104–10111.

Needham, D., and Nunn, D. S., 1990, Elastic deformation and failure of lipid bilayer membranes containing cholesterol, *Biophys. J.* **58**:997–1009.

Okwu, A. K., Xu, X.-X., Shiratori, Y., and Tabas, I., 1994, Regulation of the threshold for lipoprotein-induced ACAT stimulation in macrophages by cellular sphingomyelin content, *J. Lipid Res.* **35**:644–655.

Pagano, R. E., 1990, Lipid traffic in eucaryotic cells: mechanisms for intracellular transport and organelle-specific enrichment of lipids, *Curr. Opin. Cell Biol.* **2**:652–663.

Papahadjopoulos, D., Nir, S., and Ohki, S., 1971, Permeability properties of phospholipid membranes: effect of cholesterol and temperature, *Biochim. Biophys. Acta* **266**:561–583.

Patton, S., 1970, Correlative relationship of cholesterol and sphingomyelin in cell membranes, *J. Theor. Biol.* **29**:489–491.

Phillips, M. C., and Finer, E. G., 1974, The stoichiometry and dynamics of lecithin-cholesterol clusters in bilayer membranes, *Biochim. Biophys. Acta* **356**:199–206.

Pörn, M. I., and Slotte, J. P., 1990, Reversible effects of sphingomyelin degradation on cholesterol distribution and metabolism in fibroblasts and transformed neuroblastoma cells, *Biochem. J.* **271**:121–126.

Pörn, M. I., and Slotte, J. P., 1995, Localization of cholesterol in sphingomyelinase-treated fibroblasts, *Biochem. J.* **308**:269–274.

Pörn, M. I., Åkerman, K. E. O., and Slotte, J. P., 1991a, HDL induce a rapid and transient release of Ca^{2+} in cultured fibroblasts, *Biochem. J.* **279**:29–33.

Pörn, M. I., Tenhunen, J., and Slotte, J. P., 1991b, Increased steroid hormone secretion in mouse Leydig tumor cells after induction of cholesterol translocation by sphingomyelin degradation, *Biochim. Biophys. Acta* **1093**:7–12.

Pörn, M. I., Ares, M., and Slotte, J. P., 1993, Degradation of plasma membrane phosphatidylcholine appears not to affect the cellular cholesterol distribution, *J. Lipid Res.* **34**:1385–1392.

Ridgway, N. D., 1995, 25-Hydroxycholesterol stimulates sphingomyelin synthesis in Chinese hamster ovary cells, *J. Lipid Res.* **36**:1345–1358.

Rothblat, G. H., and Phillips, M. C., 1991, Cholesterol efflux from arterial wall cells, *Curr. Opin. Lipidology.* **2**:288–294.

Rothblat, G. H., Mahlberg, F. H., Johnson, W. J., and Phillips, M. C., 1992, Apolipoproteins, membrane cholesterol domains, and the regulation of cholesterol efflux, *J. Lipid Res.* **33:** 1091–1097.

Shiratori, Y., Okwu, A., and Tabas, I., 1994, Free cholesterol loading of macrophages stimulates phosphatidylcholine biosynthesis and up-regulation of CTP:phosphatidylcholine cytidylyltransferase, *J. Biol. Chem.* **269:**11337–11348.

Slotte, J. P., 1992, Enzyme-catalyzed oxidation of cholesterol in mixed phospholipid monolayers reveals the stoichiometry at which free cholesterol clusters disappear, *Biochemistry* **31:** 5472–5477.

Slotte, J. P., and Bierman, E. L., 1988, Depletion of plasma membrane sphingomyelin rapidly alters the distribution of cholesterol between plasma membranes and intracellular cholesterol pools in cultured fibroblasts, *Biochem. J.* **250:**653–658.

Slotte, J. P., and Lundberg, B., 1983, Effects of cholesterol surface transfer on cholesterol and phosphatidylcholine synthesis in cultured rat arterial smooth muscle cells, *Medical Biology* **61:** 223–227.

Slotte, J. P., Oram, J. F., and Bierman, E. L., 1987, Binding of HDL to cell receptors promotes translocation of cholesterol from intracellular membranes to the cell surface, *J. Biol. Chem.* **262:** 12904–12907.

Slotte, J. P., Chait, A., and Bierman, E. L., 1988, Cholesterol accumulation in aortic smooth muscle cells exposed to low density lipoproteins, *Arteriosclerosis* **8:**750–758.

Slotte, J. P., Hedström, G., Rannstrom, S., and Ekman, S., 1989, Effect of sphingomyelin degradation on cholesterol oxidizability and steady-state distribution between the cell surface and the cell interior, *Biochim. Biophys. Acta* **985:**90–96.

Slotte, J. P., Härmälä, A.-S., Jansson, C., and Pörn, M. I., 1990, Rapid turn-over of plasma membrane sphingomyelin and cholesterol in BHK cells after exposure to sphingomyelinase, *Biochim. Biophys. Acta* **1030:**251–257.

Smaby, J., Brockman, H. L., and Brown, R. B., 1994, Cholesterol's interfacial interaction with sphingomyelins and phosphatidylcholines: hydrocarbon chain structure determine the magnitude of condensation, *Biochemistry* **33:**9135–9142.

Smutzer, G., and Yeagle, P. L., 1985, Phase behavior of DMPC-cholesterol mixtures—a fluorescence anisotropy study, *Biochim. Biophys. Acta* **814:**274–280.

Spence, M. W., Clarke, J. T. R., and Cook, H. W., 1983, Pathways of sphingomyelin metabolism in cultured fibroblasts from normal and sphingomyelin lipidosis subjects, *J. Biol. Chem.* **258:** 8595–8600.

Stein, O., Ben-Naim, M., Dabach, Y., Hollander, G., and Stein, Y., 1992, Modulation of sphingomyelinase-induced cholesterol esterification in fibroblasts, CaCo-2 cells, macrophages and smooth muscle cells, *Biochim. Biophys. Acta* **1126:**291–298.

Stockton, B. W., and Smith, I. C. P., 1976, A deuterium NMR study of the condensing effect of cholesterol on egg phosphatidylcholine bilayer membranes, *Chem. Phys. Lipids* **17:**251–263.

Straume, M., and Litman, B. J., 1987, Influence of cholesterol on equilibrium and dynamic bilayer structure of unsaturated acyl chain phosphatidylcholine vesicles as determined from high order analysis of fluorescence anisotropy decay, *Biochemistry* **26:**5121–5126.

Szabo, G., 1974, Dual mechanism for the action of cholesterol on membrane permeability, *Nature* **252:**47–49.

Tabas, I., Roscoff, W. J., and Boykow, G. C., 1988, ACAT in macrophages utilizes a cellular pool of cholesterol oxidase-accessible cholesterol as substrate, *J. Biol. Chem.* **263:**1266–1272.

Urbani, L., and Simoni, R. D., 1990, Cholesterol and vesicular stomatitis virus G protein take separate routes from the endoplasmic reticulum to the plasma membrane, *J. Biol. Chem.* **265:**1919–1923.

van Echten, G., Iber, H., Stotz, H., Takatsuki, A., and Sandhoff, K., 1990, *Eur. J. Cell Biol.* **51:**135–139.

van Meer, G., 1989, Lipid traffic in animal cells, *Annu. Rev. Cell Biol.* **5:**247–275.

Voelker, D. R., 1991, Organelle biogenesis and intracellular lipid transport in eucaryotes, *Microbiol. Rev.* **55:**543–560.

Voelker, D. R., and Kennedy, E. P., 1982, Cellular and enzymic synthesis of sphingomyelin, *Biochemistry* **21:**2753–2759.

Vos, J. P., Giudici, M. L., van der Bijl, P., Magni, P., Marchesini, S., van Golde, L. M., and Lopes-Cardozo, M., 1995, Sphingomyelin is synthesized at the plasma membrane of oligodendrocytes and by purified myelin membranes: a study with fluorescent- and radio-labelled ceramide analogues, *FEBS. Lett.* **368:**393–396.

Warnock, D. E., Roberts, C., Lutz, M. S., Blackburn, W. A., Young, W. W., and Baenziger, J. U., 1993, Determination of plasma membrane lipid mass and composition in cultured Chinese hamster ovary cells using high gradient magnetic affinity chromatography, *J. Biol. Chem.* **268:** 10145–10153.

Xu, X.-X., and Tabas, I., 1991, Lipoproteins activate ACAT in macrophages only after cellular cholesterol pools are expanded to a critical threshold level, *J. Biol. Chem.* **266:**17040–17045.

Yeagle, P. L., Martin, R. B., Lala, A. K., Lin, H. K., and Bloch, K., 1977, Differential effects of cholesterol and lanosterol on artificial membranes, *Proc. Natl. Acad. Sci. USA* **74:**4924–4926.

Chapter 11

Mechanisms and (Patho)Physiological Significance of Biliary Cholesterol Secretion

Folkert Kuipers, Ronald P. J. Oude Elferink, Henkjan J. Verkade, and Albert K. Groen

1. INTRODUCTION

Fundamental and clinical research on the physiology and metabolism of cholesterol (Ch) is mainly related to the role of this sterol in two disorders that affect millions of people, i. e., atherosclerosis and gallstones. Although the etiologies of these disorders at first sight have very little in common, both are related to the proper functioning of uptake, processing, and disposition from the body of excess Ch by the liver. As nature has not provided the body with an enzymic machinery able to catabolize the Ch molecule, excretory mechanisms are required to remove excess of this ubiquitous sterol. Disposition into bile, followed by fecal loss, represents the only pathway by which quantitatively important amounts of Ch are removed. As such, the hepatobiliary pathway is of crucial importance for the maintenance of Ch homeostasis. Secretion from the liver cell into the bile occurs either in the form

---

Folkert Kuipers and Henkjan J. Verkade Groningen Institute for Drug Studies, Laboratory of Nutrition and Metabolism, Academic Hospital Groningen, Groningen, The Netherlands. **Ronald P. J. Oude Elferink and Albert K. Groen** Department of Gastrointestinal and Liver Diseases, Academic Medical Center, Amsterdam, The Netherlands.

Subcellular Biochemistry, Volume 28: *Cholesterol: Its Functions and Metabolism in Biology and Disease,* edited by Robert Bittman. Plenum Press, New York, 1997.

of free Ch or after its conversion to bile salts (BS). Both routes contribute about equally to daily fecal sterol excretion. Values for fecal neutral sterol loss (including both biliary Ch and Ch from desquamated intestinal cells) in humans on a Western type diet range from 300–1200 mg/day and those for acidic sterols (i.e., BS) range from 200–800 mg/day. Given a dietary Ch intake in the order of 200–600 mg/day, these values imply that at least about half of the daily sterol output must originate from endogenous synthesis. In this chapter we review the current knowledge of mechanisms involved in secretion of Ch into the aqueous environment of hepatic bile and briefly discuss some pathological consequences of disturbances in this process. To put the available data into a comprehensive framework, we begin the chapter with a condensed overview of current concepts of hepatic bile formation. For more details on this subject, see Coleman (1987) and Nathanson and Boyer (1991).

2. BILE FORMATION

2.1. The Liver as a Secretory Epithelium

The liver can be considered as a secretory epithelium and, as such, catalyzes vectorial transport of solutes. This secretory function is confined to the liver parenchymal cells or hepatocytes. These cells are able to transport compounds from their sinusoidal, blood-facing side, to their canalicular pole for secretion into the bile. The canalicular membrane corresponds to the apical pole of a typical epithelial cell and is separated from the basolateral membrane by tight junctions. In contrast to the situation in typical epithelia, the apical membrane of the hepatocyte is present in the form of a belt-like structure surrounding the cell (Figure 1). The canaliculi thus formed between adjacent hepatocytes give rise to a "chicken wire" luminal network that drains into bile ductules that, in turn, combine to form bile ducts. In spite of its limited size, with an estimated contribution of only 10% of total cell surface, the canalicular plasma membrane domain is of crucial importance for bile formation and hepatobiliary secretion processes.

Bile secretion of a given compound may involve a number of steps, i.e., uptake by specialized carrier proteins or by (facilitated) diffusion; intracellular transport; in some cases, phase I and phase II biotransformation reactions; and, finally, secretion across the canalicular membrane into the bile canalicular lumen (Oude Elferink *et al.*, 1995a). Separate ATP-dependent transport systems for BS and for non-BS organic anions (OA) such as bilirubin conjugates (cMOAT = canalicular multispecific organic anion transporter) have been functionally characterized in rat and human liver (Oude Elferink *et al.*, 1995a; Zimniak and Awashti, 1993). So far, only cMOAT has been characterized at the molecular level (Paulusma *et al.*, 1995). Transporters for secretion of reduced gluthathione (GSH) and GSH-complexes have also been identified at the bile canalicular membrane (Dutczak

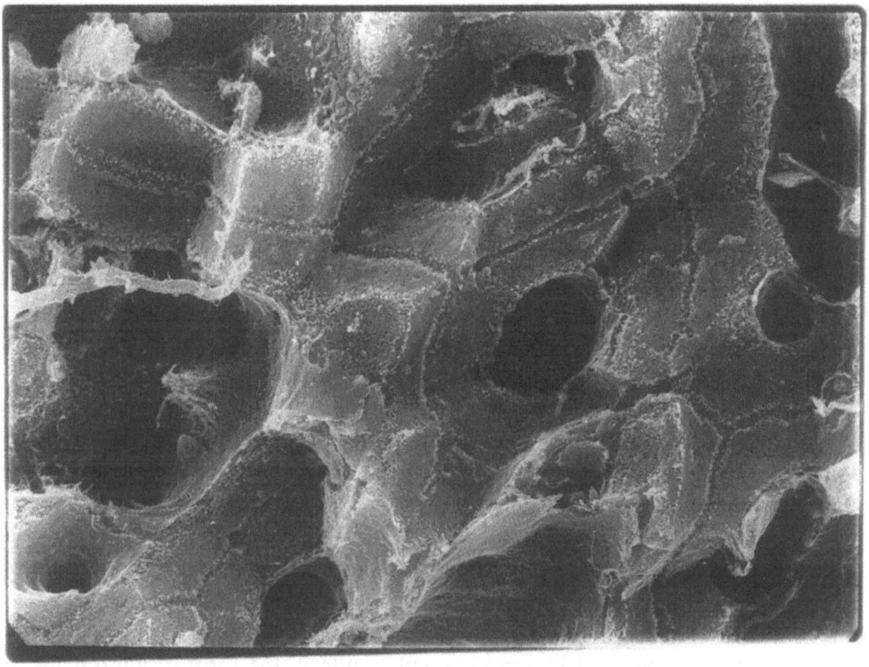

FIGURE 1. Freeze-fracture transmission electron micrograph of rat liver, showing the chicken-wire network of bile canaliculi. Magnification 2300 x. (The photograph was obtained from Dr. G.M.M. Groothuis, University Center for Pharmacy, Groningen, The Netherlands.)

and Ballatori, 1994; Fernandez-Checa *et al.,* 1993). The murine mdr1a/mdr1b P-glycoproteins (Pgp) and their human homologue MDR1 Pgp, members of the superfamily of A(TP)B(inding)C(assette)-containing transport ATPases, are involved in disposition of organic cations (Oude Elferink *et al.,* 1995a). Surprisingly, the closely related mdr2 (MDR3) Pgp was recently found to play an important role in phospholipid (PL) and Ch secretion (Smit *et al.,* 1993; Oude Elferink *et al.,* 1995b). Identification and characterization of these transporters have led to important new viewpoints in the past few years; in particular, the identification of mdr2 Pgp as a canalicular protein involved in lipid transport has changed our way of thinking about hepatobiliary lipid secretion, as discussed in section 4.

2.2. Bile Composition

Bile salts are the major organic constituents of bile, with a contribution of up to 60–70 mol% in humans and most animal species. Biliary BS consist of a mixture of primary and secondary species, in humans conjugated to either taurine or glycine. In addition to BS, bile contains millimolar concentrations of other water-soluble compounds, such as dipeptides and amino acids. Equally high concentrations of the water-insoluble compounds Ch and PLs are present in bile in micellar (with BS) or vesicular form. Bile also contains a variety of proteins, which are derived from blood (e.g., IgA and albumin), intracellular compartments (e.g., lysosomal enzymes), and the canalicular membrane (e.g., aminopeptidase N). Finally, a host of other endogenous compounds, such as bilirubin conjugates, protoporphyrins, vitamins, steroid hormone metabolites, and xenobiotics, are present in bile.

With respect to the composition of biliary lipids, two important features should be emphasized. Firstly, Ch is exclusively present in bile in its unesterified form. Secondly, biliary PL has a specific composition that is very different from that of the bile canalicular membrane and, in fact, of any intracellular membranous compartment in the liver cell (Hardison and Evans, 1986). Bile PL consists predominantly of phosphatidylcholine (PC) of a specific, relatively hydrophilic fatty acid composition (Hay *et al.*, 1993). The *sn*-1 position of biliary PC usually is occupied by palmitate (16:0), whereas the *sn*-2 position almost invariably contains an unsaturated species, predominantly oleate (18:1) or linoleate (18:2). Only minor amounts of sphingomyelin and phosphatidylethanolamine, both major constituents of the canalicular hepatocytic plasma membrane, are found in bile. Thus, biliary PL secretion does not occur in a random fashion but rather involves an advanced level of "sorting."

Ch and (bile-specific) PLs are present in bile in the form of aggregates, of which size, shape, and composition are primarily determined by their concentrations relative to those of BS, as well as by the physicochemical characteristics of PL and BS present. A large body of evidence is available to indicate that simple micelles (consisting of BS alone or Ch and BS), mixed micelles (containing Ch, PL, and BS), and unilamellar vesicles (composed of Ch, PL, and trace amounts of BS) can coexist as biliary lipid carriers in a dynamic, i.e., interchangeable, fashion. In addition, multilamellar vesicles (liquid crystals) may be present in supersaturated biles (see also section 4).

2.3. Bile Formation Is an Osmotic Process

The liver produces substantial amounts of bile, i.e., in the order of 0.5–1 liters/day in humans (Erlinger, 1982) and about 40 ml/day in rats (Kuipers *et al.*, 1985). In humans and most animals, primary or "hepatic" bile formed at the level of the canaliculi is modified during passage through the ductules, which are lined by a separate type of epithelial cells, and concentrated in the gallbladder before it is expelled into the intestine in response to a meal. Some animal species, like the rat, do not have a gallbladder; in this case only ductular modification takes place. The generation of bile flow is an osmotically driven process. Secretion of osmotically active substances provides an osmotic gradient between the lumen of the canaliculus and plasma, giving rise to flow of water and small ions through the tight junctions. It was established more than 50 years ago that BS provide the main driving force for bile formation, the so-called BS-dependent fraction of bile flow (Erlinger, 1982; Nathanson and Boyer, 1993). In rodents, GSH secretion appears to be the major determinant of the remaining BS-independent fraction (Ballatori and Truong, 1989), which, in principle, combines the actions of all non-BS compounds in the bile. In humans, there appears to be very little BS-independent flow.

2.4. Enterohepatic Circulation/Reabsorption of Biliary Lipids

Although often considered as an exclusively "excretory pathway," which may be true for certain hormone and drug metabolites, most biliary constituents are in fact (partly) reabsorbed from the intestine to be reutilized by the body. Bile salts (BS), in particular, are very effectively conserved in the so-called enterohepatic circulation (Hofmann, 1984). This circulation enables the body to maintain effectively high concentrations of these valuable detergents at the sites of their physiological actions, i.e., liver, biliary tree, and intestinal lumen. The obligatory "leak" of BS from this circulation into the feces serves as an important route for Ch removal from the body.

Robins (1975) showed that biliary PC also is very efficiently absorbed from the intestine in rats, but, in contrast to BS, these molecules are extensively hydrolyzed in the intestinal lumen and remodeled after their absorption. In fact, only the choline moiety was found to be reutilized to a significant but small extent for biliary PC synthesis and secretion. In accordance with this observation, we found that bile-type PC, rapidly introduced into the liver of bile-diverted rats by means of a liposomal carrier, was only minimally secreted into the bile in intact form, but was rather extensively metabolized (Verkade *et al.,* 1991).

Because of its potential impact on Ch homeostasis, reabsorption of biliary Ch has been studied in some detail, both in humans and in experimental animals. In this respect, it is of importance to note that at least 60–70% of the Ch present in the intestinal lumen of humans is of endogenous origin (bile, desquamated enterocytes) and that only a relatively small fraction comes from dietary sources. Consequently, in order to understand the relationship between Ch absorption and plasma Ch levels, it is essential to understand the handling of Ch from the different sources. A number of studies using intestinal perfusion techniques in humans (Simmonds *et al.,* 1967; Samuel and McNamara, 1983) have suggested that biliary Ch actually is absorbed more efficiently than dietary Ch in the upper intestine. This effect has been ascribed to the fact that biliary Ch is delivered to the gut in a form readily available for absorption, i.e., in BS/PL micelles. Dietary Ch, on the other hand, must first be transferred to the micellar form from its oily phase (with dietary triglycerides) before absorption can take place. However, as reviewed extensively by Wilson and Rudel (1994), it appears from *in vivo* studies in experimental animals that, while the absorption of biliary Ch is potentially greater, the time available for fat/Ch digestion, emulsification, and micellization in the intestinal lumen is sufficient to obtain essentially complete mixing of Ch from endogenous and exogenous sources. As a result, the percentage absorption of Ch from both sources should be equal when the entire intestinal tract is taken into consideration. However, as the amount of endogenous Ch typically is much larger than the dietary load, its importance in the maintenance of Ch homeostasis is evident and should be taken into account when evaluating the significance of Ch absorption in development of hypercholesterolemia.

3. THE ORIGIN OF BILIARY CHOLESTEROL

Considerable efforts have been made to define the origin of biliary Ch. An important issue in this respect concerns the extent to which *de novo* synthesis contributes to biliary Ch secretion. Animal studies employing tritiated water as a precursor of Ch yielded values on the order of 7–16% (Turley and Dietschy, 1981; Robins and Brunengraber, 1982; Schreibner *et al.*, 1993) under control conditions in rats and hamsters. Based on the observation that the specific activity of biliary Ch is considerably higher than that of hepatic Ch in this type of experiments, it has been suggested that newly synthesized Ch is preferentially secreted into bile. Robins *et al.* (1993) provided evidence that, in fact, the difference in specific activities may reflect secretion of (*de novo* formed) Ch from a specific subpopulation of hepatocytes. It is known that Ch biosynthesis predominantly occurs in periportal hepatocytes representing only 25–30% of total cell mass (Singer *et al.*, 1984). The same population of cells transports the majority of BS during their enterohepatic cycling (Groothuis *et al.*, 1982). Since BS stimulate biliary Ch secretion (see section 4), the flux of BS through Ch-synthesizing cells may contribute to a relatively high proportion of newly formed Ch molecules in bile in comparison to total hepatic Ch.

It should be realized that the collection of bile required for execution of this type of experiment may significantly affect the results, since interruption of the enterohepatic circulation of BS markedly affects on hepatic Ch metabolism. Among other effects, Ch synthesis rapidly increases under these conditions while, at the same time, the flux of BS decreases (Smit *et al.*, 1990). Therefore, we have recently performed studies to quantify Ch biosynthesis and secretion of newly synthesized Ch into bile by making use of a rat model that allows collection of small bile samples while the enterohepatic circulation remains intact (Kuipers *et al.*, 1985). In these studies (Bandsma *et al.*, unpublished results), we used ^{13}C-labeled acetate combined with the recently developed method of mass isotopomer distribution analysis (Neese *et al.*, 1993) to quantify Ch biosynthesis. We found that 8.0 $\pm$ 1.5% of biliary Ch is derived from *de novo* synthesis during the light period in chow-fed rats. The contribution of newly synthesized Ch increases to 18.9 $\pm$ 4.2% in chronically bile-diverted rats, in which Ch biosynthesis is strongly upregulated (Smit *et al.*, 1990). Thus, even when hepatic Ch biosynthesis is high, the majority of biliary Ch still is derived from "preformed" pools, possibly including Ch of hepatocytic membranes, Ch released from cholesteryl ester (ChE) pools, and Ch/ChE entering the liver after uptake of plasma lipoproteins.

Numerous studies in animals and humans, extensively reviewed by Botham and Bravo (1995), have addressed the issue to what extent Ch and ChE from various lipoprotein classes contribute to biliary Ch secretion. Although interpretation of these studies, usually making use of radiolabeled Ch and/or ChE, is hampered by the extensive exchange and equilibration of these molecules, available data suggest that lipoprotein Ch(E) replenishes a single hepatic Ch pool that turns over rapidly and comprises the major source of biliary Ch (and BS). This pool appears to exchange mainly with plasma high density lipoprotein (HDL) (Schwartz *et al.*, 1993). Preferential transfer of HDL-Ch to bile is supported by recent data showing rapid secretion of HDL-associated sitostanol, a plant sterol that can be used to trace free Ch (Robins *et al.*, 1995a), into bile in isolated perfused rat livers (Robins *et al.*, 1995b). There are data to suggest that free Ch associated with HDL may be transported directly from blood through the plasma membrane of the hepatocyte to the canalicular membrane without first entering the interior of the cell and mixing with newly synthesized Ch (Robins *et al.*, 1985). It is very tempting to speculate that such "express" handling of HDL-Ch is related to the phenomenon known as selective uptake of HDL-Ch(E), first described by Pittman *et al.* (1987). In this process, lipid-laden HDL particles associate with as yet undefined hepatic receptor systems to deliver their lipid content without being internalized. Direct and rapid disposition of HDL-Ch would match perfectly with the function of this lipoprotein class in reverse-Ch transport, i.e., removal of excess Ch from the periphery to the liver for disposition from the body.

4. REGULATION OF BILIARY CHOLESTEROL SECRETION

As outlined previously, it is of great clinical interest to define the factors that determine the quantity of biliary Ch secretion, both in absolute terms as well as relative to BS and PL. Data published during the past three decades have unequivocally established that secretion of Ch and PL are coupled; in general, fluctuations in PL secretion result in parallel changes in Ch secretion. It is also evident that the secretion of both lipids is, in part, governed by BS. Recent studies by our groups have generated a novel concept of how and where in the sequence of events leading to secretion of Ch these PL and BS exert their actions. This concept will be described in the next section and related to existing models of biliary lipid secretion. It is also well known that under certain experimental conditions deviations in the close coupling between Ch on the one hand and PL and BS on the other may occur. Potential mechanisms of this "uncoupling" will be discussed shortly.

4.1. A New Concept for the Bile Salt-Dependent Regulation of Biliary Cholesterol and Phospholipid Secretion

Table I summarizes the rates of biliary lipid secretion and ratios between the different biliary components in bile of diverse species. Clearly, there is an enormous species-to-species variation in biliary outputs of BS, PL, and Ch. With respect to BS, it is important to note that the type of BS may vary: Bile of some species is very rich in hydrophobic BS, whereas mice and pigs, for instance, have very hydrophilic BS in their biles. The reason for this variability is not clear at present but probably relates to the species-specific expression of the various enzymes involved in BS biosynthesis. Apparently, evolutionary selection pressure is not very high for BS composition. The difference in BS hydrophobicity can explain in part the variation in output of PL and Ch. Many studies have demonstrated a direct positive relation between BS hydrophobicity and lipid secretion (Gurantz and Hofmann, 1984; Nakagaki *et al.*, 1984; Kitani *et al.*, 1985; Rahman and Coleman, 1987; Cohen *et al.*, 1992; Coleman and Rahman, 1992). Inspection of Table I reveals, however, that BS hydrophobicity is not the sole factor controlling lipid secretion. For instance, guinea pigs and rabbits have a very hydrophobic BS composition but secrete very little lipid into their bile. Clearly, other factors besides BS hydrophobicity must be involved as well. Until recently it was assumed that the rate of PC synthesis determines maximal lipid output. This assumption was mainly based on the observation that feeding rats a choline-deficient diet greatly reduced PL secretion (Robins and Armstrong, 1976). A concomitant de-

Table I
Bile Composition in Various Mammalian Species[a]

Species	BS	PL	Ch	BS/PL	PL/Ch	H.I.	Bile Flow
Rat[b]	37	7	0.6	5.3	12	−0.31	65
Mouse[b]	83	22	4	3.8	5.5	−0.45	70
Hamster[c]	105	22	9	4.8	2.4	0.22	50
Rabbit[d]	35	0.07	0.04	500	2	0.52	85
Dog[e]	42	27	1	1.6	27	0.14	6
Guinea Pig[f]	10	0.1	0.004	100	25	0.52	160
Chicken[g]	35	6.3	0.33	5.6	19	0.36	20
Pig[h]	190	26	2	7.3	13	0.05	9
Ox[h]	134	14	0.6	9.6	23	0.17	nd
Prairie Dog[i]	82	18	3.3	4.6	5.5	0.09	10
Monkey[j]	152	46	20	3.3	2.3	0.26	10
Human[k]	150–220	46–58	5.3–20	3.3–3.8	2.3–11	0.32	6

[a] Representative values were taken from the literature. BS, PL, and Ch concentrations are given in mmol/l. Bile flow is given as ml/kg/day. nd = not determined. The hydrophobicity index (H.I.) of biliary BS was calculated according to Heuman, 1989. References: [b] Caldwell *et al.*, 1965; [c] Cohen *et al.*, 1993; [d] Coleman *et al.*, 1979; [e] Klaassen and Watkins, 1984; [f] Holzbach *et al.*, 1971; [g] Clausen *et al.*, 1965; [h] Nakayama, 1969; [i] Brenneman *et al.*, 1972; [j] Osuga and Portman, 1972; [k] Nakayama and VanderLinden, 1971.

crease in Ch secretion was observed in line with the close coupling between PL and Ch output seen in general (Robins and Armstrong, 1976).

Mazer and Carey (1984) presented a mathematical model of biliary lipid secretion that incorporated these hypotheses. Their model accurately predicted the hyperbolic relationships between BS output on the one hand and PL and Ch secretion on the other. In this model the maximal rate of PL output is fully determined by the synthesizing capacity, which is in line with the decreased PL secretion observed during choline-deficiency (Robins and Armstrong, 1976). A coupling constant was introduced to explain the coupling between Ch and PL output. At the time the model was constructed, very little was known about mechanisms underlying the different processes, and therefore these constants lacked a solid mechanistic basis. This is probably the reason that, despite its predictive success, the model of Mazer and Carey did not receive the full appreciation it deserved.

Theories on the mechanism of biliary lipid secretion changed after the discovery of the presence of Ch/PL vesicles in bile (Somjen and Gilat, 1983; Somjen and Gilat, 1985; Pattinson, 1985; Ulloa *et al.,* 1987; Lee *et al.,* 1987). Small unilamellar vesicles with diameters ranging from 60–100 nm were first demonstrated to exist in gallbladder bile but soon afterwards their presence was also confirmed in hepatic bile (Cohen *et al.,* 1989). It was hypothesized that these vesicles could be the means by which Ch and PL are secreted by the hepatocytes. Indeed, Ulloa *et al.* (1987) presented electron microscopic evidence for their presence in the canalicular space between two hepatocytes and Cohen *et al.* (1989) used glass-fiber-mediated quasi-elastic light scattering to demonstrate the presence of vesicles in bile from prairie dogs with a bile fistula. Their approach also generated dynamic information, because soon after secretion the vesicles disappeared due to solubilization by BS micelles (Cohen *et al.,* 1989).

Coleman (1987) synthesized these findings into a new model of biliary lipid secretion. In this model BS monomers or micelles induce vesiculization of the canalicular membrane. To account for the specificity of PL species it was assumed that intracellular PL is transported to specialized regions of the canaliculus by means of intracellular tranport vesicles. These vesicles subsequently fuse with the membrane and generate microdomains specifically prone to undergo "budding" induced by BS. The model did not explain the mechanism by which the BS control lipid secretion.

This point has been a matter of debate for many years (see Verkade *et al.,* 1995, for review). The controversy has centered around the question of whether BS regulate lipid secretion via a "push" or "pull" mechanism. Because increased BS-driven biliary lipid secretion does not lead to appreciable changes in the intracellular lipid pools, both transport and synthesis of the lipid must be strictly controlled. The push mechanism implies that BS driven transport of intracellular, bile-destined lipid to the canalicular membrane is rate-controlling. In this case, canalicular BS have only a permissive role in lipid solubilization. In the pull scenario, on the other hand, the major controlling step is the BS-membrane interaction in the canaliculus (Coleman, 1987; Verkade *et al.,* 1995; Coleman and Rahman, 1992; Coleman *et al.,* 1989).

Convincing evidence for the latter mechanism comes from studies with cholephilic organic anions. Hydrophilic organic anions such as ampicillin and dibromosulphthalein inhibit secretion of both PL and Ch (Apstein, 1984; Apstein and Russo, 1985; Tazuma *et al.*, 1988; Verkade *et al.*, 1990; Yamashita *et al.*, 1992; Verkade *et al.*, 1993a; Verkade *et al.*, 1993b; Yamashita *et al.*, 1993). This inhibition requires the presence of the organic anions in the canalicular space; in mutant GY/TR$^-$ rats, which have a defect in biliary organic anion secretion, the inhibition does not occur (Verkade *et al.*, 1990; Verkade *et al.*, 1993a; Yamashita *et al.*, 1993). Interestingly, only organic anions showing affinity for BS micelles inhibit lipid secretion (Verkade *et al.*, 1995). Organic anions with affinity for the vesicular phase in bile have no effect, suggesting that partitioning of a molecule in the membrane does not influence lipid secretion. The picture that emerges from these studies is that hydrophilic organic anions occupy sites in BS micelles that play a role in the interaction of the micelles with the membrane (Verkade *et al.*, 1995).

Using a model system of sonicated bile type unilamellar Ch/PL vesicles, we investigated the effect of OA on BS-induced membrane damage. Surprisingly, hydrophilic OA did not affect the interaction of BS with vesicles. The conclusion must therefore be that the interaction of BS with the canalicular membrane is more complicated than a system of direct micellization either from the membrane or from specialized microdomains in it. In principle, the composition of the model vesicles should mirror the composition of the hypothetical specialized domains. Apparently, little direct micellization at these sites occurs or some other mechanism is operational.

Studies with a transgenic mouse model have recently shed new light on these phenomena and indicated that substantial refinements of existing models of biliary lipid secretion are required. By serendipity, we found that mice with a disrupted gene encoding mdr2 P-glycoprotein showed a defect in biliary lipid secretion (Smit *et al.*, 1993). The mdr2 Pgp is a member of the multidrug-resistance-related P-glycoprotein family. It shares 73% homology to mdr 1, the Pgp responsible for multidrug resistance in many tumor cells. Homozygous (mdr2 −/−) mice secrete almost no PL and Ch in their bile, indicating a role for mdr2 Pgp in the export machinery. Heterozygous (+/−) mice show reduced PL output but a normal Ch secretion (Smit *et al.*, 1993). This led to the conclusion that the mdr2 Pgp is involved in transmembrane PL transport. Subsequently, direct evidence for this assumption was provided in two studies. Ruetz and Gros (1994) showed that NBD-labeled PC was flipped from the inner to the outer leaflet of the membrane of vesicles isolated from a Sec mutant yeast strain transfected with mdr2. The extent of the effect, however, was rather limited. Recently, these results were confirmed for isolated rat liver canalicular membrane vesicles (Nies *et al.*, 1995). Smith *et al.* (1994) used fibroblasts isolated from a mouse in which the human MDR 3, the homologue of mouse mdr 2, was overexpressed. These investigators used radiolabeled PC as a probe and found that exchange of PC to acceptor vesicles was stimulated in cells overexpressing MDR 3. Whether Ch also flips in a protein-dependent way is not clear at present (see Chapter 9 by Johnson *et al.*).

The presence of a flippase has, of course, direct consequences for the concept of the lipid secretion process. Figure 2 shows a schematic representation of the processes involved in biliary lipid secretion. There is no obvious role for a flippase in a model where both leaflets of the membrane are secreted simultaneously, as proposed by Coleman. Since PL output is almost completely abrogated in mdr2 (−/−) mice, one has to conclude that lipid secretion can only proceed from the outer leaflet of the membrane. The simplest mechanism would be direct micellization of PL and Ch in mixed micelles. However, such a scenario does not explain the well-established presence of biliary vesicles. The most convincing evidence for lipid secretion in the form of vesicles came from very recent morphologic studies of Crawford *et al.* (1995), who used an ultrarapid slam-freeze fixation technique to demonstrate actual vesiculisation of the canalicular membrane. Budding of the membrane was not observed but, in accordance with activity of a flippase, the vesicle seems to form on the outer leaflet and the membrane seems thickened at the contact site. Whether the vesicles are indeed formed by direct vesiculation of the outer leaflet or from a supersaturated mixed micellar phase close to the membrane is not clear.

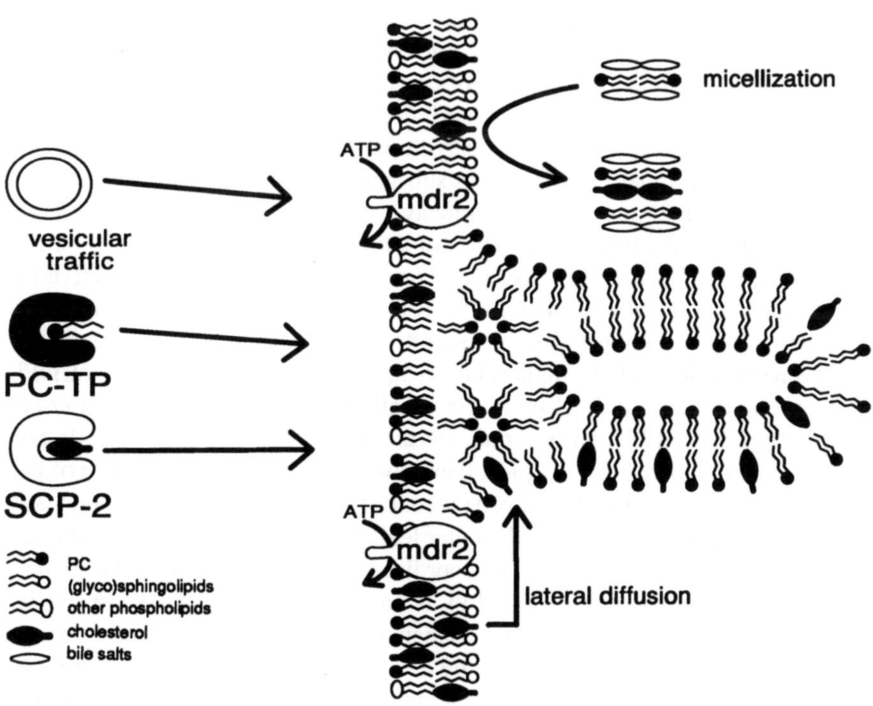

FIGURE 2. Schematic representation of the processes involved in hepatobiliary lipid secretion. See section 4 for details. PC-TP, phosphatidylcholine-transfer protein; SCP-2, sterol carrier protein-2.

Crawford *et al.* (1995) also applied the same technique on the mdr2 mouse model. As expected, a greatly reduced number of vesicular structures was observed in (−/−) mice compared with (+/+) animals. Interestingly, in (+/−) mice the amount of vesicles observed was about half of that in (+/+) controls. The speculative model (Figure 2) accounts for these observations. We speculate that mdr2 Pgp activity creates unstable domains in the membrane by establishing an excess of PL in the outer leaflet. This, either directly or indirectly via the influence of BS, creates inverted micelles that donate PL to a growing bilayer structure. Two important questions arise: (i) is this the sole pathway by which biliary lipid is secreted? and (ii) where does Ch enter the process? Whether only vesiculation occurs or whether direct micellization also occurs probably depends on the type of BS present in bile. Particularly hydrophobic BS rapidly solubilize membrane PL and Ch to form mixed micelles. The relative contribution of vesicular secretion and direct micellization may vary for different types of BS (and thus for different animal species; see Table I), micellization becoming more important with more hydrophobic BS. This may explain why Crawford *et al.* (1995) found no relation between the amount of vesicular structures at the canalicular membrane and the hydrophobicity index of BS, whereas hydrophobic BS induced considerably more PC output per mole of BS compared with hydrophilic BS.

Evidence for direct micellization of Ch alone came from studies with mdr2 (−/−) mice. As described above, these mice do not secrete Ch under normal conditions. When the endogenous BS pool consisting mainly of muricholate is replaced by feeding the mice cholate, Ch is indeed secreted into bile (Oude Elferink *et al.*, 1995b). Likewise, infusion of the hydrophobic BS deoxycholate restored Ch output to almost normal levels without substantial induction of PL output, suggesting that the direct micellization pathway can proceed at a relatively high rate (Oude Elferink *et al.*, 1995b). Secretion of Ch in the absence of PL does not, of course, prove that such a pathway is also operative in the presence of active PL output. As shown in Table I, the biliary Ch/PL ratio varies strongly from species to species; cats and dogs, for example, secrete very little Ch into their biles. Moreover, even in species with the highest biliary lipid output, the biliary ratio of Ch/PL is lower than the ratio prevailing in the canalicular membrane (Barnwell *et al.*, 1987; Yousef *et al.*, 1987; Evans *et al.*, 1976; Graham *et al.*, 1989). Thus, despite the strict coupling between the lipids, there always is preferential secretion of PL.

How can this apparent paradox be reconciled? The simplest explanation is to assume lateral diffusion of Ch in the outer leaflet into the PL-rich microdomains, as depicted in Figure 2. This ensures coordinate secretion of both lipids. In most species the coupling between Ch and PL is not constant. Interestingly, in species with a relatively high Ch output, such as man, the ratio of Ch/PL decreases at high output rates, indicating rate limitation by a step involved in transport of Ch to the outer leaflet of the canalicular membrane. In contrast, in species with low Ch output, such as mice, the Ch/PL ratio increases at high rates, indicating that in these species a step in PL transport (e.g., mdr2 Pgp) becomes rate limiting. When in man the rate of Ch diffusion into the PL-rich microdomains controls Ch output, a step involved in this process must become limiting at high flux. More than 70% of the pool of free Ch resides in the plasma membrane. Lateral diffusion of Ch has been shown to be very fast. The local Ch pool in the canalicular membrane can be replenished from the bulk pool in the basolateral domains of the membrane. There is, however, one obstacle for this pathway. In the outer leaflet lateral lipid diffusion is blocked at the tight junctions, sealing the canaliculus from the blood compartment. Hence, flux to the canalicular domain must proceed via the inner leaflet, and rapid flipping to the outer leaflet is required. Whether active or passive, this step is a good candidate to become limiting at high flux. This scenario fully explains the coupling between Ch and PL output.

Feeding animals BS such as cholate increases Ch and PL secretion. Mdr2 Pgp almost fully controls the rate of PL output under control conditions in mice. Therefore, one almost has to assume that dietary cholate will upregulate the flippase. Frijters *et al.* (1995) demonstrated that this is indeed the case. The mRNA encoding mdr2 Pgp increased by about 50% when normal mice were fed a diet supplemented with 0.1% cholate. Such a mechanism makes physiological sense. The endogenous BS pool of mice is very hydrophilic. About 60% is tauromuricholate and the remainder is taurocholate. Cholate feeding changes the BS pool fully to taurocholate, making this pool considerably more hydrophobic. Since biliary PL may serve to neutralize BS toxicity, enhanced PL secretion may serve to protect the ductular epithelial cell lining and/or the lining of the gallbladder. In line with these considerations, dietary ursodeoxycholate, a very hydrophilic BS, does not increase mdr2 Pgp expression. Interestingly, the coupling between Ch and PL secretion remains intact under the different dietary regimens mentioned.

4.2. Experimental Models of Cholesterol Hypersecretion; How Does This Fit with the Concept?

In addition to the marked species differences in lipid to BS ratios summarized in Table I, it is possible to dissociate Ch output from that of PL and/or BS within a single species by experimental means. Most studies reporting such effects have been performed with rats; consequently, all studies referred to in this section concern studies with this animal species. A first interesting phenomenon worth mentioning concerns the hypersecretion of both Ch and PL relative to BS that has been reported in mutant GY/TR$^-$ rats (Verkade *et al.,* 1993b). These rats express deficient organic anion transport activity due to a point deletion in the cMOAT gene (Paulusma *et al.,* 1995). The BS secretion rate, BS pool composition, and hepatic BS synthesis rates in these animals are indistinguishable from that in control Wistar (NW) rats. In spite of this, bile flow is significantly reduced in the mutant animals, caused by a 47% reduction of the BS-independent flow. Low bile flow in the mutants is probably related to their inability to secrete reduced glutathione into bile. We found that the concentrations of both Ch and PL are markedly higher in

GY/TR⁻ rats than in control animals over a wide range of biliary BS concentrations (Figure 3). We have attributed this effect to the prolonged exposure of the GY/TR⁻ bile canalicular membrane to micellar BS concentrations (exposure time theory), as there were also no differences found in canalicular membrane composition between the strains. In our model (Figure 2), this would imply either more rapid vesiculization or a larger contribution of direct micellization to the secretion process. The first possibility should be associated with increased mdr2 Pgp activity; this issue is currently under investigation.

In addition to "canalicular events" related to mdr2 Pgp activity and BS concentration, there are indications that certain intrahepatic factors may coregulate biliary Ch disposition. Thus, a number of reports have appeared in the literature that show isolated hypersecretion of Ch relative to both PL and BS under different experimental conditions. A marked, isolated increase in biliary Ch secretion resulted from treatment with orotic acid (Stone and Evans, 1992), progesterone (Nervi *et al.*, 1983), or 25-hydroxycholesterol (Stone *et al.*, 1985); cessation of lovastatin administration (Bilharz *et al.*, 1989); and feeding of diets supplemented with diosgenin (Thewles *et al.*, 1993; Holland *et al.*, 1993), niacin (Holland *et al.*, 1993), or fish oil (Smit *et al.*, 1991; 1994). Feeding rats a Ch-enriched diet had the opposite effect (Smit *et al.*, 1993). These effects have been ascribed to alterations in the so-called "recruitable" pool of Ch, caused by changes in Ch synthesis, Ch esterification by affecting ACAT activity, and/or VLDL production. It is of interest to note that for diosgenin (Thewles *et al.*, 1993) and fish oil (Smit *et al.*, 1994), it has been demonstrated that under the condition of hypersecretion, secretion of Ch still is controlled by BS. Likewise, impaired Ch secretion in Ch-fed rats still closely follows changes in BS output (Smit *et al.*, 1993). In the model shown in Figure 2, this would mean that the entry of Ch into vesiculating domains should be increased or decreased, respectively, under the conditions mentioned. Alternatively, more or less Ch may be available for direct micellization under the conditions mentioned. In this respect it is of importance to note that BS composition is more hydrophobic in rats than in mice (Table I) and that BS composition is readily changed under different dietary conditions in rats. We found a small increase in "steady state" Ch content of total hepatic plasma membranes of fish oil-fed rats (Smit *et al.*, 1991), but to our knowledge no data are available on Ch content of canalicular membranes under conditions mentioned above. It should be realized that completely pure canalicular membranes are impossible to obtain from rat liver; consequently, contamination with intracellular membranes will always hamper the interpretation of results obtained from such studies.

5. PATHOLOGY OF BILIARY CHOLESTEROL SECRETION

The disorder most prominently related to biliary Ch secretion is, of course, gallstone disease. The etiology of this disease in connection with Ch hyper-

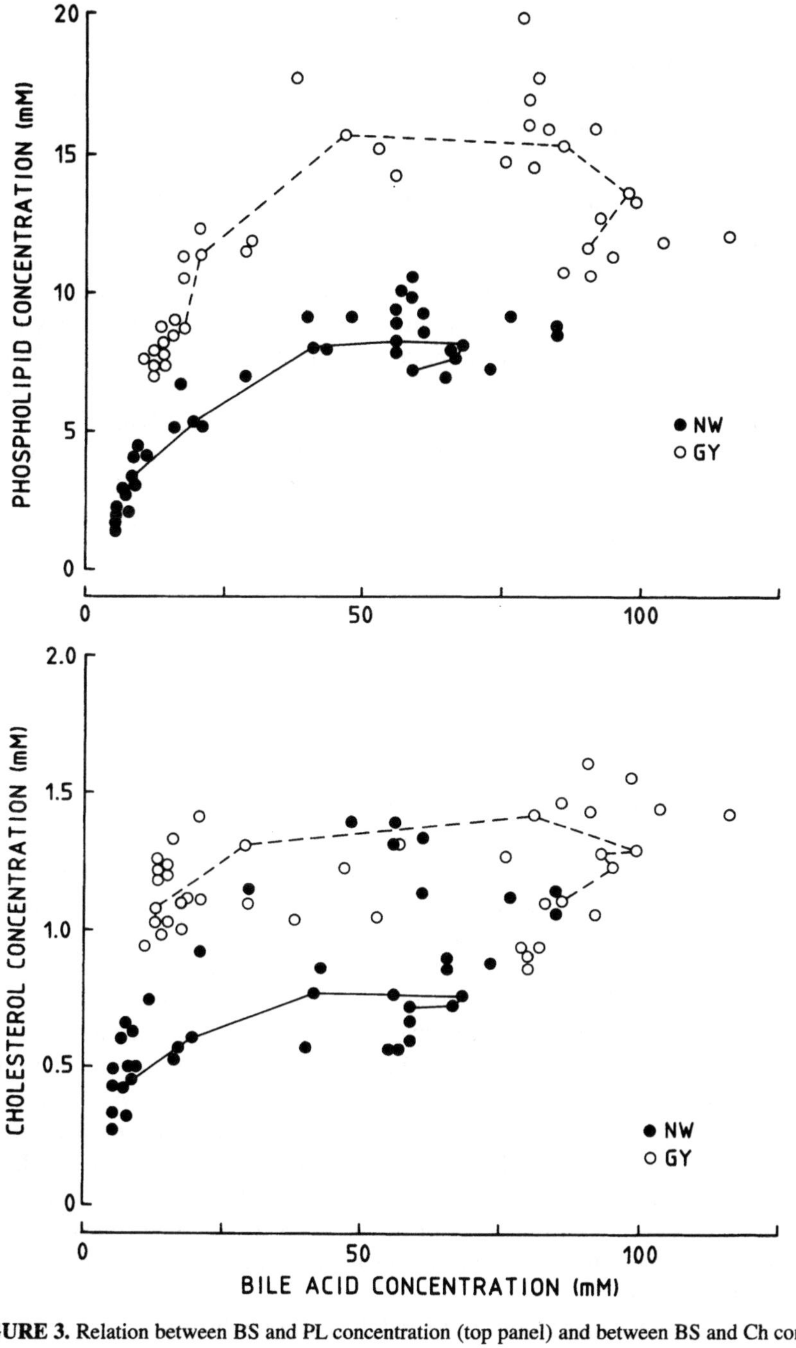

FIGURE 3. Relation between BS and PL concentration (top panel) and between BS and Ch concentration (bottom panel) during the first 5 h after interruption of the enterohepatic circulation in unanaesthetized control (NW) and GY/TR$^-$ rats ($n = 5$ in both groups). Rats were equipped with permanent catheters in the bile duct and duodenum as described by Kuipers *et al.* (1985). At time zero, catheters were disconnected and bile was collected in 30-min fractions. The secretion pattern of one characteristic animal is depicted as a solid (NW) or dotted (GY) line. During the bile collection period, the BS concentration in bile decreases, i.e., the secretion pattern moves from right to left in the figure. Reproduced from Verkade *et al.* (1993b) with permission.

secretion will be discussed shortly. In addition, Ch hyposecretion may have some important pathological implications, both for the regulation of plasma lipid and in the etiology of certain forms of liver disease.

5.1. Cholesterol Hypersecretion in Humans: Relation to Gallstone Disease

Hypersecretion of Ch into bile is a very common condition in humans (Holzbach *et al.*, 1973; Sedaghat and Grundy, 1980). Ch supersaturated bile can crystallize the excess Ch and thereby initiate the formation of Ch gallstones, a disease that affects up to 10% of the population in the Western hemisphere. To what extent human bile is supersaturated with Ch has been a matter of controversy for many years. Ch solubilization in bile is a complex process involving BS, PL, and water. In the 1970s a number of groups constructed phase diagrams to describe the interactions of BS, PL, Ch, and water (Admirand and Small, 1968; Carey and Small, 1978; Carey, 1978; Hegardt and Dam, 1971). These studies were carried out in model systems, in most cases using the BS taurocholate and egg yolk PC in addition to Ch. Carey showed that the boundary of Ch solubility could be adequately fitted with a fifth-power polynomial (Carey, 1978). The critical tables of Carey, published in 1978, are used to date to calculate the degree of supersaturation of human bile. There is evidence that differences in the BS mixture and the different types of PL that may be present in human bile can give rise to erroneous values for the Ch saturation index (CSI) when the tables of Carey are used. Some groups demonstrated crystallization of Ch in biles that were undersaturated according to these tables (Whiting and Watts, 1985; Gallinger *et al.*, 1985; Gollish *et al.*, 1983).

The major problem, however, with application of this "equilibrium approach" is the fact that human bile is a non-equilibrium system. Bile secretion is a dynamic process, and since crystallization of Ch is a rather slow event *in vivo,* the kinetics of the process, rather than the thermodynamics, determine the outcome. Inspection of the interactions between the biliary lipids that take place during passage through the biliary tree is revealing in this respect. In section 4 we have seen that Ch and PL are secreted by the hepatocyte mainly in vesicular form, whereas BS monomers are transported separately by a specific ATP-dependent transporter. In the canalicular space the BS aggregate and form micelles. Once formed, the micelles will attack the vesicles and solubilize their content. The velocity of this process under the conditions prevailing in the canalicular space is not clear. It is, however, on a timescale of minutes rather than of seconds, since the presence of the vesicles has been documented in freshly secreted hepatic bile (Cohen *et al.*, 1989).

The initial vesicular Ch/PL ratio and the hydrophobicity of the secreted BS probably determine the lifetime of the vesicles. This is due to the preference of BS micelles for PL. At the concentrations present in the biliary tree, BS take up 4–5 times more PL than Ch. If the initial ratio in the vesicles is higher than about 0.25, the BS will not be able to solubilize all of the vesicles. In bile of most species, the Ch/PL ratio is low and the vesicles are rapidly solubilized by BS micelles present in the canaliculus. These biles are always undersaturated and crystals never form. In man, however, the situation is different. Although the composition of hepatic bile has not been studied in great detail, it can be inferred when extrapolating from the lipid composition in gallbladder bile that the initial vesicular Ch/PL ratio in subjects without Ch stones is about 0.3. Hence, on average, even bile from control subjects is supersaturated with Ch and BS and will not be able to solubilize all of the vesicular lipids. In biles from Ch gallstone patients the Ch/PL ratio is higher than 0.3 and, consequently, in these biles a significant amount of Ch and PL is carried in vesicles with a high Ch/PL ratio. Vesicles with a Ch/PL ratio of higher than unity are unstable and, depending on the time of residence in the gallbladder, Ch can crystallize. The crystals can aggregate on a template of mucin, calcium salts, and calcium binding proteins, and thereby develop into a stone. This relatively simple sequence of events probably forms the basic mechanism of stone formation. It does not, however, explain all of the available data. For instance, as discussed above, many patients without stones appear to have bile supersaturated with Ch, yet their bile does not crystallize. Hence either CSI overestimates the true degree of supersaturation or the bile contains factors that slow down the rate of crystallization.

Several proteins have been isolated from bile that show crystallization-inhibiting activity when added to model systems. This class of proteins includes apolipoprotein A-I, IgA, and an as yet not fully characterized protein of 128 kDa (Holzbach *et al.,* 1984; Kibe *et al.,* 1984; Busch *et al.,* 1991). The physiological relevance of these proteins has not been firmly established. In contrast to the inhibitors, an even greater number of crystallization-promoting proteins has been described (Groen *et al.,* 1990; Harvey and Strasberg, 1990; Upadhya *et al.,* 1990; Harvey *et al.,* 1991; Offner *et al.,* 1991; Nuutinen *et al.,* 1993; Abei *et al.,* 1994; Yamashita *et al.,* 1995). These proteins all share one property: They all bind to the lectin concanavalin A. The promoters consist of two main groups: immunoglobulins and acute phase proteins. Both types of proteins have amphiphatic properties that probably play a role in the crystallization-promoting activity. The activity of the promoters has been demonstrated solely in model systems and, similarly to the inhibitors, their physiological relevance has not been established. On the contrary, most of these factors probably lack an essential feature. Pattinson and Willis (1991) showed that the nucleation time of native bile is insensitive to proteolytic activity. In these studies, both fast-and slow-crystallizing biles were incubated with an excess of pronase. Despite the fact that almost all protein was degraded, no effect on the nucleation time was observed. In an earlier study, we demonstrated pronase resistance of the main nucleating promoting activity in bile (Groen *et al.,* 1990). The pronase resistance provides a powerful discriminating feature to evaluate the relevance of a putative promoter or inhibitor. Unfortunately, almost all groups in the field have failed to apply this criterion, which has generated unnecessary confusion. It is beyond doubt that certain biliary proteins may regulate cholesterol crystallization. Their physiological importance is, however, still not settled.

5.2. Cholesterol Hyposecretion

Hyposecretion of Ch into bile is seldom encountered in human disease or animal models. An obvious reason for this scarceness is that Ch hyposecretion by itself probably does not cause evident pathology. As far as we know, the only disease in which hyposecretion of Ch at normal PL and BS secretion rates has been detected is sitosterolemia, a condition characterized by a defect in intestinal Ch uptake and increased uptake of plant sterols (see Chapter 15 by Salen *et al.*). Interestingly, the level of serum Ch is normal in these patients, suggesting that the biliary defect has something to do with transport of Ch through the bilayer of the canalicular membrane. It would be very interesting to characterize the basic defect underlying this disease further.

Hyposecretion of Ch, in combination with hyposecretion of PL but normal BS, leads to severe liver disease. Such a condition is found in the mdr2 knockout mice; so far, no human counterpart has been found showing similar pathology. In the absence of mdr2 Pgp activity, BS micelles cannot take up lipid from the canalicular membrane and, since the BS are potent detergents, other membranes are attacked, leading to cell damage and inflammatory reactions (Smit *et al.*, 1993). The pathology in the mdr2 (−/−) mice strongly depends on the composition of the BS pool. When fed lab chow, mdr2 (−/−) mice have a rather hydrophilic BS pool, consisting mainly of tauromuricholate. Under these conditions the onslaught in the biliary tree is relatively limited; progressive liver disease slowly develops, leading ultimately to hepatocellular carcinoma, but the lifespan of the animals is not significantly affected (Mauad *et al.*, 1994). If the BS pool is replaced by the slightly more hydrophobic BS taurocholate, the animals develop massive liver pathology. Apparently, biliary lipid has a crucial role in protecting biliary epithelium against the detergent action of BS. In addition to this function, biliary Ch and PL also play a role in controlling serum levels of Ch. In mdr2 (−/−) mice, the total serum Ch was decreased by about 50% and the lipoprotein profile differed as well. HDL, the predominant Ch carrier in mice, was strongly decreased, whereas VLDL was increased. The underlying mechanism is not yet clear but probably relates to the packaging of Ch in lipoproteins in liver and intestine.

The role of biliary lipid in body Ch homeostasis is only beginning to be appreciated. Clearly much further work is required to elucidate the underlying mechanisms.

6. CONCLUSIONS AND PERSPECTIVES

Recent findings, in particular those obtained with mdr2 knockout mice, have generated a novel concept of the mechanism(s) involved in hepatobiliary lipid transport. It should be realized, however, that the concept still is speculative and that more studies are needed to firmly establish its validity. Although available evidence strongly indicates that the process of biliary lipid secretion is regulated in quantitative terms at the level of the bile canalicular membrane, questions concerning the supply of these lipids to their site of exit remain to be answered. In particular, the potential roles of vesicular lipid transport, PC transfer protein, and sterol carrier proteins need to be defined. Further studies in our laboratories will also be aimed at further characterization of the (patho)physiological significance of bile lipids. From data available so far, it appears that biliary secretion not only represents an exit pathway for excess Ch from the body, but also may fulfill important functions, such as hepatobiliary transport of hydrophobic organic compounds (e.g., protoporphyrins), absorption of dietary lipids by providing material needed for chylomicron formation, and protection of epithelial cell linings from the cytotoxic effects of detergent BS. The mdr2 knockout and mdr2 transgenic mouse models will be of great practical importance in this respect.

ACKNOWLEDGMENT. This work is supported by the Netherlands Organization of Scientific Research NWO (Grant 902-23-097). Folkert Kuipers is an established investigator of the Netherlands Heart Foundation.

7. REFERENCES

Abei, M., Nuutinen, H., Kawczak, P., Schwarzendrube, J., Pillay, S. P., and Holzbach, R. T., 1994, Identification of human biliary alpha(l)-acid glycoprotein as a cholesterol crystallization promoter, *Gastroenterology* **106:**231–238.

Admirand, W. H., and Small, D. M., 1968, The physicochemical basis of cholesterol gallstone formation in man, *J. Clin. Invest.* **47:**1043–1052.

Apstein, M. D., 1984, Inhibition of biliary phospholipid and cholesterol secretion by bilirubin in the Sprague–Dawley and Gunn rat, *Gastroenterology* **87:**634–638.

Apstein, M. D., and Russo, A. R., 1985, Ampicillin inhibits biliary cholesterol secretion, *Dig. Dis. Sci.* **30:**253–256.

Ballatori, N., and Truong, A. T., 1989, Relation between biliary glutathione excretion and bile acid-independent bile flow, *Am. J. Physiol.* **265:**G1–G7.

Barnwell, S. G., Tuchweber, B., and Yousef, I. M., 1987, Biliary lipid secretion in the rat during infusion of increasing doses of unconjugated bile acids, *Biochim. Biophys. Acta* **922:**221–233.

Bilharz, L. E., Spady, D. K., and Dietschy, J. M., 1989, Inappropriate hepatic cholesterol synthesis expands the cellular pool of sterol available for recruitment by bile acids in the rat, *J. Clin. Invest.* **84:**1181–1187.

Botham, K. M., and Bravo, E., 1995, The role of lipoprotein cholesterol in biliary steroid secretion. Studies with in vivo experimental models, *Prog. Lipid Res.* **34:**71–97.

Brenneman, D. E., Connor, W. E., Forker, E. L., and DenBesten, L., 1972, The formation of abnormal bile and cholesterol gallstones from dietary cholesterol in the prairie dog, *J.Clin. Invest.* **51:**1495–1503.

Busch, N., Matiuck, N., Sahlin, S., Pillay, S. P., and Holzbach, R. T., 1991, Inhibition and promotion of cholesterol crystallization by protein fractions from normal human gallbladder bile, *J. Lipid Res.* **32:**695–702.

Caldwell, F. T., Levitsky, K., and Rosenberg, B., 1965, Dietary production and dissolution of cholesterol gallstones in the mouse, *Am. J. Physiol.* **209:**473–478.

Carey, M. C., 1978, Critical tables for calculating the cholesterol saturation of native bile, *J. Lipid Res.* **19:**945–955.

Carey, M. C., and Small, D. M., 1978, The physical chemistry of cholesterol solubility in bile, *J. Clin. Invest.* **61:**998–1026.

Clausen, J., Kruse, I., and Dam, H., 1965, Fractionation and characterization of proteins and lipids in bile, *Scand. J. Clin. Lab. Invest.* **17:**325–335.

Cohen, B. I., Miki, S., Mosbach, E. H., Ayyad, N., Stenger, R. J., Mikami, T., Yoshii, M., Kihira, K., and Hoshita, T., 1993, Bile acid sulfonates alter cholesterol gallstone incidence in hamsters, *Hepatology* **17:**103–110.

Cohen, D. E., Angelico, M., and Carey, M. C., 1989, Quasielectric light scattering evidence for vesicular secretion of biliary lipids, *Am. J. Physiol.* **257:**G1–G8.

Cohen, D. E., Leighton, L. S., and Carey, M. C., 1992, Bile salt hydrophobicity controls vesicle secretion rates and transformation in native bile, *Am. J. Physiol.* **263:**G386–G395.

Coleman, R., 1987, Biochemistry of bile secretion, *Biochem. J.* **24:**249–261.

Coleman, R., and Rahman, K., 1992, Lipid flow in bile formation, *Biochim. Biophys. Acta* **1125:** 113–133.

Coleman, R., Iqbal, S., Godfrey, P. P., and Billington, D., 1979, Membranes and bile formation, *Biochem. J.* **178:**201–208.

Coleman, R., Rahman, K., Bellringer, M. E., Kan, K., and Hamlin, S., 1989, Secretion of biliary lipids and its control, in: *Trends in Bile Acid Research* (G. Paumgartner, A. Stiehl, and W. Gerok, eds.), pp. 161–176, Kluwer Academic Publishers, Dordrecht.

Crawford, J. M., Crawford, A. R., Hatch, V. C., Stearns, R. C., Barnes, S., and Godleski, J. J., 1995, Hepatocellular secretion of biliary lipid:bile salt-induced vesiculation of the canalicular membrane outer leaflet, in: *Bile Acids in Gastroenterology. Basic and Clinical Advances* (A. F. Hofmann, G. Paumgartner, and A. Stiehl, eds.), pp. 254–257, Kluwer Academic Publishers, Dordrecht.

Dutczak, W. J., and Ballatori, N., 1994, Transport of glutathione-methylmercury complex across liver canalicular membranes on reduced glutathione carriers, *J. Biol. Chem.* **269:**9746–9751.

Erlinger, S., 1982, Bile flow, in: *The Liver: Biology and Pathobiology* (I. Arias, H. Popper, D. Schachter, and D. A. Shafritz, eds.), pp. 407–427, Raven Press, New York.

Evans, W., Kremmer, T., and Culvenor, J., 1976, Role of membranes in bile formation. Comparison of the composition of bile and a liver-canalicular plasma membrane subfraction, *Biochem. J.* **154:**589–595.

Fernandez-Checa, J., Yi, Y., Garcia-Ruiz, C., Knezic, Z., Ahara, S., and Kaplowitz, N., 1993, Expression of rat liver GSH transport in *Xenopus laevis* oocytes, *J. Biol. Chem.* **268:**2324–2328.

Frijters, C. M. A., Ottenhoff, R., van Wijland, M. J. A., van Nieuwkerk, C., Groen, A. K., and Oude Elferink, R. O. E., 1995, Regulation of mdr2 P-glycoprotein expression by bile salts, *Hepatology* **22:**322A.

Gallinger, S., Taylor, R. D., Harvey, P. R. C., Petrunka, C. N., and Strasberg, S. M., 1985, Effect of mucous glycoprotein on the nucleation time of human bile, *Gastroenterology* **89:**648–658.

Gollish, S. H., Burnstein, M. D., Ilson, R. G., Petrunka, C. N., and Strasberg, S. M., 1983, Nucleation of cholesterol monohydrate crystals from hepatic and gallbladder bile of patients with cholesterol gallstones, *Gut* **24:**836–844.

Graham, J., Ahmed, H., and Northfield, T. C., 1989, Unidirectional pathway for vesicular cholesterol transport, in: *Trends in Bile Acid Research* (G. Paumgartner, A. Stiehl, and W. Gerok, eds.), pp. 177–188, Kluwer Academic Publishers, Dordrecht.

Groen, A. K., Noordam, C., Drapers, J. A., Egbers, P., Jansen, P. L. M., and Tytgat, G. N., 1990, Isolation of a potent cholesterol nucleation-promoting activity from human gallbladder bile: role in the pathogenesis of gallstone disease, *Hepatology* **11:**525–533.

Groothuis, G. M. M., Hardonk, M. J., Keulemans, K. P. T., Nieuwenhuis, P., and Meijer, D. K. F., 1982, Autoradiographic and kinetic demonstration of acinar heterogeneity of taurocholate transport, *Am. J. Physiol.* **243:**G455–G462.

Gurantz, D., and Hofmann, A. F., 1984, Influence of bile acid structure on bile flow and biliary lipid secretion in the hamster, *Am. J. Physiol.* **247:**G736–G748.

Hardison, W. G. M., and Evans, H. E., 1986, Postobstructive subcellular organelle and biliary lipid composition in the rat, *J. Hepatol.* **3:** 318–327.

Harvey, P. R. C., and Strasberg, S. M., 1990, Biliary proteins and their role as nucleating inhibitors/promotors, in: *Gallstone Disease. Pathophysiology and Therapeutic Approaches* (W. Swobodnik, H. Ditschuneit, and R. D. Soloway, eds.), pp. 55–66, Springer-Verlag, Berlin.

Harvey, P. R. C., Upadhya, G. A., and Strasberg, S. M., 1991, Immunoglobulins as nucleating proteins in the gallbladder bile of patients with cholesterol gallstones, *J. Biol. Chem.* **266:**13996–14003.

Hay, D. W., Cahalane, M. J., Timofeyeva, N., and Carey, M. C., 1993, Molecular species of lecithin in human gallbladder bile, *J. Lipid Res.* **34:**759–768.

Hegardt, F. G., and Dam, H., 1971, The solubility of cholesterol in aqueous solution of bile salts and lecithin, *Z. Ernahr.* **10:** 223–233.

Heinemann, T., Axtmann, G., and Vonbergmann, K., 1993, Comparison of intestinal absorption of cholesterol with different plant sterols in man, *Eur. J. Clin. Invest.* **23:**827–831.

Heman, D. M., 1989, Quantitative estimation of the hydrophilic–hydrophobic balance of mixed bile salt solutions, *J. Lipid Res.* **30:**719–730.

Hofmann, A. F., 1984, Chemistry and enterohepatic circulation of bile acids, *Hepatology* **4:**4S–14S.

Holland, R. E., Rahman, K., Morris, A. I., Coleman, R., and Billington, D., 1993, Effects of niacin on biliary lipid output in the rat, *Biochem. Pharmacol.* **45:**43–49.

Holzbach, R. T., Marsh, M. E., and Hallberg, M. C., 1971, The effect of pregnancy on lipid composition of guinea pig gallbladder bile, *Gastroenterology* **60:**288–293.

Holzbach, R. T., Marsh, M., Oleszewski, M., and Holan, K., 1973, Cholesterol solubility in bile. Evidence that supersaturated bile is frequent in healthy man, *J. Clin. Invest.* **52:**1467–1479.

Holzbach, R. T., Kibe, A., Thiel, E., Howell, J. H., Marsh, M., and Hermann, R. E., 1984, Biliary proteins. Unique inhibitors of cholesterol crystal nucleation in human gallbladder bile, *J. Clin. Invest.* **73:**35–45.

Kibe, A., Holzbach, R. T., LaRusso, N. F., and Mao, S. J. T., 1984, Inhibition of cholesterol crystal formation by apolipoproteins in supersaturated model bile, *Science* **225:**514–516.

Kitani, K., Ohta, M., and Kanai, S., 1985, Tauroursodeoxycholate prevents biliary protein excretion by other bile acids in the rat, *Am. J. Physiol.* **248:**G407–G417.

Klaassen, C. D., and Watkins III, J. B., 1984, Mechanisms of bile formation, hepatic uptake and biliary excretion, *Pharmacol. Rev.* **36:**1–67.

Kuipers, F., Havinga, R., Bosschieter, H., Toorop, G. P., Hindriks, F. R., and Vonk, R. J., 1985, Enterohepatic circulation in the rat, *Gastroenterology* **88:**403–411.

Lee, S. P., Park, H. Z., Mandani, H., and Kaler, E. W., 1987, Partial characterization of a nonmicellar system of cholesterol solubilization in bile, *Am. J. Physiol.* **252:**G374–G383.

Mauad, T. H., van Nieuwkerk, C. M. J., Dingemans, K. P., Smit, J. J. M., Schinkel, A. H., Notenboom, R. G. E., van den Bergh Weerman, M. A., Verkruisen, R. P., Groen, A. K., Oude Elferink, R. P. J., Van der Valk, M. A., Borst, P., and Offerhaus, G. J. A., 1994, Mice with homozygous disruption of the mdr2 P-glycoprotein gene: a novel animal model for studies of nonsuppurative inflammatory cholangitis and hepatocarcinogenesis, *Am. J. Pathol.* **145:**1237–1245.

Mazer, N. A., and Carey, M. C., 1984, Mathematical model of biliary lipid secretion: a quantitative analysis of physiological and biochemical data from man and other species, *J. Lipid Res.* **25:** 932–953.

Nakagaki, M., Danzinger, R. G., Hofmann, A. F., and DiPietro, R. A., 1984, Biliary secretion and hepatic metabolism of taurine-conjugated 7α-hydroxy and 7β-hydroxy bile acids in the dog. Defective hepatic transport and bile hyposecretion, *Gastroenterology* **87:**647–659.

Nakayama, F., 1969, Composition of gallstone and bile: Species difference, *J. Lab. Clin. Med.* **73:**623–629.

Nakayama, F., and VanderLinden, W., 1971, Bile composition: Sweden versus Japan, *Am. J. Surgery* **122:**8–12.

Nathanson, N. H., and Boyer, J. L., 1991, Mechanisms and regulation of bile secretion, *Hepatology* **14:**551–566.

Neese, R. A., Faix, D., Kletke, C., Wu, K., Wang, A. C., Shackleton, C. M. C., and Hellerstein, M. K., 1993, Measurement of endogenous synthesis of plasma cholesterol in rats and humans using MIDA, *Am. J. Physiol.* **264:**E136–E147.

Nervi, F., del Pozo, R., Covarrubias, C. F., and Ronco, B. O., 1983, The effect of progesterone on the regulatory mechanisms of biliary cholesterol secretion, *Hepatology* **3:**360–367.

Nies, A. T., Gatmaitan, Z., and Arias, I. M., 1995, The bile canalicular membrane contains an ATP-dependent phosphatidylcholine translocator which is regulated by taurocholate, *Hepatology* **22:**295A.

Nuutinen, H., Jüngst, D., Lange, V., Abei, M., Schwarzendrube, J., Corradini, S. G., Williams, C., Chandler, K., and Holzbach, R. T., 1993, Alpha-1-acid glycoprotein (AAG) as a promoter of cholesterol crystallization in gallstone disease-correlation between the concentration of AAG and nucleation time, *Gastroenterology* **104:**A966.

Offner, G. D., Gong, D., and Afdhal, N. H., 1991, Aminopeptidase N is a 130 kD pronucleating glycoprotein in human hepatic and gallbladder bile, *Hepatology* **14:**180.

Osuga, T., and Portman, O. W., 1972, Relationship between bile composition and gallstone formation in squirrel monkey, *Gastroenterology* **63:**122–133.

Oude Elferink, R. P. J., Meijer, D. K. F., Kuipers, F., Jansen, P. L. M., Groen, A. K., and Groothuis, G. M. M., 1995a, Hepatobiliary secretion of organic compounds; molecular mechanisms of membrane transport, *Biochim. Biophys. Acta* **1241:** 215–268.

Oude Elferink, R. P. J., Ottenhoff, R., Van Wijland, M. J. A., Smit, J. J. M., Schinkel, A. H., and Groen, A. K., 1995b, Regulation of biliary lipid secretion by mdr2-P-glycoprotein in the mouse, *J. Clin. Invest.* **95:**31–38.

Pattinson, N. R., 1985, Solubilisation of cholesterol in human bile, *FEBS Lett.* **181:**339–342.

Pattinson, N. R., and Willis, K. E., 1991, Nucleation of cholesterol crystals from native bile and the effect of protein hydrolysis, *J. Lipid Res.* **32:**215–221.

Paulusma, C. C., Bosma, P. J., Bakker, C. T. M., Otter, M., Zaman, G. J. R., and Oude Elferink, R. P. J., 1995, Cloning of a liver-specific MRP-homologue which is deficient in the TR-rat, *Hepatology* **22:**295A.

Pittman, R. C., Knecht, T. P., Rosenbaum, M. S., and Taylor, C. A., 1987, A non-endocytotic mechanism for the selective uptake of high density lipoprotein-associated cholesterol esters, *J. Biol. Chem.* **262:**2443–2450.

Rahman, K., and Coleman, R., 1987, Biliary lipid secretion and its control. *Biochem. J.* **245:**531–536.

Robins, S. J., 1975, Recirculation and reutilization of micellar bile lecithin, *Am. J. Physiol.* **229:** 598–602.

Robins, S. J., and Armstrong, M. J., 1976, Biliary lecithin secretion. 11. Effects of dietary choline and biliary lecithin synthesis, *Gastroenterology* **70:**397–402.

Robins, S. J., and Brunengraber, H., 1982, Origin of biliary cholesterol and lecithin in the rat: contribution of new synthesis and preformed hepatic stores, *J. Lipid Res.* **23:**604–608.

Robins, S. J., Fasulo, J. M., Collins, M. A., and Patton, G. M., 1985, Evidence for separate pathways for transport of newly synthesized and preformed cholesterol into bile, *J. Biol. Chem.* **260:** 6511–6513.

Robins, S. J., Fasulo, J. M., Lessard, P. D., and Patton, G. M., 1993, Hepatic cholesterol synthesis and the secretion of newly synthesized cholesterol in bile, *Biochem. J.* **289:**41–44.

Robins, S. J., Fasulo, J. M., Pritzker, C. R., and Patton, G. M., 1995a, Hepatic transport of unesterified cholesterol in the rat is traced by the plant sterol, sitostanol, *Hepatology* **22:**240A.

Robins, S. J., Fasulo, J. M., Salen, G., and Patton, G. M., 1995b, Preferential sterol transport from HDL to bile, *Hepatology* **22:**240A.

Ruetz, S., and Gros, P., 1994, Phosphatidylcholine translocase: a physiological role for the mdr2 gene, *Cell* **77:**1071–1082.

Samuel, P., and McNamara, D. J., 1983, Differential absorption of exogenous and endogenous cholesterol in man, *J. Lipid Res.* **24:**265–276.

Schreibner, J., Fuchs, M., Schiemann, M., Tauber, G., Hormann, E., and Stange, E. F., 1993, Bile acid synthesis from newly synthesized vs. preformed cholesterol precursor pools in the rat, *Hepatology* **17:**1095–1102.

Schwartz, C. C., Zech, L. A., VandenBroek, J. M., and Cooper, P. S., 1993, Cholesterol kinetics in subjects with bile fistula. Positive relationship between size of the bile acid precursor pool and bile acid synthesis rate, *J. Clin. Invest.* **91:**923–938.

Sedaghat, A., and Grundy, S. M., 1980, Cholesterol crystals and the formation of cholesterol gallstones, *N. Engl. J. Med.* **302:**1274–1277.

Simmonds, J. W., Hofmann, A. F., and Theodor, E., 1967, Absorption of cholesterol from a micellar solution: intestinal perfusion studies in man, *J. Clin. Invest.* **46:**874–890.

Singer, I. I., Kawka, D. W., Kazasis, D. W., Alberts, D. W., Chen, J. S., Huff, J. W., and Ness, G. C., 1984, Hydroxymethylglutaryl-coenzymeA reductase-containing hepatocytes are distributed periportally in normal and mevinolin-treated rat livers, *Proc. Natl. Acad. Sci. USA* **81:**5556–5560.

Smit, J. J. M., Schinkel, A. H., Oude Elferink, R. P. J., Groen, A. K., Wagenaar, E., Van Deemter, L., Mol, C. A. A. M., Ottenhoff, R., Van der Lugt, N. M. T., van Roon, M. A., Van der Valk, M. A., Offerhaus, G. J. A., Berns, A. J. M., and Borst, P., 1993, Homozygous disruption of the murine mdr2 P-glycoprotein gene leads to a complete absence of phospholipid from bile and to liver disease, *Cell* **75:**451–462.

Smit, M. J., Temmerman, A. M., Havinga, R., Kuipers, F., and Vonk, R. J., 1990, Short-and long-term effects of biliary drainage on hepatic cholesterol metabolism in the rat, *Biochem. J.* **269:**781–788.

Smit, M. J., Temmerman, A. M., Wolters, H., Kuipers, F., Beijnen, A. C., and Vonk, R. J., 1991, Dietary fish oil-induced changes in intrahepatic cholesterol transport and bile acid synthesis in rats, *J. Clin. Invest.* **88:**943–951.

Smit, M. J., Kuipers, F., Vonk, R. J., Temmerman, A. M., Jaeckle, S., and Windler, E. T. T., 1993, Effects of dietary cholesterol on bile formation and hepatic processing of chylomicron remnant cholesterol in the rat, *Hepatology* **17:**445–454.

Smit, M. J., Verkade, H. J., Havinga, R., Vonk, R. J., Scherphof, G. L., In't Veld, G., and Kuipers, F., 1994, Dietary fish oil potentiates bile acid-induced cholesterol secretion into bile in rats, *J. Lipid Res.* **35:**301–310.

Smith, A. J., Timmermans-Hereijgers, J. L. P. M., Roelofsen, B., Wirtz, K. W. A., van Blitterswijk, W. J., Smit, J. J. M., Schinkel, A. H., and Borst, P., 1994, The human MDR3 P-glycoprotein promotes translocation of phosphatidylcholine through the plasma membrane of fibroblasts from transgenic mice, *FEBS Lett.* **354:**263–266.

Somjen, G. J., and Gilat, T., 1983, A non-micellar mode of cholesterol transport in human bile, *FEBS Lett.* **61:**265–268.

Somjen, G. J., and Gilat, T., 1985, Contribution of vesicular and micellar carriers to cholesterol transport in human bile, *J. Lipid Res.* **26:**699–704.

Stone, B. G., and Evans, C. D., 1992, Evidence for a common biliary cholesterol and VLDL cholesterol precursor pool in rat liver, *J. Lipid Res.* **33:**1665–1675.

Stone, B. G., Erickson, S. K., Graig, W. Y., and Cooper, A. D., 1985, Regulation of rat biliary cholesterol secretion by agents that alter intrahepatic cholesterol metabolism, *J. Clin. Invest.* **76:**1773–1781.

Tazuma, S., Barnhart, R. L., Reeve, L. E., Tokumo, H., and Holzbach, R. T., 1988, Biliary secretion of organic anions in the dog: association with defined lipid particles, *Am. J. Physiol.* **255:** G745–G751.

Thewles, A., Parslow, R. A., and Coleman, R., 1993, Effect of diosgenin on biliary cholesterol transport in the rat, *Biochem. J.* **291**:793–798.
Turley, S. D., and Dietschy, J. M., 1981, The contribution of newly synthesized cholesterol to biliary cholesterol in the rat, *J. Biol. Chem.* **256**:2438–2446.
Ulloa, N., Garrido, J., and Nervi, F., 1987, Ultracentrifugal isolation of vesicular carriers of biliary cholesterol in native human and rat bile, *Hepatology* **7**:235–244.
Upadhya, G. A., Yuan, J., Harvey, P. R. C., and Strasberg, S. M., 1990, Immunoglobulin (Ig) content of biliary Con A binding glycoproteins, *Hepatology* **12**:1007A (Abstract 678).
Verkade, H. J., Wolbers, M. J., Havinga, R., Uges, D. R., Vonk, R. J., and Kuipers, F., 1990, The uncoupling of biliary lipid from bile acid secretion by organic anions in the rat, *Gastroenterology* **99**:1485–1492.
Verkade, H. J., Derksen, J. T. P., Gerding, A., Scherphof, G. L., Vonk, R. J., and Kuipers, F., 1991, Differential hepatic processing and biliary secretion of head-group and acyl chains of liposomal phosphatidylcholines, *Biochem. J.* **275**:139–144.
Verkade, H. J., Havinga, R., Gerding, A., Vonk, R. J., and Kuipers, F., 1993a, Mechanism of bile acid-induced biliary lipid secretion in the rat-effect of conjugated bilirubin, *Am. J. Physiol.* **264**: G462–G469.
Verkade, H. J., Wolters, H., Gerding, A., Havinga, R., Fidler, V., Vonk, R. J., and Kuipers, F., 1993b, Mechanism of biliary lipid secretion in the rat: a role for bile acid-independent bile flow? *Hepatology* **17**:1074–1080.
Verkade, H. J., Vonk, R. J., and Kuipers, F., 1995, New insights into the mechanism of bile acid-induced biliary lipid secretion, *Hepatology* **21**:1174–1189.
Whiting, M. J., and Watts, J. M. C. K., 1985, Cholesterol gallstone pathogenesis: a study of potential nucleating agents for cholesterol crystal formation in bile, *Clin. Sci.* **68**:589–596.
Wilson, M. D., and Rudel, L. L., 1994, Review of cholesterol absorption with emphasis on dietary and biliary cholesterol, *J. Lipid Res.* **35**:943–955.
Yamashita, G., Tazuma, S., and Kajiyama, G., 1992, Effects of organic anions on biliary lipid secretion in rats. Importance of association with biliary secretion of biliary lipids, *Am. J. Physiol.* **257**:G1–G8.
Yamashita, G., Tazuma, S., Horikawa, K., Aihara, N., Ochi, H., Teramen, K., Yamashita, Y., Sasaki, M., Ohya, T., and Kajiyama, G., 1993, Partial characterization of mechanism(s) by which sulphobromophthalein reduces biliary lipid secretion, *Biochem. J.* **291**:173–177.
Yamashita, G., Corradini, S. G., Secknus, R., Takabayashi, A., Williams, C., Hays, L., Chernosky, A. L., and Holzbach, R. T., 1995, Biliary haptoglobin, a potent promoter of cholesterol crystallization at physiological concentrations, *J. Lipid Res.* **36**:1325–1333.
Yousef, I. M., Barnwell, S., Gratton, F., Tuchweber, B., Weber, A., and Roy, C. C., 1987, Liver cell membrane solubilization may control maximum secretory rate of cholic acid in the rat, *Am. J. Physiol.* **252**:G84–G91.
Zimniak, P., and Awashti, Y. C., 1993, ATP-dependent transport systems for organic anions, *Hepatology* **17**:330–339.

Chapter 12

Cholesterol Deposition in Atherosclerotic Lesions

Howard S. Kruth

1. INTRODUCTION

Cholesterol is a molecule of great biomedical significance. We cannot live without it, and often, we cannot live with it. Cholesterol accumulation in blood vessels leads to the development of atherosclerotic plaque-like lesions. Heart attacks and some types of strokes occur when these plaques rupture and blood clots form on their surfaces. In the United States, heart attacks and strokes kill more than 500,000 people each year (American Heart Association, 1995).

In 1910 Windaus established the significance of cholesterol as a quantitatively important constituent of atherosclerotic lesions. A few years later, Anitschkow and Chalatow (1913) showed the importance of cholesterol in the blood as a factor in the development of atherosclerotic lesions. These investigators demonstrated that feeding cholesterol to rabbits caused atherosclerotic lesions to form. Later, it was recognized that cholesterol is carried in the blood as part of complex lipoprotein particles, with most of the cholesterol occurring in the esterified, rather than unesterified, form (reviewed by Gotto *et al.*, 1986).

The Framingham study established that an elevated level of cholesterol [if carried in so-called atherogenic lipoproteins, such as low density lipoprotein

---

Howard S. Kruth Section of Experimental Atherosclerosis, National Heart, Lung, and Blood Institute, National Institutes of Health, Bethesda, Maryland 20892.

Subcellular Biochemistry, Volume 28: Cholesterol: Its Functions and Metabolism in Biology and Medicine, edited by Robert Bittman. Plenum Press, New York, 1997.

(LDL)] is a risk factor for the development of atherosclerotic cardiovascular disease (Kannel *et al.,* 1979; Castelli, 1984). However, atherosclerotic cardiovascular disease occurs frequently in individuals who do not have elevated levels of atherogenic lipoproteins (Kannel and Wilson, 1992). Also, atherosclerotic lesions (similar to those that occur in humans) can develop in arteries of swine, even when these animals are fed a cholesterol-free diet their entire lives (French *et al.,* 1963; Kruth and Fry, 1984), and even though cholesterol levels in swine are generally lower than the level considered normal in humans. These observations suggest that although elevated levels of plasma cholesterol can lead to more cholesterol entering blood vessels (Niehaus *et al.,* 1977; Nordestgaard *et al.,* 1992), additional factors must influence the degree to which plasma cholesterol is taken up and accumulated by the vessel wall.

2. UPTAKE OF PLASMA LIPOPROTEINS INTO THE VESSEL WALL

The size of lipoprotein particles is a factor influencing their entry into the vessel wall. The larger the lipoprotein particle, the less efficiently does it enter the vessel wall (Stender and Zilversmit, 1981; Stender and Hjelms, 1988; Nordestgaard and Zilversmit, 1989; Nordestgaard *et al.,* 1990; Nordestgaard *et al.,* 1992; Mamo and Wheeler, 1994). Indeed, chylomicrons, the largest lipoproteins (diameter > 75 nm), do not appear to enter the vessel wall, and elevation of these lipoproteins is not associated with the development of atherosclerosis (Nordestgaard and Zilversmit, 1988). However, remnants of chylomicrons and other smaller lipoproteins do enter the vessel wall.

Significant lipoprotein accumulation in the vessel wall depends on the presence of an intact endothelium. The endothelium lines the vessel wall where it contacts the blood. After physical removal of the endothelium in rabbits, lipoproteins and lipid preferentially accumulate where the endothelium has regenerated, not where the endothelium is absent (Falcone *et al.,* 1980; Alavi *et al.,* 1983; Roberts *et al.,* 1983; Day *et al.,* 1985; Chang *et al.,* 1992; Chang *et al.,* 1993; Olsson *et al.,* 1995). Thus, the endothelium may play an active role in transporting lipoproteins into the vessel wall. At the same time, the endothelium can serve as a barrier to removal of lipoproteins that have accumulated in the subendothelial space. Additionally, it has been suggested that an altered connective tissue matrix underlying the regenerated endothelium may bind lipoproteins (Alavi *et al.,* 1989).

Transport of lipoproteins across the endothelium reflects the property of polarized cells to carry out the process of transcytosis. In this process, molecules or particles taken up on one side of the cell are transported and released, generally at the opposite side of the cell. With endothelial cells, uptake of lipoproteins into both smooth-surfaced and coated vesicles occurs (reviewed by Simionescu, 1988). Some lipoprotein taken up is delivered to lysosomes, where the lipoprotein is degraded. The fate of cholesterol derived from this lipoprotein remains to be determined. Other lipoprotein is transported across the endothelial cell and released into the subendothelial space where the lipoprotein and its contained cholesterol may accumulate. It is unknown whether uptake of lipoproteins destined for transcytosis is mediated by specific endothelial cell receptors. Paracellular transport of lipoproteins between endothelial cells may also contribute to lipoprotein accumulation, such as when endothelial cells transiently separate during cell division (Weinbaum and Chien, 1993).

3. EVIDENCE THAT EXTRACELLULAR ESTERIFIED CHOLESTEROL (EC)-RICH DROPLETS IN ATHEROSCLEROTIC LESIONS ARE DERIVED MAINLY FROM DEPOSITED LIPOPROTEINS

3.1. Chemical Analysis of Atherosclerotic Lesions

Data from a number of laboratories indicates that build-up of cholesterol from deposited lipoproteins accounts for most of the extracellular cholesteryl ester in human atherosclerotic lesions. This assertion is in contrast to the generally accepted theory—that most extracellular cholesteryl ester in lesions is derived from lipid droplets released from lipid-filled cells (called foam cells) when these cells die in lesions. Studies of atherosclerosis in experimental animals, particularly the cholesterol-fed rabbit, have reinforced this idea. In this animal model of atherosclerosis, lipid accumulation in lesions occurs predominantly in foam cells (Scott *et al.*, 1976).

Each point of view about the origin of extracellular lipid leads to a different conceptualization about the mechanism of lipid accumulation in atherosclerotic lesions. In one case, blood-derived lipoproteins first accumulate in the extracellular space, and only then are secondarily taken up by cells, such as macrophages, within the vessel wall. In the other case, it is presumed that macrophage uptake of lipoproteins is the driving force for lipid accumulation in the vessel wall. That is, if macrophages did not take up the lipoproteins that enter the vessel wall, then these lipoproteins would not accumulate, but rather, the lipoproteins would be free to leave the vessel wall.

Smith and Slater (1972; reviewed by Smith, 1974) first pointed out that extracellular deposition of lipid is a primary locus of lipid accumulation in human lesions and is not secondary to release of lipid from foam cells. These investigators noted that extracellular lipid could accumulate in regions of the vessel wall where there were no foam cells and no evidence of cell death. They also carried out detailed chemical analysis of areas of atherosclerotic lesions where most of the lipid was contained in foam cells, or where most of the lipid was present as extracellular droplets. One particular chemical finding indicates a dissimilarity between the origin of the lipid that composes extracellular and intracellular lipid deposits in atherosclerotic lesions (Smith *et al.*, 1968). This is the pattern of fatty acids esterified with the deposited cholesterol.

Areas of human atherosclerotic lesions with extracellular lipid deposits show cholesterol esterified predominantly with linoleate. This is in contrast to areas of lesions that contain foam cells. These areas show cholesterol esterified predominantly with oleate. This is an important finding (confirmed by Künnert and Krug, 1971; Rapp *et al.*, 1983; Guyton and Klemp, 1994) because most cholesterol in plasma lipoproteins is esterified with linoleate, while most cholesterol stored in cellular lipid droplets is esterified with oleate (Smith and Slater, 1973). Thus, the fatty acid pattern of extracellular cholesteryl ester suggests that this lipid is derived directly from deposited plasma lipoproteins, rather than from released cellular lipid droplets.

On the other hand, if plasma lipoproteins deposit and are then the source of accumulated cholesteryl ester in foam cells, the predominance of oleate fatty acid in intracellular cholesteryl esters cannot be accounted for. The studies of Brown, Goldstein, and their coworkers (reviewed by Brown and Goldstein, 1983) later explained the predominance of oleate in intracellular cholesteryl ester lipid droplets. They showed that the fatty acid profile of lipoprotein-derived cholesteryl ester changes during processing of lipoprotein cholesteryl ester by cells. Specifically, lipoprotein cholesteryl esters are hydrolyzed in lysosomes; then, unesterified cholesterol released from lysosomes is preferentially re-esterified with oleate and stored in cellular lipid droplets. This explains how the fatty acid pattern of plasma lipoprotein cholesteryl ester changes to that pattern of intracellular lipid droplet cholesteryl ester. However, release of lipid droplets from foam cells cannot account for the fatty acid pattern of extracellular cholesteryl esters.

3.2. Detection of Extracellular EC-Rich Lipid Droplets in Atherosclerotic Lesions

Structural studies show that extracellular cholesteryl ester in human atherosclerotic lesions is contained within lipid droplets and that these extracellular lipid droplets are different in size from intracellular lipid droplets. By examining tissue sections of lesions stained with lipid-soluble dyes, Smith *et al.* (1967) observed that extracellular lipid droplets were smaller than intracellular lipid droplets. Using techniques that preserve lipid, Guyton and coworkers visualized the extracellular lipid droplets in lesions by electron microscopy and confirmed this observation (Bocan *et al.*, 1986; Guyton and Klemp, 1988; Guyton and Klemp, 1989). Their ultrastructural studies showed that the extracellular lipid droplets ranged in size from 60–200 nm, while the intracellular lipid droplets were greater than 400 nm in diameter (see Figure 8 in Bocan *et al.*, 1986).

Kruth (1984a) directly demonstrated the presence of cholesteryl ester in the extracellular lipid droplets using the fluorescent cholesterol probe, filipin (see Chapter 16 by Severs). Filipin labels these lipid particles after treatment of lesion tissue sections with cholesterol esterase. Cholesterol esterase treatment hydrolyzes the cholesteryl ester of the extracellular and intracellular lipid droplets, thus providing unesterified cholesterol for filipin to bind (filipin does not bind to esterified cholesterol). Filipin staining also indicates that the extracellular esterified cholesterol (EC)-containing lipid droplets are smaller than the intracellular EC-containing lipid droplets.

Recent studies of lipid deposition in the human cornea also support the conclusion that extracellular EC-rich lipid droplets are derived from lipoproteins, and not from release of foam cell lipid droplets (Gaynor *et al.,* 1996). The cornea and blood vessels are similar in that both are composed of dense connective tissue with embedded cells (keratocytes and smooth muscle cells, respectively). Cholesterol accumulates in the peripheral cornea (adjacent to the vessels that nourish the cornea) and forms a circular opacity called arcus lipoides, the presence of which can be associated with the development of cardiovascular disease (Chambless *et al.,* 1990). In both the cornea and vessel wall, cholesterol deposition increases with age and is accelerated by elevated levels of atherogenic lipoproteins such as LDL. EC-rich lipid droplets accumulate in the extracellular space of the cornea and are similar to the extracellular EC-rich lipid droplets that deposit in atherosclerotic lesions. In the cornea, the EC-rich lipid droplets accumulate in the absence of foam cells or of lipid droplet accumulation by corneal cells. This is further evidence that extracellular EC-rich lipid droplets can be derived from direct deposition of plasma lipoproteins, rather than from lipid droplets of cells.

3.3. Isolation of Extracellular EC-Rich Lipid Droplets from Atherosclerotic Lesions

3.3.1. Experimental Considerations when Isolating Lipid Particles from Vascular Tissue

It is desirable that lipid particles be quantitatively extracted from the vessel wall and not be changed during their extraction. In this regard, a number of protease inhibitors, as well as antioxidants, can be added to extraction buffers to help maintain the integrity of protein and lipid components of lipid particles (Ylä-Herttuala *et al.,* 1990). However, their presence does not ensure that changes will not occur to proteins especially susceptible to breakdown, such as apolipoprotein B (apo B) (Rapp *et al.,* 1994).

Homogenization of vascular tissue increases the yield of extracted lipid particles, but may produce artifactual changes in the lipid particles being isolated (Hoff and Clevidence, 1987). Homogenization has the potential to convert larger lipid particles (such as the extracellular EC-rich lipid droplets and intracellular lipid droplets) into smaller lipid particles, and also to cause the formation of artifactual lipid–protein aggregates (Smith and Slater, 1970). Our laboratory has found that $d < 1.01$ g/ml lipid particles isolated from extracts of homogenized human aortas are smaller (assessed by electron microscopy) than $d < 1.01$ g/ml lipid particles isolated from extracts of minced aortas (unpublished data). With size assessed by gel-filtration chromatography, Hoff *et al.* (1991) found that apo B-containing lipid particles isolated from minced human aortas are the size of LDL. In contrast, apo B-containing lipid particles isolated from homogenized aortas were composed of LDL-sized, and also a fraction of larger, lipid particles. Electron microscopic examination of apo B-containing lipid particles isolated from homogenized aortas of WHHL rabbits showed them to be smaller than LDL (the smallest apo B-containing lipoprotein) isolated from plasma of the same rabbits (Daugherty *et al.*, 1988). All these findings indicate that homogenization disrupts lipid droplet particles, which can result in reassociation of homogenized lipid with protein or lipoproteins. Thus, homogenization causes artifacts and should not be used to isolate lipid particles from tissue. The nature of the artifact (i.e., smaller than normal particles or formation of larger than normal lipid complexes) could depend on the extent of homogenization and the type of homogenizer. Buffer extraction of vascular tissue minced finely with scissors appears to release lipid particles with the least artifacts, although much less lipid is released.

Enzyme dissociation of vascular tissue using elastase, collagenase, and chondroitinase ABC is also effective in releasing lipid particles (Hoff and Gaubatz, 1979; and unpublished data from our laboratory). Besides releasing lipid particles physically trapped within the connective tissue of the vessel wall, enzymatic dissociation has the potential to release lipid particles attached to the connective tissues. Although elastin is an important connective tissue of the vessel wall, use of prolonged elastase treatment to enhance tissue dissociation could convert LDL to larger lipid particles. This is because the apo B component of LDL significantly degrades during treatment of tissue with this protease (see section 3.4).

Direct electrophoresis of lipid particles from pieces of vessel wall is another method that has been used to recover lipid particles from this tissue (Smith, 1974; Hoff *et al.*, 1983; Hoff *et al.*, 1986; Spring and Hoff, 1989). The largest lipid particles can be separated from smaller lipid particles such as LDL and high density lipoprotein (HDL) with this method. This is because lipid particles above a certain size do not enter most gels used for electrophoresis. However, it should be determined whether the failure of lipid particle fractions to enter a gel is due to the presence of very large lipid particles or is due to the presence of aggregates of small lipid particles. The same should be considered when lipid particle size is assessed by gel-filtration chromatography. For example, lipid particles may seem to be large because they elute with the column void volume. Such lipid particles could, in fact, be aggregates of small particles that would elute later than the void volume if these lipid particles were not aggregated.

Electron microscopy of negatively stained lipid particles is helpful in assessing lipid particle size, aggregation, and purity (Forte and Nordhausen, 1986). Knowing whether the isolated fraction is really a mixture of two types of lipid particles, such as liposomes and droplets, is important. For example, co-isolation from lesions of unesterified cholesterol-rich liposomes (described in section 4) with LDL accounts for some of the unusual chemical characteristics reported in some studies for lesion-derived LDL (Shaikh *et al.,* 1988; Daugherty *et al.,* 1988; Ylä-Herttuala *et al.,* 1989). Because they have the same density, these two types of lipid particles cannot be separated by density gradient centrifugation, a procedure often used to purify LDL isolated from lesions.

3.3.2. Characterization of Extracellular EC-Rich Lipid Droplets Isolated from Atherosclerotic Lesions

The presence of LDL in grossly normal and lesioned vessel walls was first indicated by immunostaining and extraction studies that showed the presence of apo B in the vessel walls (Kao and Wissler, 1965; Walton and Williamson, 1968; Smith and Slater, 1970; Hoff *et al.,* 1974; Hoff *et al.,* 1977; Hoff and Gaubatz, 1977; Hoff *et al.,* 1978a; Hoff *et al.,* 1978b). Hoff *et al.* (1979a), Hoff and Gaubatz (1982), and Hoff and Morton (1985) isolated and extensively characterized the apo B-containing LDL-sized particles that accumulate in human atherosclerotic lesions. These lipoproteins showed varying degrees of breakdown of their apo B component, and a slightly lower mean density and an increased electrophoretic mobility compared to plasma LDL (Morton *et al.,* 1986).

Additional evidence for the existence in atherosclerotic lesions of an EC-rich lipid particle larger than LDL but smaller than intracellular lipid droplets was obtained when these EC-rich lipid droplets were isolated from atherosclerotic lesions (Hollander *et al.,* 1979; Hoff *et al.,* 1979a; Hoff and Gaubatz, 1982; Chao *et al.,* 1990). Chao *et al.* (1990) isolated and purified the extracellular EC-rich lipid droplets from the other types of lipid particles present in extracts of minced human aortas. Filtration through a 0.45 μm (pore-size) filter separated lipid droplets [which have been shown to be greater than 0.4 μm, (Hata *et al.,* 1974; Bocan *et al.,* 1986)] and cholesterol crystals from other smaller lesion lipid particles. Then, LDL and HDL were removed from the aortic extracts by gel-filtration chromatography through 2% agarose (Figure 1). Lipid particles that eluted in the column void volume were then separated into two distinct fractions by centrifugation in a continuous density gradient (Figures 2 and 3). One fraction of density < 1.01 g/ml was composed of EC-rich lipid particles in which greater than 75% of cholesterol was esterified (Table I). These lipid particles were solid droplets except for the presence of some pitted regions on their surfaces (Figure 4). The size of the lipid droplets ranged from 40–200 nm in diameter, similar to the size of the extracellular lipid droplets determined *in situ* with electron microscopy (Bocan *et al.,* 1986).

Due to the larger size of intracellular lipid droplets compared with extracellular EC-rich lipid droplets, Lang and Insull (1970) were able to isolate the intracellular lipid droplets from other lipid particles in atherosclerotic lesions by differential flotation. Comparison of the chemical characteristics of extracellular EC-rich lipid droplets with plasma LDL and intracellular lipid droplets isolated

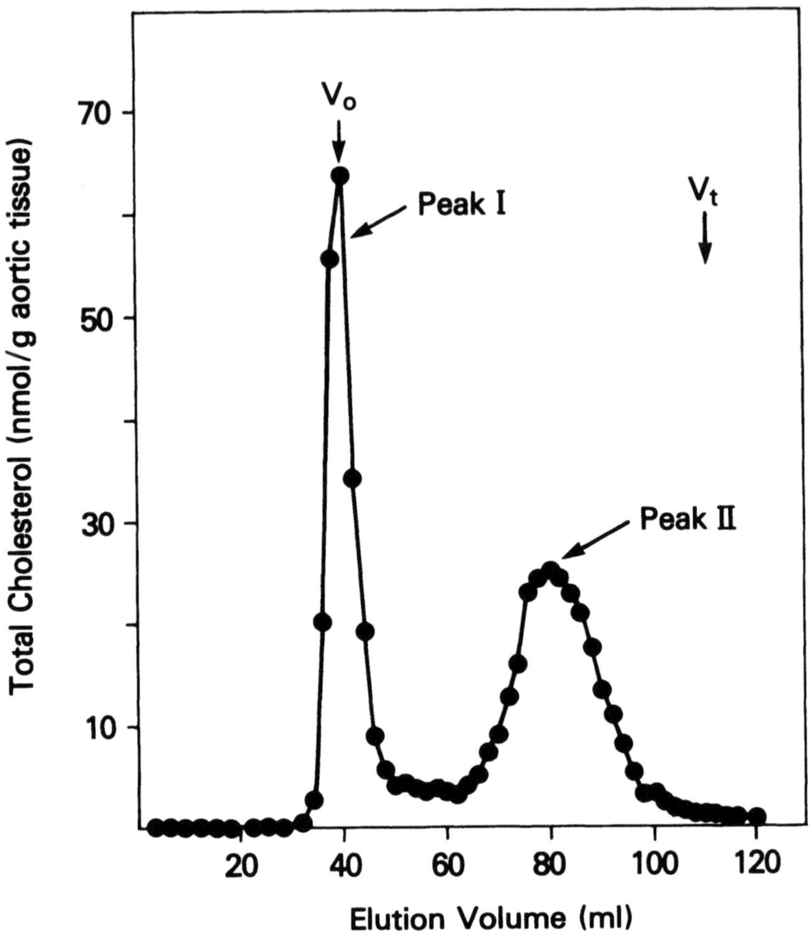

FIGURE 1. Gel-filtration chromatography of a buffer extract of minced human aorta. Cellular lipid droplets and cholesterol crystals were removed from this and all other extracts by filtration through a 0.45 μm pore-size filter prior to gel-filtration through 2% agarose. Plasma LDL eluted at the same position as peak II. Approximately 30% of the surface of the aortic tissue used to prepare this extract was covered with atherosclerotic lesions. From Chao *et al.* (1988).

from atherosclerotic lesions is shown in Tables II and III. The findings for these EC-rich lipid particles isolated from lesions confirm what Smith and coworkers found by chemical analysis of microdissected defined regions of lesions (Smith *et al.,* 1968; Smith and Slater, 1973). Mainly, oleate is the major fatty acid esterified to cholesterol in cellular lipid droplets, while linoleate is the major fatty acid esterified to cholesterol in extracellular EC-rich lipid droplets (similar to the linoleate predominance in plasma LDL). In addition, the percent cholesterol that is

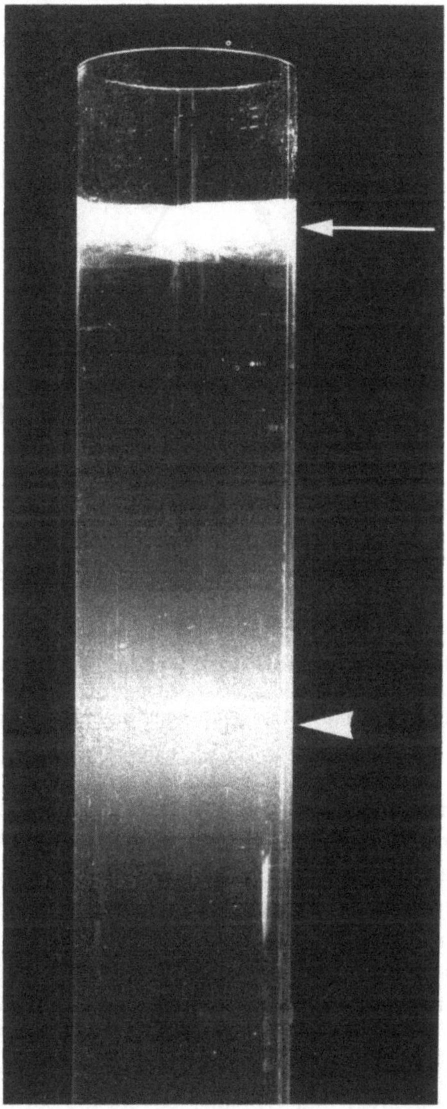

FIGURE 2. Photograph of separated EC-rich lipid droplet and UC-rich liposome fractions. These fractions were obtained following isopycnic NaCl density gradient centrifugation of void volume fractions prepared from human aortic extracts gel-filtered as shown in Figure 1. Band I (arrow) contains EC-rich lipid droplets with a $d < 1.01$ g/ml. Band II (arrowhead) contains UC-rich liposomes in a broad opaque zone distributed between d 1.03 to 1.05 g/ml. From Chao *et al.* (1990).

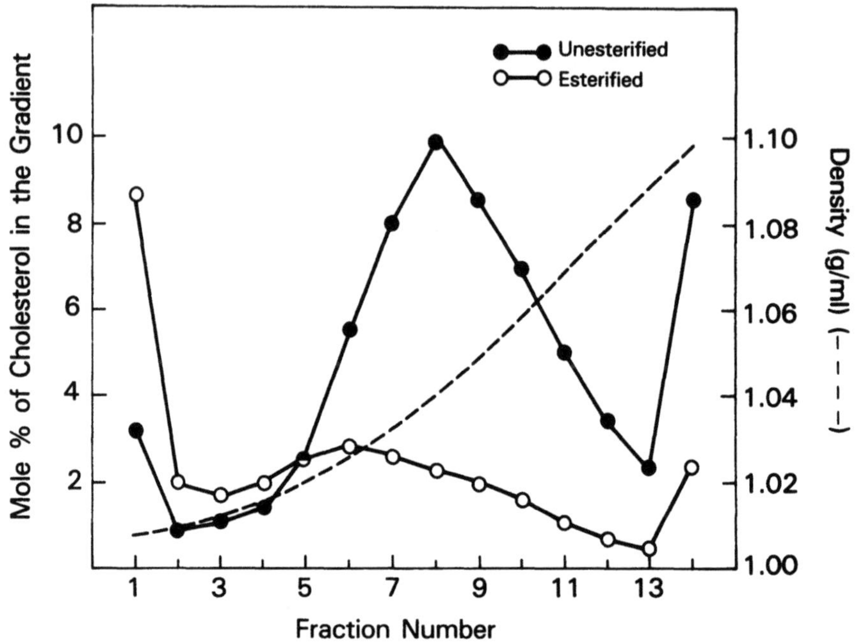

FIGURE 3. Typical cholesterol profile of separated EC-rich lipid droplets and UC-rich liposomes following NaCl density gradient centrifugation as described in Figure 2. Esterified cholesterol peaks at the top of the gradient (fraction #1) ($d < 1.01$ g/ml), while unesterified cholesterol peaks at density 1.04 g/ml. From Chao *et al.* (1990).

esterified in the extracellular EC-rich lipid droplets and plasma LDL is about 75%, whereas greater than 95% of cholesterol is esterified in intracellular lipid droplets.

On the other hand, the unesterified cholesterol to phospholipid molar ratio of plasma LDL and intracellular lipid droplets is about 1:1, while this ratio is greater than 2:1 for the extracellular EC-rich lipid droplets. The high cholesterol to phospholipid molar ratio in extracellular EC-rich lipid droplets may reflect the fact that substantial unesterified cholesterol also accumulates extracellularly in lesions (Buck and Rossiter, 1951; Bötcher and Woodford, 1962; Insull and Bartsch, 1966; Kruth, 1983; Kruth, 1984a; Kruth and Fry, 1984). This cholesterol could saturate the phospholipid monolayer that typically covers the surface of lipid droplets. Overall, extracellular EC-rich lipid droplets show a greater chemical similarity to LDL than to intracellular lipid droplets.

3.4. Extracellular EC-Rich Lipid Droplets May Be Derived from LDL That Fuses

Although extracellular lipid droplets visualized in lesions are smaller than intracellular lipid droplets, nevertheless, the *in situ* and isolated extracellular lipid

Table I
Lipid Composition of LDL-Sized Lipid Particles, Extracellular EC-Rich Lipid Droplets, and Extracellular UC-Rich Liposomes Isolated from Human Atherosclerotic Aortas[a]

Lipid constituent	LDL-sized lipid particles[b] (4)	EC-rich lipid droplets[c] (6)	UC-rich liposomes[c] (6)	
Total cholesterol (TC)	78 ± 2.7	91 ± 0.5[d]	76 ± 1.7	Mole %
unesterified (UC)	28 ± 0.8	22 ± 1.0[d]	59 ± 0.9	Mole %
esterified (EC)	50 ± 3.2	69 ± 1.3[d]	17 ± 1.5	Mole %
Phospholipid (PL)	22 ± 2.7	9 ± 0.5[d]	24 ± 1.7	Mole %
UC/PL	1.3 ± 0.1	2.5 ± 0.2	2.6 ± 0.2	molar ratio
% EC/TC	64 ± 3.1	76 ± 1.2[d]	22 ± 1.7	

[a]Values are means ± SEM. The number of separate aorta preparations is shown in parentheses.
[b]Adapted from Chao *et al.* (1988). Data are for lipid particles isolated in peak II as shown in Figure 1 herein.
[c]From Chao *et al.* (1990). Data are for lipid particles isolated as shown in the upper band ($d < 1.01$ g/ml) and lower band (peak d at 1.04 g/ml) of Figures 2 and 3 herein. Intracellular lipid droplets are larger than the EC-rich lipid droplets that accumulate in the extracellular space of lesions. Lipid particles the size of intracellular lipid droplets were removed by microfiltration during purification of the extracellular EC-rich lipid droplets. Thus, the isolated EC-rich fractions are primarily composed of extracellular EC-rich lipid droplets. The triglyceride content of EC-rich lipid droplets and UC-rich liposomes is less than 4% of their total lipid.
[d]Significant difference between EC-rich lipid droplets and UC-rich liposomes, $P < 0.001$.

droplets are about 5–10 times larger than LDL particles (Smith *et al.*, 1967; Bocan *et al.*, 1986; Chao *et al.*, 1990). Fusion of LDL particles is a mechanism that could explain the origin of these extracellular EC-rich lipid droplets. LDL is a stabilized emulsion of cholesteryl ester oil in water, in which protein (apo B) and lipid (phospholipid and unesterified cholesterol) surface components stabilize the 22-nm spherical LDL particles. Consequently, disruption of these stabilizing surface components may lead to aggregation and fusion of LDL particles. Indeed, it was shown some time ago that enzymatic modification of LDL protein or phospholipids with proteases or lipases, respectively, alters the physical properties of solutions of LDL (Horlick, 1954; Bernfeld and Kelley, 1964; Margolis and Langdon, 1966). LDL solutions treated with these enzymes become turbid, indicating that LDL particles fuse or aggregate, thereby altering how the solution of LDL scatters light.

More recently, Kovanen and coworkers (Kokkenen and Kovanen, 1985; Kokkenen *et al.*, 1986; Kovanen and Kokkenen, 1991) showed that certain proteases contained in granules released from mast cells *in vitro* (a cell present in small numbers within blood vessels) can induce fusion of LDL particles. Piha *et al.* (1995) and our laboratory (unpublished studies) found that a sufficient degree of apo B degradation is required to produce LDL fusion. Only certain proteases can induce fusion of LDL particles. Trypsin, pronase, and elastase produce

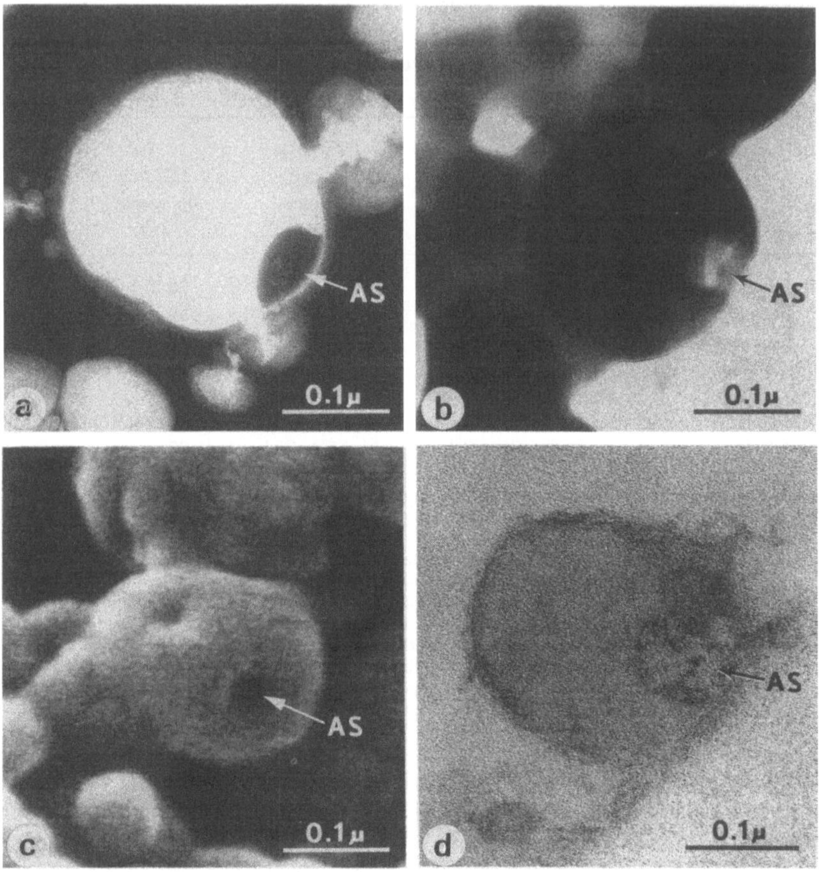

FIGURE 4. Ultrastructural analysis of isolated, extracellular EC-rich lipid droplets. (a), (b), and (d) were viewed with a transmission electron microscope; (c) was viewed with a scanning electron microscope. (a) EC-rich lipid droplets stained with phosphotungstic acid and viewed as a whole mount. (b) EC-rich lipid droplets fixed with OsO_4 and viewed as a whole mount. (c) EC-rich lipid droplets prepared with a gold coating and viewed as a whole mount. (d) EC-rich lipid droplets fixed with OsO_4, thin-sectioned, and stained with lead hydroxide. AS indicates an aqueous space in each of the droplets. From Chao *et al.* (1990).

LDL fusion, while collagenase, thrombin, and plasmin do not produce LDL fusion, even with prolonged (48 h) protease treatment of LDL (Piha *et al.*, 1995; and unpublished data from our laboratory). Limited treatment of LDL with trypsin (2 h) does not alter LDL size, consistent with previous reports (Chapman *et al.*, 1978; Chao *et al.*, 1992). However, prolonged treatment of LDL with trypsin (24 h) sufficiently degrades LDL such that some LDL fuses and forms larger particles (Figure 5). It remains to be determined what vessel wall proteases [e.g., chymase and carboxypeptidase A in mast cells or matrix metalloproteinases in smooth

Table II
Lipid Composition of Plasma LDL, Lesion Extracellular EC-Rich Lipid Droplets, and Lesion Intracellular Lipid Droplets[a]

Lipid constituent	Plasma LDL[b] (3)	Extracellular EC-rich lipid droplets[c] (6)	Intracellular lipid droplets[d] (21)	
Total cholesterol (TC)	80	91	96	Mole %
unesterified cholesterol (UC)	22	22	3	Mole %
esterified cholesterol (EC)	58	69	93	Mole %
Phospholipid (PL)	20	9	4	Mole %
UC/PL	1.1	2.5	0.8	molar ratio
% EC/TC	73	76	97	

[a]Values are means. The number of donors for plasma LDL or number of separate human aorta preparations used to isolate lipid particles is shown in parentheses.
[b]Unpublished data from our laboratory.
[c]Data from Chao *et al.* (1990).
[d]Data adapted from Lang and Insull (1970).

Table III
Fatty Acid Composition of Cholesteryl Esters in Plasma LDL, Lesion Extracellular EC-Rich Lipid Droplets, and Lesion Intracellular Lipid Droplets[a]

Fatty acid	Plasma LDL[b] (3)	Extracellular EC-rich lipid droplets[c] (4)	Intracellular lipid droplets[d] (21)
14:0	1	2	1
16:0	15	21	9
16:1	8	3	6
18:0	5	6	2
18:1	21	25	50
18:2	44	35	15
20:3	1	0	7
20:4	6	7	7

[a]Values are mean percentages. The number of donors for plasma LDL or number of separate human aorta preparations used to isolate lipid particles is shown in parentheses.
[b]Unpublished data from our laboratory.
[c]Adapted from *Chao et al.* (1990).
[d]From Lang and Insull (1970).

muscle cells and macrophages] (Kokkonen *et al.*, 1986; Galis *et al.*, 1994) can degrade apo B *in vivo.* It is important to note that apo B is also susceptible to breakdown due to lipoprotein-associated proteases as well as oxidation (Chung *et al.*, 1986; Fong *et al.*, 1987; Lee and Singh, 1988; Chulkova and Tertov, 1993).

Evidence from many studies shows the loss of immunoreactive apo B from lipoprotein particles in lesions. Colocalization of apo B and cholesteryl ester in

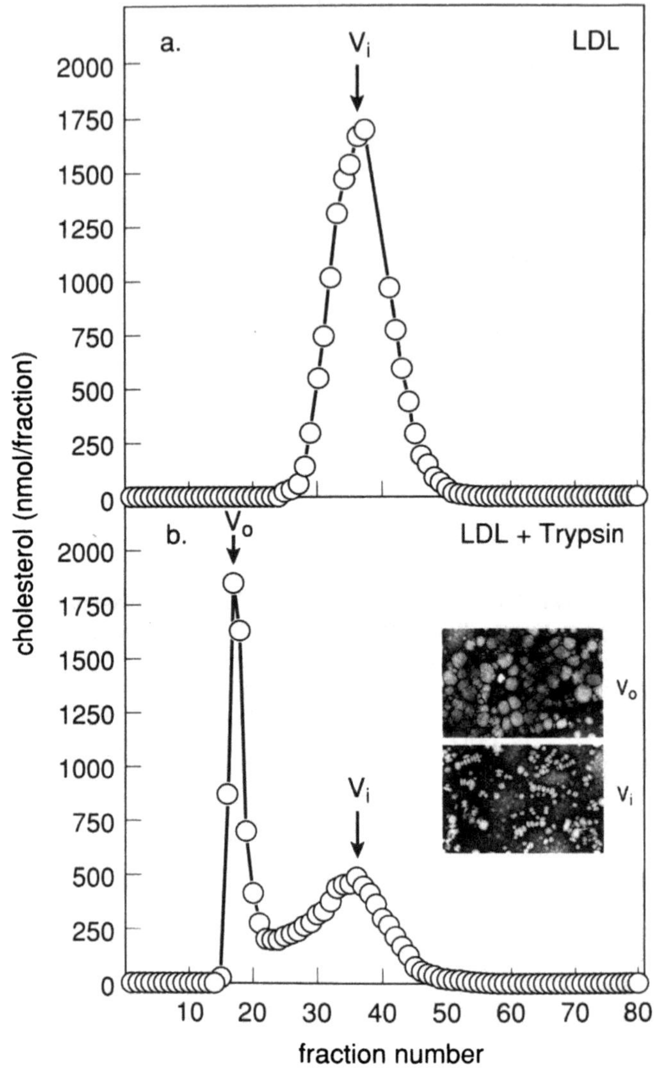

FIGURE 5. Fusion of plasma LDL induced by prolonged treatment of LDL with protease. Human plasma LDL (5 mg/ml) was incubated with bovine pancreas trypsin (1800 U/ml)(code TRTPCK, Worthington) for 24 h at 37°C and then gel-filtered through 2% agarose. Some of the protease-treated LDL elutes in the void volume (V_o) (b). These LDL particles have fused to form lipid particles up to three times the diameter of normal LDL particles (V_i). Control LDL incubated without trypsin does not fuse (a). (Unpublished data from our laboratory.)

human atherosclerotic lesions shows that extracellular EC-rich lipid particles found deep within the intima lack apo B (Kruth and Shekhonin, 1994). Apo B of LDL isolated from normal aortic tissue is mostly intact (Ylä-Herttuala *et al.*, 1988), while apo B of LDL isolated from lesioned aortic tissue is often degraded (Morton *et al.*, 1986; Steinbrecher and Lougheed, 1992; Rapp *et al.*, 1994). This suggests that changes in lesions predispose to apo B breakdown. The amount of apo B extractable from grossly normal aortic tissue can account for most of the cholesterol present in this tissue. On the other hand, even in lesions that lack foam cells (and thus also lack intracellular deposits of cholesterol), the amount of apo B present can only account for about 10% of the cholesterol accumulated in these lesions (Smith, 1981; Hoff *et al.*, 1979b). All these findings suggest that apo B is lost from LDL in lesions and that larger lipid particles form from this LDL.

Chylomicron remnants are plasma lipoproteins derived from lipolysis of chylomicrons after their release by the intestine. Chylomicron remnants are transiently elevated after eating and occur in certain genetic dyslipidemias (Patsch *et al.*, 1989; Mahley *et al.*, 1991). These lipoproteins have been implicated in the development of atherosclerosis in many studies (Zilversmit, 1979; reviewed in Havel, 1994). The extracellular EC-rich lipid droplets isolated from lesions are similar to chylomicron remnants in size, density (<1.01 g/ml), and enrichment with cholesteryl ester ($>75\%$ of cholesterol is esterified) (Sata *et al.*, 1972; Fainaru *et al.*, 1982). Thus, chylomicron remnants may be an additional source of some extracellular EC-rich lipid droplets found in lesions.

4. LIPOSOMES RICH IN UNESTERIFIED CHOLESTEROL (UC) ALSO ACCUMULATE IN ATHEROSCLEROTIC LESIONS

4.1. Identification and Characterization of UC-Rich Liposomes in Atherosclerotic Lesions

A second lipid particle, with unusual staining characteristics, has been identified within the extracellular space of human and animal atherosclerotic lesions (Kruth, 1983; Kruth, 1984b; Kruth and Fry, 1984). These lipid particles are spherical and label with filipin directly, without cholesterol esterase treatment, indicating that the particles contain unesterified cholesterol. The particles do not stain with lipid-soluble dyes such as oil red-O, which shows an absence of an oily cholesteryl ester core that is characteristic of the extracellular EC-rich lipid droplets. The staining characteristics of these filipin positive, oil red-O negative particles are consistent with the presence of unesterified cholesterol (UC)-rich liposomes in lesions. Indeed, ultrastructural studies of atherosclerotic lesions often reveal the presence of small extracellular liposomal-like structures, although no specific origin for these particles was suggested from these studies (Geer and Guidry, 1964;

Table IV
Phospholipid Composition of UC-Rich Liposomes[a]

Phospholipid class	Cholesterol-fed NZW rabbits 11 wks (3)	16 wks (3)	30 wks (2)	Humans > 40 yrs (4)	
Phosphatidylcholine	53	52	41	29	Mole %
Sphingomyelin	29	36	44	57	
Lysophosphatidylcholine	8	5	11	9	
Other	9	7	4	5	
PC/SPM	1.8	1.4	0.9	0.5	molar ratio
UC/PL	2.4	2.0	2.5	2.0	

[a]Rabbits were fed 1% cholesterol for the indicated number of weeks. The number of aortas shown in parentheses were pooled for isolation of UC-rich liposomes from rabbits. Values for humans are means of UC-rich liposomes isolated from 4 separate aortas. PC, phosphatidylcholine; SPM, sphingomyelin; UC, unesterified cholesterol; PL, phospholipid. Rabbit data adapted from Chao *et al.* (1994); human data adapted from Chao *et al.* (1990).

Moss and Benditt, 1970; Tucker *et al.*, 1971; Hoff, 1972; Trillo and Prichard, 1979; Toda *et al.*, 1980; Yamada *et al.*, 1988).

Following the identification of UC-rich liposomes in atherosclerotic lesions, these particles were isolated from lesions (Chao *et al.*, 1988). The UC-rich liposomes have a >2:1 molar ratio of unesterified cholesterol to phospholipid (Table I), are enriched in sphingomyelin (Table IV), range in size from 40–200 nm in diameter, and have densities of 1.03–1.05 g/ml (Figures 2 and 3). The structure of the liposomes isolated from lesions shows variation with respect to whether the liposomes contain phospholipids organized in a lamellar or hexagonal phase (Bangham, 1963), and whether the liposomes have few (oligolamellar) or many (multilamellar) membranes (Figure 6). These differences in structure suggest the possibility that not all UC-rich liposomes in lesions share the same origin.

4.2. Analysis of UC-Rich Liposomes Accumulating in Developing Atherosclerotic Lesions of Rabbits

Investigators have visualized both extracellular UC-rich liposomes and EC-rich lipid droplets in the extracellular space of early developing atherosclerotic lesions in cholesterol-fed rabbits (Kruth, 1985a; Simionescu *et al.*, 1986; Guyton and Klemp, 1992; Simionescu *et al.*, 1993). While UC-rich liposomes progressively accumulate in aortas of cholesterol-fed rabbits as lesions develop (Figure 7), very little EC-rich lipid droplets accumulate in these aortas (Stehbens and Ludatscher, 1983; Simionescu, 1986; Chao *et al.*, 1994). Possible reasons for this are that EC-rich droplets accumulating in lesions are rapidly converted to UC-rich liposomes, or that macrophages rapidly take up the EC-rich lipid droplets in

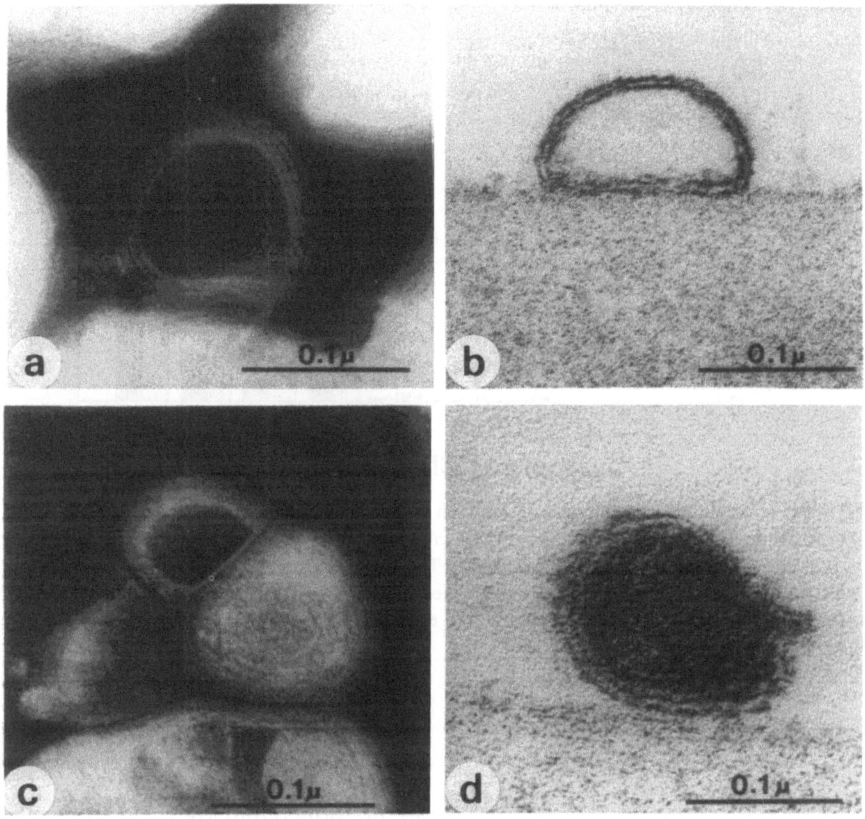

FIGURE 6. Ultrastructural analysis of isolated UC-rich liposomes. (a) and (c) are negatively stained whole mount preparations of particles. (b) and (d) are thin-sectioned particles fixed with OsO_4 and stained with lead hydroxide. (a) An oligolamellar liposome is shown. The aqueous center of the liposome is demonstrated by the dark (electron opaque) image created by phosphotungstic acid stain that has penetrated into the liposome. (b) Thin section of an oligolamellar liposome. (c) Oligolamellar liposome with central aqueous space next to a solid multilamellar liposome. (d) Thin section of a solid multilamellar liposome. From Chao *et al.* (1990).

lesions. On the other hand, in human aortas, the ratio of total cholesterol in EC-rich lipid droplets relative to UC-rich liposomes varies greatly between 0.4–2.7 (Chao *et al.*, 1990).

The earliest detectable UC-rich liposomes isolated from aortas of cholesterol-fed rabbits show a high molar ratio of unesterified cholesterol to phospholipid (>2:1) characteristic of the UC-rich liposomes isolated from human aortas (Tables I and IV) (Chao *et al.*, 1994). This suggests that UC-rich liposomes are not derived from progressive cholesterol enrichment of cellular membranes that usually have an unesterified cholesterol to phospholipid molar ratio of <1:1. Rather,

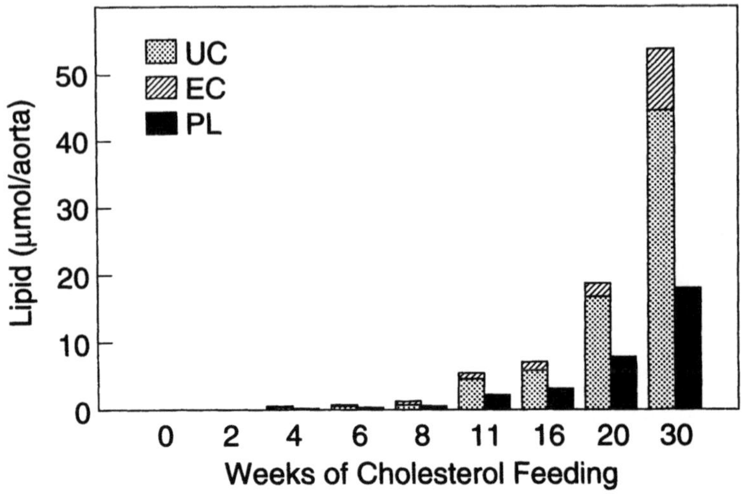

FIGURE 7. Progressive accumulation of UC-rich liposomes in aortas of cholesterol-fed New Zealand White rabbits. UC-rich liposomes of $d = 1.04$ g/ml were isolated by gel-filtration chromatography and density-gradient centrifugation procedures. The values shown were determined from the peak density fraction of the isolated UC-rich liposomes, and thus underestimate the total amount of UC-rich liposomes. UC, unesterified cholesterol; EC, esterified cholesterol; PL, phospholipid. From Chao *et al.* (1994).

the findings suggest that UC-rich liposomes are formed initially with a high unesterified cholesterol to phospholipid molar ratio.

Besides cholesterol, sphingomyelin also accumulates with development of atherosclerotic lesions in blood vessels (Rouser and Solomon, 1968; Smith, 1974 and contained references). UC-rich liposomes are one locus of this accumulating sphingomyelin and also other glycosphingolipids (Mukhin *et al.,* 1995). In fact, sphingomyelin is the major phospholipid of UC-rich liposomes isolated from human aortas (Table IV). UC-rich liposomes formed early in developing lesions of cholesterol-fed rabbits contain phosphatidylcholine as their major phospholipid (Chao *et al.,* 1994). However, with increasing times of cholesterol feeding, sphingomyelin replaces phosphatidylcholine as the predominant phospholipid in these UC-rich liposomes. The switch in phosphatidylcholine to sphingomyelin predominance in the accumulating UC-rich liposomes may partly reflect the higher affinity of sphingomyelin compared with other phospholipids for cholesterol (see Chapter 10 by Slotte).

Experimental evidence suggests that sphingomyelin accumulation in lesions results not just from deposition of plasma-derived lipoproteins (lipoproteins contain much more phosphatidylcholine than sphingomyelin), but also comes from synthesis of sphingomyelin within the vessel wall (McCandless and Zilversmit,

1956; Seth and Newman, 1975). This is because the phosphatidylcholine to sphingomyelin molar ratio in serum of cholesterol-fed rabbits is >2.5:1, even when UC-rich liposomes isolated from the aortas of these rabbits show a phosphatidylcholine to sphingomyelin molar ratio of <1:1 (Chao *et al.,* 1994). Moreover, cholesterol stimulates synthesis of sphingomyelin in vascular cells, including macrophages (Schmitz *et al.,* 1990; Greemland and Savion, 1993; Okwu *et al.,* 1994). Also, cholesterol accumulation in tissues of individuals with Niemann–Pick type C disease is accompanied by what seems to be a compensatory increase in tissue sphingomyelin (Vanier, 1983 and see Chapter 14 by Pentchev *et al.*). Thus, it is likely that the shift in phospholipid predominance from phosphatidylcholine to sphingomyelin in atherosclerotic lesions reflects not only the high affinity of sphingomyelin for cholesterol, but also the metabolism of the vessel wall in response to deposited cholesterol.

4.3. The Origin of UC-Rich Liposomes in Atherosclerotic Lesions

The mutually exclusive distribution of extracellular EC-rich lipid droplets and UC-rich liposomes in lesions suggests that one type of extracellular lipid particle converts into the other type of lipid particle (Kruth, 1984b; Kruth and Fry, 1984). Chao *et al.* (1992) demonstrated a mechanism by which EC-rich lipid particles can convert to UC-rich liposomes that show a >2:1 unesterified cholesterol to phospholipid molar ratio. Hydrolysis of the cholesteryl ester core of LDL with cholesterol esterase converts this 22-nm oil droplet into larger UC-rich liposomes (65–160 nm in diameter) (Figure 8). These LDL-derived liposomes are similar in size to the extracellular UC-rich liposomes found in atherosclerotic lesions. Likewise, EC-rich lipid droplets isolated from human atherosclerotic lesions also transform into UC-rich liposomes following enzymatic hydrolysis of their cholesteryl ester (unpublished data from our laboratory). This hydrolytic mechanism of formation of UC-rich liposomes can explain the finding that the earliest detectable UC-rich liposomes in lesions show a >2:1 unesterified cholesterol to phospholipid molar ratio, rather than the liposomes showing a progressive increase in this ratio (see section 4.2).

Cholesterol esterases found in plasma (Zolfaghari *et al.,* 1992) and in cells of the vessel wall (Hajjar *et al.,* 1983) potentially could hydrolyze the extracellular EC-rich lipid particles in lesions. Liposome formation from extracellular EC-rich lipid particles (e.g., LDL particles and EC-rich lipid droplets) could conceivably occur following cell death and release of cellular cholesterol esterase in lesions. Advanced stages of lesion development typically show focal regions of cell death, usually in the deepest regions of the plaque where a lipid core forms. Release of cellular cholesterol esterases in the lipid core could transform EC-rich lipid particles into the UC-rich liposomes found in the lipid cores of human and WHHL rabbit atherosclerotic lesions (WHHL is a type of rabbit with spontaneous, genetically determined hypercholesterolemia) (Figure 6 of Guyton and Klemp, 1988;

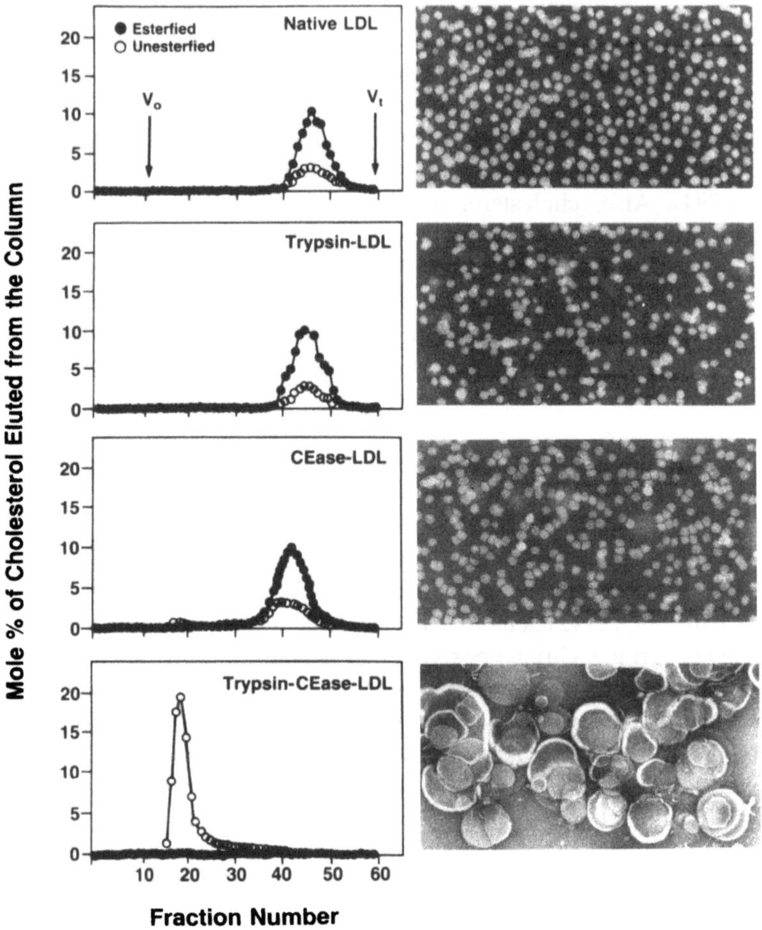

FIGURE 8. Formation of UC-rich liposomes from plasma LDL. LDL was incubated at 37°C with buffer (i.e., native LDL), trypsin, cholesterol esterase, or trypsin followed by cholesterol esterase (2 h for each treatment). The treated LDL samples were gel-filtered through 2% agarose. Native LDL, trypsin-treated LDL, and cholesterol esterase-treated LDL all eluted substantially in the internal volume of the column. Only trypsin-cholesterol esterase-treated LDL eluted exclusively with the void volume of the column, due to the conversion of LDL into UC-rich liposomes. Negatively stained lipoprotein particles for each treatment are shown to the right of each elution pattern. From Chao *et al.* (1992).

Figure 2 of Guyton and Klemp, 1989; unpublished observations of WHHL lesions from our laboratory). The pitting of the surface of some extracellular EC-rich lipid droplets may represent sites where cholesterol esterase has acted on these lipid particles (Guyton and Klemp, 1989; Chao *et al.*, 1990).

Besides occurring in lesion cores, UC-rich liposomes often surround foam cells in areas of lesions that do not show evidence of cell death. This suggests a link between formation of these UC-rich liposomes and the process of foam cell formation (Geer and Guidry, 1964; Tucker *et al.*, 1971; Toda *et al.*, 1980; Kruth, 1983; Kruth, 1984b; Yamada *et al.*, 1988). One possibility is that monocyte-macrophages take up extracellular EC-rich lipid droplets and lipoproteins and then hydrolyze the cholesteryl ester in these lipid particles. Evidence for this possibility comes from studies of Schmitz and Robenek (Schmitz *et al.*, 1988; Robenek and Schmitz 1988) who showed that, under certain conditions, cultured mouse peritoneal macrophages produce UC-rich liposomes during processing of acetylated LDL.

Macrophages also excrete unesterified cholesterol in discoidal cholesterol–phospholipid complexes following uptake of acetylated LDL (Basu *et al.*, 1982; Kruth *et al.*, 1994). The amphipathic protein, apo E, stabilizes these discoidal cholesterol–phospholipid complexes. Treatment of these discoidal particles with the protease chymotrypsin converts the discs into liposomes (Figure 9). This occurs presumably because of loss of the stabilizing effect of apo E. If small discoidal particles are transformed into large liposomes in lesions, this could contribute to the occurrence of UC-rich liposomes around foam cells. Also, conversion of small discs to large liposomes should hinder transport of cholesterol out of lesions by the discoidal lipoproteins (discussed in section 8).

Still, other sources of UC-rich liposomes in atherosclerotic lesions have been suggested. Activated platelets release cholesterol-containing liposomes (Sandberg *et al.*, 1985; Skarlatos *et al.*, 1988). The platelet liposomes promote clot formation by providing phospholipid, a procoagulant. At the same time that the platelet lipid particles function in hemostasis, the particles could be a source of lesion cholesterol if they should enter the vessel wall. Cultured macrophages and smooth muscle cells take up the platelet lipid particles, which causes significant cholesterol accumulation in these cells (Kruth, 1985b; Curtiss *et al.*, 1987; Mendelsohn and Loscalzo, 1988). Lastly, it has been proposed that surface lipids (mainly unesterified cholesterol and phospholipid) from triglyceride-rich lipoproteins (e.g., chylomicrons and very low density lipoproteins) are another possible source of UC-rich liposomes found in lesions (Chung *et al.*, 1994). This lipid is released as the triglyceride-rich lipoproteins undergo lipolysis at the endothelial surface.

Interestingly, no UC-rich liposomes accumulate in arcus lipoides of the cornea. Two major differences between lipid deposition in the cornea and vessel wall are that cell necrosis and macrophages are absent in the mature cornea. Thus, in cornea, there are no dying cells to release cholesterol esterases, and there are

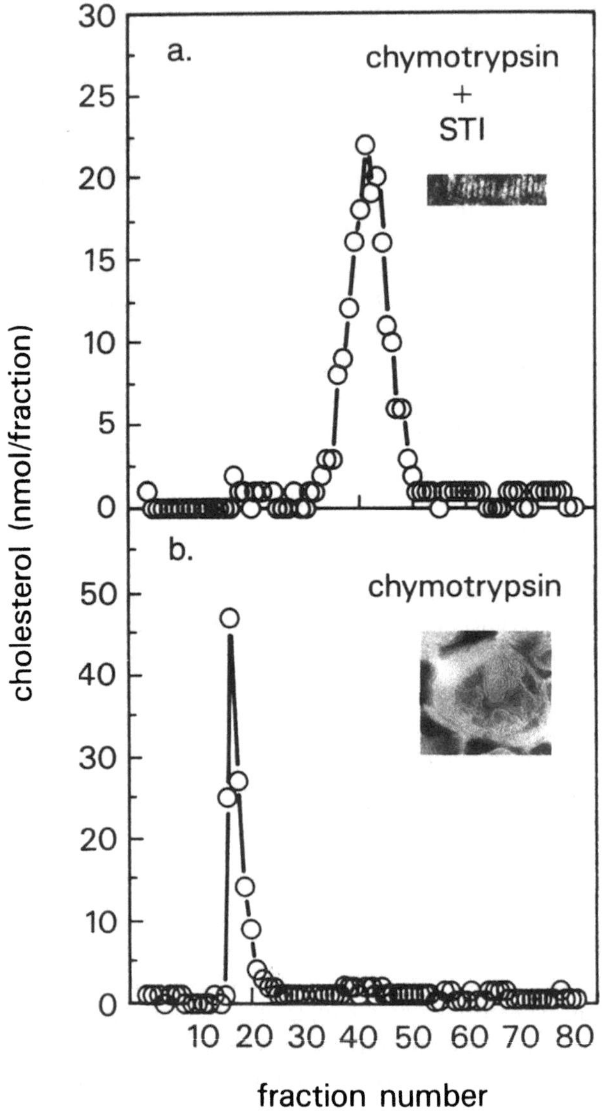

FIGURE 9. Formation of UC-rich liposomes from macrophage-derived discoidal HDL. Apo E-containing discoidal HDL particles, produced by cultured cholesterol-enriched human monocyte-derived macrophages as described in Kruth *et al.* (1995), were incubated for 4 h at 37°C with (a) bovine pancreas chymotrypsin (35 U/ml)(cat. no. 230832, Calbiochem) that was inhibited with soybean trypsin inhibitor (2.3 mg/ml)(code SI, Worthington) or (b) chymotrypsin alone. Gel-filtration of the incubated mixtures through 2% agarose shows that chymotrypsin converts the small discoidal particles into larger liposomes that elute in the column void volume. Insets show negatively stained lipid particles following incubations. (Unpublished data from our laboratory.)

no macrophages that could hydrolyze the cholesteryl ester in extracellular EC-rich lipid particles. This may explain why UC-rich liposomes do not occur in the cornea.

5. CRYSTALLINE FORMS OF CHOLESTEROL IN ATHEROSCLEROTIC LESIONS

Cholesterol in atherosclerotic lesions occurs not only in droplets and liposomes, but also in crystals that may be outside or inside cells, appearing free within the cytoplasm or within lysosomes (Stary and Malinow, 1982; Lupu *et al.,* 1987; Nakamura and Ohtsubo, 1992). Unesterified cholesterol is very insoluble in water and would normally crystallize at body temperature (Renshaw *et al.,* 1983). However, association of unesterified cholesterol with phospholipid (such as in liposomes or cell membranes) prevents unesterified cholesterol from crystallizing (Small and Shipley, 1974; Small, 1988). In addition, conversion of unesterified cholesterol to an oil through esterification of cholesterol with a fatty acid can also prevent crystallization. This is because most cholesteryl esters do not crystallize at body temperature (Stewart, 1961; Small and Shipley, 1974; Lundberg, 1985). On the other hand, lipase-mediated hydrolysis of phospholipid (associated with unesterified cholesterol in membranes of cells and liposomes) or cholesteryl ester (in lipoproteins, extracellular EC-rich lipid droplets, and intracellular EC-rich lipid droplets) both produce free cholesterol that, in sufficient concentration, should crystallize within lesions.

Release of lipases (e.g., cholesterol esterases and phospholipases) from dying cells in the lesion core (as discussed in section 4.3) can explain the formation of the flat plate-shaped cholesterol crystals found in the lesion core (Figure 10a) (Small and Shipley, 1974; Katz *et al.,* 1976; Katz and Small, 1980; Small, 1988). Cholesterol crystals can occur not only as flat plates but also as rods, depending on the conditions of crystallization (Garti *et al.,* 1981). This may explain the occurrence of a second type of cholesterol crystal—rod-shaped—that can be isolated from atherosclerotic lesions (Figure 10b). Similar crystals accumulate in lysosomes of hepatocytes from mice that suffer from a cholesterol storage disease similar to Niemann-Pick type C disease in humans (Shio *et al.,* 1982). The mouse and human diseases demonstrate defective trafficking of cholesterol that results in buildup of unesterified cholesterol in lysosomes.

Normally, uptake and hydrolysis of lipoprotein-derived cholesteryl esters do not result in formation of intracellular cholesterol crystals. However, it has been recently shown that uptake of cholesteryl ester lipid droplets into macrophages results in formation of intralysosomal cholesterol crystals (Tangirala *et al.,* 1994). This occurs as the accumulated cholesteryl ester undergoes hydrolysis within the lysosomes. The rate and extent to which cells accumulate cholesterol, and also the degree to which a cell can solubilize accumulated unesterified cholesterol (by its

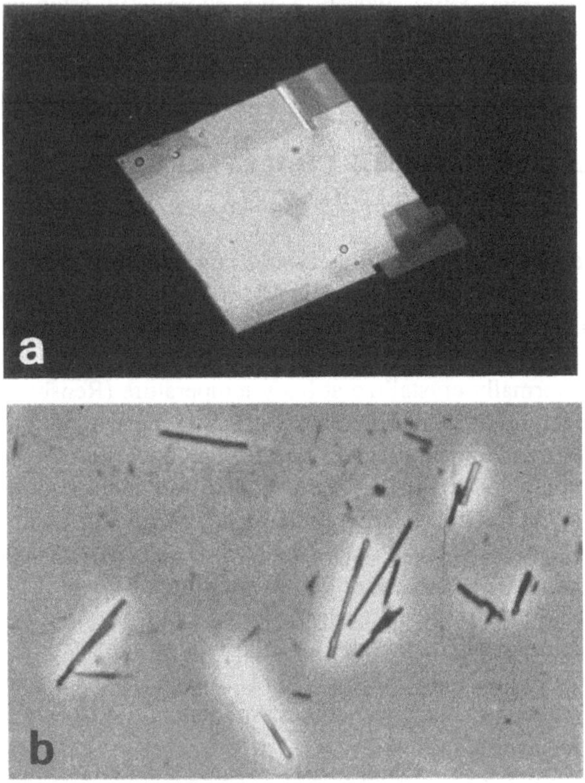

FIGURE 10. Photomicrographs of plate (a) and rod-shaped (b) cholesterol crystals isolated from human atherosclerotic lesions. (Unpublished data from our laboratory.)

association with phospholipid or conversion to an oil) should determine whether accumulated cholesterol crystallizes in cells.

6. THE NATURE OF LIPID IN EARLY DEVELOPING ATHEROSCLEROTIC LESIONS

Determining the nature of the lipid particles that first accumulate within the subendothelial space of beginning lesions has been difficult. Studies show droplet and liposomal lipid particles and also apo B within early developing lesions (Simionescu *et al.*, 1986; Mora *et al.*, 1987; Nistor *et al.*, 1987; Guyton and Klemp, 1992; Tirziu *et al.*, 1995). However, it has not been possible to determine whether plasma lipoproteins first enter the subendothelial space and then transform into larger EC-rich lipid droplets, or whether lipoproteins form EC-rich lipid droplets (in the plasma or during transport across the endothelium) before they enter the subendothelial space. As discussed above, although unesterified cholesterol-rich liposomes are frequent in early developing lesions, their prevalence could reflect a rapid conversion of lipoproteins or EC-rich lipid droplets into these liposomes within the subendothelial space.

In any case, subendothelial lipid accumulates before accumulation of subendothelial foam cells. Early developing lesions in cholesterol-fed rabbits show subendothelial accumulation of lipid by week two, before cellular accumulation of lipid occurs, beginning at week four (Kruth, 1985a; Simionescu *et al.*, 1986). In one intriguing study, injection of LDL into the bloodstream of a rabbit resulted in the subendothelial accumulation of lipid particles (the same size or larger than LDL) within 2 hours of the injection (Nievelstein *et al.*, 1991). These observations of early developing lesions are consistent with the proposal that extracellular lipid results in foam cell formation, rather than the reverse.

7. ACCUMULATION OF EXTRACELLULAR LIPID BY MACROPHAGES

7.1. Plasma Lipoproteins and Macrophage Cholesterol Accumulation

Whether cellular accumulation of extracellular lipid in lesions helps to clear this lipid from lesions or further traps this lipid remains controversial. In either case, there has been great interest in examining the interaction of plasma lipoproteins and lesion-derived lipid particles with vascular wall cells, such as macrophages. Of the plasma lipoproteins considered atherogenic, only very low density lipoproteins that exhibit β-mobility (i.e., β-VLDL) cause substantial cholesterol accumulation in macrophages (Mahley *et al.*, 1980; Fainaru *et al.*, 1982; Rotheneder and Kostner, 1986; Gianturco and Bradley, 1988; Skarlatos *et al.*, 1993a). These β-VLDL include chylomicron remnants and cholesteryl ester-rich lipoproteins synthesized by the liver. On the other hand, the major atherogenic lipoprotein in plasma, LDL, induces little or no cholesterol accumulation in macrophages. Thus, it is not clear how native LDL could cause cholesterol accumulation in macrophages in atherosclerotic lesions.

A possible explanation of how LDL can cause cholesterol accumulation in macrophages is that native LDL becomes modified. It has been shown that LDL can be modified (by e.g., acetylation, oxidation, glycation, or desialylation) such that it is bound and taken into macrophages by receptors that do not ordinarily bind native LDL (Goldstein *et al.*, 1979; Mahley *et al.*, 1979; Fogelman *et al.*, 1980; Henriksen *et al.*, 1981; Shechter *et al.*, 1981; Henriksen *et al.*, 1983; Lopes-Virella *et al.*, 1991; Hoff *et al.*, 1989; Orekhov *et al.*, 1991a; Orekhov *et al.*, 1992). With many of these modifications, modified LDL that has aggregated induces greater cholesterol accumulation in macrophages than modified LDL that remains unaggregated (Tertov *et al.*, 1989; Hoff *et al.*, 1989; Hoff and O'Neil, 1991; Hoff and Cole, 1991; Hoff *et al.*, 1992; Tertov *et al.*, 1992; Kawabe *et al.*, 1994). Aggregation of LDL particles induced by vortexing or treatment with either phospholipase C or sphingomyelinase (Khoo *et al.*, 1988; Suits *et al.*, 1989; Heinecke *et al.*, 1991; Xu and Tabas, 1991) also converts LDL to a form that can induce cholesterol accumulation in macrophages. Additionally, complexing of LDL with glycosaminoglycans or antibodies promotes macrophage cholesterol accumulation, possibly mediated by receptors that recognize the glycosaminoglycans or the antibodies, in addition to the LDL (Falcone *et al.*, 1984; Orekhov *et al.*, 1987; Orekhov *et al.*, 1989; Tertov *et al.*, 1990; Hurt *et al.*, 1990; Lopes-Virella, 1991; Orekhov *et al.*, 1991b; Khoo *et al.*, 1992; Vijayagopal *et al.*, 1992; Viyayagopal *et al.*, 1993; Morganelli *et al.*, 1995).

What most of these modifications of LDL have in common is that they convert monomeric LDL to aggregated, polymeric LDL. Aggregated LDL is multivalent and has the potential to cross-link the receptors to which it binds. Crosslinking of receptors is essential to triggering many cellular responses, including some types of endocytosis (Tabas *et al.*, 1991; Miyamoto *et al.*, 1995). Multivalency of ligand may be a factor in the uptake of β-VLDL into surface-connected tubules of mouse peritoneal macrophages and acetylated LDL into surface-connected compartments of human monocyte-derived macrophages (Myers *et al.*, 1993; Kruth *et al.*, 1995). Lipoprotein uptake into these tubules and compartments seems to influence the rate of processing and fate of these lipoproteins within the macrophage.

7.2. Do Lipid Particles Isolated from Atherosclerotic Lesions Induce Cholesterol Accumulation in Human Monocyte-Macrophages?

LDL-sized lipoproteins and some larger lipid particles isolated from lesions and purified by gel-filtration chromatography can induce cholesterol accumulation in macrophages (Clevidence *et al.*, 1984; Morton *et al.*, 1986; Hoff and Clevidence, 1987; Ylä-Herttuala *et al.*, 1989; Hoff and O'Neil, 1991; Steinbrecher and Lougheed, 1992; Vijayagopal *et al.*, 1992; Ismail *et al.*, 1994). It has been suggested that the tendency of lesion LDL to aggregate (from oxidative changes to this LDL or interaction of the LDL with connective tissue components) contributes to the capacity of lesion LDL to promote cellular cholesterol accumulation.

Ultrastructural studies show liposomes within vacuoles of lesion macrophages, consistent with uptake of some liposomes from the extracellular space (Moss and Benditt, 1970; Hoff, 1972; Haley *et al.*, 1977). However, we have found that neither UC-rich liposomes nor EC-rich lipid droplets isolated from human atherosclerotic lesions induce cholesterol accumulation in human monocyte-macrophages (unpublished data from our laboratory). UC-rich liposomes formed *in vitro* from LDL (as described in section 4.3) induce massive cholesterol accumulation in human monocyte-macrophages (unpublished data from our laboratory). Possibly, this does not occur when apo B in these liposomes is extensively degraded. UC-rich liposomes and EC-rich lipid droplets detected in, and isolated from, atherosclerotic lesions lack apo B (Mora *et al.*, 1990; Kruth and Shekhonin, 1995, and unpublished data from this laboratory).

In some studies, large lipid particles lacking apo B isolated from lesions appear to induce cholesterol accumulation in macrophages *in vitro*. This may be due to the presence of small apo B-containing lipoproteins aggregated with the larger lipid particles (Hoff and Clevidence, 1987; Steinbrecher and Lougheed, 1992). Besides cholesterol in droplets and liposomes, macrophages have been incubated with cholesterol crystals. Artificially produced cholesterol crystals induce substantial cholesterol accumulation in macrophages (Werb and Cohn, 1972; Koren *et al.*, 1991; Skarlatos *et al.*, 1993b). Cholesterol crystals isolated from lesions have not yet been tested for their ability to enrich cultured vascular cells with cholesterol.

8. PATHWAYS BY WHICH LIPIDS CAN LEAVE THE VESSEL WALL

Cholesterol that accumulates in normal regions of the vessel wall is much less than would be predicted based on the rate of influx of plasma lipoprotein cholesterol into the vessel wall (Stender and Hjelms, 1984). Thus, there must be processes that normally remove lipid from the vessel wall as well as atherosclerotic lesions. One potential mechanism for removal of cholesterol from the vessel wall is through the action of HDL particles that circulate in the blood and are also present in the vessel wall (Heideman and Hoff, 1982; Smith, 1990).

Many studies suggest that HDL functions as an anti-atherogenic lipoprotein (reviewed in Pieters *et al.*, 1994). HDL particles could infiltrate the intima from the blood stream or be produced *de novo* within the intima by monocyte-macrophages (discussed in section 4.3). These HDL particles have the potential to pick up cholesterol from cells (Ho *et al.*, 1980; Chapter 9 by Johnson *et al.*) and the extracellular space, and to carry that cholesterol out of the vessel wall. Only HDL particles are small enough to filter through the internal elastic membrane that separates the intima (the luminal layer of the vessel wall where lipid accumulates) from the rest of the vessel wall (Shimamoto, 1975; Smith and Staples, 1980; Smith and Ashall, 1983; Nordestgaard *et al.*, 1990). Passage through the internal elastic membrane is necessary for lipoproteins to reach the lymphatic drainage of the vessel wall.

HDL can solubilize cholesterol crystals (Adams and Abdulla, 1978; Abdulla and Adams, 1978) and also take up cholesterol (Jonas *et al.*, 1978; Gordon *et al.*, 1983; Koo *et al.*, 1985). Preliminary data from our laboratory suggests that HDL can also partially solubilize UC-rich liposomes and EC-rich lipid droplets. While cholesterol enrichment of artificial liposomes generally makes these liposomes resistant to disruption by HDL (Finkelstein and Weissmann, 1979; Scherphof, 1979; Kirby *et al.*, 1980; Tall *et al.*, 1980; Allen, 1981), some apparent breakdown of lesion UC-rich liposomes occurs (Chao *et al.*, 1991).

Lipoproteins larger than HDL that are too large (e.g., LDL) to filter through an intact internal elastic membrane can, presumably, only leave the intima by moving back across the endothelium into the bloodstream. Much investigation of lipoprotein transport from the blood into the vessel wall has been carried out. However, only a few studies have examined the potential for cell-mediated transport of lipoproteins and lipid out of the vessel wall (Cotran, 1965; Robinson and Apkarian, 1991). Build up of subendothelial lipid could drive such a process in which endothelial cells sense and transport lipid back into the blood. Also, monocyte-macrophages can leave lesions with their accumulated lipid by migrating between endothelial cells (Gerrity, 1981; Joris *et al.*, 1983). However, the quantitative significance of cell-mediated or the other possible mechanisms of cholesterol removal from lesions has not been determined.

9. WHY DOES LIPID ACCUMULATE IN THE VESSEL WALL?

Lipid could accumulate in the vessel wall because of both increased uptake into the wall and decreased removal from it (Table V). An elevated concentration of lipoproteins in the blood is a risk factor for the development of atherosclerosis, presumably because this provides more lipid that can potentially enter the vessel wall. Importantly, accumulation of lipid and development of lesions in the vessel wall are not random processes but occur at selective sites (McGill *et al.*, 1957; Gerrity *et al.*, 1977; Cornhill *et al.*, 1990). Some studies show that increased uptake of plasma lipoproteins and other colloidal substances by endothelial cells

Table V
Possible Factors Influencing Lipid Accumulation in Atherosclerotic Lesions

Increased Entry of Lipid into Lesions

1. Elevated concentration of lipid in plasma.
2. Increased uptake and transport of lipid into subendothelial space by endothelial cells.
3. Increased uptake of lipid through intercellular spaces of endothelium.

Decreased Removal of Lipid from Lesions

1. Lipid becomes trapped in intima because it cannot pass through internal elastic membrane.
2. Endothelium fails to transport intimal lipid back into the plasma.
3. Lipid sequestered in foam cells that fail to migrate out of lesion.
4. Lipid binds to connective tissue components.
5. Lipid particles enlarge in lesion and are physically trapped by connective tissue components.
6. Insufficient removal of lipid by HDL.

occurs at these sites (Duff *et al.*, 1954; Shimamoto *et al.*, 1976; Stemerman *et al.*, 1986; Wiklund *et al.*, 1987; Nielsen *et al.*, 1992). The presence of endothelial cells with specialized properties is a possible factor determining the occurrence of such sites. Some endothelial cells show characteristics of macrophages, such as phagocytosis (Ryan *et al.*, 1986) and scavenger type receptors that function in uptake of modified lipoproteins (Stein and Stein, 1980; Baker *et al.*, 1984). Thus, modifications of lipoproteins that enhance lipoprotein uptake by macrophages could also stimulate lipoprotein uptake by those endothelial cells with macrophage-like characteristics.

Physical trapping within the connective tissue matrix, adherence to the connective tissue matrix, and uptake by macrophages and smooth muscle cells in the vessel wall cells all could contribute to vessel wall retention of lipoproteins and lipid particles (reviewed by Williams and Tabas, 1995). In fact, one study in rabbits suggests that a slower rate of efflux of plasma lipoproteins, rather than a faster rate of influx, differentiates lesion-prone from lesion-susceptible sites within the aorta (Schwenke and Carew, 1989a, 1989b). Size should be an important parameter influencing lipid particle retention. This is because the greater the size of the lipid particle, the greater is the likelihood that it would be trapped within the connective tissue matrix of the vessel wall (Nordestgaard *et al.*, 1995). Indeed, the rate at which lipoproteins leave the vessel wall is inversely proportional to their size.

It has been suggested that lipoprotein binding to connective tissue components (e.g., elastin, collagen, glycosaminoglycans) or fibrin in lesions (Srinivasin *et al.*, 1975; Smith *et al.*, 1976; Walton and Morris, 1977; Mawhinney *et al.*, 1978; Nakashima *et al.*, 1985; Camejo *et al.*, 1985; Linden *et al.*, 1986; Smith and Cochran, 1990; Berenson *et al.*, 1990) could account for lipoprotein retention in lesions. Interestingly, besides a loose bound fraction of apo B (the protein marker of atherogenic lipoproteins such as LDL) a tight bound fraction of apo B occurs in atherosclerotic lesions but is not found in the normal vessel wall (Hoff *et al.*, 1977; Hoff *et al.*, 1978a; Hoff *et al.*, 1978b). Moreover, apo(a), the protein marker for the atherogenic lipoprotein Lp(a), is mostly tightly bound within atherosclerotic plaques (Pepin *et al.*, 1991). Further, lipoprotein particles that have lost their surface protein show even greater binding to connective tissue (Paananen and Kovanen, 1994; Paananen *et al.*, 1995).

Lipid may not only become trapped in the extracellular space, but also can be trapped within cells such as macrophages. If macrophages that have accumulated cholesterol fail to transport this cholesterol out of lesions or transfer the cholesterol to HDL, then this cholesterol will also be retained within lesions. Prolonged storage of lipid in lesions may eventually result in its oxidation and polymerization, forming an insoluble material, called ceroid or lipofuscin (Slater and Smith, 1972). It is unlikely that cells can eliminate any ceroid or lipofuscin they have accumulated (Ball *et al.*, 1987).

10. CONCLUSIONS

A predominant point of view has been that trapping of lipoprotein cholesterol within macrophage foam cells is the cause of lipid accumulation in atherosclerotic lesions. Extracellular lipid is believed to accumulate when the foam cells die and release their contained lipid. This lipid then forms the lipid core of the plaque. It follows from this conceptualization of lipid accumulation that, if macrophage uptake of lipoproteins could be blocked, then cholesterol accumulation and atherosclerosis would be prevented.

However, this explanation for the accumulation of cholesterol in lesions does not explain why most of the accumulated extracellular cholesteryl ester has a fatty acid pattern different from that of cellular lipid droplets, and is contained within particles much smaller than lipid droplets (discussed in section 3). The foam cell hypothesis of lipid accumulation also does not explain the occurrence of lesions that show no evidence of foam cell formation. The findings reviewed here suggest that even if uptake of lipid by macrophages in the vessel wall did not occur, lipid would nevertheless continue to accumulate in the vessel wall.

Is it better for lipid to be extracellular or intracellular in atherosclerotic lesions? To the extent that extracellular or intracellular lipid resists mobilization through the interaction of extracellular lipid particles and foam cells with HDL, both extracellular and intracellular lipid may continue to increase in the vessel wall. Sufficient accumulation of extra-or intracellular lipid could prove toxic to the cells that are surrounded by or contain this lipid, respectively (Peng *et al.*, 1979; Evensen *et al.*, 1983; Morel *et al.*, 1983; Hessler *et al.*, 1983; Naseem and Heald, 1987; Smith and Johnson, 1989; Guyton *et al.*, 1990; Chisolm *et al.*, 1994; Carpenter *et al.*, 1995).

If most foam cells could leave the vessel wall or release their accumulated cholesterol to HDL, then foam cell formation should contribute to reverse cholesterol transport from lesions. Macrophage foam cells are more frequent in lesion-resistant regions of the vessel wall than in lesion-prone regions of the vessel wall (Wissler, 1994). This is consistent with a role of macrophages in helping to clear lesion lipid. Moreover, buildup of extracellular lipid could be more harmful than buildup of intracellular lipid because progressive extracellular lipid accumulation predisposes plaques to rupture (Falk, 1992; Fuster *et al.*, 1992; Davies *et al.*, 1993). These observations suggest that, overall, macrophage accumulation of lipid could be helpful rather than harmful with respect to lesion development.

On the other hand, once an atherosclerotic plaque with a lipid core covered by a thin cap has already formed, cholesterol accumulation by macrophages may become harmful. This is because cholesterol accumulation by macrophages stimulates their secretion of proteases and procoagulants. This could contribute to cap rupture and formation of obstructing blood clots on lesions (Hartung *et al.*, 1985; Falcone and Ferene, 1988; Schuff-Werner *et al.*, 1989; Rouis *et al.*, 1990; Lesnik *et al.*, 1992; Galis *et al.*, 1994; Galis *et al.*, 1995).

Obviously, relationships between the lipid deposition process and the development of atherosclerotic lesions are complex. Additional research is needed to assess mechanisms of lipid uptake and removal from the vessel wall, and also to determine which of these processes are rate limiting for vessel wall lipid accumulation. Further examination of the inter-relationships between plasma-derived lipoproteins and lesion droplet, crystalline, and liposomal pools of cholesterol will increase our understanding of the lipid deposition process in atherosclerotic lesions.

ACKNOWLEDGMENTS. I would like to thank my colleagues who have contributed to studies from this laboratory, Dr. Henry Hoff and Dr. Paulette Gaynor for helpful suggestions concerning the manuscript, and Carol Kosh for help in preparation of the manuscript.

11. REFERENCES

Abdulla, Y. H., and Adams, C. W. M., 1978, The action of human high density lipoprotein on cholesterol crystals. Part 2. Biochemical observations, *Atherosclerosis* **31**:473–480.

Adams, C. W. M., and Abdulla, Y. H., 1978, The action of human high density lipoproteins on cholesterol crystals. Part 1. Light-microscopic observations, *Atherosclerosis* **31**:465–471.

Alavi, M., Dunnett, C. W., and Moore, S., 1983, Lipid composition of rabbit aortic wall following removal of endothelium by balloon catheter, *Arteriosclerosis* **3**:413–419.

Alavi, M. Z., Richardson, M., and Moore, S., 1989, The *in vitro* interactions between serum lipoproteins and proteoglycans of the neointima of rabbit aorta after a single balloon catheter injury, *Am. J. Pathol.* **134**:287–294.

Allen, T. M., 1981, A study of phospholipid interactions between high-density lipoproteins and small unilamellar vesicles, *Biochim. Biophys. Acta* **640**:385–397.

American Heart Association, 1995, *American Heart Association Heart and Stroke Facts:* Statistical Supplement, Dallas, TX.

Andreeva, E. R., Rekhter, M. D., Romanov, Y. A., Antonova, G. M., Antonov, A. S., Mironov, A. A., and Orekhov, A. N., 1992, Stellate cells of aortic intima, II: arborization of intimal cells in cells in culture, *Tissue Cell* **24**:697–704.

Anitschkow, N., and Chalatow, S., 1913, Uber experimentelle cholesterinsteatose und ihre bedeutung fur die entstehung einiger pathologischer prozesse, *Zentralbl. Allg. Pathol.* **24**:1–9.

Baker, D. P., Van Lenten, B. J., Fogelman, A. M., Edwards, P. A., Kean, C., and Berliner, J. A., 1984, LDL, scavenger, and β-VLDL receptors on aortic endothelial cells, *Arteriosclerosis* **4**:248–255.

Ball, R. Y., Carpenter, K. L. H., and Mitchinson, M. J., 1987, What is the significance of ceroid in human atherosclerosis? *Arch. Pathol. Lab. Med.* **111**:1134–1140.

Bangham, A. D., 1963, Physical structure and behavior of lipids and lipid enzymes, *Adv. Lipid Res.* **1**:65–104.

Basu, S. K., Ho, Y. K., Brown, M. S., Bilheimer, D. W., Anderson, R. G. W., and Goldstein, J. L., 1982, Biochemical and genetic studies of the apoprotein E secreted by mouse macrophages and human monocytes, *J. Biol. Chem.* **257**:9788–9795.

Berenson, G. S., Radhakrishnamurthy, B., Srinivasa, S. R., Vijayagopal, P., and Dalferes, E. R., 1990, Nature and importance of proteoglycans in the atherosclerotic plaque, in: *Pathobiology of the Human Atherosclerotic Plaque* (S. Glagov, W. P. Newman, and S. A. Schaffer, eds.), pp. 189–208, Springer-Verlag, New York.

Bernfeld, P., and Kelley, T. F., 1964, Proteolysis of human serum β-lipoprotein, *J. Biol. Chem.* **239**:3341–3346.

Bocan, T. M. A., Schifani, T. A., and Guyton, J. R., 1986, Ultrastructure of the human aortic fibrolipid lesion. Formation of the atherosclerotic lipid-rich core, *Am. J. Pathol.* **123**:413–424.

Bötcher, C. J. F., and Woodford, F. P., 1962, Chemical changes in the arterial wall associated with atherosclerosis, *Fed. Proc.* **21**:15–19.

Brown, M. S., and Goldstein, J. L., 1983, Lipoprotein metabolism in the macrophage: Implications for cholesterol deposition in atherosclerosis, *Annu. Rev. Biochem.* **52**:223–261.

Buck, R. C., and Rossiter, R. J., 1951, Lipids of normal and atherosclerotic aortas, *Arch. Pathol.* **51**:224–237.

Camejo, G., Hurt, E., and Romano, M., 1985, Properties of lipoprotein complexes isolated by affinity chromatography from human aorta, *Biomed. Biochim. Acta* **44**:389–401.

Carpenter, K. L. H., Taylor, S. E., Van der Veen, C., Williamson, B. K., Ballantine, J. A., and Mitchinson, M. J., 1995, Lipids and oxidized lipids in human atherosclerotic lesions at different stages of development, *Biochim. Biophys. Acta* **1256**:141–150.

Castelli, W. P., 1984, Epidemiology of coronary heart disease: The Framingham Study, *Am. J. Med.* **76**:4–12.

Chambless, L. E., Fuchs, F. D., Linn, S., Kritchevsky, S. B., Larosa, J. C., Segal, P., Rifkind, B. M., 1990, The associaiton of corneal arcus with coronary heart disease and cardiovascular disease mortality in the lipid research clinics mortality follow-up study, *Am. J. Public Health* **80**:1200–1204.

Chang, M. Y., Lees, A. M., and Lees, R. S., 1992, Time course of ^{125}I-labeled LDL accumulation in the healing, balloon-deendothelialized rabbit aorta, *Arterioscler. Thromb.* **12**:1088–1098.

Chang, M. Y., Lees, A. M., and Lees, R. S., 1993, Low density lipoprotein modification and arterial wall accumulation in a rabbit model of atherosclerosis, *Biochemistry* **32**:8518–8524.

Chao, F.-F., Blanchette-Mackie, E. J., Skarlatos, S. I., Gamble, W., Resau, J. H., Mergner, W. T., and Kruth, H. S., 1988, Unesterified cholesterol-rich lipid particles in atherosclerotic lesions of human and rabbit aortas, *Am. J. Pathol.* **131**:73–83.

Chao, F.-F., Blanchette-Mackie, E. J., Chen, Y.-J., Dickens, B. F., Berlin, E., Amende, L. M., Skarlatos, S. I., Gamble, W., Resau, J. H., Mergner, W. T., and Kruth, H. S., 1990, Characterization of two unique cholesterol-rich lipid particles isolated from human atherosclerotic lesions, *Am. J. Pathol.* **136**:169–179.

Chao, F.-F., Rifai, N., Blanchette-Mackie, E. J., Resau, J. H., Amende, L. M., and Kruth, H. S., 1991, *In vitro* solubilization of aortic unesterified cholesterol-rich lipid particles by high density lipoproteins, *Clin. Chem.* **37**:921 (abstract).

Chao, F.-F., Blanchette-Mackie, E. J., Tertov, V. V., Skarlatos, S. I., Chen, Y.-J., and Kruth, H. S., 1992, Hydrolysis of cholesteryl ester in low density lipoprotein converts this lipoprotein to a liposome, *J. Biol. Chem.* **267**:4992–4998.

Chao, F.-F., Blanchette-Mackie, E. J., Dickens, B. F., Gamble, W., and Kruth, H. S., 1994, Development of unesterified cholesterol-rich lipid particles in atherosclerotic lesions of WHHL and cholesterol-fed NZW rabbits, *J. Lipid Res.* **35**:71–83.

Chapman, M. J., Goldstein, S., and Mills, G. L., 1978, Limited tryptic digestion of human serum low-density lipoprotein: isolation and characterization of the protein-deficient particle and of its apoprotein, *Eur. J. Biochem.* **87**:475–488.

Chisolm, G. M., Ma, G., Irwin, K. C., Martin, L. L., Gunderson, K. G., Linberg, L. F., Morel, D. W., and DiCorleto, P. E., 1994, 7β-hydroperoxycholest-5-en-3β-ol, a component of human atherosclerotic lesions, is the primary cytotoxin of oxidized human low density lipoprotein, *Proc. Natl. Acad. Sci. USA* **91**:11452–11456.

Chulkova, T. M., and Tertov, V. V., 1993, Degradation of human apolipoprotein B-100 by apolipoprotein (a), *FEBS Lett.* **336**:327–329.

Chung, B.-H., Im, J. H., and Bowdon, H. R., 1986, Lipolysis-induced degradation of apolipoproteins B and E of human very low density lipoprotein, *J. Biol. Chem.* **261**:2960–2967.

Chung, B.-H., Tallis, G., Yalamoori, V., Anantharamaiah, G. M., and Segrest, J. P., 1994, Liposome-like particles isolated from human atherosclerotic plaques are structurally and compositionally similar to surface remnants of triglyceride-rich lipoproteins, *Arterioscl. Thromb.* **14:** 622–635.

Clevidence, B. A., Morton, R. E., West, G., Dusek, D. M., and Hoff, H. F., 1984, Cholesterol esterification in macrophages. Stimulation by lipoproteins containing apo B isolated from human aortas, *Arteriosclerosis* **4:**196–207.

Cornhill, J. F., Herderick, E. E., and Stary, H. C., 1990, Topography of human aortic sudanophilic lesions, *Monogr. Atheroscler.* **15:**13–19.

Cotran, R. S., 1965, Endothelial phagocytosis: An electron-microscopic study, *Exp. Mol. Pathol.* **4:**217–231.

Curtiss, L. K., Black, A. S., Takagi, Y., and Plow, E. F., 1987, New mechanism for foam cell generation in atherosclerotic lesions, *J. Clin. Invest.* **80:**367–373.

Daugherty, A., Zweifel, B. S., Sobel, B. E., and Schonfeld, G., 1988, Isolation of low density lipoprotein from atherosclerotic vascular tissue of Watanabe Heritable hyperlipidemic rabbits, *Arteriosclerosis* **8:**768–777.

Davies, M. J., Richardson, P. D., Woolf, N., Katz, D. R., and Mann, J., 1993, Risk of thrombosis in human atherosclerotic plaques: role of extracellular lipid, macrophage, and smooth muscle cell content, *Br. Heart J.* **69:**377–381.

Day, A. J., Alavi, M., and Moore, S., 1985, Influx of [^{3}H, ^{14}C]cholesterol-labelled lipoprotein into re-enthelialised and de-endothelialised areas of ballooned aortas in normal-fed and cholesterol-fed rabbits, *Atherosclerosis* **55:**339–351.

Duff, G. L., McMillan, G. C., and Lautsch, E. V., 1954, The uptake of colloidal thorium dioxide by the arterial lesions of cholesterol atherosclerosis in the rabbit, *Am. J. Pathol.* **30:**941–955.

Evensen, S. A., Galdal, K. S., and Nilsen, E., 1983, LDL-induced cytotoxicity and its inhibition by anti-oxidant treatment in cultured human endothelial cells and fibroblasts, *Atherosclerosis* **49:**23–30.

Fainaru, M., Mahley, R. W., Hamilton, R. L., and Innerarity, T. L., 1982, Structural and metabolic heterogeneity of β-very low density lipoproteins from cholesterol-fed dogs and from humans with Type III hyperlipoproteinemia, *J. Lipid Res.* **23:**702–714.

Falcone, D. J., and Ferenc, M. J., 1988, Acetyl-LDL stimulates macrophage-dependent plasminogen activation and degradation of extracellular matrix, *J. Cell Physiol.* **135:**387–396.

Falcone, D. J., Hajjar, D. P., and Minick, C. R., 1980, Enhancement of cholesterol and cholesteryl ester accumulation in re-endothelialized aorta, *Am. J. Pathol.* **99:**81–104.

Falcone, D. J., Mated, N., Shio, H., Minick, C. R., and Fowler, S. D., 1984, Lipoprotein-heparin-fibronectin-denatured collagen complexes enhance cholesteryl ester accumulation in macrophages, *J. Cell Biol.* **99:**1266–1274.

Falk, E., 1992, Why do plaques rupture? *Circulation* **86:**III-30-III-42.

Finkelstein, M. C., and Weissmann, G., 1979, Enzyme replacement via liposomes. Variations in lipid composition determine liposomal integrity in biological fluids, *Biochim. Biophys. Acta* **587:**202–216.

Fogelman, A. M., Shechter, I., Seager, J., Hokom, M., Child, J. S., and Edwards, P. A., 1980, Malondialdehyde alteration of low density lipoproteins leads to cholesteryl ester accumulation in human monocyte-macrophages, *Proc. Natl. Acad. Sci. USA* **77:**2214–2218.

Fong, L. G., Parthasarathy, S., Witztum, J. L., and Steinberg, D., 1987, Nonenzymatic oxidative cleavage of peptide bonds in apoprotein B-100, *J. Lipid Res.* **28:**1466–1477.

Forte, T. M., and Nordhausen, R. W., 1986, Electron microscopy of negatively stained lipoproteins, *Methods Enzymol.* **128:**442–457.

French, J. E., Jennings, M. A., Poole, J. C. F., Robinson, D. S., and Florey, S. H., 1963, Intimal changes in the arteries of ageing swine, *Proc. R. Soc. London [Biol]* **158:**24–42.

Fuster, V., Badimon, J. J., and Badimon, L., 1992, Clinical-pathological correlations of coronary disease progression and regression, *Circulation* **86:**III 1–11.

Galis, Z. S., Sukhova, G. K., Lark, M. W., and Libby, P., 1994, Increased expression of matrix metalloproteinases and matrix degrading activity in vulnerable regions of human atherosclerotic plaques, *J. Clin. Invest.* **94:**2493–2503.

Galis, Z. S., Sukhova, G. K., Kranzhofer, R., Clark, S., and Libby, P., 1995, Macrophage foam cells from experimental atheroma constitutively produce matrix-degrading proteinases, *Proc. Natl. Acad. Sci. USA* **92:**402–406.

Garti, N., Karpuj, L., and Sarig, S., 1981, Correlation between crystal habit and the composition of solvated and nonsolvated cholesterol crystals, *J. Lipid Res.* **22:**785–791.

Gaynor, P. G., Zhang, W.-Y., Salehizadeh, B., Hunsberger, S., Pettiford, B., and Kruth, H. S., 1996, Cholesterol accumulation in human cornea: Evidence that extracellular cholesteryl ester-rich lipid particles deposit independently of foam cells, *J. Lipid Res.* **37:**1849–1861.

Geer, J. C., and Guidry, M. A., 1964, Cholesteryl ester composition and morphology of human normal intima and fatty streaks, *Exp. Mol. Pathol.* **3:**485–499.

Gerrity, R. G., 1981, The role of the monocyte in atherogenesis. II. Migration of foam cells from atherosclerotic lesions, *Am. J. Pathol.* **103:**191–200.

Gerrity, R. G., Richardson, M., Somer, J. B., Bell, F. P., and Schwartz, C. J., 1977, Endothelial cell morphology in areas of *in vivo* Evans Blue uptake in the aorta of young pigs. II. Ultrastructure of the intima in areas of differing permeability to proteins, *Am. J. Pathol.* **89:** 313–334.

Gianturco, S. H., and Bradley, W. A., 1988, Lipoprotein-mediated cellular mechanisms for atherogenesis in hypertriglyceridemia, *Semin. Thromb. Hemost.* **14:**165–169.

Goldstein, J. L., Ho, Y. K., Basu, S. K., and Brown, M. S., 1979, Binding site on macrophages that mediates uptake and degradation of acetylated low density lipoprotein, producing massive cholesterol deposition, *Proc. Natl. Acad. Sci. USA* **76:**333–337.

Gordon, V., Innerarity, T. L., and Mahley, R. W., 1983, Formation of cholesterol- and apoprotein E-enriched high density lipoproteins *in vitro*, *J. Biol. Chem.* **258:**6202–6212.

Gotto, A. M., Jr., Pownall, H. J., and Havel, R. J., 1986, Introduction to the plasma membrane, *Methods Enzymol.* **128:**3–41.

Greemland, M., and Savion, N., 1993, Sphingomyelin biosynthesis and efflux correlates with cholesterol metabolism and is higher in vascular endothelial cells than in smooth muscle cells, *Eur. Heart J.* **14:**687–691.

Guyton, J. R., and Klemp, K. F., 1988, Ultrastructural discrimination of lipid droplets and vesicles in atherosclerosis: Value of osmium-thiocarbohydrazide-osmium and tannic acid-paraphenylenediamine techniques, *J. Histochem. Cytochem.* **36:**1319–1328.

Guyton, J. R., and Klemp, K. F., 1989, The lipid-rich core region of human atherosclerotic fibrous plaques. Prevalence of small lipid droplets and vesicles by electron microscopy, *Am. J. Pathol.* **134:**705–717.

Guyton, J. R., and Klemp, K. F., 1992, Early extracellular and cellular lipid deposits in aorta of cholesterol-fed rabbits, *Am. J. Pathol.* **141:**925–936.

Guyton, J. R., and Klemp, K. F., 1994, Development of the atherosclerotic core region. Chemical and ultrastructural analysis of microdissected atherosclerotic lesions from human aorta, *Arterioscl. Thromb.* **14:**1305–1314.

Guyton, J. R., Black, B. L., and Seidel, C. L., 1990, Focal toxicity of oxysterols in vascular smooth muscle cell culture, *Am. J. Pathol.* **137:**425–434.

Hajjar, D. P., Minick, C. R., and Fowler, S., 1983, Arterial neutral cholesteryl esterase. A hormone-sensitive enzyme distinct from lysosomal cholesteryl esterase, *J. Biol. Chem.* **258:** 192–198.

Haley, N. J., Shio, H., and Fowler, S., 1977, Characterization of lipid-laden aortic cells from cholesterol-fed rabbits. I. Resolution of aortic cell populations by metrizamide density gradient centrifugation, *Lab. Invest.* **37:**278–296.

Hartung, H.-P., Kladetzky, R. G., and Hennerici, M., 1985, Chemically modified low density lipoproteins as inducers of enzyme release from macrophages, *FEBS Lett.* **186:**211–214.

Hata, Y., Hower, J., and Insull, W., Jr., 1974, Cholesteryl ester-rich inclusions from human aortic fatty streak and fibrous plaque lesions of atherosclerosis. I. Crystalline properties, size and internal structure, *Am. J. Pathol.* **75:**423–454.

Havel, R. J., 1994, Postprandial hyperlipidemia and remnant lipoproteins, *Curr. Opin. Lipidol.* **5:**102–109.

Heideman, C. L., and Hoff, H. F., 1982, Lipoproteins containing apolipoprotein A-I extracted from human aortas, *Biochim. Biophys. Acta* **711:**431–444.

Heinecke, J. W., Suits, A. G., Aviram, M., and Chait, A., 1991, Phagocytosis of lipase-aggregated low density lipoprotein promotes macrophage foam cell formation. Sequential morphological and biochemical events, *Arterioscl. Thromb.* **11:**1643–1651.

Henriksen, T., Mahoney, E. M., and Steinberg, D., 1981, Enhanced macrophages degradation of low density lipoprotein previously incubated with cultured endothelial cells: Recognition by receptors for acetylated low density lipoproteins, *Proc. Natl. Acad. Sci. USA* **78:**6499–6503.

Henriksen, T., Mahoney, E. M., and Steinberg, D., 1983, Enhanced macrophage degradation of biologically modified low density lipoprotein, *Arteriosclerosis* **3:**149–159.

Hessler, J. R., Morel, D. W., Lewis, L. J., and Chisolm, G. M., 1983, Lipoprotein oxidation and lipoprotein-induced cytotoxicity, *Arteriosclerosis* **3:**215–222.

Ho, Y.-K., Brown, M. S., and Goldstein, J. L., 1980, Hydrolysis and excretion of cytoplasmic cholesteryl esters by macrophages: stimulation by high density lipoprotein and other agents, *J. Lipid Res.* **21:**391–398.

Hoff, H. F., 1972, Human intracranial atherosclerosis. A histochemical and ultrastructural study of gross fatty streak lesions, *Am. J. Pathol.* **69:**421–437.

Hoff, H. F., and Clevidence, B. A., 1987, Uptake by mouse peritoneal macrophages of large cholesteryl ester-rich particles isolated from human atherosclerotic lesions, *Exp. Mol. Pathol.* **46:**331–344.

Hoff, H. F., and Cole, T. B., 1991, Macrophage uptake of low-density lipoprotein modified by 4-hydroxynonenal. An ultrastructural study, *Lab. Invest.* **64:**254–265.

Hoff, H. F., and Gaubatz, J. W., 1977, Ultrastructural localization of apolipoprotein B in human aortic and coronary atherosclerotic plaques, *Exp. Mol. Pathol.* **26:**214–227.

Hoff, H. F., and Gaubautz, J. W., 1979, Residual apo B in aortic plaques extracted with hydrolytic enzymes, *Artery* **6:**89–107.

Hoff, H. F., and Gaubatz, J. W., 1982, Isolation, purification, and characterization of a lipoprotein containing apo B from the human aorta, *Atherosclerosis* **42:**273–297.

Hoff, H. F., and Morton, R. E., 1985, Lipoproteins containing apoB extracted from human aortas. Structure and function, *Ann. N.Y. Acad. Sci.* **454:**183–194.

Hoff, H. F., and O'Neil, J., 1991, Lesion-derived low density lipoprotein and oxidized low density lipoprotein share a lability for aggregation, leading to enhanced macrophage degradation, *Arterioscl. Thromb.* **11:**1209–1222.

Hoff, H. F., Jackson, R. L., Mao, S. J. T., and Gotto, A. M., Jr., 1974, Localization of low-density lipoproteins in atherosclerotic lesions from human normolipemics employing a purified fluorescent-labelled antibody, *Biochim. Biophys. Acta* **351:**407–415.

Hoff, H. F., Heideman, C. L., Gotto, A. M., Jr., and Gaubatz, J. W., 1977, Apolipoprotein B retention in the grossly normal and atherosclerotic human aorta, *Circ. Res.* **41:**684–690.

Hoff, H. F., Heideman, C. L., Gaubatz, J. W., Scott, D. W., and Gotto, A. M., Jr., 1978a, Detergent extraction of tightly-bound apoB from extracts of normal aortic intima and plaques, *Exp. Mol. Pathol.* **28:**290–300.

Hoff, H. F., Heideman, C. L., Gaubatz, J. W., Scott, D. W., Titus, J. L., and Gotto, A. M., Jr., 1978b, Correlation of apolipoprotein B retention with the structure of atherosclerotic plaques from human aortas, *Lab. Invest.* **38:**560–567.

Hoff, H. F., Bradley, W. A., Heideman, C. L., Gaubatz, J. W., Karagas, M. D., and Gotto, A. M., Jr., 1979a, Characterization of low density lipoprotein-like particle in the human aorta from grossly normal and atherosclerotic regions, *Biochim. Biophys. Acta* **573:**361–374.

Hoff, H. F., Karagas, M., Heideman, C. L., Gaubatz, J. W., and Gotto, A. M., Jr., 1979b, Correlation in the human aorta of apoB fractions with tissue cholesterol and collagen content, *Atherosclerosis* **32:**259–268.

Hoff, H. F., Gerrity, R. G., Naito, H. K., and Dusek, D. M., 1983, Methods in laboratory investigation. Quantitation of apolipoproteins B in aortas of hypercholesterolemic swine, *Lab. Invest.* **48:**492–504.

Hoff, H. F., Dusek, D. M., and Lynn, M. P., 1986, Methods in laboratory investigation. Spatial distribution and accumulation of low density lipoproteins in the abdominal aorta of swine: determination by a novel electrotransfer procedure, *Lab. Invest.* **55:**377–386.

Hoff, H. F., O'Neil, J., Chisolm, G. M. III, Cole, T. B., Quehenberger, O., Esterbauer, H., and Jurgens, G., 1989, Modification of low density lipoprotein with 4-hydroxynonenal induces uptake by macrophages, *Arteriosclerosis* **9:**538–549.

Hoff, H. F., O'Neil, J., and Cole, T. B., 1991, Macrophage degradation of LDL extracted from human aortic plaques: Effect of isolation conditions, *Exp. Mol. Pathol.* **54:**72–86.

Hoff, H. F., Whitaker, T. E., and O'Neil, J., 1992, Oxidation of low density lipoprotein leads to particle aggregation and altered macrophage recognition, *J. Biol. Chem.* **267:**602–609.

Hollander, W., Paddock, J., and Colombo, M., 1979, Lipoproteins in human atherosclerotic vessels. I. Biochemical properties of arterial low density lipoproteins, very low density lipoproteins, and high density lipoproteins, *Exp. Mol. Pathol.* **30:**144–171.

Horlick, L., 1954, Serum lipoprotein stability in atherosclerosis, *Circulation* **10:**30–42.

Hurt, E., Bondjers, G., and Camejo, G., 1990, Interaction of LDL with human arterial proteoglycans stimulates its uptake by human monocyte-derived macrophages, *J. Lipid Res.* **31:**443–454.

Insull, W., Jr., and Bartsch, G. E., 1966, Cholesterol, triglyceride, and phospholipid content of intima, media, and atherosclerotic fatty streak in human thoracic aorta, *J. Clin. Invest.* **45:**513–523.

Ismail, N. A. E., Alavi, M. Z., and Moore, S., 1994, Isolation of lipoprotein-proteoglycan complexes from balloon catheter deendothelialized aortas and the uptake of these complexes by blood monocyte-derived macrophages, *Pathology* **26:**145–153.

Jonas, A., Hesterberg, L. K., and Drengler, S. M., 1978, Incorporation of excess cholesterol by high density serum lipoproteins, *Biochim. Biophys. Acta* **528:**47–57.

Joris, I., Zand, T., Nunnari, J. J., Krolikowski, F. J., and Majno, G., 1983, Studies on the pathogenesis of atherosclerosis. I. Adhesion and emigration of mononuclear cells in the aorta of hypercholesterolemic rats, *Am. J. Pathol.* **113:**341–358.

Kannel, W. B., and Wilson, P. W., 1992, Efficacy of lipid profiles in prediction of coronary disease, *Am. Heart J.* **124:**768–774.

Kannel, W. B., Castelli, W. P., and Gordon, T., 1979, Cholesterol in the prediction of atherosclerotic disease, *Ann. Intern. Med.* **90:**85–91.

Kao, V. C. Y., and Wissler, R. W., 1965, A study of the immunohistochemical localization of serum lipoproteins and other plasma proteins in human atherosclerotic lesions, *Exp. Mol. Pathol.* **4:**465–479.

Katz, S. S., and Small, D. M., 1980, Isolation and partial characterization of the lipid phases of human atherosclerotic plaques, *J. Biol. Chem.* **255:**9753–9759.

Katz, S. S., Shipley, G. G., and Small, D. M., 1976, Physical chemistry of the lipids of human atherosclerotic lesions. Demonstration of a lesion intermediate between fatty streaks and advanced plaques, *J. Clin. Invest.* **58:**200–211.

Kawabe, Y., Cynshi, O., Takashima, Y., Suzuki, T., Ohba, Y., and Kodama, T., 1994, Oxidation-induced aggregation on rabbit low-density lipoprotein by Azo initiator, *Arch. Biochem. Biophys.* **310:**489–496.

Khoo, J. C., Miller, E., McLoughlin, P., and Steinberg, D., 1988, Enhanced macrophage uptake of low density lipoprotein after self-aggregation, *Arteriosclerosis* **8:**348–358.

Khoo, J. C., Miller, E., Pio, F., Steinberg, D., and Witztum, J. L., 1992, Monoclonal antibodies against LDL further enhance macrophage uptake of LDL aggregates, *Arterioscl. Thromb.* **12:**1258–1266.

Kirby, C., Clarke, J., and Gregoriadis, G., 1980, Cholesterol content of small unilamellar liposomes controls phospholipid loss to high density lipoproteins in the presence of serum, *FEBS Lett.* **111:**324–328.

Kokkonen, J. O., and Kovanen, P. T., 1985, Low density lipoprotein degradation by rat mast cells. Demonstration of extracellular proteolysis caused by mast cell granules, *J. Biol. Chem.* **260:**14756–14763.

Kokkonen, J. O., Vartiainen, M., and Kovanen, P. T., 1986, Low density lipoprotein degradation by secretory granules of rat mast cells. Sequential degradation of apolipoprotein B by granule chymase and carboxypeptidase A, *J. Biol. Chem.* **261:**16067–16072.

Koren, E., Koscec, M., McConathy, W. J., and Fugate, R. D., 1991, Possible role of macrophages in regression of atherosclerosis, *Prog. Lipid Res.* **30:**237–243.

Koo, C., Innerarity, T. L., and Mahley, R. W., 1985, Obligatory role of cholesterol and apolipoprotein E in the formation of large cholesterol-enriched and receptor-active high density lipoproteins, *J. Biol. Chem.* **260:**11934–11943.

Kovanen, P. T., and Kokkonen, J. O., 1991, Modification of low density lipoproteins by secretory granules of rat serosal mast cells, *J. Biol. Chem.* **266:**4430–4436.

Kruth, H. S., 1983, Filipin-positive, oil red O-negative particles in atherosclerotic lesions induced by cholesterol feeding, *Lab. Invest.* **50:**87–93.

Kruth, H. S., 1984a, Histochemical detection of esterified cholesterol within human atherosclerotic lesions using the fluorescent probe filipin, *Atherosclerosis* **51:**281–292.

Kruth, H. S., 1984b, Localization of unesterified cholesterol in human atherosclerotic lesions. Demonstration of filipin-positive, oil red O-negative particles, *Am. J. Pathol.* **114:**201–208.

Kruth, H. S., 1985a, Subendothelial accumulation of unesterified cholesterol. An early event in atherosclerotic lesion development, *Atherosclerosis* **57:**337–341.

Kruth, H. S., 1985b, Platelet-mediated cholesterol accumulation in cultured aortic smooth muscle cells, *Science* **227:**1243–1245.

Kruth, H. S., and Fry, H. S., 1984, Histochemical detection and differentiation of free and esterified cholesterol in swine atherosclerosis using filipin, *Exp. Mol. Pathol.* **40:**288–294.

Kruth, H. S., and Shekhonin, B., 1995, Evidence for loss of apo B from LDL in human atherosclerotic lesions: extracellular cholesteryl ester lipid particles lacking apo B, *Atherosclerosis* **105:**227–234.

Kruth, H. S., Skarlatos, S. I., Gaynor, P. M., and Gamble, W., 1994, Production of cholesterol-enriched nascent high density lipoproteins by human monocyte-derived macrophages is a mechanism that contributes to macrophage cholesterol efflux, *J. Biol. Chem.* **269:**24511–24518.

Kruth, H. S., Skarlatos, S. I., Lilly, K., Chang, J., and Ifrim, I., 1995, Sequestration of acetylated LDL and cholesterol crystals by human monocyte-derived macrophages, *J. Cell Biol.* **129:**133–145.

Kunnert, B., and Krug, H., 1971, Composition of cholesterol esters in fatty streaks atherosclerotic plaques of the human aorta. Chromatographic investigations, *Atherosclerosis* **13:**93–101.

Lang, P. D., and Insull, W., Jr., 1970, Lipid droplets in atherosclerotic fatty streaks of human aorta, 1970, *J. Clin. Invest.* **49:**1479–1488.

Lee, D. M., and Singh, S., 1988, Degradation of apolipoprotein B-100 in human chylomicrons, *Biochim. Biophys. Acta* **960:**148–156.

Lesnik, P., Rouis, M., Skarlatos, S., Kruth, H. S., and Chapman, M. J., 1992, Uptake of exogenous free cholesterol induces upregulation of tissue factor expression in human monocyte-derived macrophages, *Proc. Natl. Acad. Sci. USA* **89:**10370–10374.

Linden, T., Wiklund, O., Fager, G., Olofsson, S.-O., and Bondjers, G., 1986, A new microimmuno assay for apolipoprotein B in arterial tissue. Studies on peroperative human biopsies, *Atherosclerosis* **62:**227–237.

Lindgren, F. T., 1975, Preparative ultracentrifugal laboratory procedure and suggestions for lipoprotein analysis, in: *Analysis of Lipid and Lipoproteins* (E. G. Perkin, ed.), pp. 205–224, American Oil Chemical Society, New York.

Lopes-Virella, M. F., Griffith, R. L., Shunk, K. A., and Virella, G. T., 1991, Enhanced uptake and impaired intracellular metabolism of low density lipoprotein complexed with anti-low density lipoprotein antibodies, *Arterioscl. Thromb.* **11:**1356–1367.

Lundberg, B., 1985, Chemical composition and physical state of lipid deposits in atherosclerosis, *Atherosclerosis* **56:**93–110.

Lupu, F., Danaricu, I., and Simionescu, N., 1987, Development of intracellular lipid deposits in the lipid-laden cells of atherosclerotic lesions. A cytochemical and ultrastructural study, *Atherosclerosis* **67:**127–142.

Mahley, R. W., Innerarity, T. L., Weisgraber, K. H., and Oh, S. Y., 1979, Altered metabolism (in vivo and in vitro) of plasma lipoproteins after selective modification of lysine residues of the apoproteins, *J. Clin. Invest.* **64:**743–750.

Mahley, R. W., Weisgraber, K. H., Innerarity, T. L., and Rall, S. C., Jr., 1991, Genetic defects in lipoprotein metabolism. Elevation of atherogenic lipoproteins caused by impaired catabolism, *Genetic Defects* **265:**78–83.

Mamo, J. C. L., and Wheeler, J. R., 1994, Chylomicrons or their remnants penetrate rabbit thoracic aorta as efficiently as do smaller macromolecules, including low-density lipoprotein, high-density lipoprotein, and albumin, *Coronary Artery Dis.* **5:**695–705.

Margolis, S., and Langdon, R. G., 1966, Studies on human serum β_1-lipoprotein. III. Enzymatic modifications, *J. Biol. Chem.* **241:**485–493.

Mawhinney, T. P., Augustyn, J. M., and Fritz, K. E., 1978, Glycosaminoglycan-lipoprotein complexes from aortas of hypercholesterolemic rabbits. Part 1—Isolation and characterization, *Atherosclerosis* **31:**155–167.

Mendelsohn, M. E., and Loscalzo, J., 1988, Role of platelets in cholesteryl ester formation by U-937 cells, *J. Clin. Invest.* **81:**62–68.

McCandless, E. L., and Zilversmit, D. B., 1956, The effect of cholesterol on the turnover of lecithin, cephalin and sphingomyelin in the rabbit, *Arch. Biochem. Biophys.* **62:**402–410.

McGill, H. C., Geer, J. C., and Holman, R. L., 1957, Sites of vascular vulnerability in dogs demonstrated by Evans Blue, *Arch. Pathol.* **64:**303–311.

Miyamoto, S., Akiyama, S., and Yamada, K. M., 1995, Synergistic roles for receptor occupancy and aggregation in integrin transmembrane function, *Science* **267:**883–885.

Mora, R., Lupu, F., and Simionescu, N., 1987, Prelesional events in atherogenesis. Colocalization of apolipoprotein B, unesterified cholesterol and extracellular phospholipid liposomes in the aorta of hyperlipidemic rabbit, *Atherosclerosis* **67:**143–154.

Mora, R., Simionescu, M., and Simionescu, N., 1990, Purification and partial characterization of extracellular liposomes isolated from the hyperlipidemic rabbit aorta, *J. Lipid Res.* **31:**1793–1807.

Morel, D. W., Hessler, J. R., and Chisolm, G. M., 1983, Low density lipoprotein cytotoxicity induced by free radical peroxidation of lipid, *J. Lipid Res.* **24:**1070–1076.

Morganelli, P. M., Rogers, R. A., Kitzmiller, T. J., and Bergeron, A., 1995, Enhanced metabolism of LDL aggregates mediated by specific human monocyte IgG Fc receptors, *J. Lipid Res.* **36:**714–724.

Morton, R. E., West, G. A., and Hoff, H. F., 1986, A low density lipoprotein-sized particle isolated from human atherosclerotic lesions is internalized by macrophages via a non-scavenger-receptor mechanism, *J. Lipid Res.* **27:**1124–1134.

Moss, N. S., and Benditt, E. P., 1970, The ultrastructural of spontaneous and experimentally induced arterial lesions. III. The cholesterol-induced lesions and the effect of a cholesterol and oil diet on the preexisting spontaneous plaque in the chicken aorta, *Lab. Invest.* **23:**521–535.

Mukhin, D. N., Chao, F.-F., and Kruth, H. S., 1995, Glycosphingolipid accumulation in the aortic wall is another feature of human atherosclerosis, *Arterioscler. Thromb. Vasc. Biol.* **15:**1607–1615.

Myers, J. N., Tabas, I., Jones, N. L., and Maxfield, F. R., 1993, β-very low density lipoprotein is sequestered in surface-connected tubules in mouse peritoneal macrophages, *J. Cell Biol.* **123:**1389–1402.

Nakamura, H., and Ohtsubo, K.-I., 1992, Ultrastructure appearance of atherosclerosis in human and experimentally-induced animal models, *Electron Microsc. Rev.* **5:**129–170.

Nakashima, Y., Matsushima, T., Takahara, K., Kuroiwa, A., and Nakamura, M., 1985, The analysis of lipids and glycosaminoglycans of low-density lipoprotein-glycosaminoglycans complexes isolated from normal, fatty streaks, and fibrous plaques of human aortic intima, *Inter. Angiol.* **4:**487–493.

Naseem, S. M., and Heald, F. P., 1987, Cytotoxicity of cholesterol oxides and their effects on cholesterol metabolism in cultured human aortic smooth muscle cells, *Biochem. Int.* **14:**71–84.

Niehaus, C. E., Nicoll, A., Wootton, R., Williams, B., Lewis, J., Goltart, D. J., and Lewis, B., 1977, Influence of lipid concentrations and åge on transfer of plasma lipoprotein into human arterial intima, *Lancet* **2:**469–471.

Nielsen, L. B., Nordestgaard, B. G., Stender, S., and Kjeldsen, K., 1992, Aortic permeability to LDL as a predictor of aortic cholesterol accumulation in cholesterol-fed rabbits, *Arterioscl. Thromb.* **12:**1402–1409.

Nievelstein, P. F. E. M., Fogelman, A. M., Mottino, G., and Frank, J. S., 1991, Lipid accumulation in rabbit aortic intima 2 hours after bolus infusion of low density lipoprotein. A deep-etch and immunolocalization study of ultrarapidly frozen tissue, *Arterioscl. Thromb.* **11:**1795–1805.

Nistor, A., Bulla, A., Filip, D. A., and Radu, A., 1987, The hyperlipidemic hamster as a model of experimental atherosclerosis, *Atherosclerosis* **68:**159–173.

Nordestgaard, B. G., and Zilversmit, D. B., 1988, Large lipoproteins are excluded from the arterial wall in diabetic cholesterol-fed rabbits, *J. Lipid Res.* **29:**1491–1500.

Nordestgaard, B. G., and Zilversmit, D. B., 1989, Comparison of arterial intimal clearances of LDL from diabetic and nondiabetic cholesterol-fed rabbits: differences in intimal clearance explained by size differences, *Arteriosclerosis* **9:**176–183.

Nordestgaard, B. G., Hjelms, E., Stender, S., and Kjeldsen, K., 1990, Different efflux pathways for high and low density lipoproteins from porcine aortic intima, *Arteriosclerosis* **10:**477–485.

Nordestgaard, B. G., Tybjaerg-Hansen, A., and Lewis, B., 1992, Influx of *in vivo* of low density, intermediate density, and very low density lipoproteins into aortic intimas of genetically hyperlipidemic rabbits: roles of plasma concentration, extent of aortic lesions, and lipoprotein particle size as determinants, *Arterioscler. Thromb.* **12:**6–18.

Okwu, A. K., Xu, X. X., Shiratori, Y., and Tabas, I., 1994, Regulation of the threshold for lipoprotein-induced acyl-CoA:cholesterol O-acetyltransferase stimulation in macrophages by cellular sphingomyelin content, *J. Lipid Res.* **35:**644–655.

Olsson, G., Wiklund, O., and Bondjers, G., 1995, Effects of injury on apo B kinetics and concentration in rabbit aorta, *Arterioscl. Thromb. Vasc. Biol.* **15:**930–936.

Orekhov, A. N., Tertov, V. V., Mukhin, D. N., Koteliansky, V. E., Glukhova, M. A., Khashimov, K. A., and Smirnov, V. N., 1987, Association of low-density lipoprotein with particulate connective tissue matrix components enhances cholesterol accumulation in cultured subendothelial cells of human aorta, *Biochim. Biophys. Acta* **928:**251–258.

Orekhov, A. N., Tertov, V. V., Mukhin, D. N., Koteliansky, V. E., Glukhova, M. A., Frid, M. G., Sukhova, G. K., Khashimov, K. A., and Smirnov, V. N., 1989, Insolubilization of low density lipoprotein induces cholesterol accumulation in cultured subendothelial cells of human aorta, *Atherosclerosis* **79:**59–70.

Orekhov, A. N., Tertov, V. V., and Mukhin, D. N., 1991a, Desialylated low density lipoprotein-naturally occurring modified lipoprotein with atherogenic potency, *Atherosclerosis* **86:**153–161.

Orekhov, A. N., Tertov, V. V., Kabakov, A. E., Adamova, I. Y., Pokrovsky, S. N., and Smirnov, V. N., 1991b, Autoantibodies against modified low density lipoprotein. Nonlipid factor of blood plasma that stimulates foam cell formation, *Arterioscl. Thromb.* **11:**316–326.

Orekhov, A. N., Tertov, V. V., Sobenin, I. A., Smirnov, V. N., Via, D. P., Guevara, J., Gotto, A. M. Jr., and Morrisett, J. D., 1992, Sialic acid content of human low density lipoproteins affects their interaction with cell receptors and intracellular lipid accumulation, *J. Lipid Res.* **33:**805–817.

Paananen, K., and Kovanen, P. T., 1994, Proteolysis and fusion of low density lipoprotein particles independently strengthen their binding to exocytosed mast cell granules, *J. Biol. Chem.* **269:**2023–2031.

Paananen, K., Saarinen, J., Annila, A., and Kovanen, P. T., 1995, Proteolysis and fusion of low density lipoprotein particles strengthen their binding to human aortic proteoglycans, *J. Biol. Chem.* **270:**12257–12262.

Patsch, W., Patsch, J. R., and Gotto, A. M., Jr., 1989, The hyperlipoproteinemias, *Hepatic Diseases* **73:**859–893.

Peng, S.-K., Tham, P., Taylor, C. B., and Mikkelson, B., 1979, Cytotoxicity of oxidation derivatives of cholesterol on cultured aortic smooth muscle cells and their effect on cholesterol biosynthesis, *Am. J. Clin. Nutr.* **32:**1033–1042.

Pepin, J. M., O'Neil, J. A., and Hoff, H. F., 1991, Quantification of apo[a] and apoB in human atherosclerotic lesions, *J. Lipid Res.* **32:**317–327.

Pieters, M. N., Schouten, D., and Van Berkel, T. J. C., 1994, *In vitro* and *in vivo* evidence for the role of HDL in reverse cholesterol transport, *Biochim. Biophys. Acta* **1225:**125–134.

Piha, M., Lindstedt, L., and Kovanen, P. T., 1995, Fusion of proteolyzed low-density lipoprotein in the fluid phase: A novel mechanism generating atherogenic lipoprotein particles, *Biochemistry* **34:**10120–10129.

Rapp, J. H., Connor, W. E., Lin, D. S., Inahara, T., and Porter, J. M., 1983, Lipids of human atherosclerotic plaques and xanthomas: clues to the mechanism of plaque progression, *J. Lipid Res.* **24:**1329–1335.

Rapp, J. H., Lespine, A., Hamilton, R. L., Colyvas, N., Chaumeton, A. H., Tweedie-Hardman, J., Kotite, L., Kunitake, S. T., Havel, R. J., and Kane, J. P., 1994, Triglyceride-rich lipoproteins isolated by selected-affinity anti-apolipoprotein B immunosorption from human atherosclerotic plaque, *Arterioscl. Thromb.* **14:**1767–1774.

Renshaw, P. F., Janoff, A. S., and Miller, K. W., 1983, On the nature of dilute aqueous cholesterol suspensions, *J. Lipid Res.* **24:**47–51.

Ridgway, N. D., 1995, 25-Hydroxycholesterol stimulates sphingomyelin synthesis in Chinese hamster ovary cells, *J. Lipid Res.* **36:**1345–1358.

Robenek, H., and Schmitz, G., 1988, Ca^{++} antagonists and ACAT inhibitors promote cholesterol efflux from macrophages by different mechanisms. II. Characterization of intracellular morphologic changes, *Arteriosclerosis* **8:**57–67.

Roberts, A. B., Lees, A. M., Lees, R. S., Strauss, H. W., Fallon, J. T., Taveras, J., and Kopiwoda, S., 1983, Selective accumulation of low density lipoproteins in damaged arterial wall, *J. Lipid Res.* **24:**1160–1167.

Robinson, K. A., and Apkarian, R. P., 1991, Ultrastructure of coronary arterial endothelium in atherosclerotic swine suggests lipid retro-endocytosis, *Scanning Microsc.* **5:**533–539.

Rotheneder, M., and Kostner, G. M., 1986, Cholesterol esterification in mouse peritoneal macrophages in the presence of pathological human plasma lipoproteins, *Biol. Chem. Hoppe Seyler* **367:**1219–1222.

Rouis, M., Nigon, F., Lafuma, C., Hornebeck, W., and Chapman, M. J., 1990, Expression of elastase activity by human monocyte-macrophages is modulated by cellular cholesterol content, inflammatory mediators, and phorbol myristate acetate, *Arteriosclerosis* **10:**246–255.

Rouser, G., and Solomon, R. D., 1968, Changes in phospholipid composition of human aorta with age, *Lipids* **4:**232–234.

Ryan, U. S., Maxwell, G., Olazabal, B., Hart, M., and Mayfield, L., 1986, Endothelial cell phagocytosis and activation, *Fed. Proc.* **45:**101–108.

Sandberg, H., Bode, A. P., Dombrose, F. A., Hoechli, M., and Lentz, B. R., 1985, Expression of coagulant activity in human platelets: Release of membranous vesicles providing platelet factor 1 and platelet factor 3, *Thromb. Res.* **39:**63–79.

Sata, T., Havel, R. J., and Jones, A. L., 1972, Characterization of subfractions of triglyceride-rich lipoproteins separated by gel chromatography from blood plasma of normolipemic and hyperlipemic humans, *J. Lipid Res.* **13:**757–768.

Schechter, I., Fogelman, A. M., Haberland, M. E., Seager, J., Hokom, M., and Edwards, P. A., 1981, The metabolism of native and malondialdehyde-altered low density lipoproteins by human monocyte-macrophages, *J. Lipid Res.* **22:**63–71.

Scherphof, G., Morselt, H., Regts, J., and Wilschut, J. C., 1979, The involvement of the lipid phase transition in the plasma-induced dissolution of multilamellar phosphatidylcholine vesicles, *Biochim. Biophys. Acta* **556:**196–207.

Schmitz, G., Robenek, H., Beuck, M., Krause, R., Schurek, A., and Niemann, R., 1988, Ca^{++} antagonists and ACAT inhibitors promote cholesterol efflux from macrophages by different mechanisms. I. Characterization of cellular lipid metabolism, *Arteriosclerosis* **8:**46–56.

Schmitz, G., Beuck, M., Fischer, H., Nowicka, G., and Robenek, H., 1990, Regulation of phospholipid biosynthesis during cholesterol influx and high density lipoprotein-mediated cholesterol efflux in macrophages, *J. Lipid Res.* **31:**1741–1752.

Schuff-Werner, P., Claus, G., Armstrong, V. W., Kostering, H., and Seidel, D., 1989, Enhanced procoagulatory activity (PCA) of human monocytes/macrophages after *in vitro* stimulation with chemically modified LDL, *Atherosclerosis* **78:**109–112.

Schwenke, D. C., and Carew, T. E., 1989a, Initiation of atherosclerotic lesions in cholesterol-fed rabbits. I. Focal increases in arterial LDL concentration precede development of fatty streak lesions, *Atherosclerosis* **9:**895–907.

Schwenke, D. C., and Carew, T. E., 1989b, Initiation of atherosclerotic lesions in cholesterol-fed rabbits. II. Selective retention of LDL vs. selective increases in LDL permeability in susceptible sites of arteries, *Arteriosclerosis* **9:**908–918.

Scott, R. F., Daoud, A. S., and Florentin, R. A., 1976, Animal models in atherosclerosis, in: *Atherosclerosis in Primates* (J. Strong, ed.), pp. 120–145, S. Karger, New York.

Seth, S. K., and Newman, H. A. I., 1975, Sphingomyelin and other phospholipid metabolism in the rabbit atheromatous and normal aorta, *Circ. Res.* **36:**294–299.

Shaikh, M., Martini, S., Quiney, J. R., Baskerville, P., LaVille, A. E., Browse, N. L., Duffield, R., Turner, P. R., and Lewis, B., 1988, Modified plasma-derived lipoproteins in human atherosclerotic plaques, *Atherosclerosis* **69:**165–172.

Shimamoto, T., 1975, Hyperreactive arterial endothelial cells in atherogenesis and cyclic AMP phosphodiesterase inhibitor in prevention and treatment of atherosclerotic disorders, *Jpn. Heart J.* **16:**76–97.

Shimamoto, T., Hidaka, H., Moriya, K., Kobayashi, M., Takahashi, T., and Numano, F., 1976, Hyperreactive arterial endothelial cells: A clue for the treatment of atherosclerosis, *Ann. N.Y. Acad. Sci.* **275:**266–285.

Shio, H., Fowler, S., Bhuvaneswaran, C., and Morris, M. D., 1982, Lysosome lipid storage disorder in NCTR-BALB/c mice. II. Morphologic and cytochemical studies, *Am. J. Pathol.* **108:**150–159.

Simionescu, N., 1988, Prelesional changes of arterial endothelium in hyperlipoproteinemic atherogenesis, in: *Endothelial Cell Biology in Health and Disease* (N. Simionescu and M. Simionescu, eds.), pp. 385–429, Plenum Press, New York.

Simionescu, N., Vasile, E., Lupu, F., Popescu, G., Simionescu, M., 1986, Prelesional events in atherogenesis. Accumulation of extracellular cholesterol-rich liposomes in the arterial intima and cardiac valves of the hyperlipidemic rabbit, *Am. J. Pathol.* **123:**109–125.

Simionescu, N., Sima, A., Dobrian, A., Tirziu, D., and Simionescu, M., 1993, Pathobiochemical changes of the arterial wall at the inception of atherosclerosis, *Curr. Top. Pathol.* **87:**1–45.

Skarlatos, S. I., Amende, L. M., Chao, F.-F., Blanchette-Mackie, E. J., Gamble, W., and Kruth, H. S., 1988, Biochemical characterization of isolated cholesterol-phospholipid particles continuously released from rat and human platelets after activation, *Lab. Invest.* **59:**344–352.

Skarlatos, S. I., Dichek, H. L., Fojo, S. S., Brewer, H. B., and Kruth, H. S., 1993a, Absence of triglyceride accumulation in lipoprotein lipase-deficient human monocyte-macrophages incubated with human very low density lipoprotein, *J. Clin. Endocrinol. Metab.* **76:**793–796.

Skarlatos, S. I., Rouis, M., Chapman, M. J., and Kruth, H. S., 1993b, Heterogeneity of cellular cholesteryl ester accumulation by human monocyte-derived macrophages, *Atherosclerosis* **99:**229–240.

Slater, R. S., and Smith, E. B., 1972, The microdissection of large atherosclerotic plaques to give morphologically and topographically defined fractions for analysis. Part 2. Studies on "Nile Blue" cells, *Atherosclerosis* **15:**57–69.

Small, D. M., 1988, Progression and regression of atherosclerotic lesions. Insights from lipid physical biochemistry, *Arteriosclerosis* **8:**103–129.

Small, D. M., and Shipley, G. G., 1974, Physical-chemical basis of lipid deposition in atherosclerosis. The physical state of the lipids helps to explain lipid deposition and lesion reversal in atherosclerosis, *Science* **185:**222–229.

Smith, E. B., 1974, The relationship between plasma and tissue lipids in human atherosclerosis, *Adv. Lipid Res.* **12:** 1–49.

Smith, E. B., 1981, Relationship between lipids and atherosclerosis, in: *Haemostasis and Thrombosis* (A. L. Bloom, and D. P. Thomas, eds.), pp. 554–574, Churchill Livingstone, New York.

Smith, E. B., 1990, Transport, interactions and retention of plasma proteins in the intima: the barrier function of the internal elastic lamina, *Eur. Heart J.* **11:**72–81.

Smith, E. B., and Ashall, C., 1983, Low-density lipoprotein concentration in interstitial fluid from human atherosclerotic lesions. Relation to theories of endothelial damage and lipoprotein binding, *Biochim. Biophys. Acta* **754:**249–257.

Smith, E. B., and Cochran, S., 1990, Factors influencing the accumulation in fibrous plaques of lipid derived from low density lipoprotein. II. Preferential immobilization of lipoprotein (a) Lp(a), *Atherosclerosis* **84:**173–181.

Smith, E. B., and Slater, R., 1970, The chemical and immunological assay of low density lipoproteins extracted from human aortic intima, *Atherosclerosis* **11:**417–438.

Smith, E. B., and Slater, R. S., 1972, The microdissection of large atherosclerotic plaques to give morphologically and topographically defined fractions for analysis. Part I. The lipids in the isolated fractions, *Atherosclerosis* **15:**37–56.

Smith, E. B., and Slater, R., 1973, Relationship between plasma lipids and arterial tissue lipids, *Nutr. Metabol.* **15:**17–26.

Smith, E. B., and Staples, E. M., 1980, Distribution of plasma proteins across the human aortic wall. Barrier functions of endothelium and internal elastic lamina, *Atherosclerosis* **37:**579–590.

Smith, E. B., Evans, P. H., and Downham, M. D., 1967, Lipid in the aortic intima. The correlation of morphological and chemical characteristics, *J. Atheroscl. Res.* **7:**171–186.

Smith, E. B., Slater, R. S., and Chu, P. K., 1968, The lipids in raised fatty and fibrous lesions in human aorta. A comparison of the changes at different stages of development, *J. Atheroscl. Res.* **8:**399–419.

Smith, E. B., Massie, I. B., and Alexander, K. M., 1976, The release of an immobilized lipoprotein fraction from atherosclerotic lesions by incubation with plasmin, *Atherosclerosis* **25:**71–84.

Smith, L. L., and Johnson, B. H., 1989, Biological activities of oxysterols, *Free Radic. Biol. Med.* **7:**285–332.

Spring, P. M., and Hoff, H. F., 1989, LDL accumulation in the grossly normal human iliac bifurcation and common iliac arteries, *Exp. Mol. Pathol.* **51:**179–185.

Srinivasan, S. R., Dolan, P., Radhakrishnamurthy, B., Pargaonkar, P. S., and Berenson, G. S., 1975, Lipoprotein-acid mucopolysaccharide complexes of human atherosclerotic lesions, *Biochim. Biophys. Acta* **388:**58–70.

Srinivasan, S. R., Vijayagopal, P., Dalferes, E. R. Jr., Abbate, B., Radhakrishnamurthy, B., and Berenson, G. S., 1986, Low density lipoprotein retention by aortic tissue. Contribution of extracellular matrix, *Atherosclerosis* **62:**201–208.

Stary, H. C., and Malinow, M. R., 1982, Ultrastructure of experimental coronary artery atherosclerosis in cynomolgus macaques. A comparison with the lesions of other primates, *Atherosclerosis* **43:**151–175.

Stehbens, W. E., and Ludatscher, R. M., 1983, The susceptibility of renal arterial forks in rabbits to dietary-induced lipid deposition, *Pathology* **15:**475–485.

Stein, O., and Stein, Y., 1980, Bovine aortic endothelial cells display macrophage-like properties towards acetylated [^{125}I]-labelled low density lipoprotein, *Biochim. Biophys. Acta* **620:** 631–635.

Steinbrecher, U. P., and Lougheed, M., 1992, Scavenger receptor-independent stimulation of cholesterol esterification in macrophages by low density lipoprotein extracted from human aortic intima, *Arterioscl. Thromb.* **12:**608–625.

Stemerman, M. B., Morrel, E. M., Burke, K. R., Colton, C. K., Smith, K. A., and Lees, R. S., 1986, Local variation in arterial wall permeability to low density lipoprotein in normal rabbit aorta, *Arteriosclerosis* **6:**64–69.

Stender, S., and Hjelms, E., 1984, *In vivo* influx of free and esterified plasma cholesterol into human aortic tissue without atherosclerotic lesions, *J. Clin. Invest.* **74:**1871–1881.

Stender, S., and Hjelms, E., 1988, *In vivo* transfer of cholesterol ester from high and low density plasma lipoproteins into human aortic tissue, *Arteriosclerosis* **8:**252–262.

Stender, S., and Zilversmit, D. B., 1981, Transfer of plasma lipoprotein components and of plasma proteins into aortas of cholesterol-fed rabbits: molecular size as a determinant of plasma lipoprotein influx, *Arteriosclerosis* **1:**38–49.

Stewart, G. T., 1961, Mesomorphic forms of lipid in the structure of normal and atheromatous tissues, *J. Path. Bact.* **81:**385–393.

Suits, A. G., Chait, A., Aviram, M., and Heinecke, J. W., 1989, Phagocytosis of aggregated lipoprotein by macrophages: low density lipoprotein receptor-dependent foam-cell formation, *Proc. Natl. Acad. Sci. USA* **86:**2713–2717.

Tabas, I., Myers, J. N., Innerarity, T. L., Xu, X.-X., Arnold, K., Boyles, J., and Maxfield, F. R., 1991, The influence of particle size and multiple apoprotein E-receptor interactions on the endocytic targeting of β-VLDL in mouse peritoneal macrophages, *J. Cell Biol.* **115:** 1547–1560.

Tall, A. R., 1980, Studies on the transfer of phosphatidylcholine from unilamellar vesicles into plasma high density lipoproteins in the rat, *J. Lipid Res.* **21:**354–363.

Tangirala, R. K., Jerome, W. G., Jones, N. L., Small, D. M., Johnson, W. J., Glick, J. M., Mahlberg, F. H., and Rothblat, G. H., 1994, Formation of cholesterol monohydrate crystals in macrophage-derived foam cells, *J. Lipid Res.* **35:**93–104.

Tertov, V. V., Sobenin, I. A., Gabbasov, Z. A., Popov, E. G., and Orekhov, A. N., 1989, Lipoprotein aggregation as an essential condition of intracellular lipid accumulation caused by modified low density lipoprotein, *Biochem. Biophys. Res. Commun.* **163:**489–494.

Tertov, V. V., Orekhov, A. N., Kacharava, A. G., Sobenin, I. A., Perova, N. V., and Smirnov, V. N., 1990, Low density lipoprotein-containing circulating immune complexes and coronary atherosclerosis, *Exp. Mol. Pathol.* **52:**300–308.

Tertov, V. V., Orekhov, A. N., Sobenin, I. A., Gabbasov, Z. A., and others, 1992, Three types of naturally occurring modified lipoproteins induce intracellular lipid accumulation due to lipoprotein aggregation, *Circ. Res.* **71:**218–228.

Thompson, T. E., and Tillack, T. W., 1985, Organization of glycosphingolipids in bilayers and plasma membranes of mammalian cells. *Annu. Rev. Biophys. Biophys. Chem.* **14:**361–386.

Tirziu, D., Dobrian, A., Tasca, C., Simionescu, M., and Simionescu, N., 1995, Intimal thickening of human aorta contain modified reassembled lipoproteins, *Atherosclerosis* **112:**101–114.

Toda, T., Leszczynski, D., McGibbon, W. H., and Kummerow, F. A., 1980, Coronary arterial lesions in sexually mature non-layers, layers, and roosters, *Virchows Arch. A.* **388:**123–135.

Trillo, A. A., and Prichard, R. W., 1979, Early endothelial changes in experimental primate atherosclerosis, *Lab. Invest.* **41:**294–302.

Tucker, C. F., Catsulis, C., Strong, J. P., and Eggen, D. A., 1971, Regression of early cholesterol-induced aortic lesions in rhesus monkeys, *Am. J. Pathol.* **65:**493–514.

Vanier, M. T., 1983, Biochemical studies in Niemann-pick disease. I. Major sphingolipids of liver and spleen, *Biochim. Biophys. Acta* **750:**178–184.

Vijayagopal, P., Srinivasan, S. R., Radhakrishnamurthy, B., and Berenson, G. S., 1992, Lipoprotein-proteoglycan complexes from atherosclerotic lesions promote cholesteryl ester accumulation in human monocytes/macrophages, *Arterioscl. Thromb.* **12:**237–249.

Vijayagopal, P., Srinivasan, S. R., Radhakrishnamurthy, B., and Berenson, G. S., 1993, Human monocyte-derived macrophages bind low-density-lipoprotein-proteoglycan complexes by a receptor different from the low-density-lipoprotein receptor, *Biochem. J.* **289:**837–844.

Walton, K. W., and Morris, C. J., 1977, Studies on the passage of plasma proteins across arterial endothelium in relation to atherogenesis, *Prog. Biochem. Pharmacol.* **14:**138–152.

Walton, K. W., and Williamson, N., 1968, Histological and immunofluorescent studies on the evolution of the human atheromatous plaque, *J. Atheroscl. Res.* **8:**599–624.

Watson, K. E., Bostrom, K., Ravindranath, R., Lam, T., Norton, B., and Demer, L. L., 1994, TGF-1 and 25-hydroxycholesterol stimulate osteoblast-like vascular cells to calcify, *J. Clin. Invest.* **93:** 2106–2113.

Weinbaum, S., and Chien, S., 1993, Lipid transport aspects of atherogenesis, *J. Biomech. Eng.* **115:** 602–610.

Werb, Z., and Cohn, Z. A., 1972, Cholesterol metabolism in the macrophage. 3. Ingestion and intracellular fate of cholesterol and cholesterol esters, *J. Exp. Med.* **135:**21–44.

Wiklund, O., Bjornheden, T., Olofsson, S.-O., and Bondjers, G., 1987, Influx and cellular degradation of low density lipoproteins in rabbit aorta determined in an *in vitro* perfusion system, *Arteriosclerosis* **7:**565–571.

Williams, K. J., and Tabas, I., 1995, The response-to-retention hypothesis of early atherogenesis, *Arterioscl. Thromb. Vasc. Biol.* **15:**551–561.

Windaus, A., 1910, Über den gehalt normaler und atheromatöser aorten an cholesterin und cholesterin und cholesterinestern, *Hoppe-Seyl. Z.* **67:**174–176.

Wissler, R. W., 1994, New insights into the pathogenesis of atherosclerosis as revealed by PDAY, *Atherosclerosis* **108:**S3–S20.

Xu, X.-X., and Tabas, I., 1991, Sphingomyelin enhances low density lipoprotein uptake and ability to induce cholesteryl ester accumulation in macrophages, *J. Biol. Chem.* **266:**24849–24858.

Yamada, T., Press, M., Vesselinovitch, D., and Wissler, R. W., 1988, Quantitative ultrastructural analysis of coronary atherosclerotic involvement in two macaque species, *Exp. Mol. Pathol.* **48:**1–23.

Ylä-Herttuala, S., Jaakkola, O., Ehnholm, C., Tikkanen, M. J., Solakivi, T., Sarkioja, T., and Nikkari, T., 1988, Characterization of two lipoproteins containing apolipoproteins B and E from lesion-free human aortic intima, *J. Lipid Res.* **29:**563–572.

Ylä-Herttuala, S., Palinski, W., Rosenfeld, M. E., Parthasarathy, S., Carew, T. E., Butler, S., Witztum, J. L., and Steinberg, D., 1989, Evidence for the presence of oxidatively modified low density lipoprotein in atherosclerotic lesions of rabbit and man, *J. Clin. Invest.* **84:**1086–1095.

Ylä-Herttuala, S., Palinski, W., Rosenfeld, M. E., Steinberg, D., and Witztum, J. L., 1990, Lipoproteins in normal and atherosclerotic aorta, *Eur. Heart J.* **11:**Suppl. E, 88–99.

Zilversmit, D. B., 1979, Atherogenesis: a postprandial phenomenon, *Circulation* **60:**473–485.

Zolfaghari, R., Harrison, E. H., Han, J. H., Rutter, W. J., and Fisher, E. A., 1992, Tissue and species differences in bile salt-dependent neutral cholesteryl ester hydrolase activity and gene expression, *Arterioscler. Thromb.* **12:**295–301.

Chapter 13

Cholesterol Metabolism and Tumor Cell Proliferation

Peter S. Coleman, Li-Chuan Chen, and
Laura Sepp-Lorenzino

1. INTRODUCTION

The functions assumed by free cholesterol in mammalian cells are diverse, ranging from that as a starting substrate for the synthesis of steroid hormones and bile salts to its less specifically understood but critical role as a principal lipid component of every cell membrane. Cholesterol biosynthesis begins with metabolically supplied cytoplasmic acetyl CoA that also serves as the common precursor for *de novo* fatty acid synthesis, and can be expressed by the reaction: 18 acetyl CoA + $\frac{1}{2}O_2$ + 10 $H^+ \rightarrow$ cholesterol + 9 CO_2 + 18 CoA-SH. During the synthesis, the genesis of mevalonic acid (MVA) from 3-hydroxy-3-methylglutaryl CoA (HMG-CoA), catalyzed by HMG-CoA reductase (HMGR), the third and principal rate-controlling enzyme of the pathway, leads to the production of diverse polyisoprenoid intermediates. Metabolism of MVA demonstrates considerable versatility through ensuing branching reactions that yield a host of terpene compounds used in the

Peter S. Coleman The Boston Biomedical Research Institute, Laboratory of Metabolic Regulation, Boston, MA 02114. **Li-Chuan Chen** Clinical Pharmacology Branch, National Cancer Institute, National Institutes of Health, Bethesda, MD 20892. **Laura Sepp-Lorenzino** Memorial Sloan-Kettering Cancer Center, Program in Cell Biology and Genetics and Department of Medicine, New York, NY 10021.

Subcellular Biochemistry, Volume 28: Cholesterol: Its Functions and Metabolism in Biology and Medicine, edited by Robert Bittman. Plenum Press, New York, 1997.

construction of an assortment of required metabolites [dolichol, ubiquinone (Coenzyme Q), isopentenyl adenine, heme a/a_3], in addition to cholesterol itself. This chapter focuses on associations that link cholesterol and mitochondrial citrate export, as well as the metabolism of citrate-derived isoprenoid precursor metabolites *en route* to cholesterol synthesis, with deregulated cell proliferation (i.e., cancer).

1.1. Cholesterol and Cancer: Correlations through a Glass, Darkly

Reflecting the diversity of its varied and often nonspecific cellular roles, cholesterol's relationship to human disease, in general, appears tenuous or inexplicit, principally because clinical diagnosis of a disease is made when the patient presents macroscopic symptoms that disturb the whole organism's normative physiology; molecular-level mechanisms that underlie and provoke the symptoms are initially veiled. The deposition of lipid-filled structures within the endothelial cell luminal lining of the arterial wall's smooth muscle layer, for example, is universally diagnostic of atherosclerosis in its most insidious form—coronary artery disease, which is responsible for more than 500,000 deaths annually in the U.S. alone. Cholesterol-based atherogenesis evolves from lesions in this luminal endothelium, and, upon smooth muscle migration through these lesions, eventually leads to formation of gelatinous and fibrous plaques or thickenings (see Chapter 12 by Kruth). Together with progressive smooth muscle cell proliferation and recruitment of platelets, such thickenings turn into cholesterol-rich, calcified deposits or streaks that protrude, narrow, and ultimately block the channel of such critical vessels as coronary arteries to the point where, due to the reduced blood flow, oxygen starvation becomes the decisive element in sustaining normal cardiac muscle function. Yet, the underlying molecular mechanisms that induce the progressive arterial accumulation of cholesterol-rich deposits in the first place are not clearly understood. Nevertheless, due to the coalescence of tight statistical correlations between circulating plasma cholesterol levels, the pathogenesis of atherosclerosis, and clinical cardiovascular events, treatment to lower the total plasma cholesterol level is a common strategy currently prescribed by national and international concensus panels and medical organizations (see Chapter 17 by Biessen *et al.*).

Malignant cell growth, or cancer, occurs as a result of abnormal gene function, which in turn gives rise to various metabolic aberrations: altered signal transduction, different modes of transcription, changes in cell cycle regulation, reduced capacity for DNA repair, anomalous DNA replication, and the mismanagement of the programs controlling normal cell senescence (apoptosis). These undesirable pathologies all become possible upon recruitment of either mutant or incorrectly expressed enzyme proteins whose functions can alter metabolic processes responsible for the homeostasis of the differentiated cell and promote untimely cell proliferation instead. For decades there has been serious speculation that cholesterol and aspects of its complex cellular metabolism are implicated in carcinogenesis. Clinical epidemiological studies, conducted in the U.S. and elsewhere, have attempted to discern statistical correlations between human serum (or ascitic fluid) cholesterol concentration and the ability to predict cancer risk (Keys *et al.*, 1985; Hiatt and Fireman, 1986; Sherwin *et al.*, 1987; Knekt *et al.*, 1988; Kromhout *et al.*, 1988; Mortensen *et al.*, 1988; Schatzkin *et al.*, 1988). The conclusions of these and numerous other such studies remain controversial, despite indications that preclinical malignancy generally appears to be inversely related to the serum cholesterol level across nearly all subject populations. Such statistical correlation investigations were motivated, in part, from early reports based on animal dietary, cellular, and enzyme-specific studies (Dietschy and Siperstein, 1967; Sabine, 1975; Kandutsch and Chen, 1977), all of which pointed to the involvement of serum or cellular cholesterol, and/or the cholesterogenesis pathway itself, in modulating the rate of cell proliferation.

1.2. The Current Climate of Research Emphasis: Isoprenoid Metabolic Intermediates and Prenylated Proteins

More than a decade of accumulated data accentuate the tight connection between uncontrolled cell proliferation and a specific, early metabolic branch off the main cholesterol synthesis pathway, arising after the synthesis of MVA from HMG-CoA. Correlations between the metabolism and utilization of certain MVA-derived polyisoprenoid intermediates and cell proliferation have stirred excitement not only insofar as cancer cell growth, but increasingly in the area of arterial smooth muscle cell proliferation that accompanies atherogenic disease. This heightened interest in cholesterol pathway-derived polyisoprenoids (particularly C_{15} farnesyl and C_{20} geranylgeranyl moieties) and its connection to cell proliferation devolves upon production of a diverse class of proteins that become post-translationally modified with covalently bound farnesyl or geranylgeranyl isoprenoids—the prenylated proteins. Commensurate with reports that continue to characterize the properties of, and identify, the ever-increasing numbers of prenylated proteins, has been an accumulation of information about their role in intracellular molecular signaling. Because of the close correlations between intracellular signaling, the involvement of prenylated proteins, and the latter's reliance on the cholesterol pathway, references made a decade ago to "pleiotropic effects of cholesterogenesis on cell proliferation" now appear to be somewhat more clearly defined through identification of certain farnesylated proteins (such as the activated Ras GTPase) that serve as critical information transmission molecules due to their participation as early components of interlocking signaling cascades. Current research, however, generates reports on almost a daily basis about protein–protein associations of increasing complexity. Clearly, our understanding of the roles of cholesterol and its MVA-derived precursors in regulating cell proliferation by means of such signaling is still far from complete.

1.3. Previous Reviews and Chapter Prospective

Several previous reviews have attempted to collate direct or indirect connections linking various functions assigned to molecular cholesterol, the regulation of its synthesis, and tumor cell proliferation (Chen *et al.*, 1978; Coleman and Lavietes, 1981; Chen, 1984; Siperstein, 1984; Coleman and Sepp-Lorenzino, 1990; Coleman, 1991). Short recapitulations of some of these correlations recently appeared (Rao, 1995; Siperstein, 1995). This chapter will draw on selected references from these past reviews and summarize ideas that remain relevant according to current knowledge. Those lingering concepts will be extended with recent findings on the role of MVA-derived intermediates that appear to be intimately involved in cell proliferation signaling, particularly regarding the functions of farnesylated Ras. Recent information will also be discussed on peroxisomally compartmentalized cholesterol (and other lipid) metabolism. In addition, current studies will be documented on inhibitors of main-line cholesterol biosynthesis reactions and their potential as antitumor agents. Finally, ongoing work and speculations about the inhibition of specific protein prenylation enzymes and cell proliferation will be addressed as a potentially promising avenue for the advent of future tumoristatic drugs.

2. RAMIFICATIONS OF MEMBRANE CHOLESTEROL CONTENT

2.1. Subcellular Membrane Cholesterol Distribution

In addition to the fact that eukaryotic cell membranes contain either cholesterol or a close structural relative, virtually every major metabolic pathway, at some point during turnover, must directly involve, or be the result of, interactions with membranes. The relationship between tumor growth and cholesterol *per se* appears indirect partly because of this sterol's imprecisely defined role as a ubiquitous lipid element inserted, to varying extents, in the bilayer of different types of membranes. Since parallel phospholipid leaflet bilayers form the structural template of membranes, functions exhibited by proteins inserted into this bilayer differentially reflect the immediate lipid environment of membrane proteins (Spector and Yorek, 1985). Accordingly, localized membranes in the cell (plasma, mitochondrial, nuclear, etc.) constitute functionally specific organelles in their own right, with each type of membrane being responsible for, and controlling, discrete metabolic reaction sequences.

It is not obvious why mammalian cell membranes contain cholesterol. However, as all mammalian cell membranes contain free cholesterol (Yeagle, 1987, 1991, 1992), the major physico-chemical effect of its presence is an increased order or rigidity in the acyl-chain core of the membrane phospholipids and, thus, a lower degree of membrane fluidity. Membrane cholesterol also appears to be spatially organized into clusters rather than exhibiting a homogeneously uniform (or random) distribution amidst the diverse bilayer phospholipids. The concept of cholesterol-rich domains or patches (Jain, 1975; Jain and White, 1977) has been recently re-emphasized (Rothblat *et al.*, 1992).

Different mammalian membranes contain different amounts of free cholesterol, and in normal cells such membrane-specific cholesterol levels are fairly well maintained, although the homeostatic mechanisms involved in sustaining membrane-specific cholesterol levels remain unclear (Table I). The cholesterol content assessed as characteristic of specific cell membranes—plasma, nuclear, mitochondrial, microsomal, etc.—has been found to be altered in diverse disease states, such as cancer (Coleman and Sepp-Lorenzino, 1990; Coleman, 1991). Quantitation of membrane lipids, particularly cholesterol, remains difficult and controversial due to: (i) the propensity for varying degrees of cross-contamination between subcellular membrane fractions accompanying isolation; (ii) the variety of methods used to evaluate membrane cholesterol content; and (iii) the broad diversity of cell types studied. For example, the plasma membrane of intact cultured fibroblasts has been reported to contain greater than 90% of the cell cholesterol when assessed by the cholesterol oxidase method (Lange *et al.*, 1989; Lange, 1992), although this high value or a generalization of it applicable to all cell types has been vigorously disputed (van Meer, 1987). Notwithstanding these variances, the example for normal rat liver given in Table I indicates that the plasma membrane, and lysosomal membranes that derive from it, invariably possess the highest cholesterol levels, and mitochondria the lowest.

The question arises how cholesterol becomes distributed among the different cell membranes subsequent to its synthesis in the cytoplasmic/ER environment.

Table I
Rat Liver Membranes: General Lipid, Protein, Phospholipid, and Cholesterol Content

Membrane type	Protein[a] (% dry wt)	Lipid[a] (% dry wt)	P-lipid[b] (μg/mg protein)	Chol[b] (μg/mg protein)	Chol/P-lipid[b] (%)	Chol[c] (mol %)
Plasma	58	42	672	128	19	38
Lysosome	—	—	156	38	24	48
Golgi	—	—	825	78	9	19
Microsome (av.)	67	33	374	14	4	7
Nuclear	66	34	500	38	8	15
Mitochondria	52 (outer) 76 (inner)	48 (outer) 24 (inner)	155[d]	3[d]	2	4

[a]Data essentially from Guidotti (1972).
[b]Data from Daum (1985); Table I.
[c]Calculated from Daum (1985); cholesterol MW = 386.6, P-lipid $MW_{(av)}$ = 770.
[d]Average of both membranes. These values fluctuate widely depending on the tissue from which mitochondria were derived; see Daum (1985), Table II.

With regard to the plasma membrane, its relatively high cholesterol content (whether assessed as $<50\%$ or $>90\%$ of total), compared with intracellular membranes, may be a function of the level of membrane sphingomyelin. As with other phospholipids (Op den Kamp, 1979), sphingomyelin appears to be distributed asymmetrically, the outer leaflet of the plasma membrane being sphingomyelin-enriched relative to the inner (Slotte *et al.*, 1989). It was shown that sphingomyelin helps maintain the cholesterol homeostasis of at least the plasma membrane by modulating the activity of cytoplasmic acyl-CoA:cholesterol acyl transferase, which catalyzes the formation of cholesteryl esters from the unesterified cholesterol (Chen *et al.*, 1993; Okwu *et al.*, 1994; see Chapter 10 by Slotte). The formation of cholesterol-derived cholesteryl esters, which themselves are not membrane lipid components but rather are compartmentalized as storage forms of the sterol, can thus effect a redistribution of at least plasma membrane cholesterol under circumstances where it would rise beyond some normal threshold level (Liscum and Dahl, 1992; Liscum and Underwood, 1995). As emphasized above, the underlying basis for an empirically determined "normal" level of free cholesterol in any of the cell's membranes, whether of metabolic or physicochemical origin, is not clear. This statement is also probably as applicable to the phospholipid heterogeneity found within any of the membranes of cells (Wirtz, 1991).

Apart from its role as a ubiquitous membrane structural element, unesterified cholesterol has no known specific function in any cellular reaction, except as a metabolic substrate or a proven modulator of water and cation (Na^+, H^+) transport across membranes (Haines, 1993). The literature contains, on the one hand, recurrent observations of distally related associations between the amount of membrane cholesterol and metabolic phenomena linked to cell proliferation, and on the other hand, increasingly focused molecular-level conjectures on the more direct involvement of MVA-derived intermediary metabolites in the control of signaling processes that lead either to cellular differentiation or proliferation. These two themes comprise points of emphasis for the remainder of this chapter.

2.2. Tumors: Diminished Control of Cholesterogenesis and Membrane Cholesterol Enrichment

Studies up through the mid-1970s on the linkage between cholesterol and cell growth specifically emphasized this sterol's requirement as fulfilling the need for creation of new membranes; the requirement could be met either by *de novo* biosynthesis or by means of dietary or cell culture media supplementation. However, tumor tissues, and certain membranes derived from them, were found to contain a higher free cholesterol content compared with their normal cell progenators, and esterified cholesterol levels in tumors were similarly increased [see, for example, van Hoeven and Emmelot (1972); Brenneman *et al.* (1975)]. By that time, it had also been learned that the enzymology of cholesterol synthesis, whose regulation in normal cells and tissues is stringently adjusted by multiple molecular mechanisms, including feedback downregulation at the MVA-generating step catalyzed by HMGR (Edwards *et al.,* 1984; Rudney and Sexton, 1986), demonstrated absence of sterol-metabolite feedback control at this locus in tumors, resulting in high continuous rates of tumor cholesterogenesis (see review by Coleman and Lavietes, 1981, and references therein).

With tumors, reports on the loss of ordinarily tight regulation of cholesterogenesis found in normal cells and tissues, mainly at the HMGR step, were seen to provide a consistent hallmark of neoplasia (Siperstein, 1970; Sabine, 1975; Pedersen, 1978). Furthermore, it became firmly established that proliferating cells in culture, grown in medium containing cholesterol-depleted serum, compensate for the lack of this required exogenous "nutrient" by actively synthesizing cholesterol *de novo* (Faust *et al.,* 1982). When grown in culture, frank tumor cells, in which cholesterogenesis is rapid, constant, and lacking in feedback control at the HMGR locus, mirror whole tumor tissue growth profiles by exhibiting a higher cholesteryl ester content, while yielding isolated membrane fractions either enriched in cholesterol or manifesting altered cholesterol/phospholipid membrane compositions, as compared with the membrane lipids of well-differentiated, slow-, or non-dividing cells. Depending on the tissue or proliferating cell line examined, as well as the particular membrane fraction scrutinized, and generally correlating with the rates of cell growth *in vivo* or in culture, the membrane cholesterol enrichment of tumors ranges widely, from moderate to many-fold, compared to non-tumorigenic cells (van Hoeven and Emmelot, 1972; Dessi *et al.,* 1992) but is invariably higher than normal. Moreover, cholesterol-enriched plasma membranes of many tumors were found to actively exfoliate cholesterol-rich, tumor antigen-containing microvesicles, serving not only to "decoy" the host's immune surveillance away from the proliferating neoplasm, but also perhaps as a homeostatic strategy to reestablish a more "normal" membrane cholesterol content (see van Blitterswijk *et al.,* 1977; van Blitterswijk *et al.,* 1979; Boctor and Bystryn, 1982; van Blitterswijk *et al.,* 1982; van Blitterswijk *et al.,* 1984; and references cited in Coleman and Lavietes, 1981). By the same token, the loss of control over cholesterogenesis *de novo* at the HMGR locus of the pathway, conceded as a phenotypic characteristic of tumors (Brown *et al.,* 1978; Chen *et al.,* 1978; Brown and Goldstein, 1980; Chen, 1984; Siperstein, 1984; Goldstein and Brown, 1990), stimulated attempts to understand more about the interplay of metabolic routes that connect with cholesterogenesis in tumors (e.g., Azrolan and Coleman, 1989).

Finally, the role of cholesterogenesis, when studied intensively with specific inhibitors of HMGR (see section 4.1), proved critical to verifying that there exists a definitive link between cholesterogenesis *de novo* and cell proliferation, and that this linkage is universal. Thus, all dividing cells (differentiated as well as transformed) adhere to this principle, including normal smooth muscle cells, atrial myocytes, and endothelial, fibroblast, and neuronal cells, each of which displays retarded growth upon HMGR inhibition, similar to what is observed with more blatantly proliferating tumor cell systems. Supporting observations continue to accumulate to the present day (Habenicht *et al.*, 1980; Roussillon *et al.*, 1983; Langan and Volpe, 1987a; Langan and Volpe, 1987b; Cox *et al.*, 1988; Stange *et al.*, 1988; Tertov *et al.*, 1988; Corsini *et al.*, 1995).

2.3. High Carbon Flux through Cholesterogenesis: MVA-derived Isoprenoids in G1 Induce DNA Replication and Cell Proliferation

Many of the aforementioned phenomenological observations (e.g., Kandutsch and Hancock, 1971; Kandutsch and Chen, 1977; Kandutsch *et al.*, 1977) evolved from correlations made between the active process of cholesterogenesis *de novo* and the temporal characteristics of the cell cycle. The fact that the cholesterogenesis pathway in tumors lacks the metabolic controls common in normal cells and tissues, thereby resulting in a high flux of carbon precursor incorporation into cholesterol, strengthened even further the linkage between rates of cholesterogenesis *de novo* and the cholesterol-enriched membranes of tumors (Spady, 1991).

Regarding these direct correlations with cell division, the report by Chen *et al.* (1975), which revealed that in synchronized cell cultures, the specific inhibition of cholesterogenesis prior to S-phase led to G1 arrest, remains one of that era's most direct demonstrations. The conclusions drawn from this study reinforced the primary requirement of an actively operating, high precursor-carbon flux through the cholesterogenic pathway in order for cells to proliferate, over and above that necessary to produce sufficient cholesterol for the construction of membranes. Later reports (Perkins *et al.*, 1982; Langan and Volpe, 1987; Tabacik and Aliau, 1989) continued to emphasize the fact that non-sterol/non-cholesterol precursors derived during active cholesterogenesis are essential to stimulate DNA synthesis and proliferation. By the early 1980s, it was verified that inhibitors of cholesterogenesis that also blocked proliferation did so only when added prior to the decline of cholesterol stimulation, that is, during early-to-mid-G1.

Questions on the temporal involvement of sterol synthesis during the cell cycle thus highlighted the relation between specific segments of the cholesterogenesis pathway and cell proliferation. Such studies, done initially by Siperstein's laboratory (Quesney-Huneeus *et al.*, 1979; Quesney-Huneeus *et al.*, 1983), revealed that pathway intermediates arising between formation of MVA and the cyclization of squalene to yield lanosterol, served in some undefined capacity as critical triggers—or molecular signals—that promoted transit from the G1-to S-phase of the cell cycle. Therefore, it became clear that the cell's need to acquire cholesterol during growth occurred by *de novo* synthesis mainly in the early G1-phase of the cell cycle, which presumably fulfilled demands for, at least, new membrane synthesis required during duplication of cell organelles upon mitosis.

But free cholesterol (whether supplied extracellularly through LDL receptor-mediated uptake, or by mobilizing intracellular stores of cholesteryl esters) showed no clear effect on induction of DNA replication.

Genetic and metabolic research on the cell cycle had, by then, concluded that once a critical G1/S-phase boundary point is reached, cells become committed to complete their entire division cycle without requiring further intervention by extracellular factors (see reviews by Pardee, 1989; Norbury and Nurse, 1992; Peeper *et al.,* 1994). Thus, in the early-to-mid 1980s, an important conclusion of various laboratories was the time-restricted requirement for creation of some MVA-derived metabolite(s) that helped to stimulate DNA synthesis and ensuing cell proliferation, with such an intermediate acting during a relatively brief period corresponding to late G1, just prior to reaching the G1/S-phase boundary (Fairbanks *et al.,* 1984; Fairbanks *et al.,* 1986). These observations, usually made with transformed or tumor cell populations, were immediately expanded to include proliferation of minimally transformed, non-neoplastic cells, thereby generalizing the phenomenon. For example, in partially hepatectomized rats, 8 h post-hepatectomy (i.e., just prior to the onset of DNA synthesis in the surgically challenged liver) the activity of HMGR increases concomitant with decreases in squalene and cholesterol synthesis in the regenerating liver tissue (Trentalance *et al.,* 1984). This, again, supported the concept that before DNA replication in S-phase, MVA-derived intermediates might be diverted away from the more distal reactions of the cholesterogenic pathway, reflecting the S-phase dependence on certain non-sterol-producing MVA metabolites. These and other studies confirmed that among numerous signals proposed to help initiate transit through the cell cycle by allowing entry of proliferating cells into the S-phase, those arising from MVA metabolism generated seemingly specific (but at that time undefined) cholesterol pathway-derived isoprenoid intermediates as molecular signals that functioned only during the time-restricted late G1-phase. MVA-derived dolichyl phosphate (precursor for *N*-linked glycoprotein synthesis) was proposed as one of these critical regulators of transit through the cell cycle (Doyle *et al.,* 1991) whereas other common isoprenoid derivatives (ubiquinone, isopentenyl adenine) proved less encouraging in this regard (Maltese and Aprille, 1985).

To summarize, all cells continuously engaged in the cell cycle (e.g., tumors), have lost the expected feedback downregulation of cholesterogenesis at the HMGR locus (Siperstein, 1995). As a result, their membranes are often found to be cholesterol-rich relative to normal cell counterparts. All proliferating cells (*in vivo* and *in vitro*; embryonic and normal) manifest a high carbon flux through the cholesterogenesis pathway during the G1-phase, whereas terminally differentiated, non-dividing cells display minimal cholesterogenic pathway activity in G1. Finally, there exists a strict temporal correlation between MVA-derived isoprenoid biosynthesis during the G1-phase and subsequent DNA replication in S-phase, such that for new DNA synthesis to occur, specific MVA-derived isoprenoid elements of the cholesterogenic pathway must be expressed prior to the G1/S boundary point, during mid-to-late G1.

2.4. Tumor Mitochondrial Membrane Cholesterol, Citrate Transport, and the Supply of Acetyl Carbons for Increased Tumor Cholesterogenesis

All nucleated mammalian cells have the capacity for cholesterogenesis *de novo*. The liver manifests among the highest rates because, in addition to its large mass, it generates cholesterol for export to supply other tissues. Mammalian tissues, such as liver, that assume responsibility for extensive biosynthesis, exhibit commensurately high aerobic respiration and possess numerous mitochondria. Therefore, liver mitochondria-based reactions are pre-eminent not only by serving as the furnace of bioenergetics, but also because mitochondria control cellular anabolism due to their compartmentalized channeling of precursor anabolites required by cytoplasmic synthetic machinery. Mitochondria assume such ancillary responsibilities by delivering substrate metabolites, created in the matrix space, across the double membrane of the organelle and into the cytoplasmic mileu, a process catalyzed by numerous inner membrane-associated anion substrate transport proteins (LaNoue and Schoolwerth, 1979; LaNoue and Schoolwerth, 1984). Mitochondria can occupy as much as 40% of the volume of some mammalian cells, and their membranes comprise a substantial portion of the total cellular membrane surface area, particularly when taking into account the contribution provided by the highly folded inner mitochondrial membrane (Blouin *et al.*, 1977; Weibel, 1984). It is noteworthy that profiles (both ultrastructural and metabolic) accepted as characteristic of normal cell mitochondria have been reported to be different in tumors, including the general observation that tumor cells are deficient in their cellular mitochondrial content (see review by Pedersen, 1978; also Posner and Coleman, unpublished observations).

A well-documented characteristic of mitochondria isolated from diverse tumors is the enrichment of their membranes with cholesterol, relative to mitochondria derived from normal progenitor tissues (Feo *et al.*, 1973; Pani *et al.*, 1973; Farias *et al.*, 1975; Feo *et al.*, 1975; Morton *et al.*, 1976; Coleman *et al.*, 1978; Pedersen, 1978). Since it has been ascertained repeatedly that the level of membrane cholesterol can affect the activity of membrane-integrated enzymes and transport proteins (e.g., Kaschnitz *et al.*, 1976; Kaplan *et al.*, 1982; Krämer, 1982; Yeagle, 1991, and many others), studies were reported describing various perturbations of inner mitochondrial membrane enzymology and metabolism (rates of electron transport oxidation, uncoupler-stimulated ATPase, adenine nucleotide transport, etc.) as a function of experimental manipulation of cholesterol content or the cholesterol/phospholipid ratio of mitochondrial membranes (Graham and Green, 1970; Farias *et al.*, 1975; Coleman *et al.*, 1978; Hackenbrock *et al.*, 1980; Schneider *et al.*, 1980; Hackenbrock, 1981; Krämer, 1982). In a series of detailed physicochemical studies employing liposomes fused with specific lipid vesicle systems constructed from inner mitochondrial membranes (mitoplasts), electron transfer rates were shown to correlate with the degree of lateral diffusion of redox components (Schneider *et al.,* 1980; Schneider *et al.,* 1982; Gupte *et al.,* 1984; Chazotte and Hackenbrock, 1988; Chazotte and Hackenbrock, 1991). Lateral diffusibility of membrane proteins is, of course, modulated by the degree of membrane cholesterol, among other variables. Collectively, the data from Hackenbrock's laboratory revealed that mitochondrial redox reactions occur by random collisional, long-range diffusion between redox partners, and that rates of electron transfer can be modulated by non-productive "obstructive" collisions that depend, in part, on the protein concentration of the inner membrane, and on the relative cholesterol content of the bilayer (see Chazotte and Hackenbrock, 1991, for review).

Similarly, altered tumor mitochondrial membrane-mediated metabolism was investigated in some depth by Coleman's laboratory as a function of the degree of experimentally manipulated cholesterol-enrichment of mitochondrial membranes, with specific emphasis on tricarboxylic acid (TCA or Krebs) cycle respiration coupled to effects on the electroneutral exchange-transport of mitochondrially generated citrate from the matrix compartment to the cytoplasm by the inner membrane's citrate transport protein (CTP) carrier (Kaplan *et al.,* 1982; Parlo and Coleman, 1984; Coleman, 1986; Kaplan *et al.,* 1986; Parlo and Coleman, 1986). The observations from these studies were viewed as directly relevant to the aforementioned high rates of cholesterogenesis *de novo* noted in tumors, and they correlated as well with the requirement of a high flux of carbon precursor metabolites through the cholesterogenesis pathway for cell cycle transit, particularly the segment between MVA and squalene. Because of the unique relevance of mitochondrial citrate transport to general lipogenesis, these studies will be briefly recapitulated.

On isolation, normal liver mitochondria are poor in total cholesterol ($\leq$6 μg/mg protein; Table I) whereas mitochondria from malignant tissues (e.g., rapidly growing experimental hepatomas), when carefully prepared so as to retain their functional coupling in oxidative phosphorylation (maximal respiratory control ratios upon ADP addition), exhibit a striking cholesterol enrichment, often several-fold higher (Pedersen, 1978; Coleman, 1991; also Posner and Coleman, unpublished observations). While ADP-stimulated respiratory rates fueled by most mid-entry TCA cycle substrates (e.g., α-ketoglutarate, even glutamate) are similar for both tumor and normal mitochondria despite differences in membrane cholesterol content, Coleman's group unexpectedly observed that respiration promoted by pyruvate and citrate, which are glycolysis-derived immediate-entry TCA cycle substrates, was dramatically reduced in mitochondria from the graded growth-rate series of Morris hepatomas. That is, pyruvate or citrate exhibited virtually no state-3 O_2 uptake with highly cholesterol-rich mitochondria derived from the rapidly proliferating Morris hepatoma 3924A (Figure 1). Analysis revealed that instead of undergoing the molecular transformations expected with normal mitochondria for the sequence of TCA cycle oxidations and decarboxylations, the carbons of citrate and its precursor pyruvate were preferentially exported (as citrate) from the matrix of cholesterol-rich

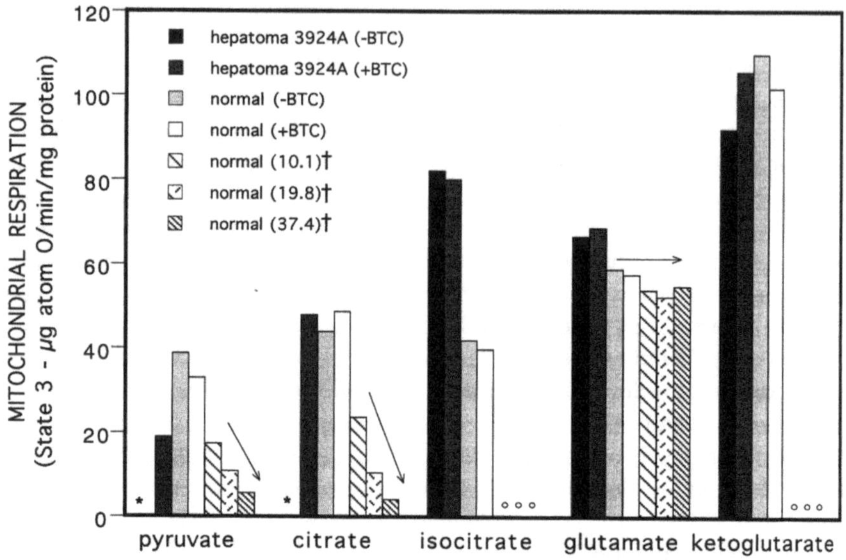

FIGURE 1. Rates of mitochondrial O_2 consumption by Morris hepatoma 3924A, normal liver, and cholesterol-enriched normal liver mitochondria, ±150 μM BTC. In addition to the key, other symbols are: *, ADP-stimulated respiration did not occur; ∘, respiration with these substrates not determined after artificial cholesterol-enrichment; † numbers in parentheses give mitochondrial cholesterol content (μg/mg protein) upon experimental enrichment of normal mitochondrial preparations to three increasing levels of cholesterol (see text). The Morris hepatoma 3924A mitochondria possessed a total cholesterol content of ~26 μg/mg protein, while those from normal liver yielded ~5 μg/mg protein. Thus, the degree of artificial cholesterol enrichment of normal mitochondria (shown by the last 3 bars of each substrate set) was: ~2-fold (10.1 μg/mg protein), ~4-fold (19.8 μg/mg protein) and ~7.5-fold (37.4 μg/mg protein). The decreasing trend for pyruvate and citrate respiration with increasing cholesterol enrichment is shown by the slanting arrows; note that for glutamate the respiratory rate is virtually independent of mitochondrial membrane cholesterol content. Data were compiled from the original references, which should be consulted for experimental methods (Coleman and Lavietes, 1981; Parlo and Coleman, 1984; Coleman, 1991).

hepatoma mitochondria (Parlo and Coleman, 1984). Rates of mitochondrial citrate efflux were assessed as > four-fold higher with tumor than with normal control organelles, and the steady-state level of tumor mitochondrial matrix citrate was reduced to ~14% that of control (Parlo and Coleman, 1984; Kaplan *et al.*, 1986). It was ascertained that the cause of these apparent metabolic anomalies could be attributed directly to the level of mitochondrial membrane cholesterol. This was accomplished by artificially enriching normal liver mitochondria with cholesterol using a solid-state molecule-transfer method (Coleman *et al.*, 1978), to levels equal to or exceeding those of tumor mitochondria (see Figure 2 of review by Coleman and Lavietes, 1981), and measuring metabolic parameters in such model preparations. The resulting cholesterol-enriched normal mitochondria accurately mimicked the pattern of depressed rates of pyruvate/citrate-driven respiration, while evidencing accelerated citrate efflux profiles that had been displayed by actual tumor mitochondria (Parlo and Coleman, 1984; Kaplan *et al.*, 1986; Coleman, 1991).

Additional support for such an altered route of citrate carbon flow in cholesterol-rich tumor mitochondria was acquired by evaluating the effects of the specific inhibitor of the inner membrane-affiliated CTP carrier, 1,2,3-benzenetricarboxylate (BTC) (Bryla, 1981). Upon BTC addition to either tumor or cholesterol-enriched normal mitochondria, thus blocking mitochondrially generated citrate export, pyruvate and citrate reacquired their capacity to fuel respiration (see Figure 1), manifesting ADP-stimulated state-3 O_2 consumption approaching normal values (Parlo and Coleman, 1984; Coleman, 1991). As a result, BTC inhibited the export of citrate carbons via the CTP carrier, apparently restricting their utilization to the respiratory reactions of the TCA cycle. Overall, the results implied that cholesterol enrichment of the inner mitochondrial membrane could induce a stimulation of the CTP carrier activity, resulting in the preferential export of mitochondrially generated citrate to the extra-mitochondrial space.

These phenomena were verified subsequently at the whole-cell and viable tissue level. Rat hepatoma tissue incubations exhibited suppressed *pyruvate*-fueled CO_2 evolution concomitant with a high flux of pyruvate carbons into newly synthesized cholesterol relative to normal liver tissue (Coleman, 1986a, 1986b, Parlo and Coleman, 1986). The addition of BTC to these viable tumor tissue systems reestablished pyruvate respiration-derived CO_2 evolution, while inhibiting cholesterogenesis. Therefore, observations made at the whole-cell level corroborated the results with isolated tumor mitochondria and linked the latter data to the enhanced cholesterogenesis observed with rapidly proliferating tumors. Also, the cellular and tissue level data indirectly implied that BTC must be readily capable of crossing the cell membrane to exert its effect on mitochondria. These findings were taken as strong experimental support for a hypothesis stating that mitochondria of tumor cells appear to operate by means of an abbreviated or truncated Krebs cycle (Parlo and Coleman, 1984; Parlo and Coleman, 1986; Coleman and Sepp-Lorenzino, 1990; Coleman, 1991).

2.5. The Truncated Krebs Cycle, Mitochondrially Supplied Citrate, and Cell Growth

The feasibility of a truncated Krebs cycle was originally formulated as a means to explain accumulated phenomenological observations relating deregulated tumor cholesterogenesis and cell proliferation (Lavietes and Coleman, 1980; Coleman and Lavietes, 1981). Briefly, operation of the proposed truncation in effect removes pyruvate carbons (derived from preceding anerobic carbohydrate catabolism), along with its mitochondrially generated acetyl CoA, from direct participation in the TCA cycle. Instead, the known presence of abundant general transaminase activity within the mitochondrial matrix would allow for nonketogenic amino acids (e.g., glutamate) to fuel an abbreviated cycle (Moreadith and Lehninger, 1984), with entry into it via specific anion exchange transport carriers (see Kaplan, 1996, for a recent review) at the level of α-ketoglutarate, in order to regenerate sufficient cycle-sustaining oxaloacetate. Non-limiting oxaloacetate would thereby fulfill dual roles as: (i) transaminase co-substrate (to perpetuate the truncation scheme), and (ii) condensation co-substrate together with pyruvate-derived acetyl CoA to yield citrate, which would then be preferentially exported by cholesterol-rich tumor mitochondria (Parlo and Coleman, 1984; Parlo and Coleman, 1986). The underlying principle of such a proposal relies on the fact that the enhanced rate of cholesterogenesis observed in tumors (e.g., the Morris hepatoma 3924A, in which the rate is many times higher than normal liver) demands a high and constantly replenished supply of cytoplasmic acetyl CoA to support the equally high flux of precursor carbons through synthetic lipogenesis pathways, particularly that which generates MVA. Mitochondria are the fundamental repositories of metabolically generated citrate, and in cholesterogenically active cells the export of mitochondrial citrate to the cytoplasm fulfills the requirement of being the major source of that compartment's acetyl CoA, by way of the intervening cytosolic ATP–citrate lyase reaction (Srere, 1975). Thus, the hypothesis of a truncated Krebs cycle stipulates that tumors adjust the pattern of their mitochondrial respiratory metabolism so as to compensate for the cell's demand for a high cholesterogenic flux by utilizing alternative bioenergetic fuels (e.g., glutamate; see Moreadith and Lehninger, 1984), while simultaneously expelling the cytoplasm's acetyl CoA precursor (mitochondrial pyruvate-generated citrate) to the extramitochondrial milieu where cholesterogenesis occurs.

One prediction arising from the proposed operation of such a truncated Krebs cycle, but at the cellular level, would be the demonstration that proliferating cells in culture, which depend on the active operation of the cholesterogenic pathway (i.e., the MVA-derived, DNA synthesis-stimulating element(s), described in section 2.3), actually require, rather than merely tolerate, the export of pyruvate-derived citrate from mitochondria to sustain cell division. Subsequent studies (Rao and Coleman, 1989) found that when mitochondrial citrate export was blocked with BTC during the G1-phase of the cell cycle in pyruvate-supplied murine lymphoma cell suspension cultures, cholesterogenesis, DNA synthesis, and cell proliferation were all inhibited. This inhibition of cell growth, correlative with the fueling of cholesterogenesis via carbon transit through mitochondria, could be completely reversed upon removal of the BTC from the culture medium. However, in order to establish further support for the truncated Krebs cycle concept, it was necessary to directly demonstrate all of the following: i) that BTC is readily taken up by, and capable of release from, cells—i.e., that its transport across the plasma membrane is not hindered; ii) that the inhibitory effect of BTC on the mitochondrial CTP carrier is specific for that particular membrane protein by not manifesting significant interference with the activity of cytoplasmic ATP-citrate lyase; and iii) that the rate of carbon flux through cholesterogenesis, starting from citrate itself—rather than from acetyl CoA—retains the aberrantly decontrolled pattern observed with rapidly proliferating tumor cell systems compared with nondividing normal ones. The first objective was met by the use of [^{3}H]BTC (Rao and Coleman, 1989). The second and third criteria were demonstrated by experiments with post-mitochondrial supernatants (PMS), focusing particularly on rates of citrate-fueled cholesterogenesis in tumor and normal tissue (Azrolan and Coleman, 1989).

While PMS systems retain the capacity for lipogenesis, they must be supplemented with ATP and other cofactors normally provided by mitochondrial metabolism. Compared with such systems from normal rat liver, the PMS from rapidly proliferating hepatoma tissue was found to exhibit, calculated on a per-cell basis, a nine-fold increase in ATP-citrate lyase activity and the amount and activity of HMGR. These increases correlated with an observed 9–10-fold increase in the rate of citrate carbon incorporation into newly synthesized cholesterol (Azrolan and Coleman, 1989; Coleman, 1991). Insofar as the relative inhibitory effect of BTC, its interaction with the CTP carrier of mitochondria obviously cannot be a factor in mitochondria-free PMS systems. The data shown in Table II (Azrolan and Coleman, unpublished observations) address the site of BTC inhibition of cholesterogenesis by verifying that BTC is only ~1.5% as effective an inhibitor of ATP-citrate lyase activity with the PMS derived from either normal or tumor cells

Table II
Insensitivity of Post-Mitochondrial Supernatant ATP-Citrate Lyase Activity to Inhibition by BTC with Normal and Hepatoma Preparations[a,b]

PMS source	[BTC], mM	ATP-citrate lyase (nmol malate or OAA/min/10^9 cells)	% inhibition
Normal rat liver	0	2.14 ± 0.23	—
	25	2.05 ± 0.26	4
	50	1.69 ± 0.18	21
	100	0.98 ± 0.08	54
Hepatoma 3924A	0	18.31 ± 1.42	—
	25	17.21 ± 1.56	6
	50	12.82 ± 1.02	30
	100	9.70 ± 0.87	47

[a]Azrolan and Coleman, unpublished observations.
[b]PMS preparations (100 μl) from normal rat liver or Morris hepatoma 3924A were mixed with 880 μl of assay buffer (10 mM K^+-citrate, 10 mM $MgCl_2$, 10 mM dithiothreitol, 0.5 mM NADH, 0.6 units of malate dehydrogenase, 0.33 mM coenzyme A, 100 mM Tris-HCl, pH 7.5), ± the indicated final concentrations of BTC, and the reaction was initiated by addition of 20 μl of 250 mM ATP. ATP-citrate lyase activity was calculated by monitoring the oxidation of NADH at 340 nm. This ATP-citrate lyase activity was normalized to a per-cell basis with data obtained from dispersed cell preparations from each tissue source counted under a microscope with a hemocytometer, whose DNA content/cell was determined versus calf thymus DNA standards. Data are from 5 individual PMS preparations assayed in duplicate ±SEM. See Azrolan and Coleman (1989) for details.

(~50% inhibition at 100 mM BTC) than when employed as an inhibitor of the CTP carrier in isolated mitochondrial incubations, where maximal citrate transport inhibition was achieved with ≤150 μM BTC (Parlo and Coleman, 1984; Parlo and Coleman, 1986).

Although the CTP carrier is considered organ-specific—that is, expressed predominantly in the mitochondria of tissues that carry out the more specialized metabolic function of lipogenesis, such as liver and kidney (Kaplan, 1996)—the data obtained with the cultured murine lymphoma (Rao and Coleman, 1989) suggest that its tissue distribution may be more general than previously believed.

Summarizing sections 2.4 and 2.5: Cholesterol-rich mitochondria are deficient in their ability to oxidize pyruvate and pyruvate-derived citrate by the TCA cycle, rapidly exporting mitochondrially generated citrate instead via the CTP carrier. Direct BTC-inhibition of the CTP carrier re-establishes pyruvate- and citrate-fueled respiration in cholesterol-enriched mitochondria, implying that mitochondrial membrane enrichment with cholesterol stimulates CTP activity, while at the same time lending experimental support for the operation of a truncated Krebs cycle in tumor mitochondria. At the cellular level, viable hepatoma tissue incubations manifest a much larger pyruvate → cholesterol carbon flux and a commensurately lower rate of CO_2 evolution than normal liver, whereas the presence of BTC abolishes these metabolic profiles. BTC, which rapidly penetrates into cells, inhibits cholesterogenesis, DNA synthesis, and cell proliferation in pyruvate-fueled, G1-synchronized murine lymphoma suspension cultures, and wash-out of the BTC reverses the inhibition of all three phenomena. At the subcellular (cytoplasmic) level, citrate-fueled cholesterol synthesis in tumor PMS incubations is ≥nine-fold higher than in normal liver PMS, an increase paralleled in the tumor both with regard to the amount and activity of HMGR as well as in the activity of ATP–citrate lyase. At concentrations many times greater than that which proved effective for inhibiting the mitochondrial CTP carrier, BTC minimally inhibits the ATP–citrate lyase activity of the PMS system, implying that the ability of BTC to inhibit pyruvate-fueled cholesterogenesis in cells specifically is due to its effect on the mitochondrial CTP carrier.

2.6. Unresolved Questions

Results based on mitochondrial cholesterol enrichment appear to directly implicate the stimulation of CTP carrier activity in order to sustain a high and constant cytoplasmic level of pyruvate-derived citrate. This elevated CTP-mediated citrate transport activity must therefore be a direct function of the mitochondrial inner membrane cholesterol content. Unequivocal data on inner membrane cholesterol content of tumor versus normal mitochondria do not seem to be available, and given the difficulties in performing creditable membrane cholesterol assays, as alluded to previously (section 2.1), collection of such data would almost surely arouse controversy. However, the purification, reconstitution, subsequent cloning, and overexpression of the rat liver CTP carrier has recently been achieved (Kaplan *et al.,* 1990; Kaplan and Mayor, 1993; Kaplan *et al.,* 1993; Xu *et al.,* 1995). It is thus of interest to ask whether the carrier's kinetic profile responds to the more readily performed manipulation of cholesterol content of the standard asolectin liposomal vesicle preparation used in CTP transporter reconstitution assays, as was indeed hinted at from the data on the citrate transport experiments with intact hepatoma mitochondria (Kaplan *et al.,* 1982).

Nevertheless, it would still remain to be determined how cholesterol moves from the outer to the inner mitochondrial membrane, only by which means would it then be capable of modulating CTP activity. Existence of periodic "contact points" between the outer and inner membranes, indicated by electron microscopy, affords one possible mechanism allowing for such intermembrane communication and ensuing molecule transfer (see review by Hartl *et al.,* 1989). Alternatively, the formation of intermembrane space cholesterol-rich lipid vesicles might serve as direct physical conduits (Martinez *et al.,* 1988; Echegoyen *et al.,* 1993). In steroid hormone synthesizing tissues, sterol carrier proteins (e.g., either SCP-2, or a fatty acid binding protein of 11.3 kDa, or a 2.2 kDa steroidogenesis activator peptide, SAP) have all been proposed by various investigators to perform this role (see Nestler *et al.,* 1990, for review). However, a new and intriguing possibility arises upon invoking the recently discovered 30-kDa steroidogenic acute regulatory (StAR) protein as a vehicle for delivering cholesterol from the outer to the inner mitochondrial membrane (Lin *et al.,* 1995; see Chapter 8 by Martínez and Strauss). Indeed, the mechanism by which extramitochondrial cholesterol is translocated to the inner mitochondrial membrane in steroidogenic (e.g., adrenal) tissue for reaction with the matrix-located cytochrome P-450 side chain cleavage enzyme in order to generate steroid hormones has been a long-standing puzzle (Privalle *et al.,* 1983). The solution to this dilemma may lie in the proposed role of StAR protein (Waterman, 1995), which appears necessary to conduct extramitochondrially derived cholesterol to the inner membrane site of interaction with P-450. Expanding this proposal, and although an unexplored conjecture, StAR protein might serve to effect movement of cholesterol between outer and inner mitochondrial membranes of many, if not all, tissues, and could thus be responsible for inner membrane cholesterol enrichment in tumors, even if they do not normally generate the steroid hormones of more specialized tissues. However, since StAR protein does not seem to be expressed in every steroidogenically active tissue (Lin *et al.,* 1995), other means for moving cholesterol into the inner mitochondrial membrane may exist in such tissues, and thus the mechanism remains an important area for future research.

Further, as described in section 2.3, more than a decade of research with proliferating cell cultures has established that, during G1, a prerequisite for continued cell cycling is the metabolism of MVA through the segment of cholesterogenesis prior to squalene. However, what is lacking is the ability to quantitatively correlate the extent of the deregulated, high rates of cholesterogenesis observed in tumors (i.e., the degree or "fold" of enhanced HMGR activity and carbon flux through the pathway), with some intracellular threshhold concentration of any of the MVA-derived isoprenoid intermediates. In the cell, of course, different branchpoint pathways of MVA metabolism from farnesyl pyrophosphate (FPP), which yield the known spectrum of polyisoprenoid metabolites (ubiquinone, dolichol, cholesterol itself, etc.), can each manifest individual rates and diverse modes of regulation of carbon flux through them, based on the kinetic characteristics of the rate-limiting enzyme for the particular pathway branch. But it is not clear how to relate, in some meaningful stoichiometric manner, the MVA-derived isoprenoid prerequisite for cell growth to the many-fold increase in continuous carbon flux through the deregulated HMGR locus that generates the MVA, as observed with proliferating tumors (Azrolan and Coleman, 1989). In other words, while the phenomenology appears to be clear, the molecular stoichiometry upon which the phenomenology is founded is not.

Additional complexity has arisen from the fact that the cholesterogenic pathway, whose localization was restricted to the combined soluble cytosolic *plus* ER membrane compartments since the 1950s, more recently has been revealed to involve the significant (some would claim predominant, if not exclusive) participation of peroxisomes. Added to this are reports indicating that centrally important metabolites derived from MVA (e.g., FPP) may also be generated within mitochondria (see Grünler *et al.*, 1994, for review). Consequently, the cellular locus of cholesterogenesis, and thereby of the MVA-generated intermediates critical for cell growth, is a puzzle undergoing current decoding.

3. CELL COMPARTMENTALIZATION OF CHOLESTEROGENESIS

3.1. Peroxisomes and Isoprenoid Synthesis

Peroxisomes are bounded by a single membrane; conduct well-known oxido-reductions involving lipids, amino acids, and purines; and constitute the compartmentalized source of catalase-mediated H_2O_2-dissipating reactions. Moreover, many of the chief cholesterogenic enzymes, as well as of sterol carrier protein 2 (SCP-2) and a newly discovered farnesylated protein called PxF are all localized to peroxisomes (Krisans *et al.*, 1980; Krisans *et al.*, 1988; Krisans, 1992; Stamellos *et al.*, 1992; Stamellos *et al.*, 1993; Biardi *et al.*, 1994; James *et al.*, 1994; Krisans *et al.*, 1994; Krisans, 1996). Thus, new avenues arise for research on the utilization of citrate-derived acetyl CoA in the cytoplasm and the manner of control over cholesterogenesis.

The varied metabolism carried out by peroxisomes, and their roles in specific diseases (including cancer), were addressed in a previous review (Goldfisher and

Reddy, 1984). Recent evidence now shows that peroxisomes possess, in addition to their lipid β-oxidizing machinery, all of the cholesterogenic enzymes: that is, acetoacetyl CoA thiolase (Thompson and Krisans, 1990), HMG CoA synthase (Krisans *et al.*, 1988), HMGR (Keller *et al.*, 1985), mevalonate kinase (Stamellos *et al.*, 1992; Biardi *et al.*, 1994), mevalonate phosphate (MVAP) kinase, mevalonate pyrophosphate decarboxylase (MVAPPD) (Krisans, 1996), and farnesyl diphosphate (FPP) synthase (Krisans *et al.*, 1994), as well as enzymes for the synthesis of squalene (Stamellos *et al.*, 1993), and for the conversion of lanosterol to cholesterol (Appelkvist *et al.*, 1990; Appelkvist *et al.*, 1995). Moreover, peroxisomes are also known to contain the *cis* prenyltransferase activity that serves as the committed step in dolichol synthesis (Ericsson *et al.*, 1993) and, as well, apolipoprotein E, which participates in cholesterol transport, helping to maintain cholesterol homeostasis (Hamilton *et al.*, 1990; see Chapter 9 by Johnson, *et al.*). Enlarging the cholesterol metabolic profile of peroxisomal responsibility is the presence of the 13.5-kDa SCP-2 and a 58-kDa protein, termed SCPx. The latter's derived amino acid sequence reveals the presence of tandem activities, as it contains the complete SCP-2 sequence, as well as that for 3-oxoacyl CoA thiolase (Seedorf *et al.*, 1994). Among the more established roles of peroxisome metabolism is the capacity to catalyze the β-oxidation of medium, long, and very long chain fatty acids; medium and long chain dicarboxylic fatty acids; the carboxyl side chains of prostaglandins and bile acid intermediates; and the fatty acyl side chains of some xenobiotics (Reddy and Mannaerts, 1994). In view of their established cholesterogenic ability, the acetyl CoA generated via such peroxisomal fatty acid β-oxidation might serve to provide the substrate precursor required for active cholesterogenesis to proceed within these organelles (Schepers *et al.*, 1988; Hayashi and Miwa, 1989; Schepers *et al.*, 1990). This possibility receives support from clinically relevant data showing that patients suffering from peroxisomal deficiency diseases have low serum cholesterol levels, and cells with a reduced complement of peroxisomes reveal a similarly reduced capacity to synthesize cholesterol (Krisans, 1992).

In view of the very recent appreciation for the expanded metabolic roles of peroxisomes, it is not surprising that peroxisomal cholesterogenesis in tumor cells has not yet been explored. It has already been suggested, however, that the functions of peroxisomes are important for development and cell differentiation. For example, there are claims of a decrease in the number of diaminobenzidine-positive (catalase histochemical stain) peroxisomes and corresponding catalase activity during neoplastic transformation (Keller *et al.*, 1993). Analysis of carcinoma cells from human breast and colon revealed significantly lower activity of peroxisomal enzymes, such as catalase, urate oxidase, and fatty acyl CoA oxidase, than in the adjacent healthy mucosa. Moreover, an inverse relationship was reported between such enzyme activities and tumor grade (Keller *et al.*, 1993). Whether the decrease in peroxisome number during neoplastic transformation is accompanied by a quantitatively related decrease in the isoprenoid and cholesterol synthesis enzyme machinery in peroxisomes, with a compensatory increase in other compartments, remains unknown.

3.2. Peroxisome Proliferators as Modulators of Cholesterogenesis

Considerations of peroxisomal lipid (fatty acid and cholesterol) metabolism require mention of a group of compounds termed rodent "peroxisome proliferators". These compounds have diverse structures and exert pleiotropic effects on hepatocytes, primarily in rats and mice (but not humans); typical effects are an induction of peroxisome number and an increase in peroxisomal enzyme activity related to fatty acid β-oxidation (Reddy and Lalwani, 1983). Peroxisome proliferators comprise a wide array of structurally unrelated compounds, including: (i) hypolipidemic/hypocholesterolemic agents of the clofibrate type, such as gemfibrozil, ciprofibrate, and benzafibrate, or agents with similar biological effects but with structures far different from that of clofibrate; (ii) plasticizing agents, such as di(2-ethylhexyl)phthalate (DEHP) and other like agents; (iii) phenoxy herbicides, such as 2,4-dichloro-and 2,4,5-trichlorophenoxyacetic acid; (iv) common anti-inflammatory drugs, such as acetylsalicyclic acid and ibuprofen; (v) certain leukotriene antagonists; (vi) perfluorinated fatty acids; (vii) naturally generated steroid hormones, such as dehydroepiandrosterone (DHEA); and (viii) several other miscellaneous compounds, such as valproic acid (Reddy and Lalwani, 1983; Reddy and Mannaerts, 1994). The pleiotropic effects on rodent hepatocytes caused by treatment with these agents are believed to be mediated via intracellular elements known as peroxisome proliferator-activated receptors, or PPARs, a few of which have recently been cloned in varied animal species (Isserman and Green, 1990).

Since peroxisomes possess the capacity for significant involvement in cellular cholesterogenesis, and since treatment of rodent hepatocytes with peroxisome proliferators raises the peroxisome number and enzyme content, it has been of interest to determine whether the MVA pathway is responsive to treatment with such agents in cells from non-rodent species. In fact, peroxisome proliferators have been reported to affect the activities of cholesterogenic enzymes, but the overall results have been contradictory and confusing. For example, clofibrate treatment yields conflicting data with regard to the activities of both HMGR and a subsequent pathway enzyme, MVAPPD. Reports have indicated that clofibrate inhibits chicken liver HMGR and MVAPPD activities *in vivo* (Castillo *et al.*, 1990) and both chicken liver and brain HMGR and MVAPPD activities *in vitro* (Zafra *et al.*, 1987; Castillo *et al.*, 1990). These conclusions coincide with previous results obtained with rat liver microsomal HMGR (Cohen *et al.*, 1974). Other work, however, claimed that clofibric acid does not inhibit MVAPPD, MVA kinase, or MVAP kinase activities *in vitro* (Shama Bhat and Ramasarma, 1979b). Yet, the incorporation of ^{14}C-MVA into cholesterol was shown, even earlier, to decrease in rat liver homogenates upon treatment of rats *in vivo* with clofibrate (Azarnoff *et al.*, 1965), reflecting the more recently reported decrease in HMGR and MVAPPD activities *in vivo* (Castillo *et al.,* 1990). Others, however, found that clofibrate treatment *in vivo* did not affect microsomal HMGR activity, whereas a related compound, benzafibrate, inhibited it. These conflicting data are difficult to reconcile when other results claim that gemfibrozil, related to clofibrate, induced HMGR activity in both peroxisomal and microsomal fractions (Hashimoto *et al.,* 1995), a result consistent with that of a previous report showing increased incorporation of ^{14}C-acetate into cholesterol in rat liver by gemfibrozil treatment *in vivo* (Maxwell *et al.,* 1983).

The principal cellular location of MVA kinase, as with HMGR, adds fuel to the confusion over peroxisome proliferator effects. MVA kinase, once thought to be exclusively a cytosolic enzyme, more recently has been demonstrated in several reports as predominantly (or entirely) localized to liver peroxisomes (Stamellos *et al.,* 1992; Biardi *et al.,* 1994). The MVA kinase activity formerly seen in the cytosol has been claimed to be due to artifacts arising from subcellular fractionation, which results in the leakage of MVA kinase from peroxisomes (Stamellos *et al.,* 1992). A recent study concluded that except for gemfibrozil, three other peroxisome proliferators—clofibrate, fenofibrate, and the steroid DHEA—all decreased MVA kinase activity and the enzyme protein level (Biardi *et al.,* 1994). If MVA kinase and HMGR are assumed to be affected similarly by peroxisome proliferators, one could then infer that clofibrate should decrease peroxisomal HMGR activity as it does MVA kinase activity. With such contrasting results, it is impossible to draw conclusions about the effects of clofibrate-type serum cholesterol-lowering compounds on peroxisomal and/or microsomal HMGR and affiliated pathway enzymes. It may be that such effects result from different peroxisome proliferators exerting a spectrum of influences on cholesterogenic pathway enzymes as a function of the compartment in which cholesterogenesis occurs, and that such effects may therefore result from a differential accessibility of various peroxisome proliferators to these cholesterogenically active compartments.

Further complicating the matter is the well-established characteristic of the HMGR diurnal cycle. The highest specific activity of HMGR in rat liver microsomes occurs midway through the dark (feeding) cycle, whereas the highest activity in peroxisomal HMGR activity in rat liver was found at the second hour of the light cycle (Rusnak and Krisans, 1987). It appears, then, that further work aimed at determining the effects of peroxisome proliferators on cholesterogenesis must consider these diurnal HMGR activity factors as well, or else the confusions accompanying this area of interest will persist.

On the other hand, apart from the aforementioned contradictory or unexplained effects regarding peroxisomal phenomena and the known capacity to inhibit liver cholesterol biosynthesis, clofibrate displays intriguing consequences on mitochondrial structure, and recently was found to elevate the activities of selective TCA cycle enzymes (Prager *et al.,* 1993). Prager *et al.* (1993) reported that upon clofibrate feeding, a higher metabolic flux occurred through the rat liver mitochondrial TCA cycle, commensurate with enhanced CO_2 production. Such findings are strikingly reminiscent (but in the reverse) of metabolic patterns revealed by experiments that probed the truncated TCA cycle in mitochondria from tumors where cholesterogenesis is elevated (Parlo and Coleman, 1984; Parlo and Coleman, 1986). On the basis of such observations, there appears to be further indirect phenomenologic support for a linkage or correlation between acetyl carbon flux through mitochondrial respiratory metabolism (see Figure 1) versus extramitochondrial cholesterogenesis, and factors that contribute to the regulation of cell proliferation.

Despite the fact that there also is clear experimental support that peroxisomes contain cholesterogenic enzymes capable of synthesizing FPP, the subsequent fate of peroxisomal FPP remains another unresolved issue. Whether peroxisomal FPP is utilized for protein prenylation within these organelles, is exported to other compartments (mitochondria, cytosol), or is used to support the branch pathways that generate other more common isoprenoids (ubiquinone, dolichol, etc.) has yet to be determined. One fact is beyond dispute, however; the MVA-derived intermediates FPP and geranylgeranyl-PP (GGPP) each behave as integral precursor metabolites necessary for the manufacture of many important prenyl-modified signal-transducing proteins.

4. MVA-DERIVED FARNESYL- AND GERANYLGERANYL-MODIFIED PROTEINS: SIGNAL ELEMENTS FOR CELL PROLIFERATION OR DIFFERENTIATION

4.1. Prenylated Proteins: Alternative Roles for Farnesyl-PP

Discovery of the requirement of MVA metabolite intermediates for progression through the cell cycle (during G1), as previously discussed in sections 2.3 and 2.5, relied heavily on the experimental use of fungal metabolites called statins (Endo *et al.*, 1976; Alberts *et al.*, 1980; Endo, 1992). Statins (see Figure 2) penetrate cells and competitively inhibit HMGR, which leads to depletion of intracellular MVA, starving the remainder of the cholesterogenic pathway of substrate and thereby reducing cholesterol production. As a result of their unique site of action, statins effect synchronization of dividing cell populations to early G1, the time when maximal carbon flux through the cholesterogenesis pathway normally occurs by virtue of HMGR displaying its highest activity within proliferating cells (Jakobisiak *et al.*, 1991; Keyomarsi *et al.*, 1991; Sepp-Lorenzino *et al.*, 1991), thus corroborating the requirement for MVA metabolite production in order to complete the cell cycle. Statin inhibition of tumor cell proliferation and signaling is considered later in section 5.3; see also Chapter 17 by Biessen *et al.*

In 1984, a statin-type inhibitor (mevinolin/lovastatin) was employed to deplete endogenous MVA stores in Swiss 3T3 cells, following which the addition of radiolabeled MVA revealed, for the first time, the covalent incorporation of then-unidentified MVA-metabolite(s) into a seemingly vast array of cellular proteins with a range of M_r from <15 kDa–>60 kDa (Schmidt *et al.*, 1984).

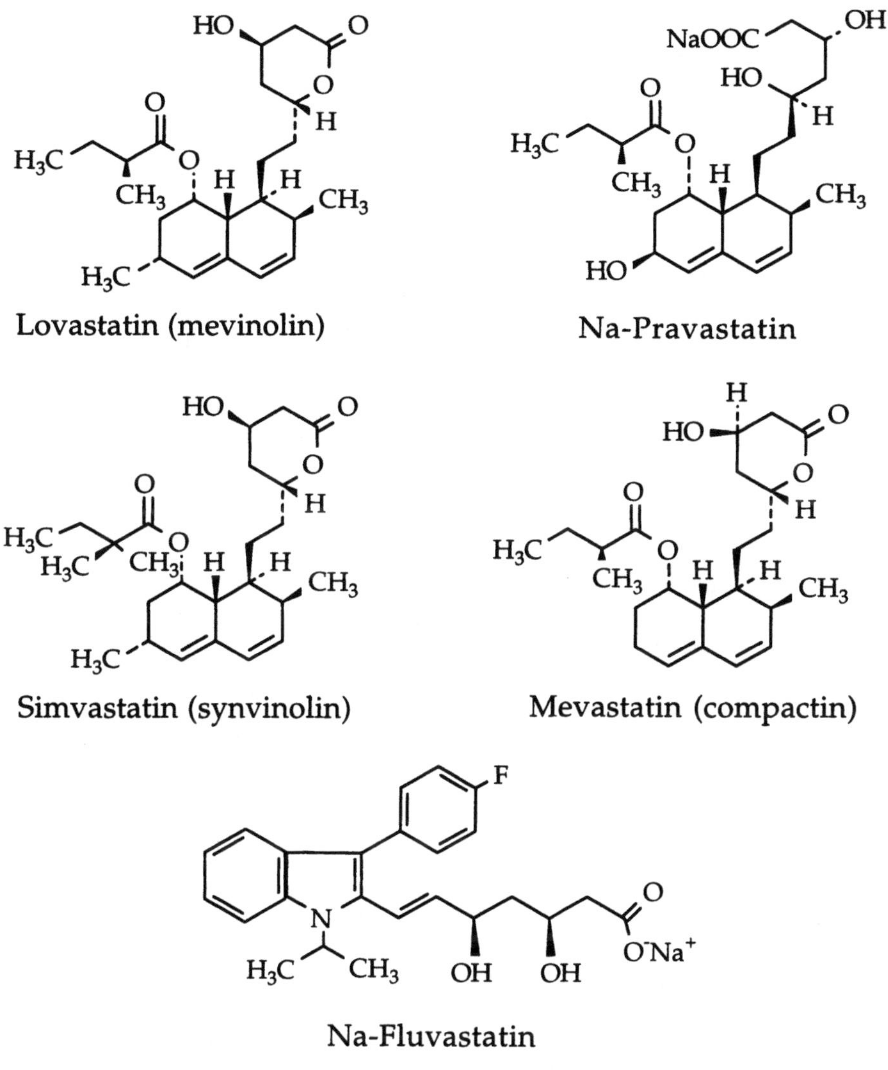

FIGURE 2. Molecular structures of several commercially produced and clinically prescribed statin-type drugs. All statins are serum cholesterol-lowering drugs that act as competitive inhibitors of HMGR, the major rate-limiting enzyme of the cholesterol synthesis pathway. Fluvastatin (Sandoz), a substituted indole heptenoate derivative, is an entirely synthetic HMGR inhibitor (rather than the natural product of fungal metabolism) and is structurally distinct from the other statins shown.

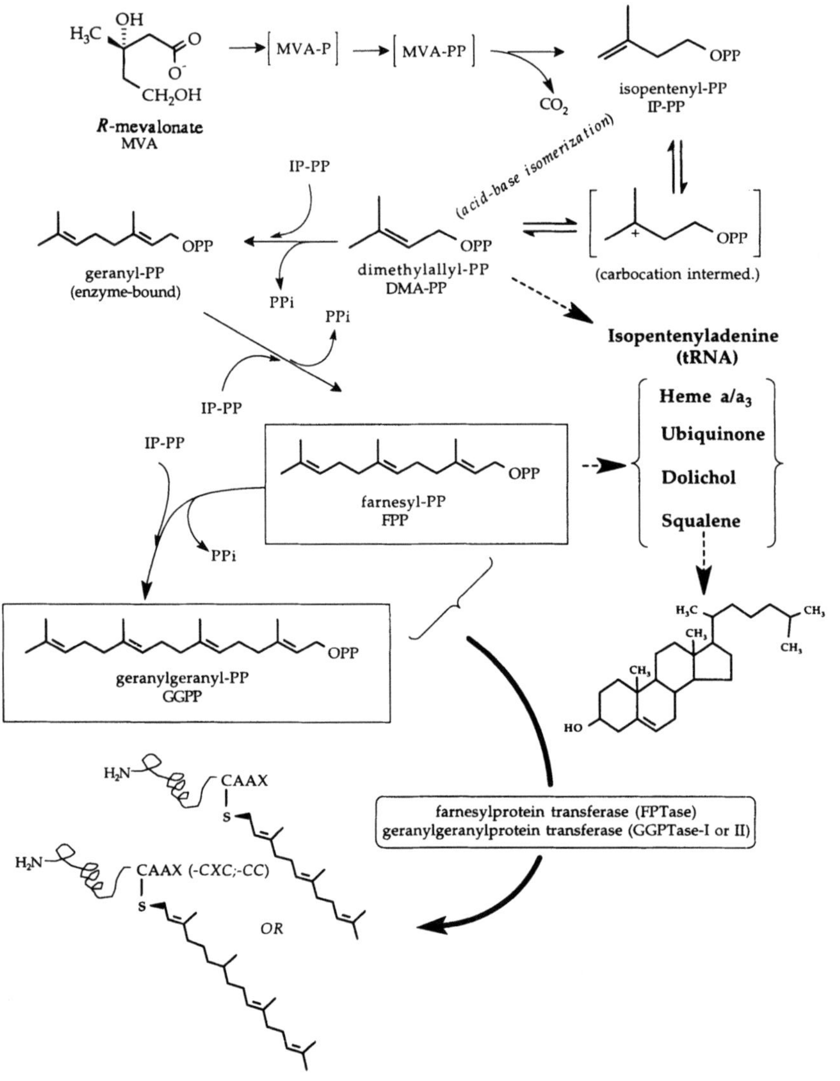

FIGURE 3. Metabolic branch pathway for covalent prenylation of proteins by MVA-derived farnesyl and geranylgeranyl intermediates.

Incorporation of MVA metabolite(s) into proteins occurred in the presence of cycloheximide or chloramphenicol, indicating that the modification is a posttranslational event. The covalent, posttranslational modification of proteins was an unexpected finding, and it was ultimately revealed that only two modifying terpene moieties were involved: all-*trans*-farnesyl and a C_{20} derivative, all-*trans*-geranylgeranyl (Bruenger and Rilling, 1986; Farnsworth *et al.* a, 1990; Rilling *et al.*, 1990), which are covalently attached by a thioether bond only to cysteine residues (see Figure 3).

An impressively large number of prenyl-modified proteins was soon (and continues to be) identified in diverse animal and human cell lines. Included in the growing list are the γ subunits of several membrane-bound heterotrimeric (α,β,γ) G-proteins, the α subunit of cGMP phosphodiesterase, classical Ras proteins together with diverse members of the large Ras-related family of small GTP-binding proteins, two nuclear lamin proteins (pre-lamin A and lamin B), and even a certain species of myelin-specific cyclic nucleotide phosphodiesterase (CNPase). Although, due to their diversity, classification of the complete profile of functions for all prenylated proteins discovered is not yet possible, it is clear that many such proteins perform as intermediary signal transducing elements, which is why so much recent excitement has accompanied their discovery. Numerous papers and reviews have appeared that focus on identifying, characterizing, and assigning putative functions to the vast array of prenylated proteins, and these should be consulted for the pertinent references they cite (Glomset *et al.*, 1990; Maltese, 1990; Adamson *et al.*, 1992; Glomset *et al.*, 1992; Giannakouros and Magee, 1993; Sepp-Lorenzino *et al.*, 1994; Casey, 1995; Sanford *et al.*, 1995; Wedegaertner *et al.*, 1995; Zhang and Casey, 1996). Certainly, additional as-yet-unidentified prenylated proteins will continue to be revealed in time; e.g., the peroxisomal membrane-affiliated PxF that was recently described (James *et al.*, 1994).

The consensus sequence for protein prenylation is a conserved carboxyl-terminus consisting of a 4-residue motif in the form of CAAX (C = cysteine; A = aliphatic; X = [Met, Ser, Ala] forming one class, or [Leu or Phe] forming a second class of terminal X residues), with prenylation always occurring on the Cys (Clarke *et al.*, 1988; Stimmel *et al.*, 1990). Other Ras-family G-proteins (e.g., Rab), which were found to be exclusively geranylgeranyl-modified, revealed an alternative C-terminus motif, viz., CC/CXC, (Cys-Cys or Cys-[various]-Cys). The recognition element for determining whether the farnesyl or geranylgeranyl moiety would be directed to the CAAX-motif depends heavily on the identity of the terminal (X) residue; when X is Met, Ser or Ala farnesylation is specified, whereas when it is Leu or Phe geranylgeranylation occurs (Khosravi-Far *et al.*, 1991; Moores *et al.*, 1991; Khosravi-Far *et al.*, 1992; Kinsella and Maltese, 1992; Schafer and Rine, 1992). Parallel research disclosed that the C-terminal CAAX motif of the extensive array of small G-proteins, as well as other proteins with higher masses, not only undergoes prenyl modification at the cysteine, but is subsequently processed in a two-step reaction sequence involving first, proteolytic cleavage and removal of the-AAX terminus, and second, methyl esterification of the exposed carboxyl group of the prenylated Cys residue (Maltese, 1990; Casey, 1995). The resulting electrically neutral C-terminus takes on a more hydrophobic character, suggesting that Cys prenylation and methyl esterification, yielding the mature modified protein species, would encourage its association with membranes (Black, 1992; Clarke, 1992).

Membrane anchoring indeed seems to be a prerequisite for the expression of cellular function for many prenylated proteins (Farnsworth *et al.*, 1990b; Glomset *et al.*, 1990; Hancock *et al.*, 1990; Braun *et al.*, 1991; Hancock *et al.*, 1991; Nigg *et al.*, 1992; Sepp-Lorenzino *et al.*, 1994; Soldati *et al.*, 1994). On the other hand, the prenyl modification might be necessary for specific protein–protein association at topologically apposing non-polar domains rather than (or in addition to) membrane localization (Shirataki *et al.*, 1992; Kuroda *et al.*, 1993; Marshall, 1993). This possibility is particularly intriguing since, through application of a cell-free/membrane-free, Ras-dependent signal transduction assay, it has recently been reported that farnesylation of Ras is the sole requirement for exhibiting an *in vitro* membrane-free signal transduction that mimicks the type of protein kinase information transfer observed in whole cell systems (McGeady *et al.*, 1995).

Prenylated proteins seem to exhibit a multi-compartment subcellular distribution, with the majority of the smaller species (those displaying $M_r < 30$ kDa) found in the post-nuclear (cytosol plus membrane-vesicle/-fragment) fraction (Maltese and Sheridan, 1987). Larger prenylated proteins with $M_r \approx 50–70$ kDa pellet with the nuclear fraction, the major component of which seems to be lamin B (68 kDa), found associated with insoluble nuclear matrix-intermediate filaments (Sepp-Lorenzino *et al.*, 1989). There also is some evidence for the phosphorylation of prenylated proteins (Sepp-Lorenzino *et al.*, 1989). Although an intriguing observation, the role of phosphorylation, perhaps in regulating the biological function of various prenylated proteins [e.g., lamin B (Hornbeck *et al.*, 1988)], needs further exploration.

Since production of MVA-derived metabolites is prerequisite to transit from the G1-to S-phase of the cell cycle (see section 2.3), the question arose of whether the prenylation of proteins correlates with discrete segments of the cycle, and has been addressed to some degree in at least three reports. Maltese's group (Maltese and Sheridan, 1988) reported that the prenyl modification of proteins is metabolically stable since no evidence was found for turnover of preformed [^{3}H]MVA-labeled prenylated proteins over an 8 h post-labeling chase in cultured cells. A subsequent study (Sepp-Lorenzino *et al.*, 1991), however, revealed that the process of protein prenylation in synchronized, MVA-starved Hep G2 cultures, upon [^{3}H]MVA addition, appears to be linked to different phases of the cycle, and that peak farnesyl (assayed as farnesol) production correlates with the G1/S interphase. This implies, however, that despite the stability of the prenylation of previously modified proteins, protein synthesis coupled to cell division exposes new proteins that require prenyl modification as a function of cell proliferation. Recent studies with regenerating rat liver (Bruscalupi *et al.*, 1994) appear to support such a concept since the data indicated an enhanced incorporation of [^{3}H]MVA into prenylated proteins commensurate with a 20-fold increase in the production of C_{15} (farnesyl) and C_{20}-derived isoprenoids, which occurred just prior to the onset of S-phase (16 h post partial hepatectomy) in the regenerating lobe of the liver.

The importance of the question of the fate of prenylated proteins subsequent to their formation, and the linkage with cell proliferation, demands further exploration since it is difficult to reconcile the well-established diversion of MVA-derived FPP during the critical late G1-phase, presumably to serve as a modifier of protein structure and function, if all of its protein targets are already stably prenylated. Of course, cells may possess a means of disposing of such stably modified, signal-transducing prenylated proteins during each round of the cell division cycle, perhaps by exporting them through specific carriers across the plasma membrane to the extracellular milieu from where they can be excreted (this possibility is raised again in section 5.4). On the other hand, one can speculate that other modes of MVA-linked metabolism, apart from generating FPP-derived prenylated proteins, may be required for cell cycle transit and cell proliferation (Sawamura *et al.*, 1993; Forman *et al.*, 1995), in which case substantive fluctuations in protein prenylation commensurate with the G1/S checkpoint would not be expected to occur. Two informative reviews on protein prenylation, with overlapping perspectives, are worth consulting (Maltese, 1990; Schafer and Rine, 1992).

4.2. Protein Prenylation Enzymes and Neoplastic Transformation: The Focus on *S*-Farnesylated p21^{ras}

Three enzyme activities have been found to catalyze protein prenylation: (i) a farnesyl protein transferase (FPTase) that operates on CAAX-motif proteins with terminal X residues principally of Met or Ser (or Ala; Phe is also tolerated); (ii) a geranylgeranylprotein transferase-I (GGPTase-I) that selects CAAX proteins where the strongly preferred X residue is Leu; and (iii) a second type of GGPTase (GGPTase-II), which is specified by proteins with the CC (or possibly CXC) terminus motif, and for which evidence implies that both cysteines may become prenylated (Moores *et al.*, 1991; Reiss *et al.*, 1991; Khosravi-Far *et al.*, 1992; Moomaw and Casey, 1992; Seabra *et al.*, 1992; Yokoyama and Gelb, 1993). Upon purification of the enzymes from various sources, rapid progress ensued in cloning and expressing them in amounts convenient for characterizing their kinetic and substrate-specific properties (Pompliano *et al.*, 1992; Chen *et al.*, 1993; Omer *et al.*, 1993; Zhang *et al.*, 1994a). Characteristics of substrate binding suggests a random order mechanism; either substrate may bind first, followed by the second (Pompliano *et al.*, 1992). Photoaffinity labeling studies support the idea that the FPTase catalytic site lies in a pocket at the interface between the α and β subunits of this heterodimeric enzyme (Ying *et al.*, 1994).

Knowledge of the presence of a C-terminal CAAX motif on p21*ras* proteins (Willumsen *et al.*, 1984), and the fact that whether they are mutationally oncogenic or wild-type, they all carry this prenylation recognition sequence, was advanced further upon verification that p21*ras* proteins encoded by all transforming *ras* genes (H-, K-, and N-*ras*) become *S*-farnesylated on the terminal cysteine (Cys^{186}) as prerequisite to the enzymatic cleavage of the-AAX tripeptide and the subsequent methylation of the residual farnesyl-Cys-COOH (Hancock *et al.*, 1989). Mutationally "activated" Ras proteins, which are found in many human cancers, depend upon farnesylation for their transforming capacity, and undergo further C-terminus processing (Gutierrez *et al.*, 1989; Downward, 1990; Cox and Der, 1992). The Cys-farnesylation and subsequent carboxymethylation, together with additional covalent lipid modifications in certain cases (e.g., palmitoylation of H-and N-but not K-Ras; Newman and Magee, 1993) is understood as the means for targeting Ras to the inner leaflet of the plasma membrane (Glomset *et al.*, 1990; Hancock *et al.*, 1990; Hancock *et al.*, 1991). It has been repeatedly verified that only by means of such membrane affiliation will Ras be able to manifest its biological function (Marshall, 1993). Generally speaking, posttranslational covalent modification of proteins with lipid moieties other than, or in addition to, the isoprenoid intermediates generated by the cholesterogenesis pathway, has become recognized as a mode of assuring membrane recruitment, and an important means by which such modified proteins may intervene as signal transducing elements to mediate extra- to intracellular information transfer. A summary and a comprehensive review of protein prenylation have recently appeared (Casey, 1995; Zhang and Casey, 1996).

4.3. Inhibition of Farnesyl-Ras-Mediated Cell Proliferation Signaling

It has become axiomatic that the numerous *ras*-like genes encode 21–26-kDa guanine nucleotide binding proteins, and these are involved in signal transduction which, in turn, regulates proliferation and differentiation programs. Thus, mutational activation of *ras*-encoded proteins was shown to confer oncogenicity on a class of sarcoma viruses and to contribute to the transformation of a variety of cell lines (Barbacid, 1987; Bishop, 1991).

Among the increasingly complex variety of signal transduction pathways, aspects of the receptor tyrosine kinase, Ras-dependent, mitogen-activated protein (MAP) kinase transduction route [that is, the Ras/Raf/MEK/ERK pathway; see Marshall (1995)] have recently become somewhat better understood, relative to other Ras-independent pathways (Hall, 1990; Ridley and Hall, 1992; Blenis, 1993; Aronheim *et al.*, 1994; Hall, 1994; Leevers *et al.*, 1994; Quilliam *et al.*, 1994; Stokoe *et al.*, 1994). Nevertheless, there are issues of considerable importance that remain unclarified, such as postulated (unidentified) components believed to operate as parallel or upstream elements of differentially activated signaling routes, all of whose paths ultimately seem to converge at the MAP kinase locus (see Figure 4). As of mid-1995, the simplified diagram shown in Figure 4 was rendered

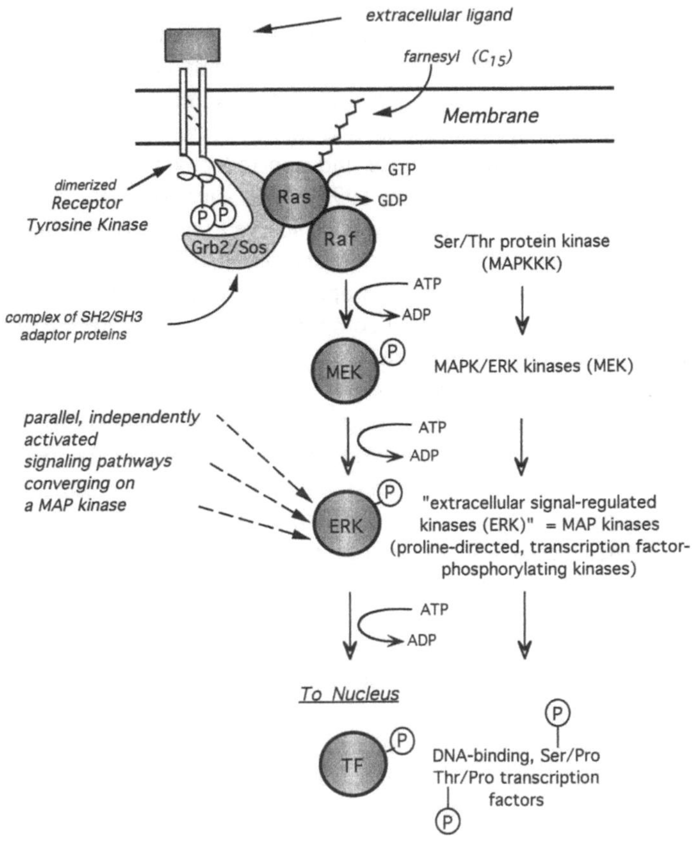

FIGURE 4. Ras/Raf-mediated, protein kinase-activated signaling pathway.

partly obsolete by evidence for at least two Ras/Raf-activated signaling pathways (one MAP-kinase-dependent and the other MAP-kinase-independent). Currently, it appears that each such Ras/Raf-activated pathway can assume responsibility for differential phosphorylation of transcription factors c-fos and c-jun (Agarwal *et al.,* 1995). One must assume that further modifications of the detailed routes of signaling through Ras will continue to be exposed.

Since activating mutations of the *ras* gene have been identified in a large percentage of human tumors, especially pancreatic, lung, colorectal, and bladder carcinomas (Medema and Bos, 1993), and given the pivotal role performed by Ras as a required component of one of the principal signal transduction cascades even in those tumors in which it is not mutationally altered, the function of Ras might be capable of hyperstimulation if upstream elements are inappropriately activated. For example, functional Ras was shown to be required for signaling and cell transformation by tyrosine kinases, inasmuch as neutralizing antibodies or dominant negative *ras* mutants can block these events (Smith *et al.*, 1986; Blenis, 1993; Boguski and McCormick, 1993; Medema and Bos, 1993; Ruderman, 1993). This scenario is especially apparent in some common human cancers, such as breast, esophageal, prostate, and small cell lung carcinoma, in which constitutive activation of both receptor and non-receptor tyrosine kinases often accompanies transformation (Rosen, 1995). With such logic, the development and application of anti-Ras therapies was conceived, not only for those cancers in which mutational activation of Ras occurs, but also for an even wider variety of neoplasias in which a functional (as opposed to a mutational) activation of wild type Ras is found to be an integral component of the transforming phenotype.

A current strategy for inhibiting Ras signaling, and thereby tumor cell proliferation, has targeted those enzymes involved directly in posttranslational Ras farnesylation (Gibbs, 1992; Gibbs *et al.*, 1994; Prendergast and Gibbs, 1994). In fact, the absolute requirement of farnesylation for acquisition of Ras biological activity was convincingly demonstrated by studies with *S. cerevisiae,* since deletion of FPTase activity in this yeast, either by inactivating mutations of the FPTase β-subunit (RAM1 locus) or by means of a knockout approach, gave rise to cells that were incapable of supporting the phenotype of an activated RAS2 allele (Powers *et al.*, 1986; Goodman *et al.*, 1988; Chen *et al.*, 1991; Kohl *et al.*, 1991; Schafer and Rine, 1992). Subsequent posttranslational modifications of farnesylated $p21^{ras}$—the proteolytic removal of the last three amino acid residues and carboxylmethylation of the farnesylated *C*-terminal Cys—were shown to be non-essential for transforming biological activity, and thus became less appealing as targets for the development of specific inhibitors of $p21^{ras}$ function.

Before the specific inhibition of $p21^{ras}$ farnesylation with farnesylprotein transferase-targeted proto-drugs is discussed, it is necessary to return to the effects on cell proliferation and protein prenylation signaling events that have been observed upon blocking the early steps of main-line cholesterogenesis itself.

5. PHARMACOLOGIC INHIBITION OF CHOLESTEROGENESIS AND PROTEIN PRENYLATION IN THE CONTROL OF TUMOR PROLIFERATION

5.1. Inhibitors of Cholesterogenesis as Antitumor Agents

Because tumor cells consistently exhibit increased rates of cholesterol synthesis, and since specific MVA-derived isoprenoids of the cholesterogenic pathway must be generated during mid-to-late G1 phase for new DNA synthesis to occur, the blocking of this pathway, particularly the segment containing the enzymes HMGR to FPP synthase, has been suggested as a means for suppressing the growth and spread of human cancer (Buchwald, 1992; Elson and Yu, 1994). In recognition of this fact, a variey of cholesterogenesis inhibitors that possess antiproliferative ability have been discovered by screening their capacity to induce non-cytotoxic cytostasis.

5.2. Statins

Treating proliferating cells with statin-type drugs (see section 4.1 and Figure 2) depletes them of endogenously produced MVA, the product of the HMGR reaction and the precursor for synthesis of all non-steroidal and steroidal isoprenoids. Concomitant with the inhibition of HMGR, however, exposure to statins leads to a compensatory upregulation of HMGR mRNA transcription and a subsequent increase in the amount of HMGR per cell, despite manifestation of a reduced carbon flux through the MVA pathway (Brown *et al.*, 1978; Watson *et al.*, 1985).

In a variety of cell systems, regardless of whether they possess wild type or mutated Ras, statins inhibit cell proliferation accompanied by G1 arrest (Maltese and Sheridan, 1988; Jakobisiak *et al.*, 1991; Keyomarsi *et al.*, 1991; Sepp-Lorenzino *et al.*, 1991). All these effects can be reversed by the addition of exogenous MVA, but not by cholesterol. Since statins block isoprenoid biogenesis, protein prenylation is similarly inhibited and the biological function of the unprocessed proteins becomes impaired. For example, following lovastatin treatment, Ras protein substantially accumulates in the cytoplasm rather than affiliating with the plasma membrane fraction (Jakobisiak *et al.*, 1991) and is thus incapable of transmitting signals from activated EGF, IGF-1, and insulin receptors in fibroblasts (Vincent *et al.*, 1991). Although many studies have demonstrated the antiproliferative effect and inhibition of protein prenylation by statins in cultured tumor cells, only two animal studies *in vivo* have shown that lovastatin inhibits the growth of tumor explants in nude mice: one study with neuroblastoma (Maltese *et al.*, 1985) and a second with *ras*-transformed cells (Sebti *et al.*, 1991). However, the lovastatin concentrations required for inhibition of tumor growth or cell proliferation (>25 μM) are much (~100-fold) greater than those required to inhibit cholesterol synthesis (Chin *et al.*, 1982; Skalnik *et al.*, 1985). The presence of such high statin concentrations necessary to arrest cell proliferation results in toxicity and apoptosis (Pérez-Sala and Mollinedo, 1994), implying a requirement for isoprenoid intermediates in critical cellular processes other than those needed to sustain cholesterol synthesis itself.

The early experiments by Quesney-Huneeus *et al.* (1983) described in section 2.3 established that one or several non-steroidal isoprenoid moieties were required for proliferation. On this basis it was later conjectured that due to the apparently central role of farnesylated Ras in signaling and cell transformation, the observed cytostatic effect of lovastatin was probably based on an inhibition of Ras farnesylation. However, subsequent research indicated that such a view is, at best, an oversimplification, since it has proven difficult to certify that Ras behaves as the single and unique prenylated protein responsible for signaling in cell transformation and proliferation. Thus, DeClue and coworkers compared the effects of lovastatin treatment on cells transformed by various oncogenes whose activities were either dependent on, or independent of, *ras* gene function (DeClue *et al.*, 1991). They observed that NIH-3T3 cells transformed by v-*src* or v-*ras,* which are known to be dependent on Ras farnesylation, were no more susceptible to growth inhibition by lovastatin than those cultures transformed by v-*raf* or by a myristoylated form of v-*ras,* both of which are independent of FPTase activity and, thereby, Ras farnesylation (see Figure 4; note that Raf lies immediately downstream from, and its activity is therefore independent of, Ras processing). Similarly, from the standpoint of statin-induced growth inhibition, tumors possessing activating *ras* mutations, such as certain bladder carcinoma cell lines (EJ and T24; Jakobisiak *et al.*, 1991; Sebti *et al.*, 1991) or pancreatic carcinoma lines (human-CAV, MIA Paca2, CAPAN2 and PANC1; Sumi *et al.*, 1992), were found to be no more sensitive to lovastatin than tumor cell lines with the wild type *ras* gene, such as Hep G2 hepatoma cells and the 70Z/3 murine lymphoma (Sepp-Lorenzino *et al.*, 1989; Sepp-Lorenzino *et al.*, 1991). Based on such results, one might infer that the antiproliferative effect of HMGR-inhibiting statin drugs is not proximally tied to the direct inhibition of Ras farnesylation, if indeed inhibition of Ras farnesylation is considered the sole cause of cessation of transforming cell growth.

5.3. Cyclic Monoterpenes (e.g., d-Limonene) Inhibit HMGR Synthesis and Protein Prenylation

Certain natural terpeneoids have also been found to decrease HMGR activity by suppressing the synthesis of this enzyme (Clegg *et al.*, 1982; Pearce *et al.*, 1992; Parker *et al.*, 1993), again, in contrast to the direct enzyme inhibition mode of action elicited by statins. These HMGR suppressors are often plant isoprenoid derivatives and include mono-and bi-cyclic monoterpenes as well as the farnesylated benzopyran tocotrienol compounds (Figure 5). Administration of such metabolites *in vivo* to rats at doses of 3 μmol/g body weight resulted in a decrease in liver HMGR activity and in serum cholesterol (Middleton *et al.*, 1979; Qureshi *et al.*, 1988). Of those monoterpenoids tested in such a fashion, their scale of relative potency in lowering HMGR activity and serum cholesterol levels was: menthol > cineole > borneol > menthone > d-limonene (Clegg *et al.*, 1980). Antiproliferative effects of the more potent of these HMGR suppressor monoterpenoids have not been studied in detail, whereas considerable attention has been drawn to only one of them, d-limonene.

The monocyclic monoterpene metabolite d-limonene, found in essential oils from citrus fruits, is a metabolic product derived from the C_{10} geranyl-PP isoprenoid in plants. It has been shown to have both chemopreventive and chemotherapeutic activity on a variety of chemically induced rodent tumors, including mammary carcinomas (reviewed in Crowell and Gould, 1994). In 7,12-dimethylbenz(a)anthracene (DMBA)-induced carcinogenesis, for instance, d-limonene

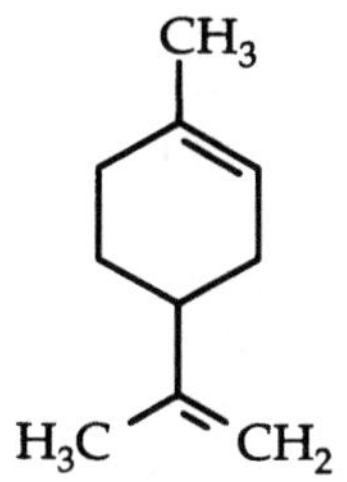

Limonene

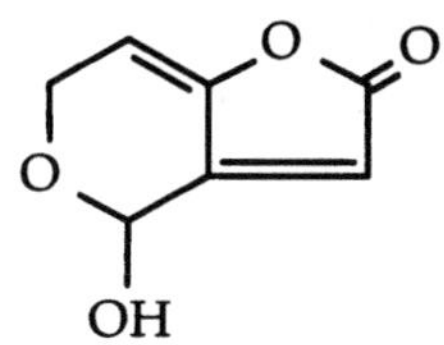

Patulin

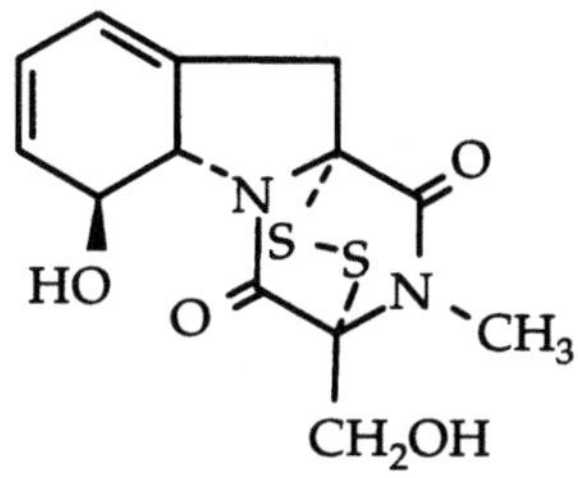

Gliotoxin

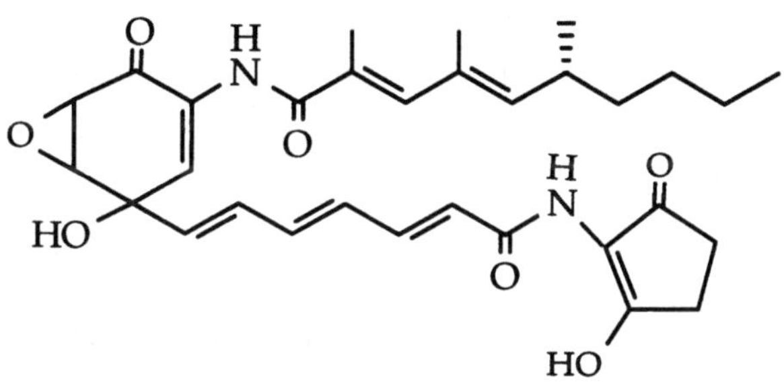

Manumycin

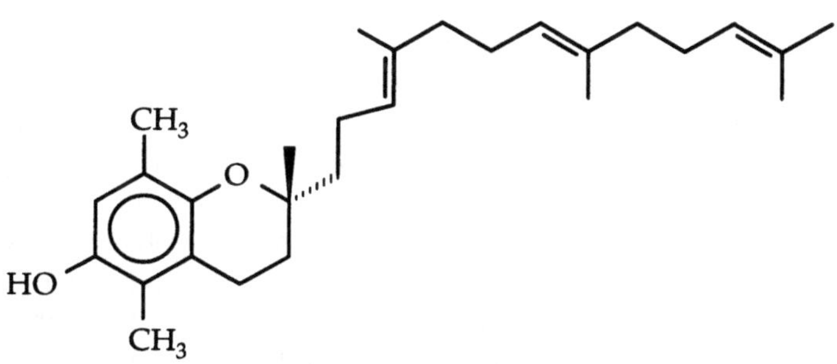

β-Tocotrienol

FIGURE 5. Molecular structures of other inhibitors of cholesterogenesis, protein prenylation and cell proliferation.

was effective as a chemopreventive agent both during the initiation, as well as the promotion/progression, stages of multi-stage carcinogenesis. Its effect was associated with an induction of "detoxification" enzymes that rendered the metabolites of DMBA less carcinogenically active and more easily excretable. Similarly, d-limonene was found to be capable of preventing development of mammary carcinomas induced *in situ* upon direct transfer of the v-Ha-*ras* gene with a retroviral vector (Gould *et al.,* 1994). Moreover, in rats bearing frank DMBA-induced mammary carcinomas, d-limonene treatment caused tumor regression by cytostasis, not cytotoxicity (Haag *et al.,* 1992).

Despite recent aggressive study, the mechanism of the growth inhibitory effects of d-limonene is not clear. It is known to be metabolically converted into the more biologically potent derivatives perillic acid, dihydroperillic acid, and their methyl esters (Crowell *et al.,* 1992), which, when supplied to tissue culture systems, are powerful inhibitors of cell proliferation of both untransformed (NIH-3T3) and transformed (HT29 colon carcinoma) cells (Crowell *et al.,* 1992; Crowell *et al.,* 1994). The effects of d-limonene and its metabolites on cholesterogenesis and protein prenylation were investigated, where it was found that at concentrations (mM range) capable of reducing proliferation of cultured cells, the overall rate of cholesterol synthesis appeared to be unaffected; however, there was observed to be a partial inhibition of protein prenylation, particularly of small GTP-binding proteins (Crowell *et al.,* 1991). Similar results were obtained by Schulz *et al.* (1994) with investigations on the effects of d-limonene, perillic acid, and perillic alcohol on the proliferation of human lymphocytes. These compounds inhibited phytohemagglutinin-stimulated peripheral lymphocyte proliferation, resulting in G1 arrest, in a concentration-dependent manner correlating with inhibition of prenylation of the 21–26-kDa GTP-binding protein species. Yet, these effects could not be overcome by supplementation with MVA, nor by the addition of isopentenyl adenine, ubiquinone, or dolichol, indicating that while the critical target of d-limonene action most probably lies downstream of the HMGR locus, it appears to bypass, in some fashion, formation of the currently identified non-steroidal isoprenenoid metabolites of the classical cholesterogenesis metabolic pathway.

It is noteworthy that d-limonene and its derivatives appear to inhibit specifically prenylation of the 21-26 kDa protein species (Crowell *et al.,* 1991), as this group includes not only $p21^{ras}$ (which is farnesylated) but also the less well-defined group of proteins that are geranylgeranylated and that constitute the major class of prenylated G-proteins. The blocking of protein farnesylation by active metabolites of d-limonene has been demonstrated by inhibition of FPTase in an *in vitro* enzyme study (Gelb *et al.,* 1995). Yet, perhaps due to differential intracellular microcompartmentalization, it is conceivable that these anti-proliferative monoterpenes preferentially target some aspect of the metabolism of cytosolic GTP-binding proteins, not merely by affecting protein prenylation *per se,* but possibly by interfering with a more subtle processing of Ras and Ras-like proteins prior to prenylation. If, for example, the effectiveness of these compounds as antiproliferative agents were merely dependent on the exclusive inhibition of Ras processing, then cells transformed by v-*ras* should be exquisitely sensitive to d-limonene growth inhibition, whereas cells whose transformation is independent of *ras* gene product function should be refractory. This concept was tested by comparing the effects of d-limonene and several other monoterpenes on the clonogenicity of epithelial cells, either untransformed or transformed by v-*ras* or v-*raf*. As mentioned above in discussing the effects of statin-type agents (see Figure 4), v-*raf* is an oncogene whose product (Raf) is engaged downstream of Ras in the signal transduction pathway, and whose transforming potential would thus be independent of *ras* gene function (Ruch and Sigler, 1994). As with the experiments of DeClue *et al.* (1991), it was observed that v-*ras*-transformed cells were not any more susceptible to growth inhibition by d-limonene than any of the other activated Ras-independent transformed cell lines. This result, together with the absence of Ras accumulation exclusively in the cytosolic fraction following treatment with the d-limonene-type monoterpenes, makes it difficult to support a proposed mechanism for these compounds as being due merely to suppression of Ras farnesylation. Similar observations by Gould *et al.* (1994) indicated that with *N*-nitroso-*N*-methylurea-induced mammary carinomas, d-limonene was extremely effective in causing tumor regression, but was independent of the presence of activating *ras* mutations in these tumors. Regardless of the currently veiled molecular mechanism of action of d-limonene, it may turn out to depend not directly on the inhibition of Ras processing, but rather on the presence of still-unidentified Ras-like G-proteins, perhaps modified by geranylgeranylation, against which d-limonene acts by preventing participation of such prenylated proteins in non-Ras signaling cascades.

A recent clue about the manner in which d-limonene and one of its metabolic derivatives, perillyl alcohol, exert their antiproliferative effects, relative to the mode of action of lovastatin, was revealed by the observation that the monoterpenes diminish the level of total cellular Ras (immunochemically determined), as well as cause a decrease of [^{35}S]Met incorporation into Ras; lovastatin does neither (Hohl and Lewis, 1995). Moreover, the reduction in the level of cellular Ras by perillyl alcohol, but not that of the G_α subunit (which is not subject to prenylation) points to a functional specificity for the monoterpenes toward Ras (or Ras-like) proteins. Hohl and Lewis (1995) suggested that monoterpenes act either by inhibiting manufacture (i.e., transcription/translation) of Ras, by stimulation of Ras proteolysis, or by a combination of both processes, whereas statin-type inhibitors of cell proliferation merely deplete cells of the prenyl intermediates generated via cholesterogenesis, which are required for Ras modification and subsequent localization of farnesylated Ras to the membrane.

Another isoprenoid-derived natural product with potential antitumor activity, which can be briefly mentioned here because of its topological resemblance to d-limonene, is the molecule named canventol. Canventol is a synthetic compound structurally related to sarcophytol A, a membrane-type diterpene isolated from coral. Because sarcophytol A is not abundant in nature and possesses a structure that has proven difficult to synthesize *de novo,* analogs have been created and tested for their antitumor effects. Canventol was found to be one such potent analog, since both it and sarcophytol A inhibited tumor promotion in several rodent carcinogenesis models. Due to the structural similarities between canventol and d-limonene, the former was tested for its effect on protein prenylation. Komori and coworkers found that canventol treatment inhibited MVA incorporation into a restricted class of proteins with M_r of 13, 17, and 22 kDa (Komori *et al.,* 1993). However, in contrast to d-limonene, which seems to prevent generalized protein prenylation, the prenylation of the majority of Ras-like G-proteins did not appear to be affected by caventol treatment. As with d-limonene, the antitumor effect of caventol might be attributed to its ability to inhibit geranylgeranylation of yet-unspecified small G-proteins apart from Ras.

The tocotrienols (Figure 5) are farnesylated benzopyrans generated in barley and palm oils. They exhibit hypocholesterolemic activity when administered *in vivo.* Tocotrienols reduce cholesterol synthesis in cultured rat hepatocytes and Hep G2 cells at low μM concentrations, apparently by inhibiting transcription while concomitantly enhancing degradation of the HMGR enzyme (Pearce *et al.,* 1992; Parker *et al.,* 1993; Bradfute and Simoni, 1994), thereby reducing the cell's optimal capacity for carbon flux through cholesterogenesis. Several analogs of the naturally occurring tocotrienols have been synthesized and found to elicit even greater potency for decreasing serum cholesterol (Pearce *et al.,* 1994). It has been observed that γ-tocotrienol (50 μM) and α-tocotrienol (100 μM) inhibited the growth of cultured human H69 and HeLa cells by 50%, whereas α-tocopherol (vitamin E) was without effect (Guthrie *et al.,* 1993). In addition, the survival times of mice bearing sarcoma 180 Ehrlich carcinomas and IMC carcinomas were shown to be significantly increased after intraperitoneal inocculation with α-and γ-tocotrienols (but not with α-tocopherol) (Komiyama *et al.,* 1989). The effect of tocotrienols on protein prenylation has not yet been reported, and therefore any linkage between the antitumor activity of these terpenoids and the inhibition of prenylated protein formation awaits future disclosure.

5.4. Competitive Inhibitors of MVAPPD

MVA-pyrophosphate decarboxylase, MVAPPD, catalyzes the decarboxylation of MVA-pyrophosphate (MVAPP) to isopentenyl pyrophosphate (IPP) (see Figure 3). The gene for MVAPPD has not yet been cloned, and therefore much less is known about the regulation and control of this enzyme compared with HMGR. Attempts to purify this enzyme from chicken, rat, and pig liver have been made (Toth *et al.,* 1995). However, detailed kinetic and protein structural studies are still lacking. Several compounds have been shown to inhibit MVAPPD activity, including 6-fluoromevalonate, catabolites of phenylalanine, and many plant phenolic compounds.

By inhibiting the activity of MVAPPD, 6-fluoromevalonate blocks formation of the first true isoprenoid of cholesterogenesis, IPP (Leonard *et al.,* 1990). It has been shown to prevent conversion of both acetate and MVA to cholesterol in perfused rat liver, and, as well, to inhibit the synthesis of juvenile hormone in insect larvae. The true proximal inhibitor of MVAPPD apparently is 6-fluoromevalonate-pyrophosphate, to which 6-fluoromevalonate is converted by sequential MVA kinase-mediated phosphorylations, and which yields a competitive K_i of ~37 nM for the enzyme, far below the K_m (10 μM) with its substrate MVAPP (Nave *et al.,* 1985). Upon treatment *in vitro* or *in vivo* of cholesterogenically capable systems with 6-fluoromevalonate, accumulation of MVA and its metabolites is observed, due to the inhibition of MVAPPD (Nave *et al.,* 1985; Reardon and Abeles, 1987). This accumulation has been claimed to be the principal cause of growth inhibition found with a variety of transformed and malignant T cell, B cell, and myeloid cell lines (Cuthbert and Lipsky, 1991), as well as of *ras*-transformed cells (Cuthbert and Lipsky, 1995). With *ras*-transformed cells, 6-fluoromevalonate not only blocked prenylation of Ras, but also was reported to decrease the expression of Ras mRNA and, therefore, of immunodetectible Ras protein mass, making these *ras*-transformed cells very sensitive to treatment with this agent (Cuthbert and Lipsky, 1995). Although active as an antiproliferative drug, 6-fluoromevalonate shares with other classes of HMGR inhibitors a lack of biologically (as distinguished from chemically) responsive specificity, inasmuch as the formation of both steroidogenic and non-steroidogenic isoprenoids is inhibited in treated cells.

Several natural product phenolic derivatives have also been reported to inhibit MVAPPD directly. These include phenylacetate (PA) and cinnamic acid with its hydroxy analogs, ferulic, isoferulic, and coumaric acids (Shama Bhat and Ramasarma, 1979a; Castillo *et al.,* 1991a; Castillo *et al.,* 1991b). Moreover, a number of catabolites of phenylalanine (in addition to PA) that serve as metabolic markers of phenylketonuria, such as phenylpyruvate, phenyllactate, and their hydroxyl products, are also inhibitors of MVAPPD (Castillo *et al.,* 1991a, 1991b). Cinnamic acid derivatives behave as more potent competitive inhibitors of the enzyme than PA (Shama Bhat and Ramasarma, 1979a), and, together with isoferulic acid, have been reported to act also as competitive inhibitors of the upstream cholesterol pathway enzyme MVAP kinase (Shama Bhat and Ramasarma, 1979b).

Castillo *et al.* (1991a, 1991b) showed that, when administered in mM concentrations to cell homogenates, PA, phenylpyruvate, phenyllactate, and their hydroxyl products all inhibit MVAPPD. The effects of PA and its analogs on inhibition of MVAPPD and cholesterol biosynthesis, as well as their capacity to bring about phenotypic reversion of transformed cells, have been linked to the inhibition of isoprenylation of cellular proteins and cell proliferation in various tumor lines and *ras*-transformed cells (Samid *et al.,* 1994; Hudgins *et al.,* 1995; Liu *et al.,* 1995).

When employed *in vitro,* the antiproliferative, growth inhibitory and differentiation-inducing effects of PA and its analogs encompass a wide variety of tumor cells, including glioblastoma, neuroblastoma, rhabdomyosarcoma, melanoma, hormone-dependent and-refractory prostate cancer, HL-60 leukemia, and B-chronic lymphoeytic leukemia cells (Samid *et al.*, 1992a; Samid *et al.*, 1992b; Cinatl *et al.*, 1993; Samid *et al.*, 1993; Call *et al.*, 1994; Cinatl *et al.*, 1994; Liu *et al.*, 1994; Samid *et al.*, 1994; Hudgins *et al.*, 1995; Liu *et al.*, 1995). Of all the PA analogs tested, the more potent inhibitors of MVAPPD also turn out to be the more potent inhibitors of protein isoprenylation and manifest higher antiproliferative effects *in vitro* (Hudgins *et al.*, 1995; Liu *et al.*, 1995). Similar to observations with lovastatin, *ras*-transformed cells revealed greater sensitivity to PA-effected inhibition of cell growth than their parental lines (Miller and Samid, 1994). Samid's group has also reported that *ras*-transformed cells exhibit a greater resistance than their parental lines to oxidative stress, radiation, and treatment with doxorubicin (Adriamycin), whereas the presence of lovastatin reversed this pattern of resistance (Samid *et al.*, 1991; Miller *et al.*, 1993; Miller and Samid, 1994). Furthermore, PA treatment rendered these *ras*-transformed cells capable of regaining responsiveness to doxorubicin and radiation (Shack *et al.*, 1995). Findings from human tumor and plant embryonic cells show them responding to the cytostatic effect of PA and analogs in the same concentration range and rank order, indicating that these compounds may act through common mechanisms to inhibit the growth of divergent, undifferentiated cells.

Although the concentrations of PA required to induce differentiation in tumor cells are rather high (mM), such concentrations are pharmacologically achievable in humans and test animals. The results from phase I clinical trials suggest that PA exerts antiproliferative activity against glioma (Thibault *et al.*, 1994). Such antitumor activity observed in clinical trials with PA is also supported by results from animal transplant studies, which revealed that PA inhibited both the growth of glioma cells (Ram *et al.*, 1994) as well as human prostate cancer cells (Samid, personal communication) in recipient mice.

Finally, current studies by Samid's group imply that several multidrug resistant (MDR) cells derived from breast, colon and pancreatic cancer appear more sensitive than their respective parental cells to PA and the related compound phenylbutyrate. Whether this effect is due to an enhanced reliance of MDR cells on prenylprotein synthesis warrants further investigation. However, recent findings regarding the substrate characteristics required by the overexpressed plasma membrane-associated MDR transporter of tumors (which is also called P-glycoprotein) may be of central importance. First, it has been shown that the mammalian MDR transporter, as well as its yeast homolog (Ste6), can actively transport the yeast mating pheromone called **a**-factor, which is a farnesylated, methyl-esterified peptide containing a C-terminal-CAAX motif (Sapperstein *et al.*, 1994). Casey's laboratory (Zhang *et al.*, 1994) has since reported that synthetic prenylcysteine methyl esters (i.e., *S*-farnesyl-or *S*-geranylgeranyl-Cys-$COCH_3$) bind to and thereby stimulate the ATPase activity of the MDR transporter, whether the transporter is expressed from a baculovirus system or derived from drug-resistant human breast cancer cells. The structural determinants required for MDR transporter ligand binding, examined in some detail (Zhang *et al.,* 1995), have been found to include a free N-terminal nitrogen, the prenyl group and the Cys carboxymethyl function. The broader implications of such findings extend, among other things, to speculations for a mechanism that the cell could employ to rid itself of various stably prenylated (low turnover rate) proteins (Maltese and Sheridan, 1988), which otherwise might accumulate and interfere with cell proliferation signal transduction pathways (section 4.1).

5.5. Endogenous Steroids as Antiproliferative Agents

Dehydroepiandrosterone (DHEA) is an endogenously produced steroidal intermediate in the pathway of androgen and estrogen synthesis. It is interesting that its metabolic conversion to 7-oxo derivatives serves to direct DHEA away from androgen/esterogen synthesis, and that these oxygenated DHEA metabolites behave as powerful inducers of certain mitochondrial enzymes (Lardy *et al.,* 1995). DHEA is also a peroxisome proliferator (section 3.2). Of concern to the subject of this chapter is that, among other of its manifold properties, DHEA exhibits apparent chemopreventive features, and it also seems to cause inhibition of protein prenylation (Schultz and Nyce, 1991; Schultz *et al.,* 1992). DHEA, and its sulfated metabolite DHEAS, are relatively abundant adrenal steroids originally shown to protect against natural and chemically induced tumor formation (Schwartz *et al.,* 1988). Interestingly, the human serum levels of DHEA fluctuate according to age, reaching a peak (5–15 μM) between the ages of 20–25 years and declining thereafter to levels ~10% of its peak value. Due to its putative antiproliferative/antitumor effect, it was proposed that DHEA may confer protection against tumorigenesis at a young age, and its decline in serum concentration might be associated with the increasing susceptibility to cancer during the aging process (Regelson and Kalimi, 1994). Its antiproliferative effects were recently verified by causing the induction of differentiation of B16 murine melanoma cells (Kawai *et al.,* 1995). In attempts to understand its molecular role, a human carcinoma cell line was treated with increasing concentrations of this compound (Schultz *et al.,* 1992), which led to growth arrest at G1 in a time-and dose-dependent fashion. This effect could be prevented by the addition of MVA, or with MVA *plus* cholesterol, if the cells had been pretreated with DHEA for an extended period (more than 48 h). Analysis of the pattern of prenylation indicated that incubation with DHEA, but not DHEAS, reduced [^{3}H]MVA incorporation into all classes of prenylated proteins, including Ras, in HT29 cells. In spite of the fascinating multiple properties exhibited differentially by DHEA and DHEAS, the mechanism for the antiproliferative and antitumor effect of these steroid derivatives remains largely unexplained (Regelson and Kalimi, 1994), but it is clear that by subverting protein prenylation, thereby interrupting the normal biological activities of certain target proteins, cells become impaired in their proliferative response.*

5.6. Other Protein Prenylation Inhibitors

Several groups have screened natural product repositories for compounds potentially capable of inhibiting protein prenylation (Tamanoi, 1993). By such high-efficiency screening methods a host of diverse natural products that inhibit protein farnesylation has been identified (Figure 5), including the following: patulin (Miura *et al.*, 1993); gliotoxin and its derivative acetylgliotoxin (van der Pyl *et al.*, 1992); 10′-desmethoxy-streptonigrin (Liu *et al.*, 1992); pepticinnamins (Omura *et al.*, 1993; Shiomi *et al.*, 1993); and manumycin (Hara *et al.*, 1993; Shu *et al.*, 1994). The gliotoxins, whose structures bear no features immediately resembling isoprenoid intermediates, inhibit protein prenylation by an unknown mechanism, since they behave as neither substrates nor competitive inhibitors of the prenylprotein transferase enzymes. Manumycin, produced by a *Streptomyces,* has been shown to be active *in vivo* by inhibiting tumor formation in nude mice of a fibrosarcoma cell line transformed by the activated K-*ras* gene product (Hara *et al.*, 1993). Hara and Han (1995) recently extended the analysis of FPTase and mammalian tumor cell growth inhibition by manumycin and gliotoxin with the development of a Ras/Raf-kinase/MAP kinase-dependent signaling pathway assay based on *C. elegans* vulval differentiation. Interestingly, the *C. elegans* homolog of mammalian Ras, a *let-60* gene product, is directly involved in receptor tyrosine kinase signaling of MAP kinase, and plays a key role in the control of such differentiation events as vulval cell development in this organism. The responses to treatment with manumycin and gliotoxin were found to mimic, in the *C. elegans* assay, precisely those encountered by treatment of mammlian tumors with synthetic CAAX-type FPTase inhibitors that are discussed at length in section 6; that is, these natural antitumor compounds blocked the action of activated (mutated) *let-60* protein, but not of normal *let-60* protein nor of mutations in certain upsteam or downstream signaling pathway elements prior to MAP kinase that are homologous with those mammalian factors depicted in Figure 4.

*It may be worth noting, as an aside, that not all current biological research relies on experimental strategies reduced to the scale of genetic engineering or molecule-specific manipulation. Thus, DHEA also has achieved an expanding halo of notoriety as an "anti-aging" or even an "age-reversing" steroid, in which guise it has become a minor *cause célèbre,* in addition to its experimental inhibition of cell growth *in vitro* (Bilger, 1995).

6. RECENT APPROACHES TOWARD RATIONAL DRUG DESIGN TO INHIBIT p21^{ras} FARNESYLATION

With Ras farnesylation having been established as a crucial prerequisite for enabling its biological function, the early 1990s introduced the possibility of developing specific FPTase inhibitors as potential antitumor therapeutic agents as a most obvious and promising approach (Gibbs, 1991). This, coupled with the common presence of mutationally activated Ras in many types of human cancer, has continued to impel the search by many labs for inhibitors of FPTase (referred to herein as FTI; farnesylprotein transferase inhibitors) that possess nontoxic anti-proliferative/anti-neoplastic potential.

Characterization of purified FPTase revealed that FPP seemed to be the uniquely utilized cholesterogenic pathway-derived prenyl moiety co-substrate, and that catalysis *in vitro* occurs merely upon addition of FPP plus a minimal recognition tetrapeptide sequence, consisting of the CAA[Met/Ser/Ala] domain (Reiss *et al.,* 1990; Pompliano *et al.,* 1992). Therefore, FPP and CAAX-type tetrapeptides were deemed well suited as models for the design of potential inhibitors of Ras farnesylation. As mentioned in section 4.2, catalytic constraints pertaining to the FPTase enzyme mechanism are fairly well established (Reiss *et al.,* 1991; Reiss *et al.,* 1991; Pompliano *et al.,* 1992; Reiss *et al.,* 1992; Omer *et al.,* 1993; Stradley *et al.,* 1993; Ying *et al.,* 1994). Such detailed information promoted efforts aimed towards designing molecules that could selectively target and inhibit the Ras farnesylation reaction inside cells, with the limited goal in mind of interfering directly with Ras protein function in the signaling cascade of protein–protein interactions and kinase-directed phosphorylations that lead to cell transformation and proliferation. Since the Ras farnesylation reaction involves two substrates, the design of FTI analogs has focused on synthesizing one of three types of pro-drug antimetabolites: either (i) analogs of the isoprenoid farnesyl moiety; (ii) CAAX-type peptide analog (peptidomimetic) inhibitors; or (iii) pseudo-transition-state compounds that contain both of the substrate moieties together in one molecule. The structures of some of these synthetic FTI agents are shown in Figure 6.

6.1. FPP-Based Inhibitors

The search for inhibitors that would interfere with Ras farnesylation catalyzed by FPTase recognized that in order to obtain an isoprenoid analog that would bind to and inhibit FPTase, the number of C_5 isoprene units, as well as the presence of a negatively charged (Mg^{2+}-binding?) region, were both required structural elements. Among the synthetic lipoidal inhibitors that have been found to target FPTase is a series of polyprenoids constructed such that either the pyrophosphate group is modified (Cohen *et al.,* 1995), or stereochemical changes have been introduced in the prenyl tail (Das and Allen, 1991). The FPP analog

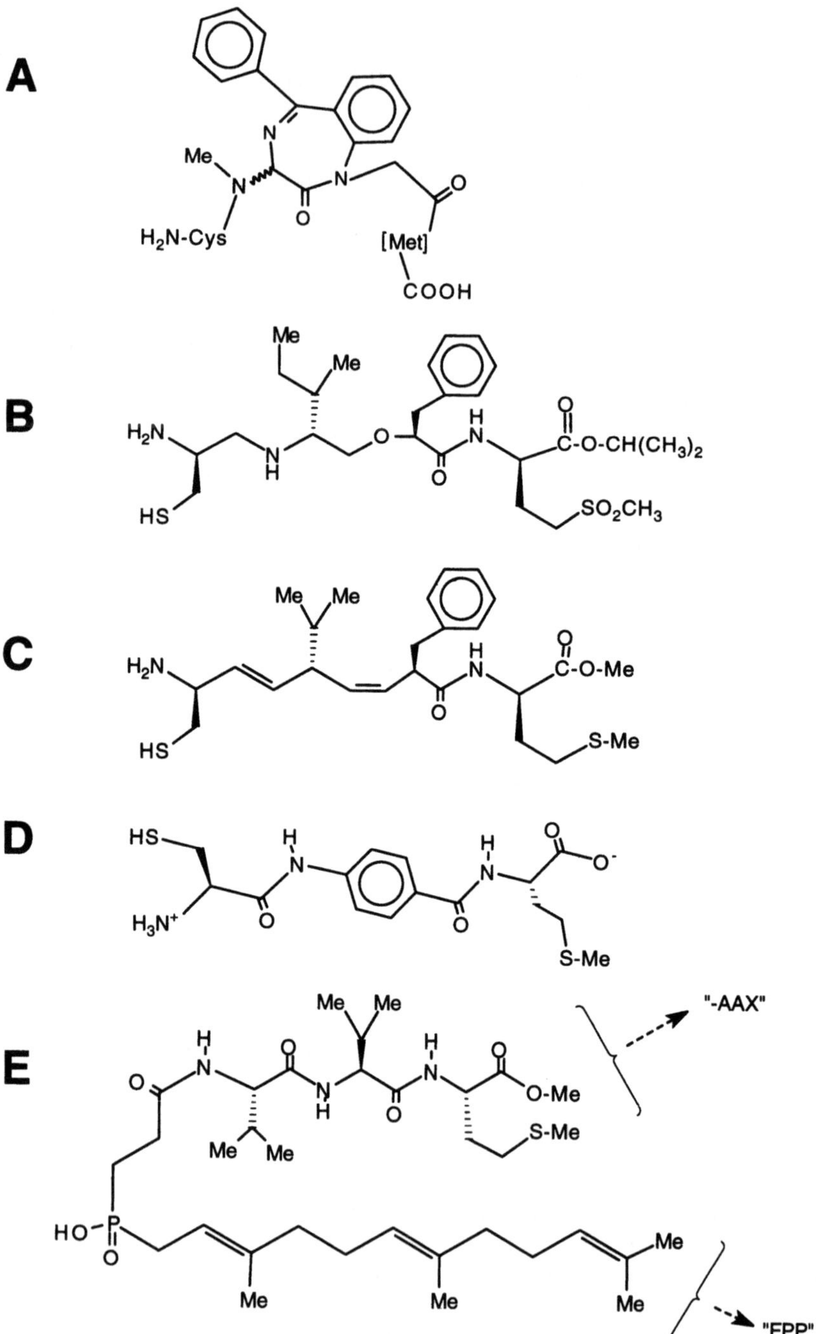

FIGURE 6. Farnesylprotein transferase inhibitor (FTI) pro-drugs that block Ras-transformed cell proliferation. **A,** see James *et al.* (1993); Marsters *et al.*, (1994); **B,** see Graham *et al.*, (1994); **C,** see Cox *et al.* (1994); Nagasu *et al.* (1995); **D,** see Vogt *et al.*, (1995); **E,** see Manne *et al.* (1995); Patel *et al.*, (1995).

(α-hydroxyfarnesyl)phosphonic acid was found to be a potent FPTase inhibitor both *in vitro* and *in vivo.* This compound, together with the recently recognized natural terpenoid products chaetomellic and zaragozic acids (Singh *et al.,* 1993), appear to behave as competitive inhibitors of FPTase *in vitro,* with respect to FPP (Gibbs *et al.,* 1993). However, only (α-hydroxyfarnesyl)phosphonic acid was active *in vivo,* as it was shown to block v-Ha-Ras processing in *ras*-transformed fibroblasts at low (1 μM) concentration (Gibbs *et al.,* 1993). The chaetomellic and zaragozic acids, despite exhibiting K_i values in the low nM range *in vitro,* were without effect *in vivo,* perhaps because of the presence of multiple CO_2H functions on each species, yielding highly charged structures that might render them incapable of penetrating the cell membrane (Tamanoi, 1993). More recent studies with synthetic farnesyl-and geranylgeranyl-like photoaffinity analogs [e.g., the prenyl-PP structural mimics 2-diazo-3,3,3-trifluoropropionyloxy-farnesyl (or -geranylgeranyl) diphosphate] indicated that they covalently label the β subunit of both FPTase, as well as GGPTase-I (Omer *et al.,* 1993; Bukhtiyarov *et al.,* 1995). These prenyl diphosphate photoreactive analogs were found to bind to the enzymes in a competitive manner versus FPP or GGPP, similar to the general binding characteristics revealed with the photoaffinity labeling CAAX peptide analogs reported by Ying *et al.* (1994). It seems fair to conclude, however, that as of mid-1995 there has been less emphasis in pursuing the application of prenyl moiety inhibitors than in seeking peptidomimetic CAAX-type FTI compounds. Nevertheless, the current availability of catalytic site-seeking photoreactive, covalently binding CAAX-type, as well as prenyl diphosphate-type, ligands for prenylprotein transferases, should now serve to encourage future studies for mapping the discrete ligand binding sites of enzymes that utilize either the peptide or the prenyl diphosphate as substrate (or both, as co-substrates), thereby allowing one to define critical amino acid residues responsible for the catalytic mechanisms of such enzymes.

6.2. CAAX-Based FPTase Inhibitors

In-depth studies have supported the general belief that the CAAX motif on the protein co-substrate possesses the principal recognition element for specifying modification either by a farnesyl or a geranylgeranyl moiety with FPTase or GGPTase-I, respectively (Reiss *et al.,* 1990; Pompliano *et al.,* 1992). Therefore, it was not surprising to find that a series of different CAAX-motif tetrapeptides could serve as competitive inhibitors of Ras farnesylation *in vitro* (Reiss *et al.,* 1990; Brown *et al.,* 1992; Pompliano *et al.,* 1992). However, since such peptides, inside a cell, would also behave as substrate analogs and undergo farnesylation themselves, their unmodified concentration would continue to decrease over time, obviating their role as competitive ligands for FPTase and preventing long-term intracellular inhibition. It also was observed that if the penultimate C-terminal residue (i.e., A_2) was substituted for a Phe, and if a positive charge was simultaneously retained by the N-terminal Cys, then such a CAAX-motif tetrapeptide no longer served as a farnesylatable substrate but behaved exclusively as a pure FPTase inhibitor (Goldstein *et al.*, 1991; Brown *et al.*, 1992). Nevertheless, peptides have limited potential for use with *in vivo* systems inasmuch as they are readily degraded by non-specific extra-and intracellular proteases, and, more importantly perhaps, are not efficiently taken up into cells due to their ionized character. These undesirable characteristics led several research groups to embark on the design of peptidomimetic molecules that would overcome the drawbacks possessed by true CAAX peptide inhibitors.

Therefore, based on the known structural and biochemical requirements of FPTase, including its coordination with Zn^{2+} and its requirement for binding a CAAX motif peptide possessing a turn-like conformation at the FPTase binding domain (Stradley *et al.*, 1993), one class of peptidomimetics was designed in which the correct geometry was conserved by complete replacement of the central A_1A_2 aliphatic amino acids with a benzodiazepine moiety (James *et al.*, 1993; Marsters *et al.*, 1994). The conformation of the benzodiazepine group sufficiently mimicked a dipeptide β-turn, which then allowed the N-terminal Cys and the carboxyl terminal residue (e.g., Met) of the CAAX analog to assume the appropriate topology for binding to the FPTase. A family of benzodiazepine (BZA) peptidomimetic inhibitors against FPTase was thus constructed (see Figure 6), each member of which was found to selectively inhibit farnesylation of Ras *in vitro*, with differing IC_{50} values (in the nM range) depending on the identity and degree of modification of the C-terminal residue (Met or Ser versus Leu). Experiments *in vivo* reported that geranylgeranylation of appropriate FPTase-directed acceptor peptides was unaffected by these BZA inhibitors, whereas MVA incorporation into Ras and lamins A and B was completely inhibited, signifying a restricted inhibition of protein farnesylation. Moreover, the BZA analogs inhibited cell growth and induced reversion of the morphology of Rat1 fibroblasts transformed by v-Ha-*ras* (James *et al.*, 1993).

FPTase inhibitors developed by a group at Merck were also derived from the basic CAAX motif, but a different synthetic scheme was employed from that of the BZA peptidomimetics of James *et al.* (1993). During their investigations, Merck workers observed that incubations of true CAAX tetrapeptides with cytosolic fractions derived from NIH 3T3 fibroblasts led to the rapid destruction of these FPTase competitive ligands, apparently by means of non-specific aminopeptidase proteolytic activity present in the cell. Indeed, a membrane-associated, Zn^{2+}-requiring endopeptidase activity has been reported capable of degrading farnesylated Ras-CAAX peptide analogs (Akopyan *et al.*, 1994). Rapid peptide bond hydrolysis, together with notoriously poor cellular uptake of peptides in general, presumably accounts for the unimpressive anti-proliferative (cytostatic) capacity of such true peptide FPTase inhibitors when tested *in vivo*, regardless of their *in vitro* potency in purified FPTase assays. This fact indicated the need for modification of the peptide bonds in the design of a pro-drug suitable for subsequent use *in vivo*. Among the FTI compounds synthesized by Merck and found to show promise both *in vitro* and *in vivo,* are those based on the C-terminal tetrapeptide of p21^{Kras}, Cys-Ile-Ile-Met. The parental tetrapeptide was subjected to reduction of the first two peptide bonds and further modified by replacement of the terminal methionine by a homoserine to yield the pseudopeptide analog Cys[ΨCH_2NH]Ile[ΨCH_2NH]Ile-homoserine. Upon lactonization of the C-terminal homoserine this derivative was converted to a more membrane-permeable pro-drug taken up into cells (Graham *et al.,* 1994).

In vitro prenylation assays indicated that several of such pseudopeptide analogs were potent as specific inhibitors of FPTase, exhibiting >5,000-fold selectivity towards FPTase over GGPTase-I or II (their IC_{50} for the C_{20} prenylation-specific enzymes was >100 μM; Kohl *et al.,* 1993). Although requiring higher concentrations when employed *in vivo* (IC_{50} ~100 μM) than *in vitro* (IC_{50} in the nM range), the homoserine lactone pro-drug derivative was found to inhibit processing and subcellular membrane localization of viral Ha-Ras in NIH 3T3 v-*ras* transformants, whereas the processing of a chimeric Ras protein, specifically engineered to be subjected to geranylgeranylation (Ras-CVLL) rather than farnesylation, was not affected. Thus, this type of pseudopeptide analog displays FPTase selectivity. In order to characterize the *in vivo* consequences of protein farnesylation inhibition on cell proliferation, Kohl and coworkers studied the effect of this type of FTI by assessing the soft agar growth of Rat1 cells transformed by oncogenic versions of *ras, raf,* and *mos* genes (Kohl *et al.,* 1993). The logic of such comparative experiments, described previously (section 5.2, Figure 4) is based on the fact that while Raf and Mos proteins are members of signal transduction cascades, their transforming potential, unlike that involving Ras, is independent of farnesylation for acquisition of biological activity. The results of this inquiry showed that treatment with the Merck homoserine-lactone FTI led to a dose-dependent inhibition of anchorage-independent growth in cells transformed by v-*ras,* but not by v-*raf* or v-*mos.* By comparison, lovastatin inhibited soft agar growth of all three transformants.

In simultaneous studies, another series of membrane-permeable peptidomimetic FTI pro-drugs was developed by the Eisai Research Institute (García *et al.,* 1993; Cox *et al.,* 1994; Nagasu *et al.,* 1995). These are derived from the parental CAAX tetrapetide CVFM modified by a strategy similar to that of the Merck group, wherein the first two peptide bonds are reduced, thereby making the resulting analogs particularly resistant to aminopeptidase proteolysis *in vivo.* Despite some subtle structural distinctions between the Eisai and Merck pro-drugs, their latest versions are quite similar to each other (see Figure 6). Eisai's FTI analogs reportedly inhibit FPTase over GGPTase I with a high *in vitro* selectivity that is retained *in vivo* since, at μM concentrations, these analogs prevent Ras and lamin B processing without affecting that of Rap1 (which is geranylgeranylated). Morevover, biological Ras function was found to be blocked by these FTI pro-drugs, evaluated as inhibition of v-*ras*-induced oocyte maturation and transformation of NIH-3T3 cells (Cox *et al.*, 1994). In transformation assays, they blocked the anchorage-independent growth induced by oncogenic, farnesylated Ras but not that by geranylgeranylated or myristoylated versions of Ras, or by oncogenically activated Raf. The Eisai pro-drugs also appear capable of preventing oncogenically activated, Ras-mediated signaling of a mitogen-activated protein kinase (MAPK), and the subsequent activation of the specific transcription factors Jun and Elk-1. Such observations imply that in the presence of these FTI, the Ras-dependent signaling cascade of sequential protein kinase-mediated phosphorylations is interrupted by prevention of Ras farnesylation (Cox *et al.*, 1994).

Still another effort in the development of FTI pro-drugs is represented by the synthesis of a set of compounds by Sebti's group (Nigam *et al.*, 1993; Qian *et al.*, 1994), in which the central dipeptide, A_1A_2, of the CAAX motif once again is completely replaced by spacer groups of different lengths and conformations (see Figure 6). Here, instead of a benzodiazepine moiety (James *et al.*, 1993; Marsters *et al.*, 1994), the spacers are 4-aminobenzoic acid (4-ABA) and 3-aminomethylbenzoic acid (3-AMBA), resulting in the creation of the peptidomimetic species Cys-[4-ABA]-Met and Cys-[3-AMBA]-Met. Both compounds were found to selectively inhibit FPTase *in vitro* relative to GGPTase I (IC_{50} = 50 and 100 nM for FPTase, respectively, versus >100 μM for GGPTase-I) without becoming farnesylated by the enzyme. This selectivity was also observed *in vivo* as these compounds blocked Ras processing without affecting the geranylgeranylation of Rap1 (Nigam *et al.*, 1993; Qian *et al.*, 1994; Vogt *et al.*, 1995).

6.3. Bisubstrate Analog FPTase Inhibitors

The objective of rational drug design for development of FTI agents has also led to the creation of bisubstrate analogs that mimic the putative structure of the FPTase transition state or intermediate. These are compounds that contain components of both co-substrate partners of the enzyme as part of the same molecule. The inhibitors designed by Bristol-Myers Squibb retain the farnesyl moiety, together with the last three residues of the Ras family CAAX motif, i.e.-AAX, but omit Cys, the site of Ras farnesylation. The tripeptide and farnesyl portions are connected via a phosphonic or phosphinic acid linker (Manne *et al.*, 1995; Patel *et al.*, 1995), whose length was chosen so as to allow the two substratelike domains to adopt a geometry reminiscent of an intermediate or transition state. The thioether bond formed between the farnesyl moiety and the Cys thiol of the actual CAAX motif is replaced in this tripeptide analog by either the phosphonic or phosphinic acid linkage, since there is no Cys sulfhydryl group. Bisubstrate inhibitors of this type (see example in Figure 6) were demonstrated as being 1000-fold more potent as inhibitors of FPTase than GGPTase-I *in vitro* ($IC_{50} \approx 5$ nM versus >10 μM, respectively), to inhibit the anchorage-dependent and -independent growth of oncogenic *ras*-transformed NIH-3T3 cells, and to block the malignant neurofibromatosis type-1 phenotype (Yan *et al.,* 1995).

6.4. FTI as Potential Chemotherapeutic Agents: Bright Promise and Unsolved Problems *in Vivo*

The various synthetic peptidomimetic FTI pro-drugs (Figure 6) display exceptional potency *in vitro* as FPTase inhibitors and may hold great promise as new molecular prototypes in the chemical arsenal of tumor therapeutics. Thus, recent whole animal *in vivo* studies have been reported by both Merck and Eisai groups, employing their independently designed, but rather structurally similar, FTI prodrugs (Kohl *et al.,* 1995; Nagasu *et al.,* 1995). The Merck peptidomimetic (compound B, Figure 6) was tested in a transgenic mouse line defined as bearing the v-Ha-*ras* oncogene under the control of the mouse mammary tumor virus long terminal repeat (MMTV) (Kohl *et al.,* 1995). These transgenic mice are known to exhibit a high incidence of spontaneous mammary and salivary tumors. The results obtained were highly encouraging. Animals exhibiting palpable tumors (50-350 mm^3 in volume) and treated daily with 40 mg/kg/day of the Merck FTI for up to two weeks, displayed a dramatic regression in tumor size. Kohl *et al.* (1995) reported that in 100% (7/7) of the animals, tumors became undetectable after treatment, whereas in the vehicle-treated (control) group, tumors continued to increase in size. Furthermore, in contrast to the FTI-treated animals, the control litter developed new tumors as the experiment progressed. The overall results were dramatized further by comparison with parallel doxorubicin (Adriamycin) treatment. Doxorubicin, a chemotherapeutic agent commonly used to treat mammary tumors, when administered at the maximally tolerated dose (2 mg/kg/day), yielded only 2/7 complete, and 3/7 partial, reversions. Furthermore, in those animals not responding to doxorubicin, subsequent treatment with the Merck FTI indicated that complete remission occurred.

The Merck whole animal studies also revealed the important mechanistic result that continued exposure of the tumor to the FTI pro-drug was required for complete tumor remission. In other words, the effect of the FTI is only transient; when the prodrug was no longer present due to systemic excretion, the *ras*-transformed tumors reappeared. Moreover, the cytostatic rather than cytotoxic action of this pro-drug is supportive of the fact that constant inhibition of oncogenic Ras farnesylation is necessary for inhibition of cell proliferation. Despite long term exposure of animals to the Merck analog at the reported dosage, no ill effects were manifest, and chronically treated animals showed no inclination toward cachexia. Histopathological analysis of tissues, including those in which farnesylated proteins are assumed to play a significant role (rapidly proliferating tissues as well as retina, skeletal muscle, etc.) showed no discernable alterations following chronic (11-week) treatment with the pro-drug.

Intact animal studies, recently reported by the Eisai group with their FTI analog (compound C, Figure 6), included combined evaluations on the growth inhibition of *ras*-transformed 3T3 cells and several human tumor lines, both of which were propagated in nude mice as xenografts. In addition, inhibition assays of colony formation in soft agar were performed (Nagasu *et al.,* 1995). From the standpoint of the future therapeutic value of such FTI pro-drugs, the results of Nagasu *et al.* (1995) are qualitatively similar to those obtained by Kohl *et al.* (1995) and Sepp-Lorenzino *et al.* (1995). Thus, the Eisai data, in general, also indicate that targeting the inhibition of protein farnesylation constitutes a potentially encouraging modality for the chemotherapeutic treatment of hyperproliferative and neoplasic diseases. Moreover, for these CAAX-type analogs, as well as for the BZA pro-drug, lack of toxicity to untransformed cell lines and normal tissues at the dosages employed (James *et al.,* 1993; Kohl *et al.,* 1993; Cox *et al.,* 1994; James *et al.,* 1994; Kohl *et al.,* 1994; Kohl *et al.,* 1995; Nagasu *et al.,* 1995; Sepp-Lorenzino *et al.,* 1995) makes such peptidomimetic FTI compounds attractive candidates for future evaluation in a clinical setting with human subjects.

Underlying all the positive indications highlighted by the recent animal trials is a bed of confusing observations. Analysis of the capacity of such FTI pro-drugs to serve as antiproliferative, antitumor agents, when assessed *in vivo* across a broad spectrum of cultured tumor cell lines rather than by whole animal treatment regimens, has revealed a sometimes puzzling array of varied responses on exposure to the FTI.

First, it should be stressed that the greatest *in vivo* successes with the various peptidomimetic FTI pro-drugs or, for that matter, with many of the other antiproliferative compounds that interrupt or inhibit isoprenoid processing of signal transduction proteins (e.g., the recently described farnesyl-amine,-acetate and -thiosalycylate derivatives (Kothapalli *et al.,* 1993; Marom *et al.,* 1995; Meigs *et al.,* 1995), have been achieved in studies of oncogenic-*ras*-transformed cell cultures. As might be expected, the activity of various FTI compounds is selective for cells transformed via Ras-mediated pathways, compared with cells utilizing alternate transduction pathways (e.g., via v-*raf* or v-*mos* transformation) that are independent of protein farnesylation. However, it has been observed that cells (such as Rat1 or NIH 3T3 fibroblasts) which have *not* been engineered to contain oncogenically activated Ras protein are seemingly insensitive to the presence of the otherwise potent FTI pro-drugs in Figure 6. That is, proliferation of untransformed cells proceeds at analog concentrations that inhibit transformed cells (Davis, 1993; García *et al.,* 1993; James *et al.,* 1993; Kohl *et al.,* 1993; Marshall, 1993; Cox *et al.,* 1994; Gibbs *et al.,* 1994; James *et al.,* 1994; Kohl *et al.,* 1994; Prendergast *et al.,* 1994; Manne *et al.,* 1995; Marshall, 1995; Patel *et al.,* 1995). It must be remembered that so-called "normal" cell growth also depends on Ras-mediated signal transduction; consequently, the distinction in the response of normal and Ras-transformed cells to FTI analogs is puzzling. Furthermore, since 1994, *in vivo* studies have indicated that while selected cell lines are susceptible to the effects of FTIs, not all cultured tumor cell lines respond equally (see, for example, Nagasu *et al.*, 1995; Sepp-Lorenzino *et al.*, 1995). Certain tumor cells, whether or not they harbor oncogenic *ras,* were found to exhibit resistance to an otherwise potent FTI pro-drug. This *in vivo* resistance to the anti-proliferative effects of the FTI was independent of, and not correlated with, the drug's proven capacity to inhibit FPTase activity *in vitro.* These revelations could suggest that a cell line would exhibit resistance if it were able to engage "alternative" signal transduction pathways that either circumvent or operate independently of the Ras-mediated route. In this regard, it was shown by Sepp-Lorenzino *et al.* (1995) that even though the FTI analog blocked growth factor-stimulated Ras activation in both sensitive as well as resistant cells, in some resistant cell lines the activation of MAP kinase was unaffected following treatment with the FTI pro-drug. This observation is corroborated by a recent finding that expression of a constitutively activated EGF receptor in NIH 3T3 cells results in the constitutive activation of MEK, and to a lesser extent MAP kinase (Montgomery *et al.*, 1995). Consequently, in these cells FTI inhibitors could be expected to be ineffective because the cells are transformed by a mechanism that bypasses the farnesylation-dependent portion of the signalling cascade.

A recent report by Sinenski's group (Dalton *et al.*, 1995) dramatically emphasizes the incompleteness of our comprehension of the heretofore "expected" role of protein farnesylation on signaling pathways that stipulate cell transformation or proliferation. Dalton *et al.* (1995) found that when cells are treated either for short or prolonged periods with 25 μM of the BZA class of FTI analog (compound A, Figure 6), the farnesylation of Ras, lamin B, and pre-lamin A was completely blocked. Yet, under such total inhibition of protein farnesylation: (i) $p21^{ras}$ still manifests membrane association; (ii) NGF-induced differentiation (neurite outgrowth) of PC-12 cells occurred unabated; (iii) viability and rate of proliferation of CHO-K1 cells were uneffected; and (iv) despite accumulation of the unfarnesylated lamin proteins in the nucleus of BZA treated cells, the structure of the nuclear lamina appeared to remain unperturbed. These workers put forth the noncontroversial hypothesis that "alternate mechanisms exist" permitting normally farnesylated CAAX-motif proteins a means of processing that is independent of farnesylation. Similarly unfathomable in terms of our current understanding is the report from Simoni's laboratory (Meigs *et al.*, 1995) demonstrating that cells treated with farnesyl acetate, a presumed naturally generated intermediate derived from MVA metabolism, manifest inhibition of DNA replication and cease proliferating, but remain otherwise intact and viable. Furthermore, upon prolonged (>6 h) exposure to farnesyl acetate, cells become irreversibly blocked from progressing from G1 to S, an observation that distinguishes this metabolic inhibitor from the actions of statins, which, upon removal, allow completion of the remainder of the cell cycle and re-establishment of growth. Farnesyl acetate was also found to inhibit protein prenylation to an extent comparable to the BZA type of

CAAX peptidomimetic. The conclusion expressed by these authors is that farnesyl acetate "acts differently" from the statins as an antiproliferative agent, and they suggest that whatever its mode of action, it is inconsistent with an interference with standard MVA metabolism.

Discrepancies on the effects of other FTI analogs with cultured cell systems are also evident. The Merck peptidomimetic inhibitor (compound B, Figure 6) was tested on a panel of human cell lines derived from a variety of malignancies. Anchorage-dependent and-independent growth of more than 70% of all tumor cell lines tested was inhibited, despite the presence of other genetic abnormalities involving positive and negative regulators of growth (Sepp-Lorenzino *et al.*, 1995). Interestingly, no clear correlation was observed between the presence of activating *ras* mutations and sensitivity towards the farnesylation inhibitor. In fact, several tumor cell lines possessing wild type *ras* were nevertheless exquisitely sensitive to the FTI analog in terms of inhibition of proliferation, the capacity for soft agar growth, and in preventing activation of Ras-mediated MAP kinases following growth factor challenge. Among these cell lines were those harboring activated and/or overexpressed tyrosine kinases, a commonly observed characteristic present in many breast, prostate, and colon carcinomas (Rosen, 1995). Because Ras function is crucial to the transduction of signals originating through activated tyrosine kinases (Medema and Bos, 1993), it was not completely unexpected to find that tumor cell lines with inherently activated receptor tyrosine kinase were sensitive to the FTI, inasmuch as Ras function would be tonically stimulated and continuously required for the maintenance of transformation. More difficult to explain from the results with the Merck analog is the fact that several cell lines possessing activating *ras* mutations manifested resistance to the pro-drug (Sepp-Lorenzino *et al.*, 1995). However, FPTase activity from resistant cell lines was not refractory to *in vitro* inhibition by the Merck FTI. Differential drug uptake rates or metabolism by the resistant versus sensitive cell lines were dismissed as potential causes for the selective lack of sensitivity to the FTI. Of course, it should not be overlooked that the presence of a *ras* mutation does not preclude the existence of other genetic lesions that could account for the resistant phenotype. Rather, it might be surmised that a cell line would exhibit resistance if it were able to engage "alternative" signal transduction pathways that either circumvent or operate independently of the Ras-mediated route (see Figure 4). It was shown that even though the FTI analog blocked growth-factor-stimulated Ras activation in both sensitive as well as resistant cells, the activation of MAP kinases was unaffected in some resistant cell lines following treatment with the FTI pro-drug (Sepp-Lorenzino *et al.*, 1995). These observations appear to resemble the data obtained with engineered cell lines for which growth in soft agar, or proliferation of v-*ras*-transformed fibroblasts propagated in nude mice, were both slowed by one or another synthetic FPTase inhibitor, but for which the tumorigenicity of downstream (or Ras-independent) elements derived from v-*raf* and v-*mos* transformed cells was unaffected (James *et al.*, 1993; Cox *et al.*, 1994; James *et al.*, 1994; Kohl *et al.,* 1994). In addition to the complexities alluded to above, and because of the important role in cell proliferation assumed by farnesylated proteins such as Ras and the nuclear lamins, it also remains puzzling that normal tissues are relatively unaffected by treatment with FTI analogs.

Is it Ras (plus-or-minus the nuclear lamins), or some other as-yet unknown prenylated protein(s), that behaves as the critical target(s) principally responsible for the profound inhibition of proliferation effected by these FTI compounds? As indicated above, most screenings of potential antitumor pro-drugs believed to be capable of interfering either with some aspect of cholesterogenically linked isoprenoid synthesis or, in the case of the FTI analogs, with protein farnesylation, evaluated their effectiveness with the use of specifically engineered v-*ras*-transformed cell lines. The logic of such experiments was to seek data indicating that in the presence of the added Ras-modification inhibitor, transformation by v-*ras* would become blocked, but that no effect would be observed if cell transformation was due, instead, to the presence of oncogenic proteins that are independent of farnesylation for activity, such as v-*raf* or v-*mos,* or even a myristoylated v-*ras* (James *et al.,* 1993; Kohl *et al.,* 1993; Cox *et al.,* 1994). However, analysis of the morphological reversion by FTI of v-*ras*-transformed fibroblasts as a function of time revealed a lack of correlation between the effect of the FTI on Ras processing and the effects on cellular morphology (Prendergast *et al.,* 1994). For example, more than 7 days were required for v-*ras*-transformed Rat-1 cells to regain morphological transformation upon removal of the inhibitor, whereas Ras was completely localized to the membrane within 48 h. Moreover, as shown in the Sepp-Lorenzino *et al.* study (1995), a survey of human tumor cell lines failed to demonstrate a correlation between *ras* mutation and sensitivity to inhibition by the FTI employed. All these problems reinforce the warning that when Ras-effected signaling is considered, the specificity of Ras interactions with both downstream as well as upstream effectors must be known and taken into account in order to develop a coherent understanding of the phenomena (Wittinghofer and Herrmann, 1995).

Several possibilities can be raised as hypotheses to help explain the various experimental discrepencies mentioned, but none can be fully verified at present. For example, Trueblood *et al.* (1993) reported that FPTase knockout mutants in yeast nevertheless maintained their viability because Ras geranylgeranylation was able to substitute for Ras farnesylation. Since GGPTase-I is over 1000-fold more resistant than FPTase to inhibition by many of the FTI shown in Figure 6, one might suggest that geranylgeranylation is relatively unaffected by the inhibitor, permitting potential target proteins to become biologically activated as a result of an alternative posttranslational prenylation. In this regard, it is recognized that Ki-RasB, but not Ha-Ras, behaves as a substrate for GGPTase-I *in vitro* (Moores *et al.,* 1991; Yokoyama *et al.,* 1991; Armstrong *et al.,* 1995), and this might account for sustained Ras function in the absence of farnesylation. However, this alternative pathway for processing does not adequately take into consideration other normally farnesylated proteins, such as the nuclear lamins. In recent FTI experiments (Sepp-Lorenzino *et al.*, 1995) the pattern of [^{3}H]MVA incorporation, as well as data from posttranslational protein processing assays, indicate that the lamins remain *un*prenylated following treatment with the farnesylation inhibitor, so the question remains unsolved at present. It must be admitted, of course, that the resistance of normal cells may only be an apparent one, and in reality may reflect a drug dosage effect. Oncogenic *ras*-transformed tumors are heavily dependent on Ras modification for their proliferative phenotype, and the tumorigenic phenotype may thus respond to a lower dose of FTI than would normal cells. In fact, Kohl *et al.* (1995) alluded to the observation that if the effective dosage (40 mg/kg/day) of the Merck FTI was increased nearly three-fold (~110 mg/kg/day), some toxicity to cells and tissues did indeed occur.

As described above, the increasing volume of studies with FTI analogs has demonstrated, in part, the ability of these pro-drugs to disrupt MAP kinase-mediated signaling in *ras*-transformed cells (Cox *et al.*, 1994; James *et al.*, 1994), but not in non-transformed cells—e.g., Rat1 cells, which possess the wild-type *ras* gene (James *et al.*, 1994; Sepp-Lorenzino *et al.*, 1995)—and these observations are difficult to reconcile with current knowledge of Ras-mediated signaling.

Nevertheless, the possibility that in normal cells GGPTase-I might display cross-prenylation activity and the capacity to farnesylate normal Ras, as was shown with RhoB (Moores *et al.*, 1991; Yokoyama *et al.*, 1991; Armstrong *et al.*, 1995) was mentioned above as a possible means to explain the apparent resistance of normal cells to FTI analog growth inhibition relative to *ras*-transformed cells. Unexpected experimental corroboration of this hypothesis to justify the apparent resistance phenomenon in normal cells came from the recent discovery of a new *ras* gene, TC21 (Chan *et al.*, 1994), the existence of which therefore increases the N-, H-and K-Ras transforming-potential protein family by one additional member. The TC21 *ras* gene is ubquitously present in all human tissues, and in support of the cross-prenylation hypothesis, the TC21 protein proved to be an excellent substrate *in vitro* for both purified FPTase and GGPTase-I (Carboni *et al.*, 1995). Most significantly, however, the Bristol-Myers Squibb bisubstrate FTI analog (compound E, Figure 6; see section 6.3), which was capable of totally blocking H-Ras farnesylation, did not affect the processing of TC21. Furthermore, the bisubstrate analog proved incapable of reverting the phenotypic characteristics of NIH 3T3 cells transformed by TC21 (inhibition of anchorage-dependent and-independent growth, reversal of transformed morphology, etc.), although it did so with cells transformed by H-*ras* (Carboni *et al.*, 1995). Consequently, signal transduction via Ras is not identical to that via TC21, despite both G-proteins belonging to the same general Ras superfamily. Thus, TC21 may indeed compensate for Ras function in normal cells, permitting them to circumvent the effects exhibited by Ras-specific FTI pro-drugs.

The suspicion that cross-prenylation may occur under certain circumstances *in vivo* (e.g. overexpression of different types of Ras/Ras-like proteins) accumulates support upon taking into consideration that: (1) K-RasB and TC21/R-Ras2 tumors, but not those dependent on the more commonly studied H-Ras oncoprotein (Leftheris *et al.*, 1996), prove resistant to FTI peptidomimetics (Carboni *et al.*, 1995; James *et al.*, 1995; Casey and Seabra, 1996; James *et al.*, 1996); (2) K-RasB tumors comprise the most common type of Ras-dependent human neoplasms; and (3) K-RasB and TC21/R-Ras2 are substrates for GGPTase-I and thus capable of an alternative prenylation should the farnesylating activity of FPTase become unavailable (Trueblood *et al.*, 1993; Armstrong *et al.*, 1995; Carboni *et al.*, 1995; Casey and Seabra, 1996). It must nevertheless be emphasized that, except for the results of one report (Lerner *et al.*, 1996), peptidomimetic inhibitors that can exhibit restricted specificity for GGPTase-I over FPTase are not available. Indeed, even the criteria for FPTase substrate specificity, based on Met, Ser, or Ala as the C-terminus ("X") residue of the CAAX motif (Moores *et al.*, 1991; Reiss *et al.*, 1991b; Casey and Seabra, 1996; Zhang and Casey, 1996) recently were reported to exhibit a plasticity when cadmium rather than zinc was substituted as the required metal ion. With cadmium, leucine-terminated CAAX peptides become farnesylated by FPTase instead of behaving, as might have been expected, as substrates specific for GGPTase-I (Zhang *et al.*, 1996).

Despite such attempts to rationalize anomalous findings, it is quite obvious that the mechanism of action of the various FTI analogs (Figure 6) and their intracellular signal transduction target(s) involves a much more complex picture than was previously (naively, perhaps) thought to exist. Nevertheless, the data so far accumulated are certainly encouraging, inasmuch as drugs of this CAAX peptidomimetic class might represent an exploitable window of opportunity into an important new therapeutic modality, at least on the basis of their specificity toward Ras farnesylation inhibition, their absence of cytotoxicity, and the potentially high therapeutic index observed to date.

7. SUMMARY AND PROSPECTUS

Research literature for this chapter was reviewed until the last quarter of 1995, with a few pertinent references from 1996 added in proof. Inevitably, some relevant studies might have been inadvertantly overlooked, for which the authors apologize. The compiled references are meant to provide the reader with at least some perspective of the complex landscape that encompassess the topic of cholesterol and its role in tumorigenesis, which casts a broad net across multiple subdisciplines. The emerging picture of where things stand as of this writing prompts us to say, "*Se non e vero, e ben trovato,*"* regarding the existence of biologically meaningful connections drawn between cholesterol as the molecule itself, the branched MVA metabolic pathway leading to cholesterol synthesis, and the

*"If it's not true, it might as well be."

uncontrolled cell proliferation characteristic of cancer. Our purpose here has been to extend the message of previous reviews on this topic by emphasizing the abundance of strong evidence in support of these correlations at the cell and molecular level.

First, for more than two decades, data have accumulated showing that proliferating cells manifest a higher flux through the MVA pathway than when they are in cell cycle arrest, and that frank tumors exhibit deregulated cholesterogenesis at the principal rate controlling locus of the pathway—the reaction catalyzed by HMGR.

Then, there is the tantalizing evidence that in order to fuel the enhanced flux through the MVA pathway of proliferating tumor cells, increasing amounts of substrate, in the form of acetyl units, become accessible to the pathway by means of metabolic adjustments in mitochondrial metabolism—perhaps through a membrane cholesterol level-based "feedback" mechanism that encourages the export of mitochondrial citrate, which, in turn, reveals a truncated TCA cycle phenotype in tumors.

Finally, it has also been ascertained that the MVA segment of cholesterol biosynthesis participates intimately in helping to control cell proliferation by generating specific prenyl intermediates as part of one branch off the central metabolic route of cholesterogenesis. These specific isoprenoids, the farnesyl and geranylgeranyl moieties, serve to covalently modify, and thereby enable the biological activity of, signal-transducing proteins like those of the Ras superfamily.

Prenylated Ras-mediated signal transduction pathways provide much of the molecular information needed to trigger cell proliferation. Since protein prenylation is centrally significant, prenylprotein transferase enzymes (e.g., FPTase) represent inviting targets for the design of chemotherapeutic drugs that would interrupt (non-toxically) such signaling events and stop tumor cell proliferation. Results have just begun to appear from studies *in vitro* and *in vivo* (at both the cell culture and whole animal level) employing the first generation of such anti-proliferative FTI pro-drugs.

Yet, there remain serious gaps in our understanding of fundamentals that cloud the molecular details of the cholesterol–cancer connection. Perhaps foremost among several pressing issues is the basic question of what governs the regulation of cholesterogenesis *per se,* since it is now known that the entire multi-step pathway may occur not in one compartment (as previously thought) but in multiple intracellular loci (cytosolic/microsomal, mitochondrial, and peroxisomal), each of which may exert different modes of control.

From the standpoint of the cholesterogenesis/cell proliferation linkage, it is important to stress the universal observation that a direct inhibition of HMGR by competitive antimetabolites (like the statins) always halts cell proliferation in late G1. This observation is to be cautiously distinguished from the effects of compounds (such as d-limonene, tocotrienols, etc.) that inhibit cell proliferation by blocking the MVA pathway through a lowering of the HMGR enzyme protein level itself (i.e., suppression of its synthesis at the transcription or translation level and/or stimulation of its proteolytic degradation). Whereas MVA addition overcomes statin inhibition of both cholesterogenesis (as befits an inhibitor acting directly on the HMGR enzyme molecule itself) and progression through the cell division cycle, MVA does not seem as readily able to reverse the inhibition caused by various other natural products (d-limonene and its metabolites, gliotoxin, manumycin) that interfere with cell proliferation and/or protein prenylation. On the other hand, for some of the more recently screened antiproliferative compounds, this question is still open to further study.

Clearly, the most prominent leitmotif for the past 20+ years has been the requirement for operation of the branched MVA metabolic pathway, which produces intermediate(s) necessary for cell proliferation. As intricacies of MVA metabolism have become more apparent with time, it remains possible that all MVA pathway intermediates (and even *their* derivatives) generated by mammalian cells may not yet have been identified. We also are unable to assure ourselves at the present time about the completeness of our knowledge on the total biological function of those intermediates that we already know about. But, together with the rapidly evolving landscape of complex signal transducing protein–protein associations that direct the tempo of cell proliferation versus differentiation, future work on the various roles assumed by isoprenoid metabolism will inspire a nexus of research areas that will guarantee a more complete understanding about what makes cells divide.

ACKNOWLEDGMENTS. In assembling this review and being involved in a small portion of the work it addresses, we owe much to those dear to us who constantly nourish our humanness—especially Jane and Dan, and Augusto and Manuel.

8. REFERENCES

Adamson, P., Marshall, C. J., Hall, A., and Tilbrook, P. A., 1992, Post-translational modifications of p21 rho proteins, *J. Biol. Chem.* **267**:20033–20038.

Agarwal, S., Corbley, M. J., and Roberts, T. M., 1995, Reconstitution of signal transduction from the membrane to the nucleus in a baculovirus expression system: activation of Raf-1 leads to hypermodificatio of c-jun and c-fos via multiple pathways, *Oncogene* **11**:427–438.

Akopyan, T. N., Couedel, Y., Orlowski, M., Fournie-Zaluski, M. C., and Roques, B. P., 1994, Proteolytic processing of farnesylated peptides: assay and partial purification from pig brain membranes of an endopeptidase which has the characteristics of E.C. 3.4.24.15, *Biochem. Biophys. Res. Commun.* **198**:787–794.

Alberts, A., Chen, J., Kuron, G., Hunt, V., Huff, J., Hoffman, C., Rothrock, J., Lopez, M., Joshua, H., Harris, E., Patchett, A., Monaghan, R., Currie, S., Stapley, E., Albers-Schonberg, G., Hensens, O., Hirshfield, J., Hoogsteen, K., Liesch, J., and Springer, J., 1980, Mevinolin: a highly potent competitive inhibitor of 3-hydroxy-3-methylglutaryl coenzyme A reductase and a cholesterol-lowering agent, *Proc. Natl. Acad. Sci. USA* **77**:3957–3961.

Appelkvist, E. L., Reinhart, M., Fischer, R., Billheimer, J., and Dallner, G., 1990, Presence of individual enzymes of cholesterol biosynthesis in rat liver peroxisomes, *Arch. Biochem. Biophys.* **282**: 318–325.

Appelkvist, E. L., Grünler, J., and Dallner, G., 1995, Isoprenoid biosynthesis in peroxisomes, *FASEB J.* **9**:A1312.

Armstrong, S. A., Hannah, V. C., Goldstein, J. L., and Brown, M. S., 1995, CAAX geranylgeranyl transferase transfers farnesyl as efficiently as geranylgeranyl to RhoB, *J. Biol. Chem.* **270:**7864–7868.

Aronheim, A., Engelberg, D., Li, N., Al-Alawi, N., Schlessinger, J., and Karin, M., 1994, Membrane targeting of the nucleotide exchange factor Sos is sufficient for activating the Ras signaling pathway, *Cell* **78:**949–961.

Azarnoff, D. L., Tucker, D. R., and Barr, G. A., 1965, Studies with ethyl chlorophenoxyisobutyrate (clofibrate), *Metabolism* **14:**959–965.

Azrolan, N. I., and Coleman, P. S., 1989, A discoordinate increase in the cellular amount of 3-hydroxy-3-methylglutaryl-CoA reductase results in the loss of rate-limiting control over cholesterogenesis in a tumor cell-free system, *Biochem. J.* **258:**421–425.

Barbacid, M., 1987, *ras* genes, *Annu. Rev. Biochem.* **56:**779–827.

Biardi, L., Sreedhar, A., Zokaei, A., Vartak, N. B., Bozeat, R. L., Shackelford, J. E., Keller, G. A., and Krisans, S. K., 1994, Mevalonate kinase is predominantly localized in peroxisomes and is defective in patients with peroxisome deficiency disorders, *J. Biol. Chem.* **269:**1197–1205.

Bilger, B., 1995, Forever young, *The Sciences* **35:**26–30.

Bishop, J. M., 1991, Molecular themes in oncogenesis, *Cell* **64:**235–248.

Black, S. D., 1992, Development of hydrophobicity parameters for prenylated proteins, *Biochem. Biophys. Res. Commun.* **186:**1437–1442.

Blenis, J., 1993, Signal transduction via MAP kinases: proceed at your own RSK, *Proc. Natl. Acad. Sci. USA* **90:**5889–5892.

Blouin, A., Bolender, R. P., and Weibel, E. R., 1977, Distribution of organelles and membranes between hepatocytes and nonhepatocytes in the rat liver parenchyma, *J. Cell Biol.* **72:** 441–455.

Boctor, A. M., and Bystryn, J.-C., 1982, Degradation of tumor-associated antigens shed by human melanoma cells in culture, *Cancer Res.* **42:**2121–2125.

Boguski, M. S., and McCormick, F., 1993, Proteins regulating Ras and its relatives, *Nature* **366:** 643–654.

Bradfute, D. L., and Simoni, R. D., 1994, Non-sterol compounds that regulate cholesterogenesis: analogues of farnesyl pyrophosphate reduce 3-hydroxy-3-methylglutaryl-coenzyme A reductase levels, *J. Biol. Chem.* **269:**6645–6650.

Braun, P. E., De Angelis, D., Shtybel, W. W., and Bernier, L., 1991, Isoprenoid modification permits 2′,3′-cyclic nucleotide 3′-phosphodiesterase to bind to membranes, *J. Neurosci. Res.* **30:**540–544.

Brenneman, D. E., Mathur, S. N., and Spector, A. A., 1975, Characterization of the hyperlipidemia in mice bearing the Ehrlich ascites tumor, *Eur. J. Cancer* **11:**225–230.

Brown, M. S., and Goldstein, J. L., 1980, Multivalent feedback regulation of HMG CoA reductase, a control mechanism coordinating isoprenoid synthesis and cell growth, *J. Lipid Res.* **21:**505–517.

Brown, M. S., Faust, J. R., Goldstein, J. L., Kaneko, I., and Endo, A., 1978, Induction of 3-hydroxy-3-methylglutaryl coenzyme A reductase activity in human fibroblasts incubated with compactin (ML-236B), a competitive inhibitor of the reductase, *J. Biol. Chem.* **253:**1121–1128.

Brown, M. S., Goldstein, J. L., Paris, K. J., Burnier, J. P., and Masters, J. C., Jr., 1992, Tetrapeptide inhibitors of protein:farnesyltransferase: amino-terminal substitution in phenylalanine-containing tetrapeptides restores farnesylation, *Proc. Natl. Acad. Sci. USA* **89:**8313–8316.

Bruenger, E., and Rilling, H. C., 1986, Prenylated proteins from kidney, *Biochem. Biophys. Res. Commun.* **139:**209–214.

Bruscalupi, G., Castellano, F., Sacco, S., and Trentalance, A., 1994, Prenylated proteins in regenerating rat liver, *Biochem. Biophys. Res. Commun.* **200:**1713–1720.

Bryla, J., 1981, Inhibitors of mitochondrial anion transport, *Inhibitors of Mitochondrial Function* (M. Erecinska and D. F. Wilson, eds.), pp. 211–259, Pergamon Press, New York.

Buchwald, H., 1992, Cholesterol inhibition, cancer, and chemotherapy, *Lancet* **339:**1154–1156.

Bukhtiyarov, Y. E., Omer, C. A., and Allen, C. M., 1995, Photoreactive analogues of prenyl diphosphates as inhibitors and probes of human protein farnesyltransferase and geranylgeranyltransferase type I, *J. Biol. Chem.* **270:**19035–19040.

Call, T. G., Stenson, M. J., and Witzig, T. E., 1994, Effects of phenylacetate on cells from patients with B-chronic lymphocytic leukemia, *Leukemia Lymphoma* **14:**145–149.

Carboni, J. M., Yan, N., Cox, A. D., Bustelo, X., Graham, S. M., Lynch, M. J., Weinmann, R., Seizinger, B. R., Der, C. J., Barbacid, M., and Manne, V., 1995, Farnesyltransferase inhibitors are inhibitors of Ras but not R-Ras2/TC21 transformation, *Oncogene* **10:**1905–1913.

Casey, P. J., 1995, Protein lipidation in cell signaling, *Science* **268:**221–225.

Casey, P. J., and Seabra, M. C., 1996, Protein Prenyltransferases, *J. Biol. Chem.* **271:** 5289–5292.

Castillo, M., Burgos, C., Rodriques-Vico, R., Zafra, M. F., and Garcia-Peregrin, E., 1990, Effects of clofibrate on the main regulatory enzymes of cholesterogenesis, *Life Sci.* **46:**397–403.

Castillo, M., Iglesias, J., Zafra, M. F., and Garcia-Peregrin, E., 1991a, Inhibition of chick brain cholesterogenic enzymes by phenyl and phenolic derivatives of phenylalanine, *Neurochem. Int.* **18:** 171–174.

Castillo, M., Martinez-Cayuela, M., Zafra, M. F., and Garcia-Peregrin, E., 1991b, Effect of phenylalanine derivatives on the main regulatory enzymes of hepatic cholesterogenesis, *Mol. Cell. Biochem.* **105:**21–25.

Chan, A. M.-L., Miki, T., Meyers, K., and Aaronson, S. A., 1994, A human oncogene of the RAS superfamily unmasked by expression cDNA cloning, *Proc. Natl. Acad. Sci. USA* **91:**7558–7562.

Chazotte, B., and Hackenbrock, C. R., 1988, The multicollisional, obstructed, long-range diffusional nature of mitochondrial electron transport, *J. Biol. Chem.* **263:**14359–14367.

Chazotte, B., and Hackenbrock, C. R., 1991, Modulation and analysis of structure and function of the inner mitochondrial membrane through the application of phospholipid enrichment and membrane fusion techniques, *Membrane Fusion* (J. Wilschut and D. Hoekstra, eds.), pp. 803-844, Marcel Dekker, New York.

Chen, H., Born, E., Mathur, S. N., and Field, F. J., 1993, Cholesterol and sphingomyelin syntheses are regulated independently in cultured intestinal cells, CaCo-2: role of membrane cholesterol and sphingomyelin content, *J. Lipid Res.* **34:**2159–2167.

Chen, H. W., 1984, Role of cholesterol metabolism in cell growth, *Fed. Proc.* **43:**126–130.

Chen, H. W., Heiniger, H.-J. and Kandutsch, A. A., 1975, Relationship between sterol synthesis and DNA synthesis in phytohemagglutinin-stimulated mouse lymphocytes, *Proc. Natl. Acad. Sci. USA* **72:** 1950–1954.

Chen, H. W., Kandutsch, A. A., and Heiniger, H.-J., 1978, The role of cholesterol in malignancy, *Prog. Exp. Tumor Res.* (D. F. H. Wallach, ed.), pp. 275–316, S. Karger, Basel.

Chen, W.-J., Andres, D., Goldstein, J. L., Russell, D., and Brown, M., 1991, cDNA cloning and expression of the peptide-binding β subunit of rat p21ras farnesyltransferase, the counterpart of yeast DPR1/RAM1, *Cell* **66:**327–334.

Chen, W.-J., Moomaw, J. F., Overton, L., Kost, T. A., and Casey, P., 1993, High level expression of mammalian protein farnesyltransferase in a baculovirus system, *J. Biol. Chem.* **268:** 9675–9680.

Chin, C. J., Luskey, K. L., Anderson, R. G. W., Faust, J. R., Goldstein, J. L., and Brown, M. S., 1982, Appearance of crystalloid endoplasmic reticulum in compactin-resistant Chinese hamster cells with a 500-fold increase in 3-hydroxy-3-methylglutaryl-coenzyme A reductase, *Proc. Natl. Acad. Sci. USA* **79:**1185–1189.

Cinatl, J., Cinatl, J., Mainke, M., Weissflog, A., Rabenau, H., Kornhuber, B., and Doerr, H. W., 1993, In vitro differentiation of human neuroblastoma cells by sodium phenylacetate, *Cancer Lett.* **70:**15–24.

Cinatl, J., Cinatl, J., Herneiz, P., Rabenau, H., Novak, M., Benda, R., Gümbel, H., Kornhuber, B., and Doerr, H. W., 1994, Induction of myogenic differentiation in a human rhabdomyosarcoma cell line by phenylacetate, *Cancer Lett.* **78:**41–48.

Clarke, S., 1992, Protein isoprenylation and methylation at carboxy-terminal cysteine residues, *Annu. Rev. Biochem.* **61:**355–386.

Clarke, S., Vogel, J., Deschenes, R., and Stock, J., 1988, Post-translational modification of the H-ras oncogene protein: evidence for a third class of protein carboxyl methyltransferase, *Proc. Natl. Acad. Sci. USA* **85:**4643–4647.

Clegg, R. J., Middleton, B., Bell, G. D., and White, D. A., 1980, Inhibition of hepatic cholesterol synthesis and 3-hyroxy-3-methylglutaryl coenzyme A reductase by mono and bicyclic monoterpenes administered in vivo, *Biochem. Pharmacol.* **29:**2125–2127.

Clegg, R. J., Middleton, B., Bell, G. D., and White, D. A., 1982, The mechanism of cyclic monoterpene inhibition of 3-hydroxy-3-methylglutaryl coenzyme A reductase in vivo in the rat, *J. Biol. Chem.* **257:**2294–2299.

Cohen, B. I., Raicht, R. F., Shefer, S., and Mosbach, E. H., 1974, Effects of clofibrate on sterol metabolism in the rat, *Biochim. Biophys. Acta* **369:**79–85.

Cohen, L. H., Valentijn, A. R. P. M., Roodenburg, L., van Leeuwen, R. E. W., Huisman, R. H., Lutz, R. J., van der Marel, G. A., and van Boom, J. H., 1995, Different analogs of farnesyl pyrophosphate inhibit squalene synthase and protein:farnesyltransferase to different extents, *Biochem. Pharmacol.* **49:**839–845.

Coleman, P. S., 1986a, Membrane cholesterol and tumor bioenergetics, *Ann. NY Acad. Sci.* **488:** 451–467.

Coleman, P. S., 1986b, The mitochondrial citrate transport carrier and the role of deregulated cholesterogenesis in tumor cell proliferation, *EBEC Reports* **4:**77.

Coleman, P. S., 1991, Deterministic coupling between cellular bioenergetics, cholesterol synthesis, cell proliferation and cancer, *Chemical Carcinogenesis 2: Modulating Factors* (A. Columbano, F. Feo, R. Pascale, and P. Pani, eds.), pp. 265–288, Plenum Press, New York.

Coleman, P. S., and Lavietes, B. B., 1981, Membrane cholesterol, tumorigenesis, and the biochemical phenotype of neoplasia, *CRC Crit. Rev. Biochem.* **11:**341–393.

Coleman, P. S., and Sepp-Lorenzino, L., 1990, The role of the cholesterol synthesis pathway during tumor cell proliferation, *Adv. Cholesterol Res.* (M. Esfahani and J. Swaney, eds.), pp. 201–270, Telford Press, NJ.

Coleman, P. S., Lavietes, B., Born, R., and Weg, A., 1978, Cholesterol enrichment of normal mitochondria in vitro: a model system with properties of hepatoma mitochondria, *Biochem. Biophys. Res. Commun.* **84:**202–207.

Corsini, A., Verri, D., Raiteri, M., Quarato, M., Paoletti, R., and Fumagalli, R., 1995, Effects of 26-aminocholesterol, 27-hydroxycholesterol and 25-hydroxycholesterol on proliferation and cholesterol homeostasis in arterial monocytes, *Arterio. Throm. Vasc. Biol.* **15:**420–428.

Cox, A. D., and Der, C. J., 1992, The ras/cholesterol connection: implications for ras oncogenicity, *Crit. Rev. Oncogen.* **3:**365–400.

Cox, D. C., Comai, K., and Goldstein, A. L., 1988, Effects of cholesterol and 25-hydroxycholesterol on smooth muscle cell and endothelial cell growth, *Lipids* **23:**85–88.

Cox, A. D., Garcia, A. M., Westwick, J. K., Kowalczyk, J. J., Lewis, M. D., Brenner, D. A., and Der, C. J., 1994, The CAAX peptidomimetic compound B581 specifically blocks farnesylated, but not geranylgeranylated or myristoylated, oncogenic ras signaling and transformation, *J. Biol. Chem.* **269:**19203–19206.

Crowell, P. L., and Gould, M. N., 1994, Chemoprevention and therapy of cancer by d-limonene, *Crit. Rev. Oncogenesis* **5:**1–22.

Crowell, P. L., Chang, R., Ren, Z., Elson, C., and Gould, M., 1991, Selective inhibition of isoprenylation of 21-26-kDa proteins by the anticarcinogen d-limonene and its metabolites, *J. Biol. Chem.* **266:**17679–17685.

Crowell, P. L., Lin, S., Vedejs, E., and Gould, M. N., 1992, Identification of circulating metabolites of the antitumor agent d-limonene capable of inhibiting protein isoprenylation and cell growth, *Cancer Chemoth. Pharmacol.* **31:**205–212.

Crowell, P. L., Ren, Z., Lin, S., Vedejs, E., and Gould, M. N., 1994, Structure-activity relationships among monoterpene inhibitors of protein isoprenylation and cell proliferation, *Biochem. Pharmacol.* **47:**1405–1415.

Cuthbert, J. A., and Lipsky, P. E., 1991, Negative regulation of cell proliferation by mevalonate or one of the mevalonate phosphates, *J. Biol. Chem.* **266:**17966–17971.

Cuthbert, J. A., and Lipsky, P. E., 1995, Suppression of the proliferation of ras-transformed cells by fluoromevalonate, an inhibitor of mevalonate metabolism, *Cancer Res.* **55:**1732–1740.

Dalton, M. B., Fantle, K. S., Bechtold, H. A., DeMaio, L., Evans, R. M., Krystosek, A., and Sinensky, M., 1995, The farnesyl protein transferase inhibitor BZA-5B blocks farnesylation of nuclear lamins and p21ras but does not affect their function or localization, *Cancer Res.* **55:** 3295–3304.

Das, N. P., and Allen, C. M., 1991, Inhibition of farnesyl transferases from malignant and non-malignant cultured human lymphocytes by prenyl substrate analogues, *Biochem. Biophys. Res. Commun.* **181:**729–735.

Daum, G., 1985, Lipids of mitochondria, *Biochim. Biophys. Acta* **822:**1–42.

Davis, R. J., 1993, The mitogen-activated protein kinase signal transduction pathway, *J. Biol. Chem.* **268:**14553–14556.

DeClue, J. E., Vass, W. C., Papageorge, A. G., Lowy, D. R., and Willumsen, B. M., 1991, Inhibition of cell growth by lovastatin is independent of *ras* function, *Cancer Res.* **51:**712–717.

Dessi, S., Batetta, B., Anchisi, C., Pani, P., and Costelli, P., 1992, Cholesterol metabolism during the growth of a rat ascites hepatoma (Yoshida AH-130), *Br. J. Cancer* **66:**787–793.

Dietschy, J. M., and Siperstein, M. D., 1967, Effect of cholesterol feeding and fasting on sterol synthesis in seventeen tissues of the rat, *J. Lipid Res.* **8:**97–104.

Downward, J., 1990, The ras superfamily of small GTP-binding proteins, *Trends Biochem. Sci.* **15:** 469–472.

Doyle, J., Kabakoff, B., and Kandutsch, A. A., 1991, Dolichyl phosphate as a regulator of cell growth, *Chemical Carcinogenesis-2: Modulating Factors* (A. Columbano, F. Feo, R. Pascale, and P. Pani, eds.), pp. 289–297, Plenum Press, New York.

Echegoyen, S., Oliva, E. B., Sepulveda, J., Diaz-Zagoya, J. C., Espinosa-Garcia, M. T., Pardo, J. P., and Martinez, F., 1993, Cholesterol increase in mitochondria: its effect on inner-membrane functions, submitochondrial localization and ultrastructure morphology, *Biochem. J.* **289:** 703–708.

Edwards, P. A., Lan, S.-F., and Fogelman, A. M., 1984, High density lipoproteins and lecithin dispersions increase the activity of 3-hydroxy-3-methylglutaryl coenzyme A reductase by increasing the rate of synthesis and decreasing the rate of degradation of the enzyme, *J. Biol. Chem.* **259:** 8190–8194.

Elson, C. E., and Yu, S. G., 1994, The chemoprevention of cancer by mevalonate-derived constituents of fruits and vegetables, *J. Nutrit.* **124:**607–614.

Endo, A., 1992, The discovery and development of HMG-CoA reductase inhibitors, *J. Lipid Res.* **33:**1569–1582.

Endo, A., Kuroda, M., and Tanazawa, K., 1976, Competitive inhibition of 3-hydroxy-3-methylglutaryl CoA reductase by ML-236A and ML-236B, fungal metabolites having hypocholesterolemic activity, *FEBS Lett.* **72:**323–325.

Ericsson, J., Appelkvist, E. L., Runquist, M., and Dallner, G., 1993, Biosynthesis of dolichol and cholesterol in rat liver peroxisomes, *Biochimie* **75:**167–173.

Fairbanks, K. P., Witte, L. D., and Goodman, D. S., 1984, Relationship between mevalonate and mitogenesis in human fibroblasts stimulated with platelet-derived growth factor, *J. Biol. Chem.* **259:**1546–1551.

Fairbanks, K. P., Barbu, V. D., Witte, L. D., Weinstein, I. B., and Goodman, D. S., 1986, Effects of mevinolin and mevalonate on cell growth in several transformed cell lines, *J. Cell. Physiol.* **127:**216–222.

Farias, R. N., Bloj, B., Morero, R. D., Sineriz, F., and Trucco, R. E., 1975, Regulation of allosteric membrane-bound enzymes through changes in membrane lipid composition, *Biochim. Biophys. Acta* **415:**231–254.

Farnsworth, C. C., Gelb, M. H., and Glomset, J. A., 1990, Identification of geranylgeranyl-modified proteins in HeLa cells, *Science* **247:**320–322.

Farnsworth, C. C., Yamane, H., Gelb, M., and Glomset, J. A., 1990, A 23 kDa GTP-binding protein from bovine brain membranes is geranylgeranylated, *FASEB J.* **4:**870a.

Faust, J. R., Lusky, K. L., Chin, D. J., Goldstein, J. L., and Brown, M. S., 1982, Regulation of synthesis and degradation of 3-hydroxy-3-methylglutaryl-coenzyme A reductase by low density lipoprotein and 25-hydroxycholesterol in UT-1 cells, *Proc. Natl. Acad. Sci. USA* **79:**5205–5209.

Feo, F., Canuto, R. A., Bertone, G., Garcea, R., and Pani, P., 1973, Cholesterol and phospholipid composition of mitochondria and microsomes isolated from Morris hepatoma 5123 and rat liver, *FEBS Lett.* **33:**229–232.

Feo, F., Canuto, R.A., Garcea, R., and Gabriel, L., 1975, Effect of cholesterol content on some physical and functional properties of mitochondria isolated from adult rat liver, fetal liver, cholesterol-enriched liver, and hepatomas AH-130, 3924A and 5123, *Biochim. Biophys. Acta* **413:**116–134.

Forman, B. M., Goode, E., Chen, J., Oro, A. E., Bradley, D. J., Perlman, T., Noonan, D. J., Burke, L. T., McMorris, T., Lamph, W. W., Evans, R. M., and Weinberger, C., 1995, Identification of a nuclear receptor that is activated by farnesol metabolites, *Cell* **81:**687–693.

García, A. M., Rowell, C., Ackermann, K., Kowalczyk, J. J., and Lewis, M. D., 1993, Peptidomimetic inhibitors of Ras farnesylation and function in whole cells, *J. Biol. Chem.* **268:**18415–18418.

Gelb, M. H., Tamanoi, F., Yokoyama, K., Ghomashchi, F., Esson, K., and Gould, M. N., 1995, The inhibition of protein prenyltransferases by oxygenated metabolites of limonene and perillyl alcohol, *Cancer Lett.* **91:**169–175.

Giannakouros, T., and Magee, A. I., 1993, Protein prenylation and associated modifications, *Lipid Modification of Proteins* (M. J. Schlesinger, eds.), pp. 136–162, CRC Press, Boca Raton, Florida.

Gibbs, J. B., 1991, Ras C-terminal processing enzymes–new drug targets? *Cell* **65:**1–4.

Gibbs, J. B., 1992, Pharmacologic probes of Ras function, *Cancer Biol.* **3:**383–390.

Gibbs, J. B., Pompliano, D. L., Mosser, S. D., Rands, E., Lingham, R. B., Singh, S. B., Scolnick, E. M., Kohl, N. E., and Oliff, A., 1993, Selective inhibition of farnesyl-protein transferase blocks processing in vivo, *J. Biol. Chem.* **268:**7617–7620.

Gibbs, J. B., Oliff, A., and Kohl, N. E., 1994, Farnesyltransferase inhibitors: Ras research yields a potential cancer therapeutic, *Cell* **77:**175–178.

Glomset, J. A., Gelb, M. H., and Farnsworth, C. C., 1990, Prenyl proteins in eukaryotic cells: a new type of membrane anchor, *Trends Biochem. Sci.* **15:**139–142.

Glomset, J., Gelb, M., and Farnsworth, C., 1992, Geranylgeranylated proteins, *Biochem. Soc. Trans.* **20:**479–484.

Goldfisher, S., and Reddy, J. K., 1984, Peroxisomes (microbodies) in cell pathology, *Intl. Rev. Exp. Pathol.* **26:**45–84.

Goldstein, J. L., and Brown, M. S., 1990, Regulation of the mevalonate pathway, *Nature* **343:**425–430.

Goldstein, J. L., Brown, M. S., Stradley, S. J., Reiss, Y., and Gierasch, L. M., 1991, Nonfarnesylated tetrapeptide inhibitors of protein farnesyltransferase, *J. Biol. Chem.* **266:**15575–15578.

Goodman, L. E., Perou, C. M., Fujiama, A., and Tamanoi, F., 1988, Structure and expression of yeast DPR1, a gene essential for the processing and intracellular localization of ras proteins, *Yeast* **4:** 271–281.

Gould, M. N., Moore, C. J., Zhang, R., Wang, B., Kennan, W. S., and Haag, J. D., 1994, Limonene chemoprevention of mammary carcinoma induction following direct in situ transfer of v-Ha-ras, *Cancer Res.* **54:**3540–3543.

Graham, J. M., and Green, C., 1970, The properties of mitochondria enriched *in vitro* with cholesterol, *Eur. J. Biochem.* **12:**58–66.

Graham, S. L., deSolms, S. J., Giuliani, E. A., Kohl, N. E., Mosser, S. D., Oliff, A. I., Pompliano, D. L., Rands, E., Breslin, M. J., Deana, A. A., Garsky, V. M., Scholz, T. H., Gibbs, J. B., and Smith, R. L., 1994, Pseudopeptide inhibitors of Ras farnesyl-protein transferase, *J. Med. Chem.* **37**:725–732.

Grünler, J., Ericsson, J., and Dallner, G., 1994, Branch-point reactions in the biosynthesis of cholesterol, dolichol, ubiquinone and prenylated proteins, *Biochim. Biophys. Acta* **1212**:259–277.

Guidotti, G., 1972, Membrane proteins, *Annu. Rev. Biochem.* **41**:731–752.

Gupte, S., Wu, E.-S., Hoechli, L., Hoechli, M., Jacobson, K., Sowers, A. E., and Hackenbrock, C. R., 1984, Relationship between lateral diffusion, collisional frequency, and electron transfer of mitochondrial inner membrane oxidation-reduction components, *Proc. Natl. Acad. Sci. USA* **81**: 2606–2610.

Guthrie, N., Nesaretnam, K., Chambers, A. F., and Carroll, K. K., 1993, Inhibition of breast cancer cell growth by tocotrienols, *FASEB J.* **7**:A7.

Gutierrez, L., Magee, A., Marshall, C., and Hancock, J., 1989, Post-translational processing of p21ras is two-step and involves carboxyl-methylation and carboxy-terminal proteolysis, *EMBO J.* **8**:1093–1098.

Haag, J. D., Lindstrom, M. J., and Gould, M. N., 1992, Limonene-induced regression of mammary carcinomas, *Cancer Res.* **52**:4021–4026.

Habenicht, A. J. R., Glomset, J. A., and Ross, R., 1980, Relation of cholesterol and mevalonic acid to the cell cycle in smooth muscle and Swiss 3T3 cells stimulated to divide with platelet-derived growth factor, *J. Biol. Chem.* **255**:5134–5140.

Hackenbrock, C. R., 1981, Lateral diffusion and electron transfer in the mitochondrial inner membrane, *Trends Biochem. Sci.* **6**:151–154.

Hackenbrock, C. R., Schneider, H., Lemasters, J. J., and Hochli, M., 1980, Diffusional activity of redox components related to electron transfer in the mitochondrial energy transducing membrane, *EBEC Reports* **1**:23–24.

Haines, T. H., 1993, Water transport across biological membranes, *FEBS Lett.* **346**:115–122.

Hall, A., 1990, The cellular functions of small GTP-binding proteins, *Science* **249**:635–640.

Hall, A., 1994, A biochemical function for Ras-at last, *Science* **264**:1413–1414.

Hamilton, R. L., Wong, J. S., Guo, L. S., Krisans, S. K., and Havel, R. J., 1990, Apolipoprotein E localization in rat hepatocytes by immunogold labeling of cryo-thin sections, *J. Lipid Res.* **31**: 1589–1604.

Hancock, J. F., Magee, A. I., Childs, J. E., and Marshall, C. J., 1989, All ras proteins are polyisoprenylated but only some are palmitoylated, *Cell* **57**:1167–1177.

Hancock, J. F., Paterson, H., and Marshall, C. J., 1990, A polybasic domain or palmitoylation is required in addition to the CAAX motif to localize p21-ras to the plasma membrane, *Cell* **63**: 133–139.

Hancock, J. F., Cadwallader, K., and Marshall, C. J., 1991, Methylation and proteolysis are essential for efficient membrane binding of prenylated p21(K-ras(B)), *EMBO J.* **10**:641–646.

Hara, M., Akasaka, K., Akinaga, S., Okabe, M., Nakano, H., Gomez, R., Wood, D., Uh, M., and Tamanoi, F., 1993, Identification of Ras farnesyltransferase inhibitors by microbial screening, *Proc. Natl. Acad. Sci. USA* **90**:2281–2285.

Hartl, F.-U., Pfanner, N., Nicholson, D. W., and Neupert, W., 1989, Mitochondrial protein import, *Biochim. Biophys. Acta* **988**:1–45.

Hashimoto, F., Ishikawa, T., Hamada, S., and Hayashi, H., 1995, Effect of gemfibrozil on lipid biosynthesis from acetyl-CoA derived from peroxisomal β-oxidation, *Biochem. Pharmacol.* **49**: 1213–1221.

Hayashi, H., and Miwa, A., 1989, The role of peroxisomal fatty acyl CoA β-oxidation in bile acid synthesis, *Arch. Biochem. Biophys.* **274**:582–589.

Hiatt, R. A., and Fireman, B. H., 1986, Serum cholesterol and the incidence of cancer in a large cohort, *J. Chronic Dis.* **39**:861–870.

Hohl, R. J., and Lewis, K., 1995, Differential effects of monoterpenes and lovastatin on RAS processing, *J. Biol. Chem.* **270:**17508–17512.

Hornbeck, P., Huang, K.-P., and Paul, W., 1988, Lamin B is rapidly phosphorylated in lymphocytes after activation of protein kinase C, *Proc. Natl. Acad. Sci. USA* **85:**2279–2283.

Hudgins, W. R., Shack, S., Myers, C. E., and Samid, D., 1995, Antitumor activity of phenylacetate and derivatives: Correlation with lipophilicity and inhibition of protein prenylation, *Biochem. Pharmacol.* **50:**1273–1279.

Isserman, I., and Green, S., 1990, Activation of a member of the steroid hormone receptor superfamily by peroxisome proliferators, *Nature* **347:**645–650.

Jain, M. K., 1975, Role of cholesterol in biomembranes and related systems, *in: Curr. Topics Memb. Transport* (F. Bronner and A. Kleinzeller, eds.), pp. 1–84, Academic Press, NY.

Jain, M. K., and White, H. B., III, 1977, Long-range order in biomembranes, *Adv. Lipid Res.* **15:**1–60.

Jakobisiak, M., Bruno, S., Skierski, J., and Darzynkiewicz, Z., 1991, Cell cycle-specific effects of lovastatin, *Proc. Natl. Acad. Sci. USA* **88:**3628–3632.

James, G., Goldstein, J. L. and Brown, M. S., 1996, Resistance of K-RasBV12 proteins to farnesyltransferase inhibitors in Rat1 cells, *Proc. Natl. Acad. Sci. USA* **93:** 4454–4458.

James, G. L., Goldstein, J. L., Brown, M. S., Rawson, T. E., Somers, T. C., McDowell, R. S., Crowley, C. W., Lucas, B. K., Levinson, A. D., and Marsters, J. C. J., 1993, Benzodiazepine peptidomimetics: potent inhibitors of Ras farnesylation in animal cells, *Science* **260:**1937–1942.

James, G. L., Brown, M. S., Cobb, M. H., and Goldstein, J. L., 1994, Benzodiazepine peptidomimetic BZA-5B interrupts the MAP kinase activation pathway in H-ras transformed Rat-1 cells, but not in untransformed cells, *J. Biol. Chem.* **269:**27705–27714.

James, G. L., Goldstein, J. L., Pathak, R. K., Anderson, R. G. W., and Brown, M. S., 1994, PxF, a prenylated protein of peroxisomes, *J. Biol. Chem.* **269:**14182–14190.

James, G. L., Goldstein, J. L. and Brown, M. S., 1995, Polylysine and CVIM sequences of K-RasB dictate specificity of prenylation and confer resistence to benzodiazepine peptidomimetic in vitro *J. Biol. Chem.* **270:** 6221–6226.

Kandutsch, A. A., and Chen, H. W., 1977, Consequences of blocked sterol synthesis in culture cells: DNA synthesis and membrane composition, *J. Biol. Chem.* **252:**409–415.

Kandutsch, A. A., and Hancock, R. L., 1971, Regulation of the rate of sterol synthesis and the level of β-hydroxy-β-methyglutaryl coenzyme A reductase activity in mouse liver and hepatomas, *Cancer Res.* **31:**1396–1400.

Kandutsch, A. A., Heiniger, H.-J., and Chen, H. W., 1977, Effects of 25-hydroxycholesterol and 7-ketocholesterol inhibitors, *Biochim. Biophys. Acta* **486:**260–272.

Kaplan, R. S., 1996, Mitochondrial transport processes, *in: Molecular Biology of Membrane Transport Disorders* (S. G. Schultz, T. Andreoli, A. Brown *et al*, eds.), pp. 277–302, Plenum Press, NY.

Kaplan, R. S., Mayor, J. A., Johnston, N., and Oliveira, D. L., 1990, Purification and characterization of the reconstitutively active tricarboxylate transporter from rat liver, *J. Biol. Chem.* **265:** 13379–13385.

Kaplan, R. S., and Mayor, J. A., 1993, Structure, function and regulation of the tricarboxylate transport protein from rat liver mitochondria, *J. Bioener. Biomemb.* **25:**503–514.

Kaplan, R. S., Mayor, J. A., and Wood, D. O., 1993, The mitochondrial tricarboxylate transport protein. cDNA cloning, primary structure, and comparison with other mitochondrial transport proteins, *J. Biol. Chem.* **268:**13682–13690.

Kaplan, R. S., Parlo, R. A., and Coleman, P. S., 1986, Measurement of citrate transport in tumor mitochondria, *Meth. Enzymol.* **125:**671–691.

Kaplan, R. S., Morris, H. P., and Coleman, P. S., 1982, Kinetic characteristics of citrate influx and efflux with mitochondria from Morris hepatomas 3924A and 16, *Cancer Res.* **42:**4399–4407.

Kaschnitz, R. M., Morris, H. P., and Hatefi, Y., 1976, Oxidative phosphorylation properties of mitochondria isolated from transplanted hepatoma, *Biochim. Biophys. Acta* **449:**224–235.

Kawai, S., Yahata, N., Nishida, S., Nagai, K., and Mizushima, Y., 1995, Dehydroepiandrosterone inhibits B16 mouse melanoma cell growth by induction of differentiation, *Anticancer Res.* **15**:427–431.

Keller, G.-A., Barton, M. C., Shapiro, D. J., and Singer, S. J., 1985, 3-Hydroxy-3-methylglutaryl coenzyme A reductase is present in peroxisomes in normal rat liver cells, *Proc. Natl. Acad. Sci. USA* **82**:770–774.

Keller, G.-A., Cable, S., El Bouhtoury, F., Heusser, S., Scotto, C., Armbruster, L., Ciolek, E., Colin, S., Schilt, J., and DauAa, M., 1993, Peroxisomes through cell differentiation and neoplasia, *Biol. Cell* **77**:77–88.

Keyomarsi, K., Sandoval, L., Band, V., and Pardee, A. B., 1991, Synchronization of tumor and normal cells from G1 to multiple cell cycles by lovastatin, *Cancer Res.* **51**:3602–3609.

Keys, A., Aravanis, C., Blackburn, H., Buzina, R., Dontas, A. S., Fidanza, F., Karvonen, M. J., Menotti, A., Nedeljkovic, S., and Punsar, S., 1985, Serum cholesterol and cancer mortality in the Seven Countries study, *Am. J. Epidemiol.* **121**:870–883.

Khosravi-Far, R., Lutz, R. J., Cox, A. D., Conroy, L., Bournes, J. R., Sinensky, M., Balch, W. E., Buss, J. E., and Der, C. J., 1991, Isoprenoid modification of rab proteins terminating in CC or CXC motifs, *Proc Natl. Acad. Sci. USA* **88**:6264–6268.

Khosravi-Far, R., Clark, G. J., Abe, K., Cox, A. D., McLain, T., Lutz, R. J., Sinensky, M., and Der, C. J., 1992, Ras (CXXX) and Rab (CC/CXC) prenylation signal sequences are unique and functionally distinct, *J. Biol. Chem.* **267**:24363–24368.

Kinsella, B., and Maltese, W., 1992, rab GTP-binding proteins with three different carboxyl-terminal cysteine motifs are modified in vivo by 20-carbon isoprenoids, *J. Biol. Chem.* **267**:3940–3945.

Knekt, P., Reunanen, A., Aromaa, A., Heliovaara, M., Hakulinen, T., and Hakama, M., 1988, Serum cholesterol and risk of cancer in a cohort of 39,000 men and women, *J. Clin. Epidemiol.* **41**:519–530.

Kohl, N., Diehl, R., Shaber, M., Rands, E., Soderman, D., He, B., Moores, S., Pompliano, D., Ferro-Novick, S., Powers, S., Thomas, K., and Gibbs, J. B., 1991, Structural homology among mammalian and *Saccharomyces cerevisiae* isoprenyl-protein transferases, *J. Biol. Chem.* **266**: 18884–18888.

Kohl, N. E., Mosser, S. D., deSolms, J., Guiliani, E. A., Pompliano, D. L., Graham, S. L., Smith, R. L., Scolnick, E. M., Oliff, A., and Gibbs, J. B., 1993, Selective inhibition of ras-dependent transformation by a farneslytransferase inhibitor, *Science* **260**:1937–1943.

Kohl, N. E., Wilson, F. R., Mosser, S. D., Giuliani, E. A., deSolmes, S. J., Conner, M. W., Anthony, M. J., Holtz, W. J., Gomez, R. P., Lee, T.-J., Smith, R. L., Graham, S. L., Hartman, G. D., Gibbs, J. B., and Oliff, A., 1994, Protein farnesyl transferase inhibitors block the growth of ras-dependent tumors in nude mice, *Proc. Natl. Acad. Sci. USA* **91**: 9141–9145.

Kohl, N. E., Omer, C. A., Conner, M. W., Anthony, N. J., Davide, J. P., deSolms, S. J., Guiliani, W. A., Gomez, R. P., Graham, S. L., Hamilton, K., Handt, L. K., Hartman, G. D., Koblan, K. S., Kral, A. M., Miller, P. J., Mosser, S. D., O'Neill, T. J., Rands, E., Schaber, M. D., Gibbs, J. B., and Oliff, A., 1995, Inhibition of farnesyltransferase induces regression of mammary and salivary carcinomas in ras transgenic mice, *Nature Medicine* **1**:792–797.

Komiyama, K., Lizuka, K., Yamaoka, M., Watanabe, H., Tsuchiya, N., and Umezawa, I., 1989, Studies on the biological activity of tocotrienols, *Chem. Pharm. Bull.* **37**:1369–1371.

Komori, A., Suganuma, M., Okabe, S., Zou, X., Tius, M. A., and Fujuki, H., 1993, Caventrol inhibits tumor promotion in CD-1 mouse skin through inhibition of tumor necrosis factor α release and protein isoprenylation, *Cancer Res.* **53**:3462–3464.

Kothapalli, R., Guthrie, N., Chambers, A. F., and Carroll, K. K., 1993, Farnesylamine: an inhibitor or farnesylation and growth of ras-transformed cells, *Lipids* **28**:969–973.

Krämer, R., 1982, Cholesterol as activator of ADP-ATP exchange in reconstituted liposomes and in mitochondria, *Biochim. Biophys. Acta* **693**:296–304.

Krisans, S. K., 1992, The role of peroxisomes in cholesterol metabolism, *Am. J. Respir. Cell Mol. Biol.* **7:**358–364.

Krisans, S. K., 1996, Cell compartmentalization of cholesterol biosynthesis, *in: Peroxisomes: Biology and Role in Toxicology and Disease* (J. K. Reddy, T. Suga, G. P. Mannaerts, S. Subramani, and P. B. Lazerow, eds.) *Annals New York Acad. Sci.* **804:** 142–164.

Krisans, S. K., Motensen, R. M., and Lazarow, P. B., 1980, Acyl CoA synthetase in rat liver peroxisomes: computer-assisted analysis of cell fractionation experiments, *J. Biol. Chem.* **255:** 9599–9607.

Krisans, S. K., Rusnak, N., Keller, G.-A., and Edwards, P. A., 1988, Localization of 3-hydroxy-3-methylglutaryl coenzyme A synthase in rat liver peroxisomes, *J. Cell Biol.* **107:**122a.

Krisans, S. K., Ericsson, J., Edwards, P. A., and Keller, G. A., 1994, Farnesyl-diphosphate synthase is localized in peroxisomes, *J. Biol. Chem.* **269:**14165–14169.

Kromhout, D., Bosschieter, E. B., Dreijver, M., and de Lezenne Coulander, C., 1988, Serum cholesterol and 25-year incidence of and mortality from myocardial infarction and cancer: the Zutphen study, *Arch. Intern. Med.* **148:**1051–1055.

Kuroda, Y., Suzuki, N., and Kataoka, T., 1993, The effect of posttranslational modifications on the interaction of Ras2 with adenylyl cyclase, *Science* **259:**683–686.

Langan, T. J., and Volpe, J. J., 1987, Cell cycle-specific requirement for mavalonate, but not for cholesterol, for DNA synthesis in glial primary cultures, *J. Neurochem.* **49:**513–521.

Langan, T. J., and Volpe, J. J., 1987, Oligodendroglial differentiation in glial primary cultures: requirement for mevalonate, *J. Neurochem.* **48:**1804–1808.

Lange, Y., 1992, Tracking cell cholesterol with cholesterol oxidase, *J. Lipid Res.* **33:**315–321.

Lange, Y., Swaisgood, M. H., Ramos, B. V., and Steck, T. L., 1989, Plasma membranes contain half the phospholipid and 90% of the cholesterol and sphingomyelin in cultured human fibroblasts, *J. Biol. Chem.* **264:**3786–3793.

LaNoue, K. F., and Schoolwerth, A. C., 1979, Metabolite transport in mitochondria, *Annu. Rev. Biochem.* **48:**871–922.

LaNoue, K. F., and Schoolwerth, A. C., 1984, Metabolite transport in mammmalian mitochondria, *in: Bioenergetics* (L. Ernster, ed.), pp. 221–268, Elsevier, NY.

Lardy, H., Partridge, B., and Wei, Y., 1995, Ergosteroids: induction of thermogenic enzymes in liver of rats treated with steroids derived from dehydroepiandrosterone, *Proc. Natl. Acad. Sci. USA* **92:** 6617–6619.

Lavietes, B., and Coleman, P. S., 1980, The role of lipid metabolism in neoplastic differentiation, *J. Theor. Biol.* **85:**523–542.

Leevers, S. J., Paterson, H. F., and Marshall, C. J., 1994, Requirement for Ras in Raf activation is overcome by targeting Raf to the plasma membrane, *Nature* **369:**411–414.

Leftheris, K., Kline, T., Vite, G. D., Cho, Y. H., Bhide, R. S., Patel, D. V., Patel, M. M., Schmidt, R. J., Weller, H. N., Andahazy, M. L., Carboni, J. M., Gullo-Brown, J. L., Lee, F. Y., Ricca, C., Rose, W. C., Yan, N., Barbacid, M., Hunt, J. T., Meyers, C. A., Seizinger, B. R., Zahler, R., and Manne, V., 1996, Development of highly potent inhibitors of Ras farensyltransferase possessing cellular and in vivo activity, *J. Med. Chem.* **39:** 224–236.

Leonard, S., Beck, L., and Sinensky, M., 1990, Inhibition of isoprenoid biosynthesis and the post-translatinal modification of pro-p21*ras*, *J. Biol. Chem.* **265:**5157–5160.

Lerner, E. C., Qian, Y., Hamilton, A. D., and Sebti, S. M., 1996, Disruption of oncogenic K-Ras4B processing and signaling by a potent geranylgeranyltransferase I inhibitor *J. Biol. Chem.* **270:** 26770–26773.

Lin, D., Sugawara, T., Strauss III, J. F., Clark, B. J., Stocco, D. M., Saenger, P., Rogol, A., and Miller, W. L., 1995, Role of steroidogenic acute regulatory protein in adrenal and gonadal steroidogenesis, *Science* **267:**1828–1831.

Liscum, L., and Dahl, N. K., 1992, Intracellular cholesterol transport, *J. Lipid Res.* **33:**1239–1254.

Liscum, L., and Underwood, K. W., 1995, Intracellular cholesterol transport and compartmentation, *J. Biol. Chem.* **270:**15443–15446.

Liu, W. C., Barbacid, M., Bulgar, M., Clark, J. M., Crosswell, A. R., Dean, L., Doyle, T. W., Fernandes, P. B., Huang, S., Manne, V., Pirnik, D. M., Wells, J. S., and Meyers, E., 1992, 10′-desmethoxystreptonigrin, a novel analog of streptonigrin, *J. Antibiot.* **45:**454–457.

Liu, L., Shack, S., Stetler-Stevenson, W., Hudgins, W. R., and Samid, D., 1994, Differentiation of cultured human melanoma cells induced by the aromatic fatty acids phenylacetate and phenylbutyrate, *J. Invest. Dermatol.* **103:**335–340.

Liu, L., Hudgins, W. R., Shack, S., Yin, M. Q., and Samid, D., 1995, Cinnamic acid: a natural product with potential use in cancer intervention, *Int. J. Cancer* **62:**345–350.

Maltese, W. A., 1990, Posttranslational modification of proteins by isoprenoids in mammalian cells, *FASEB J.* **4:**3319–3328.

Maltese, W. A., and Aprille, J. R., 1985, Relation of mevalonate synthesis to mitochondrial ubiquinone content and respiratory function in cultured neuroblastoma cells, *J. Biol. Chem.* **260:** 11524–11529.

Maltese, W. A., and Sheridan, K. M., 1987, Isoprenylated proteins in cultured cells: subcellular distribution and changes related to altered morphology and growth arrest induced by mevalonate deprivation, *J. Cell Physiol.* **133:**471–481.

Maltese, W. A., and Sheridan, K. M., 1988, Isoprenoid synthesis during the cell cycle, *J. Biol. Chem.* **263:**10104–10110.

Maltese, W. A., and Sheridan, K. M., 1988, Studies of 3-hydroxy-3-methylglutaryl coenzyme A synthease and reductase and isoprenoid labeling in cells synchronized by centrifugal elutriation, *J. Biol. Chem.* **263:**10104–10110.

Maltese, W. A., Defendi, R., Green, R. A., Sheridan, K. M., and Donley, D. K., 1985, Suppression of murine neuroblastoma growth in vivo by mevinolin, a competitive inhibitor of 3-hydroxy-3-methylglutaryl-coenzyme A reductase, *J. Clin. Invest.* **76:**1748–1754.

Manne, V., Yan, N., Carboni, J. M., Tuomari, A. V., Ricca, C. S., Brown, J. G., Andahazy, M. L., Schmidt, R. J., Patel, D., Zahler, R., Weinmann, R., Der, C. J., Cox, A. D., Hunt, J. T., Gordon, E. M., Barbacid, M., and Seizinger, B. R., 1995, Bisubstrate inhibitors of farnesyltransferase: a novel class of specific inhibitors of ras transformed cells, *Oncogenes* **10:**1763–1779.

Marom, M., Haklai, R., Ben-Baruch, G., Marciano, D., Egozi, Y., and Kloog, Y., 1995, Selective inhibition of Ras-dependent cell growth by farnesylthiosalicylic acid, *J. Biol. Chem.* **270:** 22263–22270.

Marshall, C. J., 1993, Protein prenylation: a mediator of protein-protein interactions, *Science* **259:**1865–1866.

Marshall, C. J., 1995, Specificity of receptor tyrosine kinase signaling: transient versus sustained extracellular signal-regulated kinase activation, *Cell* **80:**179–185.

Marshall, M. S., 1993, The effector interactions of p21ras, *Trends Biochem. Sci.* **18:**250–254.

Marsters, J. C. J., McDowell, R. S., Reynolds, M. E., Oare, D. A., Somers, T. C., Stanley, M. S., Rawson, T. E., Struble, M. E., Burdick, D. J., and Chan, K. S., 1994, Benzodiazepine peptidomimetic inhibition of farnesyltransferase, *Bioorg. Med. Chem.* **2:**949–957.

Martinez, F., Eschegoyen, S., Briones, R., and Cuellar, A., 1988, Cholesterol increase in mitochondria: a new method of cholesterol incorporation, *J. Lipid Res.* **29:**1005–1011.

Maxwell, R. E., Nawrocki, J. W., and Uhlendorf, P. D., 1983, Some comparative effects of gemfibrozil, clofibrate, bezafibrate, cholestyramine and compactin on sterol metabolism in rats, *Atherosclerosis* **48:**195–203.

McGeady, P., Kuroda, S., Shimizu, K., Takai, Y., and Gelb, M. H., 1995, The farnesyl group of H-Ras facilitates the activation of a soluble upstream activator of mitogen-activated protein kinase, *J. Biol. Chem.* **270:**26347–26351.

Medema, R. H., and Bos, J. L., 1993, The role of p21ras in receptor tyrosine kinase signaling, *Crit. Rev. Oncogenesis* **4:**15–61.

Meigs, T. E., Sherwood, S. W., and Simoni, R. D., 1995, Farnesyl acetate, a derivative of an isoprenoid of the mevalonate pathway, inhibits DNA replication in hamster and human cells, *Exp. Cell Res.* **219:**461–470.

Middleton, A., Middleton, B., White, D. A., and Bell, G. D., 1979, The effects of monocyclic terpenes on hepatic S-3-hydroxy-3-methylglutaryl coenzyme A reductase in vivo, *Biochem. Soc. Trans.* **7**:407–408.

Miller, A. C., and Samid, D., 1994, Tumor resistance to oxidative stress: association with ras oncogene expression and reversal by lovastatin, an inhibitor of P21 ras isoprenylation, *Int. J. Cancer* **59**:1–6.

Miller, A. C., Kariko, K., Myers, C. E., Clark, E. P., and Samid, D., 1993, Increased radioresistance of EJras-transformed human osteosarcoma cells and its modulation by lovastatin, an inhibitor of p21 ras isoprenylation, *Int. J. Cancer* **53**:302–307.

Miura, S., Hasumi, K., and Endo, A., 1993, Inhibition of protein prenylation by patulin, *FEBS Lett.* **318**:88–90.

Montgomery, R. B., Moscatello, D. K., Wong, A. J., Cooper, J. A., and Stahl, W. L., 1995, Differential modulation of mitogen-activated protein (MAP) kinase/extracellular signal-related kinase kinase and MAP kinase activities by a mutant epidermal growth factor receptor, *J. Biol. Chem.* **270**: 30562–30566.

Moomaw, J., and Casey, P., 1992, Mammalian protein geranylgeranyltransferase, *J. Biol Chem.* **267**: 17438–17443.

Moores, S. L., Schaber, M., Mosser, S., Rands, E., O'Hara, M., Garsky, V., Marshall, M., Pompliano, D., and Gibbs, J., 1991, Sequence dependence of protein isoprenylation, *J. Biol. Chem.* **266**:14603–14610.

Moreadith, R. W., and Lehninger, A. L., 1984, The pathways of glutamate and glutamine oxidation by tumor cell mitochondria, *J. Biol. Chem.* **259**:6215–6221.

Mortensen, P. B., Kristensen, S. D., Bloch, A., Jacobsen, B. A., and Rasmussen, S. N., 1988, Diagnostic value of ascitic fluid cholesterol levels in the prediction of malignancy, *Scand. J. Gastroenterol.* **23**:1085–1088.

Morton, R., Cunningham, C., Jester, R., Waite, M., Miller, N., and Morris, H. P., 1976, Alteration of mitochondrial function and lipid composition in Morris 7777 hepatoma, *Cancer Res.* **36**: 3246–3254.

Nagasu, T., Yoshimatsu, K., Rowell, C., Lewis, M. D., and Garcia, A. M., 1995, Inhibition of human tumor xenograft growth by treatment with farnesyl transferase inhibitor B956, *Cancer Res.* **55**:5310–5314.

Nave, J.-F., d'Orchymont, H., Ducep, J.-B., Piriou, F., and Jung, M., 1985, Mechanism of the inhibition of cholesterol biosynthesis by 6-fluoromevalonate, *Biochem. J.* **227**:247–254.

Nestler, J. E., Takagi, K., and Strauss III, J. F., 1990, Lipoprotein and cholesterol metabolism in cells that synthesize steroid hormones, *in: Advances in Cholesterol Research* (M. Esfahani and J. Swaney, eds.), pp. 133–169, Telford Press, Caldwell, NJ.

Newman, C. M. H., and Magee, A. I., 1993, Posttranslational processing of the ras superfamily of small GTP-binding proteins, *Biochim. Biophys. Acta* **1155**:79–96.

Nigam, M., Seong, C.-M., Qian, Y., Hamiton, A. D., and Sebti, S. M., 1993, Potent inhibition of human tumor $p21^{ras}$ farnesyltransferase by A_1A_2-lacking $p21^{ras}CA_1A_2X$ peptidomimetics, *J. Biol. Chem.* **269**:20695–20698.

Nigg, E. A., Kitten, G. T., and Vorburger, K., 1992, Targeting lamin proteins to the nuclear envelope: the role of CaaX box modifications, *Biochem. Soc. Trans.* **20**:500–504.

Norbury, C., and Nurse, P., 1992, Animal cell cycles and their control., *Annu. Rev. Biochem.* **61**: 441–470.

Okwu, A. K., Xu, X.-X., Shiratori, Y., and Tabas, I., 1994, Regulation of the threshold for lipoprotein-induced acyl-CoA:cholesterol O-acyltransferase stimulation in macrophages by cellular sphingomyelin content, *J. Lipid Res.* **35**:644–655.

Omer, C.A., Kral, A.M., Diehl, R.E., Prendergast, G.C., Powers, S., Allen, C.M., Gibbs, J.B., and Kohl, N.E., 1993, Characterization of recombinant human farnesyl-protein transferase: cloning, expression, farnesyl diphosphate binding and functional homology with yeast prenyl-protein transferases, *Biochemistry* **32**:5167–5176.

Omura, S., van der Pyl, D., Inokoshi, J., and Takashima, H., 1993, Pepticinnamins, new farnesyl-protein tranferase inhibitors produced by an actinomycete. I. Producing strain, fermentation, isolation and biological activity, *J. Antibiot.* **46:**222–228.

Op den Kamp, J. A. F., 1979, Lipid asymmetry in membranes, *Annu. Rev. Biochem.* **48:**47–71.

Pani, P., Canuto, R., Garcea, R., and Feo, F., 1973, Lipid composition of subcellular particles isolated from rat liver and tumor cells, *Biochem. Soc. Trans.* **1:**972–976.

Pardee, A. B., 1989, G1 events and regulation of cell proliferation, *Science* **246:**603–608.

Parker, R. A., Pearce, B. C., Clark, R. W., Gordon, D. A., and Wright, J. J. K., 1993, Tocotrienols regulate cholesterol production in mammalian cells by post-transcriptional suppression of 3-hydroxy-3-methylglutaryl-coenzyme A reductase, *J. Biol. Chem.* **268:**11230–11238.

Parlo, R. A., and Coleman, P. S., 1984, Enhanced rate of citrate export from cholesterol-rich hepatoma mitochondria, *J. Biol. Chem.* **259:**9997–10003.

Parlo, R. A., and Coleman, P. S., 1986, Continuous pyruvate carbon flux to newly synthesized cholesterol and the suppressed evolution of pyruvate-generated CO_2 in tumors: further evidence for a persistent truncated Krebs cycle in hepatomas, *Biochim. Biophys. Acta* **886:** 169–176.

Patel, D. V., Gordon, E. M., Schmidt, R. J., Weller, H. N., Young, M. G., Zahler, R., Barbacid, M., Carboni, J. M., Gullo-Brown, J. L., Hunihan, L., Ricca, C., Robinson, S., Seizinger, B. R., Tuomari, A. V., and Manne, V., 1995, Phosphinyl acid-based bisubstrate analog inhibitors of Ras farnesyl protein transferase, *J. Med. Chem.* **38:**435–442.

Pearce, B. C., Parker, R. A., Deason, M. E., Qureshi, A. A., and Wright, J. J. K., 1992, Hypocholesterolemic activity of synthetic and natural tocotrienols, *J. Med. Chem.* **35:**3595–3606.

Pearce, B. C., Parker, R. A., Deason, M. E., Dischino, D. D., Gillespie, E., Qureshi, A. A., Volk, K., and Wright, J. J. K., 1994, Inhibitors of cholesterol biosynthesis. 2. Hypocholesterolemic and antioxidant activities of benzopyran and tetrahydronaphthalene analogues of the tocotrienols, *J. Med. Chem.* **37:**526–541.

Pedersen, P. L., 1978, Tumor mitochondria and the bioenergetics of cancer cells, *in: Prog. Exp. Tumor Res.* (D. F. H. Wallach, ed.), pp. 190–274, S. Karger, Basel.

Peeper, D. S., van der Eb, A. J., and Zantema, A., 1994, The G1/S cell-cycle checkpoint in eukaryotic cells, *Biochim. Biophys. Acta* **1198:**215–230.

Pérez-Sala, D., and Mollinedo, F., 1994, Inhibition of isoprenoid biosynthesis induces apoptosis in human promyelocytic HL-60 cells, *Biochem. Biophys. Res. Commun.* **199:**1209–1215.

Perkins, S. L., Ledin, S. F., and Stubbs, J. D., 1982, Linkage of the isoprenoid biosynthetic pathway with induction of DNA synthesis in mouse lymphocytes. Effects of compactin on mitogen-induced lymphocytes in serum-free medium, *Biochim. Biophys. Acta* **711:**83–89.

Pompliano, D., Rands, E., Schaber, M., Mosser, S., Anthony, N., and Gibbs, J., 1992, Steady-state kinetic mechanism of Ras farnesyl:protein transferase, *Biochemistry* **31:**3800–3807.

Powers, S., Michaelis, S., Broek, D., S., A.-A., Field, J., Herskowitz, I., and Wigler, M., 1986, RAM, a gene of yeast required for a functional modification of RAS proteins and for production of mating pheromone a-factor, *Cell* **47:**413–422.

Prager, C., Schön, H. J., Nikfardjam, M., Schmid, D., Untersalmberger, M., Kremser, K., and Kramar, R., 1993, Clofibrate elevates enzyme activities of the tricarboxylic acid cycle in rat liver, *J. Lipid Res.* **34:**359–364.

Prendergast, G. C., and Gibbs, J. B., 1994, Ras regulatory interactions: Novel targets for anti-cancer intervention?, *BioEssays* **16:**187–191.

Prendergast, G. C., Davide, J. P., deSolms, S. J., Giuliani, E. A., Graham, S. L., Gibbs, J. B., Oliff, A., and Kohl, N. E., 1994, Farnesyltransferase inhibition causes morphological reversion of ras-transformed cells by a complex mechansim that involves regulation of the actin cytoskeleton, *Mol. Cell Biol.* **14:**4193–4202.

Privalle, C. T., Crivello, J. F., and Jefcoate, C. R., 1983, Regulation of intramitochondrial cholesterol transfer to side-chain cleavage cytochrome P-450 in rat adrenal gland, *Proc. Natl. Acad. Sci. USA* **80:**702–706.

Qian, Y., Blaskovich, M. A., Saleem, M., Seong, C.-M., Wathen, S. P., Hamilton, A. D., and Sebti, S. M., 1994, Design and structural requirements of potent peptidomimetic inhibitors of p21ras farnesyltransferase, *J. Biol. Chem.* **269:**12410–12413.

Quesney-Huneeus, V., Wiley, M. H., and Siperstein, M. D., 1979, Essential role for mevalonate synthesis in DNA replication, *Proc. Natl. Acad. Sci. USA* **76:**5056–5060.

Quesney-Huneeus, V., Galick, H. A., Siperstein, M. D., Erickson, S. K., Spencer, T. A., and Nelson, J. A., 1983, The dual role of mevalonate in the cell cycle, *J. Biol. Chem.* **258:**378–385.

Quilliam, L. A., Huff, S. Y., Rabun, K. M., Wei, W., Park, W., Broek, D., and Der, C. J., 1994, Membrane-targeting potentiates guanine nucleotide exchange factor CDC25 and SOS1 activation of Ras transforming activity, *Proc. Natl. Acad. Sci. USA* **91:**8512–8516.

Qureshi, A. A., Mangels, A. R., Din, Z. Z., and Elson, C. E., 1988, Inhibition of hepatic mevalonate biosynthesis by the monoterpene, *D*-limonene., *J. Agric. Food Chem.* **36:**1220–1224.

Ram, Z., Samid, D., Walbridge, S., Oshiro, E. M., Viola, J. J., Tao-Cheng, J.-H., Shack , S., Thibault, A., Meyers, C. E., and Oldfield, E. H., 1994, Growth inhibition, tumor maturation, and extended survival in experimental brain tumors in rats treated with phenylacetate, *Cancer Res.* **54:** 2923–2927.

Rao, K. N., 1995, The significance of the cholesterol biosynthetic pathway in cell growth and carcinogenesis, *Anticancer Res.* **15:**309–314.

Rao, S., and Coleman, P. S., 1989, Control of DNA replication and cell growth by inhibiting the export of mitochondrially derived citrate, *Exp. Cell Res.* **180:**341–352.

Reardon, J. E., and Abeles, R. H., 1987, Inhibition of cholesterol biosynthesis by fluorinated mevalonate analogues, *Biochemistry* **26:**4717–4722.

Reddy, J. K., and Lalwani, N. D., 1983, Carcinogenesis by hepatic peroxisome proliferators: evaluation of the risk of hypolipidemic drugs and industrial plasticizers to humans, *Crit. Res. Toxicol.* **12:**1–58.

Reddy, J. K., and Mannaerts, G. P., 1994, Peroxisomal lipid metabolism, *Annu. Rev. Nutr.* **14:**343–370.

Regelson, W., and Kalimi, M., 1994, Dehydroepiandrosterone (DHEA)—the multifunctional steroid. II. Effects on the CNS, cell proliferation, metabolic and vascular, clinical and other effects. Mechanism of action? *Ann. NY Acad. Sci.* **719:**564–575.

Reiss, Y., Goldstein, J. L., Seabra, M. C., Casey, P. J., and Brown, M. S., 1990, Inhibition of purified p21ras farnesyl:protein transferase by Cys-AAX tetrapeptides, *Cell* **62:**81–88.

Reiss, Y., Seabra, M., Armstrong, S., Slaughter, C., Goldstein, J., and Brown, M., 1991, Non-identical subunits of p21ras farnesyltransferase: peptide binding and farnesyl pyrophosphate carrier functions, *J. Biol Chem.* **266:**10672–10677.

Reiss, Y., Stradley, S., Gierasch, L. M., Brown, M. S., and Goldstein, J. L., 1991, Sequence requirement for peptide recognition by rat brain p21ras protein farnesyltransferase, *Proc. Natl. Acad. Sci. USA* **88:**732–736.

Reiss, Y., Brown, M. S., and Goldstein, J. L., 1992, Divalent cation and prenyl pyrophosphate specificities of the protein farnesyltransferase from rat brain, a zinc metalloenzyme, *J. Biol. Chem.* **267:**6404–6408.

Ridley, A. J., and Hall, A., 1992, The small GTP binding protein rho regulates the assembly of focal adhesions and actin stress fibers in response to growth factors, *Cell* **70:**389–399.

Rilling, H. C., Bruenger, E., Epstein, W. W., and Crain, P. F., 1990, Prenylated proteins: the structure of the isoprenoid group, *Science* **247:**318–320.

Rosen, N., 1995, Oncogenes, *in: The Molecular Basis of Cancer* (J. Mendelsohn, P. M. Howley, M. A. Israel, and L. A. Liotta, eds.), pp. 105–116, W. B. Saunders Company, Philadelphia.

Rothblat, G. H., Mahlberg, F. H., Johnson, W. J., and Phillips, M. C., 1992, Apolipoproteins, membrane cholesterol domains and the regulation of cholesterol efflux, *J. Lipid Res.* **33:** 1091–1097.

Roussillon, S., Astruc, M., Defay, R., Tabacik, C., Descomps, B., and Crastes de Paulet, A., 1983, DNA and cholesterol biosynthesis in synchronized embryonic rat fibroblasts. Temporal relationships between HMG-CoA reductase activity, sterol biosynthesis and thymidine incorporation into DNA, *Biochim. Biophys. Acta* **763:**1–10.

Ruch, R. J., and Sigler, K., 1994, Growth inhibition of rat liver epithelial tumor cells by monoterpenes does not involve Ras plasma membrane association, *Carcinogenesis* **15:**787–789.

Ruderman, J. V., 1993, MAP kinase and the activation of quiescent cells, *Curr. Opin. Cell Biol.* **5:** 207–213.

Rudney, H., and Sexton, R. C., 1986, Regulation of cholesterol biosynthesis, *Annu. Rev. Nutr.* **6:** 245–272.

Rusnak, N., and Krisans, S. K., 1987, Diurnal variation of HMG-CoA reductase activity in rat liver peroxisomes, *Biochem. Biophys. Res. Commun.* **148:**890–896.

Sabine, J. R., 1975, Defective control of lipid biosynthesis in cancerous and precancerous liver, *Prog. Biochem. Pharmacol.* **10:**269–307.

Samid, D., Miller, A. C., Rimoldi, D., Gafner, J., and Clark, E. P., 1991, Increased radiation resistance in transformed and non-transformed cells with elevated ras proto-oncogene expression, *Rad. Res.* **126:**244–250.

Samid, D., Shack, S., and Sherman, L. T., 1992a, Phenylacetate: a novel nontoxic inducer of tumor cell differentiation, *Cancer Res.* **52:**1988–1992.

Samid, D., Yeh, A., and Prasanna, P., 1992b, Induction of erythroid differentiation and fetal hemoglobin production in human leukemic cells treated with phenylacetate, *Blood* **80:** 1576–1581.

Samid, D., Shack, S., and Meyers, C. E., 1993, Selective growth arrest and phenotypic reversion of prostate cancer cells in vitro by nontoxic pharmacological concentrations of phenylacetate, *J. Clin. Invest.* **91:**2288–2295.

Samid, D., Ram, Z., Hudgins, W. R., Shack, S., Liu, L., Walbridge, S., Oldfield, E. H., and Myers, C. E., 1994, Selective activity of phenylacetate against malignant gliomas: resemblance to fetal brain damage in phenylketonuria, *Cancer Res.* **54:**891–895.

Sanford, J. C., Foster, L., Kapadia, Z., and Wessling-Resnick, M., 1995, Analysis of the stoichiometry of Rab protein prenylation, *Anal. Biochem.* **224:**547–556.

Sapperstein, S., Berkower, C., and Michaelis, S., 1994, Nucleotide sequence of the yeast STE14 gene which encodes an isoprenylcysteine carboxyl methyltransferase and demonstration of its essential role in a-factor export, *Mol. Cell Biol.* **14:**1438–1449.

Sawamura, M., Li, N., Nara, Y., and Yamori, Y., 1993, Proliferative effect of mevalonate metabolites other than isoprenoids on cultured vascular smooth muscle cells, *Clin. Exp. Pharm. Physiol.* **20:** 509–514.

Schafer, W. R., and Rine, J., 1992, Protein prenylation: genes enzymes, targets and functions, *Annu. Rev. Genet.* **26:**209–237.

Schatzkin, A., Hoover, R. N., Taylor, P. R., Zeigler, R. G., Carter, C. L., Albanes, D., Larson, D. B., and Licitra, L. M., 1988, Site-specific analysis of total serum cholesterol and incident cancer in the National Health and Nutrition Examination Survey I Epidemiologic Follow-up Study, *Cancer Res.* **48:**452–458.

Schepers, L., Casteels, M., Vamecq, J., Parmentier, G., van Veldhoven, P. P., and Mannaerts, G. P., 1988, β-Oxidation of the carboxyl side-chain of prostaglandin E2 in rat liver peroxisomes and mitochondria, *J. Biol. Chem.* **263:**2724–2731.

Schepers, L., van Veldhoven, P. P., Casteels, M., Eyssen, H., and Mannaerts, G. P., 1990, Presence of three acyl-CoA oxidases in rat liver peroxisomes, *J. Biol. Chem.* **265:**5242–5246.

Schmidt, R. A., Schneider, C. L., and Glomset, J. A., 1984, Evidence for post-translational incorporation of a product of mevalonic acid into Swiss 3T3 cell proteins, *J. Biol. Chem.* **259:**10175–10180.

Schneider, H., Lemasters, J. J., Höchli, M., and Hackenbrock, C. R., 1980, Fusion of liposomes with mitochondrial inner membranes, *Proc. Natl. Acad. Sci. USA* **77:**442–446.

Schneider, H., Lemasters, J. J., Höchli, M., and Hackenbrock, C. R., 1980, Liposome-mitochondrial inner membrane fusion, *J. Biol. Chem.* **255:**3748–3756.

Schneider, H., Höchli, M., and Hackenbrock, C. R., 1982, Relationship between the density distribution of intramembrane particles and electron transfer in the mitochondrial inner membrane as revealed by cholesterol incorporation, *J. Cell Biol.* **94:**387–393.

Schultz, S., and Nyce, J. W., 1991, Inhibition of protein isoprenylation and p21ras membrane association by dehydroepiandrosterone in human colonic adenocarcinoma cells in vitro, *Cancer Res.* **51**:6563–6567.

Schultz, S., Klann, R. C., Schonfeld, S., and Nyce, J. W., 1992, Mechanisms of cell growth inhibition and cell cycle arrest in human colonic adenocarcinoma cells by dehydroepiandrosterone: role of isoprenoid biosynthesis, *Cancer Res.* **52**:1372–1376.

Schwartz, A. G., Whitcomb, J. M., Nyce, J. W., Lewbart, M. L., and Pashko, L. L., 1988, Dehydroepiandrosterone and structural analogs: a new class of cancer chemopreventive agents, *Adv. Cancer Res.* **51**:391–424.

Seabra, M., Goldstein, J., Sudhof, M., and Brown, M., 1992, Rab geranylgeranyl transferase, *J. Biol. Chem.* **267**:14497–14503.

Sebti, S. M., Tkalavic, G. T., and Jani, J. P., 1991, Lovastatin, a cholesterol biosynthesis inhibitor, inhibits the growth of human H-ras oncogene transformed cells in nude mice, *Cancer Commun.* **3**:141–147.

Seedorf, U., Brysch, P., Engel, T., Schrage, K., and Assmann, G., 1994, Sterol carrier protein X is peroxisomal 3-oxoacyl CoA thiolase with intrinsic sterol carrier and lipid transfer activity, *J. Biol. Chem.* **269**:21277–21283.

Sepp-Lorenzino, L., Azrolan, N. I., and Coleman, P. S., 1989, Cellular distribution of cholesterogenesis-linked, phospho-isoprenylated proteins in proliferating cells, *FEBS Lett.* **245**:110–116.

Sepp-Lorenzino, L., Rao, S., and Coleman, P. S., 1991, Cell-cycle-dependent, differential prenylation of proteins, *Eur. J. Biochem.* **200**:579–590.

Sepp-Lorenzino, L., Coleman, P. S., and Larocca, J., 1994, Isoprenylated proteins in myelin, *J. Neurochem.* **62**:1539–1545.

Sepp-Lorenzino, L., Ma, Z., Kohl, N. E., Rands, E., Gibbs, J. B., Oliff, A., and Rosen, N., 1995, A peptidomimetic inhibitor of farnesyl:protein transferase inhibits the anchorage-dependent and-independent growth of human tumor cell lines, *Cancer Res.* **55**:5302–5309.

Shack, S., Chen, L.-C., Miller, A. C., Danesi, R., and Samid, D., 1995, Increased susceptibility of ras-transformed cells to phenylacetate is associated with inhibition of p21ras isoprenylation and phenotypic reversion, *Intl. J. Cancer* **63**:124–129.

Shama Bhat, C., and Ramasarma, T., 1979a, Effect of phenyl and phenolic acids on mevalonate-5-phosphate kinase and mevalonate-5-pyrophosphate decarboxylase of the rat brain, *J. Neurochem.* **32**:1531–1537.

Shama Bhat, C., and Ramasarma, T., 1979b, Inhibition of rat liver mevalonate pyrophosphate decarboxylase and mevalonate phosphate kinase by phenyl and phenolic compounds, *Biochem. J.* **181**:143–151.

Sherwin, R. W., Wentworth, D. N., Cutler, J. A., Hulley, S. B., Kuller, L. H., and Stamler, J., 1987, Serum cholesterol levels and cancer mortality in 361,662 men screened for the Multiple Risk Factor Intervention Trial, *JAMA* **257**:943–948.

Shiomi, K., Yang, H., Inokoshi, J., van der Pyl, D., Nakagawa, A., Takeshima, H., and Omura, S., 1993, Pepticinnamins, new farnesyl-protein tranferase inhibitors produced by an actinomycete. II. Structural elucidation of pepticinnamin E, *J. Antibiot.* **46**:229–234.

Shirataki, H., Kaibuchi, K., Yamaguchi, T., Wada, K., Horiuchi, H., and Takai, Y., 1992, A possible target protein for smg-25A/rab3A small GTP-binding protein, *J. Biol. Chem.* **267**: 10946–10949.

Shu, Y. Z., Huang, S., Wang, R. R., Lam, K. S., Klohr, S. E., Volk, K. J., Pirnik, D. M., Wells, J. S., Fernandes, P. B., and Patel, P. S., 1994, Manumycins E, F, and G, new members of manumycin class antibiotics, from *Streptomyces sp., J. Antibiot.* **47**:324–333.

Singh, S. B., Zink, D. L., Liesch, J. M., Goetz, M. A., Jenkins, R. G., Nallin-Olmstead, M., Silverman, K. C., Bills, G. F., Mosley, R. T., Gibbs, J. B., Albers-Schonberg, G., and Lingham, R. B., 1993, Isolation and structure of chaetomellic acids A and B from *Chaetomella acutiseta*: farnesyl pyrophosphate mimic inhibitos or Ras farnesyl-protein transferase, *Tetrahedron* **49**:5917–5926.

Siperstein, M. D., 1970, Regulation of cholesterol biosynthesis in normal and malignant tissues, *in: Current Topics in Cell Regulation* (B. L. Horecker and E. R. Stadtman, eds.), pp. 65–100, Academic Press, New York.

Siperstein, M. D., 1984, Role of cholesterogenesis and isoprenoid synthesis in DNA replication and cell growth, *J. Lipid Res.* **25:**1462–1468.

Siperstein, M. D., 1995, Cholesterol, cholesterogenesis and cancer, *Adv. Exper. Med. Biol.* **369:** 155–166.

Skalnik, D. G., Brown, D. A., Borwn, P. C., Friedman, R. L., Hardeman, E. C., Schimke, R. T., and Simoni, R. D., 1985, Mechanisms of 3-hydroxy-3-methylglutaryl coenzyme A reductase over-accumulation in three compactin-resistant cell lines, *J. Biol. Chem.* **260:**1991–1994.

Slotte, J. P., Hedström, G., Rannström, S., and Ekman, S., 1989, Effects of sphingomyelin degradation on cell cholesterol oxidizability and steady-state distribution between the cell surface and the cell interior, *Biochim. Biophys. Acta* **985:**90–96.

Smith, M. R., De Gudicibus, S. J., and Stacey, D. W., 1986, Requirement for c-ras proteins during viral oncogene transformation, *Nature* **320:**540–543.

Soldati, T., Shaprio, A. D., Svejstrup, A. B. D., and Pfeffer, S. R., 1994, Membrane targeting of the small GTPase Rab9 is accompanied by nucleotide exchange, *Nature* **369:**76–78.

Spady, D. K., 1991, Regulation of cholesterol metabolism in normal and malignantly transformed tissue in vivo, *in: Chemical Carcinogenesis-2: Modulating Factors* (A. Columbano, F. Feo, R. Pascale, and P. Pani, eds.), pp. 299–309, Plenum Press, NY.

Spector, A. A., and Yorek, M. A., 1985, Membrane lipid composition and cellular function, *J. Lipid Res.* **26:**1015–1035.

Srere, P., 1975, The enzymology of the formation and breakdown of citrate, *Adv. Enzymol.* **43:**57–101.

Stamellos, K. D., Shackelford, J. E., Tanaka, R. D., and Krisans, S. K., 1992, Mevalonate kinase is localized in rat liver peroxisomes, *J. Biol. Chem.* **267:**5560–5568.

Stamellos, K. D., Shackelford, J. E., Shechter, I., Jiang, G., Conrad, D., Keller, G. A., and Krisans, S. K., 1993, Subcellular localization of squalene synthase in rat hepatic cells. Biochemical and immunochemical evidence, *J. Biol. Chem.* **268:**12825–12836.

Stange, E. F., Preclik, G., Schneider, A., and Reimann, F., 1988, The role of enterocyte cholesterol metabolism in intestinal cell growth and differentiation, *Scand. J. Gastroent.* **151:**79–85.

Stimmel, J. B., Deschenes, R., Volker, C., Stock, J., and Clarke, S., 1990, Evidence for an *S*-farnesyl-cysteine methyl ester at the carboxyl terminus of the Saccharomyces cerevisiae RAS2 protein, *Biochemistry* **29:**9651–9659.

Stokoe, D., Macdonald, S. G., Cadwallader, K., Symons, M., and Hancock, J. F., 1994, Activation of Raf as a result of recruitment to the plasma membrane, *Science* **264:**1463–1467.

Stradley, S. J., Rizo, J., and Gierasch, L. M., 1993, Conformation of a heptapeptide substrate bound to protein farnesyltransferase, *Biochemistry* **32:**12586–12590.

Sumi, S., Beauchamp, D., Townsend, C. M., Jr., Uchida, T., Murakami, M., Rajaraman, S., Ishizuka, J., and Thompson, J. C., 1992, Inhibition of pancreatic adenocarcinoma cell growth by lovastatin, *Gastroenterology* **103:**982–989.

Tabacik, C., and Aliau, S., 1989, Cholesterogenesis and cell division in phytohemagglutinin-stimulated human lymphocytes: a comparative study with several inhibitors, *Biochim. Biophys. Acta* **1011:** 149–157.

Tamanoi, F., 1993, Inhibitors of Ras farnesyltransferases, *Trends Biochem. Sci.* **18:**349–353.

Tertov, V. V., Orekhov, A. N., Ryong, L. H., and Smirnov, V. N., 1988, Intracellular cholesterol accumulation is accompanied by enhanced proliferative activity of human aortic intimal cells, *Tissue Cell.* **20:**849–854.

Thibault, A., Cooper, M. R., Figg, W. D., Venzon, D. J., Sartor, A. O., Tompkins, A. C., Weinberger, M. S., Headlee, D. J., McCall, N. A., Samid, D., and Meyers, C. E., 1994, A phase I and pharmacokinetic study of intraveous phenylacetate in patients with cancer, *Cancer Res.* **54:** 1690–1694.

Thompson, S. L., and Krisans, S. K., 1990, Rat liver peroxisomes catalyze the initial step in cholesterol synthesis: the condensation of acetyl CoA units into acetoacetyl CoA, *J. Biol. Chem.* **265:** 5731–5735.

Toth, M. J., Park, J., and Huwyler, L., 1995, Purification of rat liver mevalonate pyrophosphate decarboxylase, *FASEB J.* **9:**A1293.

Trentalance, A., Leoni, S., Mangiantini, M. T., Spagnuolo, S., Feingold, K., Hughes-Fulford, M., Siperstein, M. D., Cooper, A. D., and Erickson, S. K., 1984, Regulation of 3-hydroxy-3-methylglutaryl coenzyme A reductase and cholesterol synthesis and esterification during the first cell cycle of liver regeneration, *Biochim. Biophys. Acta* **794:**142–151.

Trueblood, C. E., Ohya, Y., and Rine, J., 1993, Genetic evidence for in vivo cross-specificity of the CaaX-box protein prenyltransferases farnesyltransferase and geranylgeranyltransferase-I in *Saccharomyces cerevisiae, Mol. Cell Biol.* **13:**4260–4275.

van Blitterswijk, W. J., Emmelot, P., Hilkmann, H. A. M., Oomenmeulmans, E. P. M., and Inbar, M., 1977, Differences in lipid fluidity among isolated plasma membranes of normal and leukemic lymphocytes and membranes exfoliated from their cell surface, *Biochim. Biophys. Acta* **467:** 309–320.

van Blitterswijk, W. J., Emmelot, P., Hilkmann, H. A. M., Higers, J., and Feltkamp, C. A., 1979, Rigid plasma membrane-derived vesicles, enriched in tumor-associated surface antigens (MLr) occurring in the ascites fluid of a murine leukemia (GRSL), *Intl. J. Cancer* **23:**62–70.

van Blitterswijk, W. J., De Veer, G., Krol, J. H., and Emmelot, P., 1982, Comparative lipid analysis of purified plasma membranes and shed extracellular membrane vesicles from normal murine thymocytes and leukemic GRSL cells, *Biochim. Biophys. Acta* **688:**495–504.

van Blitterswijk, W. J., Hilkmann, H., and Hengeveld, T., 1984, Differences in membrane lipid composition and fluidity of transplanted GRSL lymphoma cells, depending on their site of growth in the mouse, *Biochim. Biophys. Acta* **778:**521–529.

van der Pyl, D., Inokoshi, J., Shiomi, K., Yang, H., Takeshima, H., and Omura, S., 1992, Inhibition of farnesyl:protein transferase by gliotoxin and acetylgliotoxin, *J. Antibiot.* **45:**1802–1805.

van Hoeven, R., and Emmelot, P., 1972, Studies on plasma membranes. XVIII. Lipid class composition of plasma membranes isolated from rat and mouse liver and hepatomas, *J. Membr. Biol.* **9:**105–126.

van Meer, G., 1987, Plasma membrane cholesterol pools, *Trends Biochem. Sci.* **12:**375–376.

Vincent, T., Wulfert, E., and Merler, E., 1991, Inhibition of growth factor signaling pathways by lovastatin, *Biochem. Biophys. Res. Commun.* **180:**1284–1289.

Vogt, A., Qian, Y., Blaskovich, M. A., Fossum, R. D., Hamilton, A. D., and Sebti, S. M., 1995, A non-peptide mimetic of Ras-CAAX: selective inhibition of farnesyltransferase and Ras processing, *J. Biol. Chem.* **270:**660–664.

Waterman, M. R., 1995, A rising StAR: an essential role in cholesterol transport, *Science* **267:** 1780–1781.

Watson, J. A., Havel, C. M., Lobos, D. V., Baker, F. C., and Morrow, C. J., 1985, Isoprenoid synthesis in isolated embryonic Drosophila cells, *J. Biol. Chem.* **260:**14083–14091.

Wedegaertner, P. B., Wilson, P. T., and Bourne, H. R., 1995, Lipid modifications of trimeric G proteins, *J. Biol. Chem.* **270:**503–506.

Weibel, E. R., 1984, *The Pathway for Oxygen: Structure and Function in the Mammalian Respiratory System,* Harvard University Press, Cambridge, MA.

Willumsen, B. M., Norris, K., Papageorge, A. G., Hubbert, N. L., and Lowy, D. R., 1984, Harvey murine sarcoma virus p21 ras protein: biological and biochemical significance of the cysteine nearest the carboxy terminus, *EMBO J.* **3:**2581–2585.

Wirtz, K. W. A., 1991, Phospholipid transfer proteins, *Annu. Rev. Biochem.* **60:**73–99.

Wittinghofer, A., and Herrmann, C., 1995, Ras-effector interactions, the problem of specificity, *FEBS Lett.* **369:**52–56.

Xu, Y., Mayor, J. A., Gresme, D., Wood, D., and Kaplan, R. S., 1995, High-yield bacterial expression, purification, and functional reconstitution of the tricarboxylate transport protein from rat liver mitochondria, *Biochem. Biophys. Res. Commun.* **207:**783–789.

Yan, N., Ricca, C., Fletcher, J., Glover, T., Seizinger, B. R., and Manne, V., 1995, Farnesyltransferase inhibitors block the neurofibromatosis type I (NFI) malignant phenotype, *Cancer Res.* **55:** 3569–3575.

Yeagle, P., 1987, *The Membranes of Cells,* Academic Press, NY.

Yeagle, P., 1992, The dynamics of membrane lipids, *in: The Structure of Biological Membranes* (P. Yeagle, ed.), pp. 157–174, CRC Press, Boca Raton, FL.

Yeagle, P. L., 1991, Modulation of membrane function by cholesterol, *Biochimie* **73:**1303–1310.

Ying, W., Sepp-Lorenzino, L., Cai, K., Aloise, P., and Coleman, P. S., 1994, Photoaffinity-labeling peptide substrates for farnesyl-protein transferase and the intersubunit location of the active site, *J. Biol. Chem.* **269:**470–477.

Yokoyama, K., and Gelb, M. H., 1993, Purification of a mammalian protein geranylgeranyltransferase, *J. Biol. Chem.* **268:**4055–4060.

Yokoyama, K., Goodwin, G., Ghomashchi, F., Glomset, J. A., and Gelb, M., 1991, A protein geranylgeranyltransferase from bovine brain: implications for protein prenylation specificity, *Proc. Natl. Acad. Sci. USA* **88:**5302–5306.

Zafra, M. F., Riquelme, S., Castillo, M., and Garcia-Peregrin, E., 1987, Effect of clofibrate on brain mevalonate-5-pyrophosphate decarboxylase, *Neurochem. Res.* **12:**787–791.

Zhang, F., and Casey, P. J., 1996, Protein prenylation: molecular mechanisms and functional consequences, *Annu. Rev. Biochem.* **65:**241–269.

Zhang, F. L., Diehl, R. E., Kohl, N. E., Gibbs, J. B., Giros, B., Casey, P. J., and Omer, C. A., 1994a, cDNA cloning and expression of rat and human protein geranylgeranyltransferase type-I, *J. Biol. Chem.* **269:**3175–3180.

Zhang, L., Sachs, C. W., Fine, R. L., and Casey, P. J., 1994b, Interaction of prenylcysteine methyl esters with the multidrug resistance transporter, *J. Biol. Chem.* **269:**15973–15976.

Zhang, L., Sachs, C. W., Fu, H.-W., Fine, R. L., and Casey, P. J., 1995, Characterization of prenylcysteines that interact with P-glycoprotein and inhibit drug transport in tumor cells, *J. Biol. Chem.* **270:**22859–22865.

Zhang, F. L., Fu, H. W., Casey, P. J., and Bishop, W. R., 1996, Substitution of cadmium for zinc in farnesyl:protein transferase alters its substrate specificity, *Biochemistry* **35:**8166–8171.

Chapter 14

Biological Implications of the Niemann–Pick C Mutation

Peter G. Pentchev, E. Joan Blanchette-Mackie, and Laura Liscum

1. INTRODUCTION

The unique and varied membranes of a cell permit the compartmentalization of cellular functions. The synthesis of these distinct membrane domains is largely achieved through the protein and lipid sorting machinery of a cell. While an extensive experimental technology has evolved to study intracellular protein sorting (Rothman, 1994), cellular lipid topology has remained more a phenomenological science, with less progress made in elucidating specific molecular mechanisms. The relative paucity of experimental techniques available to study intracellular lipid trafficking has made the delineation of inherited metabolic defects particularly critical for delineating metabolic pathways of transport.

The cholesterol metabolic pathway has provided some prime examples of gene mutations that have played key roles in delineating lipid pathways. The seminal biological clue that led to a molecular understanding of the uptake and

Peter G. Pentchev National Institute of Neurological Disorders and Stroke, Developmental and Metabolic Neurology Branch, National Institutes of Health, Bethesda, Maryland 20892. **E. Joan Blanchette-Mackie** Laboratory of Cell Biochemistry and Biology, National Institute of Diabetes and Digestive and Kidney Diseases, National Institutes of Health, Bethesda, Maryland 20892. **Laura Liscum** Department of Physiology, Tufts University School of Medicine, Boston, Massachusetts 02111.

Subcellular Biochemistry, Volume 28: Cholesterol: Its Functions and Metabolism in Biology and Medicine, edited by Robert Bittman. Plenum Press, New York, 1997.

metabolic utilization of cholesterol came from work on an inherited metabolic disorder, familial hypercholesterolemia (FH) (Brown and Goldstein, 1986). Studies with cultured fibroblasts from patients revealed a defect in the internalization of exogenous cholesteryl esters complexed within lipoproteins, thus suggesting the presence of a plasma membrane receptor (Goldstein *et al.,* 1995). Once endocytosed via this receptor, lipoprotein particles are translocated to lysosomes for hydrolytic processing. The free cholesterol generated within the lysosomes is mobilized and enriches other cellular compartments, initiating a cascade of metabolic responses. The trafficking of endocytosed cholesterol to postlysosomal targets has, however, remained largely an uncharacterized process. This lack of study reflected both an assumption of spontaneous intracellular cholesterol distribution and the absence of viable experimental approaches. Elucidation of the cellular lesion responsible for Niemann–Pick C (NP-C) disease has provided both strong experimental support, as well as heuristic impetus, for a lysosomal cholesterol transport pathway that is protein-mediated. In this chapter we will review the progress made in our understanding of intracellular cholesterol transport that can be directly linked to the NP-C mutation.

2. NIEMANN–PICK C DISEASE

NP-C disease is an autosomal-recessive neurovisceral lipid storage disorder (Pentchev *et al.,* 1995). Despite a common classification (Crocker and Farber, 1958), NP-C is clearly different phenotypically and genotypically from the Niemann-Pick A and B disorders, which are primary metabolic lesions in sphingomyelin catabolism (Schuchman and Desnick, 1995). The typical NP-C clinical presentation begins in late childhood and is marked by progressive ataxia, dystonia, and dementia, with death occurring in the second decade (Pentchev *et al.,* 1995). Lipid-filled foam cells infiltrate many tissues. Unesterified cholesterol, sphingomyelin, and glycolipids accumulate in reticuloendothelial tissues. This lipid accumulation is reflected in an enlargement of the liver and spleen. Only glycolipids are reported to be elevated in the brain. The disease is panethnic and quite rare in the general population, with an estimated frequency of 1 in 1,000,000.

3. CELLULAR LESION

3.1. Historical Evolution

Following its long and misleading classification as a sphingomyelinosis, NP-C was eventually recognized as a cellular cholesterol lipidosis primarily because of serendipitous studies carried out in a murine lipid storage disorder. First, it was found that high cholesterol diets induced a phenotype reminiscent of NP-C (i.e., hepatosplenomegaly and an anomalous storage of unesterified cholesterol in liver and spleen) in a neuroviscerally affected mutant mouse strain of unknown etiology (Pentchev *et al.,* 1984). Next, experiments with fibroblast cultures derived from the affected animals provided *in vitro* evidence that LDL uptake caused cellular accumulation of free cholesterol without an induction of cholesterol esterification (Pentchev *et al.,* 1984). Evidence for the close association of the murine mutation to the lesion in cholesterol metabolism was provided by the finding that cell cultures derived from heterozygous mutant mice displayed an intermediary deficiency in the processing of exogenously derived cholesterol (Pentchev *et al.,* 1986). Given the similar lipid storage profiles, fibroblasts derived from NP-C patients were examined for a possible cholesterol processing error and the same lesion was found (Pentchev *et al.,* 1986). These results provided the first experimental grounds to suggest that the NP-C mutation affects cellular cholesterol metabolism.

3.2. Lysosomal Sequestration of Endocytosed Cholesterol

This newly found defect in cellular cholesterol metabolism provided a fertile venue for further characterization of the NP-C mutation. Receptor-mediated uptake of LDL labeled with [^{3}H]cholesteryl linoleate by NP-C fibroblasts was shown by subcellular fractionation to lead to an abnormal lysosomal accumulation of endocytosed [^{3}H]cholesterol that was linked to a deficient rate of transfer to the plasma membrane (Liscum *et al.,* 1989). Analysis by filipin fluorescence microscopy of the intracellular distribution of cholesterol after LDL incubation also showed abnormal sequestration of cholesterol in lysosomes and deficient enrichment of the plasma membrane (Sokol *et al.,* 1986). As anticipated from this abnormal lysosomal sequestration of cholesterol, induction of homeostatic responses normally associated with cellular cholesterol enrichment was shown to be uniformly delayed in mutant cells (Liscum and Faust, 1987; Pentchev *et al.,* 1987). The defect in cholesterol movement was found to be specific for exogenously derived cholesterol, since movement of endogenously synthesized cholesterol to the plasma membrane is normal in NP-C cells (Liscum *et al.,* 1989).

3.3. Cholesterol Distribution within the Golgi Complex

Excessive accumulation of endocytosed cholesterol in lysosomes and subsequent delayed relocation to the plasma membrane strongly suggested a mutation localized to the lysosomes. Partial delay in this intracellular trafficking of lysosomal cholesterol in cells derived from obligate NP-C heterozygotes reinforced this concept (Kruth *et al.,* 1986). However, in spite of the extensive lysosomal involvement in cholesterol storage, the specific intracellular location of the transport block remains to be resolved.

Most recently, examination of filipin-stained NP-C cells has turned our attention to the Golgi apparatus (Figure 1). Upon internalization of LDL by NP-C cells, cholesterol-filled lysosomes migrate to the perinuclear region adjacent to the Golgi complex (Blanchette-Mackie *et al.,* 1988). During the early phase of LDL uptake (4 h), filipin fluorescence is seen not only in NP-C lysosomes but also in the adjacent Golgi area, while normal cells show no such foci of filipin fluorescence. By 24 h of LDL incubation, NP-C (but not normal) lysosomes are intensely and uniformly fluorescent, and the Golgi-associated filipin fluorescence is now comparable in NP-C and normal cells.

Freeze-fracture electron microscopy of filipin-treated samples was employed to probe more precisely the Golgi distribution of cholesterol (Coxey *et al.,* 1993). In normal cells, LDL uptake led to a sterol enrichment of all Golgi cisternae with an increasing *cis* to *trans* gradient. In NP-C cells, uptake of LDL led to an anomalous enrichment of *trans* Golgi cisternae with cholesterol, while adjacent *cis*-medial cisternae and *trans* Golgi transport vesicles were cholesterol deficient relative to normal cells.

These experiments collectively suggested that the Golgi is involved in the intracellular distribution of endocytosed cholesterol and that the NP-C mutation affects the participation of both the lysosomes and Golgi complex in the process (Figure 2). A current challenge is to determine whether the mutation more directly targets cholesterol transport (out of lysosomes) or sorting (Golgi processing). The argument for a defect targeted directly to lysosomes finds conceptual support in studies that have documented facilitated movement of metabolites from lysosomes (Gahl, 1989). The idea for a primary defect in Golgi targeting is supported by recent studies suggesting that the Golgi participates in a number of processes related to the intracellular trafficking of cholesterol (see section 5.4). If the protein affected by the NP-C mutation is involved in cholesterol sorting in the Golgi complex, this molecule must not sort and shuttle *de novo* cholesterol from the endoplasmic reticulum to plasma membrane, as those pathways are normal in NP-C cells (Liscum *et al.,* 1989).

4. MOLECULAR LESION

In addition to the human disorder, three separate animal models of NP-C disease have been reported—in the mouse (Pentchev *et al.,* 1986; Ohno *et al.,* 1992), cat (Lowenthal *et al.,* 1990), and dog (Kuwamura *et al.,* 1993). Genetic analyses have been carried out on the murine and human disorders. First, the murine NP-C mutation was linked to mouse chromosome 18 (Sakai *et al.,* 1991). The positional cloning of the human mutation was then provided with an expeditious start by the finding that the murine disorder was complemented by human chromosome 18 (Kurimasa *et al.,* 1993) and by the strong syntenic relationship between mouse and human chromosomes 18. The human mutation was initially

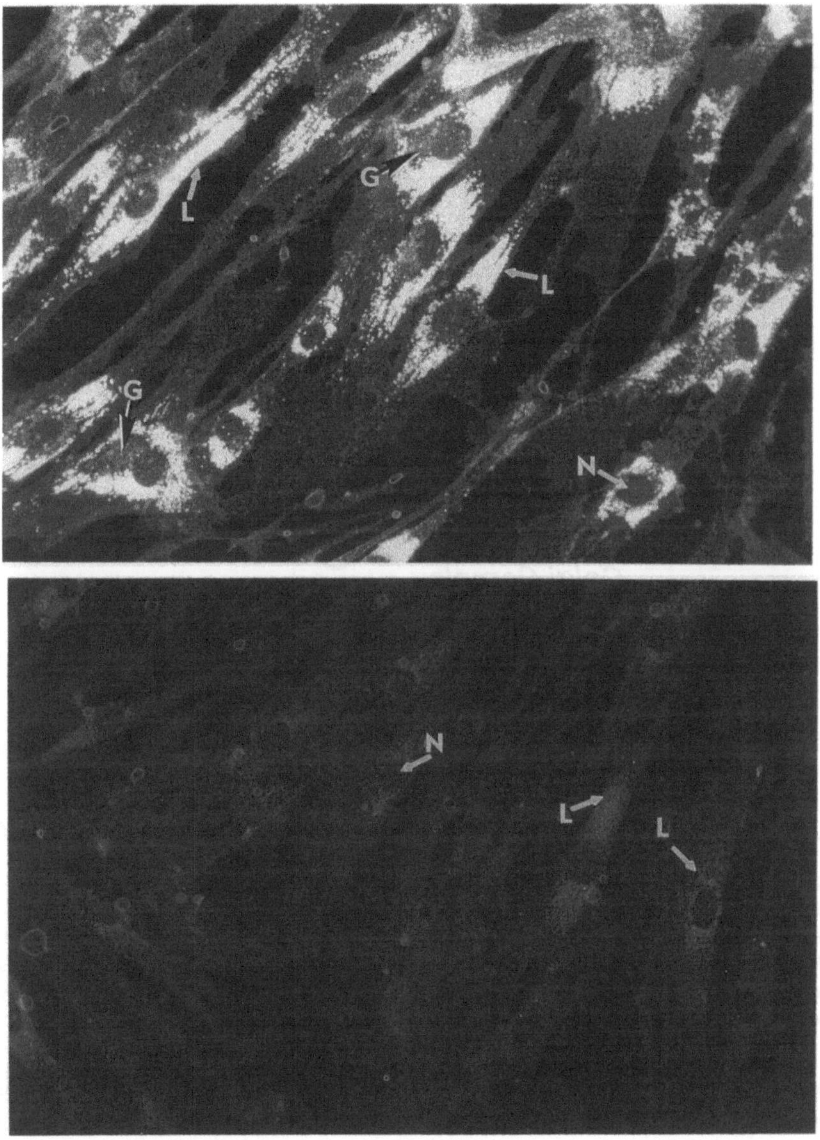

FIGURE 1. Cytochemical analysis of intracellular cholesterol distribution by filipin fluorescence microscopy. Cultured Niemann-Pick C (top) and normal fibroblasts (bottom) were incubated with LDL for 24 h followed by fixation and staining with the cholesterol-specific marker filipin. There is little intracellular filipin fluorescence in normal cells and intense filipin fluorescence in Niemann-Pick C cells. In mutant cells, fluorescent filipin-cholesterol complexes are located in lysosomes (L) and in the Golgi complex (G) situated in the perinuclear area. Nucleus, N 280X.

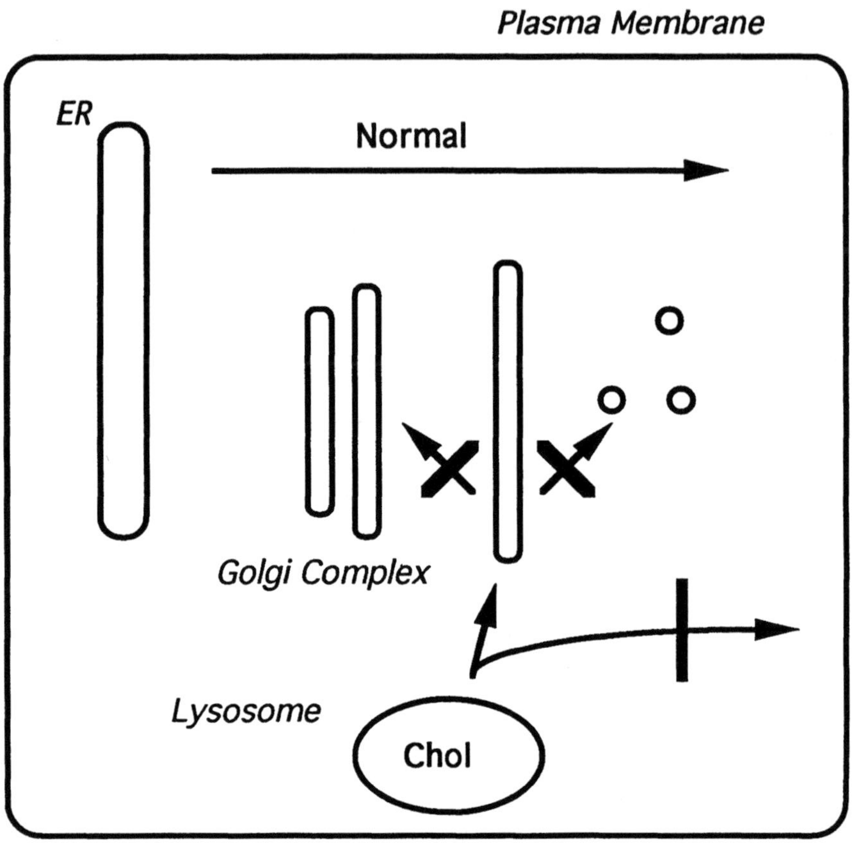

FIGURE 2. Pathways of cholesterol movement in NP-C fibroblasts. The movement of cholesterol between the endoplasmic reticulum and plasma membrane is normal in NP-C fibroblasts. However, LDL-derived cholesterol accumulates in lysosomes and in the *trans* cisternae of the Golgi complex. LDL-cholesterol transfer to other cellular compartments and mediation of homeostatic responses is greatly delayed in NP-C cells. The route and mechanism of LDL-cholesterol transport are not known, but current evidence suggests that the Golgi complex plays a role in sterol sorting.

linked to the pericentromeric region of chromosome 18 (Carstea *et al.*, 1993) and more detailed microsatellite mapping then localized it to the 18q11 region, in an interval spanning approximately 2 CM (Carstea *et al*, 1997). Recent cell fusion studies have revealed two separate complementation groups within approximately 100 NP-C fibroblast cultures examined (Steinberg *et al.*, 1994; Vanier *et. al.*, 1996). The major complementation group (encompassing greater than 90% of cultures analyzed) is linked to the defined chromosome 18 region, while the minor complementation group in the one multiplex family was shown not to link to this chromosome 18 region.

5. INFLUENCES OF THE NP-C MUTATION ON THE STUDY OF INTRACELLULAR CHOLESTEROL TRANSPORT: A HEURISTIC PROCESS

5.1. Primary Human Fibroblasts

Several recent reviews have dealt with the characterization of intracellular cholesterol transport (Billheimer and Reinhart, 1990; Schroeder *et al.,* 1991; Lange, 1992; Liscum and Dahl, 1992; Liscum and Faust, 1994; Liscum and Underwood, 1995). Our current aim is to consider how knowledge and appreciation of the NP-C lesion have influenced research strategies.

The NP-C mutation exhibits a clear gene dosage effect; that is, the lysosomal cholesterol accumulation and homeostatic responses in fibroblasts from obligate heterozygotes are intermediate between that of normal and homozygotes. This finding suggests that the process of cholesterol egress from lysosomes is protein-mediated and argues in favor of a protein with a structural (receptor or transporter), rather than enzymatic, function (Pentchev *et al.,* 1994a). The existence of at least two NP-C complementation groups indicates that several proteins are involved in this trafficking pathway. Furthermore, since rapid progress is being made in positional cloning of the NP-C gene, protein candidates will soon be at hand.

5.2. Somatic Cell Mutants

Another approach for identifying cholesterol transport molecules is to isolate somatic cell mutants that express single gene defects in cholesterol's transport or cellular distribution. One advantage of this strategy is that complementation analysis of a large family of selected mutants has the potential of revealing multiple alleles that produce the desired phenotype. That is, a somatic cell genetic approach will identify genes that, in addition to the NP-C gene(s), play a role in cholesterol movement. Chinese hamster ovary (CHO) cells have proved to be an excellent model system for such an analysis; they are an immortalized cell line that is functionally haploid for many genes. Exogenous DNA can readily be introduced into mutant CHO cells, thus facilitating the cloning of both complementing and suppressing genes. Two laboratories have independently isolated cholesterol transport-defective CHO cells (Cadigan *et al.,* 1990; Dahl *et al.,* 1992). Complementation analysis has revealed that six mutant lines are in the same complementation class (i.e., express defects in the same gene) and exhibit the NP-C biochemical phenotype. A second mutant class exhibits a defect in a separate cholesterol transport pathway (Dahl *et al.,* 1994) (Figure 3). Complementation analysis between the CHO mutants and NP-C fibroblasts is in progress to determine conclusively whether either mutant class shares the same gene defect as NP-C cells.

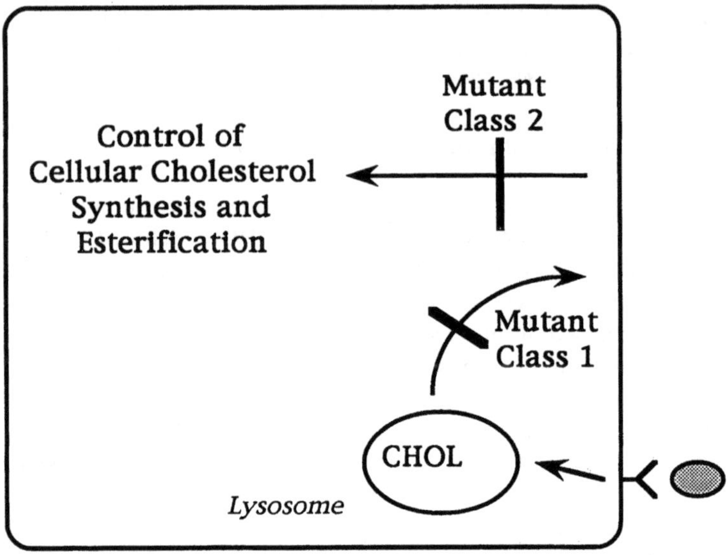

FIGURE 3. Somatic cell mutants in intracellular cholesterol transport. In normal cells, LDL-derived cholesterol is transported to the plasma membrane and other cellular sites. The rising cholesterol levels cause suppression of cellular cholesterol synthesis and activation of cholesterol esterification. Mutant class 1 exhibits defective movement of LDL-derived cholesterol from lysosomes to the plasma membrane and greatly delayed regulation of cellular cholesterol homeostasis. Mutant class 2 shows normal movement of cholesterol to the plasma membrane but defective transport to the cell interior and abnormal control of homeostatic responses.

5.3. Pharmacological Agents that Perturb Cholesterol Movement between Organelles

Another major field of interest centers on the action of a class of compounds called hydrophobic amines, such as U18666A and imipramine. Early studies of such compounds centered on the effects of U18666A on cholesterol synthesis (reviewed by Liscum and Dahl, 1992). We found, however, that the characteristics of a U18666A- or imipramine-treated cell mimicked the NP-C phenotype (Liscum and Faust, 1989; Rodriguez-Lafrasse *et al.*, 1990). LDL-cholesterol accumulates in the lysosomes of U18666A- or imipramine-treated cells, resulting in impaired movement of LDL-cholesterol to the plasma membrane and delayed homeostatic responses. The family of hydrophobic amines with this action also includes stearylamine, RV-538, and sphinganine (Roff *et al.*, 1991) (Figure 4). The mechanism(s) of action of these agents is unknown. Hydrophobic amines might interfere with cholesterol movement by directly or indirectly inhibiting a cholesterol transport

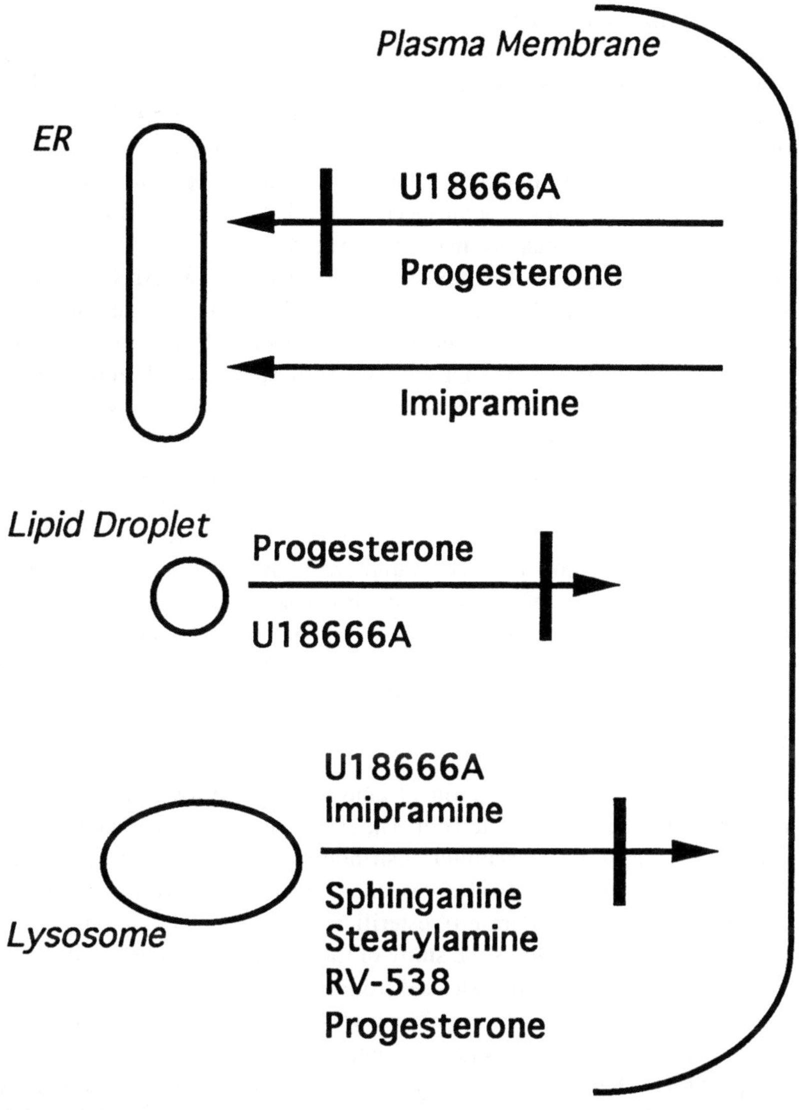

FIGURE 4. Pharmacological agents that inhibit intracellular cholesterol transport pathways. Movement of LDL-derived cholesterol out of lysosomes is inhibited by hydrophobic amines (U18666A, imipramine, sphinganine, stearylamine, and RV-538) and the steroid progesterone. Transport of cholesterol from the plasma membrane to the endoplasmic reticulum and mobilization of cholesterol from lipid droplets is prevented by U1866A and progesterone.

molecule, or perhaps by altering membrane biophysical characteristics. The hypothesis that they work through a common mechanism to inhibit the action of a lysosomal component is most likely too simplistic, since U18666A also inhibits the movement of cholesterol from the cell surface to the cell interior while imipramine does not (Harmala *et al.*, 1994; Lange, 1994). Further elucidation of the mechanism of hydrophobic amine action will be valuable, since a hydrophobic amine binding site may represent a part of the putative cholesterol transport molecule or may reveal information on the transport mechanism itself.

A second major class of drugs found to perturb cholesterol transport are steroids. An earlier study of the processing of cholesteryl esters in macrophages (McGookey and Anderson, 1983) led to a study of the effect of progesterone on the intracellular distribution of LDL-derived cholesterol in cultured fibroblasts; progesterone was found to induce a phenotype strikingly similar to NP-C (Butler *et al.*, 1992). Progesterone was also shown to cause cholesterol enrichment of *trans* Golgi cisternae, after LDL uptake, in a manner resembling NP-C (Blanchette-Mackie *et al.*, 1997). Among a number of other steroids tested, only pregnenolone induced similar effects. A more comprehensive structure-function analysis found that a number of compounds with an oxo group at C_{17} or C_{20} induced an accumulation of endocytosed cholesterol in lysosomes (Aikawa *et al.*, 1994). Such a structural specificity suggests the targeting of a specific protein involved in interactions with sterol (Pentchev *et al.*, 1994b). The possibility that the NP-C mutation and progesterone target the same protein can be further explored by attempting to isolate CHO cell mutants that have developed resistance to progesterone action.

The drug-induced sequestration of cholesterol in lysosomes is readily reversible when the drugs are washed out. The subsequent release of a large bolus of lysosomal cholesterol for intracellular distribution can be monitored by: (i) transfer to extracellular acceptors, (ii) fluorescence microscopy of filipin-stained cells, and (iii) measurement of cholesterol esterification. The reversible block of lysosomal cholesterol egress allows the study of mechanisms that control the transfer process. When CHO cells were released from a U18666A block, transfer of lysosomal cholesterol to the plasma membrane was not affected by agents disturbing lysosomal pH, cytoskeletal organization, energy production, or protein synthesis (Liscum, 1990). When mouse macrophages were released from a progesterone block, cholesterol mobilization was affected by pH perturbations with both NH_4Cl and chloroquine, as well as by bafilomycin A, which inhibits the vacuolar-type H^+-ATPase (Furuchi *et al.*, 1993). When human fibroblasts were released from progesterone treatment, cholesterol trafficking was retarded by monensin and NH_4Cl but was not affected by either brefeldin A or cytoskeletal poisons (Blanchette-Mackie *et al.*, 1997). These variable findings may reflect differences in cell types and/or method of analysis. The development of a repertoire of molecular and chemical means of modulating intracellular cholesterol transport bodes well for future experimental exploration.

5.4. Mechanisms Involved in Removal of Cholesterol from the Cell

Cellular levels of free cholesterol are under tight control. Cholesterol biosynthetic and esterification enzymes and LDL receptor activity are all exquisitely sensitive to cellular levels of free cholesterol. Upon internalization of LDL, cholesterol must exit the lysosomes and move throughout the cell in order to mediate the transcriptional regulation, enzyme activation, and protein degradation necessary to limit the further acquisition of cholesterol. This tight control is necessary because unregulated accumulation of cholesterol is cytotoxic. As a further defense mechanism against excess free cholesterol, mechanisms exist to shuttle cholesterol to the plasma membrane and translocate it to extracellular acceptors, such as HDL. This "reverse cholesterol transport" is an important research topic because of its involvement in the pathophysiology of atherosclerosis (Fielding and Fielding, 1995; also Chapter 11 by Johnson *et al.*).

5.4.1. Cholesterol Crystals

NP-C fibroblasts and atherogenic foam cells display many similar characteristics, and knowledge gained in one field is certain to benefit the other. In NP-C cells, excess lysosomal cholesterol accumulates because of the transport defect. Cholesterol synthesis and receptor-mediated internalization of LDL continue, despite the rising influx of cholesterol due to its metabolic sequestration in lysosomes. The storage leads to formation of cholesterol crystals in NP-C lysosomes (Pentchev *et al.,* 1994a). Similarly, macrophages are transformed into foam cells because lipoprotein uptake is not homeostatically controlled. Cholesterol crystals also form in the lysosomes of foam cells, perhaps because delivery of lipoprotein cholesterol into lysosomes exceeds their capacity to relocate the sterol to other intracellular sites (Tangirala *et al.,* 1994). This may represent an inherent kinetic limitation in the capacity of lysosomes to transport cholesterol.

5.4.2. Oxidized LDL Metabolism

J774 mouse macrophages also display an NP-C-like lysosomal sequestration of LDL-cholesterol and deficient induction of cholesterol esterification when cells are incubated with oxidized LDL, but not when cells internalized equal amounts of acetylated LDL (Roma *et al.,* 1992). The possibility that oxidized by-products of the modified LDL inhibit lysosomal cholesterol efflux remains to be explored. This idea is supported by the finding that oxysterols isolated from oxidized LDL and added to acetylated LDL cause a lysosomal accumulation of cholesterol (Maor and Aviram, 1994). Not all oxysterols cause this phenotype, however, as 25- and 26-hydroxycholesterol and 7-ketocholesterol do not alter lysosomal cholesterol flux in mouse macrophages (Aikawa *et al.,* 1994) or cultured human fibroblasts (Pentchev, unpublished data). The mechanism of lysosomal cholesterol trapping when cells endocytose oxidized LDL remains to be resolved and is of great relevance to both NP-C disease and atherosclerosis.

5.4.3. Mobilization of Cholesterol Ester Droplets

The similarities in the cholesterol lipidoses presented by NP-C and atherogenic foam cells highlight the utility of the genetic lesion in dissecting relevant intracellular pathways of cholesterol mobilization. Many of the experimental protocols developed to test other pathways of cholesterol transfer are now being applied to study the movement from cholesteryl ester lipid droplets to the plasma membrane, since that pathway is a major route for reverse cholesterol transport. Long before it was determined to cause a NP-C phenotype, progesterone was shown to disrupt the cholesteryl ester cycle of cultured mouse macrophage foam cells (McGookey and Anderson, 1983), causing cholesteryl ester droplets to disappear and unesterified cholesterol to accumulate in filipin-positive perinuclear vesicles. Based on our current understanding of progesterone's effect on intracellular cholesterol trafficking, it is possible that the perinuclear vesicles represent cholesterol-filled lysosomes. More recently, it was shown that progesterone blocks the intracellular translocation of free cholesterol derived from the cholesteryl ester pool to the plasma membrane (Mazzone *et al.*, 1995). Like progesterone, U18666A also prevents lipid droplet-derived cholesterol from being transported freely in the cell (Warner *et al.*, 1995). These findings suggest that lysosomes play a major role in routing free cholesterol to the plasma membrane for transfer to suitable extracellular acceptors, such as HDL.

The Golgi complex may also play a role in mobilization of cholesterol derived from cholesteryl ester droplets. Disruption of the Golgi complex structure and function with drugs such as monensin and brefeldin A block relocation of lipid droplet-derived cholesterol to the plasma membrane, thereby preventing HDL-mediated cellular cholesterol efflux (Mendez, 1995). The trafficking of cholesterol from cholesteryl ester pools to the mitochondria for steroid synthesis has been shown to involve a transient passage through the plasma membrane (Nagy and Freeman, 1990). Treatment of cultured rat adrenocortical and ovarian granulosa cells with okadaic acid blocked utilization of intracellular cholesteryl ester pools for steroidogenesis (Azhar *et al.*, 1994). The action of okadaic acid is postulated to be due to disruption of Golgi complex structure. Based on such studies, it has been speculated that the Golgi apparatus could be involved in targeting of cholesterol to various intracellular sites of utilization and storage (Mendez, 1995).

Thus, the Golgi complex may play as central a role in all aspects of sterol trafficking as it does in protein trafficking. Since NP-C fibroblasts display altered cholesterol distribution in the Golgi, identification of the NP-C gene product may reveal a Golgi protein involved in cholesterol trafficking. The application of such a finding may be of great importance in understanding the critical link between the intracellular transport and processing of cholesterol.

6. REFERENCES

Aikawa, K., Furuchi, T., Fujimoto, Y., Arai, H., and Inoue, K., 1994, Structure-specific inhibition of lysosomal cholesterol transport in macrophages by various steroids, *Biochim. Biophys. Acta* **1213:**127–134.

Azhar, S., Frazier, J. A., Tsai, L., and Reaven, E., 1994, Effect of okadaic acid on utilization of lipoprotein-derived cholesteryl esters by rat steroidogenic cells, *J. Lipid Res.* **35:**1161–1176.

Billheimer, J. T., and Reinhart, M. P., 1990, Intracellular trafficking of sterols, in: *Subcellular Biochemistry*, Volume 16 (H. J. Hilderson, ed.), Chapter 10, Plenum Press, New York.

Blanchette-Mackie, E. J., Dwyer, N. K., Amende, L. M., Kruth, H. S., Butler, J. D., Sokol, J., Comly, M. E., Vanier, M. T., August, J. T., Brady, R. O., and Pentchev, P. G., 1988, Type-C Niemann-Pick disease: low density lipoprotein uptake is associated with premature cholesterol accumulation in the Golgi complex and excessive cholesterol storage in lysosomes, *Proc. Natl. Acad. Sci. USA* **85:**8022–8026.

Blanchette-Mackie, E. J., Coxey, R. O., Campbell, G., Butler, J. D., Dwyer, N. K., Brady, R. O., and Pentchev, P. G., 1997, Cholesterol sequestration in Golgi and lysosomes: effect of progesterone, in preparation.

Brown, M. S., and Goldstein, J. L., 1986, A receptor-mediated pathway for cholesterol homeostasis, *Science* **232:**34–47.

Butler, J. D., Blanchette-Mackie, J., Goldin, E., O'Neill, R. R., Carstea, E. D., Roff, C. F., Patterson, M. C., Patel, S., Comly, M. E., Cooney, A., Vanier, M. T., Brady, R. O., and Pentchev, P. G., 1992, Progesterone blocks cholesterol translocation from lysosomes, *J. Biol. Chem.* **267:**23797–23805.

Cadigan, K. M., Spillane, D. M., and Chang, T. Y., 1990, Isolation and characterization of Chinese hamster ovary cell mutants defective in intracellular low density lipoprotein-cholesterol trafficking, *J. Cell Biol.* **110:**295–308.

Carstea, E. D., Polymeropoulos, M. H., Parker, C. C., Detera-Wadleigh, S. D., O'Neill, R. R., Patterson, M. C., Goldin, E., Xiao, H., Straub, R. E., Vanier, M. T., Brady, R. O., and Pentchev, P. G., 1993, Linkage of Niemann-Pick disease type C to human chromosome 18, *Proc. Natl. Acad. Sci. USA* **90:**2002–2004.

Carstea, E. D., Parker, C. C., Fandino, L. B., Overhauser, J., Weissenback, J., Vanier, M. T., and Polymeropoulus, M. H., 1997, Localization of the major gene for human Niemann-Pick type C to 18q11, in preparation.

Coxey, R. A., Pentchev, P. G., Campbell, G., and Blanchette-Mackie, E. J., 1993, Differential accumulation of cholesterol in Golgi compartments of normal and Niemann-Pick type C fibroblasts incubated with LDL: a cytochemical freeze-fracture study, *J. Lipid Res.* **34:**1165–1176.

Crocker, A. C., and Farber, S., 1958, Niemann-Pick disease: a review of eighteen patients, *Medicine* **37:**1–95.

Dahl, N. K., Reed, K. L., Daunais, M. A., Faust, J. R., and Liscum, L., 1992, Isolation and characterization of Chinese hamster ovary cells defective in the intracellular metabolism of LDL-derived cholesterol, *J. Biol. Chem.* **267:**4889–4896.

Dahl, N. K., Daunais, M. A., and Liscum, L., 1994, A second complementation class of cholesterol transport mutants with a variant Niemann-Pick type C phenotype, *J. Lipid Res.* **35:**1839–1849.

Fielding, C. J., and Fielding, P. E., 1995, Review: molecular physiology of reverse cholesterol transport, *J. Lipid Res.* **36:**211–228.

Furuchi, T., Aikawa, K., Arai, H., and Inoue, K., 1993, Bafilomycin A_1, a specific inhibitor of vacuolar-type H^+-ATPase, blocks lysosomal cholesterol trafficking in macrophages, *J. Biol. Chem.* **268:**27345–27348.

Gahl, W. A., 1989, Lysosomal membrane transport in cellular nutrition, *Annu. Rev. Nutr.* **9:**39–61.

Goldstein, J. L., Hobbs, H. H., and Brown, M. S., 1995, Familial hypercholesterolemia, in: *The Metabolic and Molecular Basis of Inherited Disease* (C. R. Scriver, A. L. Beaudet, W. S. Sly, and D. Valle, eds.), 7th ed., Vol. 2, pp. 1981–2030, McGraw-Hill, New York.

Harmala, A. S., Porn, M. I., Mattjus, P., and Slotte, J. P., 1994, Cholesterol transport from plasma membranes to intracellular membranes is inhibited by 3β-[2-(diethylamino)ethoxy]androst-5-en-17-one, *Biochim. Biophys. Acta* **1211:**317–325.

Kruth, H. S., Comly, M. E., Butler, J. D., Vanier, M. T., Fink, J. K., Wenger, D. A., Patel, S., and Pentchev, P. G., 1986, Type C Niemann-Pick disease: abnormal metabolism of low density lipoprotein in homozygous and heterozygous fibroblasts, *J. Biol. Chem.* **261:**16769–16774.

Kurimasa, A., Ohno, K., and Oshimura, M., 1993, Restoration of cholesterol metabolism in 3T3 cell lines derived from the sphingomyelinosis mouse (spm/spm) by transfer of a human chromosome 18, *Hum. Genet.* **92:**157–162.

Kuwamura, M., Awakura, T., Shimada, A., Umemura, T., Kagota, K., Kawamura, N., and Naiki, M., 1993, Type C Niemann-Pick disease in boxer dog, *Acta Neuropathol.* **85:**345–348.

Lange, Y., 1992, Review: tracking cell cholesterol with cholesterol oxidase, *J. Lipid Res.* **33:**315–321.

Lange, Y., 1994, Cholesterol movement from plasma membrane to rough endoplasmic reticulum. Inhibition by progesterone, *J. Biol. Chem.* **269:**3411–3414.

Liscum, L., 1990, Pharmacological inhibition of the intracellular transport of low-density lipoprotein-derived cholesterol in Chinese hamster ovary cells, *Biochim. Biophys. Acta* **1045:**40–48.

Liscum, L., and Dahl, N. K., 1992, Review: intracellular cholesterol transport, *J. Lipid Res.* **33:**1239–1254.

Liscum, L., and Faust J. R., 1987, Low density lipoprotein (LDL)-mediated suppression of cholesterol synthesis and LDL uptake is defective in Niemann-Pick type C fibroblasts, *J. Biol. Chem.* **262:**17002–17008.

Liscum, L., and Faust, J. R., 1989, The intracellular transport of low density lipoprotein-derived cholesterol is inhibited in Chinese hamster ovary cells cultured with 3-β-[2-(diethylamino) ethoxy]androst-5-en-17-one, *J. Biol. Chem.* **264:**11796–11806.

Liscum, L., and Faust, J. R., 1994, Compartmentation of cholesterol within the cell, *Curr. Opin. Lipidology* **5:**221–226.

Liscum, L., and Underwood, K. W., 1995, Minireview: intracellular cholesterol transport and compartmentation, *J. Biol. Chem.* **270:**15443–15443.

Liscum, L., Ruggiero, R. M., and Faust, J. R., 1989, The intracellular transport of low density lipoprotein-derived cholesterol is defective in Niemann-Pick type C fibroblasts, *J. Cell Biol.* **108:**1625–1636.

Lowenthal, A. C., Cummings, J. F., Wenger, D. A., Thrall, M. A., Wood, P. A., and de Lahunta, A., 1990, Feline sphingolipidosis resembling Niemann-Pick disease type C, *Acta Neuropathol.* **81:**189–197.

Maor, I., and Aviram, M., 1994, Oxidized low density lipoprotein leads to accumulation of unesterified cholesterol as a result of lysosomal trapping of the lipoprotein hydrolyzed cholesteryl ester, *J. Lipid Res.* **35:**803–819.

Mazzone, T., Krishna, M., and Lange, Y., 1995, Progesterone blocks intracellular translocation of free cholesterol derived from cholesteryl ester in macrophages, *J. Lipid Res.* **36:**544–551.

McGookey, D. J., and Anderson, R. G. W., 1983, Morphological characterization of the cholesteryl ester cycle in cultured mouse macrophage foam cells, *J. Cell Biol.* **97:**1156–1168.

Mendez, A. J., 1995, Monensin and brefeldin A inhibit high density lipoprotein-mediated cholesterol efflux from cholesterol-enriched cells, *J. Biol. Chem.* **270:**5891–5900.

Nagy, L., and Freeman, D. A., 1990, Effect of cholesterol transport inhibitors on steroidogenesis and plasma membrane cholesterol transport in cultured MA-10 Leydig tumor cells, *Endocrinology* **126:**2267–2276.

Ohno, K., Nanba, E., Miyawaki, T., Kitagawa, T., and Takeshita, K., 1992, A cell line derived from sphingomyelinosis mouse shows alterations in intracellular cholesterol metabolism similar to those in type C Niemann-Pick disease, *Cell Struc. Func.* **17:**229–235.

Pentchev, P. G., Boothe, A. D., Kruth, H. S., Weintroub, H., Strivers, J., and Brady, R. O., 1984, A genetic storage disorder in BALB/C mice with a metabolic block in esterification of exogenous cholesterol, *J. Biol. Chem.* **259:**5784–5791.

Pentchev, P. G., Comly, M. E., Kruth, H. S., Patel, S., Proestel, M., and Weintroub, H., 1986, The cholesterol storage disorder of the mutant BALB/C mouse. A primary genetic lesion closely linked to defective esterification of exogenously derived cholesterol and its relationship to human type C Niemann-Pick disease, *J. Biol. Chem.* **261:**2772–2777.

Pentchev, P. G., Comly, M. E., Kruth, H. S., Tokoro, T., Butler, J., Sokol, J., Filling-Katz, M., Quirk, J. M., Marshall, D. C., Patel, S., and Vanier, M. T., 1987, Group C Niemann-Pick disease: faulty regulation of low-density lipoprotein uptake and cholesterol storage in cultured fibroblasts, *FASEB J.* **1:**40–45.

Pentchev, P. G., Blanchette-Mackie, E. J., and Davidowicz, E. A., 1994a, The Niemann-Pick C gene: a key to pathways of intracellular cholesterol transport, *Trends Cell Biol.* **4:**365–369.

Pentchev, P. G., Brady, R. O., Blanchette-Mackie, E. J., Vanier, M. T., Carstea, E. D., Parker, C. C., Goldin, E., and Roff, C. F., 1994b, Review: the Niemann-Pick C lesion and its relationship to the intracellular distribution and utilization of LDL cholesterol, *Biochim. Biophys. Acta* **1225:** 235–243.

Pentchev, P. G., Vanier, M. T., Suzuki, K., and Patterson, M., 1995, Niemann-Pick C disease type C: a cellular cholesterol lipidosis, in: *The Metabolic and Molecular Basis of Inherited Disease* (C. R. Scriver, A. L. Beaudet, W. S. Sly, and D. Valle, eds.), 7th ed., Vol. 2, pp. 2625–2639, McGraw-Hill, New York.

Rodriguez-Lafrasse, C., Rousson, R., Bonnet, J., Pentchev, P. G., Louisot, P., and Vanier, M. T., 1990, Abnormal cholesterol metabolism in imipramine-treated fibroblast cultures. Similarities with Niemann-Pick type C disease, *Biochim. Biophys. Acta* **1043:**123–128.

Roff, C. F., Goldin, E., Comly, M. E., Cooney, A., Brown, A., Vanier, M. T., Miller, S. P. F., Brady, R. O., and Pentchev, P. G., 1991, Type C Niemann-Pick disease: use of hydrophobic amines to study defective cholesterol transport, *Dev. Neurosci.* **13:**315–320.

Roma, P., Bernini, F., Fogliatto, R., Bertulli, S. M., Negri, S., Fumagalli, R., and Catapano, A. L., 1992, Defective catabolism of oxidized LDL by J774 murine macrophages, *J. Lipid Res.* **33:**819–829.

Rothman, J. E., 1994, Mechanisms of intracellular protein transport, *Nature* **372:**55–63.

Sakai, Y., Miyawaki, S., Shimizu, A., Ohno, K., and Watanabe, T., 1991, A molecular genetic map of mouse chromosome 18, including spm, Grl-1, Fim-2/c-fms, and Mbp, *Biochem. Genet.* **29:**103–113.

Schroeder, F., Jefferson, J. R., Kier, A. B., Knittel, J., Scallen, T. J., Wood, W. G., and Hapala, I., 1991, Minireview: membrane cholesterol dynamics; cholesterol domains and kinetic pools, *Proc. Soc. Exptl. Biol. Med.* **196:**235–252.

Schuchman, E. H., and Desnick, R. J., 1995, Niemann-Pick disease types A and B: acid sphingomyelinase deficiencies, in: *The Metabolic and Molecular Basis of Inherited Disease* (C. R. Scriver, A. L. Beaudet, W. S. Sly, and D. Valle, eds.), 7th ed., Vol. 2, pp. 2601–2624, McGraw-Hill, New York.

Sokol, J., Blanchette-Mackie, E. J., Kruth, H. S., Dwyer, N. K., Amende, L. M., Butler, J. D., Robinson, E., Patel, S., Brady, R. O., Comly, M. E., Vanier, M. T., and Pentchev, P. G., 1986, Type C Niemann-Pick disease: lysosomal accumulation and defective intracellular mobilization of low density lipoprotein cholesterol, *J. Biol. Chem.* **263:**3411–3417.

Steinberg, S. J., Ward, C. P., and Fensom, A. H., 1994, Complementation studies in Niemann-Pick disease type C indicate the existence of a second group, *J. Med. Genet.* **31:**317–320.

Tangirala, R. K., Jerome, W. G., Jones, N. L., Small, D. M., Johnson, W. J., Glick, J. M., Mahlberg, F. H., and Rothblat, G. H., 1994, Formation of cholesterol monohydrate crystals in macrophage-derived foam cells, *J. Lipid Res.* **35:**93–104.

Vanier, M. T., Duthel, S., Rodriguez-Lafrasse, C., Pentchev, P., and Carstea, E., D., 1996, Genetic heterogeneity in Niemann-Pick C disease. A study using somatic cell hybridization and linkage analysis, *Am. J. Hum. Genet.* **58:**118–125.

Warner, G. J., Stoudt, G., Bamberger, M., Johnson, W. J., and Rothblat, G. H., 1995, Cell toxicity induced by inhibition of acyl coenzyme A:cholesterol acyltransferase and accumulation of unesterified cholesterol, *J. Biol. Chem.* **270:**5772–5778.

Chapter 15

Sitosterolemia

Gerald Salen, Sarah Shefer, Lien Nguyen, Gene C. Ness, G. S. Tint, and Ashok K. Batta

1. INTRODUCTION

It is more than 21 years since the first report on sitosterolemia appeared (Bhattacharyya and Connor, 1974). Two sisters with tendon xanthomatosis and normal plasma cholesterol levels were found to have elevated plant sterol concentrations in the plasma. A high percentage of dietary sitosterol was absorbed from the intestine as measured by the sterol balance technique, and was believed to account for the plant sterol accumulation. Since the original report, 30 patients from 19 families have been detected (Bhattacharyya and Connor, 1974; Shulman *et al.*, 1976; Khachadurian and Salen, 1980; Miettinen, 1980; Whitington *et al.*, 1980; Kwiterovich *et al.*, 1981; Wang *et al.*, 1981; Lin *et al.*, 1983; Salen *et al.*, 1983; Beaty *et al.*, 1986; Björkhem and Skrede, 1989; Belamarich *et al.*, 1990; Hidaka *et al.*, 1990; Skrede *et al.*, 1985; Lutjohan *et al.*, 1995). The clinical presentation includes tendon xanthomas, aortic stenosis with accelerated atherosclerosis (particularly affecting males at a young age), hemolytic episodes, and arthritis and arthralgias. The risk of atherosclerosis was observed in several young male subjects who died prematurely because of acute myocardial infarctions associated

Gerald Salen, Sarah Shefer, Lien Nguyen, G. S. Tint, and Ashok K. Batta Department of Medicine and the Liver Center, University of Medicine and Dentistry—New Jersey Medical School, Newark, New Jersey 07103. **Gerald Salen and G. S. Tint** Research Service and Medical Service, Department of Veterans Affairs Medical Center, East Orange, New Jersey 07019 **Gene C. Ness** Department of Biochemical and Molecular Biology, College of Medicine, University of South Florida Medical School, Tampa, Florida 33612.

Subcellular Biochemistry, Volume 28: Cholesterol: Its Functions and Metabolism in Biology and Medicine, edited by Robert Bittman. Plenum Press, New York, 1997.

with extensive coronary and aortic arteriosclerosis. The youngest was a 13-year-old Amish male who had 4 other affected siblings (Kwiterovich *et al.,* 1981). In addition, a 17-year-old male personally followed by the authors developed angina pectoris, showed an abnormal cardiac stress test with decreased coronary artery perfusion, and died suddenly of an acute myocardial infarction while exercising (Salen *et al.,* 1985b). Examination of his coronary arteries at postmortem revealed 60% occlusion of the left anterior descending coronary artery (Figure 1). However, multiple microinfarctions were noted in the myocardium, which suggested that the atherosclerosis process had begun earlier, was chronic, and had progressed.

Sitosterolemia is inherited as a recessive trait (Beaty *et al.,* 1986). Heterozygotes are clinically and biochemically normal, although plasma sitosterol levels of some heterozygous subjects may be slightly but significantly increased over controls. These values still differ quantitatively from homozygotes by 10–20-fold (Wang *et al.,* 1981; Hidaka *et al.,* 1990; Salen *et al.,* 1992a). Of interest is the high degree of inheritance of the homozygous state. In two unrelated families symptomatic sitosterolemia was present in 4/4 and 2/4 siblings, respectively.

The purpose of the present report is to summarize our current knowledge about sitosterolemia. Emphasis will be placed on the characteristic clinical and biochemical abnormalities that underlie the diagnosis and pathogenesis of this condition, and the deficiency of HMG-CoA reductase that results in very reduced cholesterol biosynthesis, which we believe is the inherited defect.

2. BIOCHEMICAL FEATURES

2.1. Sterol Content

The hallmark biochemical feature of sitosterolemia is the demonstration of elevated concentrations of sitosterol (24α-ethylcholesterol) in the plasma (Bhattacharyya and Connor, 1974; Rao *et al.,* 1975; Shulman *et al.,* 1976; Khachadurian and Salen, 1980; Miettinen, 1980; Whitington *et al.,* 1980; Kwiterovich *et al.,* 1981; Wang *et al.,* 1981; Lin *et al.,* 1983; Salen *et al.,* 1983; Salen *et al.,* 1985a; Skrede *et al.,* 1985; Beaty *et al.,* 1986; Björkhem and Skrede, 1989; Belamarich *et al.,* 1990; Hidaka *et al.,* 1990; Lutjohan *et al.,* 1995). Actually, all dietary sterols are found in plasma (Gregg *et al.,* 1986), but since sitosterol is usually the most abundant in the diet, proportionately greater quantities are present in plasma and tissues. For this reason, the condition has been named sitosterolemia (Gregg *et al.,* 1986). In addition, the respective 5α-dihydro derivative of cholesterol (cholestanol) and the 5α-dihydro plant sterol derivatives, 5α-campestanol and 5α-sitostanol, are present in increased amounts in plasma and tissues (Dayal *et al.,* 1982; Salen *et al.,* 1985a). Since diets contain only small amounts of cholestanol, 5α-campestanol, and 5α-sitostanol, the 5α-dihydro derivatives probably are produced endogenously in larger amounts (Salen *et al.,* 1985a).

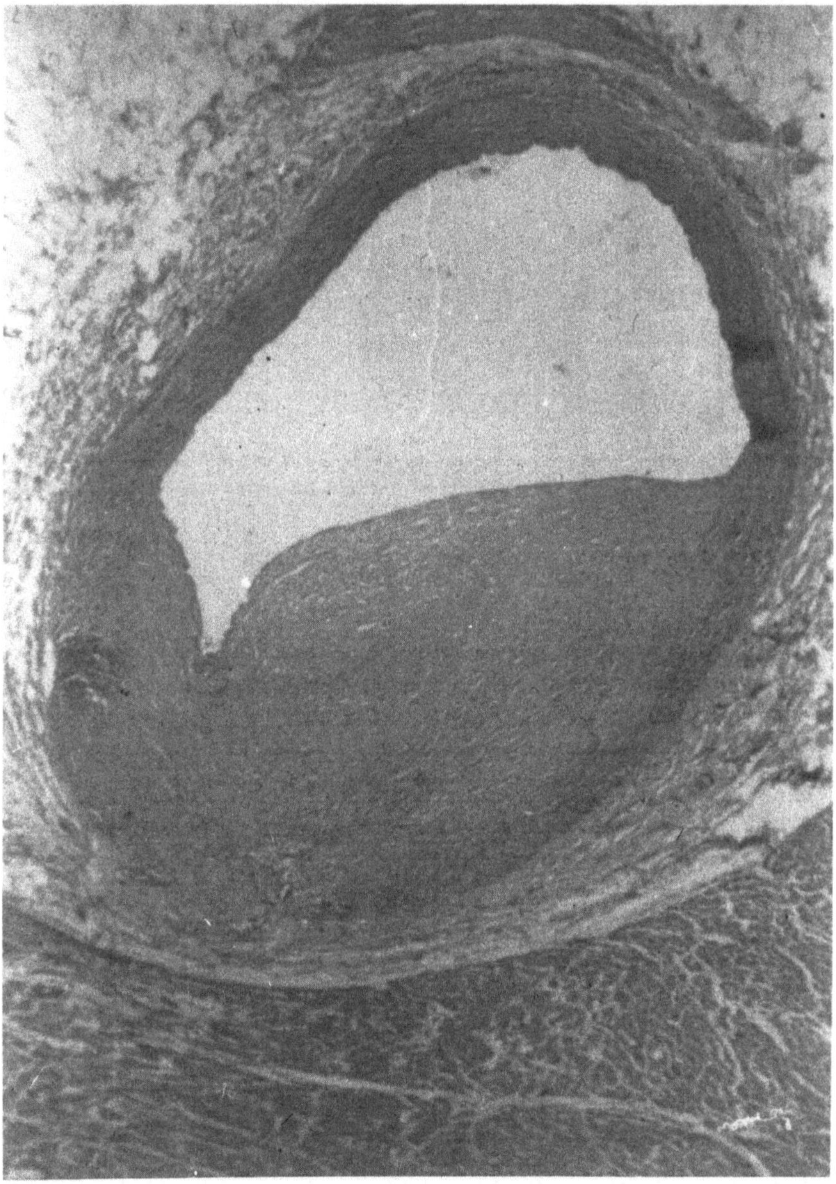

FIGURE 1. Section of coronary artery from a 17-year-old sitosterolemic male who died of an acute myocardial infarction. Sixty percent occulsion of the vessel lumen by atherosclerotic plaque and thickening of the vessel wall is observed. (Data are from Salen *et al.*, 1992a.)

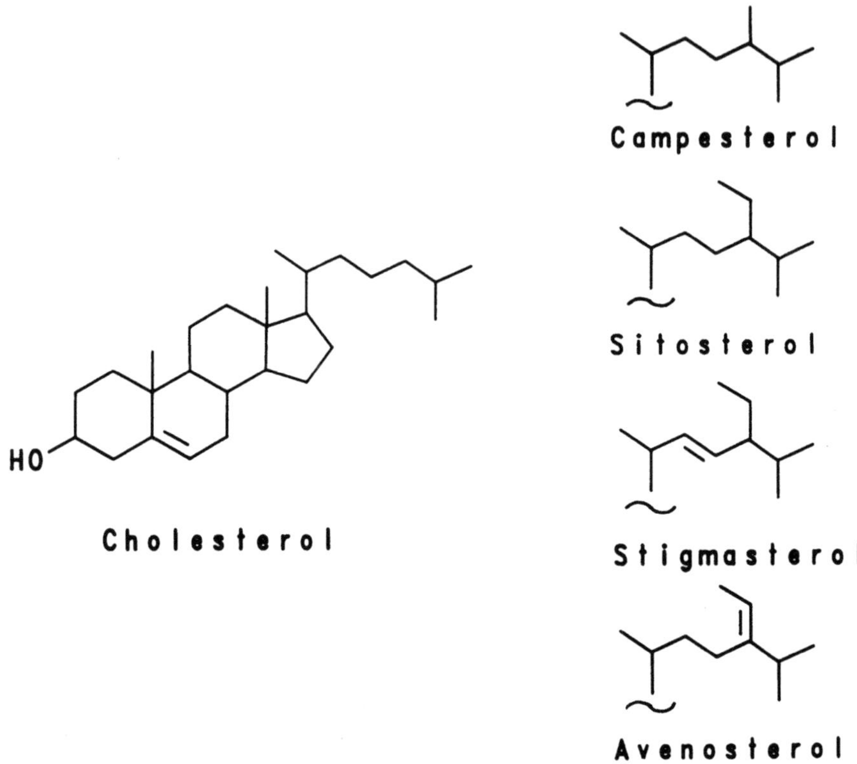

FIGURE 2. Structure of plant sterols found in plasma of sitosterolemic subjects. These sterols have the same ring nucleus as cholesterol, but differ by the addition of substituents on the side chain at carbon 24, the presence of a double bond at carbon 22 in stigmasterol, and a double bond between carbons 24 and 28 in avenosterol.

The diagnosis of sitosterolemia is established by demonstrating increased amounts of plant sterols (campesterol, sitosterol, stigmasterol, and avenosterol; see Figure 2 for structures) and 5α-stanols (see Figure 3 for structures) in plasma and tissues (Bhattacharyya and Connor, 1974; Salen *et al.*, 1985a, 1985b, 1992a). The usual colorimetric assay that depends on a double bond between carbons 5 and 6, and the enzymatic method that detects the 3α-hydroxy group do not distinguish sitosterol from cholesterol. Therefore, gas-liquid chromatography using a capillary column is usually necessary to find plant sterols and 5α-stanols and establish the diagnosis (Figure 4), although high performance liquid chromatography can also be used (Hidaka *et al.*, 1990).

In one family with four affected siblings, the unsaturated sterols represented about 16% of the total plasma sterols, with cholestanol and 5α-dihydro plant stanols making up about 4% (Khachadurian and Salen, 1980; Salen *et al.*, 1983). In other families, plant sterol and 5α-stanols may account for as little as 11% to as

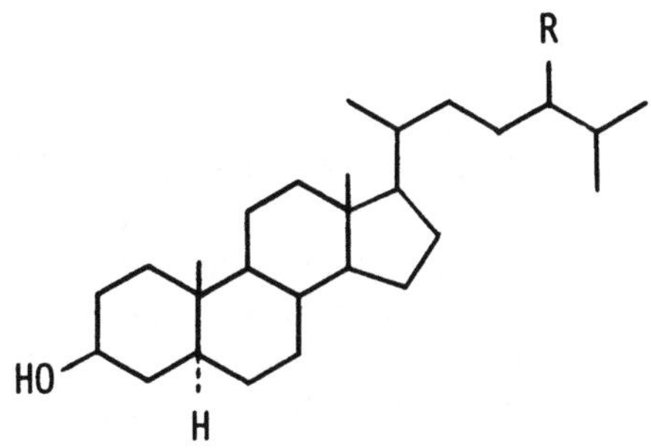

R = H Cholestanol

R = CH_3 24-methyl Cholestanol

R = C_2H_5 24-ethyl Cholestanol

FIGURE 3. Structures of 5α-saturated stanols found in plasma of sitosterolemic subjects. Cholestanol is the 5α-dihydro derivative of cholesterol; 5α-campestanol and 5α-sitostanol are the 5α-dihydro derivatives of campesterol and sitosterol, respectively.

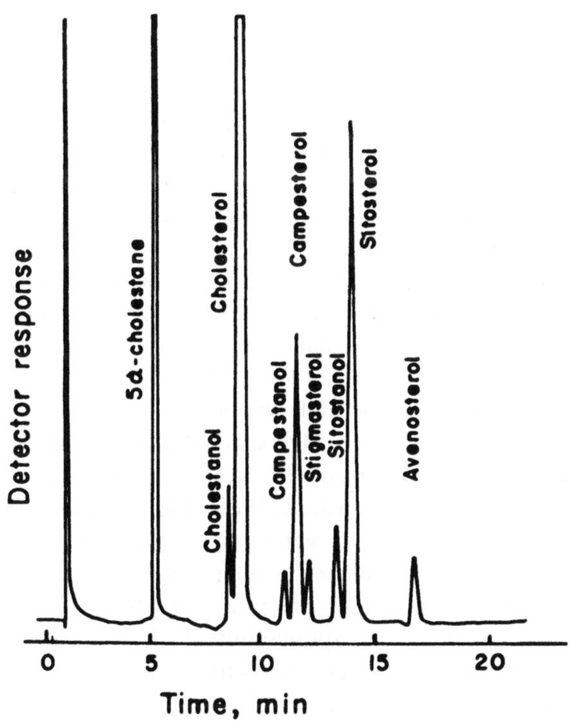

FIGURE 4. Capillary gas-liquid chromatogram of plasma sterols and 5α-stanols from a homozygous sitosterolemic subject. In normal plasma only cholesterol and trace amount (< 1%) of cholestanol and sitosterol are detected. 5α-Cholestane is added as an internal standard. The chromatogram was performed on a Chrompack CP-Wax-57CB capillary column. (Data are from Salen *et al.*, 1992a).

much as 25% of the plasma sterols (Kwiterovich *et al.*, 1981; Salen *et al.*, 1983, 1985a; Lutjohan *et al.*, 1995). Thus, cholesterol represents only about 80% of the total plasma sterols in sitosterolemic homozygotes. The concentration and distribution of sterols and stanols from a number of sitosterolemic homozygotes from five families and their heterozygous relations are presented in Table I. For controls, values from 20 healthy subjects are given, showing that cholesterol normally represents 99.6% of the total sterols, with about 0.2% cholestanol and 0.2% plant sterols. In some heterozygotes, cholestanol and sitosterol levels are similar to controls (Nye *et al.*, 1988; Nguyen *et al.*, 1990a, 1991; Salen *et al.*, 1992a). However, in several obligate heterozygotes, cholestanol and sitosterol levels were slightly but significantly higher than the control means, but still substantially less than found in their homozygous offsprings (Wang *et al.*, 1981; Hidaka *et al.*, 1990). Plasma cholesterol concentrations also may vary considerably in homozygotes. As illustrated in Table I, cholesterol levels may be low, but are usually increased over age-matched controls. However, at times in some subjects, extremely high cholesterol concentrations that resemble the levels found in LDL receptor-deficient hypercholesterolemic subjects are detected (i.e., subject LBU).

2.2. Lipoproteins

Plasma lipoproteins have been measured in homozygous sitosterolemic subjects. The increased amounts of unsaturated plant sterols and saturated 5α-stanols are distributed in about the same proportion among the various lipoprotein fractions (HDL, LDL,IDL, VLDL) (Bhattacharyya and Connor, 1974; Shulman *et al.*, 1976; Salen *et al.*, 1989). However, LDL concentrations tend to be elevated, reflecting higher total sterol concentrations compared with age- and sex-matched controls. Despite the incorporation of increased amounts of plant sterols and 5α-saturated stanols, preparative density gradients for each lipoprotein class were similar to that of controls (Salen *et al.*, 1989). The major proportion of the total low density lipoproteins was isolated in the subfraction of $d = 1.034$ g/ml. The mean particle diameter, 25.7 ± 2.8 nm, for sitosterolemic LDL was not unusual as determined by electron microscopy, and the sitosterolemic LDL was not distinguishable morphologically from normal LDL. Concentrations of HDL tended to be normal or low in homozygous sitosterolemic subjects (Salen *et al.*, 1989). Electron microscopy of HDL from a male sitosterolemic subject with severe symptomatic atherosclerosis showed that the particles were round, with a mean diameter of 8.5 ± 1.7 nm, consistent with the predominance of small, dense HDL (Salen *et al.*, 1989).

Plasma concentrations of apolipoprotein B are usually increased and apolipoprotein A-I decreased, for sitosterolemic homozygotes (Salen *et al.*, 1989). However, normal plasma apo B and A-I levels were present in heterozygotes (Nguyen *et al.*, 1990; Salen *et al.*, 1992a). Thus, the apolipoprotein pattern reflects increased LDL and usually low HDL concentrations detected in these patients as determined by analytical and preparative ultracentrifugation (Salen *et al.*, 1989).

Table I
Plasma Sterol and Stanol Concentrations

Patients	Age (years)	Cholesterol (mg/dl)	Sitosterol (mg/dl)	Campesterol (mg/dl)	5α-Cholestanol (mg/dl)	5α-Sitostanol (mg/dl)	5α-Campestanol (mg/dl)
Homozygotes							
KCN[a]($n = 10$)[b]	29	184 ± 25	15 ± 2	7.5 ± 1.1	2.1 ± 1.0	2.9 ± 1.0	1.3 ± 0.9
KC[a]($n = 10$)	23	202 ± 25	14 ± 4.1	8 ± 3.1	4.7 ± 1.0	2.2 ± 1	1.4 ± 0.2
TC[a]($n = 10$)	27	233 ± 12	21 ± 8.3	10 ± 0.5	3.8 ± 1.4	5.4 ± 1.2	1.9 ± 1.0
RC[a]($n = 10$)	18	249 ± 39	20 ± 5.5	13 ± 1.5	7.5 ± 2.4	3.9 ± 1.0	2.6 ± 0.9
GB[c]($n = 10$)	28	292 ± 8	13 ± 0.9	8 ± 0.2	1.8 ± 0.1	1.9 ± 1.0	0.8 ± 0.4
CL	42	134	27	13	1.6	3.0	2.6
JBR[d]	14	256	30	14.5	4.0	2.7	2.5
LBU	24	482	56	24.0	11.0	6.0	4.0
Heterozygotes							
AC[a]($n = 5$)	50	210 ± 26	0.95 ± 0.17	ND[e]	0.65 ± 0.21	ND	ND
VC[a]($n = 5$)	56	194 ± 14	0.36 ± 0.09	ND	0.34 ± 0.19	ND	ND
DB[c]($n = 3$)	25	204 ± 27	0.66 ± 0.05	ND	0.34 ± 0.13	ND	ND
RBR[d]	48	254	0.8	ND	1.2	ND	ND
Controls							
$n = 20$	17–62	180 ± 5	0.22 ± 0.20	ND	0.20 ± 0.20	ND	ND

[a]C-family–includes heterozygous parents.
[b]The number of samples was analyzed in parentheses.
[c]B-family–includes heterozygous sister.
[d]R-family–includes heterozygous mother.
[e]ND, not detected.

2.3. Tissue Sterol Concentrations

Tissue sterol concentrations were measured in a 17-year-old sitosterolemic homozygote male who died unexpectedly of an acute myocardial infarction, and showed about 16% plant sterols and 5α-stanols in plasma (Salen *et al.*, 1985b). The total sterol concentrations in red blood cells, liver, lung, and heart were not different from controls, but the cholesterol concentrations in these tissues were reduced, and offset by increased amounts of plant sterols and 5α-saturated stanols. Of importance, the individual plant sterols and 5α-saturated stanols were deposited in the tissues in about the same proportion that they were present in LDL. This suggested that the tissue sterols originated from plasma. In contrast, brain sterols in the sitosterolemic subject were composed almost entirely of cholesterol (Salen *et al.*, 1985b). Thus, despite the presence of large amount of plant sterols and 5α-stanols, the blood-brain barrier in sitosterolemia remains intact and is not permeable to circulating LDL. This is in contrast to cerebrotendinous xanthomatosis (CTX), another lipid storage disease, where increased amounts of cholestanol are deposited in the brain, suggesting that in CTX the blood-brain barrier is damaged and becomes more permeable to circulating LDL (Salen *et al.*, 1983, 1985b). Interestingly, atheromas in the aorta of this sitosterolemic subject contained increased amounts of esterified, sterols; about 50% of the cholesterol and sitosterol were esterified, as compared with only 10% esterified sterols in visceral organs (Salen *et al.*, 1985b)

2.4. Bile Sterols and Bile Acids

Sterol composition in bile is different from controls in sitosterolemia. Not only is less cholesterol secreted into the bile, but sitosterol appears in the same or lower proportion relative to cholesterol in bile as compared to plasma (Bhattacharyya and Connor, 1974; Shulman *et al.*, 1976; Lin *et al.*, 1983; Salen *et al.*, 1983; Bhattacharyya *et al.*, 1991). Normally the liver preferentially secretes sitosterol into bile; there is a threefold enrichment of sitosterol relative to cholesterol as compared to blood in control subjects (Salen *et al.*, 1970). Biliary bile acids, including cholic acid, deoxycholic acid, and lesser amounts of chenodeoxycholic acid were secreted into bile in amounts adequate to prevent steatorrhea (Bhattacharyya and Connor, 1974; Miettinen, 1980; Lin *et al.*, 1983; Salen *et al.*, 1983; Shefer *et al.*, 1988). No unusual biliary bile acids were detected, although it has not been established if sitosterol and other plant sterols can be converted to primary bile acids in homozygotes. Bhattacharyya *et al.* (1991) reported radioactive bile acids derived from [^{14}C]sitosterol in the feces of three sitosterolemic subjects, but the precise identification of the structure of these compounds was not carried out. In contrast, other investigators have found 21-carbon steroid acids in bile of rats derived from sitosterol (Boberg *et al.*, 1989; Lund *et al.*, 1991). However, it was noted that the large quantities of sitosterol and cholestanol in sitosterolemic liver competitively inhibited cholesterol 7α-hydroxylase, the rate-determining enzyme for bile acid synthesis, which may eventually lead to decreased bile acid production and deficient pool size (Shefer *et al.*, 1988, 1994; Boberg *et al.*, 1990). Thus, sitosterolemic liver has lost both the capacity to recognize sitosterol and the ability to preferentially secrete the 24-ethyl sterol into the bile. In addition, cholesterol secretion into bile (see Chapter 11 by Kuipers *et al.*) is markedly diminished. Also, since biliary cholesterol secretion (lithogenicity) relative to bile acids and phospholipids is decreased (Miettinen, 1980), gallstones may not form and have not been reported in sitosterolemic subjects (Salen, unpublished results).

2.5. Sterols and Monocytes

Monocytes' (mononuclear leukocytes) sterol composition has been measured in four sitosterolemic homozygotes (Nguyen *et al.*, 1988). The sterols and stanols are similar in composition (sitosterol, campesterol, cholestanol, 5α-campestanol, and 5α-sitostanol) as found in LDL, indicating that the plant sterols and stanols originate from plasma. However, total sterol concentrations in monocytes from the sitosterolemic homozygotes were 2–3 times higher than in control monocytes. Thus, monocytes, which are precursors to foam cells, contain increased quantities of cholesterol, plant sterols, and 5α-stanols derived from plasma, which may contribute to the accelerated atherosclerotic process in this disease.

3. SITOSTEROL METABOLISM

It has long been known that sitosterol is poorly absorbed from the intestine (Gould *et al.*, 1969; Salen *et al.*, 1970). The low plasma concentrations found in animals and humans fed large amounts of dietary plant sterols attest to restricted intestinal absorption (Salen *et al.*, 1970). However, early sitosterol balance studies where fecal outputs were measured and subtracted from dietary inputs gave confusingly high values (between 30–50% of intake) for absorption (Salen *et al.*, 1970).

3.1. Sitosterol Turnover

To overcome potential errors in the balance technique, sitosterol absorption has been measured by two independent isotopic methods. The isotope kinetic method estimates absorption by mathematical analysis of specific activity decay curves after intravenous pulse-labeling with a tracer dose of radioactive sitosterol (Goodman and Noble, 1968; Grundy and Ahrens, 1969; Salen *et al.*, 1970; Goodman *et al.*, 1973). In normal and hyperlipidemic subjects, the plasma specific activity decay of sitosterol is much more rapid than the specific activity decay of cholesterol when both isotopic sterols were injected intravenously together (Figure 5) (Salen *et al.*, 1970, 1989, 1992a; Lin *et al.*, 1983; Bhattacharyya *et al.*, 1991). These decay curves can be divided into two exponentials and analyzed mathematically according to the two-pool model.

Table II lists values for controls and three homozygotes sitosterolemic subjects from two unrelated families. Since sitosterol is not synthesized endogenously in normal humans and sitosterolemic subjects (Salen *et al.*, 1970; Lin *et al.*, 1983), the production rate is equivalent to absorption and in the control subjects amounted to less than 10 mg per day, or less than 5% of daily intake. The mean total sitosterol body pool size in the controls was also calculated and amounted to about 130 mg.

3.2. Sitosterol Absorption—Kinetic Method

In contrast, sitosterol turnover in the three homozygous sitosterolemic subjects was much slower compared with controls. Sitosterol production rates were 5–10 times larger than the control mean, confirming the enhanced absorption found in homozygotes with this disease. Total body pools of sitosterol were also

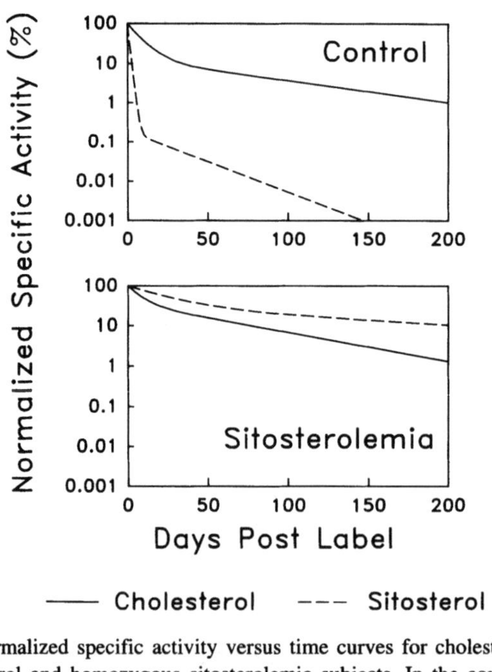

FIGURE 5. The normalized specific activity versus time curves for cholesterol and sitosterol are illustrated for a control and homozygous sitosterolemic subjects. In the control subject, sitosterol decays more rapidly than cholesterol, as contrasted with slower decay of sitosterol than cholesterol in the homozygote.

Table II
Sitosterol Turnover in Normal and Sitosterolemic Patients[a]

	Controls	CL	Sitosterolemic homozygotes[b] KCN	KEC
$t_A^{1/2}$ (days)	3.8 ± 0.2	7.0	2.7	11
$t_B^{1/2}$ (days)	13.8 ± 2.4	91	24	112
K_A (day^{-1})	0.17 ± 0.04	0.015	0.067	0.014
M_A (mg)	80 ± 36	5100	2400	1700
M_B (mg)	46 ± 22	4400	2400	1800
$M_A + M_B$ (mg)	126 ± 32	9500	4800	3500
PR_A (mg/day)	7.9 ± 2.3	80	162	52

[a]Data are from Lin *et al.* (1983); Salen *et al.* (1989); Salen *et al.* (1992a); $t_A^{1/2}$ = half life pool A; $t_B^{1/2}$ = half life pool B; K_A = coefficent of elimination; M_A = pool size A; M_B = pool size B; $M_A + M_B$ = total pool size; PR_B = production rate; equal absorption since there is no endogenous synthesis.
[b]Patient KCN and KEC are sisters and unrelated to CL.

tremendously enlarged and ranged from 3500–9500 mg. Although absorption was increased in these subjects, the extraordinary body pool sizes did not relate linearly to absorption as seen in normal individuals (Salen *et al.*, 1970, 1989, 1992a). The discrepancy could be explained by the fact that the elimination constant from pool A (K_A) was 2–10 times more rapid in control subjects than sitosterolemic homozygotes. Thus, sitosterolemic homozygotes hyperabsorb sitosterol from the intestine and also retain the plant sterol in body tissues (Lin *et al.*, 1983; Salen *et al.*, 1989, 1992a). This finding of very slow sitosterol turnover associated with increased absorption and very delayed removal has also been confirmed in three additional sitosterolemic subjects from two unrelated families (Bhattacharyya *et al.*, 1991) who received radioactive sitosterol intravenously. Body pool sizes were extraordinarily expanded as noted previously. In addition, sitosterol turnover has been studied in two sitosterolemic obligate heterozygotes (parents of homozygotes). The results showed that sitosterol absorption was increased 2–3 times over controls, but body pool sizes were not expanded beyond normal because sitosterol removal was rapid (Salen *et al.*, 1992a). Thus, heterozygotes still retain the ability to excrete sitosterol normally and do not accumulate the plant sterol, despite enhanced absorption.

3.3. Sitosterol Absorption—Plasma Dual Isotope Method

Enhanced sitosterol absorption in homozygotes and heterozygotes has been confirmed independently by absorption measurements obtained by adapting the plasma dual-isotope ratio method used to study cholesterol absorption (Table III) (Salen *et al.*, 1989, 1992a). In this technique, [^{14}C]sitosterol is fed and

Table III
Sitosterol and Cholesterol Absorption[a]

Controls	Homozygotes (n = 2)	Sitosterolemic KCN	KEC
Sitosterol absorption (%)	4.5	63	28
Cholesterol absorption (%)	46	49	69

[a]Sterol absorption was determined by the plasma dual-isotope ratio method. Data from Salen *et al.* (1989).

[22,23-^{3}H]sitosterol is administered intravenously at the same time. The ^{3}H/^{14}C ratio is then determined in plasma sitosterol and compared to the ideal ratio, which is calculated by dividing the total oral dose by total injected dose and is equivalent to 100% absorption (Salen *et al.*, 1989). In two homozygotes, 28% and 63% of dietary sitosterol were absorbed, respectively, which is in good agreement with the absorption values calculated in these subjects by the independent isotope kinetic method (Table II). Two healthy control subjects who consumed the same diet absorbed an average of 4.5% of dietary sitosterol, while two obligate heterozygotes absorbed 15 and 17% of dietary sitosterol, respectively (Salen *et al.*, 1992a). More recently, Lutjohan *et al.* (1995) found that sitosterol absorption average 16% in three unrelated subjects using the fecal balance method, in which the recovery of 5α-sitostanol and chromic oxide that are not absorbed from the intestine served as internal standards. Thus, sitosterol absorption is enhanced in homozygotes and heterozygotes, but the plant sterols accumulate in homozygotes because removal is so hindered.

4. CHOLESTEROL ABSORPTION AND TURNOVER

Cholesterol absorption as measured by the plasma dual-isotope ratio method tended to be at the high end of the normal range (59% of intake) in sitosterolemic homozygotes, as reported in Table III (Salen *et al.*, 1989; Lutjohan *et al.*, 1995). Thus, increased sitosterol absorption does not interfere with cholesterol absorption in sitosterolemic homozygotes, although it is believed that sitosterol and cholesterol share the same intestinal absorption pathway. Also, it is important to realize that cholesterol is absorbed about 10 times more efficiently than sitosterol in healthy control subjects, but that percent sitosterol absorption approaches cholesterol absorption in some sitosterolemic homozygotes (Table III) (Salen *et al.*, 1989). There was no difference in cholesterol absorption between controls and heterozygotes (Salen *et al.*, 1992a). Although it has been suggested that the limited absorption of sitosterol compared to cholesterol in normal subjects may relate to greater affinity of sitosterol for intestinal bile acid micelles (Tilvis and Miettinen, 1986), diminished intestinal sitosterol esterification (Kuksis and Huang, 1962) and reduced sitosterol enterocyte transport (Sylvén and Borgstrom, 1969) may also be responsible. The upregulation of these mechanisms seems unlikely to explain the increased absorption of sitosterol in sitosterolemic homozygotes. Perhaps, the absence of a protein or polypeptide located in the enterocyte that discriminates sitosterol from cholesterol will ultimately explain the enhanced absorption in these subjects.

An important biochemical finding observed in the sitosterolemic subjects is reduced cholesterol turnover (Table IV) (Miettinen, 1980; Lin *et al.*, 1983; Salen *et al.*, 1989). Not only is the plasma specific activity decay of cholesterol much slower in homozygotes than control subjects, but turnover (PR_A = synthesis plus absorbed cholesterol) is markedly reduced. Calculations of cholesterol turnover by the isotope kinetic method revealed values that were 50–70% smaller in homozygotes than similarly fed controls (Lin *et al.*, 1983; Salen *et al.*, 1989, 1992a; Bhattacharyya *et al.*, 1991). Moreover, since cholesterol absorption tended to be large in sitosterolemic subjects, the diminished daily production must result from decreased cholesterol synthesis. When turnover values were corrected by subtracting absorbed cholesterol, average cholesterol synthesis was about 50% lower in sitosterolemic homozygotes than healthy controls (Salen *et al.*, 1989, 1992a; Bhattacharyya *et al.*, 1991). In support, Miettinen (1980), found cholesterol synthesis as measured by the sterol balance technique was 50 and 80% lower in a homozygous sitosterolemic subject than similarly fed control subjects when studied on two occasions separated by 4 years. Lutjohan *et al.* (1995) also recently reported that cholesterol synthesis was 5 times lower in three unrelated homozygotes than

Table IV
Cholesterol Turnover[a]

	Controls (n = 4)	Sitosterolemic homozygotes KL	KCN	KEC
$t^{1/2}_A$ (days)	6.7 ± 0.6	8.6	2.4	5.0
$t^{1/2}_B$ (days)	53 ± 6	81	24	74
K_A (day^{-1})	0.045 ± 0.006	0.023	0.077	0.024
M_A (grams)	29 ± 8	31	11	21
M_B (grams)	48 ± 16	34	13	41
$M_A + M_B$ (grams)	77 ± 9	65	24	65
PR_A (mg/day)	1450 ± 560	670	860	710
Synthesis (mg/kg/day)	14.6 ± 6.0[b]	NA[c]	9.5[b]	5.9[b]

[a]Data from Miettinen (1980); Lin *et al.* (1983); and Salen *et al.* (1989).
[b]Synthesis estimated by subtracting absorbed cholesterol (Table III) from turnover (PR_A).
[c]Not available.

in controls. In contrast, cholesterol turnover and synthesis as measured by isotope kinetic method in sitosterolemic heterozygotes resembled control subjects and was not decreased (Salen *et al.*, 1992a).

5. MECHANISM OF REDUCED CHOLESTEROL SYNTHESIS

5.1. HMG-CoA Reductase Content

A major discovery from balance and isotopic turnover studies was that cholesterol synthesis in sitosterolemic homozygotes was extremely low (Table IV) (Miettinen, 1980; Lin *et al.*, 1983; Salen *et al.*, 1989, 1992a; Bhattacharyya *et al.*, 1991; Lutjohan *et al.*, 1995). In order to better understand this observation, 3-hydroxy-3-methylglutaryl coenzyme A reductase (HMG-CoA reductase), the rate-controlling enzyme for cholesterol biosynthesis, was measured in liver microsomes from four sitosterolemic homozygotes (Nguyen *et al.*, 1990b; Shefer *et al.*, 1994). For comparison, HMG-CoA reductase was measured in liver specimens obtained from 11 liver transplant donors (whose livers became available when appropriate recipients could not be located). In the control livers (Table V), the mean HMG-CoA reductase activity was 5.3 and 8.2 times greater, respectively, than the values from two sitosterolemic homozygotes' liver specimens. About 72% of the HMG-CoA reductase was expressed (active) in the sitosterolemic livers compared only to 49% in the controls.

HMG-CoA reductase protein concentrations were determined in these same microsomal specimens by immunoblotting and densitometric scanning (Table V). In the control liver microsomes, the mean relative mass of HMG-CoA reductase per mg of microsomal protein was 6.8 and 8.9 times larger, respectively, than the values for the two sitosterolemic livers. Thus, markedly reduced HMG-CoA reductase activity and enzyme protein characterize sitosterolemic liver. However, when the catalytic efficiency of HMG-CoA reductase (activity per

Table V
Hepatic Microsomal HMG CoA Reductase Activity and Mass[a]

	HMG-CoA Reductase		
Subjects	Activity (pmol/mg/min)	Mass (peak area/mg)	Catalytic efficiency (pmol/min/peak area)
Controls (n = 11)	98.1 ± 28.8	1.4 ± 0.14	68.6
Sitosterolemic homozygotes			
KCN	11.9	0.16	74.3
TC	18.4	0.21	76.2

[a]Data are from Nguyen *et al.* (1990b).

unit protein) was calculated by dividing the enzyme specific activity by the enzyme mass, no difference was detected between control and sitosterolemic livers. This suggests that although reduced quantities of HMG-CoA reductase are produced by the sitosterolemic livers, the catalytic function of the formed enzyme is normal.

5.2. HMG-CoA Reductase mRNA

To further explore the severe deficiency of HMG-CoA reductase, poly A^+ RNA was isolated from liver specimens obtained from two control subject and one homozygous sitosterolemic subject and hybridized with pRED 227 and pHRED 102, which are full-length sequence cDNA probes for hamster and human HMG-CoA reductase, respectively, and pCAT 10, a probe for human catalase mRNA. The Northern blots (Figures 6A and 6B) showed virtually no transcript for sitosterolemic HMG-CoA reductase mRNA, as contrasted with good transcripts for HMG-CoA reductase mRNA from the control specimens. Both control and sitosterolemic specimens gave transcripts for catalase mRNA that indicated that the RNA isolated from control and sitosterolemic livers was intact. Thus, the deficiency of microsomal HMG-CoA reductase in sitosterolemic livers can be attributed to the very low levels of HMG-CoA reductase mRNA that are available for enzyme translation (Nguyen *et al.*, 1990b).

5.3. Hepatic LDL Receptor Membrane Binding

LDL receptor binding was also measured in 12 control and 2 sitosterolemic liver membrane preparations. Total binding (assayed in the absence of unlabeled LDL) was 54% and 80% higher, respectively, in the two sitosterolemic liver membrane preparations than the mean for the control measurements (Table VI). Similarly, high affinity receptor-mediated LDL binding recorded as the difference between total binding and non-specific binding (assayed in the presence of abundant unlabeled LDL) was 2.2 and 3.3 times greater, respectively, in the sitosterolemic than the control livers. Therefore, sitosterolemic livers expressed increased numbers of LDL receptors, so that a much higher proportion of circulating LDL was receptor-bound, and more LDL was taken up from plasma, than by control liver membranes (Nguyen *et al.*, 1990b).

In separate experiment, Beil *et al.* (1988) measured *in vivo* LDL turnover and found greater production associated with rapid catabolism consistent with the expression of more LDL receptor in a sitosterolemic subject compared with three matched controls.

In two sitosterolemic liver specimens, lipofuscin-like pigment was distributed in the liver cytosol (Nguyen *et al.*, 1990b). The nature of this pigment has not been determined at this time (Figure 7).

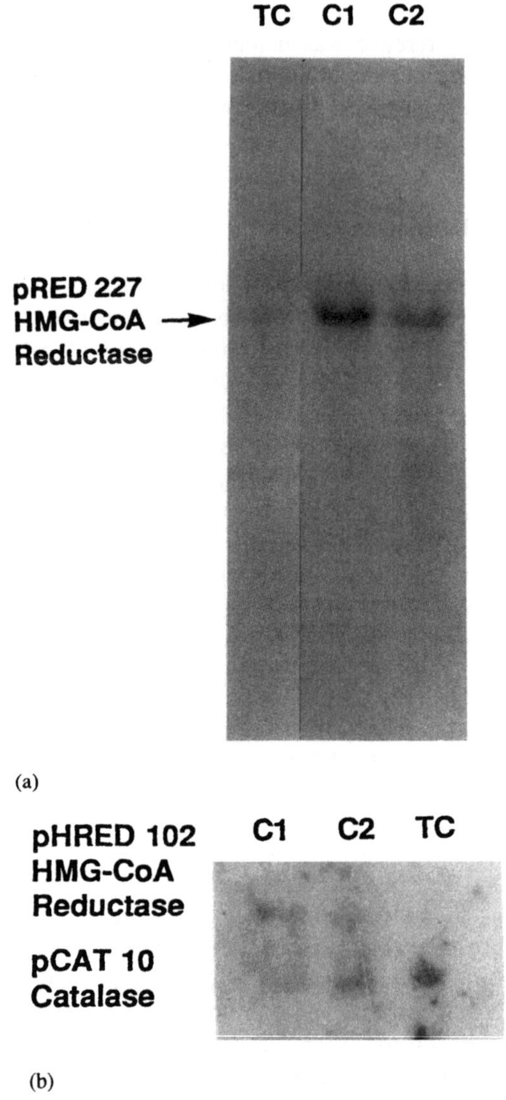

FIGURE 6. Northern blot analysis of sitosterolemic hepatic mRNA. Northern blots of liver poly A$^+$ RNA from a sitosterolemic homozygote (TC) and two control subjects who were probed with (A) pRED 227 and (B) pHRED 102 for HMG-CoA reductase mRNA and pCAT 10 for catalase mRNA. Virtually no signal from sitosterolemic HMG-CoA reductase mRNA was detected. (Data are reprinted with permission from Nguyen *et al.*, 1990b; Salen *et al.*, 1992a.)

Table VI
Hepatic LDL Receptor Binding[a]

Subjects	Total	Receptor-mediated
	(ng/mg protein)	
Control (n = 12)	204.0 ± 10.0	95 ± 8.2
Sitosterolemic homozygotes		
TC	315.3	193.2
KCN	336.8	312.8

[a]Data are from Nguyen *et al.* (1990b).

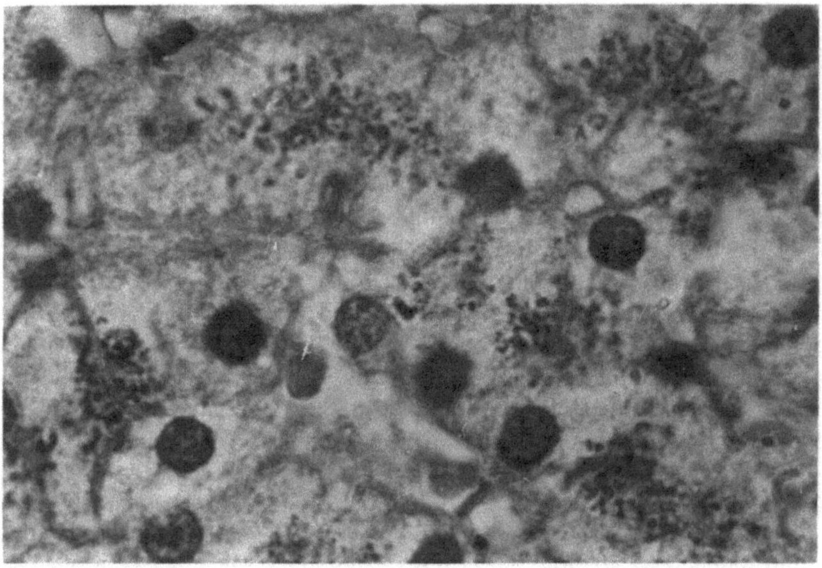

FIGURE 7. Light microscopy of liver from sitosterolemic homozygote that shows lipofuscin-like granules deposited randomly in cytosol of hepatocyte. Hemotoxylin and eosin × 250. (Data are reprinted with permission from Nguyen *et al.*, 1990b; Salen *et al.*, 1992a.)

6. TREATMENT OF SITOSTEROLEMIA

6.1. Bile Acid Malabsorption

Bile acid malabsorption produced by either binding resins (cholestyramine or colestipol) or ileal bypass surgery are effective treatments of sitosterolemia (Miettinen, 1980; Whitington *et al.*, 1980; Salen *et al.*, 1983, 1985a, 1993; Belamarich *et al.*, 1990). Plasma cholesterol concentrations declined dramatically (by

25–50%) and the percent reduction in plasma sterol concentrations obtained with these drugs or surgery is greater than when applied to hypercholesterolemic subjects. In most sitosterolemic patients, plant sterols usually decrease proportionally to cholesterol, and cholestanol and 5α-saturated stanols are virtually eliminated from the blood (Salen *et al.*, 1985a). However, it is important to realize that not all patients respond similarly, and cholestyramine treatment produced less reduction of plasma plant sterols than cholesterol concentrations in a Japanese sitosterolemic family (Hidaka *et al.*, 1990). Of note, clinical improvement including disappearance of xanthomas, diminution of aortic stenosis murmur, and decreased episodes of angina pectoris and arthritic attacks have been noted in several subjects treated with either cholestyramine or ileal bypass surgery (Whitington *et al.*, 1980; Belamarich *et al.* 1990; Nguyen *et al.*, 1990a; Salen *et al.*, 1993).

6.2. Lovastatin

Lovastatin, a competitive inhibitor of cholesterol biosynthesis that is widely used in the treatment of hypercholesterolemia, has been tried but has not been an effective treatment in sitosterolemia. Plasma cholesterol, plant sterols, or 5α-saturated stanols were not reduced in three homozygotic sitosterolemic subjects (Nguyen *et al.*, 1990a, 1991; Salen *et al.*, 1993).

The effect of the various treatments can be explained by examining cholesterol biosynthesis and LDL receptor function in freshly isolated peripheral mononuclear leukocytes (monocytes). These cells synthesize cholesterol and express HMG-CoA reductase activity and LDL receptors in parallel to the liver. In five homozygotic sitosterolemic subjects from three unrelated families, mononuclear leukocyte cholesterol synthesis, as measured by the conversion of acetate to cholesterol, was 30–70% below the mean value from 16 healthy control subjects (Nguyen *et al.*, 1990a, 1991). Subnormal monocyte cholesterol synthesis in the sitosterolemic subjects was supported by measurements of HMG-CoA reductase activity, which were 50–70% lower in the homzoygotes than the control mean. In contrast, LDL receptor function in monocytes from four of the five homozygous patients was increased 60% over the control mean. Thus, sitosterolemic mononuclear leukocytes manifest the same defect in cholesterol biosynthesis as the liver and compensate for the decrease in cholesterol synthesis by the increased expression of LDL receptors.

When the enterohepatic circulation of bile acids is interrupted by cholestyramine reducing the hepatic bile acid flux, HMG-CoA reductase activity increased 13% and LDL receptor function rose 40% in freshly isolated monocytes from healthy control subjects (Nguyen *et al.*, 1990a). In contrast, monocytes from four homozygous sitosterolemic subjects (from three unrelated families) failed to upregulate either cholesterol synthesis (conversion of acetate to cholesterol) or HMG-CoA reductase activity when treated similarly (Nguyen *et al.*, 1990a, 1991). In fact, monocytes HMG-CoA reductase activity paradoxically declined. However, monocyte LDL receptor function responded normally to bile acid malabsorption by increasing the HMG-CoA reductase activity by 20–30% in the sitosterolemic mononuclear cells (Grundy *et al.*, 1971; Shepard *et al.*, 1989, Nguyen *et al.*, 1990a, 1991).

Lovastatin treatment produced no change in plasma sterol concentrations in two unrelated sitosterolemic homozygous subjects and caused only a small rise in monocyte HMG-CoA reductase activity, compared with a 28% reduction in plasma sterol concentrations and a 38% increase in monocyte HMG-CoA reductase activity in control and hypercholesterolemic subjects (Nguyen *et al.*, 1990a, 1991). Although lovastatin competitively inhibits mevalonic acid synthesis, HMG-CoA reductase activity normally increases. Apparently, the block in cholesterol produces greater gene expression for the synthesis of HMG-CoA reductase. LDL receptor function also did not change in the homozygous sitosterolemic monocytes, compared with a 41% increase in receptor-mediated LDL binding in control cells from subjects treated with lovastatin (Nguyen *et al.*, 1990a, 1991).

These results indicate a major abnormality in cholesterol homeostasis in sitosterolemic subjects. Depressed cholesterol biosynthesis arises from a pronounced deficiency of the rate-controlling enzyme, HMG-CoA reductase, caused by the virtual absence of HMG-CoA reductase mRNA. Interruption of the enterohepatic circulation of bile acids reduces the hepatic bile acid flux and normally stimulates bile acid synthesis (Nguyen *et al.*, 1990a), but fails to increase cholesterol production in sitosterolemic homozygotes. Cholesterol 7α-hydroxylase activity rises in response to bile acid malabsorption (Shefer *et al.*, 1988) so that more bile acids are formed (Miettinen, 1980). Cholesterol biosynthesis (HMG-CoA reductase activity) should increase and more LDL receptors should be expressed to provide additional cholesterol as the substrate for bile acid synthesis. In normal subjects, the decrease in plasma cholesterol reflects the balance between input of new cholesterol (synthesis) and the removal and catabolism of LDL. Because sitosterolemic subjects cannot upregulate HMG-CoA reductase, the demand for more substrate for bile acid synthesis can only be met by the catabolism of LDL. Thus, there is greater than expected fall in plasma sterol concentrations (cholesterol and plant sterols) in sitosterolemic subjects.

The failure of lovastatin to serve as an effective treatment suggests that sitosterolemic cholesterol synthesis is so low that further inhibition of HMG-CoA reductase does not increase LDL receptor function. With this in mind, hypercholesterolemic patients who do not respond to therapeutic doses of lovastatin should have their plasma sterols tested by GLC, since the failure to respond to lovastatin may indicate sitosterolemia (Nguyen *et al.*, 1990a, 1991).

In other experiments, during a low cholesterol and low sitosterol diet, 24-h urinary mevalonate excretion that reflects cholesterol synthesis was measured in a female sitosterolemic homozygote and was less than 50% of the 24-h urinary mevalonate excretion in control subjects (Table VII). In contrast, 24-h urinary

Table VII
Urinary Mevalonate Excretion[a]

Subject/treatment	μmol/kg per 24 h
Sitosterolemic homozygote-female	
Low sterol diet	11.0
+ Lovastatin	6.5
+ Cholestyramine	7.0
Sitosterolemic heterozygote	
Low sterol diet	23.0
+ Lovastatin	13.0
+ Cholestyramine	30.0
Controls (n = 23)	23.1 ± 2.5

[a]Data are from Salen *et al.* (1993).

mevalonate output was normal in her heterozygous father. Treatment with lovastatin reduced 24-h mevalonate excretion by about half in both this subject and her father, the heterozygote. In contrast, cholestyramine, which caused bile acid malabsorption and stimulated cholesterol biosynthesis, increased urinary mevalonate excretion 30% in the heterozygote (father), but paradoxically, 24-h urinary mevalonate excretion declined 36% in this female homozygote (Salen *et al.,* 1993). These results support low cholesterol biosynthesis in homozygotes and the failure of bile acid malabsorption to increase synthesis as key mechanisms in sitosterolemic homozygotes.

7. INHERITED ABNORMALITY

At the present time, the principal inherited defect has not been established with certainty. However, three abnormal mechanisms have been linked to the pathogenesis of sitosterolemia and predisposition to atherosclerosis: hyperabsorption of sitosterol, decreased sitosterol elimination, and reduced cholesterol synthesis. The hyperabsorption of sitosterol and other dietary sterols from the intestine is well documented, but by itself will not cause the enormous sitosterol pools in this disease. Sitosterolemic heterozygotes also hyperabsorb sitosterol but do not accumulate the plant sterol (Salen *et al.,* 1992a). Not until the intestinal pathway for cholesterol absorption is elucidated will the exact mechanism for sitosterol hyperabsorption be understood.

To date, all sitosterolemic homozygotes show diminished hepatic secretion of sitosterol and cholesterol. Bile contains reduced amounts of both sitosterol and cholesterol, and biliary sterol excretion is further decreased when dietary intake is restricted (Bhattacharyya *et al.,* 1991). Clearly, the combination of decreased removal with increased absorption accounts for the giant sitosterol and other plant sterol pools.

A third key feature of the disease is abnormal regulation of cholesterol biosynthesis. We have demonstrated that reduced cholesterol synthesis results from a deficiency of HMG-CoA reductase in the liver and mononuclear cells of sitosterolemic subjects (Nguyen *et al.*, 1990a, 1990b, 1991; Salen *et al.*, 1993; Shefer *et al.*, 1995) and that HMG-CoA reductase mRNA is barely detected in the liver. LDL receptor function is enhanced in most sitosterolemic homozygotes to provide cellular sterols. Attempts to stimulate cholesterol synthesis (HMG-CoA reductase) by inducing bile acid malabsorption (cholestyramine or ileal bypass surgery) or by supplying a low sterol diet did not increase HMG-CoA reductase activity in freshly isolated mononuclear cells (Nguyen *et al.*, 1990a, 1991; Salen *et al.*, 1993). Thus, the upregulation of cholesterol synthesis is prevented in sitosterolemia.

Moreover, it is important to emphasize that HMG-CoA reductase activity and the expression of LDL receptors are normally coordinately regulated in the same direction. In other words, factors stimulating HMG-CoA reductase increase the number of LDL receptors (Grundy *et al.*, 1971; McNamara *et al.*, 1980). In contrast, sitosterolemic mononuclear cells and hepatocytes show diminished HMG-CoA reductase activity and enzyme mass in combination with increased LDL receptor expression (McNamara *et al.*, 1980; Nguyen *et al.*, 1990a, 1990b, 1991). These observations lead us to believe that the inherited defect in sitosterolemia involves an abnormality in the regulation of the HMG-CoA reductase gene.

However, at this time it is still not possible to establish which mechanism is primary. It is still possible that enhanced sitosterol absorption and accumulation are primary events. Therefore, low cholesterol synthesis and enhanced receptor function may conceivably relate to the accumulated sterols and stanols and/or an oxygenated derivative (Salen *et al.*, 1983). However, Boberg *et al.* (1989) have reported that sitosterol is not an effective feedback inhibitor of HMG-CoA reductase, and Shefer *et al.* (1984, 1994) noted that neither cholestanol nor sitosterol inhibits HMG-CoA reductase activity and mRNA in rat liver. Thus, neither cholestanol nor sitosterol are downregulators of cholesterol biosynthesis.

8. CONCLUSION

Sitosterolemia is a rare inherited lipid storage disease characterized chemically by the accumulation of plant sterols and 5α-saturated stanols in plasma and tissues. Very low cholesterol synthesis due to a deficiency of HMG-CoA reductase associated with increased intestinal plant sterol absorption and slow hepatic sterol removal are major biochemical features. Because cholesterol synthesis cannot upregulate, bile acid malabsorption mobilizes body sterols for bile acid synthesis, dramatically lowers plasma and monocyte sterol concentrations, and may halt the progression of the atherosclerotic process.

ACKNOWLEDGMENTS. The work in the authors' laboratories has been funded by Research Service, Veterans Affairs Medical Center, and U.S. Public Health Service Grants HL 17818 and DK 26756.

9. REFERENCES

Beaty, T. J., Kwiterovich, P. O., Jr., Khoury, M. J., White, S., Bachorik, P. S., Smith, H. H., Teng, B., and Sniderman, A., 1986, Genetic analysis of plasma sitosterol, apoprotein B and lipoproteins in a large Amish pedigree with sitosterolemia, *Am. J. Hum. Genet.* **38:**492–504.

Beil, U. F., Vega, G. L., Ma, P. A., Gary, A., and Grundy, S. M., 1988, Lipoprotein kinetics in beta sitosterolemia, *Arteriosclerosis* **8:**587A.

Belamarich, P. F., Deckelbaum, R. J., Starc, T., Dobrin, B. E., Tint, G. S., and Salen, G. 1990, Response to diet and cholestyramine in a patient with sitosterolemia, *Pediatrics* **86:**977–981.

Bhattachharyya, A., and Connor, W. E., 1974, β-Sitosterolemia and xanthomatosis. A newly described lipid storage disease in two sisters. *J. Clin. Invest.* **53:**1033–1043.

Bhattacharyya, A. K., Connor, W. E., Lin, D. S., McMurry, M. M., and Shulman, R. S., 1991, Sluggish sitosterol turnover and hepatic failure to excrete sitosterol into bile cause expansion of body pool of sitosterol in patients with sitosterolemia and xanthomatosis, *Atheroscler. Thromb.* **11:**1287–1294.

Björkhem, I., and Skrede, S., 1989, Familial diseases with storage of sterols other than cholesterol: cerebrotendinous xanthomatosis and phytosterolemia, in: *The Metabolic Basis of Inherited Disease* (C. R. Scriver, A. L. Beaudet, W. S. Sly, and D. Valle, eds.), pp. 1293–1302, McGraw-Hill, New York.

Boberg, K. M., Akerlund, J. E., and Björkhem, I., 1989, Effect of sitosterol on the rate-limiting enzymes in cholesterol biosynthesis and degradation, *Lipids* **24:**9–12.

Boberg, K. M., Lund, E., Olund, J., Björkhem, I., 1990, Formation of C_{21} bile acids from plant sterols in the rat, *J. Biol. Chem.* **265:**7967–7975.

Dayal, B., Tint, G. S., Batta, A. K., Speck, J., Khachadurian, A. K., Shefer, S., and Salen, G., 1982, Identification of 5α-stanols in patients with sitosterolemia and xanthomatosis: stereochemistry of the protonolysis of steroidal organoboranes, *Steroids* **40:**233–243.

Goodman, D. S., and Noble, R. P., 1968, Turnover of plasma cholesterol in man, *J. Clin. Invest.* **47:**231–241.

Goodman, D. S., Noble, R. P., and Dell, R. B., 1973, Three-pool model of the long-term turn over of plasma cholesterol in man, *J. Lipid Res.* **14:**178–188.

Gould, R. G., Jones, R. J., LeRoy, G. V., Wissler, R. W., and Talyor, G. B., 1969, Absorbability of β-sitosterol in humans, *Metabolism* **18:**652–662.

Gregg, R. E., Connor, W. E., Lin, D. S., and Brewer, H. B., Jr., 1986, Abnormal metabolism of shellfish sterols in a patient with sitosterolemia and xanthomatosis, *J. Clin. Invest.* **77:**1864–1872.

Grundy, S. M., and Ahrens, E. H., Jr., 1969, Measurements of cholesterol turnover, synthesis, and absorption in man carried out by isotope kinetic and sterol balance methods, *J. Lipid Res.* **10:**91–107.

Grundy, S. M., Ahrens, E. H., Jr., and Salen, G., 1971, Interruption of the enterohepatic circulation of bile acids in man: Comparative effects of cholestyramine and ileal exculsion on cholesterol metabolism, *J. Lab. Clin. Med.* **78:**94–121.

Hidaka, H. T., Nakamura, T. K., Aohi, T., Kojima, H., Nakajima, Y., Kosugi, K., Hatanaka, I., Harada, M., Kobayaski, M., Tamura, A., Fuju, T., and Shiegeta, Y., 1990, Increased plasma plant sterol levels in heterozygotes with sitosterolemia and xanthomatosis, *J. Lipid Res.* **31:**881–888.

Khachadurian, A. K., and Salen, G., 1980, Familial phytosterolemia: cholestanolemia and abnormal bile salt composition, *Clin. Res.* **28:**397A.

Kuksis, A., and Huang, T. C., 1962, Differential absorption of plant sterols in the dog, *Can. J. Biochem.* **40:**1493–1495.

Kwiterovich, P. O., Jr., Smith, H. H., Connor, W. E., Bachorik, P. S., McKusick, V. A., Teng, B., and Sniderman, A., 1981, Hyperapobetalipoproteinemia in two families with xanthomatosis and phytosterolemia, *Lancet* **1:**466–469.

Lin, H. J., Wang, C., Salen, G., Lam, K. C., and Chan, T. K., 1983, Sitosterol and cholesterol metabolism in a patient with sitosterolemia and xanthomatosis, *Metabolism* **32:**126–133.

Lund, E., Boberg, K. M., Bystrom, S., Oland, J., Carlstrom, K., and Björkhem, I., 1991, Formation of novel C_{21} bile acids from cholesterol in the rat: structure identification of the major di and trihydroxylated species, *J. Biol. Chem.* **266:**4929–4937.

Lutjohan, D., Björkhem, I., Beil, U. F., and von Bergmann, K., 1995, Sterol absorption and sterol balance in phytosterolemia evaluated by deuterium-labeled sterols: Effect of sitostanol treatment, *J. Lipid Res.* **36:**1763–1773.

McNamara, D. J., Davidson, N. O., and Fernandez, S., 1980, In vitro cholesterol synthesis in freshly isolated mononuclear cells of human blood: effect of in vivo administration of clofibrate and/or cholestyramine, *J. Lipid Res.* **21:**65–71.

Miettinen, T. A., 1980, Phytosterolemia, xanthomatosis and premature atherosclerotic arterial disease: a case with high plant sterol absorption, impaired sterol elimination and low cholesterol synthesis, *Eur. J. Clin. Invest.* **10:**27–35.

Nguyen, L., Shefer, S., Salen, G., Horak, I., Tint, G. S., and McNamara, D. J., 1988, The effect of abnormal plasma sterol composition on cellular sterol content and composition and low density lipoprotein uptake and degradation by monocytes and lymphocytes in sitosterolemia with xanthomatosis, *Metab. Clin. Exp.* **37:**346–351.

Nguyen, L., Salen, G., Shefer, S., Shore, V., Tint, G. S., and Ness, G., 1990a, Unexpected failure of bile acid malabsorption to stimulate cholesterol synthesis in sitosterolemia with xanthomatosis: comparison with lovastatin, *Arteriosclerosis* **10:**289–297.

Nguyen, L., Shefer, S., Salen, G., Ness, G., Tint, G. S., Zaki, F. G., and Rani, I., 1990b, Molecular defect in cholesterol synthesis in sitosterolemia with xanthomatosis, *J. Clin. Invest.* **86:** 926–931.

Nguyen, L., Salen, G., Shefer, S., Tint, G. S., Shore, V., and Ness, G., 1990c, Decreased cholesterol biosynthesis in sitosterolemia with xanthomatosis: Diminished mononuclear leukocyte 3-hydroxy-3-methylglutaryl coenzyme A reductase activity and enzyme protein associated with increased low-density lipoprotein receptor function, *Metab. Clin. Exp.* **39:**436–443.

Nguyen, L. B., Cobb, M., Shefer, S., Salen, G., Ness, G., and Tint, G. S., 1991, Regulation of cholesterol biosynthesis in sitosterolemia: Effects of lovastatin, cholestyramine and dietary sterol restriction, *J. Lipid Res.* **32:**1941–1948.

Nye, E. R., Sutherland, W. H. F., Mortimer, J. G., and Stringer, H. C. W., 1988, Sitosterolemia and heterozygous familial hypercholesterolemia in a three-year-old girl: case report, *N. Z. Med. J.* **101:**418–419.

Rao, M. K. G., Perkins, E. G., Connor, W. E., and Bhattacharyya, A. K., 1975, Identification of β-sitosterol, campesterol, and stigmasterol in human serum, *Lipids* **10:**566–568.

Salen, G., Grundy, S. M., Ahrens, E. H., Jr., 1970, The metabolism of β-sitosterol in man, *J. Clin. Invest.* **49:**952–967.

Salen, G., Shefer, S., and Berginer, V., 1983, Familial diseases with storage of sterols other than cholesterol, in: *The Metabolic Basis of Inherited Disease* (J. B. Stanbury, J. B. Wyngaarden, D. S. Fredrickson, M. S. Brown, and J. L. Goldstein, eds.), pp. 710–730, McGraw-Hill, New York.

Salen, G., Kwiterovich, P. O., Jr., Shefer, S., Tint, G. S., Horak, I., Shore, V., Dayal B., and Horak, E., 1985a, Increased plasma cholestanol and 5α-saturated plant sterol derivatives in subjects with sitosterolemia and xanthomatosis, *J. Lipid Res.* **26:**203–209.

Salen, G., Horak, I., Rothkopf, M., Cohen, J. L., Speck, J., Tint, G. S., Shore, V., Dayal, B., Chen, T., and Shefer, S., 1985b, Lethal atherosclerosis associated with abnormal plasma and tissue sterol composition in sitosterolemia with xanthomatosis, *J. Lipid Res.* **26:**1126–1133.

Salen, G., Shore, V., Tint, G. S., Forte, T., Shefer, S., Horak, I., Horak, E., Dayal, B., Nguyen, L., Batta, A. K., Lindgren, F. T., and Kwiterovich, P. O., Jr., 1989, Increased sitosterol absorption, decreased removal, and expanded body pools compensate for reduced cholesterol synthesis in sitosterolemia with xanthomatosis, *J. Lipid Res.* **30:**1319–1330.

Salen, G., Tint, G. S., Shefer, S., Shore, V., and Nguyen, L., 1992a, Increased sitosterol absorption is offset by rapid elimination to prevent accumulation in heterozygotes with sitosterolemia, *Atheroscler. Thromb.* **12:**563–568.

Salen, G., Shefer, S., Nguyen, L., Ness, G. C., Tint, G. S., and Shore, V., 1992b, Sitosterolemia, *J. Lipid Res.* **33:**945–955.

Salen, G., Cobb, P., Nguyen, L., Shefer, S., and Tint, G. S., 1993, Abnormal regulation of cholesterol biosynthesis by a low sterol diet, cholestyramine and lovastatin in sitosterolemia, in: *Bile Acids and the Hepatobiliary System* (G. Paugartner, A. Stiehl, and W. Gerok, eds.), pp. 85–91, Kluwer Academic Publishers, Lancaster, U.K.

Shefer, S., Hauser, S., Salen, G., Zaki, F. G., Bullock, J., Salgado, E., and Shevitz, J., 1984, Comparative effects of cholestanol and cholesterol on hepatic sterol and bile acid metabolism in the rat, *J. Clin. Invest.* **74:**1773–1781.

Shefer, S., Salen, G., Nguyen, L., Batta, A. K., Packin, V., Tint, G. S., and Hauser, S., 1988, Competitive inhibition of bile acid synthesis by endogenous cholestanol and sitosterol in sitosterolemia with xanthomatosis, *J. Clin. Invest.* **82:**1833–1839.

Shefer, S., Salen, G., Bullock, J., Nguyen, L. B., Ness, G. C., Zhoo, Z., Belamarich, P. F., Chowdhary, I., Lerner, S., Batta, A. K., and Tint, G. S., 1994, The effect of increased hepatic sitosterol on the regulation of 3-hydroxy-3-methyl-glutaryl-coenzyme A reductase and cholesterol 7α-hydroxylase in the rat and sitosterolemic homozygotes, *Hepatology* **20:**213–219.

Shepard, J., Packard, C. J., Bicker, S., Laurie, T. D. V., and Morgan, H. G., 1989, Cholestyramine promotes receptor-mediated low-density lipoprotein catabolism, *N. Engl. J. Med.* **302:**1219–1222.

Shulman, R. S., Bhattacharyya, A. K., Connor, W. E., and Fredrickson, D. S., 1976, β-Sitosterolemia and xanthomatosis, *N. Engl. J. Med.* **294:**472–483.

Skrede, B., Björkhem, I., Bergesen, O., Kayden, H., and Skrede, S., 1985, The presence of 5α-saturated sitostanol in the serum of a patient with phytosterolemia, and its biosynthesis from plant sterols in rats with bile fistula, *Biochim. Biophys. Acta* **836:**368–375.

Sylvén, C., and Borgstrom, B., 1969, Absorption and lymphatic transport of cholesterol and sitosterol in the rat, *J. Lipid Res.* **10:**179–182.

Tilvis, R. S., and Miettinen, T. A., 1986, Serum plant sterols and their relation to cholesterol absorption, *Am. J. Clin. Nutr.* **43:**92–97.

Wang, G., Lin, H J., Chan, T. K., Salen, G., Chan, W. C., and Tse, T. F., 1981, A unique patient with coexisting cerebrotendinous xanthomatosis and β-sitosterolemia, *Am. J. Med.* **71:**313–319.

Whitington, G. L., Ragland, J. B., Sabesin, S. M., and Kuiken, L. B., 1980, Neutral sterolemia and xanthomas, *Circulation* **59:**11–33 (abstract).

Chapter 16

Cholesterol Cytochemistry in Cell Biology and Disease

Nicholas J. Severs

1. INTRODUCTION

Cholesterol localization in membranes, cells, and tissues is a topic of substantial and continuing interest to the biomedical cell biologist. Knowing precisely where and in what amount cholesterol is present offers potentially useful insights into the functional role of this lipid as a constituent of normal membranes, and the part it plays in the pathogenesis of diseases that involve dysfunction of cholesterol homeostasis. The cholesterol content of membranes influences a range of properties, such as fluidity, stability, the activities of membrane-bound enzymes, and the adhesive and electrical properties of the cell surface. Differences in cholesterol content characterize the various specialized membrane systems of the cell; plasma membranes and the membranes of endocytic organelles are normally cholesterol-rich, whereas endoplasmic reticulum, nuclear, and mitochondrial membranes typically have much lower cholesterol levels. Furthermore, the concept that cholesterol distribution may be heterogeneous within a given membrane, reflecting a mosaiclike pattern of functional differentiation, has also been widely debated. Despite the wealth of information from *in vitro* biophysical studies,

---

Nicholas J. Severs National Heart and Lung Institute, Imperial College of Science, Technology, and Medicine, Royal Brompton Hospital, London, England.

Subcellular Biochemistry, Volume 28: Cholesterol: Its Functions and Metabolism in Biology and Medicine, edited by Robert Bittman. Plenum Press, New York, 1997.

however, there remain many gaps in our understanding of the functional significance of cholesterol distribution in relation to the diverse membrane-mediated cellular activities that are fundamental to tissue function.

The aim of this chapter is to review the various approaches to the localization of cholesterol in cells, membranes, and tissues, to provide a critical assessment of their advantages and limitations, and to examine some of their current applications in biomedical cell biology. Cholesterol localization methods fall into four basic categories. The first relies on agents that bind selectively to cholesterol, and in so doing induce distinctive electron microscopically visible alterations in membrane structure. The second involves using the fluorescence properties of the complexes formed between cholesterol and a binding agent for localization at the light microscopical level. The third approach relies on enzymes and other reagents to produce a dense reaction product at cholesterol-rich sites that is detectable by thin-section electron microscopy (EM) or light microscopy. The fourth is based on antibody labeling techniques, in particular, immunogold labeling. Although some degree of overlap exists between these categories, they nevertheless provide a convenient framework for discussion.

2. STRATEGIES FOR CHOLESTEROL LOCALIZATION

2.1. Ultrastructural Localization of Cholesterol Using Agents that Perturb Membrane Structure: Filipin and the Saponins

Techniques aimed at localizing cholesterol by using agents that, in binding to the target lipid, perturb membrane structure, were developed in the late 1970s and early 1980s. The best known and most widely used reagent of this type is the polyene antibiotic filipin (Andrews and Cohen, 1979; Elias *et al.*, 1979). Further agents that have seen substantial use in this type of approach are the saponins digitonin and tomatin. Bacterial exotoxins, such as streptolysin O and Θ-hemolysin, and other polyene antibiotics, such as nystatin and amphotericin B, have also been considered as candidate cholesterol probes, although their less discrete and less conspicuous effects on membrane structure make them unsuitable for this purpose (Severs and Robenek, 1983; Elias *et al.*, 1978; Kitajima *et al.*, 1976; Verkleij *et al.*, 1973).

Electron microscopy of the effects of the saponins and filipin on membranes has a long history. Saponins were applied in early negative stain EM investigations of lipid structure (Lucy and Glauert, 1964; Bangham and Horne, 1962; Glauert *et al.*, 1962), and digitonin was subsequently used with the specific object of retaining and localizing cholesterol for thin-section EM (Frühling *et al.*, 1971; Scallen and Dietert, 1969; Ökrös, 1968). Morphological studies on the effects of filipin on membranes developed from biophysical investigations of membrane permeability and the lytic effects of a range of polyene antibiotics conducted on unfixed cells

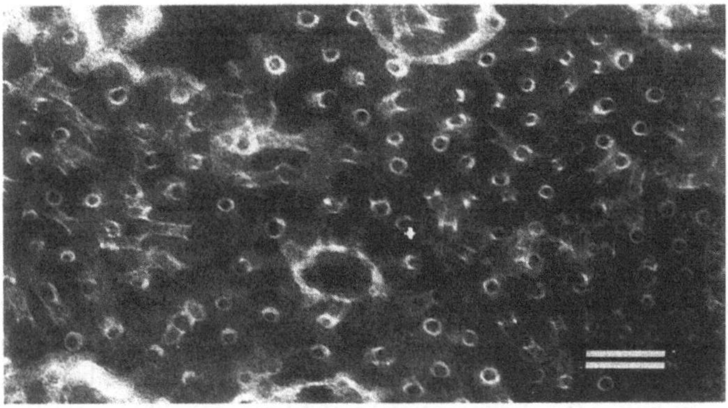

FIGURE 1. Filipin-induced lesions examined by negative staining electron microscopy appear as distinct circular rings in the membrane. This example shows a fragment of isolated plasma membrane (non-luminal) from rat urinary bladder epithelial cell that was treated with filipin before examination on a carbon-coated grid with potassium phosphotungstate staining. Scale bar = 100 nm.

(for reviews, see Bittman 1978; Norman *et al.* 1976). These studies defined the specific interaction of filipin with 3β-hydroxysterols (i.e., cholesterol in mammalian cells and ergosterol in fungi), and concurrent morphological investigations of unfixed isolated membranes examined by negative staining (Bittman *et al.*, 1974; Kinsky *et al.*, 1966) demonstrated that filipin–sterol interaction results in distinctive circular membrane lesions (Figure 1). The mechanism of filipin-sterol interaction and formation of these lesions was defined as a multistep process involving: i) penetration of filipin molecules into one monolayer of the membrane, ii) binding of filipin to cholesterol to form primary complexes (each consisting of one sterol and one filipin molecule), and iii) aggregation of primary complexes into multimolecular aggregates of a size sufficient to distort the membrane (i.e, form the ultrastructurally visible lesion) (Severs *et al.*, 1981; De Kruijff and Demel, 1974; De Kruijff *et al.*, 1974). The technique of freeze-fracture EM was first applied during these early studies to examine the effects of filipin in unfixed membranes, and revealed the lesions to be distinctive deformations 15–30-nm in diameter deformations corresponding to those seen by negative staining (Tillack and Kinsky, 1973; Verkleij *et al.*, 1973).

2.1.1. Filipin and the Saponins as Cholesterol Probes for Freeze-Fracture Electron Microscopy

From the background outlined above, the idea that the lesions induced by filipin and the saponins could be used as markers for mapping membrane cholesterol at ultrastructural resolution, in particular by freeze-fracture EM, was proposed (Andrews and Cohen, 1979; Elias *et al.*, 1979) and, subsequently, enthusiastically promoted. Methodological aspects of the use of filipin for this specific purpose were investigated in detail; glutaraldehyde fixation of the sample was recommended to minimize membrane disruption and obtain uniformly sized lesions, quantitative studies in a model system established a direct relationship between sterol content and filipin lesion number, and tomatin was reported to offer advantages over digitonin (Friend and Bearer, 1981; Robinson and Karnovsky, 1980; Elias *et al.,* 1979).

The standard methodology established from these studies for the freeze-fracture cytochemical detection of cholesterol involves incubating the biological specimen (isolated membranes, cells, or tissue samples) in filipin or one of the saponins during or after treatment in glutaraldehyde or glutaraldehyde/paraformaldehyde fixative. Filipin is poorly soluble in aqueous solutions; the agent is therefore dissolved in a droplet of dimethyl sulfoxide before addition to the fixative or buffer (final concentration of filipin, 50–200 $\mu g\ cm^{-3}$). Incubation is normally done in the dark, at room temperature with constant mixing for 2–24 hours. The samples are then cryoprotected with glycerol, rapidly frozen, and platinum–carbon freeze-fracture replicas are prepared as in the standard freeze-fracture procedure for transmission EM (for a recent review, see Severs and Shotton, 1995). Freeze fracture displays membranes in striking *en face* view, enabling the distribution and abundance of the "markers" to be readily determined. The typical appearance of the lesions induced by filipin and one of the saponins, tomatin, as viewed by freeze-fracture EM, is illustrated in Figure 2.

The spectacular images revealed by the filipin/freeze-fracture technique captured the imagination of cell morphologists during the 1980s, and a plethora of papers followed in which variations in sterol content between and within membranes were reported. In the course of this enthusiasm, many authors assumed that interpretation of the results was quite straightforward—that the distribution and abundance of filipin-induced deformations directly reflected local variations in cholesterol content in the native membrane. However, as we shall see in section 3, this interpretative concept was overly simplistic, and the literature of the period requires critical reading to discriminate fact from artifact.

2.1.2. Thin-Section Electron Microscopy of Filipin- and Saponin-Induced Membrane Deformations

Apart from freeze fracture, details of the membrane distortion induced by filipin and the saponins may also be viewed by thin-section EM (Figure 3). In this case, after incubation with the agent (given after or during aldehyde fixation), the samples are post-fixed in osmium tetroxide and embedded in epoxy resin, following conventional procedures. Information from thin sectioning complements that obtained by freeze fracture, allowing, for example, the correlation of lesion distribution with peripheral membrane protein specializations on the cytoplasmic side of the membrane. However, membranes that have been deformed by filipin frequently show a reduction in definition, and sectional profiles lack the three-

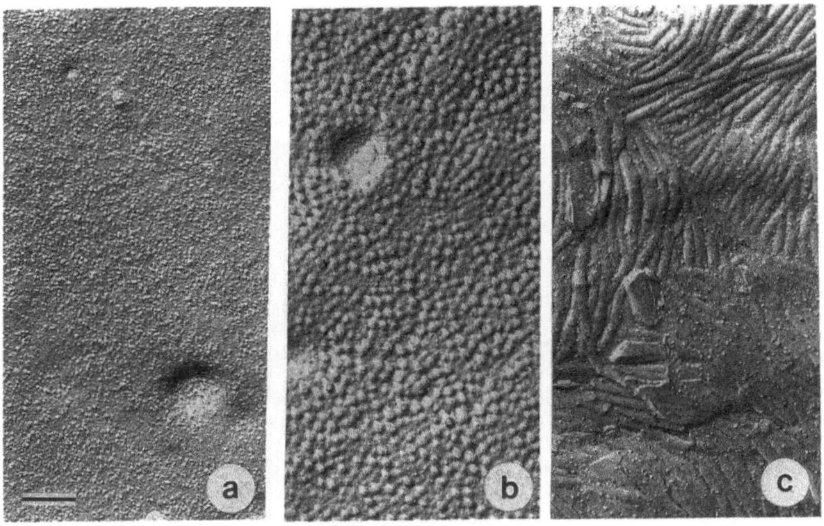

FIGURE 2. Characteristic structure of the deformations induced by filipin and tomatin in a cholesterol-rich membrane, the fibroblast plasma membrane, as revealed by freeze-fracture electron microscopy. (a) The untreated membrane presents a smooth contour, seen here in P-face view (i.e., fracture face of the half-membrane leaflet left attached to the protoplasm after freeze fracture; for further details of nomenclature see Branton *et al.*, 1975). (b) Filipin, by interacting with cholesterol, produces ~25 nm-diameter rounded deformations of uniform size in aldehyde-fixed specimens. The deformations are seen here as protrusions on the P face; complementary E faces show corresponding depressions. (The E face is defined as the fracture face of the half-membrane leaflet left attached to the extracellular space after freeze fracture.) Different types of membrane differ from one another with respect to the polarity of the induced deformations. Other morphological forms of lesion also occur, particularly in specimens treated with the agent before fixation (Luciano *et al.*, 1989; Behnke *et al.*, 1984; Severs *et al.*, 1981). (c) Tomatin complexes with membrane cholesterol to form hemitubules, 40–60 nm in width, which are typically elevated toward the side of the membrane from which the agent has penetrated. The plasma membrane thus shows elevated cylindrical structures on the P face; corresponding depressions are seen on E faces (Miller, 1984). As with filipin, the established procedure is to apply the agent to aldehyde-fixed samples. Lesion morphology may vary according to the specimen and experimental conditions (Severs and Simons, 1987; Miller, 1984; Frühling *et al.*, 1971). Scale bar = 200 nm.

dimensional information provided by freeze fracture. Thin sectioning has thus been applied largely as a supporting technique to the freeze fracture approach.

2.2. Filipin as a Fluorescent Probe for Localization of Cholesterol at the Light Microscopical Level

Filipin–cholesterol complexes have major absorption and emission bands at 357 nm and 480 nm, respectively (Bittman, 1978; Norman *et al.*, 1976; Bittman *et al.*, 1974). Ultraviolet excitation for epifluorescence light microscopy thus permits localization of unesterified cholesterol in filipin-treated samples, at a resolution

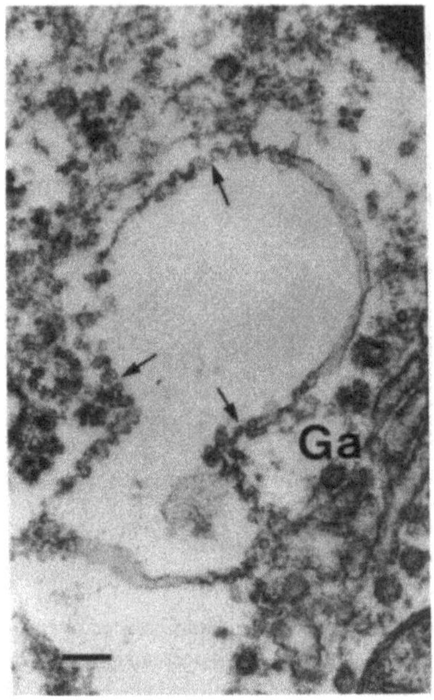

FIGURE 3. Effect of filipin as seen by thin-section EM, from a Golgi-associated membrane in a superficial epithelial cell from rat bladder. Note corrugated membrane profiles (arrows), corresponding to the lesions observed by freeze-fracture EM. GA, Golgi apparatus. Scale bar = 100 nm. (Reprinted by permission of Academic Press.)

sufficient to discriminate, for example, membrane macrodomains of differing fluorescence signal intensity. The approach may be applied to suspensions of isolated cells, cultured cell monolayers, or tissue sections (for examples of images, see section 5). For cells, the procedure typically involves exposing paraformaldehyde-fixed cells briefly to filipin prior to direct examination (e.g., El Yandouzi *et al.*, 1994; Pumplin and Bloch, 1983; Tamm and Tamm, 1983). For tissue samples, cryosections of the fixed tissue are prepared; these are usually stained with standard histological stains (e.g., hematoxylin) after the filipin treatment (Kruth, 1984a, 1984b; Kruth and Fry, 1984; Kruth, 1983). The major application of the latter approach has been in the detection of cholesterol deposition in pathological material (see section 5). The filipin/fluorescence technique can be adapted to detect cholesteryl ester (Kruth, 1984a) by first extracting unesterified cholesterol using ethanol treatment, and then incubating in cholesterol esterase. Unesterified cholesterol, derived from hydrolysis of the ester, is then accessible to detection by filipin.

The fluorescence property of filipin–cholesterol binding has also been exploited to identify cholesterol-rich lipoproteins following polyacrylamide gel electrophoresis (Smejkal and Hoff, 1994a, 1994b), and has potential for cytochemical localization at the EM level. General procedures exist for localizing fluorescent probes at the EM level by a photoconversion reaction in the presence of 3,3'-diaminobenzidine (DAB) (Sandell and Masland, 1988; Moranto, 1982). Oxygen and hydroxyl radicals produced by irradiation of the fluorescently labeled specimen oxidize DAB to give an electron-dense marker similar to that produced by the peroxidase-catalyzed reaction of DAB with H_2O_2 (see section 2.3). The applicability of this approach with filipin has yet to be evaluated.

2.3. Enzyme Cytochemistry

Ultrastructural localization of cholesterol by enzyme cytochemistry relies on the production of a highly electron-dense reaction product, oxidized DAB, as a marker for use in thin-section EM (Jones and Miyai, 1981). The technique, originally developed from a method for determining total serum cholesterol for clinical chemistry (Allain *et al.*, 1974), uses the enzymes cholesterol esterase and cholesterol oxidase. The former enzyme hydrolyzes esterified cholesterol to free cholesterol; the latter oxidizes free cholesterol to produce cholestenone and hydrogen peroxide (H_2O_2). Using horseradish peroxidase as a catalyst, the H_2O_2 is then used to oxidize DAB (i.e., create the visible marker). The same principle may be used for histochemical localization of cholesterol at the light microscopical level, an approach that predated EM localization (Emeis *et al.*, 1977).

Brief fixation of tissues in glutaraldehyde (e.g., 0.5% glutaraldehyde fixative for 30 min) is normally required to obtain optimal results. Vibratome sections of the fixed tissue are then incubated in the enzyme-detection medium (e.g., cholesterol esterase/cholesterol oxidase/DAB and horseradish peroxidase for the detection of total cholesterol) for several hours. Samples are then processed for conventional thin-section EM. Examples of detailed experimental protocols and composition of incubation media are given in Pelletier and Vitale (1994).

In contrast to the use of digitonin for localization of unesterified cholesterol in thin sections, the enzymatic method is versatile, allowing selective localization of free cholesterol, esterified cholesterol, or total cholesterol (i.e., free plus esterified). Detection of total cholesterol is achieved by the use of both enzymes, as outlined above, whereas labeling of free cholesterol only is done by omitting the cholesterol esterase step from the standard incubation procedure. Specific detection of cholesterol esters may be achieved by first removing the free cholesterol before following the standard procedure, a step accomplished either by treatment with ethanol or by preincubation in cholesterol oxidase without DAB. Figure 4, taken from the recent study of Pelletier and Vitale (1994), illustrates the application of these approaches to the localization of free and esterified cholesterol in testicular cells.

2.4. Antibodies to Cholesterol and Colloidal Gold Labeling

The introduction of colloidal gold particles as markers in the 1980s revolutionized ultrastructural cytochemistry, and the use of these markers has obvious

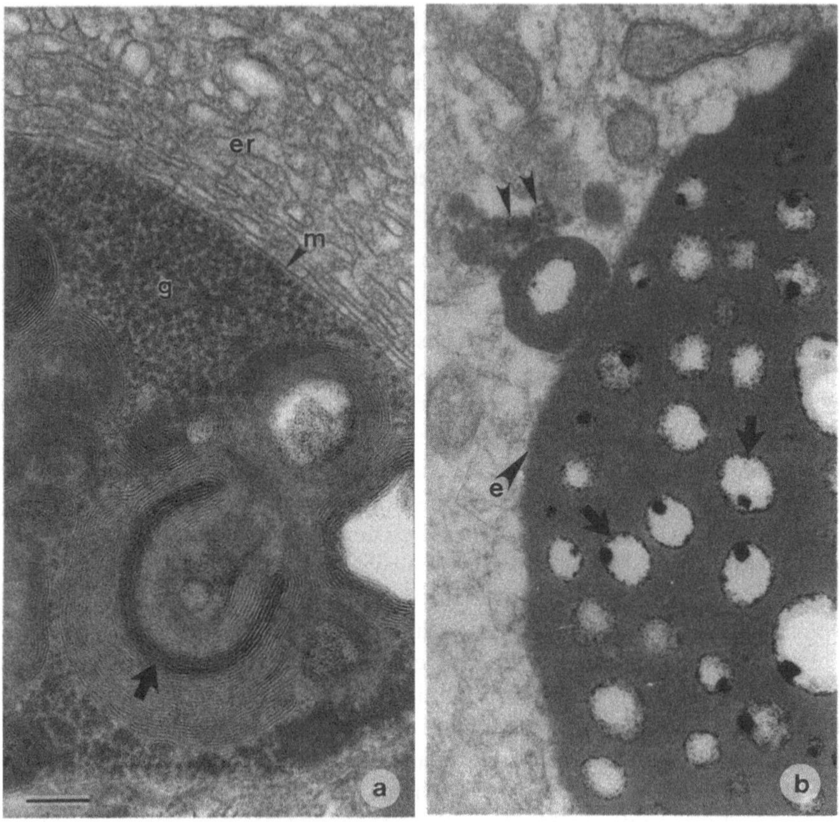

FIGURE 4. Examples of localization of free and esterified cholesterol by enzyme cytochemistry and thin-section electron microscopy in testicular cells (Pelletier and Vitale, 1994). (a) Localization of free cholesterol in a thin section from a guinea-pig Leydig cell using an incubation medium consisting of cholesterol oxidase, DAB, horseradish peroxidase, and dimethyl sulfoxide. Prominent reaction product is heterogeneously distributed within lysosomes, visible in curled stacks of membrane (arrow) and ground substance (g). The limiting membrane (m) of the lysosome and the endoplasmic reticulum membranes show no labeling. (b) Localization of cholesterol ester in a Sertoli cell lipid droplet from mink testis. Sections were pre-incubated in a medium containing cholesterol oxidase (from *Streptomyces* sp) to enzymatically oxidize the free cholesterol, before incubation in "total" medium containing cholesterol esterase, cholesterol oxidase, DAB, and horseradish peroxidase. Circular profiles of reaction product, often enclosing distinct dense granules, are apparent (arrows). Arrowheads indicate small, positively labeled lipid droplets apparently in the process of fusing; e, indicates edge of large lipid droplet. Scale bar = 200 nm. Micrographs kindly provided by R. Marc Pelletier.

attractions for the localization of cholesterol, the proteins concerned with its extracellular and intracellular transport, and the enzymes responsible for its metabolism. Because gold markers are of small and uniform size, and of high electron-scattering power, they rapidly became established in ultrastructural cytochemistry (see reviews by Hayat, 1989; Horisberger, 1983; and Roth, 1983). The basic principle of the standard approach is to use antibody-coated colloidal gold particles as a secondary probe to detect sites at which a specific primary antibody has bound to the target antigen. Alternatively, gold particles may be coated directly with the primary antibody, or with other ligands such as lipoproteins, enzymes, or lectins. Colloidal gold labeling combined with pre-embedding or post-embedding techniques for thin-section EM, and with platinum–carbon surface replicas and freeze-fracture EM, have become widely applied in cell biology for the localization of a comprehensive range of protein components. Adaptations of this approach using silver enhancement provide a highly sensitive detection method for light microscope visualization (Hacker, 1989).

Although these approaches have proven informative in the study of a range of aspects of cholesterol metabolism, notably in the investigation of the properties of lipoprotein receptors (see section 5), their application for the direct detection of cholesterol is less straightforward. Opportunities for the detection of lipids in general is restricted by the comparative lack of suitable antibody probes. Gold-conjugated phospholipase-A_2 has been used for the detection of phospholipids (Maraldi *et al.*, 1992; Coulombe *et al.*, 1988), but lipids themselves, including cholesterol, generally make poor immunogens. Nevertheless, a number of papers have reported production of antibodies to specific classes of lipid, including cholesterol (Compère *et al.*, 1995; Kan and Rauch, 1993; Maneta-Peyret *et al.*, 1992; Miyazawa *et al.*, 1992; Blanchette-Mackie *et al.*, 1989; Swartz *et al.*, 1988). Swartz *et al.* (1988) successfully raised anticholesterol monoclonal antibodies using cholesterol-rich liposomes as an immunogen in mice, and immunogold labeling in combination with these antibodies has been demonstrated in two test systems—crystalline cholesterol (Figure 5) and cholesterol-rich liposomes (Blanchette-Mackie *et al.*, 1989; Swartz *et al.*, 1988). The feasibility of applying the immunogold approach to cell fractions and to localize extracellular deposits of cholesterol in tissues is currently under investigation (E.J. Blanchette-Mackie, personal communication).

3. PITFALLS IN THE APPLICATION OF FILIPIN AND THE SAPONINS TO LOCALIZE MEMBRANE CHOLESTEROL AT THE ULTRASTRUCTURAL LEVEL

As noted in section 2.1.1, filipin enjoyed intense popularity as a cholesterol probe for freeze fracture during the first half of the 1980s. Despite a subsequent decline in interest, a steady stream of papers applying filipin to study membrane

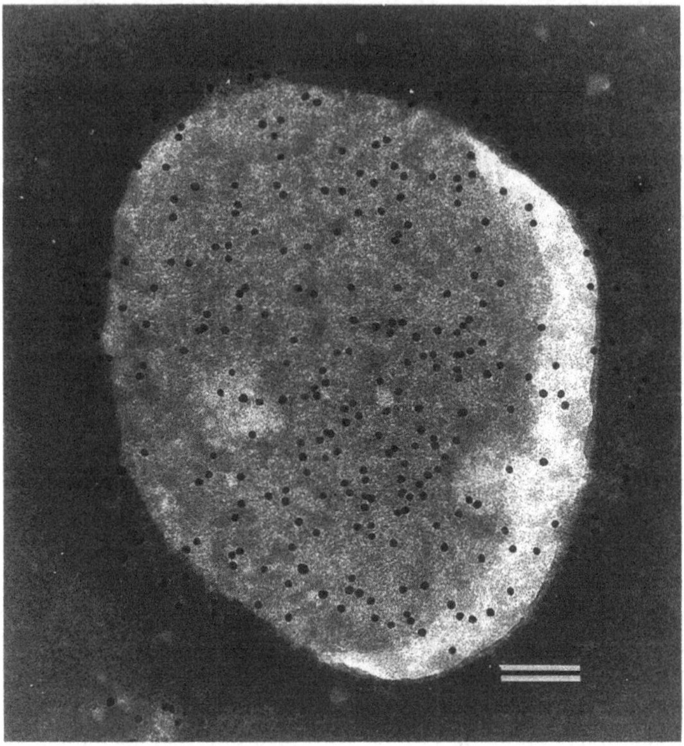

FIGURE 5. Immunogold labeling of crystalline cholesterol. Negatively stained cholesterol crystal treated with primary mouse monoclonal antibody against cholesterol followed by secondary antibody-5 nm gold complexes. The surface of the crystal is heavily labeled with gold complexes. Scale bar = 100 nm. Micrograph kindly provided by Joan Blanchette-Mackie. (From Blanchette-Mackie *et al.*, 1989; Reprinted by permission of Wiley-Liss Inc.)

cholesterol has continued. The approach merits further discussion here because of the sheer number of publications on the topic and because the technique has certain pitfalls for the unwary. For more detailed treatment of this subject, the reader is referred to other reviews (Severs, 1995; Miller, 1989; Severs, 1984a, 1984b; Severs and Robenek, 1983).

3.1. False Negative Responses to Filipin in Protein-Rich Membrane Domains

The first indication that caution is required in the interpretation of filipin-labeling patterns came from the discovery that in one structurally specialized membrane of high cholesterol content, the luminal plasma membrane of the mammalian urinary bladder, filipin and digitonin gave discordant results (Severs *et al.*, 1981; Warren and Severs, 1980). The bladder luminal membrane has an unusual structure comprising hexagonal arrays of integral membrane proteins, which are organized as a mosaic of large semi-rigid plaques, separated from one another by narrow, flexible interplaque regions (Figure 6a,c). From the high ratio of plaque to interplaque area, and the large quantity of cholesterol measured in isolated membrane fractions, it was clear that significant quantities of cholesterol are present in both domains (Severs *et al.*, 1981; Warren and Severs, 1980). However, when the membrane was treated with filipin, deformations were formed only in the narrow interplaque zones; the plaques were entirely resistant to the agent (Figure 6b,d). Digitonin, by contrast, induced lesions in both domains (Figure 7). This led to the conclusions that: i) the presence of closely packed integral membrane proteins inhibits perturbation of cholesterol-rich membrane by filipin; ii) negative responses to filipin do not always indicate a lack of cholesterol; and iii) parallel application of other methods is required to verify the significance of filipin-labeling patterns with respect to cholesterol localization (Severs *et al.*, 1981).

This warning was at first largely unheeded, but further instances of filipin-negative responses in domains containing closely packed integral membrane proteins were soon identified (for a review see Severs and Robenek, 1983). A similar lack of filipin sensitivity was found to characterize membrane domains rich in peripheral membrane proteins, i.e., domains characterized by the presence of electron-dense membrane coats visible by thin-section EM but which show no obvious intramembrane structural specialization in freeze fracture. In smooth muscle cells, for example, the membrane-associated dense bodies that attach the contractile apparatus to the cytoplasmic surface of the plasma membrane prevent formation of filipin lesions but not tomatin lesions (Severs and Simons, 1983; see section 4). Similarly, negative responses to filipin in retroviral envelopes (Feltkamp and van der Waerden, 1982), in protozoan/ectosymbiotic bacterial junctions (Tamm and Tamm, 1983), and in clathrin-coated vesicles (Steer *et al.*, 1984), were all found to be converted to positive responses after enzymatic removal of the proteins. The evidence was clear; reliable deductions about cholesterol localization could not be based solely on filipin-labeling patterns observed in freeze fracture. The reason filipin fails to detect cholesterol in protein-rich domains, despite its specificity for 3β-hydroxysterols, appears to be that densely packed proteins, whether within or closely associated with the lipid bilayer, either interfere with the aggregation of primary complexes and/or physically prevent the membrane from being deformed (Severs and Robenek, 1983; Severs and Simons, 1983; Severs *et al.*, 1981). Alternative approaches may help to identify the discrepancies, notably the use of two different agents (i.e., filipin and tomatin) for single and double labeling, and experimental removal of peripheral protein coats by proteolytic treatment prior to filipin treatment. Epifluorescence light microscopy is also useful for verification, as it may detect primary filipin–cholesterol binding even when ultrastructurally visible lesions are not formed. Although none of these methods

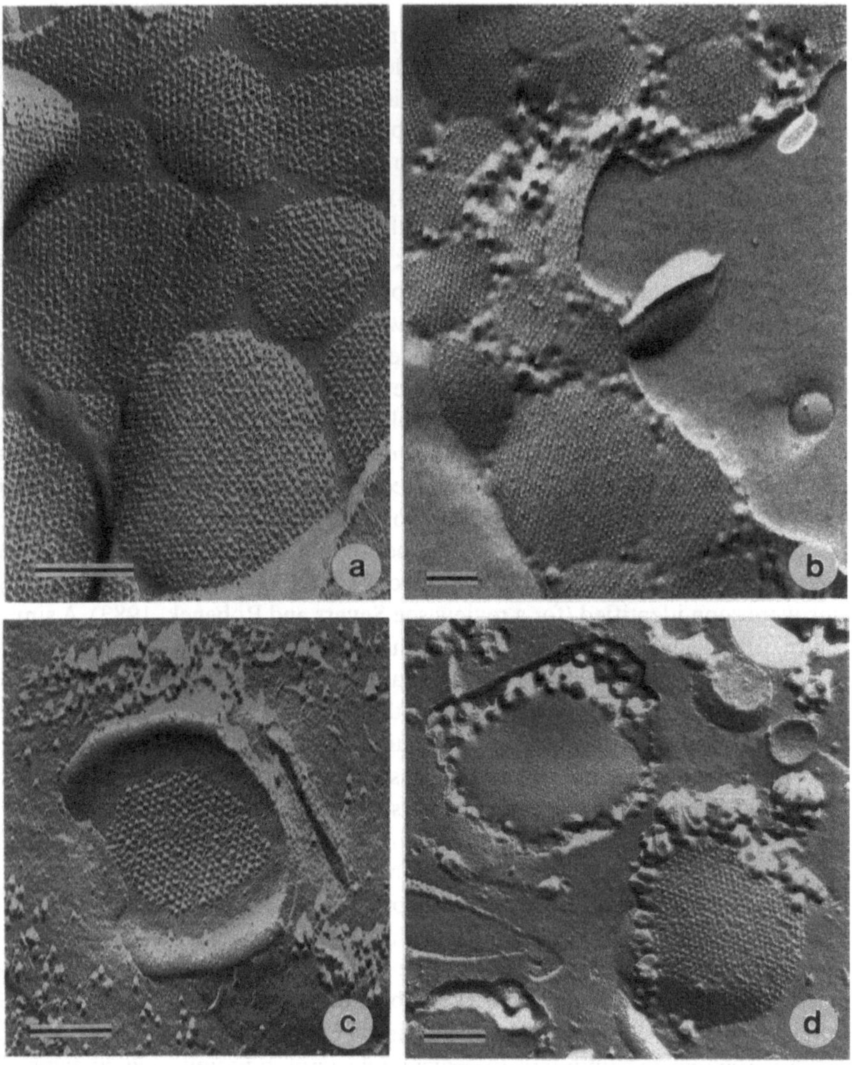

FIGURE 6. Effects of filipin on the luminal plasma membrane and cytoplasmic discoidal vesicles of rat urinary bladder. (a) The luminal membrane has an unusual mosaic structure, consisting of plaques of hexagonally ordered protein subunits, separated by smooth interplaque regions. (b) When the membrane is treated with filipin, lesions are formed only in the narrow interplaque zones, never within the plaques. (c) Discoidal vesicles, which form a cytoplasmic pool of membrane exchangeable by exocytosis and endocytosis with the luminal surface, also carry hexagonal arrays. (d) As with the luminal membrane, these subunit arrays are not affected by filipin; the surrounding non-specialized membrane is highly sensitive to the treatment. All views shown are of the E face of the membrane (i.e., the fracture face of the half-membrane sheet left attached to the extracellular space or endoplasmic space after freeze fracture), with the exception of the upper vesicle in (d), which shows the complementary view (P face). Scale bar = 200 nm. (From Severs *et al.*, 1981. Reprinted by permission of Academic Press.)

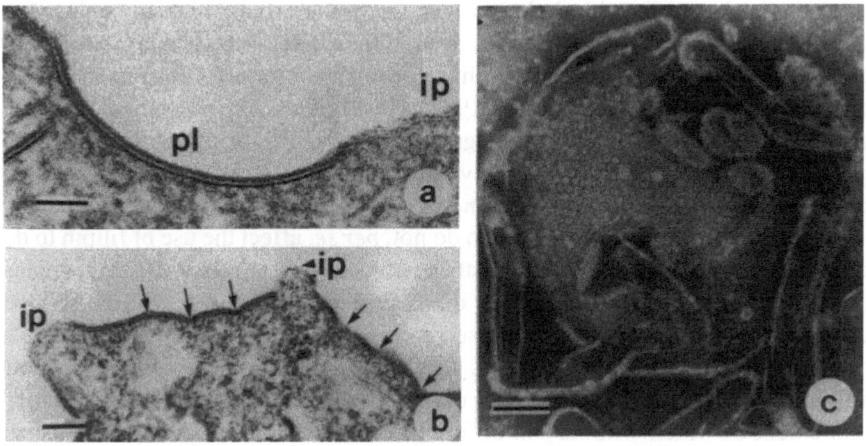

FIGURE 7. Bladder luminal membrane after digitonin treatment. (a) Thin-section view of untreated luminal membrane. Note the asymmetrically thickened unit membrane structure of the plaques (pl), which present a gently sloping concave shape at the surface. ip, interplaque region. (b) Thin-section view of the membrane after digitonin treatment. The agent has caused sharp corrugations in the membrane (arrows) (compare freeze-fracture view in Figure 6a,b). (c) Negative staining of isolated membrane fragments confirms that the plaque regions are sensitive to digitonin treatment. (From Severs *et al.*, 1981. Reprinted by permission of Academic Press.) Scale bar = 100 nm.

alone is infallible, meaningful information could evidently be obtained by using them in combination (reviewed in Severs, 1984a, and Severs and Robenek, 1983).

3.2. Limitations of Spatial Resolution

The deficiencies in filipin labeling encountered in practice refocused attention on the underlying mechanism of filipin–cholesterol complexation and lesion formation, and this raised further interpretative questions, especially with regard to the spatial resolution of localization at the ultrastructural level. If, as outlined in section 2.1, lateral mobility of primary complexes is necessary for formation of an aggregate, the very process of lesion formation will involve lateral movement of cholesterol molecules from their original locations in the membrane (Maurin *et al.*, 1988; Behnke *et al.*, 1984b; Severs and Robenek, 1983). The question then arises as to how accurately the location of the ultimately formed deformation represents the original location of its constituent sterol molecules. From the theoretical diffusion coefficient of cholesterol, and the time taken for a filipin lesion to form, Miller (1984, 1989) argued that any attempt to pinpoint the original position of cholesterol in the membrane could be thwarted from the outset. There are grounds for believing, however, that the diffusion of cholesterol in natural membranes is more limited than in model membrane systems. Protein–lipid interactions (e.g., interaction of caveolin with cholesterol; see section 4) would markedly reduce free diffusion in natural membranes, and, even in liposomes, biophysical factors such as overall cholesterol concentration have been shown to affect diffusion characteristics (Miller, 1989; Alecio *et al.*, 1982). Nevertheless, the theoretical considerations still leave an undesirable degree of uncertainty in the ability of filipin deformations, at the ultrastructural level, to pinpoint accurately the distribution of cholesterol *within the membrane plane*.

The caveats on spatial resolution do not, *per se*, affect the use of filipin to deduce other features of cholesterol distribution, such as the relative overall cholesterol contents between topologically discrete membrane systems, or transbilayer differences in cholesterol content based on the side of the membrane from which the deformations project (Orci *et al.*, 1980). However, in these applications, one still has to contend with the influence of membrane-associated proteins, the nature and extent of which will vary in different membranes. In addition, a number of other factors have been reported to influence the response observed, in particular, the molar ratio of cholesterol to filipin, and the concentration of sphingomyelin in the membrane (Behnke *et al.*, 1984a, 1984b).

3.3. Appraisal and Current Status of Techniques

When the above considerations were brought to the fore, the uncertainties and complexities of what can or cannot be said about filipin-labeling patterns in freeze-fractured membranes dampened interest in the technique. In the ensuing critical atmosphere, doubts on the reliability of digitonin, which had remained dormant in the literature (Frühling *et al.*, 1971), also resurfaced (Miller, 1989). What guidance may be offered, then, from the accumulated studies to date? Taken as a whole, the evidence shows that the labeling patterns formed by treating membranes with filipin are indeed determined by cholesterol (or other 3β-hydroxysterol) content, but only in part; a variety of other factors are also involved. Cholesterol concentration certainly does, in many instances, seem to be a major factor in determining the extent of the response observed, but membrane proteins and probably other lipids have important modifying influences that cannot be ignored. The relationship of the position and the number of lesions to the location and amount of cholesterol is at best approximate, and will vary according to the specific factors pertaining to the system under investigation. Membranes do differ in a consistent manner in their responses to filipin and other agents, and this does reflect real differences in a combination of biochemical and biophysical properties, particularly cholesterol content and distribution. However, the relative contributions of the various factors involved are usually difficult to determine.

Despite these drawbacks, filipin is too useful a probe to be dismissed out of hand. The above limitations are most acute in attempts to identify regional variations in cholesterol content within the plane of a given membrane at high resolution, but the validity of using the agent for other applications has to be assessed on a case-by-case basis, depending on the precise nature of the proposed experimental application. Alternative applications of filipin are many and varied. For example, the cholesterol binding property of the agent has been exploited for purposes other than direct cholesterol localization, such as in the investigation of the barrier function of tight junctions (Miller and Baldridge, 1985; Pelletier and Friend, 1983). The failure of filipin to label cholesterol in protein-rich domains has itself been turned to an advantage by using the agent as a probe to reveal the position of underlying dense peripheral membrane protein specializations in planar freeze-fracture views of membrane. Examples of this application are the identification of the position of dense bodies in smooth muscle cells and of clathrin-coated pits in studies on low-density-lipoprotein internalization (Davis and Shivers, 1992; Robenek and Severs, 1992). The distinctive filipin labeling patterns of different cell types may similarly be used to assign the correct cellular identification to freeze-fractured plasma membranes in situations where they cannot otherwise be easily identified (Pelletier and Byers, 1992). Filipin has also been used as an indirect probe in studies on lipid metabolism and transport; for example, the affinity of filipin for the *trans*-Golgi region, originally demonstrated in freeze fracture (Orci *et al.*, 1981; Severs, 1981), has been exploited to inhibit the transport of a ceramide analogue to this region (Pagano *et al.*, 1991). However, the most sustained applications of filipin in recent investigations have focused on two main themes; the role of cholesterol in caveolae, and cholesterol localization in lipid-related disorders. The following sections will explore these topics in more detail.

4. CHOLESTEROL AND CAVEOLAE

Caveolae are vesicular invaginations of the plasma membrane that are especially prominent in smooth muscle cells, endothelial cells, and fibroblasts. Although they have frequently been described as "non-coated" vesicles, to distinguish them from the classical clathrin-coated vesicle responsible for receptor-mediated endocytosis, caveolae possess a distinctive, though less conspicuous coat, readily visualized by ultrarapid freezing/deep etch electron microscopy. This comprises arrays of filaments and strands arranged in an irregular striped pattern, one component of which is a 21-kDa integral membrane phosphoprotein named caveolin (Rothberg *et al.*, 1992), also known as VIP21 (Kurzchalia *et al.*, 1992, 1994). In the older literature, caveolae were frequently referred to as pinocytotic vesicles, and presumed to function in fluid phase endocytosis; subsequent opinion held that they were more likely to be static inpocketings of the plasma membrane with a variety of potential functions (reviewed by Severs, 1988). The evidence now emerging is of a versatile plasma membrane organelle whose functions may encompass a variety of forms of endocytosis, carrier transport, and signal transduction.

Cholesterol was first proposed to play a role in caveolar organization from filipin/freeze-fracture studies on smooth muscle cell plasma membranes. In the smooth muscle cell, caveolae are distributed in distinctive rows and bands, running down the length of the cell (Figure 8a). The membrane zones between the bands of caveolae are the sites at which the dense bodies beneath the plasma membrane anchor the contractile apparatus and cytoskeleton to the cell surface. When the smooth muscle cell plasma membrane is treated with filipin, lesions are formed predominantly in the caveolar band regions, but not in the interband zones (Figure 8b). This led to the proposal that smooth muscle caveolae are organized in longitudinally oriented cholesterol-rich zones of membrane, and that these zones alternate with cholesterol-poor zones (Montesano, 1979). However, when the potential dangers of taking filipin-labeling patterns at face value became apparent from studies on membrane domains containing closely packed proteins (section 3.1), further investigations of the smooth muscle plasma membrane using alternative cholesterol probes were undertaken (Severs and Simons, 1983). These demonstrated that tomatin, in contrast to filipin, affects both the non-caveolar interband regions and the caveolar band regions (Figure 8c). Sequential double labeling with both agents demonstrated that, after initial treatment of the membrane with filipin, tomatin is still able to detect cholesterol in the interband regions (Figure 8d). The result with tomatin could not, therefore, be explained by lateral translocation of cholesterol. Hence, it was concluded that cholesterol distribution in the smooth muscle cell plasma membrane is uniform rather than heterogeneous, at least at the level of resolution afforded by freeze-fracture cytochemistry (Severs and Simons, 1986, 1983). Cholesterol clearly is present in the interband zones, but filipin is incapable of detecting it, the action of the agent being inhibited by the compact peripheral membrane proteins of the membrane-associated dense bodies, as discussed in section 3.1.

Application of the filipin/freeze-fracture technique to endothelial cells similarly led to divergent interpretations. In one study, the membranous rims of caveolae (endothelial vesicles) were reported to be preferentially sensitive to filipin (Simionescu *et al.,* 1983), but, just as in the smooth muscle cell, other studies questioned the general significance of this finding (Simons and Severs, 1986). There the matter appeared to rest until interest in caveolae revived with the development of new hypotheses of caveolar function, the identification of characteristic protein constituents of the caveolar membrane, and further investigation of a link between cholesterol and caveolae.

Current opinion suggests that caveolae may function not only as classical endocytic organelles and in transcytosis (i.e., endocytosis coupled to exocytosis, as in transendothelial transport), but also in potocytosis, a process proposed to involve transient closure of caveolae, as distinct from the pinching-off to form discrete intracellular vesicles that occurs in classical endocytosis (Stahl and Mueller, 1995; Anderson *et al.,* 1992; Keller *et al.,* 1992; Ying *et al.,* 1992). According to the potocytosis theory, caveolae are specialized structures that function in the transmembrane carrier transport of low molecular weight ligands (such as folate),

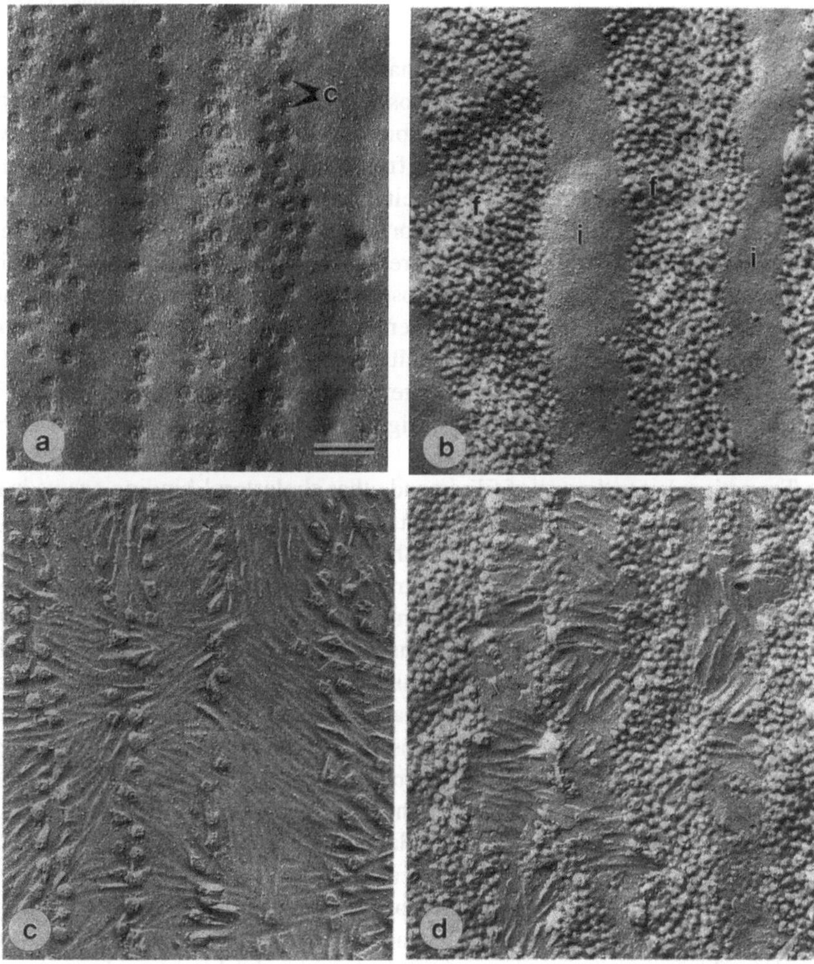

FIGURE 8. Effects of single and double labeling with filipin and tomatin on the smooth muscle cell plasma membrane. (a) In common with other cells, the plasma membrane of the smooth muscle cell contains vesicular invaginations called caveolae (c), shown here in planar freeze-fracture view of an untreated control membrane. The caveolae are seen as circular depressions in this P-face view, where the fracture plane skips across the caveolar neck, and are arranged in characteristic bands along the long axis of the cell. The non-caveolar interband regions of membrane are the sites at which dense bodies attach the contractile apparatus to the cytoplasmic surface of the plasma membrane. (b) Treatment with filipin leads to the formation of abundant deformations in the caveolar band regions while the non-caveolar interband regions are insensitive to the agent (i). In this P-face view, the filipin lesions (f) appear as ~25-nm bumps, distinguished from the caveolae, which appear as larger depressions. (c) Tomatin treatment consistently leads to the formation of corrugations in the interband (non-caveolar) regions; the caveolar bands are also affected where the spacing between the caveolae is sufficiently large to accommodate the tomatin-induced tubular structure. As this example shows an E-face view, the caveolar necks appear raised. (d) Sequential double treatment with filipin followed by tomatin. Cholesterol in the interband regions, which remains undetected by filipin, is detected by the subsequent tomatin treatment. Scale bar = 200 nm.

which bind to clusters of glycosyl-phosphatidylinositol (GPI)-anchored receptors (Rothberg *et al.,* 1990). The theory proposes that receptor–ligand complexes are concentrated within the invaginated membrane of the caveola and, when the caveolar neck closes, the ligands dissociate from the receptors and are transported across the membrane via membrane protein channels or carriers (Anderson *et al.,* 1992). Apart from GPI-anchored receptors, numerous other integral membrane proteins have been reported to be preferentially localized in caveolae. Notable among these are calcium ATPase, inositol 1,4,5-triphosphate receptor, heterotrimeric GTP-binding proteins, and members of the Src family of protein tyrosine kinases (Dietzen *et al.,* 1995; Schnitzer *et al.,* 1995c; Chang *et al.,* 1994; Fujimoto, 1993; Fujimoto *et al.,* 1992; Stefanová *et al.,* 1991), a combination indicating a potential role for caveolae in signal transduction (for reviews, see Parton and Simons, 1995; Anderson, 1993).

The rationale for the use of filipin and other cholesterol-binding agents in recent studies on caveolae differs from that of the earlier freeze-fracture studies in that the agents are used to sequester, and thereby deplete, the (unfixed) membrane of cholesterol, rather than to form ultrastructural markers. Treatment of cell cultures with low concentrations of filipin in this way results in a marked alteration in the distribution of the folate receptor; instead of appearing in clusters associated with caveolae, the receptor becomes diffusely distributed in the membrane (as determined by immunolabeling with primary antibody followed by labeled secondary antibody, at both the light microscopy and EM levels; Rothberg *et al.,* 1990). Treatment with other cholesterol-binding agents (nystatin and digitonin), or depletion of cholesterol by culturing the cells in lipoprotein-deficient medium and inhibiting cholesterol synthesis, similarly leads to increased dispersion of folate receptors. Apart from caveolae implicated in potocytosis, parallel findings appear to apply to caveolae engaged in classical endocytosis and transcytosis. In endothelial cells *in vitro* and *in situ,* extraction of cholesterol by perfusion or superfusion with low concentrations of filipin (or nystatin) has been shown to lead to disassembly of caveolae, with consequent reduction in transcellular and intracellular transport of albumin and insulin via the endothelial caveolar vesicle transport pathway (Schnitzer *et al.,* 1994; Schnitzer *et al.,* 1995a). Structural studies using the rapid-freezing/deep-etch technique applied to (unfixed) fibroblast plasma membranes reveal that filipin treatment results in rapid flattening out of caveolae and disintegration of the striped caveolin coat (Rothberg *et al.,* 1992). Moreover, when caveolar cholesterol is oxidized with cholesterol oxidase, caveolin reportedly migrates from the plasma membrane to the Golgi apparatus (Smart *et al.,* 1994). Taken together, these studies indicate that: i) cholesterol plays a key role in maintaining caveolar structure and in modulating the interaction of GPI-anchored receptors; and ii) the enriched cholesterol environment of caveolar membrane is maintained by caveolin.

These conclusions have been further strengthened as the molecular topology of caveolin and the nature of its interaction with cholesterol have been clarified.

The caveolin molecule is thought to form a hairpin loop-like structure in the membrane, with the amino and carboxy termini on the cytoplasmic side (Dupree *et al.*, 1993), and has two key properties that promote the formation of specialized membrane microdomains. First, caveolin binds directly and specifically to cholesterol (Murata *et al.*, 1995), which in turn associates with sphingolipids, creating a specialized lipid environment. Second, caveolin interacts with itself to form homo-oligomers, which apparently act as a scaffold in which the functional proteins are concentrated within this environment (Sargiacomo *et al.*, 1995).

In summary, current evidence indicates that cholesterol is enriched in caveolae and serves a key role in their structure and function. Earlier freeze-fracture cytochemical studies, despite being based on shaky foundations, may thus turn out to have pointed to a potentially important link. Our concepts of caveolar function and organization are evolving rapidly, however, and some aspects are currently under active debate. For example, the original reports that GPI-anchored proteins are clustered (i.e., localized within caveolae) under steady-state conditions were conducted using a primary antibody followed by a labeled secondary antibody; single step labeling with fluorescently labeled monoclonal antibody, on the other hand, is reported to reveal a dispersed distribution of the proteins (Mayor *et al.*, 1994). A helpful approach is to draw a distinction between morphologically defined caveolae and detergent insoluble glycosphingolipid-enriched complexes containing GPI-linked and signal transduction proteins that, according to the various viewpoints, may exist separately, at the rim, or within caveolae (Schnitzer *et al.*, 1995b; reviewed by Parton and Simons, 1995). Despite the remaining uncertainties, however, the concept that the organization of specific protein components of the caveolar membrane into functional assemblies depends on cholesterol, and that abnormal levels of cholesterol may lead to dysfunction of transport and signal transduction properties, remains a central hypothesis for current investigation (Anderson, 1993).

5. CYTOCHEMICAL LOCALIZATION OF CHOLESTEROL, LIPOPROTEINS, AND LIPOPROTEIN RECEPTORS IN DISEASE

Localization studies have made a key contribution in providing a structural basis for understanding the cholesterol-related events of atherosclerosis and other less well known "cholesterol-storage" disorders, such as Niemann–Pick type C disease (see Chapter 14 by Pentchev *et al.*) and Schnyder's corneal crystalline dystrophy. Atherosclerosis, the pathological process by which the arterial wall becomes thickened with lipid and fibromuscular material, leads to coronary artery disease and stroke, the major causes of death in industrialized countries (see Chapter 12 by Kruth). Niemann–Pick disease refers to a heterogeneous collection of genetically inherited lipid storage disorders, the type C variant of which involves a defect in the ability of the cells to esterify excess cholesterol destined for intracellular storage. Schnyder's corneal crystalline dystrophy, also an inherited condition associated with hyperlipidemia, manifests as crystalline and diffuse opacification of the cornea as a result of lipid deposition.

The technique of staining histological sections with filipin for fluorescence microscopical detection of unesterified cholesterol (section 2.2), used in parallel with oil red O staining for detection of esterified cholesterol, has proven effective for investigating the nature of intracellular and extracellular cholesterol accumulation in the arterial wall, both in animal models of disease (Chao *et al.*, 1988; Kruth and Fry, 1984; Kruth, 1983) and in human atherosclerosis (Kruth, 1984a,b). An example, demonstrating localized accumulations of cholesterol-rich lipid in the intimal layer of human coronary artery, an early event in the pathogenesis of atherosclerosis, is illustrated in Figure 9. The same

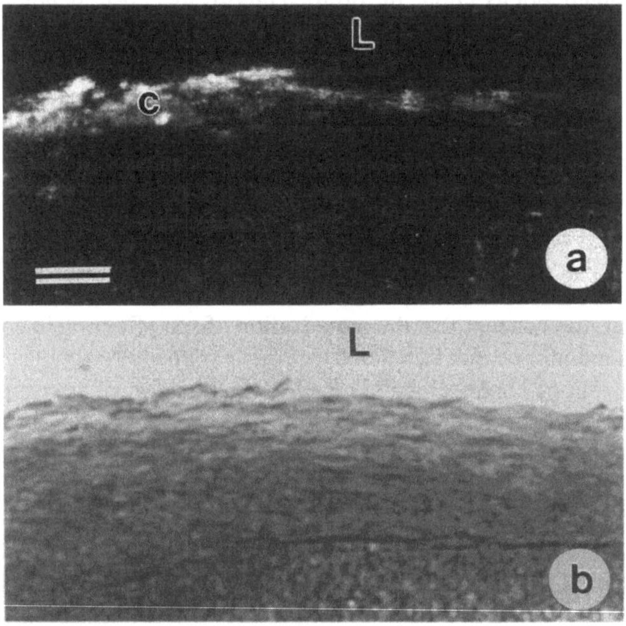

FIGURE 9. Cholesterol-rich lipid deposits in the intima of human coronary artery, as seen in a cryosection viewed by fluorescence microscopy after filipin labeling (a). A corresponding brightfield view (methylene blue staining) of the same region is shown in (b). The brightly fluorescent zone in (a) shows a small deposit of cholesterol-rich lipid (c) in the minimally thickened intima of a coronary artery from a transplant patient without symptoms of coronary artery disease. Such lipid deposits are derived from low density lipoprotein (the major carrier of cholesterol) that infiltrate into the intima, and stimulate monocytes to penetrate the arterial wall and transform into macrophages, which avidly take up the lipid and develop a foamy morphology. Death of lipid-laden macrophages ultimately results in the formation of a lipid-rich core in the atherosclerotic plaque, a zone of weakness which is prone to rupture, with consequent life-threatening thrombosis. (Coronary artery from explanted heart from patient with dilated cardiomyopathy.) L, lumen. Scale bar = 100 μm.

approach applied to Schnyder's corneal crystalline dystrophy has revealed that unesterified cholesterol is a major component of extracellular lipid deposition in the subepithelial and superficial stromal membrane in and around Bowman's layer (Figure 10), responsible for opacification of the cornea (Rodrigues *et al.*, 1990, 1987). Filipin fluorescence labeling applied to Niemann–Pick C cultured fibroblasts after exposure to LDL demonstrates massive accumulation of cholesterol in the lysosomes and Golgi region (see Figure 1 of Chapter 14), reflecting a delayed homeostatic response to exogenous cholesterol, which is characteristic of the disease (Vanier *et al.*, 1991; Rodriguez-Lafrasse *et al.*, 1990; Sokol *et al.*, 1988). From fluorescence and freeze-fracture studies using filipin, NPC fibroblasts were found to differ from normal fibroblasts in the pattern of cholesterol enrichment in the membranes of the Golgi apparatus, with accumulation in the trans cisternae, suggesting that a defect in the modulation of cholesterol transport through the Golgi might play a part in this process (Coxey *et al.*, 1993).

It is important to bear in mind that detection of cholesterol alone, in the context of the metabolic defects underlying these disorders, will provide only one facet of a complex picture. Comprehensive insights into the mechanisms by which cells control their cholesterol intake, and how different types of lipoproteins are processed by cells, have been obtained by localization of lipoprotein receptors, lipoproteins, and their apoprotein constituents in experimentally manipulated cultured cells and in diseased tissue, a field pioneered by Robenek (reviewed by Robenek, 1989). These approaches to the investigation of cholesterol metabolism

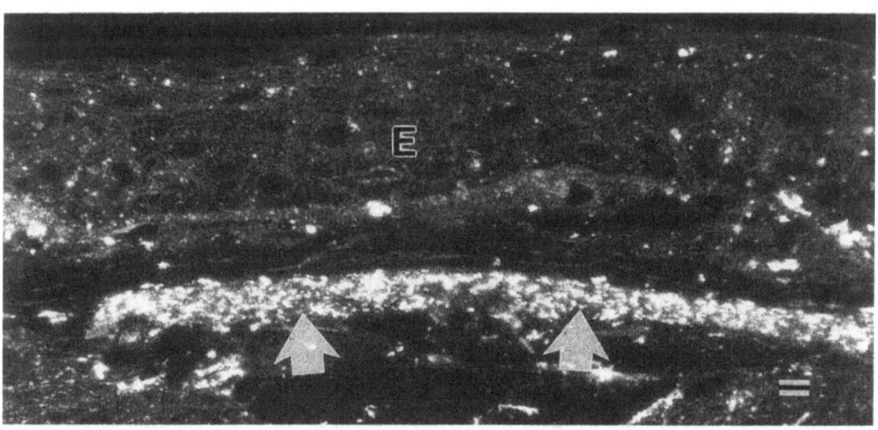

FIGURE 10. Cholesterol localization in Schnyder's corneal crystalline dystrophy, using filipin as a fluorescent probe for unesterified cholesterol in cryosections examined by light microscopy. Small deposits of cholesterol are scattered throughout the cornea, and larger accumulations (arrow) occur beneath the epithelium (E). Scale bar = 10 μm. (Micrograph kindly provided by Joan Blanchette-Mackie, from Rodrigues *et al.*, 1990).

at the cellular level have relied centrally on colloidal gold cytochemistry (section 2.4), a technique that has proven highly versatile for application in diverse EM methods. How cells organize and display their receptors at the cell surface, and the dynamic events of receptor-mediated endocytosis by which lipoproteins are taken into the cell and the receptors subsequently recycled back to the cell surface, have been directly visualized for a range of lipoproteins by means of platinum–carbon surface replicas of cultured cells. The specificities of receptors for different classes of lipoprotein in fibroblasts, smooth muscle cells, macrophages, and endothelial cells have been investigated in detail using this approach, and the intracellular processing routes after internalization of gold-tagged lipoproteins tracked by thin-section EM. Key aspects studied include the mechanisms by which macrophages accumulate, process, and release cholesterol (Kruth *et al.,* 1995; Schmitz *et al.,* 1987, 1985b), the effects of drugs on these processes (Robenek and Schmitz, 1988; Schmitz *et al.,* 1988), and the identification of a specific defect in intracellular trafficking in Tangier disease (an autosomal recessive disorder characterized by absence of high density lipoproteins) (Robenek and Schmitz, 1991; Schmitz *et al.,* 1985a). Localization studies using colloidal gold are also of proven value for tracking lipoprotein interactions in intact tissues. For example, the infiltration and deposition of low-density lipoprotein into the intima has been directly visualized by ultrarapid freeze/deep etch EM (Frank *et al.,* 1995), and detailed studies conducted on localization of lipoproteins and apoproteins in the human atherosclerotic lesion by immunogold labeling of cryosectioned specimens (Kaesberg *et al.,* 1993). For more comprehensive treatment of this subject, the reader is referred to other recent reviews (Robenek and Severs, 1993, 1992; Robenek, 1989).

6. CONCLUDING COMMENT

Cholesterol-binding agents and localization methods have had a chequered history, but with appropriate choice of technique to the topic of investigation, and with due care in interpretation of results, these methods have a useful place in research on the biological roles of cholesterol in the normal and dysfunctional cell. Localization of the protein components (lipoproteins and lipoprotein receptors) responsible for cholesterol transport and intracellular trafficking, and of the enzymes responsible for cholesterol metabolism, provide complementary dimensions to such studies, and a comprehensive set of methods of proven value is available for this type of investigation. As the intricacies of cholesterol and cellular dysfunction in relation to disease are progressively unraveled, there will continue to be a need for the perspectives that these approaches provide.

ACKNOWLEDGMENTS. I am indebted to Dr. R. Marc Pelletier (Department of Anatomy, University of Montreal, Canada) for providing the illustrations for Figure 4 from his elegant study on the enzymatic localization of cholesterol (Pelletier and Vitale, 1994), and to Dr. Joan Blanchette-Mackie (NIH, Bethesda, Maryland) for providing Figures 5 and 10 from her pioneering studies on cholesterol localization in disease (Blanchette-Mackie *et al.,* 1989; Rodrigues *et al.,* 1990). The following publishers are acknowledged for permission to reproduce previously published micrographs: Academic Press (Figures 3, 6, 7); Wiley-Liss (Figure 5). The help of Stephen Rothery in the preparation of figures, and for undertaking the experimental work for Figure 9, is much appreciated. Findings reviewed from the author's laboratory on cardiovascular cells come from projects supported by the British Heart Foundation.

7. REFERENCES

Alecio, M. R., Golan, D. E., Veatch, W. R., and Rando, R. R., 1982, Use of a fluorescent cholesterol derivative to measure lateral mobility of cholesterol in membranes, *Proc. Natl. Acad. Sci. USA* **79:**5171–5174.

Allain, C. C., Poon, L. S., Chan, C. S. G., Richmond, W., and Fu, P. C., 1974, Enzymatic determination of total serum cholesterol, *Clin. Chem.* **20:**470–475.

Anderson, R. G. W., 1993, Caveolae: where incoming and outgoing messengers meet, *Proc. Natl. Acad. Sci. USA* **90:**10909–10913.

Anderson, R. G. W., Kamen, B. A., Rothberg, K. G., and Lacey, S. W., 1992, Potocytosis: sequestration and transport of small molecules by caveolae, *Science* **255:**410–411.

Andrews, L. D., and Cohen, A. I., 1979, Freeze-fracture evidence for the presence of cholesterol in particle-free patches of basal disks and the plasma membrane of retinal rod outer segments of mice and frogs, *J. Cell Biol.* **81:**215–228.

Bangham, A. D., and Horne, R. W., 1962, Action of saponin on biological cell membranes, *Nature* **196:**952–953.

Behnke, O., Tranum-Jensen, J., and van Deurs, B., 1984a, Filipin as a cholesterol probe. I. Morphology of filipin-cholesterol interaction in lipid model systems, *Eur. J. Cell Biol.* **35:**189–199.

Behnke, O., Tranum-Jensen, J., and van Deurs, B., 1984b, Filipin as a cholesterol probe. II. Filipin-cholesterol interaction in red blood cell membranes, *Eur. J. Cell Biol.* **35:**200–215.

Bittman, R., 1978, Sterol-polyene antibiotic complexation: probe of membrane structure, *Lipids* **13:**686–691.

Bittman, R., Chen, W. C., and Anderson, O. R., 1974, Interaction of filipin III and amphotericin B with lecithin-sterol vesicles and cellular membranes. Spectral and electron microscope studies, *Biochemistry* **13:**1364–1373.

Blanchette-Mackie, J. E., Dwyer, N. K., and Amende, L. M., 1989, Cytochemical studies of lipid metabolism: immunogold probes for lipoprotein lipase and cholesterol, *Am. J. Anat.* **185:**255–263.

Branton, D., Bullivant, S., Gilula, N. B., Karnovsky, M. J., Moor, H., Muhlethaler, K., Northcote, D. H., Packer, L., Satir, B., Satir, P., Speth, V., Staehelin, L. A., Steere, R. L., and Weinstein, R. S., 1975, Freeze-etching nomenclature, *Science* **190:**54–56.

Chang, W.-J., Ying, Y., Rothberg, K. G., Hooper, N. M., Turner, A. J., Gambliel, H. A., De Gunzburg, J., Mumby, S. M., Gilman, A. G., and Anderson, R. G. W., 1994, Purification and characterization of smooth muscle cell caveolae, *J. Cell Biol.* **126:**127–138.

Chao, F-F., Amende, L. M., Blanchette-Mackie, J. E., Skarlatos, S. I., Gamble, W., Resau, J. H., Mergner, W. T., and Kruth, H. S., 1988, Unesterified cholesterol-rich lipid particles in atherosclerotic lesions of human and rabbit aortas, *Am. J. Pathol.* **131:**73–83.

Compère, P., Maneta-Peyret, L., Goffinet, G., and Cassagne, C., 1995, Immunogold labelling of fatty acyl chains, *J. Immunol. Methods* **181:**201–209.

Coulombe, P. A., Kan, F. W. K., and Bendayan, M., 1988, Introduction of a high-resolution cytochemical method for studying the distribution of phospholipids in biological tissues, *Eur. J. Cell Biol.* **46:**564–576.

Coxey, R. A., Pentchev, P. G., Campbell, G., and Blanchette-Mackie, E. J., 1993, Differential accumulation of cholesterol in Golgi compartments of normal and Niemann–Pick type C fibroblasts incubated with LDL: a cytochemical freeze-fracture study, *J. Lipid Res.* **34:**1165–1176.

Davis, E. C., and Shivers, R. R., 1992, Ordered distribution of membrane-associated dense plaques in intact quail gizzard smooth muscle cells revealed by freeze-fracture following treatment with cholesterol probes, *Anat. Rec.* **232:**385–392.

De Kruijff, B., and Demel, R. A., 1974, Polyene antibiotic-sterol interactions in membranes of *Acholeplasma laidlawii* cells and lecithin liposomes. III. Molecular structure of the polyene antibiotic-cholesterol complexes, *Biochim. Biophys. Acta* **339:**57–70.

De Kruijff, B., Gerritsen, W. J., Oerlemans, A., van Dijck, P. W. M., Demel, R. A., and van Deenen, L. L. M., 1974, Polyene antibiotic-sterol interactions in membranes of *Acholeplasma laidlawii* cells and lecithin liposomes. II. Temperature dependence of the polyene antibiotic-sterol complex formation, *Biochim. Biophys. Acta* **339:**44–56.

Dietzen, D. J., Hastings, W. R., and Lublin, D. M., 1995, Caveolin is palmitoylated on multiple cysteine residues. Palmitoylation is not necessary for localization of caveolin to caveolae, *J. Biol. Chem.* **270:**6838–6842.

Dupree, P., Parton, R. G., Raposo, G., Kurzchalia, T. V., and Simons, K., 1993, Caveolae and sorting in the *trans*-Golgi network of epithelial cells, *EMBO J.* **12:**1597–1605.

El Yandouzi, E. H., Zlatkine, P., Moll, G., and Le Grimellec, C., 1994, Cholesterol distribution in renal epithelial-cells LLC-PK1 as determined by cholesterol oxidase–evidence that glutaraldehyde fixation masks plasma-membrane cholesterol pools, *Biochemistry* **33:**2329–2334.

Elias, P. M., Goerke, J., Friend, D. S., and Brown, B. E., 1978, Freeze-fracture identification of sterol-digitonin complexes in cell and liposome membranes, *J. Cell Biol.* **78:**577–596.

Elias, P. M., Friend, D. S., and Goerke, J., 1979, Membrane sterol heterogeneity. Freeze-fracture detection with saponins and filipin, *J. Histochem. Cytochem.* **27:**1247–1260.

Emeis, J. J., Van Gent, C. M., and Van Sabben, C. M., 1977, An enzymatic method for the histochemical localization of free and esterified cholesterol separately, *Histochem. J.* **9:**197–204.

Feltkamp, C. A., and van der Waerden, A. W. M., 1982, Membrane-associated proteins affect the formation of filipin-cholesterol complexes in viral membranes, *Exp. Cell Res.* **140:**289–297.

Frank, J. S., Mottino, G., and Nievelstein-Post, P., 1995, Early events in atherosclerosis captured by the quick-freeze/deep-etch technique, in: *Rapid Freezing, Freeze Fracture and Deep Etching* (N. J. Severs and D. M. Shotton, eds.), pp. 335–345, Wiley, New York.

Friend, D. S., and Bearer, E. L., 1981, Beta-hydroxysterol distribution as determined by freeze-fracture cytochemistry, *Histochem. J.* **13:**535–546.

Frühling, J., Penasse, W., Sand, G., and Claude, A., 1971, Réactions de la digitonine avec le cholestérol et autres lipides de la corticosurrénale du rat: étude par microscopie électronique, *J. Microscopie* **12:**83–106.

Fujimoto, T., 1993, Calcium pump of the plasma membrane is localized in caveolae, *J. Cell Biol.* **120:**1147–1157.

Fujimoto, T., Nakade, S., Miyawaki, A., Mikoshiba, K., and Ogawa, K., 1992, Localization of inositol 1,4,5-trisphosphate receptor-like protein in plasmalemmal caveolae, *J. Cell Biol.* **119:**1507–1513.

Glauert, A. M., Dingle, J. T., and Lucy, J. A., 1962, Action of saponin on biological cell membranes, *Nature* **196:**953–955.

Hacker, G. W., 1989, Silver-enhanced colloidal gold for light microscopy, in: *Colloidal Gold. Principles, Methods and Applications* (M. A. Hayat, ed.), pp. 297–321, Vol. 1, Academic Press, San Diego.

Hayat, M. A., 1989, *Colloidal Gold. Principles, Methods and Applications,* Academic Press, New York.

Horisberger, M., 1983, Colloidal gold as a tool in molecular biology, *Trends Biochem. Sci.* **8:** 395–397.

Jones, H. M., and Miyai, K., 1981, Ultrastructural localization of cholesterol by enzyme histochemistry, *Histochem. J.* **13:**1017–1028.

Kaesberg, B., Harrach, B., Dieplinger, H., and Robenek, H., 1993, In situ immunolocalization of lipoproteins in human arteriosclerotic tissue, *Arterioscler. Thromb.* **13:**133–146.

Kan, F. W. K., and Rauch, J., 1993, Immunolocalization of phospholipids using freeze-fracture cytochemistry, in: *Proceedings of the 51st Annual Meeting of the Microscopy Society of America* (G. W. Bailey and C. L. Rieder, eds.), San Francisco Press, San Francisco.

Keller, G., Siegel, M. W., and Caras, I. W., 1992, Endocytosis of glycophospholipid–anchored and trans membrane forms of CD4 by different endocytic pathways, *EMBO J.* **11:**863–874.

Kinsky, S. C., Luse, S, and van Deenen, L. L. M., 1966, Interaction of polyene antibiotics with natural and artificial membrane systems, *Fed. Proc. Fed. Am. Soc. Exp. Biol.* **25:**1503.

Kitajima, Y., Sekiya, T., and Nozawa, Y., 1976, Freeze-fracture ultrastructural alterations induced by filipin, pimaricin, nystatin and amphotericin B in the plasma membranes of *epidermophyton, saccharomyces* and red blood cells. A proposal of models for polyene-ergosterol complex-induced lesions, *Biochim. Biophys. Acta* **445:**452–465.

Kruth, H. S., 1983, Filipin-positive, oil red O-negative particles in atherosclerotic lesions induced by cholesterol feeding, *Lab. Invest.* **50:**87–93.

Kruth, H. S., 1984a, Histochemical detection of esterified cholesterol within human atherosclerotic lesions using the fluorescent probe filipin, *Atherosclerosis* **51:**281–292.

Kruth, H. S., 1984b, Localization of unesterified cholesterol in human atherosclerotic lesions. Demonstration of filipin-positive, oil-red-O-negative particles, *Am. J. Pathol.* **114:**201–208.

Kruth, H. S., and Fry, D. L., 1984, Histochemical detection and differentiation of free and esterified cholesterol in swine atherosclerosis using filipin, *Exp. Mol. Pathol.* **40:**288–294.

Kruth, H. S., Skarlatos, S. I., Lilly, K., Chang, J., and Ifrim, I., 1995, Sequestration of acetylated LDL and cholesterol crystals by human monocyte-derived macrophages, *J. Cell Biol.* **129:** 133–145.

Kurzchalia, T. V., Dupree, P., and Monier, S., 1994, VIP21-caveolin, a protein of the *trans*-Golgi network and caveolae, *FEBS Lett.* **346:**88–91.

Kurzchalia, T. V., Dupree, P., Parton, R. G., Kellner, R., Virta, H., Lehnert, M., and Simons, K., 1992, VIP21, a 21-kD membrane protein is an integral component of trans-golgi-network-derived transport vesicles, *J. Cell Biol.* **118:**1003–1014.

Luciano, L., Reale, E., Konitz, H., Boseck, U., and Boseck, S., 1989, Alignment of cholesterol in the membrane bilayer, *J. Histochem. Cytochem.* **37:**1421–1425.

Lucy, J. A., and Glauert, A. M., 1964, Structure and assembly of macromolecular lipid complexes composed of globular micelles, *J. Mol. Biol.* **8:**727–748.

Maneta-Peyret, L., Picard, J-P., Bezian, J-H., and Cassagne, C., 1992, Fatty acids rendered immunogenic, *Immunol. Lett.* **31:**227–232.

Maraldi, N. M., Zini, N., Squarzoni, S., Del Coco, R., Sabatelli, P., and Manzoli, F. A., 1992, Intranuclear localization of phospholipids by ultrastructural cytochemistry, *J. Histochem. Cytochem.* **40:**1383–1392.

Maurin, L., Bancel, F., Morin, P., and Bienvenüe, A., 1988, Interactions between a paramagnetic analogue of cholesterol and filipin, *Biochim. Biophys. Acta* **939:**102–110.

Mayor, S., Rothberg, K. G., and Maxfield, F. R., 1994, Sequestration of GPI-anchored proteins in caveolae triggered by cross-linking, *Science* **264:**1948–1951.

Miller, R. G., 1984a, Interactions between digitonin and bilayer membranes, *Biochim. Biophys. Acta* **774:**151–157.

Miller, R. G., 1984b, The use and abuse of filipin to localize cholesterol in membranes, *Cell Biol. Int. Rep.* **8:**519–535.

Miller, R. G., 1989, Freeze-fracture sterol localization with filipin, in: *Freeze-Fracture Studies of Membranes* (S. W. Hui, ed.), pp. 87–102, CRC Press, Boca Raton, FL.

Miller, R. G., and Baldridge, W. H., 1985, The tight junction as a barrier to cholesterol in canine epithelial cells, *J. Ultrastruct. Res.* **90:**275–285.

Miyazawa, A., Inoue, H., Yoshioka, T., Horikoshi, T., Yanagisawa, K., Umeda, M., and Inoue, K., 1992, Monoclonal antibody analysis of phosphatidylserine and protein kinase C localizations in developing rat cerebellum, *J. Neurochem.* **59:**1547–1554.

Montesano, R., 1979, Inhomogeneous distribution of filipin-sterol complexes in smooth muscle cell plasma membrane, *Nature* **280:**328–329.

Moranto, A. R., 1982, Neuronal mapping: a photooxidation reaction makes Lucifer yellow useful for electron microscopy, *Science* **217:**953–955.

Murata, M., Peränen, J., Schreiner, R., Wieland, F., Kurzchalia, T. V., and Simons, K., 1995, VIP21/caveolin is a cholesterol-binding protein, *Proc. Natl. Acad. Sci. USA* **92:**10339–10343.

Norman, A. W., Spielvogel, A. M., and Wong, R. G., 1976, Polyene antibiotic-sterol interaction, *Adv. Lipid Res.* **14:**127–169.

Ökrös, I., 1968, Digitonin reaction in electron microscopy, *Histochemie* **13:**91–96.

Orci, L., Miller, R. G., Montesano, R., Perrelet, A., Amherdt, M., and Vassalli, P., 1980, Opposite polarity of filipin-induced deformations in the membrane of condensing vacuoles and zymogen granules, *Science* **210:**1019–1021.

Orci, L., Montesano, R., Meda, P., Malaisse-Lagae, F., Brown, D., Perrelet, A., and Vassalli, P., 1981, Heterogeneous distribution of filipin-cholesterol complexes across the cisternae of the Golgi apparatus, *Proc. Natl. Acad. Sci. USA* **78:**293–297.

Pagano, R. E., Martin, O. C., Kang, H. C., and Haugland, R. P., 1991, A novel fluorescent ceramide analogue for studying membrane traffic in animal cells: accumulation at the Golgi apparatus results in altered spectral properties of the sphingolipid precursor, *J. Cell Biol.* **113:** 1267–1279.

Parton, R. G., and Simons, K., 1995, Digging into caveolae, *Science* **269:**1398–1399.

Pelletier, R. M., and Byers, S. W., 1992, The blood testis barrier and sertoli cell junctions: structural considerations, *Microsc. Res. Tech.* **20:**3–33.

Pelletier, R. M., and Friend, D. S., 1983, The sertoli cell junctional complex: structure and permeability to filipin in the neonatal and adult guinea pig, *Am. J. Anat.* **168:**213–228.

Pelletier, R-M., and Vitale, M. L., 1994, Filipin vs enzymatic localization of cholesterol in guinea pig, mink and mallard duck testicular cells, *J. Histochem. Cytochem.* **42:**1539–1554.

Pumplin, D. W., and Bloch, R. J., 1983, Lipid domains of acetylcholine receptor clusters detected with saponin and filipin, *J. Cell Biol.* **97:**1043–1054.

Robenek, H., 1989, Topography and internalization of cell surface receptors as analyzed by affinity- and immunolabeling combined with surface replication and ultrathin sectioning techniques, in: *Electron Microscopy of Subcellular Dynamics* (H. Plattner, ed.), pp. 141–163, CRC Press, Boca Raton, FL.

Robenek, H., and Schmitz, G., 1988, Ca^{++} antagonist and ACAT inhibitors promote cholesterol efflux from macrophages by different mechanisms. II, *Arteriosclerosis* **8:**57–67.

Robenek, H., and Schmitz, G., 1991, Abnormal processing of Golgi elements and lysosomes in Tangier disease, *Arterioscler. Thromb.* **11:**1007–1020.

Robenek, H., and Severs, N. J., 1992, Endocytosis of lipoproteins and cholesterol homeostasis, in: *Cell Interactions in Atherosclerosis* (H. Robenek and N. J. Severs, eds.), pp. 261–303, CRC Press, Boca Raton, FL.

Robenek, H., and Severs, N. J., 1993, Lipoprotein receptors in macrophages and smooth muscle cells, in: *Current Topics in Pathology, Recent Progress in Atherosclerosis Research, Vol.87* (E. Vollmer and A. Roessner, eds.), pp. 73–123, Springer-Verlag, Berlin.

Robinson, J. M., and Karnovsky, M. J., 1980, Evaluation of the polyene antibiotic filipin as a cytochemical probe for membrane cholesterol, *J. Histochem. Cytochem.* **28:**161–168.

Rodrigues, M. M., Kruth, H. S., Krachmer, J. H., and Willis, R., 1987, Unesterified cholesterol in Schnyder's corneal crystalline dystrophy, *Am. J. Ophthalmol.* **104:**157–163.

Rodrigues, M. M., Kruth, H. S., Krachmer, J. H., Vrabec, M. P., and Blanchette-Mackie, J. E., 1990, Cholesterol localization in ultrathin frozen sections in Schnyder's corneal crystalline dystrophy, *Am. J. Ophthalmol.* **110:**513–517.

Rodriguez-Lafrasse, C., Rousson, R., Bonnet, J., Pentchev, P. G., Louisot, P., and Vanier, M. T., 1990, Abnormal cholesterol metabolism in imipramine-treated fibroblast cultures. Similarities with Niemann–Pick type C disease, *Biochim. Biophys. Acta* **1043:**123–128.

Roth, J. E., 1983, The colloidal gold marker system for light and electron microscope cytochemistry, in: *Techniques in Immunocytochemistry, Vol. 2* (G. R. Bullock and P. Petrusz, eds.), pp. 217–284, Academic Press, London.

Rothberg, K. G., Ying, Y-S., Kamen, B. A., and Anderson, R. G. W., 1990, Cholesterol controls the clustering of the glycophospholipid-anchored membrane receptor for 5-methyltetrahydrofolate, *J. Cell Biol.* **111:**2931–2938.

Rothberg, K. G., Heuser, J. E., Donzell, W. C., Ying, Y.-S., Glenney, J. R., and Anderson, R. G. W., 1992, Caveolin, a protein component of caveolae membrane coats, *Cell* **68:**673–682.

Sandell, J. H., and Masland, R., 1988, Photoconversion of some fluorescent markers to a diaminobenzidine product, *J. Histochem. Cytochem.* **36:**555–559.

Sargiacomo, M., Scherer, P. E., Tang, Z., Kübler, E., Song, K. S., Sanders, M. C., and Lisanti, M. P., 1995, Oligomeric structure of caveolin: implications for caveolae membrane organization, *Proc. Natl. Acad. Sci. USA* **92:**9407–9411.

Scallen, T. J., and Dietert, S. E., 1969, The quantitative retention of cholesterol in mouse liver prepared for electron microscopy by fixation in a digitonin-containing aldehyde solution, *J. Cell Biol.* **40:**802–813.

Schmitz, G., Assmann, G., Robenek, H., and Brennhausen, B., 1985a, Tangier disease: a disorder of intracellular membrane traffic, *Proc. Natl. Acad. Sci. USA* **82:**6305–6309.

Schmitz, G., Robenek, H., Lohmann, U., and Assmann, G., 1985b, Interaction of high density lipoproteins with cholesteryl ester-laden macrophages: biochemical and morphological characterization of cell surface receptor binding, endocytosis and resecretion of high density lipoproteins by macrophages, *EMBO J.* **4:**613–622.

Schmitz, G., Robenek, H., and Assmann, G., 1987, Role of the high-density lipoprotein receptor cycle in macrophage cholesterol metabolism, *Atherosclerosis Rev.* **16:**95–107.

Schmitz, G., Robenek, H., Beuk, M., Krause, R., Schurek, A., and Niemann, R., 1988, Ca^{++} antagonist and ACAT inhibitors promote cholesterol efflux from macrophages by different mechanisms. I, *Arteriosclerosis* **8:**45–56.

Schnitzer, J. E., Liu, J., and Oh, P., 1995a, Endothelial caveolae have the molecular transport machinery for vesicle budding, docking, and fusion including VAMP, NSF, SNAP, annexins, and GTPases, *J. Biol. Chem.* **270:**14399–14404.

Schnitzer, J. E., Oh, P., Pinney, E., and Allard, J., 1994, Filipin-sensitive caveolae-mediated transport in endothelium: Reduced transcytosis, scavenger endocytosis, and capillary permeability of select macromolecules, *J. Cell Biol.* **127:**1217–1232.

Schnitzer, J. E., McIntosh, D. P., Dvorak, A. M., Liu, J., and Oh, P., 1995b, Separation of caveolae from associated microdomains of GPI-anchored proteins, *Science* **269:**1435–1439.

Schnitzer, J. E., Oh, P., Jacobson, B. S., and Dvorak, A. M., 1995c, Caveolae from luminal plasmalemma of rat lung endothelium: Microdomains enriched in caveolin, Ca^{2+}-ATPase, and inositol trisphosphate receptor, *Proc. Natl. Acad. Sci. USA* **92:**1759–1763.

Severs, N. J., 1981, Localization of cholesterol in the Golgi apparatus of cardiac muscle cells, *Experientia* **37:**1195–1198.

Severs, N. J., 1984a, Freeze-fracture cytochemical methods for studying the distribution of cholesterol in heart membranes, in: *Methods in Studying Cardiac Membranes. Volume II* (N. S. Dhalla, ed.), pp. 27–44, CRC Press, Boca Raton, FL.

Severs, N. J., 1984b, Freeze-fracture cytochemistry, in: *Electron Microscopy 1984* (A. Csanády, P. Röhlich, and D. Szabó, eds.), pp. 1747–1756, Programme Committee Eighth European Congress on Electron Microscopy, Budapest.

Severs, N. J., 1988, Caveolae: static inpocketings of the plasma membrane, dynamic vesicles or plain artifact? *J. Cell Sci.* **90:**341–348.

Severs, N. J., 1995, Lipid localization by membrane perturbation: A cautionary tale, in: *Rapid Freezing, Freeze Fracture and Deep Etching* (N. J. Severs and D. M. Shotton, eds.), pp. 225–234, Wiley, New York.

Severs, N. J., and Robenek, H., 1983, Detection of microdomains in biomembranes—an appraisal of recent developments in freeze-fracture cytochemistry, *Biochim. Biophys. Acta* **737:**373–408.

Severs, N. J., and Shotton, D. M., 1995, *Rapid Freezing, Freeze Fracture, and Deep Etching,* pp. 1–372, Wiley-Liss, New York.

Severs, N. J., and Simons, H. L., 1983, Failure of filipin to detect cholesterol-rich domains in smooth muscle plasma membrane, *Nature* **303:**637–638.

Severs, N. J., and Simons, H. L., 1986, Caveolar bands and the effects of sterol-binding agents in vascular smooth muscle plasma membrane. Single and double labelling with filipin and tomatin in the aorta, pulmonary artery and vena cava, *Lab. Invest.* **55:**295–307.

Severs, N. J., and Simons, H. L., 1987, Lack of cytochemically-detectable cholesterol in rabbit vena cava endothelial plasma membrane, *J. Anat.* **151:**233–248.

Severs, N. J., Warren, R. C., and Barnes, S. H., 1981, Analysis of membrane structure in the transitional epithelium of rat urinary bladder. 3. Localization of cholesterol using filipin and digitonin, *J. Ultrastruct. Res.* **77:**160–188.

Simionescu, N., Lupu, F., and Simionescu, M., 1983, Rings of membrane sterols surround the openings of vesicles and fenestrae, in capillary endothelium, *J. Cell Biol.* **97:**1592–1600.

Simons, H. L., and Severs, N. J., 1986, Plasmalemmal vesicles and the effects of sterol-binding agents in rabbit aortic endothelium, *J. Cell Sci.* **83:**141–153.

Smart, E. J., Ying, Y.-S., Conrad, P. A., and Anderson, R. G. W., 1994, Caveolin moves from caveolae to the Golgi apparatus in response to cholesterol oxidation, *J. Cell Biol.* **127:**1185–1197.

Smejkal, G. B., and Hoff, H. F., 1994a, Cholesterol-specific probe for lipoproteins immobilized on nitrocellulose membranes, *Biotechniques* **16:**68–70.

Smejkal, G. B., and Hoff, H. F., 1994b, Filipin staining of lipoproteins in polyacrylamide gels: sensitivity and photobleaching of the fluorophore and its use in a double staining method, *Electrophoresis* **15:**922–925.

Sokol, J., Blanchette-Mackie, E. J., Kruth, H. S., Dwyer, N. K., Amende, L. M., Butler, J. D., Robinson, E., Patel, S., Brady, R. O., Comly, M. E., Vanier, M. T., and Pentchev, P. G., 1988, Type-C Niemann–Pick disease: lysosomal accumulation and defective intracellular mobilization of LDL-cholesterol, *J. Biol. Chem.* **263:**3411–3417.

Stahl, A., and Mueller, B. M., 1995, The urokinase-type plasminogen activator receptor, a GPI-linked protein, is localized in caveolae, *J. Cell Biol.* **129:**335–344.

Steer, C. J., Bisher, M., Blumenthal, R., and Steven, A. C., 1984, Detection of membrane cholesterol by filipin in isolated rat liver coated vesicles is dependent upon removal of the clathrin coat, *J. Cell Biol.* **99:**315–319.

Stefanová, I., Horejsí, V., Ansotegui, I. J., Knapp, W., and Stockinger, H., 1991, GPI-anchored cell-surface molecules complexed to protein tyrosine kinases, *Science* **254:**1016–1019.

Swartz, G. M., Gentry, M. K., Amende, L. M., Blanchette-Mackie, J. E., and Alving, C. R., 1988, Antibodies to cholesterol, *Proc. Natl. Acad. Sci. USA* **85:**1902–1906.

Tamm, S. L., and Tamm, S., 1983, Distribution of sterol-specific complexes in a continually shearing region of a plasma membrane and at procaryotic-eucaryotic cell junctions, *J. Cell Biol.* **97:**1098–1106.

Tillack, T. W., and Kinsky, S. C., 1973, A freeze-fracture study of the effects of filipin on liposomes and human erythrocyte membranes, *Biochim. Biophys. Acta* **323:**43–54.

Vanier, M. T., Rodriguez-Lafrasse, C., Rousson, R., Gazzah, N., Juge, M-C., Pentchev, P. G., Revol, A., and Louisot, P., 1991, Type C Niemann-Pick disease: spectrum of phenotypic variation in disruption of intracellular LDL-derived cholesterol processing, *Biochim. Biophys. Acta* **1096:**328–337.

Verkleij, A. J., De Kruijff, B., Gerritsen, W. F., Demel, W. F., van Deenen, L. L. M., and Ververgaert, P. H. J., 1973, Freeze-etch electron microscopy of erythrocytes, *Acholeplasma laidlawii* cells and liposomal membranes after the action of filipin and amphotericin B, *Biochim. Biophys. Acta* **291:**577–581.

Warren, R. C., and Severs, N. J., 1980, Localization of cholesterol in the bladder luminal membrane using digitonin and filipin, in: *Electron Microscopy 1980, Vol.2* (P. Brederoo and W. de Priester, eds.), pp. 326–327, Seventh European Congress on Electron Microscopy Foundation, Leiden.

Ying, Y. S., Anderson, R. G. W., and Rothberg, K. G., 1992, Each caveola contains multiple glycosyl-phosphatidylinositol-anchored membranes, *Cold Spring Harb. Symp. Quant. Biol.* **57:**593–604.

Chapter 17

Approaches for the Design of Novel Anti-Atherogenic Compounds

Erik A. L. Biessen, Leo A. J. M. Sliedregt, and
Theo J. C. Van Berkel

1. SCOPE

Atherosclerosis, a chronic disease related to the vascular system, is the most common cause of morbidity and mortality in Western society. Epidemiological studies have established that the serum cholesterol/triglyceride level is one of the major causative factors for its occurrence. At present, patients with persistent high levels of serum cholesterol and/or triglyceride are therapeutically treated with various types of drugs. Frequently applied drugs are inhibitors of a rate-limiting enzyme in the cholesterol synthesis pathway, i.e., 3-hydroxy-3-methylglutaryl-coenzyme A (HMG-CoA) reductase, or bile acid sequestrants (fibrates). These drugs lead to an enhanced expression level of low-density lipoprotein (LDL) receptors, thereby stimulating the catabolism of LDL, the main carrier of cholesterol in human blood. Despite the general therapeutic efficacy of these drugs, a significant number of patients, including those suffering from familial hypercholesterolemia and dysbetalipoproteinemia, do not respond sufficiently to treatment with these drugs. In addition, HMG-CoA reductase inhibitors may induce unwanted side effects, since

---

Erik A. L. Biessen, Leo A. J. M. Sliedregt, and Theo J. C. Van Berkel Division of Biopharmaceutics, Leiden-Amsterdam Center for Drug Research, Sylvius Laboratory, University of Leiden, 2300 RA Leiden, The Netherlands.

Subcellular Biochemistry, Volume 28: Cholesterol: Its Functions and Metabolism in Biology and Medicine, edited by Robert Bittman. Plenum Press, New York, 1997.

they not only interfere with cholesterol synthesis but also lower the synthesis of other physiologically active isoprenoids. These considerations have prompted efforts to explore new frontiers in anti-atherosclerotic therapy in order to expand the therapeutic scope for treating hyperlipidemia and hypercholesterolemia.

In this chapter, a concise overview is given of the progress in this field, which has recently gained much attention. Promising new compounds that interfere with lipoprotein biosynthesis are discussed, such as inhibitors of squalene synthase, 2,3-oxidosqualene:lanosterol cyclase, and acylcholesterol:acyl coenzyme A transferase (ACAT).

Another strategy to lower high serum cholesterol/triglyceride levels involves drugs that directly enhance the catabolism of atherogenic lipoproteins. In this respect, ansamycins and galactose-terminated glycolipids hold great promise as selective and powerful cholesterol-lowering agents. Ansamycins probably act by stimulating reverse-cholesterol transport. Glycolipids induce hepatic uptake of lipoproteins via the asialoglycoprotein receptor on parenchymal liver cells, thereby introducing another receptor-mediated uptake route for LDL.

Finally, a third route to interfere with atherogenesis is presented, which is focused on prevention of LDL oxidation. Antioxidants such as α-tocopherol, probucol, and ubiquinone may act as anti-atherosclerotic drugs by protecting lipoproteins against oxidative modification, which is thought to be an initial key event in the process of atherogenesis.

2. INTRODUCTION

Worldwide, cardiovascular diseases (CVDs) are recognized as the major cause of death (Simons, 1986). CVDs are responsible for approximately one-half of the deaths among middle-aged adults in industrialized countries. From epidemiological studies it has been recognized that genotypic as well as phenotypic factors contribute to its occurrence, with age, sex, race, diabetes, obesity, smoking habits, hypertension, and serum cholesterol/triglyceride levels being the most prominent risk factors (Thompson and Wilson, 1992; Gotto *et al.,* 1990). The serum cholesterol/triglyceride content is one of the causative factors that is responsive to drug treatment. In man, the main cholesterol/triglyceride transporting particles in the circulation are low-density lipoprotein (LDL) and high-density lipoprotein (HDL). These two lipoproteins have opposing roles in the development of atherosclerosis. High serum levels of LDL are positively correlated with an increased occurrence of arteriosclerosis, whereas high levels of serum HDL protect against atherogenesis (Wilson *et al.,* 1990; Gordon *et al.,* 1977). Homeostasis of plasma lipoprotein levels is regulated by the rate of lipoprotein synthesis, as well as by the expression level and activity of specific lipoprotein receptors. Illustratively, HDL interacts with HDL receptors in the periphery and in the liver (Oram *et al.,* 1987; Schouten *et al.,* 1990), whereas LDL receptors are responsible for the uptake and processing of LDL (Mahley and Innerarity, 1983; Dietschy and Spady, 1986; Brown and Goldstein, 1986). In serum lipid disorders, normal processing of LDL is often impaired, leading to elevated serum levels of LDL and increased risk of CVDs (Wilson *et al.*, 1990; Brown and Goldstein, 1986; Stone *et al.*, 1974; Innerarity *et al.*, 1987; Schuster *et al.*, 1990). Patients suffering from hypercholesterolemia and/or hyperlipidemia are currently subjected to dietary measures and drug therapy. The conventional lipid-lowering drugs act via a variety of mechanisms, ranging from the induction of cholesterol flux by bile-acid sequestrants to inhibition of hepatic cholesterol biosynthesis by HMG-CoA reductase inhibitors (Brown and Goldstein, 1986; Bilheimer *et al.*, 1983). Both inhibition of cholesterol synthesis and increased secretion of cholesterol will lower intracellular sterol levels, which in turn induce upregulation of the synthesis of LDL receptors, as has been demonstrated by Brown and Goldstein and coworkers (Brown and Goldstein, 1986; Bilheimer *et al.*, 1983). As a result, LDL clearance from the bloodstream will be enhanced.

Although the above strategy is generally quite effective, approximately 5% of the patients do not adequately respond (Schuster *et al.*, 1990; Yamamoto *et al.*, 1980), including those with the homozygous form of familial hypercholesterolemia (FH), who totally lack functional LDL receptors, and those with familial defective apolipoprotein (B100) (Innerarity *et al.*, 1987; Innerarity *et al.*, 1990). These patients are usually subjected to invasive and costly plasmapheresis or LDL apheresis (Thompson *et al.*, 1989; Yokoyama *et al.*, 1985). An additional drawback of HMG-CoA reductase inhibitors is that blocking of cholesterol synthesis will also interfere with the synthesis of essential intermediates and other end products of the isoprenyl biosynthetic pathway (see Chapter 4 by Guo *et al.*). Accordingly, homeostasis of these isoprenoids will be perturbed and, since they are essentially involved in various cellular functions (Cattel *et al.*, 1989), this will lead to a number of undesired side effects. The aforementioned pitfalls have prompted the development of new strategies for treating hypercholesterolemia and hyperlipidemia.

3. NOVEL INHIBITORS OF CHOLESTEROL SYNTHESIS

As discussed above, the most frequently applied therapy to lower serum cholesterol concentrations involves the use of HMG-CoA reductase inhibitors (Hunnighake, 1992; Illingworth, 1991). HMG-CoA reductase is a rate-limiting enzyme catalyzing the biosynthesis of mevalonate, a common precursor of all isoprenyl derivatives, including ubiquinones, heme A, dolichols, isopentenyl tRNA, and isoprenylated proteins (Figure 1). Endo *et al.* (1976) was the first to isolate and identify a HMG-CoA reductase inhibitor of microbial origin (*Penicillum citrinum*), compactin. Further study revealed that pravastatin, a metabolite of compactin,

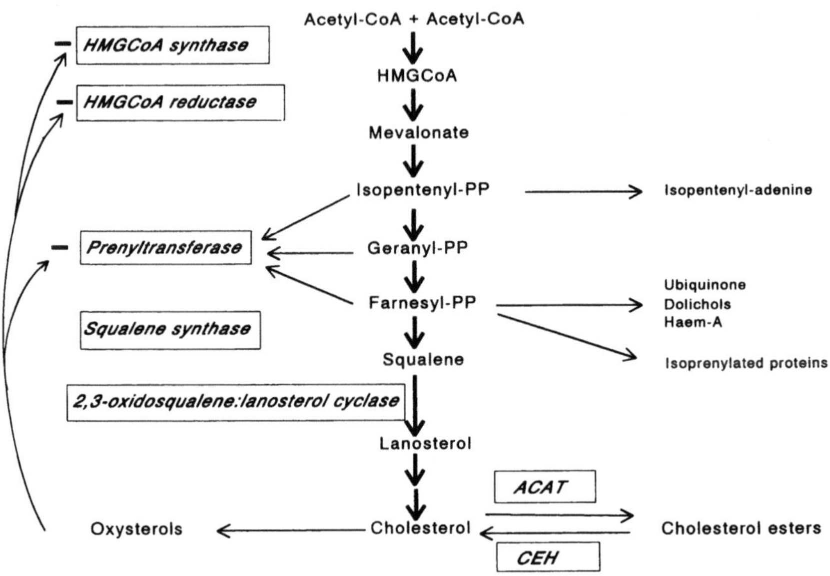

FIGURE 1. The hepatic biosynthetic pathway of cholesterol. Cholesterol homeostasis is maintained by regulation of cholesterol synthesis and dietary influx on one hand, and cholesterol catabolism on the other hand. The rate-limiting enzyme of the biosynthetic pathway is HMG-CoA reductase. The inhibitory effects of oxysterols on enzymes of this pathway are indicated by (-). Abbreviations: HMG-CoA, 3-hydroxy-3-methyl-glutaryl coenzyme-A; ACAT, acyl-CoA:cholesterol acyltransferase; CEH, cholesteryl ester hydrolase. (Adapted from Vlahcevic *et al.*, 1994, with permission.)

rather than compactin itself was the actual HMG-CoA-inhibiting compound (Tsujita *et al.*, 1986). The high hypocholesterolemic activity of these compounds triggered a whole body of pharmaceutic research on the development of novel synthetic analogues of pravastatin, leading to the design of the first fully synthetic HMG-CoA reductase inhibitor, fluvastatin (Levy *et al.*, 1993).

Clinical studies on the hypocholesterolemic effect of HMG-CoA reductase inhibitors in patients with combined hyperlipidemia or polygenic hypercholesterolemia established that a 20–40% reduction in LDL cholesterol could be achieved upon oral administration of only 20–80 mg/kg vastatin/day (the Lovastin/Pravastatin Study Group, 1993; Farmer *et al.*, 1992). Vastatins have been reported to slow the progression of atherosclerosis and to prevent the formation of new coronary lesions (Blankenhorn *et al.*, 1993; Waters *et al.*, 1994). Recent data suggest that administration of vastatins also leads to a significant reduction in cardiovascular morbidity (Scandinavian Simvastatin Survival Study Group, 1994). Interestingly, a large-scale randomized double-blind patient study established that chronic pravastatin treatment also reduced the incidence of myocardial infarction and mortality in mildly hypercholesterolemic men (>7 mM of serum cholesterol) with no history of myocardial infarction (Shepherd *et al.*, 1995).

Although HMG-CoA reductase inhibitors are quite effective in reducing the blood cholesterol levels in hypercholesterolemic patients, these inhibitors bear the major disadvantage that their target enzyme is located at an initial stage of the isoprenyl biosynthetic pathway. Accordingly, interference with mevalonate synthesis will also affect other processes that depend on the synthesis of isoprene intermediates of the cholesterol pathway. This in turn is accompanied by a number of unwanted side effects (Cohen *et al.,* 1995; Cattel *et al.,* 1995). Vastatin-induced changes in liver function and myopathy has been occasionally observed in rats (Smith *et al.,* 1991), rabbits (Nakahara *et al.,* 1992), and humans (Hunnighake, 1992; Illingworth *et al.,* 1991). Other contraindications observed upon administration of the conventional vastatins are cataract formation (Arai *et al.,* 1988) and skeletal muscle degeneration (Smith *et al.,* 1991). An effect of the vastatins on steroidogenesis has not been observed until now (Travia *et al.,* 1994). Selective inhibition of other key steps further downstream the terpene biosynthetic pathway will overcome these pitfalls.

3.1. Squalene Synthase Inhibitors

A logical target enzyme in this respect is squalene synthase, a 47-kDa membrane-bound enzyme localized on the endoplasmic reticulum of parenchymal liver cells (Sasiak and Rilling, 1988; Popjak and Agnew, 1979). Squalene synthase catalyzes the reductive head-to-head condensation of two molecules of farnesyl pyrophosphate to yield squalene (Rilling, 1978). A number of squalene synthase inhibitors and, in particular, farnesyl pyrophosphate analogues (Corey and Volante, 1976) containing a modified pyrophosphate function have been described recently (Abe *et al.,* 1994; Biller *et al.,* 1991a, 1991b; Prashad *et al.,* 1993). These inhibitors presumably inhibit squalene synthesis via competition with a reaction intermediate generated during the reductive dimerization of the farnesyl pyrophosphates. The first potent inhibitor of squalene synthase, synthesized by Biller *et al.* (1991a), had an affinity of 50 nM for the enzyme. The biphosphonic acid derivatives developed by Ciosek *et al.* (1993) appeared to be even more potent inhibitors, with affinities of up to 0.95 nM.

Although farnesyl pyrophosphate analogues are more specific inhibitors of cholesterol synthesis than HMG-CoA reductase inhibitors, they may also interfere with protein:farnesyltransferase and geranylgeranyl pyrophosphate synthase (see Chapter 3 by Ogura et al.), both key enzymes in protein isoprenylation (Cohen *et al.,* 1995; Das and Allen, 1992; Pompliano *et al.,* 1992; Gibbs *et al.,* 1993). Nevertheless, Cohen *et al.* (1995) showed that the required specificity of the farnesyl pyrophosphate analogues for squalene synthase may be realized by subtle changes in their chemical structure. FPPA-1, a farnesyl pyrophosphate analogue, displayed a 400-fold higher affinity for farnesyltransferase than for squalene synthase, whereas a methylphosphinylphosphonate analogue of FPPA-1 (for the structure see Figure 2) possessed a 10-fold higher affinity for squalene synthase.

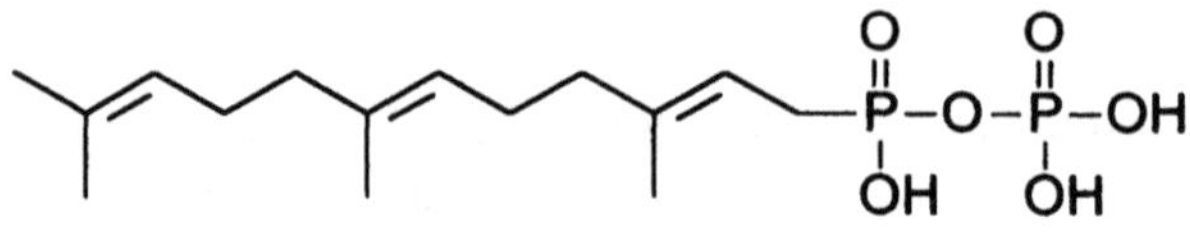

FPPA-1

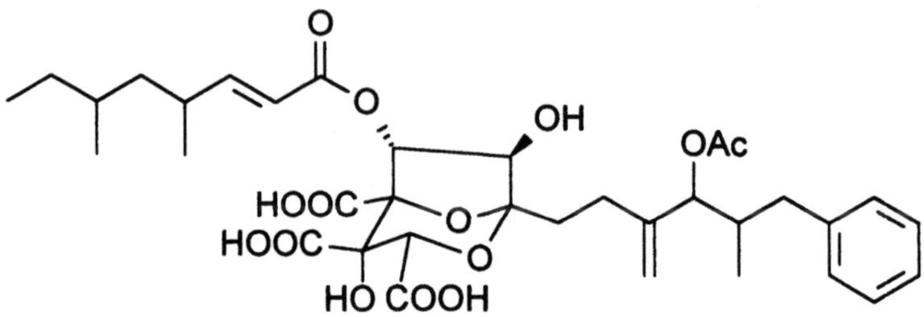

squalestatin 1

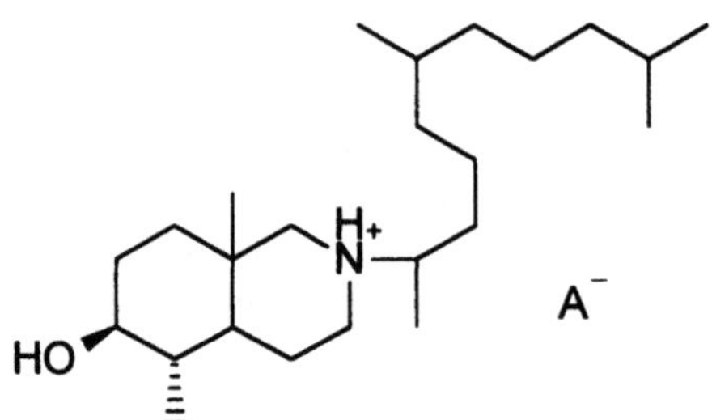

Azadecalin

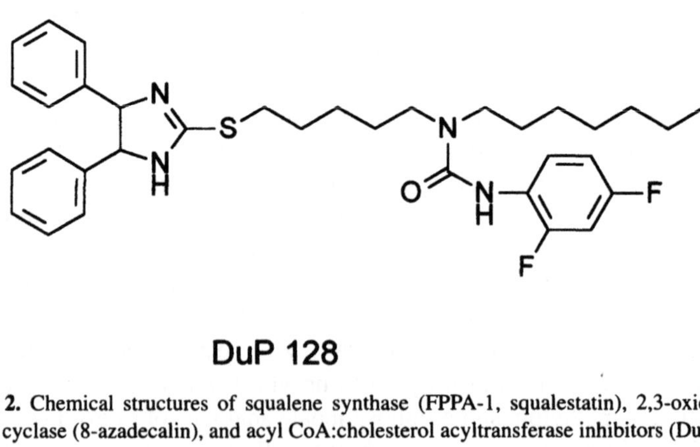

DuP 128

FIGURE 2. Chemical structures of squalene synthase (FPPA-1, squalestatin), 2,3-oxidosqualene: lanosterol cyclase (8-azadecalin), and acyl CoA:cholesterol acyltransferase inhibitors (DuP128).

Surprisingly, the most potent and selective inhibitors were not derived from logical pharmaco-chemical synthesis but by screening of fungal metabolites (Bergstrom *et al.,* 1993). Squalestatins, fungal metabolites isolated from *Phoma sp.* C2932 and *Setosphaeria khartoumensis,* appeared to be very potent inhibitors, displaying picomolar affinity for squalene synthase (Baxter *et al.,* 1992; Hasumi *et al.,* 1993). In 1992, Sidebottom and coworkers from Glaxo Group Ltd. (Sidebottom *et al.,* 1992) elucidated the chemical structures of these squalestatins (see Figure 2). A research group from Merck Research isolated structurally related compounds from *Sporormiella intermedia* and *Leptodontium elatius,* the zaragosic acids (Dufresne *et al.,* 1992). These compounds share a common structural element that is essential for inhibitory activity. It has been suggested that the inhibitory potency of the zaragosic acids results from their topological similarity to farnesyl pyrophosphate, since both compounds are characterized by triacid core moieties embedded in a lipophilic environment (Dufresne *et al.,* 1992).

Squalestatin-1 was found to inhibit cholesterol synthesis in rat brain, liver homogenates, and Chinese hamster ovary cells, leaving dolichol synthesis, unaltered (Thelin *et al.,* 1994). Apparently, squalestatin-1 is not only an effective inhibitor of cholesterol synthesis, but also a very specific one. Baxter *et al.* (1992) reported that oral administration of squalestatin-1 (10 mg/kg day) for 7 days to marmosets, a species with a lipoprotein profile similar to man, resulted in a 50% reduction in serum cholesterol content. Moreover, only apolipoprotein B100 levels appeared to be reduced, whereas apoA-1 levels remained unchanged. Intravenously administered squalestatin-1 analogues, modified at the C1 side-chain, markedly reduced the serum cholesterol level at doses as low as 0.3–1 mg/kg day (Procopiou *et al.,* 1994). In these studies, the inhibitory potency of the compounds tested did not fully correlate with their lipid-lowering activity *in vivo.* In addition, squalestatin-1 was reported to enhance ubiquinone levels by a factor of 3, probably because it activates *trans*-prenyltransferase, a key enzyme in ubiquinone biosynthesis (Thelin *et al.,* 1994). Increased levels of this antioxidant are generally considered to be beneficial (Kalèn *et al.,* 1989; Eggens *et al.,* 1990; Thelin *et al.,* 1994). The combined activity of squalestatin-1 makes it a drug with extraordinary anti-atherosclerotic features.

These promising results have considerably encouraged pharmaco-chemical research efforts on the design of synthetic squalestatin analogues (Sharatt *et al.,* 1994; Procopiou *et al.,* 1994; Koert, 1995). Recently, many efforts have been undertaken to synthesize potential key intermediates, model compounds, and the bicyclic core of squalestatin in a search for a feasible route for the total synthesis of the squalestatins (McVinish and Rizzacasa, 1994; Gurjar *et al.,* 1994; Abdel-Rahmann *et al.,* 1994; Caron *et al.,* 1994; Nicolau *et al.,* 1995). The first total syntheses of squalestatin-1 (also known as zaragosic acid A) and zaragosic acid C were reported by Nicolaou *et al.* (1994a, 1994b, 1994c) and by Carreira and Du Bois (1994) and Evans *et al.* (1994), respectively. The route used by Nicolaou *et al.* (1994a, 1994b, 1994c) required more than 30 sequential steps, with an overall yield of less than 0.01%, whereas the route used by Carreira and Du Bois (1994) was far more efficient, also entailing 30 steps but at a 2.5–3.9% overall yield. These low yields imply that still much remains to be done.

Structure–activity relationship studies are currently underway to define the essential elements of squalestatin and to enable the design of less complex squalestatin analogues (Sharatt *et al.*, 1994; Middleton *et al.*, 1995; Cox *et al.*, 1994; Kuo *et al.*, 1994; Treiber *et al.*, 1995; Procopiou *et al.*, 1993, 1996; Giblin *et al.*, 1993).

In addition, Iwasawi *et al.* (1994a) designed a new class of compounds with potent squalene synthase inhibitory activity, J-104,118. This compound possesses two characteristic features of the zaragosic acids, since it consists of a fluorobiphenyl- and dichlorophenyl-containing hydrophobic domain that is flanked by a hydrophilic dicarboxilic acid moiety. Oral administration of J-104,118 in mice appeared to result in a profound reduction of cholesterol synthesis (Iwasawi *et al.*, 1994b).

3.2. 2,3-Oxidosqualene:Lanosterol Cyclase Inhibitors

Another potentially interesting target enzyme further downstream in the cholesterol biosynthetic pathway is 2,3-oxidosqualene:lanosterol cyclase. Squalene cyclase was recently cloned and sequenced (Roessner *et al.*, 1992) and was found to consist of 728 amino acids. This enzyme is associated with the endoplasmic reticulum and catalyzes the cyclization of 2,3-oxidosqualene into lanosterol (Yamamoto *et al.*, 1969). The molecular basis of its enzymatic activity and the prerequisites for substrate recognition are only poorly understood (Grosa *et al.*, 1994). In a search for high affinity inhibitors of this enzyme, mimics of carbocation intermediates that are generated during the cyclization of 2,3-oxidosqualene have been designed (Taton *et al.*, 1986, 1992; Wannamaker *et al.*, 1992; Cerruti *et al.*, 1992; Cattel *et al.*, 1989, 1995). Acyclic azasqualene derivatives of these intermediates in which the carbocation is replaced by a nitrogen appeared to be effective in this respect (Taton *et al.*, 1986, 1992; Dodd and Oehlschlager, 1992; Duriatti *et al.*, 1985; Abe *et al.*, 1994). First generation aza derivatives such as 8-azadecalin exhibited affinities for 2,3-oxidosqualene:lanosterol cyclase of up to 1 μM (Cattel *et al.*, 1995).

Treatment of 3T3 fibroblasts *in vitro* with 8-azadecalin (for structure see Figure 2) completely blocked cholesterol synthesis (Rudney and Sexton, 1986). Aside from a direct inhibition of cholesterol synthesis, 8-azadecalin also stimulated 24,25-epoxycholesterol synthesis. After formation, 24,25-epoxycholesterol is channelled into the regulatory pathway and represses HMG-CoA reductase activity (Taylor *et al.*, 1986). Via this indirect synergistic effect, the ability of squalene cyclase inhibitors to block cholesterol synthesis is further amplified. Indeed, treatment of 3T3 fibroblasts and HepG2 cells with 8-azadecalin also resulted in a marked inhibition of HMG-CoA reductase activity (Dollis and Schuber, 1994).

Moreover, as a result of this negative feedback mechanism, cyclase inhibitors probably do not cause accumulation of isoprenyl intermediates, such as farnesyl diphosphates. Unwanted upregulation of the synthesis of ubiquinone, dolichol, heme A, and isoprenylated protein, which occurs upon treatment with squalene synthase and HMG-CoA reductase inhibitors, will be circumvented. Further *in vitro* studies confirmed the potential of cyclase inhibitors as potent (IC_{50} 100 nM) and selective blockers of cholesterol synthesis in human hepatoma cells (Boogaard *et al.*, 1987; Panini *et al.*, 1991).

Although 2,3-oxidosqualene:lanosterol cyclase seems to be a promising target for hypolipidemic therapy, it should be noted that 2,3-oxidosqualene:lanosterol cyclase is not a rate-limiting enzyme in the isoprenyl biosynthetic pathway. Therefore, only very potent inhibitors of this enzyme are applicable as hypocholesterolemic drugs. In addition, most of the acyclic azasqualene derivatives are toxic or even lethal at intravenous doses above 30 mg/kg, which is only 10-fold higher than pharmaceutically relevant doses (Cattel *et al.*, 1995).

3.3. Acylcholesterol:Acyl Coenzyme A Transferase Inhibitors

An approach that has received intensive research attention focuses on inhibition of ACAT. ACAT is a membrane-bound enzyme with a functional molecular weight of 213 kDa. It is localized in the rough endoplasmic reticulum of liver, intestinal, and ovary cells (Erickson *et al.*, 1994). Becker *et al.* (1994) recently purified, sequenced, and cloned ACAT, demonstrating that it was identical to hepatic carboxylesterase. Subtypes of ACAT have recently been identified in aorta and macrophages (Becker *et al.*, 1994; Saxena *et al.*, 1995; White *et al.*, 1994). ACAT catalyzes the intracellular esterification of cholesterol to cholesterol esters (Becker *et al.*, 1994). It was soon recognized that ACAT plays an important role in cholesterol homeostasis (Suckling and Stange, 1985). Intestinal ACAT is involved in chylomicron assembly and secretion, and thus in cholesterol absorption (Suckling and Stange, 1985; Erickson *et al.*, 1980). Similarly, hepatic ACAT mediates the acylation of cholesterol in the liver. The resulting cholesterol esters are secreted after incorporation into VLDL or are hydrolyzed by cholesterol ester hydrolase to become part of the hepatic free cholesterol pool. This pool is the major source of cholesterol-derived bile acids (Vlahcevic *et al.*, 1994), and it is also relevant for the secretion of VLDL (Carr *et al.*, 1995). Studies by Saxena *et al.* (1995) revealed that macrophage-derived ACAT regulates intracellular cholesterol ester accumulation and monocyte adhesion to the arterial wall, both crucial initial steps in the progression of atherosclerosis (Matsuda, 1994; White *et al.*, 1994). Altogether, these data imply that ACAT inhibitors will beneficially intervene with atherogenesis. ACAT inhibitors may be anti-atherosclerotic by inhibiting cholesterol absorption, by inhibiting apoB100 synthesis and VLDL secretion (Carr *et al.*, 1992; Tanaka *et al.*, 1993), and by slowing plaque formation in the arterial wall (Saxena *et al.*, 1995).

At the moment, the use and design of systemic ACAT inhibitors has progressed toward potent hypocholesterolemic drugs (Fancelli *et al.*, 1994; Higley *et al.*, 1994; Nioche *et al.*, 1995; Tawada *et al.*, 1994; Wilde *et al.*, 1995a, 1995b, Wilde and Billheimer, 1995; Picard *et al.*, 1994). Two classes of ACAT inhibitors hold great promise: the diarylimidazole analogues (Fancelli *et al.*, 1994; Higley *et al.*, 1994) and the disubstituted heterocyclic ureas (Nioche *et al.*, 1995; Tawada *et al.*, 1994; Wilde *et al.*, 1995a, 1995b, Wilde and Billheimer, 1995; Picard *et al.*, 1994). The compound with the highest *in vitro* activity yet synthesized, DuP128, had an IC_{50} value of 10 nM (for structure see Figure 2; Higley *et al.*, 1994). Low nanomolar affinities have also been reported for an aryloxythioimidazole derivative (Bani *et al.*, 1995) and a 3-imidazole benzopyrane derivative (Fancelli *et al.*, 1994), the latter claimed as being selective for aorta ACAT.

In vivo testing of the ACAT inhibitors in various animal models established the efficacy of this approach (Wrenn *et al.*, 1995; Nioche *et al.*, 1995; Carr *et al.*, 1995; Tawada *et al.*, 1994; Wilde *et al.*, 1995b; Higley *et al.*, 1994; Bani *et al.*, 1995). In African green monkeys on a cholesterol/high-fat diet, ACAT inhibitors significantly reduced the total plasma cholesterol level (Wrenn *et al.*, 1995), which could be fully attributed to a reduction of the serum LDL content. Surprisingly, and in contrast to the results of Carr *et al.* (1995), Wrenn *et al.* (1995) did not notice any effect on serum apoB100 content. Since dense cholesterol-deficient LDL is thought to be more atherogenic than cholesterol-rich LDL (Austin *et al.*, 1990), it remains to be clarified whether the overall effect of these drugs reduced atherosclerosis.

The most potent hypocholesterolemic agent appeared to be the heterocyclic urea derivative compound 21 of Nioche *et al.* (1995), which was claimed to have an ED_{25} of 0.2 mg/kg. With DuP128, CI976, YM17E, S58-035, CI277,082, CI283,546, CI283,796, and the *N*-chlorocarbonylisocyanate derivative synthesized by Picard *et al.* (1994), therapeutic doses required for a significant hypocholesterolemic response ranged from 1–20 mg/kg day (Wrenn *et al.*, 1995; Nioche *et al.*, 1995; Wilde *et al.*, 1995a, 1995, Wilde and Billheimer, 1995, Higley *et al.*, 1994, Tawada *et al.*, 1994; Picard *et al.*, 1994; Bani *et al.*, 1995).

To date, no studies have been published demonstrating anti-atherosclerotic activity of ACAT inhibitors. Studies by Bocan *et al.* (1991, 1993) showed that ACAT inhibitors not only interfere with lipoprotein assembly but also act at the arterial wall. Accumulation of cholesterol esters in foam cells in cholesterol-fed rabbits and Yucatan pigs could be significantly diminished by selective inhibitors of aorta ACAT, such as CI-976. This anti-atherosclerotic effect may already occur at doses that do not affect the serum cholesterol level, suggesting a direct interaction of this ACAT inhibitor on the vascular wall. In conclusion, each of the three interaction sites at which ACAT may stimulate atherogenesis can be selectively blocked by careful design of the ACAT inhibitors. This will enormously expand the therapeutic scope of the available ACAT inhibitors.

4. AGENTS THAT ENHANCE LIPOPROTEIN CATABOLISM

Although most of the efforts to design anti-atherosclerotic drugs are focused on compounds that interfere with lipoprotein synthesis and assembly, this is clearly not the only strategy to reduce high serum cholesterol/triglyceride levels in patients. Cholesterol/triglyceride levels can also be normalized by compounds that directly stimulate the catabolism of these lipoproteins. Two approaches to attain stimulated lipoprotein catabolism have been elaborated to date.

4.1. Ansamycins

Ansamycins, which are pivaloyl-3-piperazino derivatives of the antibiotic rifamycin, possess a high hypolipidemic activity both in non-primate and primate species. In contrast to their parent compound rifamycin, ansamycins lack antibacterial activity. Until now, four ansamycins with high cholesterol-lowering activity have been described, i. e., CGP 43371 (for structure see Figure 3), CGS 24565, CGP 33778, and CGP 23810 (Traxler *et al.*, 1990; Gibson *et al.*, 1991, 1992). Gibson *et al.* (1992, 1994) recently showed that oral administration of CGP43371 (10 mg/kg day) in chow-fed rats and hamsters reduces serum cholesterol levels by 50–55% ($P < 0.001$). The effect in cholesterol-fed rats and hamsters was less pronounced, which may be explained by a lower efficacy of CGP43371 to

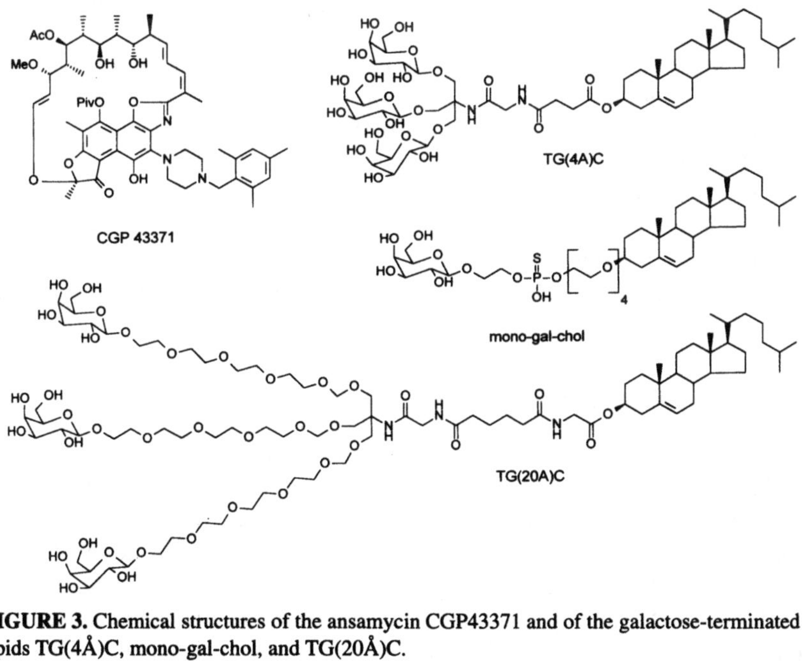

FIGURE 3. Chemical structures of the ansamycin CGP43371 and of the galactose-terminated glycolipids TG(4Å)C, mono-gal-chol, and TG(20Å)C.

reduce serum LDL and VLDL levels. Surprisingly, serum levels of apolipoproteins A1, B100, and E were not influenced after treatment with CGP43371.

Although the actual mechanism for the hypolipidemic activity is not fully identified, studies by Gibson *et al.* (1994) indicate that ansamycins enhance lipoprotein catabolism. Since ansamycins also exert their hypocholesterolemic activity in LDL receptor-deficient WHHL rabbits, a LDL receptor-mediated pathway can be excluded. Moreover, LDL receptor-related protein is probably not involved either since the *in vivo* effect of ansamycins was not influenced by a high fat load (Gibson *et al.*, 1994). When incubated with human plasma, ansamycins almost quantitatively incorporate into lipoproteins at concentrations of up to 25 μM because of their high hydrophobicity. Ansamycins preferably interact with VLDL and HDL, and to a lesser extent with LDL (Gibson *et al.*, 1995). Association is not purely a lipid-binding phenomenon since ansamycins did not associate with egg-yolk phosphatidylcholine liposomes (Leonards, 1992). Plasma clearance of the ansamycin-associated [^{14}C]-cholesterol ester labeled HDL in rats (Gibson *et al.*, 1991) and monkeys (Menear *et al.*, 1992) was slightly, but significantly, enhanced. Analysis of the tissue distribution revealed that ansamycin stimulated liver uptake of HDL by 34% ($P < 0.001$). In addition, a striking 4- to 10-fold increased accumulation of cholesterol (esters) in macrophages was observed after oral administration of CGP 43371 (Gibson *et al.*, 1994), which was not caused by a direct effect of CGP43371. Upon incorporation of ansamycins, LDL and HDL binding to HepG2 cells was increased by a factor of 2.5 and 7, respectively (Gibson *et al.*, 1995). Surprisingly, lipoprotein binding and uptake to liver cells isolated from ansamycin-treated rats was also strongly increased compared with cells from untreated rats. Since ansamycins reduce plasma HDL-derived cholesterol levels to a much greater extent than that they reduce apoA-1 levels, the most likely and intriguing hypothesis is that ansamycins act by facilitating the selective transfer of cholesterol esters from HDL to the liver (Gibson *et al.*, 1992). Both *in vivo* (Brindle and O'Buck, 1991) and *in vitro* data (Gibson *et al.*, 1994) support the hypothesis that ansamycins are resecreted by liver cells after initial uptake, and can participate in the ansamycin-mediated stimulation of reverse-cholesterol transport. To conclusively establish this hypothesis additional mechanistic studies will be required.

4.2. Galactose-Terminated Hypocholesterolemic Compounds

A second approach to enhance the catabolism of atherogenic lipoproteins involves the direct removal of these lipoproteins through alternative high capacity receptors on parenchymal liver cells. To date, this principle has been elaborated in our laboratory (Kempen *et al.*, 1984, 1987; Van Berkel *et al.*, 1985a, 1985b; Roelen *et al.*, 1991; Bijsterbosch *et al.*, 1991, 1992; Biessen *et al.*, 1994a, 1994b, 1995a, 1995b, 1995c) and by a number of other research groups (Peter *et al.*, 1990; Bernini *et al.*, 1986, 1988). A candidate receptor for this induced uptake of lipoproteins was the asialoglycoprotein receptor, which is uniquely expressed on parenchymal liver cells (Spiess, 1990). This receptor is implicated in the hepatic uptake and subsequent lysosomal processing of galactose-terminated substrates (Ashwell and Morell, 1974; Spiess, 1990). Theoretically, a bifunctional molecule, consisting of one moiety that displays a high affinity toward the asialoglycoprotein receptor and a second moiety that can associate with the atherogenic lipoprotein, should enhance the catabolism of LDL (Figure 4).

Starting from this principle, Bernini and coworkers used the specific interaction of monoclonal antibodies with apoB100, the only apolipoprotein of LDL, to create specific anchorage of a lactosylated protein to LDL (Bernini *et al.*, 1986, 1988; Bijsterbosch *et al.*, 1991). Lactosylation of immunoglobulin G yielded a compound that was avidly recognized and taken up by the asialoglycoprotein receptor (Bernini *et al.*, 1986). Administration of lactosylated IgG or Fab fragments thereof into rats, preinjected with human LDL, led to a marked reduction in the levels of LDL, whereas the HDL levels were not affected. Using [^{3}H]-cholesterol oleate-labeled LDL, Bijsterbosch *et al.* (1991) monitored the physiological fate of LDL after complexation with lactosylated antibodies. Both liver uptake and serum decay of LDL/antibody complexes were enhanced by a factor of 10 compared with uncomplexed LDL (Figure 5). Blockage of galactose receptor-mediated uptake by preinjection with Gal(NAc) drastically reduced the efficacy of lactosylated antibodies to stimulate LDL uptake, indicating that LDL was internalized via galactose-recognizing receptors on the liver. A substantial proportion of the in-

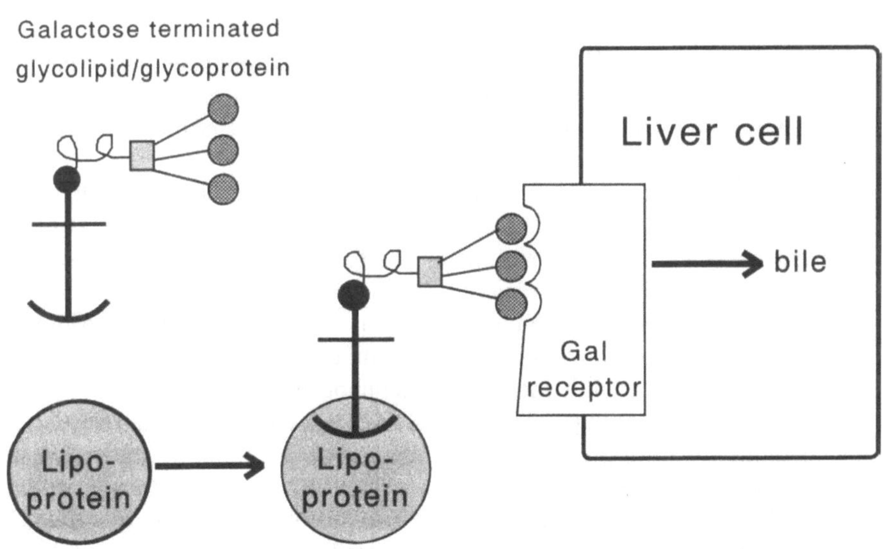

FIGURE 4. Putative mechanism for the hypocholesterolemic activity of galactose-terminated glycolipids/antibodies. Abbreviations: ASGPr, asialoglycoprotein receptor; PC, parenchymal liver cell.

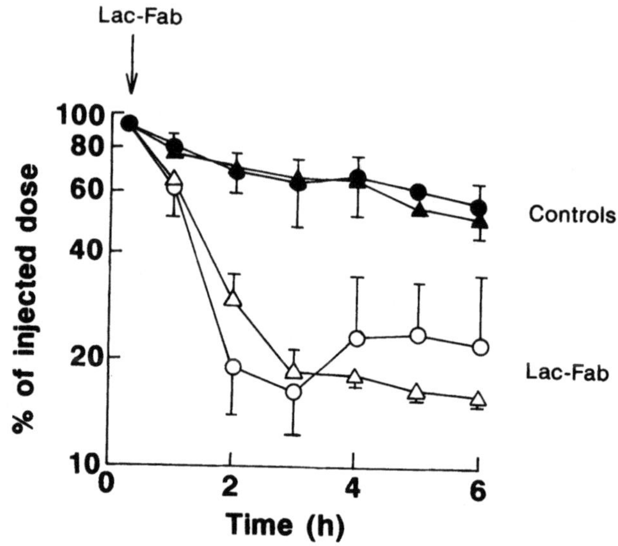

FIGURE 5. Effect of lactosylated α-(apoB100) Fab fragments on the serum decay and liver association of exogenously administered ^{125}I-LDL in rats. Rats were preinjected with human ^{125}I-LDL (23 mg/kg body weight). After 20 min, lactosylated α-(apoB100) Fab fragments (42 mg/kg body weight) (●,▲) or saline (○,△) was injected into rats. At the indicated time points, cholesterol and ^{125}I-LDL concentrations in the serum were determined. Data were normalized to serum ^{125}I-LDL and cholesterol concentrations at 3 min after injection of human LDL, and are means ±S.D. of 6–8 experiments. (Adapted from Bernini *et al.*, 1988, with permission.)

jected [^{3}H]-cholesterol oleate was excreted via the bile into the lumen of the gut, suggesting that cholesterol (esters) derived from the LDL/antibody complexes that are taken up by the liver are efficiently processed and chanelled into the biliary pathway.

Despite the efficacy and elegance of this approach to enhance the catabolism of apoB100 containing lipoproteins, clinical application is frustrated by two major disadvantages of the use of modified antibodies. Firstly, large amounts of relatively inaccessible lac-IgG are required to achieve a significant effect. The second pitfall involves the potential immunogenicity of the modified antibody. Although the immunogenicity of the IgG itself may be minimized by humanization, modification will almost certainly increase its immunogenicity.

These considerations prompted Van Berkel and coworkers to develop a new methodology for lowering serum lipoprotein levels according to the aforementioned general principle (Kempen *et al.*, 1984, 1987; Van Berkel *et al.*, 1985a, 1985b; Roelen *et al.*, 1991; Bijsterbosch *et al.*, 1992; Biessen *et al.*, 1994a,b, 1995a, 1995b, 1995c). In 1984, a cholesterylated cluster galactoside with hypocholesterolemic activity was synthesized: TG(4Å)C (for structure see Figure 3; Kempen *et al.*, 1984). Upon incubation with serum, TG(4Å)C spontaneously incorporated into LDL and HDL (Kempen *et al.*, 1984). The plasma clearance of these lipoprotein/TG(4Å)C complexes was strongly accelerated as compared to uncomplexed LDL and HDL (Van Berkel *et al.*, 1985a, 1985b). Enhanced clearance of lipoproteins also occurred after *in vivo* administration of TG(4Å)C in rats. Unfortunately, continuous intravenous infusion of high doses of TG(4Å)C was required for a significant cholesterol-lowering effect (Kempen *et al.*, 1987). Moreover, TG(4Å)C did not only induce hepatic uptake of LDL but also of HDL, which is generally thought to protect against atherogenesis (Van Berkel *et al.*, 1985b). The LDL selectivity of a monogalactosylated cholesterol derivative, mono-gal-chol (for structure see Chart 2; Roelen *et al.*, 1991) was greatly improved compared with TG(4Å)C (Bijsterbosch *et al.*, 1992).

An additional pitfall of TG(4Å)C and mono-gal-chol was the fact that both agents directed LDL to galactose-recognizing receptors (Biessen *et al.*, 1994c; Kuiper *et al.*, 1994) on Kupffer cells (Van Berkel *et al.*, 1985a; Bijsterbosch *et al.*, 1992). In contrast to the parenchymal liver cells, Kupffer cells are not directly linked to cholesterol degradation and bile secretion. It was argued that both the low lipid-lowering activity and the lack of specificity of these glycolipids to target lipoproteins to the asialoglycoprotein receptor may be caused by the moderate affinity of the galactoside moieties of the synthetic glycolipids for this receptor.

In a recent study by Biessen *et al.* (1995a, 1995b), it was shown that both the affinity and specificity of a triantennary cluster galactoside for the asialoglycoprotein receptor could be significantly improved by elongation (4Å → 20Å) of the spacer that connects the terminal galactose moieties with the branching point of the dendrite. The most selective compound, TG(20Å), displayed a 2000-fold higher affinity for the asialoglycoprotein receptor than TG(4Å). Therefore, TG(20Å) was functionalized with a cholesterol moiety to give the bifunctional compound TG(20Å)C (for structure see Figure 3; Biessen *et al.*, 1995c), and the lipid-lowering activity of TG(20Å)C was assessed. TG(20Å)C appeared to be at least 30-fold more effective in reducing serum cholesterol levels than TG(4Å)C (Biessen *et al.*, 1995a). Bolus injection of only 1.9 mg/kg of TG(20Å)C in rats led to a persistent reduction of the serum cholesterol content of 45% (Figure 6). In contrast to TG(4Å)C and mono-gal-chol, the TG(20Å)C-induced hepatic uptake of LDL was mediated by the asialoglycoprotein receptor on parenchymal liver cells, rather than by galactose receptors on Kupffer cells (Biessen *et al.*, 1994a), which may arise from the much higher affinity of this glycoside for the asialoglycoprotein receptor. In addition, it was found that after administration of TG(20Å)C, the biliary bile acid secretion was raised, while the biliary flow as such remained at a constant level (Biessen *et al.*, 1995a). Apparently, cholesterol (esters) from TG(20Å)C-treated animals are rapidly taken up by the liver and directly channelled into the biliary pathway, resulting in irreversible removal of cholesterol from the bloodstream. Although TG(20Å)C is a powerful new hypercholesterolemic agent, the modest tendency of TG(20Å)C to selectively lower serum LDL compared with HDL may be an obstacle for (pre)clinical application of

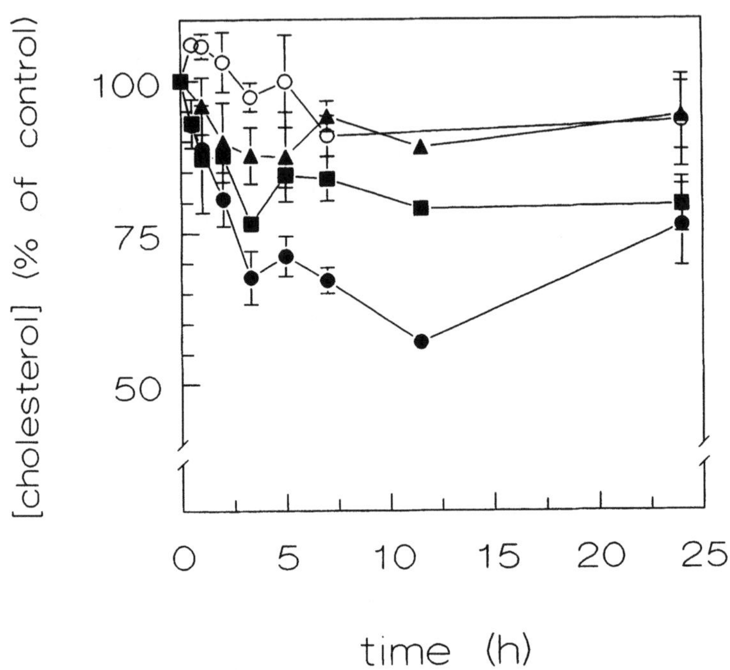

FIGURE 6. Effect of intravenous injection of TG(20Å)C on the total serum cholesterol concentration in rat. Phosphate-buffered saline (500 μl, o) or phosphate-buffered saline containing 56 μg (▲), 180 μg (■), or 560 μg of TG(20Å)C (●) was injected into the vena penis of anaesthetized rats. At the indicated times, blood samples were collected by orbital punction, and the serum cholesterol contents were determined. The total cholesterol content is plotted as % of the content at $t = 0$ (2.37 ± 0.16, 2.04 ± 0.11, 2.16 ± 0.3, 2.19 ± 0.08, and 2.28 ± 0.2, respectively). Values are means ±S.D. of three experiments. (Adapted from Biessen *et al.*, 1995, with permission.)

TG(20Å)C. Currently, studies are underway to design glycolipids that have improved selectivity toward atherogenic lipoproteins.

Recent studies have shown that TG(20Å)C and TG(4Å)C are not only effective inducers of LDL catabolism but also of Lp(a) catabolism (Biessen *et al.*, 1994b). Upon complexation of Lp(a) with TG(4Å)C or TG(20Å)C, liver uptake of Lp(a) was increased by 20-fold and 4-fold, respectively. Lp(a) is a LDL-like lipoprotein that is recognized as an independent risk factor for the occurrence of CVDs (Loscalzo, 1990). In view of the fact that an adequate Lp(a)-lowering therapy is not available to date, the glycolipid-based approach may be of value for evolving a clinically applicable therapy for treating patients with high serum Lp(a) levels.

In conclusion, glycolipids are promising hypocholesterolemics. Since the cholesterol-lowering activity of galactose-terminated glycolipids is based on an entirely new strategy that does not require the presence of functionally active LDL receptors, these compounds may be specifically applied in patients homozygous for LDL receptor deficiency (FH) or in patients with dysfunctional apoB100. Potential metabolic side effects will be circumvented since glycolipids directly remove atherogenic lipoproteins from the blood circulation through a physiological pathway.

5. ANTIOXIDANT AGENTS

Another promising point of action for the design of anti-atherogenic drugs focuses on the process by which high serum levels of LDL are supposed to induce atherogenesis. Recent studies have provided insight into the basic mechanisms behind atherogenesis in the vascular wall (Munro and Cotran, 1993; Ross, 1993). The current opinion is that oxidized LDL, rather than native LDL, initiates the atherogenic response (Witztum and Steinberg, 1991). According to this hypothesis, oxidation of LDL is the first step toward macrophage uptake and accumulation of cholesterol (esters) into cells of the vessel wall (Steinberg *et al.*, 1989). After cellular uptake, oxidized LDL triggers the synthesis of a whole body of mediators, adhesion molecules, and second messengers, leading to intracellular lipid deposition, chronic inflammatory responses, and neointima formation (Witztum and Steinberg, 1991). Immunolocalization studies established the presence of oxidized apoB100 fragments and oxidized lipids in the atherosclerotic plaque (Ylä-Herttuala *et al.*, 1989; Palinski *et al.*, 1989; Rosenfeld *et al.*, 1991). Supportive evidence for the aforementioned mechanism was also provided by epidemiological studies, showing that the oxidation resistance of LDL was associated with an increased incidence of CVDs (Rimm *et al.*, 1993). Their findings were slightly qualified by studies of Riemersma *et al.* (1991), who showed that antioxidant intake and the incidence of CVDs were only correlated in populations with low dietary intakes of antioxidants. Prevention of lipoprotein oxidation therefore is supposed to be beneficial for preventing atherosclerotic plaque formation (Steinberg *et al.*, 1989).

5.1. Antioxidant Micronutrients

The most obvious way for reducing the susceptibility of LDL to oxidation seemed to be dietary supplementation with antioxidant micronutrients, such as ascorbate, α-tocopherol (for chemical review see Figure 7), β-carotene, and flavonoids (Schäfer-Elinder and Walddius, 1994). Despite the fact that serum levels of β-carotene, ascorbate, and α-tocopherol were strongly elevated after supplementation, only α-tocopherol was able to slightly increase the resistance of lipoproteins to oxidation (Princen *et al.*, 1992; Jialal and Grundy, 1993; Reaven

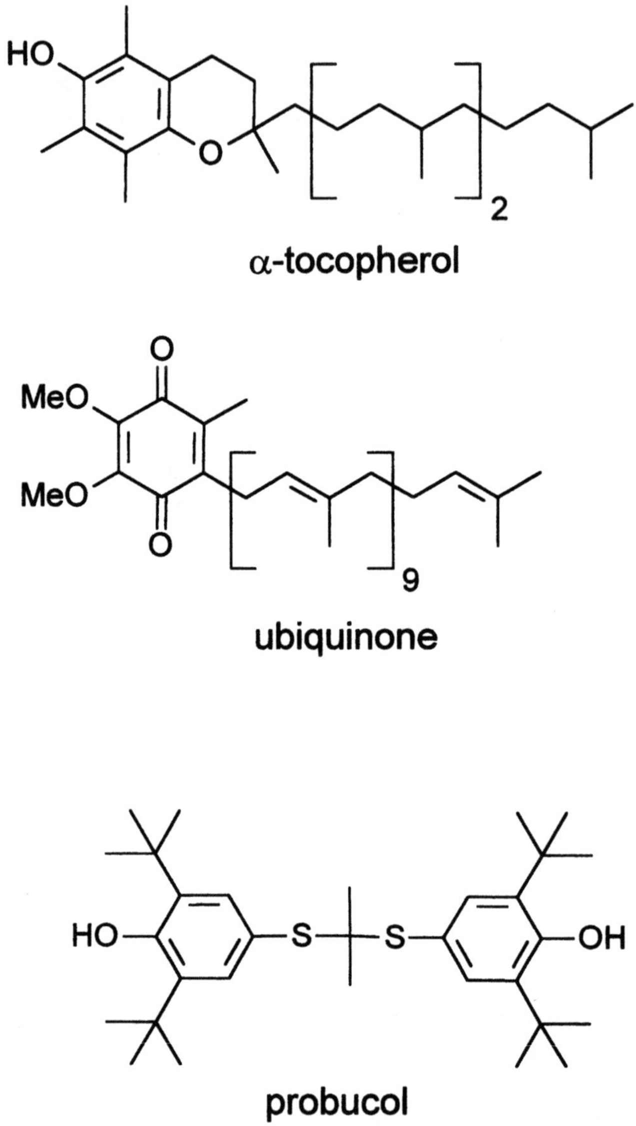

FIGURE 7. Chemical structures of the antioxidants α-tocopherol, probucol, and ubiquinone.

et al., 1993). Nevertheless, even after combined administration of large daily doses of α-tocopherol (10 g/kg) and ascorbate (0.6 g/kg) no effect on the progression of atherosclerosis in cholesterol-fed rabbits could be observed (Morel *et al.,* 1994). In contrast, Qiao *et al.* (1993) found that α-tocopherol-supplemented guinea pigs displayed significantly fewer intimal atheromatous lesions than untreated controls. Further studies in human population are needed before the anti-atherosclerotic activity of α-tocopherol can be conclusively established.

5.2. Flavonoids

Recently, flavonoids have received a great deal of attention as antioxidants with anti-atherosclerotic activity (De Whalley *et al.,* 1990). The recent interest could be attributed mainly to a Dutch epidemiological study demonstrating that flavonoid intake is inversily associated with morbidity and mortality from CVDs (Hertog *et al.,* 1993). Flavonoids are antioxidant polyphenolic compounds that abundantly occur in wine, fruits, and edible plants (Berg and Daniel 1988). Supplementary wine consumption led to a 46% decrease in the LDL susceptibility to oxidation in one study (Fuhrman *et al.,* 1995), but did not provide protection to LDL oxidation in another (Sharpe *et al.,* 1995). In short, considerable debate remains on the efficacy of dietary supplementation with micronutrients for preventing atherosclerosis. Combined with the observation of Takahashi (1995) that large doses of α-tocopherol and β-carotene are accompanied by severe hemorrhagic toxicity, the relatively low activity of these natural antioxidants has prompted various research groups to develop more powerful synthetic and endogenous substitutes.

5.3. Probucol and Analogues

A synthetic antioxidant that has aroused great expectations as an anti-atherosclerotic agent due to its high antioxidant activity is probucol [4,4′-(isopropylidenedithio)bis(2,6-di-*tert*-butylphenol)] (for structure see Figure 7). Probucol was initially introduced as an hypocholesterolemic agent. A recent patient study demonstrated that long-term treatment with probucol resulted in a 20–25% reduction in serum cholesterol over a 10 year period (Baker *et al.,* 1982). In contrast to conventional cholesterol-lowering agents, probucol was also effective in lowering serum cholesterol levels in homozygous FH patients (Baker *et al.,* 1982; Yamamoto *et al.,* 1983). Moreover, in heterozygous FH patients probucol treatment significantly accelerated regression of xanthomas (Yamamoto *et al.,* 1986).

The mechanism for its anti-atherosclerotic activity has not yet been clarified conclusively. Because of its structural similarity to the widely used antioxidant 2,6-di-*tert*-butylcresol (BHT), probucol was, however, rapidly recognized as a powerful antioxidant. Probucol, being a lipophilic compound, accumulates in lipoproteins after oral or intravenous administration (Marshall, 1982), thus protecting these lipoproteins against lipid peroxidation (Parthasarathy *et al.,* 1986). After *in vivo* administration of probucol to rabbits, primates, and rats, the resistance of LDL to oxidation was increased by a factor of 3–10 (Keaney *et al.,* 1995; Fruebis *et al.,* 1994; Sasahara *et al.,* 1994; Mambetisaeva *et al.,* 1994; Dujovne *et al.,* 1994). This was confirmed by the study of Kleinveld *et al.* (1994), in which it was demonstrated that probucol also enhanced the resistance of LDL to oxidation at doses as low as 0.025% (wt/wt). The beneficial effect of probucol on the susceptibility of LDL to oxidation led to an extensive evaluation of the anti-atherosclerotic effect of probucol. Probucol reduced the lesion progression in LDL-receptor deficient WHHL rabbits by 50% (Fruebis *et al.,* 1994; Kita *et al.,* 1987; Carew *et al.,* 1987; Braesen *et al.,* 1995) and reduced intimal lesion areas after administration in primates (Sasahara *et al.,* 1994) and Japanese quails (Bocan *et al.,* 1994). In the latter study, vascular lesion size was found to be inversily correlated to LDL susceptibility to oxidation. Chronic probucol treatment of cholesterol-fed NZW rabbits (10g/kg) also provoked a significant reduction in atherosclerosis (Morel *et al.,* 1994). Yamamoto *et al.* (1988) showed that probucol may act as an anti-atherosclerotic agent, both by suppressing cholesterol accumulation of macrophages leading to foam cell formation and by preventing oxidation of LDL. In agreement, Braesen *et al.* (1995) found that administration of probucol (1% wt/wt) for 2–4 months strongly diminished foam cell formation in plaques from WHHL rabbits and reduced plaque area and thickness, but also modulated plaque composition. Unlike untreated rabbits, plaques of treated rabbits lacked necrotic cores, contained less apoB100, and were particularly rich in vascular smooth muscle cells. The only report in which no effect of probucol on atherosclerosis was observed came from Keany *et al.* (1995).

Since the bulk of the animal studies on probucol confirmed its anti-atherosclerotic activity, an attempt was made to assess the benefits of supplementory probucol therapy in high risk hyperlipidemic patients. In general, it was found that administration of probucol dramatically decreased the susceptibility of LDL for oxidation (Dujovne *et al.,* 1994; Illingworth, 1993; Lankin *et al.,* 1993). Although the reported effects of probucol on progression and regression of atherogenesis were not fully unequivocal, the general outcome of these studies was somewhat disappointing. McDowell *et al.* (1994) reported that probucol did not have a significant effect on vascular functioning in human volunteers. The results of the extensive, prolonged, and well-defined probucol quantitative regression Swedish trial (PQRST) study were recently presented by Walldius *et al.* (1994). Surprisingly, probucol appeared to be ineffective in retarding the progression of atherosclerosis in hyperlipidemic patients. Combined treatment of probucol (1.0 g/day) and cholestyramine (8–16 g/day) appeared to be as effective in slowing the progression of atherosclerosis as a single treatment with cholestyramine. By contrast, positive results were obtained by Illingworth *et al.* (1993), who reported that probucol administration caused regression of xanthomata in severe hyperlipidemias. In addition, clinical studies of Anderson *et al.* (1995) demonstrated that additional antioxidant therapy with probucol further improved the effect of hypocholesterolemic lovastatin therapy alone on endothelium-dependent vasomotion. It can be concluded that although the anti-atherosclerotic effect of probucol remains to be decisively established, probucol may be added to conventional lipid-lowering therapy.

5.4. Ubiquinone and Analogues

Ubiquinone, or coenzyme Q_{10}, is a membrane-associated mitochondrial coenzyme with high antioxidant activity, and as such is essential for homeostasis of the mitochondrial respiratory chain (for structure see Figure 7; Olson *et al.*, 1993). Unlike the other natural antioxidants, ubiquinone is an endogenously synthesized antioxidant and is regenerated endogenously by reduction (Beyer and Ernster 1990). The presence of ubiquinone in cellular membranes is important in two respects; first, it renders membranes more fluid and second, it prevents lipid peroxidation (Jakobsson-Borin, 1994). Ubiquinone recently gained much attention because of its putative role in the congestive heart failure and some neuromuscular diseases, caused by ubiquinone deficiency-induced mitochondrial dysfunction and de-energation (Greenberg and Frishman, 1990). Illustratively, decreases in tissue levels of ubiquinone have been observed in aging, cardiomyopathy, muscle diseases oxidative stress, and carcinogenesis (Kalen *et al.*, 1989; Eggens *et al.*, 1990). In addition to its localization in tissue (between 0.06 to 0.11 mg/g), ubiquinone is also present in plasma at levels ranging from 0.57–3.03 mg/ml (Laaksonen *et al.*, 1995).

In plasma, ubiquinone associates with lipoproteins, 62% being incorporated into LDL and 35% into HDL. In lipoproteins, it presumably acts as radical scavenger and adds to the antioxidant activity of α-tocopherol and β-carotenes (Mohr *et al.*, 1992). Stocker *et al.* (1991) demonstrated that although ubiquinone is only present at very low concentrations in LDL, ubiquinone is an extremely effective antioxidant.

Since lipoprotein oxidation is generally considered to be a key event leading to cholesterol accumulation in the vessel wall and eventually to cardiovascular disease (Steinberg *et al.*, 1989; Ross, 1993), it has been proposed that a low ubiquinone to cholesterol ratio in lipoproteins leads to atherogenesis (Mohr *et al.*, 1992). The above results sugested that additional dietary supplementation of ubiquinone may prevent oxidation of LDL. Laaksonen *et al.* (1995) investigated the effect of supplementation of ubiquinone in human subjects (0–240 mg/day). It appeared that upon supplementation, the serum ubiquinone level increased dose-dependently by up to 360%. The ubiquinone-cholesterol ratio increased in all lipoprotein classes. Aside from dietary supplementation of ubiquinone, ubiquinone levels can also be raised by inhibitors of enzymes of the cholesterol synthesis pathway beyond squalene synthase (Thelin *et al.*, 1994). It remains to be established whether a higher serum ubiquinone level will slow the progression, or even induce regression, of atherosclerosis.

6. CONCLUDING REMARKS

Elevated serum cholesterol levels are commonly accepted as being an important risk factor for the development of CVDs. Moreover, epidemiological studies have revealed that morbidity and mortality rates for CVDs are significantly decreased after long-term treatment with hypocholesterolemic drugs, even in patients with moderately elevated cholesterol levels. Drug treatment of CVD patients appeared to be very cost-effective, comparing favorably with other commonly applied medical therapies (Schwartz *et al.,* 1995). Notwithstanding the success of hypocholesterolemic drugs, further research on new strategies for treating CVDs is imperative. First, a small but significant population of patients does not respond to the currently available hypocholesterolemic agents. Second, it is not quite clear to what extent the current hypocholesterolemic drugs cause adverse side effects, while chronic toxicity of these drugs may still become manifest in the remote future. Finally, as atherogenesis is a complex multifactorial disease, it is crucial to have a wide scope of anti-atherosclerotic therapies on hand for treating various classes of CVD patients.

In this review, a concise overview has been presented on recent developments in anti-atherosclerotic drug research. Of new cholesterol synthesis inhibitors, squalene synthase inhibitors seem to be very promising, although it remains to be established whether administration of these drugs will induce adverse side effects related to intracellular accumulated mevalonate, geranyl, and farnesyl pools. 2,3-Oxidosqualene:lanosterol cyclase inhibitors are also promising, but their therapeutic window seems to be rather narrow. The ACAT inhibitors are particularly interesting since they simultaneously reduce VLDL production and have anti-atherosclerotic activity on the vascular wall. A still rather unexplored approach involves the use of agents that enhance the catabolism of atherogenic lipoproteins and/or stimulate the reverse cholesterol transport, such as ansamycins and galactose-terminated hypocholesterolemic compounds. The cholesterol-lowering effect does not depend on the presence of functional lipoprotein receptors, and may therefore be of value for those CVD patients in which normal lipoprotein processing is impaired. In addition, the latter compounds may also be applied for reducing elevated serum Lp(a) levels. Although the feasibility of both classes of compounds has been demonstrated, the concepts are quite new and clinical application will require additional studies. The efficacy of antioxidant therapy in humans on atherogenesis still remains to be conclusively established. Dietary supplementation of α-tocopherol and ubiquinone may be valuable in slowing atherogenesis only at an early stage. Probucol has aroused great expectations on the basis of its powerful anti-atherosclerotic activity in primates and rabbits. The preliminary results of the extensive probucol quantitative regression Swedish trial study were, however, somewhat disappointing, in that probucol did not significantly reduce the progression of the atherosclerotic lesions. Altogether, major advances in anti-atherosclerotic drug research have recently been accomplished. Undoubtedly, one or more of the aforementioned new strategies will eventually add to, or even substitute for, the currently applied therapy.

ACKNOWLEDGMENT. This work was supported by a grant from the Dutch Heart Foundation (M93.001).

7. REFERENCES

Abdel-Rahmann, H., Adams, J. P., Boyes, A. L., Kelly, M. J., Lamont, R. B., Mansfield, D. J., Procopiou, P. A., Roberts, S. M., and Slee, D. H. 1994, Synthesis of a novel spirocyclic lactone in a potential route to squalestatin 1, *J. Chem. Soc., Perkin Trans.* 1, 1259–1262.

Abe, I., Tomesch, J. C., Wattanasin, S., and Prestwich, G. D., 1994, Inhibitors of squalene biosynthesis and metabolism: reviewing the literature published from January 1988 to September 1993, *Nat. Prod. Rep.* **11**:279–301.

Anderson, T. J., Meredith, I. T., Yeung, A. C., Frei, B., Selwyn, A. P., and Ganz, P., 1995, The effect of cholesterol-lowering and antioxidant therapy on endothelium-dependent coronary vasomotion, *New Engl. J. Med.* **332**:488–493.

Arai, M., Serzawa, N., and Terahara, S., 1988, Pravastatin sodium (CS-514), a novel cholesterol-lowering agent which inhibits HMG-CoA reductase, *Annu. Rep. Sankyo Res. Lab.* **40**:1–38.

Ashwell, G., and Morell, A. G., 1974, The role of surface carbohydrates in the hepatic recognition and transport of circulating glycoproteins, *Adv. Enzymol.* **41**:99–128.

Austin, M. A., King, M. C., Vranizan, Z. M., and Kraus, R. M., 1990, Atherogenic lipoprotein phenotype: a proposed genetic marker for coronary heart disease risk, *Circulation* **82**:495–506.

Baker, S. G., Joffe, B. I., Mendelsohn, D., and Seftel, H. C., 1982, Treatment of homozygous familial hypercholesterolemia with probucol, *S. Afr. Med. J.* **6**:7–11.

Bani, M., Bormetti, R., Ceccarelli, W., Fiocchi, R., Gobetti, M., Lombroso, M., Magnetti, S., Olgiati, V., Palladino, M., Villa, M., and Vanotti, E., 1995, Novel aryloxyalkylthioimidazoles as inhibitors of acyl-CoA: cholesterol-*O*-acyltransferase, *Eur. J. Med. Chem.* **30**:39–46.

Baxter, A., Fitzgerald, B. J., Hutson, J. L., MacCarthy, A. D., Motteram, J. M., Ross, B. C., Sapra, M., Snowden, M. A., Watson, N. S., Williams, R. J., and Wright, C., 1992, Squalestatin-1, a potent inhibitor of squalene synthase, which lowers serum cholesterol *in vivo*, *J. Biol. Chem.* **267**: 11705–11708.

Becker, A., Bottcher, A., Lackner, K. J., Fehringer, P., Notka, F., Aslanidis, C., and Schmitz, G., 1994, Purification, cloning, and expression of a human enzyme with acyl coenzyme A: cholesterol acyltransferase activity, which is identical to liver carboxylesterase, *Art. Thromb.* **14**:1346–1355.

Berg, P. A., and Daniel, P. T., 1988, Effect of flavonoid compounds on the immune response, *Prog. Clin. Biol. Res.* **280**:157–181.

Bergstrom, J. D., Kurtz, M. M., Rew, D. J., Amend, A. M., Karkas, J. D., Bostedor, R. G., Bansal, V. S., Dufresne, C., Vanmiddlesworth, F. L., Hensens, O. D., Liesch, J. M., Zink, D. L., Wilson, K. E., Onish, J., Milligan, J. A., Bills, G., Kaplan, L., Omstead, M. N., Jenkins, R. G., Huang, L., Meinz, M. S., Quinn, L., Burg, R. W., Kong, Y. L., Mochales, S., Mojena, M., Martin, I., Pelaez, F., Diez, M. T., and Alberts, A. W., 1993, Zaragosic acids: a family of fungal metabolites that are picomolar competitive inhibitors of squalene synthase, *Proc. Natl. Acad. Sci. USA* **90**:80–84.

Bernini, F., Tanenbaum, S. R., Sherrill, B. C., Gotto, A. M., Jr., and Smith, L. C., 1986, Enhanced catabolism of low density lipoproteins in rat by lactosaminated Fab fragment, *J. Biol. Chem.* **261**:9294–9299.

Bernini, F., Via, D. P., Bocan, T. M. A., Gotto, A. M., Jr., and Smith, L. C., 1988, Lactosylated Fab fragments specific for low density lipoproteins/hepatocyte targeting and hypolipoproteinemic activity, *Arteriosclerosis* **8**:825–831.

Beyer, R. E., and Ernster, L., 1990, in: *Highlights in Ubiquinone Research* (G. Lenaz, O. Barnabei, A. Rabbi, A., and M. Battino, eds.), pp. 191–213, Taylor and Francis, London.

Biessen, E. A. L., Vietsch, H., and Van Berkel, Th. J. C., 1994a, The cholesterol derivative of a new triantennary cluster galactoside directs low-and high-density lipoproteins to the parenchymal liver cell, *Biochem. J.* **302**:283–289.

Biessen, E. A. L., Vietsch, H., and Van Berkel, Th. J. C., 1994b, Cholesterol derivatized triantennary galactosides significantly enhance the catabolism of lipoprotein(a), *Circulation* **90**:I–461.

Biessen, E. A. L., Bakkeren, H. F., Beuting, D. M., Kuiper, J., and Van Berkel, Th. J. C., 1994c, Recognition of both fucose-and galactose-exposing particles by the hepatic fucose receptor depends on the particle size, *Biochem. J.* **299:**291–296.

Biessen, E. A. L., Vietsch, H., and Van Berkel, Th. J. C., 1995a, Cholesterol derivative of a new triantennary cluster galactoside lowers serum cholesterol levels and enhances secretion of bile acids in the rat, *Circulation* **91:**1847–1854.

Biessen, E. A. L., Beuting, D. M., Roelen, H. C. P. F., Van der Marel, G. A., Van Boom, J. H., and Van Berkel, Th. J. C., 1995b, Synthesis and characterization of cluster glycosides with high affinity for the hepatic asialoglycoprotein receptor, *J. Med. Chem.* **38:**1538–1546.

Biessen, E. A. L., Broxterman, H., Van Boom, J. H., and Van Berkel, Th. J. C., 1995c, The cholesterol derivative of a triantennary galactoside with high affinity for the hepatic asialoglycoprotein receptor: a potent cholesterol lowering agent, *J. Med. Chem.* **38:**1846–1852.

Bijsterbosch, M. K., Bernini, F., Bakkeren, H. F., Gotto, A. M., Jr., Smith, L. C., and Van Berkel, Th. J. C., 1991, Enhanced hepatic uptake and processing of cholesterol esters from low density lipoprotein by specific lactosaminated Fab fragments, *Art. Thromb.* **11:**1806–1813.

Bijsterbosch, M. K., Bakkeren, H. F., Kempen, H. J. M., Roelen, H. C. P. F., Van Boom, J. H., and Van Berkel, Th. J. C., 1992, A monogalactosylated cholesterol derivative that specifically induces uptake of LDL by the liver, *Art. Thromb.* **12:**1153–1160.

Bilheimer, D. W., Grundy, S. M., Brown, M. S., and Goldstein, J. L., 1983, Mevinolin and colistipol stimulate receptor-mediated clearance of low-density lipoprotein from plasma in familial hypercholesterolemia heterozygotes, *Proc. Natl. Acad. Sci. USA* **80:**4124–4128.

Biller, S. A., Forster, C., Gordon, E. M., Harrity, T., Rich, L. C., Maretta, J., and Ciosek, C. P., Jr., 1991a, Isoprenyl phosphinyl formates: new inhibitors of squalene synthase, *J. Med. Chem.* **36:**1912–1918.

Biller, S. A., Sofia, M. J., De Lange, B., Forster, C., Gordon, E. M., Harrity, T., Rich, L. C., and Ciosek, C. P., Jr., 1991b, The first potent inhibitor of squalene synthase: a profound contribution of an ether oxygen to inhibitor-enzyme interaction, *J. Am. Chem. Soc.* **113:**8522–8524.

Blankenhorn, D. H., Azen, D. Z., Kramsch, D. M., Mack, W. J., Cashine, L., Hodis, H. N., De Boer, L. W. V., Mahrer, P. R., Mastelle, M. J., and Vailos, L. I., 1993, Coronary angiographic changes with lovastatin therapy. The monitored atherosclerosis regression study (MARS), *Ann. Intern. Med.* **119:**969–976.

Bocan, T. M. A., Müller, S. B., Uhlendorf, P. D., Newton, R. S., and Krause, B. R., 1991, Comparison of CI-976, an ACAT inhibitor, and selected lipid lowering agents for anti-atherosclerotic activity in iliac-femoral and thoractic lesions, *Art. Thromb.* **11:**1830–1843.

Bocan, T. M. A., Muller, S. B., Uhlendorf, P. D., Brown, E. Q., Mazur, M. J., and Block, A. E., 1993, Inhibition of acylcoenzyme A: acyltransferase reduces the cholesteryl ester enrichment of atherosclerotic lesions in the Yucatan micropig, *Atheroscler.* **99:**175–186.

Bocan, T. M. A., Mazur, M. J., Mueller, S. B., Charlton, G., Kieft, K. A., and Krause, B. R., 1994, Atherosclerotic lesion development in hypercholesterolemic Japanese quail following probucol treatment—a biochemical and morphologic evaluation, *Pharm. Res.* **29:**65–76.

Boogaard, A., Griffioen, M., and Cohen, L. H., 1987, Regulation of 3-hydroxy-3-methylglutaryl-coenzyme A reductase in human hepatoma cell line HepG2, *Biochem. J.* **241:**345–351.

Braesen, J. H., Beisiegel, U., and Niendorf, A., 1995, Probucol inhibits not only the progression of atherosclerotic disease, but causes a different composition of atherosclerotic lesions in WHHL-rabbits, *Virch. Archiv–Int. J. Path.* **426:**179–188.

Brindle, S. D., and O'Buck, A., 1991, Tissue distribution of 14C-CGP43371 in the rat by whole body autoradiography, *Pharmacologist* **33:**208–212.

Brown, M. S., and Goldstein, J. L, 1986, A receptor-mediated pathway for cholesterol homeostasis, *Science* **232:**34–47.

Carew, T. E., Schwenke, D. C., and Steinberg, D., 1987, Antiatherogenic effect of probucol unrelated to its hypocholesterolemic effect, *Proc. Natl. Acad. Sci. USA* **84:**7725–7729.

Caron, S., McDonald, A. I., and Heathcock, C. H., 1994, Synthesis of 1-substituted and 1,4-disubstituted 2,3-di-*O*-benzyl-1,6-anhydrogalactofuranoses, *J. Org. Chem.* **60:**2780–2785.

Carr, T. P., Hamilton, R. L., and Rudel, L. L., 1995, ACAT inhibitors decrease secretion of cholesteryl esters and apolipoprotein B by perfused livers of African green monkeys, *J. Lipid Res.* **36:**25–36.

Carreira, E. M., and Du Bois, J., 1994, Synthesis of (+)-zaragosic acid C, *J. Am. Chem. Soc.* **116:**10825–10826.

Cattel, L., Ceruti, M., Balliano, G., Viola, F., Grosa, G., and Schuber, F., 1989, Drug design based on biosynthetic studies: synthesis, biological activity, and kinetics of new inhibitors of 2,3-oxidosqualene cyclase and squalene epoxidase, *Steroids* **53:**363–391.

Cattel, L., Ceruti, M., Balliano, G., Viola, F., Grosa, G., Rocco, F., and Brusa, P., 1995, 2,3-Oxidosqualene cyclase: from azasqualenes to new site-directed inhibitors, *Lipids* **30:**235–246.

Ceruti, M., Balliano, G., Viola, F., Grosa, G., Rocco, F., and Cattel, F., 1992, 2,3-epoxy-10-aza-10,11-dihydrosqualene, a high energy intermediate analogue of 2,3-oxidosqualene cyclase, *J. Med. Chem.* **35:**3050–3059.

Ciosek, C. P., Jr., Magnin, D. R., Harrity, T. W., Logan, J. V. H., Dickson, J. K., Jr., Gordon, E. M., Hamilton, K. A., Jolibois, K. G., Kunselman, L. K., Lawrence, R. M., Mookthiar, K. A., Rich, L. C., Slusarchyk, D. A., Sulsky, R. B., and Biller, S. A., 1993, Lipophilic 1,1-biphosphonates are potent squalene synthase inhibitors and orally active cholesterol lowering agents *in vivo, J. Biol. Chem.* **268:**4832–4837.

Cohen, L. H., Valentijn, A. R. P. M., Roodenburg, L., Van Leeuwen, R. E. W., Huisman, R. H., Lutz, R. J., Van der Marel, G. A., and Van Boom, J. H., 1995, Different analogues of farnesyl pyrophosphate inhibit squalene synthase and protein:farnesyltransferase to different extents, *Biochem. Pharmacol.* **49:**839–845.

Corey, E. J., and Volante, R. P., 1976, Application of unreactive analogs of terpenoid pyrophosphates to studies of multistep biosynthesis: demonstration that "presqualene pyrophosphate" is an essential intermediate on the path to squalene, *J. Am. Chem. Soc.* **98:**1291–1293.

Cox, B., Hutson, J. L., Keeling, S. E., Kirk, B. E., Srikantha, A. R. P., and Watson, N. S., 1994, The squalestatins: potent inhibitors of squalene synthase, 3-hydroxymethyl derivatives, *Bioorg. Med. Chem. Lett.* **4:**1931–1936.

Das, N. P., and Allen, C. M., 1992, Inhibition of farnesyl transferases from malignant and non-malignant cultured human lymphocytes by prenyl substrate analogues, *Biochem. Biophys. Res. Commun.* **181:**729–735.

De Rijke, Y. B., Verwey, H. F., Vogelezang, C. J. M., Princen, H. M. G., Van der Laarse, A., Bruschke, A. V. G., and Van Berkel, Th. J. C., 1992, Susceptibility of low-density lipoprotein to oxidation in coronary bypass patients, *Lancet* **340:**858–859.

De Whalley, C. V., Rankin, S. M., Hoult, J. R. S., Jessup, W., and Leake, D. S., 1990, Flavonoids inhibit the oxidative modification of low-density lipoproteins by macrophages, *Biochem. Pharmacol.* **39:**1743–1750.

Dietschy, J. M., and Spady, D. K., 1986, Regulation of low-density lipoproteins in animals and men with particular emphasis on the role of the liver, in: *Receptor Mediated Uptake in the Liver* (H. Greten, E. Windler, and U. Beisiegel, eds.), pp. 56–65, Springer Verlag, Berlin.

Dodd, D. S., and Oehlschlager, A. C., 1992, Synthesis of inhibitors of 2,3-oxidosqualene-lanosterol cyclase: conjugate addition of organocuprates to *N*-(carbobenzyloxy)-3-carbomethoxy-5,6-dihydro-4-pyridone, *J. Org. Chem.* **57:**2794–2803.

Dollis, D., and Schuber, F., 1994, Effects of a 2,3-oxidosqualene-lanosterol cyclase inhibitor, 2,3:22,23-dioxidosqualene and 24,25-epoxycholesterol on the regulation of cholesterol biosynthesis in human hepatoma cell line HepG2, *Biochem. Pharmacol.* **48:**49–58.

Dufresne, C., Wilson, K. E., Zink, D., Smith, J., Bergstrom, J. D., Kurtz, M., Rew, D., Nallin, M., Jenkins, R., Bartizal, K., Trainor, C., Bills, G., Meinz, M., Huang, L., Onishi, J., Milligan, J., Mojena, M., and Pelaez, F., 1992, The isolation and structure elucidation of zaragosic acid C, a novel potent squalene synthase inhibitor, *Tetrahedron* **48:**10221–10226.

Dujovne, C. A., Harris, W. S., Gerrond, L. L. C., Fan, J. Y. and Muzio, F., 1994, Comparison of effects of probucol versus vitamin E on *ex vivo* oxidation susceptibility of lipoproteins in hyperlipoproteinemia, *Am. J. Card.* **74**:38–42.

Duriatti, A., Bouvier-Nave, P., Benveniste, P, Schuber, F., Delprino, L., Balliano, G., and Cattel L., 1985, *In vitro* inhibition of animal and higher plants 2,3-oxidosqualene-sterol cyclases by 2-aza-2,3-dihydrosqualene and derivatives, *Biochem. Pharmacol.* **34**:2765–2777.

Eggens, I., Ekström, T. J., and Aberg, F., 1990, Biosynthesis of polyisoprenols, cholesterol and ubiquinone in highly differentiated human hepatomas, *J. Exp. Pathol.* **71**:219–232.

Endo, A., Kuroda, M., and Tsujita Y., 1976, ML-236A, ML-236B, and ML-236C. New inhibitors of cholesterogenesis produced by *Penicillum citrinum*, *J. Antibiot.* **29**:1346–1348.

Erickson, S. K., Lear, S. R., and McGreery, M. J., 1994, Functional sizes of hepatic enzymes of cholesteryl ester metabolism determined by radiation inactivation, *J. Lipid Res.* **35**:763–769.

Erickson, S. K., Shrewsbury, M. A., Brooks, C., and Meyer, D. J., 1980, Rat liver acyl-coenzyme A:cholesterol acyltransferase: its regulation *in vivo* and some of its properties *in vivo*, *J. Lipid Res.* **21**:930–941.

Evans, D. A., Barrow, J. C., Robichaud, A. J., and Sefkow, M., Asymmetric synthesis of the squalene inhibitor zaragosic acid C, 1994, *J. Am. Chem. Soc.* **116**:12111–12112.

Fancelli, D., Chiari, A., Cozzi, P., Lovisolo, P., Severino, D., and Ghiselli G., 1994, Imidazolyl benzopyrane derivatives: a new class of acyl-CoA: cholesterol acyltranferase (ACAT) inhibitors, *J. Enz. Inhib.* **8**:159–172.

Farmer, J. A., Washington, L. C., Jones, P. H., and Shapiro, D. R., 1992, Comparative effects of simvastatin and lovastatin in patients with hypercholesterolemia, *Clin. Ther.* **14**:708–717.

Fruebis, J., Steinberg, D., Dresel, H. A., and Carew, T. E., 1994, A comparison of the antiatherogenic effects of probucol and of a structural analogue of probucol in low density lipoprotein receptor-deficient rabbits, *J. Clin. Invest.* **94**:392–398.

Fuhrman, B., Lavy, A., and Aviram, M., 1995, Consumption of red wine with meals reduces the susceptibility of human plasma and low-density lipoprotein to lipid peroxidation, *Am. J. Clin. Nutrit.* **61**:549–554.

Gibbs, J. B., Pompliano, D. L., Mosser, S. D., Rands, E., Lingham, R. B., Singh, S. B., Scolnick, E. M., Kohl, N. E., and Oliff, A., 1993, Selective inhibition of farnesyl:protein transferase blocks ras processing *in vivo*, *J. Biol. Chem.* **268**:7617–7620.

Giblin, G. M. P., Bell, R., Hancock, A. P., Hartley, C. D., Inglis, G. G., Payne, J. J., Procipiou, P. A., Shingler, A. H., Smith, C., and Spooner, S. J., 1993, Semi-synthetic modifications, *Bioorg. Med. Chem. Lett.* **3**:2605–2610.

Gibson, J. C., Kothari, H. V. Kump, W., Vespa, D. B., Stucchi, A. F., and Nicolosi, R. J., 1991, Effect of a novel hypolipidemic agent CGP43371 on plasma lipoproteins of cynomolgus monkeys, *Art. Thromb.* **11**:1479a.

Gibson, J. C., Kothari, H. V., Genthe, T. M., Lee, W. H., Poirier, K. J., Sawyer, W. K., Mugrage, B., Traxler, P., Veenstra, S., Grim, M., and Kump, W., 1992, Effect of a novel series of macrocyclic hypolipidemic agents on plasma lipid and lipoprotein levels of four nonprimate species, *Atherosclerosis* **96**:147–158.

Gibson, J. C., Lee W. H., and Piccolo, J. R., 1994, The ansamycins: hypolipidemic agents stimulating cholesterol removal by nonclassical mechanisms, *J. Lipid Res.* **35**:1524–1534.

Gibson, J. C., Lee, W. H., and Stephan, Z. F., 1995, The ansamycins: a novel class of hypolipidemic agents with high affinity for lipoproteins, *Atherosclerosis* **112**:47–57.

Gordon, T., Castelli, W. P., Hjortland, M. C., Kannel, W. B., and Dawber, T. R., 1977, High-density lipoprotein as a protective factor against coronary heart disease. The Framingham study, *Am. J. Med.* **62**:707–714.

Gotto, A. M., Larosa, J. C., Hunninghake, D., Grundy, S. M., Wilson, P. W., Clarkson, T. B., and Hay, J. W., 1990, The cholesterol facts. A summary of the evidence relating dietary fats, serum cholesterol, and coronary heart disease, *Circulation* **81**:1721–1733.

Greenberg, S., and Frishman, W. H., 1990, Co-enzyme Q_{10}: a new drug for cardiovascular disease, *J. Clin. Pharmacol.* **30**:596–608.

Grosa, G., Viola, F., Ceruti M., Brusa, P., Delprino, L., Dosio, F., and Cattel, L., 1994, Synthesis and biological activity of a squalenoid maleimide and other classes of squalene derivatives as irreversible inhibitors of 2,3-oxidosqualene cyclase, *Eur. J. Med. Chem.* **29**:17–24.

Gurjar, M. K., Das, S. K., and Saha, U. K., 1994, Zaragosic acid A: interesting observations in anhydroring formation of densely functionalized carbohydrate templates, *Tetrahedron Lett.* **35**:2241–2244.

Hasumi, K., Tachikawa, K., Sakai, K., Marakawa, S., Yoshikawa, N., Kumazawa, S., and Endo, A., 1993, Competitive inhibition of squalene synthase by squalestatin 1, *J. Antibiotics* **46**:689–691.

Hertog, M. G. L., Feskens, E. J. M., Hollman, P. C. H., Katan, M. B., and Kornhout, D., 1993, Dietary antioxidant flavonoids and risk of coronary heart disease: the Zutphen elderly study, *Lancet* **342**:1007–1011.

Higley, C. A., Wilde, R. G., Maduskuie, T. P., Johnson, A. L., Pennev, P., Billheimer, J. T., Robinson, C. S., Gillies, P. J., and Wexler, R. R., 1994, Acyl CoA:cholesterol acyltransferase (ACAT) inhibitors: synthesis and structure–activity relationship studies of a new series of trisubstituted imidazoles, *J. Med. Chem.* **37**:3511–3522.

Hunninghake, D. B., 1992, HMG-CoA reductase inhibitors, *Curr. Opin. Lipidology* **3**:22–28.

Illingworth, D. R., 1991, HMG-CoA reductase inhibitors, *Curr. Opin. Lipidology* **2**:24–230.

Illingworth, D. R., 1993, The potential role of antioxidants in the prevention of atherosclerosis, *J. Nutr. Sci. Vitamin.* **39**:S43–S47.

Innerarity, T. L., Weisgraber, K. H., Arnold, K. S., Mahley, R. W., Krauss, R. M., Vega, G. L., and Grundy, S. M., 1987, Familial defective apolipoprotein B-100: low density lipoprotein with abnormal receptor binding, *Proc. Nat. Acad. Sci. USA* **84**:6919–6923.

Innerarity, T. L., Mahley, R. W., Weisgraber, K. H., Bersot, T. P., Krauss, R. M., Vega, G. L., Grundy, S. M., Friedl, W., Davignon, J., and McCarthy, B. J., 1990, Familial defective apolipoprotein B-100: a mutation of apolipoprotein B that causes hypercholesterolemia, *J. Lipid Res.* **29**:1337–1349.

Iwasawi, Y., Nonoshita, K., and Tomimoto, K., 1994a, Efficient and stereoselective synthesis of J-104,118, a novel potent inhibitor of squalene synthase, *Tetrahedron Lett.* **36**:7459–7462.

Iwasawi, Y., Hayashi, M., Nomoto, T., Shibata, J., Mitsuya, M., Yonemoto, M., Kamei, T., Miura, K., and Tomimoto, K., 1994b, Synthesis and biological activity of J-104,118, a novel inhibitor of squalene synthase, *Bioorg. Med. Chem. Lett.* **5**:1989–1994.

Jakobsson-Borin, Å., Åberg, F., and Dallner, G., 1994, Lipid peroxidation of microsomal and mitrochondrial membranes extracted with *n*-pentane and reconstituted with ubiquinol, dolichol and cholesterol, *Biochim. Biophys. Acta* **1213**:159–166.

Jialal, I., and Grundy, S. M., 1993, Effect of supplementation with alpha-tocopherol, ascorbate, and beta carotene on low-density lipoprotein oxidation, *Circulation* **88**:2780–2786.

Kalèn, A., Appelkvist, E.-L., and Dallner, G., 1989, Age-related changes in the lipid compositions of rat and human tissues, *Lipids* **24**:579–584.

Keaney, J. F., Xu, A. M., Cunningham, D., Jackson, T., Frei, B., and Vita, J. A., 1995, Dietary probucol preserves endothelial function in cholesterol-fed rabbits by limiting vascular oxidative stress, and superoxide generation, *J. Clin. Invest.* **95**:2520–2529.

Kempen, H. J. M., Hoes, C., Van Boom, J. H., Spanjer, H. H., De Lange, J., Langendoen, A., and Van Berkel, T. J. C., 1984, A water-soluble cholesteryl-containing trisgalactoside: synthesis, properties, and use in directing lipid containing particles to the liver, *J. Med. Chem.* **27**:1306–1312.

Kempen, H. J. M., Kuipers, F., Vonk, R. J., and Van Berkel, Th. J. C., 1987, Effect of infusion of "trisgalactosyl-cholesterol" on plasma cholesterol, clearance of lipoprotein cholesteryl esters, and biliary secretion in the rat, *J. Lipid Res.* **28**:659–666.

Kita, T., Nagano, Y., Yokode, M., Ishii, K., Kume, N., Ooshima, A., Yoshida, H., and Kawai, C., 1987, Probucol prevents the progression of atherosclerosis in Watanabe heritable hyperlipidemic rabbit, an animal model for familial hypercholesterolemia, *Proc. Natl. Acad. Sci. USA* **84**:5928–5931.

Kleinveld, H. A., Demacker, P. N. M., and Stalenhoef, A. F. H., 1994, Comparative study on the effect of low-dose vitamin E and probucol on the susceptibility of LDL to oxidation, and the progression of atherosclerosis in Watanabe heritable hyperlipidemic rabbits, *Art. Thromb.* **14:**1386–1391.

Koert, U., 1995, Total syntheses of zaragozic acid, *Angew. Chem. Int. Ed. Engl.* **34:**773–778.

Kuiper, J., Bakkeren, H. F., Biessen, E. A. L., and Van Berkel, Th. J. C., 1994, Characterisation of the interaction of galactose-exposing particles with rat Kupffer cells, *Biochem. J.* **299:**285–290.

Kuo, C. H., Robichaud, A. J., Rew, D. J., Bergstrom, J. D., and Berger, G. D., 1994, Design and synthesis of squalene synthase inhibitors-acyclic mimics of the zaragozic acids, *Bioorg. Med. Chem. Lett.* **4:**1591–1594.

Laaksonen, R., Riihimaki, A., Laitila, J., Martensson, K., Tikkanen, M. J., and Himberg, J. J., 1995, Serum and muscle tissue ubiquinone levels in healthy subjects, *J. Lab. Clin. Med.* **125:**517–521.

Lankin, V. Z., Revenko, V. M., Lupanov, V. P., Tikhaze, A. K., Lyakishev, A. A., and Kukharchuk, W., 1993, Impact of long term probucol administration on blood lipoprotein cholesterol levels and glutathione activity in patients with coronary atherosclerosis and hyperlipidemia, *Kardiologiya* **33:**41–44.

Leonards, K. S., Cribson, J. C., Hu, C. W., and Lee, W. H., 1992, Mechanistic studies of the hypolipidemic ansamycins CGP43371 and CGS24565, in: *Proceedings of the XIth Symposium on Drugs Affecting Lipid Metabolism,* p. 82a, Fundosione Giovanni Lorenzini, Milan, Italy.

Levy, R. I., Troendle, A. J., and Fattu, J., 1993, A quarter century of drug treatment of dyslipoproteinemia, with a focus on the new HMG-CoA reductase inhibitor fluvastatin, *Circulation* **87:**45–53.

Loscalzo, J., 1990, Lipoprotein(a). A unique risk factor for atherothrombotic disease, *Arteriosclerosis* **10:**671–679

Lovastatin/Pravastatin Study Group, 1993, A multicenter comparative trial of lovastatin and pravastatin in the treatment of hypercholesterolemia, *Am. J. Cardiol.* **71:**810–815.

Mahley, R. W., and Innerarity, T. L., 1983, Lipoprotein receptors and cholesterol homeostasis, *Biochim. Biophys. Acta* **737:**197–222.

Mambetisaeva, P. T., Kosenkov, E. I., Podrez, E. A., Samuilov, Y. D., and Kosykh, V. A., 1994, Effect of probucol and a new analog on cholesterol on lipoprotein metabolism in cultured rabbit hepatocytes, *Biochem. USSR* **59:**89–93.

Marshall, F. N., 1982, Pharmacology and toxicology of probucol, *Artery* **10:**7–21.

Matsuda, K., 1994, ACAT inhibitors as antiatherosclerotic agents—compounds and mechanisms, *Med. Res. Rev.* **14:**271–305.

McDowell, I. F. W., Brennan, G. M., McEneny, J., Young, I. S., Nicholls, D. P., McVeigh, G. E., Bruce, I., Trimble, E. R., and Johnston, G. D., 1994, The effect of probucol and vitamin E treatment on the oxidation of low-density lipoprotein and forearm vascular responses in humans, *Eur. J. Clin. Invest.* **24:**759–765.

McVinish, L. M., and Rizzacasa, M. A., 1994, Synthetic studies towards squalestatins and zaragosic acids, *Tetrahedron Lett.* **35:**923–926.

Menear, K. A., Dunnet, K. E., Hastewell, J. G., Eichler, H. G., Gibson, J. C., Kimble, E. F., Kump, W., and Taylor, P. W., 1992, Hypolipidemic activity of a novel acyclic ansamycin derivative. Synthesis and pharmacokinetic and pharmacological studies on rodents, *Art. Fors. Drug Res.* **42:**1125–1129.

Middleton, R. F., Foster, G., Cannell, R. J. P., Sidebottom, P. J., Taylor, N. L., Noble, D., Todd, M., Dawson, M. J., and Lawrence, G. C., 1995, Novel squalestatins produced by biotransformation, *J. Antibiotics* **48:**311–316.

Mohr, D., Bowry, V. W., and Stocker, R., 1992, Dietary supplementation with coenzyme Q_{10} results in increased levels of ubiquinol-10 within circulating lipoproteins and increased resistance of human low-density lipoprotein to the initiation of lipid peroxidation, *Biochim. Biophys. Acta* **1126:**247–254.

Morel, D. W., De la Llera-Moya, M., and Friday, K. E., 1994, Treatment of cholesterol-fed rabbits with dietary vitamins E and C inhibits lipoprotein oxidation but not development of atherosclerosis, *J. Nutrition* **124:**2123–2130.

Munro, J. M., and Cotran, R. S., 1988, The pathology of atherosclerosis: atherogenesis and inflammation, *Lab. Invest.* **58:**249–261.

Nakahara, K., Kuriyama, M., Yoshidome, H., Nagata, K., Nagado, T., Nakakawa, M., Arimura, K., Higuchi, I., Osame, M., and Tarumi, C., 1992, Experimental simvastatin-induced myopathy in rabbits, *J. Neurolog. Sci.* **113:**114–117.

Nicolaou, K. C., Yue, E. W., Naniwa, Y., De Riccardis, F., Nadin, A., Leresche, J. E., La Greca, S., and Yang, Z., 1994a, Zaragozic acid A/squalestatin S1: synthetic and retrosynthetic studies, *Angew. Chem. Int. Ed. Engl.* **33:**2190–2191.

Nicolaou, K. C., Yue, E. W., Naniwa, Y., De Riccardis, F., Nadin, A., Leresche, J. E., La Greca, S., and Yang Z., 1994b, Zaragozic acid A/squalestatin S1: synthetic and retrosynthetic studies, *Angew. Chem. Int. Ed. Engl.* **33:**2184–2187.

Nicolaou, K. C., Yue, E. W., Naniwa, Y., De Riccardis, F., Nadin, A., Leresche, J. E., La Greca, S., and Yang, Z., 1994c, Zaragozic acid A/squalestatin S1: synthetic and retrosynthetic studies, *Angew. Chem. Int. Ed. Engl.* **33:**2187–2190.

Nicolaou, K. C., Yue, E. W., La Greca, S., Nadin, A., Yang, Z., Leresche, J. E., Tsuri, T., Naniwa, Y., and De Riccardis, F., 1995, Synthesis of zaragozic acid A/squalestatin S1, *Chem. Eur. J.* **1:**467–494.

Nioche, J. Y., Decerprit, J., and Festal, D., 1995, Synthesis and structure–activity relationships of new ACAT inhibitors, *Eur. J. Med. Chem.* **30:**377–385.

Olson, R. E., and Rudney, H., 1993, Biosynthesis of ubiquinone, *Vitam. Horm.* **40:**1–43.

Oram, J. F., Johnson, C. J., and Aulinskas-Brown, T., 1987, Interaction of high density lipoprotein with its receptor on cultured fibroblasts and macrophages, *J. Biol. Chem.* **262:**2405–2410.

Palinski, W., Rosenfeld, M. E., Ylä-Herttuala, S., Gurtner, G. C., Socher, S. S., Butler, S. W., Parthasarathy, S., Carew, T. E., Steinberg, D., and Witztum, J. L., 1989, Low density lipoprotein undergoes oxidation *in vivo, Proc. Natl. Acad. Sci. USA* **86:**1372–1376.

Panini, S. R., Everson, G. T., and Spencer, T. A., 1991, Effect of specific inhibition of sterol biosynthesis on the uptake and utilization of low-density lipoprotein cholesterol by HepG2 cells, *J. Lipid Res.* **32:**1657–1665.

Parthasarathy, S., Printz, D. J., Boyd, D., Joy, L., and Steinberg, D., 1986, Low density lipoprotein undergoes oxidation *in vivo, Arterioscl.* **6:**505–510.

Peter, M. G., Boldt, P. C., Niederstein, Y., and Peter-Katalic, J., 1990, Synthesen von galactose-cluster-haltigen steroid derivaten, *Liebigs Ann. Chem.* 863–869.

Picard, J. A., Bousley, R. F., Lee, H. T., Hamelehle, K. L., Krause, B. R., Minton, L. L., Sliskovic, D. R., and Stanfield, R. L., 1994, Inhibitors of acyl-CoA:cholesterol *O*-acyltransferase. II. Structure-activity relationships of several series of compounds derived from *N*-(chlorocarbonyl) isocyanate, *J. Med. Chem.* **37:**2394–2400.

Pompliano, D. L., Rands, E., Schaber, M. D., Mosser, S. D., Anthony, N. J., and Gibbs, J. B., 1992, Steady state kinetic mechanism of ras farnesyl:protein transferase, *Biochemistry* **31:**3800–3807.

Popjak, G., and Agnew, W. S., 1979, Squalene synthetase, *Mol. Cell. Biochem.* **27:**97–101.

Prashad, M., Kathawala, F. G., and Scallen, T., 1993, *N*-alkyl(aryl) farnesylamines: new potent squalene synthase inhibitors, *J. Med. Chem.* **36:**1501–1509.

Princen, H. M. G., Van Poppel, G., Vogelezang, C., Buytenhek, R., and Kok, F. J., 1992, Supplementation with vitamin E but not β-carotene *in vivo* protects low-density lipoprotein from lipid peroxidation in vitro: effect of cigarette smoking, *Art. Thromb.* **12:**554–562.

Procopiou, P. A., Bailey, E. J., Bamford, M. J., Craven, A. P., Dymock, B. W., Houston, J. G., Hutson, J. L., Kirk, B. E., McCarthy, A. D., Sareen, M., Scicinski, J. J., Sharratt, P. J., Snowden, M. A., Watson, N. S., and Williams, R. J., 1994, The squalestatins: novel inhibitors of squalene synthase. Enzyme inhibitory activities and *in vivo* evaluation of C1-modified analogues, *J. Med. Chem.* **37:**3274–3281.

Procopiou, P. A., Cox, B., Kirk, B. E., Lester, M. G., McCarthy, A. D., Sareen, M., Sharratt, P. J., Snowden, M. A., Spooner, S. J., Watson, N. S., and Widdowson, J., 1996, The squalestatins: inhibitors of squalene synthase. Enzyme inhibitory activities and *in vivo* evaluation of C3-modified analogues, *J. Med. Chem.* **39:**1413–1422.

Qiao, Y., Yokoyama, M., Kameyama, K., and Asano, G., 1993, Effect of vitamin E on vascular integrity in cholesterol-fed guinea pigs, *Art. Thromb.* **13:**1885–1892.

Reaven, P. D., Khouw, A., Beltz, W. F., Parthasarathy, S., and Witztum, J. L., 1993, Effect of dietary antioxidant combinations in humans; protection of LDL by vitamin E but not β-carotene, *Art. Thromb.* **13:**590–600.

Ren, Z. B., and Gould, M. N., 1994, Inhibition of ubiquinone and cholesterol synthesis by the monoterpene perillyl alcohol, *Cancer Lett.* **76:**185–190.

Riemersma, R. A., Wood, D. A., Macintyre, C. C. A., Elton, R. A., Gey, K. F., and Oliver, M. F., 1991, Risk of angina pectoris and plasma concentrations of vitamins A, C, and E and carotene, *Lancet* **337:**1–5.

Rilling, H. C., 1966, A new intermediate in biosynthesis of squalene, *J. Biol. Chem.* **241:**3233–3236.

Rimm, E. B., Stampfer, M. J., Aschierio, A., Giovannucci, E., Colditz, G. A., and Willett, W. C., 1993, Vitamin E consumption and the risk of coronary heart disease in men, *New Engl. J. Med.* **328:**1450–1456.

Roelen, H. C. P. F., Bijsterbosch, M. K., Bakkeren, H. F., Van Berkel, Th. J. C., Kempen, H. J. M., Buytenhek, M., Van der Marel, G. A., and Van Boom, J. H., 1991, Water-soluble cholesteryl-containing phosphorothioate monogalactosides: synthesis, properties, and use in lowering blood cholesterol by directing plasma proteins to the liver, *J. Med. Chem.* **34:**1036–1042.

Roessner, J., Buntel, C. J., and Griffin, J. H., 1992, Nucleotide and deduced amino acid sequences of the oxidosqualene cyclase from *Candida albicans, J. Am. Chem. Soc.* **114:**9711–9712.

Rosenfeld, M. E., Khoo, J. C., Miller, E., Parthasarathy, S., Palinski, W. I, and Witztum, J. L., 1991, Macrophage derived foam cells freshly isolated from rabbit atherosclerotic lesions degrade modified lipoproteins, promote oxidation of low-density lipoproteins, and contain oxidation specific lipid-protein adducts, *J. Clin. Invest.* **87:**90–99.

Ross, R., 1993, The pathogenesis of atherosclerosis: a perspective for the 1990s, *Nature* **362:**801–809.

Rudney, H., and Sexton, R. C., 1986, Regulation of cholesterol biosynthesis, *Annu. Rev. Nutr.* **6:**15039–15044.

Sasahara, M., Raines, E. W., Chait, A., Carew, T. E., Steinberg, D., Wahl, P. W., and Ross, R., 1994, Inhibition of hypercholesterolemia-induced atherosclerosis in the nonhuman primate by probucol. 1. Is the extent of atherosclerosis related to resistance of LDL to oxidation? *J. Clin. Invest.* **94:**155–164.

Sasiak, K., and Rilling, H. C., 1988, Purification to homogeneity and some properties of squalene synthetase, *Arch. Biochem. Biophys.* **26:**622–627.

Saxena, U., Ferguson, E., and Newton, R. S., 1995, Acyl-coenzyme A:cholesterol-acyltransferase (ACAT) inhibitors modulate monocyte adhesion to aortic endothelial cells, *Atheroscler.* **112:**7–17.

Scandinavian Simvastatin Survival Study Group, 1994, Randomized trial of cholesterol lowering in 4444 patients with coronary heart disease: the Scandinavian simvastatin survival study (4S), *Lancet* **344:**1383–1389.

Schäfer-Elinder, L., and Walldius, G., 1994, Antioxidants and atherosclerosis progression: unresolved questions, *Curr. Opin. Lipidology* **5:**265–268.

Schouten, D., Kleinherenbrink-Stins, M. F., Brouwer, A., Knook, D. L., and Van Berkel, Th. J. C., 1990, Characterisation of the interaction of human apoE-free high density lipoprotein with human hepatocytes, *Arteriosclosis* **10:**1127–1135.

Schuster, H., Rauh, G., Korrmann, B., Hepp, T., Humphries, S., Keller, C., Wolfram, G., and Zollner, N., 1990, Familial defective apolipoprotein B-100; comparison with familial hypercholesterolemia in 18 cases detected in Munich, *Arteriosc1.* **10:**577–581.

Schwartz, J. S., Kjekshus, J., and Pedersen, T. R., 1995, Impact of cholesterol lowering on resource use and costs among patients with coronary heart disease (CHD): results from 4S, *Circulation* **92:**I510.

Sharatt, P. J., Hutson, J. L., Inglis, G. G. A., Lester, M. G., Procopiou, P. A., and Watson, N. S., 1994, Structurally simplified squalestatins-monocyclic 1,3-dioxane analogues, *Bioorg. Med. Chem. Lett.* **4:**661–666.

Sharpe, P. C., McGiath, L. T., McLean, E., Young, I. S., and Archbold, G. P. R., 1995, Effect of red wine consumption on lipoprotein(a) and other risk factors for atherosclerosis, *Q. J. Med.* **88:**101–108.

Shepherd, J., Cobbe, S. M., Ford, I., Isles, C. G., Lorimer, A. R., Macfarlane, P. W., McKillop, J. H., and Packard, C. J., 1995, Prevention of coronary heart disease with pravastatin in men with hypercholesterolemia, *New Engl. J. Med.* **333:**1301–1307.

Sidebottom, P. J., Highcock, R. M., Lane, S. J., Procopiou, P. A., and Watson, N. S., 1992, The squalestatins, novel inhibitors of squalene synthase produced by a species of Phoma II. Structure elucidation, *J. Antibiotics* **45:**648–658.

Simons, L. A., 1986, Interrelations of lipids and lipoproteins with coronary artery disease mortality in 19 countries, *Am. J. Cardiol.* **57:**5–10.

Smith, P. F., Eydelloth, R. S., Grossman, S. J., Stubbs, R. J., Schwartz, M. S., Germershausen, J. I., Vyas, K. P., Kari, P. H., and MacDonalds, J. S., 1991, HMG-CoA reductase inhibitor-induced myopathy in the rat: cyclosporine A interaction and mechanism studies, *Fharmacol. Exp. Therap.* **257:**1225–1235.

Spiess, M., 1990, The asialoglycoprotein receptor. A model for endocytic transport receptors, *Biochemistry* **29:**10009–10018.

Steinberg, D., Parthasarathy, S., Carew, T. E., Khoo, J. C., and Witztum, J. L., 1989, Beyond cholesterol. Modifications of low-density lipoprotein that increase its atherogenecity, *New Engl. J. Med.* **320:**915–924.

Stocker, R., Bowry, V. W., and Frei, B., 1991, Ubiquinol-10 protects human low-density lipoprotein more efficiently against lipid peroxidation than does α-tocopherol, *Proc. Natl. Acad. Sci. USA* **88:**1646–1650.

Stone, N. J., Levy, R. I., Frederickson, D., and Verter, J., 1974, Coronary artery disease in 116 kindreds with familial type-II hypercholesterolemia, *Circulation* **49:**49–54.

Suckling, K. E., 1994, Atherosclerosis patents: Clues to the next drug generation, *Biotechnology* **12:**1379–1380.

Suckling, K. E., and Stange, E. F., 1985, Role of acyl CoA:cholesterol acyltransferase in cellular cholesterol metabolism, *J. Lipid Res.* **26:**647–671.

Takahashi, O., 1995, Haemorrhagic toxicity of a large dose of alpha-, beta-, gamma-, and delta-tocopherols, ubiquinone, beta-carotene, retinol acetate, and L-ascorbic acid in the rat, *Food Chem. Tox.* **33:**121–128.

Tanaka, M., Jingami, H., Otani, H., Cho, M., Ueda, Y., Arai, H., Nagano, Y., Doi, T., Yokode, M., and Kita, T., 1993, Regulation of apolipoprotein B production and secretion in response to the change of intracellular cholesteryl ester contents in rabbit hepatocytes, *J. Biol. Chem.* **268:** 12713–12718.

Taton, M., Benveniste, P., and Rahier, A., 1986, *N*-[(1,5,9)-trimethyldecyl]-4a,10-dimethyl-8-aza-*trans*-decal-3β-ol, a novel potent inhibitor of 2,3-oxidosqualene cycloartenol, and lanosterol cyclases, *Biochem. Biophys. Res. Commun.* **138:**764–770.

Taton, M., Benveniste, P., Rahier, A., Johnson, W. S., Liu, H. T., and Sudhakar, A. R., 1992, Inhibition of 2,3-oxidosqualene cyclase, *Biochemistry* **31:**7892–7898.

Tawada, H., Harcourt, M., Kawamura, N., Kajino, M., Ishikawa, E., Sugiyama, Y., Ikeda, H., and Meguro K., 1994, Novel acyl-CoA:cholesterol acyltransferase inhibitors. Synthesis, and biological activity of 3-quinolylurea derivatives, *J. Med. Chem.* **37:**2079–2084.

Taylor, F. R., Kandutsch, A. A., Gayen, A. K., Nelson, J. A., Steckbeck, N. S., Phirwa, S., and Spencer, T. A., 1986, 24,25-Epoxysterol metabolism in cultured mammalian cells, and repression of 3-hydroxy-3-methylglutaryl-CoA reductase, *J. Biol. Chem.* **261:**15039–15044.

Thelin, A., Peterson, E., Hutson, J. L., McCarthy, A. D., Ericsson, J., and Dallner, G., 1994, Effect of squalestatin 1 on the biosynthesis of the mevalonate pathway lipids, *Biochim. Biophys. Acta* **1215:**245–249.

Thompson, G. R., Barbir, M., Okabayashi, K., Trayner, I., and Larkin, S., 1989, Plasmapheresis in familial hypercholesterolemia, *Arterioscl. 9 (Suppl.)* I52–I57.

Thompson, G. R., and Wilson, P. W., 1992, *Coronary Risk Factors and Their Assessment,* Science Press Ltd., London.

Travia, D., Tosi, F., Negri, C., Faccine, G., Moghetti, P., and Muggeo, M., 1994, Sustained therapy with 3-hydroxy-3-methylglutaryl-coenzyme-A reductase inhibitors does not impair steroidogenesis by adrenals and gonads, *J. Clin. Endocrinol. Metab.* **80:**836–840.

Traxler, P., Kump, W., Müller, K., and Tosch, W., 1990, Hypolipidemic activity of rifamycins derivatives, *J. Med. Chem.* **33:**552–560.

Treiber, L. R., Arison, B. H., Huang, L., MacConnel, J. G., Miller, R. R., and Stearns, R. A., 1995, Oxidative photochemical decarboxylation of zaragosic acid A, *Experientia* **51:**252–255.

Tsujita, Y., Kuroda, M., Shimada, Y., Tanzawa, K., Arai, M., Kanedo, I., Tanare, M., Masuda, H., Tarumi, C., Watanabe, Y., and Fujii, S., 1986, CS-514, a competitive inhibitor of 3-hydroxy-3-methylglutaryl coenzyme-A reductase: tissue-selective inhibition of sterol synthesis and hypolipidemic effect on various animal species, *Biochim. Biophys. Acta* **877:**50–60.

Van Berkel, Th. J. C., Kruijt, J. K., Spanjer, H. H., Nagelkerke, J. F., Harkes, L., and Kempen, H. J. M., 1985a, The effect of a water-soluble tris-galactoside-terminated cholesterol derivative on the fate of low density lipoproteins and liposomes, *J. Biol. Chem.* **260:**2694–2699.

Van Berkel, Th. J. C., Kruijt, J. K., and Kempen, H. J. M., 1985b, Specific targeting of high density lipoproteins to liver hepatocytes by incorporation of a tris-galactoside-terminated cholesterol derivative, *J. Biol. Chem.* **260:**12203–12207.

Vlahcevic, Z. R., Hylemon, P. B., and Chiang, J. Y. L., 1994, Hepatic cholesterol metabolism, in: *The Liver: Biology, and Pathobiology,* Third Edition (I. M. Arias, D. A. Schachter, and D. A. Shafritz, eds.), pp. 379–389, Raven Press, New York.

Walldius, G., Erikson, U., Olsson, A. G., Bergstrand, L., Hadell, K., Johansson, J., Kaijser, L., Lassvik, C., Molgaard, J., Nilsson, S., Schaferelinder, L., Stenport, G., and Holme, I., 1994, The effect of probucol on femoral atherosclerosis: the probucol quantitative regression Swedish trial (PQRST), *Am. J. Cardiol.* **74:**875–883.

Wannamaker, M. W., Waid, P. P., Van Sickle, W. A., McCarthy, J. R., Wilson, P., Schatzman, G. L., and Moore, W. R., 1992, *N*-(1-oxodecyl)-4a,10-dimethyl-8-aza-*trans*-decal-3β-ol: a potent competitive inhibitor of 2,3-oxidosqualene cyclase, *J. Med. Chem.* **35:**3581–3583.

Waters, D., Higginson, L., Gladstone, P., Kinball, B., Le May, M., and Boccuzzi, S. J., 1994, Effects of monotherapy with an HMG-CoA reductase inhibitor on the progression of coronary atherosclerosis as assessed by serial quantitative arteriography, the Canadian coronary atherosclerosis intervention trial, *Circulation* **89:**959–968.

White, D. A., Heffron, F., Knight, D., and Salter, A. M., 1994, The effects of two acylcoenzyme A: cholesterol acyltransferase (ACAT) inhibitors, cyclandelate, and a non-hydrolysable ether analogue, benzyl 3,3,5-trimethylcyclohexanol, on low-density lipoprotein metabolism in macrophages and hepatocytes, *Biochem. Pharmacol.* **48:**915–922.

Wilde, R. G., and Billheimer, J. T., 1995b, Acyl CoA:cholesterol acyltransferase (ACAT) inhibitors: ureas bearing two heterocyclic head groups, *Bioorg. Med. Chem. Lett.* **5:**173–176.

Wilde, R. G., Klaczkiewicz, J. D., Billheimer, J. T., Wexler, R. R., and Gillies, P. J., 1995a, Acyl CoA:cholesterol acyltransferase (ACAT) inhibitors: heterocyclic bioisosteres for the urea group in DuP 128, *Bioorg. Med. Chem. Lett.* **5:**177–180.

Wilde, R. G., Billheimer, J. T., Germain, S. J., Gillies, P. J., Higley, C. A., Kezar, H. S., Maduskuie, T. P., Shimshick, E. S., and Wexler, R. R., 1995b, Acyl CoA:cholesterol acyltransferase (ACAT) inhibitors: ureas bearing heterocyclic groups bio-isosteric for an imidazole, *Bioorg. Med. Chem. Lett.* **5:**167–172.

Wilson, P. W. F., 1990, High-density lipoprotein, low-density lipoprotein, and coronary artery disease, *Am. J. Cardiol.* **66:**7A–11A.

Witztum, J. L., and Steinberg, D., 1991, Role of oxidized low-density lipoprotein in atherogenesis, *J. Clin. Invest.* **88:**1785–1792.

Wrenn, S. M., Parks, J. S., Immermann, F. W., and Rudel, L. L., 1995, ACAT inhibitors CL 283,546 and CL 283,796 reduce LDL cholesterol without affecting cholesterol absorption in African green monkeys, *J. Lipid Res.* **36:**1199–1210.

Yamamoto, A., Sudo, H., and Endo, A., 1980, Therapeutic effects of ML-236B in primary hypercholesterolemia, *Atherosclerosis* **35:**259–266.

Yamamoto, A., Matsuzawa, Y., Kishino, B., Hayashi, R., Hirobe, K., and Kikkawa, T., 1983, Effects of probucol on homozygous cases of familial hypercholesterolemia, *Atherosclerosis* **48:**157–166.

Yamamoto, A., Matsuzawa, Y., Yokoyama, S., Funahashi, T., Yamamura, T., and Kichino, B., 1986, Effects of probucol on xanthoma regression in familial hypercholesterolemia, *Am. J. Cardiol.* **57:**49–54.

Yamamoto, A., Hara, H., Takaichi, S., Wakasugi, J. I., and Tomikawa, M., 1988, Effect of probucol on macrophages, leading to regression and atheromatous vascular lesions, *Am. J. Cardiol.* **62:**31–36.

Yamamoto, S., Lin, K., and Bloch, K., 1969, Some properties of microsomal 2,3-oxidosqualene sterol cyclase, *Proc. Natl. Acad. Sci. USA* **63:**110–117.

Ylä-Herttuala, S., Palinski, W., Rosenfeld, M. E., Parthasarathy, S., Carew, T. E., Butler, S., Witztum, J. L., and Steinberg, D., 1989, Evidence for the presence of oxidatively modified low density lipoproteins in atherosclerotic lesions of rabbit and man, *J. Clin. Invest.* **84:**1086–1095.

Yokoyama, S., Hayashi, R., Satani, M., and Yamamoto, A., 1985, Selective removal of LDL plasmapheresis in familial hypercholesterolemia, *Arterioscl.* **5:**613–622.

Index

MIX
Papier aus verantwortungsvollen Quellen
Paper from responsible sources
FSC® C105338

If you have any concerns about our products,
you can contact us on
ProductSafety@springernature.com

In case Publisher is established outside the EU,
the EU authorized representative is:
Springer Nature Customer Service Center GmbH
Europaplatz 3, 69115 Heidelberg, Germany

Printed by Libri Plureos GmbH
in Hamburg, Germany